D1583266

I G Macdonald

August 1982

# The Representation Theory of the Symmetric Group

GIAN-CARLO ROTA, *Editor*
ENCYCLOPEDIA OF MATHEMATICS AND ITS APPLICATIONS

| Volume | | Section |
|---|---|---|
| 1 | LUIS A. SANTALÓ<br>**Integral Geometry and Geometric Probability,** 1976 (2nd printing, with revisions, 1979) | Probability |
| 2 | GEORGE E. ANDREWS<br>**The Theory of Partitions,** 1976 (2nd printing, 1981) | Number Theory |
| 3 | ROBERT J. McELIECE<br>**The Theory of Information and Coding**<br>A Mathematical Framework for Communication, 1977 (2nd printing, with revisions, 1979) | Probability |
| 4 | WILLARD MILLER, Jr.<br>**Symmetry and Separation of Variables,** 1977 | Special Functions |
| 5 | DAVID RUELLE<br>**Thermodynamic Formalism**<br>The Mathematical Structures of Classical Equilibrium Statistical Mechanics, 1978 | Statistical Mechanics |
| 6 | HENRYK MINC<br>**Permanents,** 1978 | Linear Algebra |
| 7 | FRED S. ROBERTS<br>**Measurement Theory**<br>with Applications to Decisionmaking, Utility, and the Social Sciences, 1979 | Mathematics and the Social Sciences |
| 8 | L. C. BIEDENHARN and J. D. LOUCK<br>**Angular Momentum in Quantum Physics:** Theory and Application, 1981 | Mathematics of Physics |
| 9 | L. C. BIEDENHARN and J. D. LOUCK<br>**The Racah-Wigner Algebra in Quantum Theory**, 1981 | Mathematics of Physics |
| 10 | JOHN D. DOLLARD and CHARLES N. FRIEDMAN<br>**Product Integration**<br>with Application to Differential Equations, 1979 | Analysis |

GIAN-CARLO ROTA, *Editor*
ENCYCLOPEDIA OF MATHEMATICS AND ITS APPLICATIONS

---

| | | |
|---|---|---|
| 11 | WILLIAM B. JONES and W. J. THRON<br>**Continued Fractions:** Analytic Theory and Applications, 1980 | Analysis |
| 12 | NATHANIEL F. G. MARTIN and JAMES W. ENGLAND<br>**Mathematical Theory of Entropy,** 1981 | Real Variables |
| 13 | GEORGE A. BAKER, Jr., and PETER R. GRAVES-MORRIS<br>**Padé Approximants**<br>Part I: Basic Theory, 1981 | Mathematics of Physics |
| 14 | GEORGE A. BAKER, Jr., and PETER R. GRAVES-MORRIS<br>**Padé Approximants**<br>Part II: Extensions and Applications, 1981 | Mathematics of Physics |
| 15 | ENRICO. G. BELTRAMETTI and GIANNI CASSINELLI<br>**The Logic of Quantum Mechanics,** 1981 | Mathematics of Physics |
| 16 | GORDON JAMES and ADALBERT KERBER<br>**The Representation Theory of the Symmetric Group**, 1981 | Algebra |

*Other volumes in preparation*

# ENCYCLOPEDIA OF MATHEMATICS and Its Applications

GIAN-CARLO ROTA, *Editor*
**ENCYCLOPEDIA OF MATHEMATICS AND ITS APPLICATIONS**
Volume 16

Section: Algebra
P. M. Cohn and Roger Lyndon, *Section Editors*

# The Representation Theory of the Symmetric Group

**Gordon James**
Sidney Sussex College
Cambridge, Great Britain

**Adalbert Kerber**
University of Bayreuth
Bayreuth, Federal Republic of Germany

Foreword by
**P. M. Cohn**
University of London, Bedford College

Introduction by
**G. de B. Robinson**
University of Toronto

1981

**Addison-Wesley Publishing Company**
Advanced Book Program
Reading, Massachusetts

London·Amsterdam·Don Mills, Ontario·Sydney·Tokyo

**Library of Congress Cataloging in Publication Data**

James, G. D. (Gordon Douglas), 1945-
The representation theory of the symmetric group.

(Encyclopedia of mathematics and its applications;
v. 16. Section, Algebra)
Bibliography: p.
Includes index.
1. Symmetry groups. 2. Representations of groups.
I. Kerber, Adalbert. II. Title. III. Series:
Encyclopedia of mathematics and its applications;
v. 16. IV. Series: Encyclopedia of mathematics
and its applications. Section, Algebra.
QA171.J34 512'.53 81-12681
ISBN 0-201-13515-9 AACR2

American Mathematical Society (MOS) Subject Classification Scheme (1980): 20C30



Manufactured in the United States of America

ABCDEFGHIJ-HA-8987654321

# Contents

# *Editor's Statement*

A large body of mathematics consists of facts that can be presented and described much like any other natural phenomenon. These facts, at times explicitly brought out as theorems, at other times concealed within a proof, make up most of the applications of mathematics, and are the most likely to survive change of style and of interest.

This ENCYCLOPEDIA will attempt to present the factual body of all mathematics. Clarity of exposition, accessibility to the non-specialist, and a thorough bibliography are required of each author. Volumes will appear in no particular order, but will be organized into sections, each one comprising a recognizable branch of present-day mathematics. Numbers of volumes and sections will be reconsidered as times and needs change.

It is hoped that this enterprise will make mathematics more widely used where it is needed, and more accessible in fields in which it can be applied but where it has not yet penetrated because of insufficient information.

GIAN-CARLO ROTA

# Foreword

The theory of group representation has its roots in the character theory of abelian groups, which was formulated first for cyclic groups in the context of number theory (Gauss, Dirichlet, but already implicit in the work of Euler), and later generalized by Frobenius and Stickelberger to any finite abelian groups. For an abelian group all irreducible representations (over $\mathbb{C}$) are of course 1-dimensional and hence are completely described by their characters. The representation theory of finite groups emerged around the turn of the century as the work of Frobenius, Schur, and Burnside. While it applied in principle to any finite group, the symmetric group $S_n$ was a simple but important special case; —simple because its characters and irreducible representations could already be found in the rational field, important because every finite group could be embedded in some symmetric group.

Moreover, the theory can be applied whenever we have a symmetric group action on a linear space. Perhaps the simplest example is the case of a bilinear form $f(x, y)$. No theory is required to decompose $f$ into a symmetric part: $s(x, y)=f(x, y)+f(y, x)$ and an antisymmetric part: $a(x, y)=f(x, y)-f(y, x)$. These are of course just (bases for the 1-dimensional modules affording) the irreducible representations of $S_2$, 1-dimensional because $S_2$ is abelian. Taking next a trilinear form $f(x, y, z)$, we have again the symmetric and antisymmetric parts:

$$s(x, y, z)=\sum_{\sigma} f(\sigma x, \sigma y, \sigma z).$$

$$a(x, y, z)=\sum_{\sigma} \operatorname{sgn} \sigma f(\sigma x, \sigma y, \sigma z),$$

where $\sigma$ ranges over all permutations of $x, y, z$. No other linear combination of $f$'s is only multiplied by a scalar factor by the $S_3$-action (such a factor would have to be 1 or $\operatorname{sgn}\sigma$, because every permutation is a product of transpositions), but we can find pairs of linear combinations spanning a 2-dimensional $S_3$-module, e.g.

$$p=f(x, y, z)+f(y, x, z)-f(z, y, x)-f(z, x, y)$$

$$q=f(z, y, x)+f(y, z, x)-f(x, y, z)-f(x, z, y)$$

Here $p$ is obtained by 'symmetrizing' $x$, $y$ and 'antisymmetrizing' $x$, $z$, and $q$

is obtained by interchanging $x$, $z$ in $p$. If $(x, y)$ denotes the transposition of $x$, $y$ etc., then we have

$$(x,y)=p,\quad (x,z)p=q,\quad (y,z)p=-p-q,\quad (x,y)q=-p-q,$$
$$(x,z)q=p,\quad (y,z)q=q.$$

Thus we obtain the representation

$$(x,y)\rightarrow\begin{pmatrix}1 & -1\\ 0 & -1\end{pmatrix},\quad (x,z)\rightarrow\begin{pmatrix}0 & 1\\ 1 & 0\end{pmatrix},\quad (y,z)\rightarrow\begin{pmatrix}-1 & 0\\ -1 & 1\end{pmatrix},$$
$$(x,y,z)\rightarrow\begin{pmatrix}0 & -1\\ 1 & -1\end{pmatrix},\quad (x,z,y)\rightarrow\begin{pmatrix}-1 & 1\\ -1 & 0\end{pmatrix},$$

which is an irreducible 2-dimensional representation of $S_3$.

It was Alfred Young's achievement to find a natural classification of all the irreducible representations of $S_n$ in terms of 'Young tableaux', which are essentially the different ways of fully symmetrizing and antisymmetrizing. The $n$-symbols permuted are arranged in a diagram so that rows are symmetrized and columns antisymmetrized. In the above example we symmetrized $x$, $y$ and antisymmetrized $x$, $z$; this is indicated by the tableau

| | |
|---|---|
| $x$ | $y$ |
| $z$ | |

Young's derivation via tableaux was even more direct than Frobenius' and Schur's earlier method, using bialternants or S-functions, although these functions are useful in formulating combinations of representations such as plethysm.

There have been many accounts of the theory, from various points of view, and often the original sources have been hard to follow. It is good to have a general treatment, —by two authors who have both made substantial original contributions, —which combines the best of previous accounts, and systematizes and adds much that is new. After a clear exposition of Young's approach (in modern terms) they present an improved version of Specht modules giving a characteristic-free treatment and leading to a practical algorithm for estimating dimensions. The applications to combinatorics include Polya's enumeration theory, and also the less well known work of Redfield, and there is a separate chapter on the connection with representations of the general linear groups.

The comprehensive treatment, with helpful suggestions for further reading, very full references, various tables of characters, as well as the interesting historical introduction by G. de B. Robinson, will all help to make 'James-Kerber' the standard work on the subject.

P. M. Cohn

Alfred Young

# Introduction

In this introduction to the work of James and Kerber I should like to survey briefly the story of developments in the representation theory of the symmetric group. Detailed references will not be possible, but it seems worthwhile to glance at the background which has aroused so much interest in recent years.

The idea of a group goes back a long way and is inherent in the study of the regular polyhedra by the Greeks. It was Galois who systematically developed the connection with algebraic equations, early in the nineteenth century. Not long after, the geometrical relationship between the lines on a general cubic surface and the bitangents of a plane quartic curve aroused the interest of Hesse and Cayley, with a significant contribution by Schläfli in 1858 [1, Chapter IX].* Jordan in his *Traité des Substitutions*, 1870 [2], and Klein in his *Vorlesungen über das Ikosaeder*, 1884 [3], added new dimensions to Galois's work. The first edition of Burnside's *Theory of Groups of Finite Order* appeared in 1897, just at the time when Frobenius's papers in the *Berliner Sitzungsberichte* were changing the whole algebraic approach. With the appearance of *Schur's Thesis* [4] in 1901, the need for a revision of Burnside's work became apparent.

Burnside began his preface to the second edition, which appeared in 1911 [5], with the comment: "Very considerable advance in the theory of groups of finite order has been made since the appearance of the first edition of this book. In particular the theory of groups of linear substitutions has been the subject of numerous and important investigations by several writers. . . ." His preface concludes with the remark: "I owe my best thanks to the Rev. Alfred Young, M.A., Rector of Birdbrook, Essex, and former Fellow of Clare College, Cambridge, who read the whole of the book as it passed through the Press. His careful criticism has saved me from many errors and his suggestions have been of great help to me."

Alfred Young was born in 1873 and graduated from Cambridge in 1895. His first paper, "The irreducible concomitants of any number of binary quartics," appeared in the *Proceedings of the London Mathematical Society* in 1899. It had been refereed by Burnside, who told him to read the works of Frobenius and Schur; unfortunately Young knew no German, so it was not till after the war that he was able to incorporate their ideas in his important *QSA* series.

*References will be found at the end of this Introduction.

My own contact with Alfred Young began in 1929. I graduated from the University of Toronto in 1927 and was much interested in geometry, owing largely to the presence on our staff of Jacques Chapelon from Paris. I was fortunate in obtaining a small scholarship at St. John's College, Cambridge, where my first supervisor was M. H. A. Newman. Under his guidance I began to read topology. No group theory was taught in Toronto or Cambridge in those days, but its significance in topology fascinated me. Soon this became apparent to Newman, and he arranged for me to be transferred to Alfred Young as a graduate student. Young came in to Cambridge once a week to lecture. He and his wife stayed at the Blue Boar Hotel, just across the street from St. John's, where I would go to visit him. The geometrical aspects of group theory continued to interest me, and I attended Baker's tea party every week. This was where I met Donald Coxeter and several other geometers to whom I refer in the Introduction to *Young's Collected Papers*, published in Toronto [6], 1977.

After earning my Ph.D. in Cambridge in 1931, I returned to the University of Toronto. My work on the symmetric group continued, and with the cooperation of J. S. Frame and Philip Hall, yielded the dimension formulae for the irreducible representations of $S_n$ and $GL_d$ over the real field.

Richard Brauer was on staff in Toronto 1935–1948, and his interest in representation theory was responsible for much of the development which took place in those years and later, while he was at Ann Arbor 1949–1952, and Harvard 1952–1978. In 1958 I was invited to lecture at the Australian universities, and my *Representation Theory of the Symmetric Group* appeared in 1961. In 1968 my wife and I went to Christchurch, New Zealand, for three months, and it was during this period that I became interested in the application of group theory to physics. W. T. Sharp in Toronto had obtained his Ph.D. with Wigner in Princeton, and I made contact with Wybourne in New Zealand and with Biedenharn in the U.S., and attended a seminar in Bochum in West Germany in 1969. It was there that I met Adalbert Kerber and many other interesting people. Not long after, I was in touch with Gordon James, who got his Ph.D. in Cambridge with J. G. Thompson. Then when the representation theory gathering was held in Oberwolfach in 1975 I had a chance to talk with many group theorists whose writings I had read, but had never met. Afterwards my wife and I paid a brief visit to the Kerbers in historic Aachen.

It was in the autumn of 1975 that Gordon James came to spend a year at the University of Toronto. He and Kerber had begun to work on this book and we had many conversations; Kerber was largely interested in wreath products, while James had begun writing his considerable number of papers on modular theory. A number of errors had appeared in the decomposition matrices at the end of my book [7], and James has done much to improve their construction in this volume.

In April of 1976 Foata organized another gathering of group theorists in Strasbourg. He did a beautiful job, exploiting the charm of the city and its university to bring together a large number of speakers [8] on various aspects and applications of the symmetric group. Having been invited by Professor McConnell, I gave a repeat performance of my Strasbourg talk in Dublin. This was my first visit to Ireland, and it gave me much pleasure to see J. L. Synge, who had been on our staff in Toronto for many years.

It was in June 1978 that T. V. Narayana of the University of Alberta in Edmonton arranged a gathering at the University of Waterloo. He had become involved in Young's work, and his volume *Lattice Path Combinatorics with Statistical Applications* was published in our Exposition Series in 1979. The proceedings of Young Day has just appeared [9] with an introduction by J. S. Frame and a paper generalizing the hook-formulae for $O_n(2)$.

The last gathering in Oberwolfach which I attended was in January 1979. This book was well on its way but we all regretted that publication would be so long delayed. Its content contributes much to complete the picture, from the point of view of the representation theory of $S_n$, but there remains the question raised by Frame's work:—Could there be a degree formula for the irreducible representations of $S_n$ or $GL_d$ over a finite field? Future research may provide the answer.

In conclusion, let me refer to *The Theory of Partitions* [10] by Andrews, which has appeared in this series. Professor Rota's comment is worth quoting: "Professor Andrews has written the first thorough survey of this many-sided field. The specialist will consult it for the more recondite results, the student will be challenged by many deceptively simple facts, and the applied scientist may locate in it a missing identity to organize his data." When Young's *Collected Papers* came out, Andrews wrote a most interesting review of his work [11], listing 121 papers based on the original ideas of this remarkable man. The accompanying portrait of Alfred Young was sent to me by Professor Garnir.

It has given me much pleasure to work with the authors of this book, and I wish its readers every satisfaction, as well as the best of luck in further developing these ideas.

## References


1. G. Miller, H. Blichfeldt, and L. Dickson, *Finite Groups*, Dover Publications, New York, 1961.
2. C. Jordan, *Traité des Substitutions*, Paris, 1870.
3. F. Klein, *Vorlesungen über des Ikosaeder*, B. G. Teubner, Leipzig, 1884.
4. I. Schur, *Inaugural—Dissertation*, Berlin, 1901.
5. W. Burnside, *Theory of Groups of Finite Order*, Cambridge University Press, Cambridge, 1911.

6. A. Young, *Collected Papers*, University of Toronto, 1977.
7. G. de B. Robinson, *Representation Theory of the Symmetric Group*, University of Toronto, 1961.
8. D. Foata, *Combinatoire et Représentation du Group Symétrique, Strasbourg* 1976, Springer-Verlag (579), 1977.
9. T. V. Narayana, *Young-Day Proceedings—Waterloo 1978*, Marcel Dekker (57), 1980.
10. G. E. Andrews, *The Theory of Partitions*, Addison Wesley, Advanced Book Program, 1976.
11. G. E. Andrews, Bull. Am. Math. Soc., **1**, 989–997 (1979).

G. DE B. ROBINSON

# Preface

The purpose of this book is to provide an account of both the ordinary and the modular representation theory of the symmetric groups. The range of applications of this theory is vast, varying from theoretical physics through combinatorics to the study of polynomial identity algebras, and new uses are still being found. So diverse are the questions which arise that we feel justified in hoping the reader might find that some part of our text inspires him to undertake research of his own into one of the many unsolved problems in this elegant branch of mathematics.

There are several different ways of approaching symmetric group representations, and while we have tried to illuminate parts of the theory by giving more than one description of it, we have made no effort to cover every view of the subject.

The ordinary representation theory of the symmetric groups was first developed by Frobenius, but the greatest contribution to the early material came from Alfred Young. Since Young's main interest lay in quantitative substitutional analysis, it is difficult for a modern mathematican to understand his papers. The reader is referred to the book *Substitutional Analysis* by D. E. Rutherford for a pleasant account of a great part of Young's work. Both Frobenius's and Young's collected works are now available. We include an account of the group algebra and its idempotents, along the lines pursued by Young, since the symmetric group is one of the very rare cases where many aspects of general representation theory can be described explicitly. This also helps us to motivate the introduction of many combinatorial structures which turn out to be useful. The combinatorics are themselves a fascinating and fruitful basis for further study, and they continue to inspire many research papers.

The development of the modular representation theory of symmetric groups was started by T. Nakayama, who derived some $p$-modular properties of symmetric groups $S_n$, for $n<2p$, and stated a conjecture about the $p$-block structure of symmetric groups $S_n$ of arbitrary degree $n$. This conjecture was proved jointly by R. Brauer and G. de B. Robinson. Robinson and his coworkers developed these methods rapidly to study the decomposition numbers of the symmetric groups. The situation as it was in 1961 is described in Robinson's book.

Later, these methods were combined by the present authors with modular results derived from W. Specht's alternative approach to ordinary representation theory. Specht showed how to derive representations by considering submodules of a polynomial ring $F[x_1, \ldots, x_n]$, and this method yielded interesting results without referring to the characteristic of the field $F$. Using modules isomorphic to those of Specht, we explain another approach to the ordinary representations, at the same time extracting information about the modular theory.

There is a more recent approach to a characteristic-free representation theory of $S_n$ starting with a basis of a certain intertwining space which was originally derived for invariant-theoretical purposes by G.-C. Rota. We shall not present this method, as we anticipate that it will be described in another book.

The main application which we cover involves the representations of an arbitrary group, using symmetry operators on tensor space, but we also discuss combinatorics, wreath products, and permutation groups. The standard reference for symmetric functions is the book of D. E. Littlewood, but I. G. Macdonald's recent volume gives an up-to-date account of many of the results.

Both authors are greatly indebted to Professor Robinson and the University of Toronto for their generous hospitality during several visits, and they wish to record their gratitude to Professor Robinson, without whose continued enthusiasm, encouragement, and advice the book would not have been written.

In conclusion, we would like to express our thanks for the important help and criticism we received from so many colleagues and from our students, whom we have taken pleasure in working with and who have given us so much useful advice. We mention in particular H. Boerner, R. W. Carter, M. Clausen, N. Esper, H. K. Farahat, A. Golembiowski, M. Klemm, W. Lehmann, A. O. Morris, J. Neubüser, H. Pahlings, M. H. Peel, F. Sänger, D. Stockhofe, J. Tappe, K.-J. Thürlings and B. Wagner.

G. D. James
A. Kerber

# List of Symbols

| Symbol | Meaning | Definition on p. |
|---|---|---|
| $a_i(\pi)$ | The number of $i$-cycles occurring in the cycle notation of $\pi$ | 7 |
| $a(\pi)$ | The cycle type $(a_1(\pi),\ldots, a_n(\pi))$ of $\pi$ | 10 |
| $a(f; \pi)$ | The type of $(f; \pi)$ | 139 |
| $a_i^\beta(m)$ | The $p$-residue of the node which $m$ replaces in $t_i^\beta$ | 307 |
| $\mathrm{A}_n$ | The alternating group on $\{1,\ldots, n\}$ | 7 |
| $AS_\lambda$ | The alternating representation of $S_\lambda$ | 17 |
| $B_k$ | The $k$th Bell number | 228 |
| $c(\pi)$ | The number of pairwise disjoint cyclic factors of $\pi$ | 5 |
| $\mathbb{C}$ | The field of complex numbers | — |
| $C^\alpha$ | The conjugacy class of $\mathrm{S}_n$ consisting of permutations with cycle partition $\alpha$ | 12 |
| $C^{\alpha\pm}$ | The two conjugacy classes of $\mathrm{A}_n$ arising from splitting class $C^\alpha$ of $\mathrm{S}_n$ | 12 |
| $C^A(\pi)$ | The conjugacy class in $\mathrm{A}_n$ of the permutation $\pi$ | 11 |
| $C_A(\pi)$ | The centralizer in $\mathrm{A}_n$ of the permutation $\pi$ | 11 |
| $C^S(\pi)$ | The conjugacy class in $\mathrm{S}_n$ of the permutation $\pi$ | 11 |
| $C_S(\pi)$ | The centralizer in $\mathrm{S}_n$ of the permutation $\pi$ | 11 |
| $\mathrm{char}\,\mathrm{S}_n$ | The ring of generalized characters of $\mathrm{S}_n$ | 39 |
| $\mathrm{Cyc}(H)$ | The cycle index of $H$ | 170 |
| $\mathrm{Cyc}(H, D)$ | The generalized cycle index of $H$ with respect to $\chi^D$ | 171 |
| $\mathrm{Cyc}(H\|p)$ | Pólya insertion of $p$ in $H$ | 170 |
| $d_\alpha$ | The depth of $\alpha$ | 99 |
| $d^1_{\alpha,\beta}$ | A $p$-modular decomposition number of $\alpha$ | 243 |
| $d_i^\alpha(r, s)$ | The axial distance between $r$ and $s$ in $t_i^\alpha$ | 123 |
| $\mathrm{diag}\,G^*$ | The diagonal subgroup of $G^*$ | 134 |
| $\det\beta$ | The determinant of a Gram matrix for $S^\beta$ | 313 |
| $D^\beta$ | $S^\beta/(S^\beta \cap S^{\beta\perp})$ for $\beta$ $p$-regular | 299 |
| $D^1_{n,p}$ | The $p$-modular decomposition matrix of $\mathrm{S}_n$ | 244 |

| | | |
|---|---|---|
| $\left(\overset{n}{\#} D\right)^{\sim}$ | The representation of $G \operatorname{wr} H$ arising naturally from the representation $D$ of $G$ | 147 |
| $D \boxdot [\alpha]$ | The symmetrization of $D$ by $[\alpha]$ | 186 |
| $D \triangle_n D_i$ | The $n$th permutrization of $D$ by $D_i$ | 202 |
| $e^{\alpha}$ | The essentially idempotent element $\mathcal{V}^{\alpha}\mathcal{H}^{\alpha}$ of $\mathbb{Q}S_n$ arising from the tableau $t^{\alpha}$ | 31 |
| $\hat{e}^{\alpha}$ | The primitive idempotent element $(1/\kappa^{\alpha})e^{\alpha}$ of $\mathbb{Q}S_n$ arising from the tableau $t^{\alpha}$ | 106 |
| $e_i^{\alpha}$ | The essentially idempotent element arising from the standard tableau $t_i^{\alpha}$ | 106 |
| $\hat{e}_i^{\alpha}$ | The primitive idempotent element arising from the standard tableau $t_i^{\alpha}$ | 106 |
| $e_{ij}$ | A particular element of $U$ | 328 |
| $e_{ik}^{\alpha}$ | $e_i^{\alpha}\pi_{ik}^{\alpha}$ | 110 |
| $e_t$ | The polytabloid arising from $t$ | 295 |
| $E_{\alpha}$ | The projection of $\otimes^n V$ onto the homogeneous component of type $[\alpha]$ | 188 |
| $E^H$ | A certain subgroup of $G \operatorname{wr} H$ | 137 |
| $E_T$ | The element of the Weyl module corresponding to the tableau $T$ | 323 |
| $f^{\alpha}$ | The dimension of the ordinary irreducible representation $[\alpha]$ of $S_n$ | 41 |
| $f_q^{\alpha}$ | The number of different ways of reaching $[\tilde{\alpha}]$ by removing rim $q$-hooks from $[\alpha]$ | 82 |
| $(f; \pi)$ | An element of $G \operatorname{wr} H$ | 132 |
| $F$ | A field | — |
| $F^{\alpha}$ | $W^{\alpha}/(W^{\alpha} \cap W^{\alpha\perp})$ | 339 |
| $g_{\nu}(f; \pi)$ | The $\nu$th cycle product of $(f; \pi)$ | 138 |
| $G^*$ | The base group of $G \operatorname{wr} H$ | 133 |
| $G^{\mathbf{n}}$ | The set of all mappings from $\mathbf{n}$ into $G$ | 132 |
| $G \operatorname{wr} H$ | The wreath product of $G$ by $H$ | 132 |
| $[G]^H$ | The exponentiation of $G$ by $H$ | 137 |
| $G^H$ | The power of $G$ by $H$ | 137 |
| $G[H]$ | $\psi[H \operatorname{wr} G]$ | 172 |
| $G(i)$ | A certain element of $FS_n$ | 310 |
| $\mathrm{GL}(m, F)$ | The general linear group | 319 |
| $G_{m,n}(x)$ | Gaussian polynomial | 226 |
| $\mathrm{Grf}(H)$ | The group reduction function | 171 |
| $G_{X,Y}$ | A Garnir element | 301 |
| $h_i$ | A certain element of $U$ | 329 |
| $h_i^{\alpha}$ | The first-column hook length $h_{i1}^{\alpha}$ | 76 |

| | | |
|---|---|---|
| $h_{ij}^{\alpha}$ | The number of nodes in the hook $H_{ij}^{\alpha}$ | 56 |
| $h(n,k)$ | A certain "hook length" | 312 |
| $\mathcal{H}^{\alpha}$ | $\sum_{\rho\in H^{\alpha}}\rho$ | 31 |
| $H^{\alpha}$ | The horizontal group of $t^{\alpha}$ | 29 |
| $H(t^{\alpha})$ | The horizontal group of $t^{\alpha}$ | 29 |
| $H_{ij}^{\alpha}$ | The $(i,j)$-hook of $[\alpha]$ | 55 |
| $i(\ ,\ )$ | Intertwining number | 17 |
| $\mathrm{id}_{\mathbf{n}}$ | The identity mapping of $\{1,2,\dots,n\}$ | 4 |
| $I^{\alpha}$ | The simple two sided ideal of $\mathbb{Q}S_n$ corresponding to the partition $\alpha$ | 109 |
| $IS_{\lambda}$ | The identity representation of $S_{\lambda}$ | 17 |
| $l_{ij}^{\alpha}$ | The leg length of the hook $H_{ij}^{\alpha}$ | 56 |
| $L$ | $\bigoplus_{n\geqslant 0}L^{(n)}$ | 327 |
| $L^{(n)}$ | $W^{(1)}\otimes\cdots\otimes W^{(1)}$ ($n$ times) | 320 |
| $L^{\alpha}$ | $\mathcal{V}^{\alpha}L^{(n)}$ | 321 |
| $L_i^{\alpha}$ | The left ideal of $\mathbb{Q}S_n$ generated by $e_i^{\alpha}$ | 109 |
| $\mathbf{m+n}$ | $\{1,2,\dots,m+n\}$ | 45 |
| $\mathbf{m+n}\setminus\mathbf{m}$ | $\{m+1,\dots,m+n\}$ | 45 |
| $\mathbf{m}^{\mathbf{n}}$ | $\{\varphi\mid\varphi:\mathbf{n}\to\mathbf{m}\}$ | 163 |
| $M_n$ | The matrix whose $(i,k)$ entry is $(IS_{\alpha^i}\uparrow S_n,[\alpha^k])$ | 38 |
| $M^{\beta}$ | The $FS_n$-module spanned by $\beta$-tabloids | 295 |
| $\mathbf{n}$ | $\{1,2,\dots,n\}$ | 2 |
| $\mathbf{n}_{\alpha}$ | The set of dissections $\mathbf{n}^{\alpha}$ of $\mathbf{n}$ | 96 |
| $\mathbf{n}^{\lambda}$ | A dissection of $\mathbf{n}$ | 16 |
| $\mathbf{n}_i^{\lambda}$ | One of a set of disjoint subsets of $\mathbf{n}$ forming a dissection $\mathbf{n}^{\lambda}$ of $\mathbf{n}$ | 16 |
| $\mathbf{n}^{[k]}$ | The set of all $k$-subsets of $\mathbf{n}$ | 227 |
| $\mathbf{n}^{(k)}$ | The set of all $k$-tuples over $\mathbf{n}$ | 227 |
| $\mathbf{n}^{\langle k\rangle}$ | The set of all injective $k$-tuples over $\mathbf{n}$ | 228 |
| $\mathbb{N}$ | $\{1,2,3,\dots\}$ | 5 |
| $\mathbb{N}_0$ | $\{0,1,2,\dots\}$ | 2 |
| $p(k;m,n)$ | The number of proper partitions of $k$ into $\leqslant m$ parts, each one being $\leqslant n$ | 225 |
| $p(n)$ | The number of proper partitions of $n$ | 10 |
| $p'(n)$ | The number of $p$-regular proper partitions of $n$ | 243 |
| $p^{*}(n)$ | The number of $p$-regular classes of $S_n$ | 243 |
| $P(n)$ | The set of proper partitions of $n$ | 23 |
| $P^{[k]}, P^{(k)}, P^{\langle k\rangle}$ | The permutation groups on $\mathbf{n}^{[k]},\mathbf{n}^{(k)},\mathbf{n}^{\langle k\rangle}$, respectively, corresponding to the permutation group $P$ on $\mathbf{n}$ | 228 |
| $P_n^m$ | The representation of $S_n$ afforded by $\otimes^n V$ | 150 |

| | | |
|---|---|---|
| $\mathbb{Q}$ | The field of rational numbers | — |
| $r_n(g)$ | The number of $n$th roots of $g$ | 208 |
| $\mathbb{R}$ | The real number field | — |
| $R_m$ | $\sum_{i=1}^{m-1}(i, m)$ | 307 |
| $R_n(g)$ | The set of $n$th roots of $g$ | 208 |
| $R_{ij}^{\alpha}$ | The part of the rim of $[\alpha]$ associated with $H_{ij}^{\alpha}$ | 56 |
| $RS_n$ | The regular representation of $S_n$ | 36 |
| $\mathrm{st}[\alpha]$ | The set of standard $\alpha$-tableaux | 107 |
| $S^{\beta}$ | The Specht module for the partition $\beta$ | 296 |
| $S_{\lambda}$ | A Young subgroup corresponding to $\lambda$ | 16 |
| $S_n$ | The symmetric group on $\{1, 2, \ldots, n\}$ | 2 |
| $S_{\Omega}$ | The symmetric group on $\Omega$ | 2 |
| $S(k, j)$ | A Stirling number of the second kind | 231 |
| $\mathrm{SA}(n)$ | The set of self-associated partitions of $n$ | 66 |
| $\mathrm{SP}(n)$ | The set of split partitions of $n$ | 67 |
| $t^{\alpha}$ | A Young tableau | 28 |
| $t^{\alpha *}$ | The tableau obtained by removing $n$ from the standard tableau $t^{\alpha}$ | 107 |
| $t_i^{\alpha}$ | The $i$th standard $\alpha$-tableau | 107 |
| $t[\alpha]$ | The set of $\alpha$-tableaux | 107 |
| $\{t\}$ | The tabloid containing $t$ | 41 |
| $[T]$ | The equivalence class of $T$ | 324 |
| $u_{\alpha\lambda}^{i}$ | A $u$-number with respect to $p$ | 256 |
| $U$ | The universal enveloping algebra | 328 |
| $U$ | The matrix of $u$-numbers | 257 |
| $\bar{U}_F$ | The hyperalgebra | 331 |
| $V^{\alpha}$ | The vertical group of $t^{\alpha}$ | 29 |
| $V(t^{\alpha})$ | The vertical group of $t^{\alpha}$ | 29 |
| $V^{t}$ | The vertical group of $t$ | 29 |
| $\mathcal{V}^{\alpha}$ | $\sum_{\pi \in V^{\alpha}}(\mathrm{sgn}\,\pi)\pi$ | 31 |
| $\mathcal{V}^{t}$ | $\sum_{\pi \in V^{t}}(\mathrm{sgn}\,\pi)\pi$ | 31 |
| $\otimes^n V$ | $V \otimes \cdots \otimes V$ ($n$ factors) | 146 |
| $w^{\alpha}$ | The tensor corresponding to the first standard $\alpha$-tableau | 320 |
| $w_T$ | The tensor corresponding to the tableau $T$ | 322 |
| $W^{\alpha}$ | The Weyl module | 321 |
| $\mathbb{Z}$ | The ring of integers | — |
| $Z_n$ | The character table of $S_n$ | 39 |
| $\alpha'$ | The partition associated with $\alpha$ | 22 |
| $[\alpha]$ | The Young diagram associated with $\alpha$; | 22 |
| | also, the corresponding equivalence class of representations | 36 |

| | | |
|---|---|---|
| $\pi^{[k]}, \pi^{(k)}, \pi^{\langle k\rangle}$ | The permutation on $\mathbf{n}^{[k]}, \mathbf{n}^{(k)}, \mathbf{n}^{\langle k\rangle}$, respectively, corresponding to the permutation $\pi$ of $\mathbf{n}$ | 228 |
| $\sigma^{\alpha}$ | The seminormal form of $[\alpha]$ | 114 |
| $\sigma_i^{\alpha}$ | The $i$th partial sum of the parts of $\alpha$ | 26 |
| $\sigma_{ik}^{\alpha}$ | The $(i, k)$-coefficient function arising from $\sigma^{\alpha}$ | 114 |
| $\Phi(\ ,\ )$ | Bilinear form on $M^{\beta}$ | 297 |
| $\Phi_{\mathbf{Q}}(\ ,\ )$ | The natural bilinear form on $L_{\mathbf{Q}}^{(n)}$ | 330 |
| $\Phi^{\alpha}(\ ,\ )$ | A bilinear form on $L^{\alpha}$ | 331 |
| $\chi_n^D$ | The class function whose value at $g$ is $\chi^D(g^n)$ | 204 |
| $\chi^{\lambda/\mu}$ | $\sum_{\pi \in S_{\mathbf{N}}} \operatorname{sgn} \pi\, \xi^{\lambda - \mathrm{id} - (\mu - \mathrm{id}) \circ \pi}$ | 49 |
| $\omega^{\alpha}$ | The orthogonal form of $[\alpha]$ | 127 |
| $\Omega$ | A set | — |
| $\leqslant$ | Lexicographic order for partitions | 23 |
| $<$ | Total order on the set of standard $\alpha$-tableaux | 107 |
| $<$ | Total order on the set of tabloids | 303 |
| $\trianglelefteq$ | Dominance order for partitions | 23 |
| $\triangleleft$ | Is a neighbor of | 24 |
| $\simeq$ | Is isomorphic to | — |
| $\approx$ | Has the same irreducible constituents as | 244 |
| $\triangleq$ | Is similar to | 2 |
| $1_{S_n}$ | The identity mapping of $\{1, 2, \ldots, n\}$ | 4 |
| $\lvert\ \rvert$ | The cardinality of a set | — |
| $\langle\ \rangle$ | The group generated by | — |
| $\langle\ \rangle_F$ | The $F$-vector space spanned by | — |
| $\langle\langle\ \rangle\rangle_F$ | The $F$-vector space whose basis is | — |
| $\vdash$ | Is a (proper) partition of | 9 |
| $\vdash\!\dashv$ | Is a type of | 11 |
| $\models$ | Is an improper partition of | 15 |
| $\uparrow$ | The result of inducing a representation | — |
| $\downarrow$ | The restriction of a representation | — |
| $\#$ | Outer tensor product of representations | — |

# The Representation Theory of the Symmetric Group

CHAPTER 1

# *Symmetric Groups and Their Young Subgroups*

In this chapter we introduce much of the notation we shall use later on, and prove some basic results on symmetric groups. Many of the ideas concern partitions $\alpha$ of nonnegative integers $n$, and the corresponding Young subgroups $S_\alpha$ of the symmetric group $S_n$. Combinatorial structures such as Young diagrams, the diagram lattice, and Young tableaux are related to partitions, and they will help us in the next chapter to construct the ordinary irreducible representations of $S_n$.

## 1.1 Symmetric and Alternating Groups

Let $\Omega$ denote a set. A bijective mapping $\pi$ of $\Omega$ onto itself, for short

$$\pi : \Omega \rightarrowtail\!\!\!\!\twoheadrightarrow \Omega,$$

having the property that $\{\omega \mid \omega \in \Omega \text{ and } \pi(\omega) \neq \omega\}$ is finite, is called a *permutation* of $\Omega$. The order $|\Omega|$ of $\Omega$ is called the *degree* of the permutation $\pi$.

If both $\pi$ and $\rho$ are permutations of $\Omega$, then their composition is denoted by $\pi\rho$ and defined by

$$\forall \omega \in \Omega \quad (\pi\rho(\omega) := \pi(\rho(\omega))).$$

It is again bijective, keeps almost all the points fixed, and therefore is also a permutation.

If a set $P$ of permutations of $\Omega$ forms a group under this composition, we call $P$ a *permutation group* and say that $P$ is *acting* on $\Omega$. $|\Omega|$ is then called the *degree* of $P$.

ENCYCLOPEDIA OF MATHEMATICS and Its Applications, Gian-Carlo Rota (ed.). Vol. 16: G. D. James and A. Kerber, The Representation Theory of the Symmetric Group
ISBN 0-201-13515-9

The set of all the permutations of $\Omega$, i.e. the set

1.1.1 $S_\Omega := \{\pi \mid \pi : \Omega \rightarrowtail\!\!\!\twoheadrightarrow \Omega \text{ and } \pi(\omega) \neq \omega \text{ for finitely many } \omega \in \Omega\}$

is a permutation group, it is called the *symmetric group* on $\Omega$.

The elements of $\Omega$ are called *points*. If $\Omega' \subseteq \Omega$, we denote by $\pi[\Omega']$ its image under $\pi \in S_\Omega$:

$$\pi[\Omega'] := \{\pi(\omega') \mid \omega' \in \Omega'\}.$$

When defining permutation groups, the nature of the points on which they act is irrelevant in a sense to be described next.

Two permutation groups, say $P$ on $\Omega$ and $P'$ on $\Omega'$, i.e. subgroups of $S_\Omega$ and $S_{\Omega'}$ (for short: $P \leqslant S_\Omega$, $P' \leqslant S_{\Omega'}$) are called *similar* if and only if there exists a bijection $\varepsilon : \Omega \rightarrowtail\!\!\!\twoheadrightarrow \Omega'$ and an isomorphism $\phi : P \simeq P'$ such that the following holds:

1.1.2 $$\forall \pi \in P, \omega \in \Omega \quad (\phi(\pi)(\varepsilon(\omega)) = \varepsilon(\pi(\omega))).$$

(This means that by renaming the elements of $P$ by $\phi$ and the points of $\Omega$ by $\varepsilon$, we obtain $P'$.) If 1.1.2 holds, then we write

$$P \mathrel{\hat{=}} P'.$$

It is easy to check that two symmetric groups are similar if and only if they are of the same degree:

**1.1.3** $$S_\Omega \mathrel{\hat{=}} S_{\Omega'} \quad \Leftrightarrow \quad |\Omega| = |\Omega'|.$$

Hence it is only the degree that really matters.

This shows that if $n := |\Omega| \in \mathbb{N}_0 := \{0, 1, 2, 3, \ldots\}$, we may assume (up to similarity) that $\Omega = \mathbf{n}$, where

1.1.4 $$\mathbf{n} := \{1, \ldots, n\}.$$

(In particular, $\mathbf{0} = \varnothing$, the empty set.)

The symmetric group on $\mathbf{n}$ is denoted by $\mathrm{S}_n$:

1.1.5 $$\mathrm{S}_n := \{\pi \mid \pi : \mathbf{n} \rightarrowtail\!\!\!\twoheadrightarrow \mathbf{n}\}.$$

$\mathrm{S}_0$ consists of one element only (as does $\mathrm{S}_1$). An easy induction shows that the following is true:

**1.1.6** $$\forall n \geqslant 0 \quad (|\mathrm{S}_n| = n!).$$

A permutation $\pi \in \mathrm{S}_n$ is written down in full detail by putting the images

$\pi(i)$ in a row under the points $i \in \mathbf{n}$, say

$$\pi = \begin{pmatrix} 1 & \cdots & n \\ \pi(1) & \cdots & \pi(n) \end{pmatrix}.$$

This will sometimes be abbreviated by

$$\pi = \begin{pmatrix} i \\ \pi(i) \end{pmatrix}.$$

Hence, for example,

$$S_3 = \left\{ \begin{pmatrix} 1 & 2 & 3 \\ 1 & 2 & 3 \end{pmatrix}, \begin{pmatrix} 1 & 2 & 3 \\ 2 & 1 & 3 \end{pmatrix}, \begin{pmatrix} 1 & 2 & 3 \\ 3 & 2 & 1 \end{pmatrix}, \right.$$
$$\left. \begin{pmatrix} 1 & 2 & 3 \\ 1 & 3 & 2 \end{pmatrix}, \begin{pmatrix} 1 & 2 & 3 \\ 2 & 3 & 1 \end{pmatrix}, \begin{pmatrix} 1 & 2 & 3 \\ 3 & 1 & 2 \end{pmatrix} \right\}.$$

The points $1, \ldots, n$ which form the first row need not be written in their natural order; e.g.

$$\begin{pmatrix} 1 & 2 & 3 \\ 2 & 3 & 1 \end{pmatrix} = \begin{pmatrix} 2 & 1 & 3 \\ 3 & 2 & 1 \end{pmatrix}.$$

With this in mind, we call a permutation of the form

1.1.7 $$\begin{pmatrix} i_1 i_2 \cdots i_{r-1} & i_r & i_{r+1} \cdots i_n \\ i_2 i_3 \cdots i_r & i_1 & i_{r+1} \cdots i_n \end{pmatrix}$$

*cyclic* or a *cycle*. In order to emphasize $r$, we also call 1.1.7 an *r-cycle*.

A shorter notation for the cycle 1.1.7 is

1.1.8 $$(i_1, \ldots, i_r)(i_{r+1}) \cdots (i_n),$$

where the points which are *cyclically permuted* are put together in round brackets. For example

$$\begin{pmatrix} 1 & 2 & 3 \\ 1 & 3 & 2 \end{pmatrix} = (2,3)(1).$$

Commas which separate the points may be omitted if no confusion can arise (e.g. if $n \leqslant 10$), and 1-cycles can be left out, if it is clear which $n$ is meant. Hence we may write just

1.1.9 $$(i_1 \cdots i_r)$$

instead of 1.1.7 or 1.1.8. This cycle can also be expressed as

1.1.10 $$\left(i_1\pi(i_1)\pi^2(i_1)\cdots\pi^{r-1}(i_1)\right).$$

The identity mapping $\mathrm{id}_{\mathbf{n}}$, which consists of 1-cycles only, will be denoted by 1 or $1_{S_n}$:

$$1_{S_n}:=\begin{pmatrix}1\cdots n\\1\cdots n\end{pmatrix}=\mathrm{id}_{\mathbf{n}}.$$

Thus e.g.

$$S_3=\{1,(1\quad 2),(1\quad 3),(2\quad 3),(1\quad 2\quad 3),(1\quad 3\quad 2)\}.$$

The notation 1.1.9 for the cyclic permutation 1.1.7 is not uniquely determined, for the following is true:

**1.1.11** $$(i_1\cdots i_r)=(i_2\cdots i_r i_1)=\cdots=(i_r i_1\cdots i_{r-1}).$$

This means that a cycle which arises from 1.1.9 by cyclically permuting the points, describes the same permutation as does 1.1.9.

2-cycles, i.e. permutations which just interchange two points, are called *transpositions*. $S_3$ for example contains the transpositions (1 2), (1 3), and (2 3).

The *order* of a cycle $(i_1\cdots i_r)$, i.e. the order of the cyclic subgroup

$$\langle(i_1\cdots i_r)\rangle$$

generated by the cycle, is equal to its length:

**1.1.12** $$|\langle(i_1\cdots i_r)\rangle|=r.$$

The inverse of a cycle is easily obtained by reversing the sequence of the points:

**1.1.13** $$(i_1\cdots i_r)^{-1}=(i_r i_{r-1}\cdots i_1).$$

Two cycles $\pi=(i_1\cdots i_r)$ and $\rho=(j_1\cdots j_s)$ are called *disjoint* if the two sets of points which are not left fixed by $\pi$ and $\rho$ are disjoint sets. Disjoint cycles $\pi$ and $\rho$ are commuting permutations, i.e. $\pi\rho=\rho\pi$. Each permutation can be written as a product of pairwise disjoint cycles, e.g.

$$\begin{pmatrix}1&2&3&4&5&6&7&8\\7&5&2&6&3&4&1&8\end{pmatrix}=(1\quad 7)(2\quad 5\quad 3)(6\quad 4)(8).$$

The set of disjoint cycles is uniquely determined by the given permutation.

For a general $\pi \in S_n$ let $c(\pi)$ be the number of disjoint cyclic factors including 1-cycles, let $l_\nu$ $(1 \leqslant \nu \leqslant c(\pi))$ be their lengths, and choose for each $\nu$ an element $j_\nu$ of the $\nu$-th cyclic factor. Then

1.1.14
$$\pi = \prod_{\nu=1}^{c(\pi)} \left( j_\nu \pi(j_\nu) \cdots \pi^{l_\nu - 1}(j_\nu) \right).$$

This notation becomes uniquely determined if we choose the $j_\nu$ so that the following holds:

1.1.15
$$\begin{aligned} &\text{(i) } \forall 1 \leqslant \nu \leqslant c(\pi),\ s \in \mathbb{Z} \qquad (j_\nu \leqslant \pi^s(j_\nu)),\\ &\text{(ii) } \forall 1 \leqslant \nu < c(\pi) \quad (j_\nu < j_{\nu+1}). \end{aligned}$$

If this is true for 1.1.14, then 1.1.14 is called the *cycle notation* for $\pi$.

So much for notation. We now consider subsets which generate $S_n$. Because of

**1.1.16**
$$(i_1 \cdots i_r) = (i_1 i_2)(i_2 i_3) \cdots (i_{r-1} i_r),$$

each cycle, and therefore each $\pi \in S_n$, too, can be written as a product of transpositions. Hence $S_n$ is generated by its subset of transpositions (if no transpositions occur in $S_n$, then $n \leqslant 1$, but $S_0$ and $S_1$ are generated by $\varnothing$). But except for $n \leqslant 2$, we do not need every transposition. For if $1 \leqslant j < k < n$, we have

$$(j, k+1) = (k, k+1)(j, k)(k, k+1),$$

so that $(j, k+1)$ can be obtained from $(j, k)$ with the aid of the transposition $(k, k+1)$ of successive points. This shows that $S_n$ is generated by the transpositions $(k, k+1)$ of successive points, $1 \leqslant k < n$.

Another system of generators of $S_n$ is $\{(12), (1 \cdots n)\}$. This is true because, for $0 \leqslant r \leqslant n-2$, we have

$$(1 \cdots n)^r (12)(1 \cdots n)^{-r} = (r+1, r+2).$$

Hence we have proved the following:

**1.1.17** LEMMA.

$$\begin{aligned} S_n &= \langle (12), (23), \ldots, (n-1, n) \rangle \\ &= \langle (12), (1 \cdots n) \rangle. \end{aligned}$$

We would now like to introduce the sign of a permutation. In order to do this we define for $n \in \mathbb{N} := \{1, 2, \ldots\}$ the *difference product* $\Delta_n$ by

$$\Delta_1 := 1 \in \mathbb{Z},$$

and for $n \geqslant 2$ we put

$$\Delta_n := \prod_{1 \leqslant i < j \leqslant n} (j-i) \in \mathbb{Z}.$$

and define an action of $\pi \in S_n$ on $\pm\Delta_n$ by

$$\pi\Delta_n := \prod(\pi(j)-\pi(i)) \quad \text{and} \quad \pi(-\Delta_n) := -\pi(\Delta_n).$$

Since $\pi \in S_n$ is a bijection of **n**, we have for $n \geqslant 2$

$$\pi\Delta_n = \pm\Delta_n.$$

Defining the *sign* of $\pi$ by

1.1.18 $$\operatorname{sgn}\pi := \frac{\pi\Delta_n}{\Delta_n},$$

if $n \geqslant 2$, and putting

$$\operatorname{sgn} 1_{S_0} := \operatorname{sgn} 1_{S_1} := 1_{\mathbb{Z}},$$

we have for each permutation $\pi$

$$\operatorname{sgn}\pi \in \{1,-1\}.$$

Since for each $\pi, \rho \in S_n$ we have

$$(\pi\rho)\Delta_n = \pi(\rho\Delta_n),$$

the symmetric group $S_n$ acts on $\Omega := \{\Delta_n, -\Delta_n\}$ in such a way that for each $\omega \in \Omega$ we have $1\omega = \omega$ and $\pi(\rho\omega) = (\pi\rho)\omega$. This shows that the action defines a permutation representation of $S_n$ on $\Omega$, i.e. a homomorphism $S_n \to S_\Omega$. Since $\pi\Delta_n = -\Delta_n$ when $\pi$ is a transposition, the image of this permutation representation is $S_\Omega$ if and only if $n \geqslant 2$, while it is $\{1_{S_\Omega}\}$ if and only if $n < 2$. This proves the following lemma:

**1.1.19** LEMMA. *The mapping* $\operatorname{sgn}: \pi \mapsto \operatorname{sgn}\pi$ *is a homomorphism of* $S_n$ *into the multiplicative group* $\{1,-1\}$. *It is surjective if and only if* $n \geqslant 2$.

The representation of $S_n$ afforded by this homomorphism is called the *alternating representation*. The kernel of the homomorphism sgn is denoted by $A_n$:

1.1.20 $$A_n := \ker \operatorname{sgn} = \{\pi \mid \pi \in S_n, \operatorname{sgn}\pi = 1\}.$$

For example $A_3 = \{1, (1\ \ 2\ \ 3), (1\ \ 3\ \ 2)\}$.

This subgroup $A_n$ of $S_n$ is called the *alternating group* on **n**. The permutations $\pi \in S_n$, which are elements of $A_n$, are called *even* permutations, while the elements of $S_n \backslash A_n$ are called *odd* permutations.

The homomorphism theorem yields

**1.1.21**
$$\text{(i) } |A_0| = |A_1| = 1,$$
$$\text{(ii) } \forall n \geq 2 (|A_n| = n!/2).$$

For cycles $(i_1 \ldots i_r)$ we have because of 1.1.16

**1.1.22.** $(i_1 \ldots i_r) \in A_n \Leftrightarrow$ *$r$ is odd.*

More generally, $\pi \in S_n$ is even if and only if the number of cyclic factors of $\pi$ which are of even length is even. Thus, if $\pi \in S_n$, we have

**1.1.23.** $\pi \in A_n \Leftrightarrow n - c(\pi)$ *is even.*

For if we denote by $a_i(\pi)$ the number of $i$-cycles occurring in the cycle-notation of $\pi$, then $n = \sum_i i a_i(\pi)$, while $c(\pi) = \sum_i a_i(\pi)$, so that

$$\begin{aligned} n - c(\pi) &= \sum_i (i-1) a_i(\pi) \\ &\equiv \sum_j (j-1) a_j(\pi) \quad \text{(modulo 2)} \\ &\equiv \sum_j a_j(\pi) \qquad\qquad \text{(modulo 2)} \end{aligned}$$

if the last two sums are taken over those $j$ where $j$ is even. This yields 1.1.23 by an application of 1.1.22. We can rephrase this as follows: If $\pi \in S_n$, then

**1.1.24**
$$\operatorname{sgn} \pi = (-1)^{n - c(\pi)}.$$

Before we leave the alternating group in order to consider the conjugacy classes of $S_n$, we should make the following remark:

**1.1.25.** $A_n$ *is the commutator subgroup of* $S_n$. *Even more*: *each element of* $A_n$ *is itself a commutator.*

A commutator in $S_n$ is an element of the form $\pi\rho\pi^{-1}\rho^{-1}$. Hence, as $\operatorname{sgn} \pi = \operatorname{sgn} \pi^{-1}$ and $\operatorname{sgn} \rho = \operatorname{sgn} \rho^{-1}$ each commutator is contained in $A_n$. Therefore the commutator subgroup $S_n'$ of $S_n$, which is generated by all these elements, is also contained in $A_n$: $S_n' \leq A_n$. In order to prove $A_n \leq S_n'$ we verify the second half of the statement 1.1.25. To this end we note that for each $i$ such that $2i+1 \leqslant n$ we have

$$(1,\ldots,2i+1)=(1,\ldots,i+1)(i+1,\ldots,2i+1)$$

so that $(1,\ldots,2i+1)$ is of the form $\rho\sigma\rho^{-1}\sigma^{-1}$ where $\rho:=(1,\ldots,i+1)$ and $\sigma$ is a suitable element of $S_{2i+1}$ (apply 1.2.1). Also, for $i \leqslant j$

$$(1,\ldots,2i)(2i+1,\ldots,2i+2j)=(1,\ldots,i+j+1)(2i,i+j+1,\ldots,2i+2j),$$

so this element is a commutator in $S_{2i+2j}$. Similar remarks hold for arbitrary cycles of odd length and for each pair of disjoint cycles of even lengths.

Thus each even permutation, since it consists of cycles with odd lengths together with an even number of cycles with even lengths (so that we can pair them off), is of form $\rho\sigma\rho^{-1}\sigma^{-1}$ and is therefore a commutator. This completes the proof of 1.1.25.

Since for any finite group $G$ and its commutator subgroup $G'$, the commutator factor $G/G'$ is isomorphic to the group of one-dimensional characters of $G$ over $\mathbb{C}$, we get as an immediate corollary of 1.1.25:

**1.1.26.** *The only homomorphisms of* $S_n$ *into the multiplicative group of* $\mathbb{C}$ *are* $\pi \mapsto 1_{\mathbb{C}}$ *and* $\pi \mapsto \operatorname{sgn}\pi$.

## 1.2 The Conjugacy Classes of Symmetric and Alternating Groups

We shall describe the conjugacy classes of $S_n$. In order to do this, we first of all note how $\rho\pi\rho^{-1}$ is obtained from $\pi$. Since

**1.2.1** $$\rho\pi\rho^{-1}=\begin{pmatrix} i \\ \rho(i) \end{pmatrix}\begin{pmatrix} i \\ \pi(i) \end{pmatrix}\begin{pmatrix} \rho(i) \\ i \end{pmatrix}=\begin{pmatrix} \rho(i) \\ \rho\pi(i) \end{pmatrix},$$

we get $\rho\pi\rho^{-1}$ from $\pi$ by an application of $\rho$ to the points in the cyclic factors of the same $\pi$. For if

$$\pi=\cdots(\cdots i\pi(i)\cdots)\cdots,$$

then by 1.2.1 we have

$$\rho\pi\rho^{-1}=\cdots(\cdots\rho(i)\rho\pi(i)\cdots)\cdots.$$

We notice that under this process of applying $\rho$ to the points, the brackets remain invariant, so that the cyclic factors of $\rho\pi\rho^{-1}$ (in cycle notation) are of the same lengths as those of $\pi$.

On the other hand, let $\pi$ and $\sigma$ be permutations which are both products of $c(\pi)$ cyclic factors of the same lengths $l_\nu$, $1 \leqslant \nu \leqslant c(\pi)$, say

$$\pi = \prod_{\nu=1}^{c(\pi)} \left( j_\nu \cdots \pi^{l_\nu - 1}(j_\nu) \right),$$

while

$$\sigma = \prod_{\nu=1}^{c(\pi)} \left( i_\nu \cdots \sigma^{l_\nu - 1}(i_\nu) \right).$$

We now put

$$\rho := \begin{pmatrix} \cdots j_\nu \pi(j_\nu) \cdots \pi^{l_\nu - 1}(j_\nu) \cdots \\ \cdots i_\nu \sigma(i_\nu) \cdots \sigma^{l_\nu - 1}(i_\nu) \cdots \end{pmatrix}.$$

Then by 1.2.1 we obtain

$$\sigma = \rho \pi \rho^{-1}.$$

This shows that two permutations are conjugate if and only if they have the same cycle structure.

In order to make this more precise, we introduce the notion of a partition of $n$. A sequence of nonnegative integers

$$\alpha = (\alpha_1, \alpha_2, \ldots)$$

is called a (*proper*) *partition* of $n$ if and only if it satisfies

1.2.2

$$\text{(i)}\ \forall i \geqslant 1 \quad (\alpha_i \geqslant \alpha_{i+1}),$$

$$\text{(ii)}\ \sum_{i=1}^{\infty} \alpha_i = n.$$

The $\alpha_i$ are called the *parts* of $\alpha$. The fact that $\alpha$ is a partition of $n$ is abbreviated by

$$\alpha \vdash n.$$

If $\alpha \vdash n$, then by 1.2.2 (ii), there is an $h$ such that $\alpha_i = 0$ for all $i > h$. We may take the liberty of shortening $\alpha$ as follows:

$$\alpha = (\alpha_1, \ldots, \alpha_h)$$

(normally, we choose $h$ such that $\alpha_h > 0$, $\alpha_{h+1} = 0$). We list below the partitions of the first few nonnegative integers, using this convention:

$$\begin{array}{ll} n=0 & (0) \\ n=1 & (1) \\ n=2 & (2),(1,1) \\ n=3 & (3),(2,1),(1,1,1) \\ n=4 & (4),(3,1),(2,2),(2,1,1),(1,1,1,1). \end{array}$$

The number $p(n)$ of partitions of $n$ grows rapidly with $n$. E.g.

$$p(0)=1,\ p(1)=1,\ p(4)=5,\ p(10)=42,$$
$$p(20)=627,\ p(50)=204226,\ p(100)=190569292.$$

A table for $p(n)$ $(n \leqslant 100)$ can be found in the book of G. E. Andrews [1976].

The following notation is useful in the case when several nonzero parts of $\alpha$ are equal, say $a_i$ parts are equal to $i$, $1 \leqslant i \leqslant n$:

$$\alpha = \left(n^{a_n}, (n-1)^{a_{n-1}}, \ldots, 1^{a_1}\right).$$

If $a_i = 0$, then $i^{a_i}$ is usually left out. For example

$$(3,2,1^2) = (3,2,1,1,0,0,\ldots).$$

If now $\pi$ is an element of $S_n$, then the ordered lengths $\alpha_i(\pi)$, $1 \leqslant i \leqslant c(\pi)$, of the cyclic factors of $\pi$ in cycle notation form a uniquely determined partition of $n$, which we call the *cycle partition* of $\pi$, and which we denote by $\alpha(\pi)$:

1.2.3 $$\alpha(\pi) := \left(\alpha_1(\pi), \ldots, \alpha_{c(\pi)}(\pi)\right).$$

The corresponding $n$-tuple consisting of the multiplicities of parts of $\alpha(\pi)$, i.e.

1.2.4 $$a(\pi) := (a_1(\pi), \ldots, a_n(\pi))$$

is called the *cycle type* of $\pi$.

Correspondingly we call $a := (a_1, \ldots, a_n)$ a *type* of $n$ if and only if

1.2.5 $$\begin{array}{ll} \text{(i)} & \forall 1 \leqslant i \leqslant n \quad (a_i \in \mathbb{N}_0), \\ \text{(ii)} & \sum_i i a_i = n. \end{array}$$

This will be abbreviated by

$$a \vdash\!\dashv n.$$

At the beginning of this section we showed that two elements $\pi$ and $\rho$ of $S_n$ are conjugate if and only if they have the same cycle structure. Using the notion of cycle partition and cycle type and denoting by

$$C^S(\pi)$$

the conjugacy class of $\pi \in S_n$, this reads as follows:

**1.2.6 Lemma.** $C^S(\pi)=C^S(\sigma) \Leftrightarrow \alpha(\pi)=\alpha(\sigma) \Leftrightarrow a(\pi)=a(\sigma)$.

From

$$(i_1 \cdots i_r)^{-1}=(i_r \cdots i_1)$$

we obtain

**1.2.7** $$\alpha(\pi)=\alpha(\pi^{-1}),$$

for each $\pi \in S_n$. This together with 1.2.6 shows that every permutation in $S_n$ is conjugate to its inverse. Groups in which each element is a conjugate of its inverse are called *ambivalent*, so that we have proved the following

**1.2.8 Lemma.** $S_n$ *is ambivalent.*

This is not true for the alternating groups. $A_0$, $A_1$, and $A_2$ are ambivalent, but $A_3$ is not ambivalent. There are, in fact, only a very few alternating groups which are ambivalent. We would like to determine these ambivalent alternating groups, for ambivalent groups have a real character table. This problem leads us to a description of the conjugacy classes of $A_n$.

Since $A_n$ is a normal subgroup of $S_n$, a conjugacy class $C^S$ of $S_n$ is either contained in $A_n$, or contained in $S_n \backslash A_n$. 1.1.22 helps us to determine whether $C^S$ is contained in $A_n$ or not, and hence it remains to check whether the $S_n$*-class* $C^S \subseteq A_n$ is also an $A_n$*-class* or splits into several $A_n$-classes. Now the order of a conjugacy class $C^S(\pi)$ is equal to the index of the centralizer $C_S(\pi)$ of $\pi$ in $S_n$. Since for the centralizer $C_A(\pi)$ of $\pi$ in $A_n$ we have

1.2.9 $$C_A(\pi)=C_S(\pi)\cap A_n,$$

which is a subgroup of index $\leqslant 2$ in $C_S(\pi)$, either $C^S(\pi) \subseteq A_n$ is an $A_n$-class itself, i.e.

$$C^S(\pi)=C^A(\pi),$$

or $C^S(\pi)$ splits into exactly two $A_n$-classes of the same order $|C^S(\pi)|/2$. Furthermore $C^S(\pi)$ splits (for $n>1$) if and only if $C_A(\pi)=C_S(\pi)$, i.e. if and only if $C_S(\pi)$ contains no odd permutations.

$C_S(\pi)$ contains odd permutations if $\pi$ contains a cyclic factor of even length (which is then an odd element of $C_S(\pi)$) or if $\pi$ contains two cyclic factors of the same odd length, say

$$\pi=\cdots(i_\nu\cdots\pi^r(i_\nu))(i_\mu\cdots\pi^r(i_\mu)),$$

so that

$$\rho:=(i_\nu i_\mu)(\pi(i_\nu),\pi(i_\mu))\cdots(\pi^r(i_\nu),\pi^r(i_\mu))\in C_S(\pi)\backslash A_n.$$

Conversely, if $\pi$ consists of cyclic factors of pairwise different odd lengths, then by 1.2.1 $C_S(\pi)$ is generated by these cyclic factors, and hence is contained in $A_n$.

This proves the following to be true:

**1.2.10** LEMMA. *$C^S(\pi)$ splits into two $A_n$-classes of equal order if and only if $n>1$ and the nonzero parts of $\alpha(\pi)$ are pairwise different and odd. In all other cases $C^S(\pi)$ does not split.*

We denote the conjugacy class of elements with cycle partition $\alpha$ by

$$C^\alpha.$$

If this class splits into two conjugacy classes of $A_n$, we denote these by

$$C^{\alpha\pm},$$

where the notation is fixed by assuming that

1.2.11 $$(1\cdots\alpha_1)(\alpha_1+1,\ldots,\alpha_1+\alpha_2)\cdots(\cdots n)\in C^{\alpha+}.$$

In order to complete the classification of ambivalent alternating groups, we must determine which $C^{\alpha\pm}$ are ambivalent, i.e. contain the inverse of each of their elements.

**1.2.12** LEMMA. *$A_0$, $A_1$, $A_2$, $A_5$, $A_6$, $A_{10}$, and $A_{14}$ are the only ambivalent alternating groups.*

*Proof.* Let

$$\pi=(i_1\cdots i_r)\cdots(j_1\cdots j_s)$$

be a product of disjoint cycles of odd lengths in cycle notation. We can form

$$\xi := (i_2 i_r)(i_3 i_{r-1}) \cdots (j_2 j_s)(j_3 j_{s-1}) \cdots,$$

and call this permutation the *standard conjugator*. It satisfies

$$\xi \pi \xi^{-1} = \pi^{-1}.$$

(i) We notice first that $\xi$ is even if and only if the number of cyclic factors of $\pi$ whose lengths are congruent 3 modulo 4 is even.

(ii) If the standard-conjugator is even for each element in a splitting class, then $A_n$ is ambivalent.

The only splitting classes of $A_n$, $n \in \{0,1,2,5,6,10,14\}$, are the classes corresponding to the partitions

$$(5) \vdash 5, \quad (5,1) \vdash 6, \quad (7,3) \vdash 10, \quad (9,1) \vdash 10,$$
$$(13,1) \vdash 14, \quad (11,3) \vdash 14, \quad (9,5) \vdash 14.$$

In each of them the number of components congruent to 3 modulo 4 is 0 or 2, so this number is even. Hence the standard conjugator is even in each case, by (i). This shows that the groups mentioned in the statement are in fact ambivalent.

(iii) If the standard conjugator $\xi$ of $\pi$ is odd, then the $A_n$-class $C^A(\pi)$ in question is not ambivalent. For if $\sigma \in A_n$ satisfies $\sigma \pi \sigma^{-1} = \pi^{-1}$, then

$$(\sigma^{-1}\xi)\pi(\sigma^{-1}\xi)^{-1} = \pi,$$

which contradicts $C_S(\pi) = C_A(\pi)$, since $\sigma^{-1}\xi \in S_n \backslash A_n$.

(iv) It remains to show that for each $n \notin \{0,1,2,5,6,10,14\}$ there are partitions $\alpha$ with pairwise different and odd parts $\alpha_i \neq 0$ such that the number of $\alpha_i$ satisfying $\alpha_i \equiv 3$ (4) is odd.

(a) $n = 4k$, $k \in \mathbb{N}$: $(4k-1, 1)$ has the desired properties.
(b) $n = 4k+1$, $2 \leq k \in \mathbb{N}$: $(4k-3, 3, 1)$ satisfies the conditions.
(c) $n = 4k+2$, $4 \leq k \in \mathbb{N}$: Consider $(4(k-1)-3, 5, 3, 1)$.
(d) $n = 4k+3$, $k \in \mathbb{N}_0$: Take $(4k+3)$. ■

**1.2.13** COROLLARY. *The $A_n$-classes $C^{\alpha\pm}$ are ambivalent if and only if the number of parts $\alpha_i$ of $\alpha$ with the property $\alpha_i \equiv 3$ (4) is even.*

We conclude this section with an evaluation of the order of $\pi \in S_n$ and of its conjugacy class $C^S(\pi)$.

If in cycle notation

$$\pi = \prod_\nu \left( j_\nu \cdots \pi^{l_\nu - 1}(j_\nu)\right)$$

and if $r \in \mathbb{N}$, then since disjoint cycles commute, we have

$$\pi^r = \prod_\nu \left( j_\nu \cdots \pi^{l_\nu - 1}(j_\nu)\right)^r,$$

so that the order of $\pi$, i.e. the order $|\langle\pi\rangle|$ of the cyclic subgroup $\langle\pi\rangle$ of $S_n$ which is generated by $\pi$, is the least common multiple of the lengths $l_\nu$ of the cyclic factors of $\pi$. In terms of the cycle partition $\alpha(\pi)$ of $\pi$ this reads

**1.2.14** $$|\langle\pi\rangle| = \operatorname{lcm}\{\alpha_i(\pi) \mid 1 \leqslant i \leqslant c(\pi)\}.$$

The conjugacy class $C^\alpha$ which consists of the elements with cycle-partition $\alpha$ and cycle-type $a$ is obtained as follows. Its elements arise from a system of $a_i$ empty cycles of length $i$, $1 \leqslant i \leqslant n$, i.e. from

$$\underbrace{\ldots(\cdot\ldots\cdot)\ldots(\cdot\ldots\cdot)\ldots}_{\substack{|\leftarrow i \rightarrow| \quad |\leftarrow i \rightarrow| \\ a_i \text{ cycles}}}$$

by inserting the points $1, \ldots, n$ in all the $n!$ possible ways. Since cyclic permutation of points does not change a cycle and since disjoint cycles commute, each element of $C^\alpha$ arises just $\prod_i i^{a_i} a_i!$ times in this process.

This proves that for $\pi \in S_n$:

**1.2.15 Lemma.**
(i) $|C^S(\pi)| = n!/\prod_i i^{a_i(\pi)} a_i(\pi)!$,
(ii) $|C_S(\pi)| = \prod_i i^{a_i(\pi)} a_i(\pi)!$.

We postpone a detailed description of the centralizer $C_S(\pi)$ of $\pi$, since it requires the concept of wreath product.

The preceding results on conjugacy classes allow us to draw the first conclusions concerning the representation theory of $S_n$ and $A_n$. 1.2.8 for example implies that all the ordinary characters of $S_n$ are real-valued, while by 1.2.12 the following is true for alternating groups:

**1.2.16 Theorem.** $A_0$, $A_1$, $A_2$, $A_5$, $A_6$, $A_{10}$, *and* $A_{14}$ *are the only alternating groups, the character table of which is a matrix over the real number field* $\mathbb{R}$.

But the character table of $S_n$ is even a matrix over the ring $\mathbb{Z}$ of integers. For there exists a general theorem (see e.g. Huppert [1967, I, Chapter V, (13.7)]) which says that if $G$ is a finite group and for every $g \in G$ and $m \in \mathbb{N}$ which satisfies $(m, |G|)=1$, $g^m$ is conjugate to $g$, then the character table of $G$ is a matrix over $\mathbb{Z}$. $S_n$ satisfies the conditions of this theorem; it is clear that for every $\pi \in S_n$ and each $m \in \mathbb{N}$ which satisfies $(m, n!)=1$,

$$\alpha(\pi^m)=\alpha(\pi),$$

since the $m$th power of each $k$-cycle, $k \leqslant n$, is a $k$-cycle again. Thus we obtain

**1.2.17** THEOREM. *The character table of* $S_n$ *is a matrix over* $\mathbb{Z}$, *for each* $n \in \mathbb{N}_0$.

This follows also from the fact that there exists a $\mathbb{Z}$-basis of permutation characters for the character ring of $S_n$. These permutation characters are induced from the so-called Young subgroups of $S_n$, which will be introduced next.

## 1.3 Young Subgroups of $S_n$ and Their Double Cosets

We now introduce the so-called Young subgroups of $S_n$. These subgroups are named after Alfred Young (1873–1940), to whom we are indebted for large parts of the ordinary representation theory of $S_n$. The permutation representations induced from Young subgroups play a vital role in both the ordinary and the modular theory of $S_n$.

We have seen that the conjugacy classes of $S_n$ are indexed by partitions, so it is to be expected that subgroups of $S_n$ useful in representation theory should also be related to partitions. Indeed, we shall obtain a Young subgroup of $S_n$ for each partition of $n$, but we want to define a wider class of Young subgroups. In order to do this we first introduce a concept which generalizes the notion of a partition.

A sequence of nonnegative integers

$$\lambda=(\lambda_1, \lambda_2, \ldots)$$

is called an *improper partition* of $n$ if and only if

1.3.1 $$\sum_{i=1}^{\infty} \lambda_i = n.$$

The fact that $\lambda$ is an improper partition of $n$ will be abbreviated by

$$\lambda \models n.$$

Note that we require the parts of $\lambda$ to belong to $\mathbb{N}_0$. Furthermore, 1.3.1 tells us that $\lambda_i$ is eventually zero. We adopt the convention of writing

$$\lambda=(\lambda_1,\lambda_2,\ldots,\lambda_h)\vDash n$$

when $\lambda_i=0$ for all $i>h$.

The only difference between a proper and an improper partition is that we do not require the parts of an improper partition to be nonincreasing.

By $|\lambda|$ we indicate the natural number of which $\lambda$ is a proper or improper partition:

$$|\lambda|:=\sum_{i=1}^{\infty}\lambda_i.$$

Let $\lambda=(\lambda_1,\lambda_2,\ldots)$ be such an improper partition of $n$, and let

$$\mathbf{n}_i^\lambda,\qquad 1\leqslant i,$$

denote subsets of **n** which are pairwise disjoint, and which satisfy

$$\forall 1\leqslant i\quad \left(|\mathbf{n}_i^\lambda|=\lambda_i\right),$$

so that we have the following *dissection* of **n**:

$$\mathbf{n}=\dot{\bigcup_{i=1}^{\infty}}\mathbf{n}_i^\lambda.$$

If now $S_i^\lambda$ denotes the subgroup of $S_n$ consisting of the $\lambda_i!$ elements which leave each point of $\mathbf{n}\backslash\mathbf{n}_i^\lambda$ fixed, then the product

1.3.2 $$S_\lambda:=\prod_{i=1}^{\infty}S_i^\lambda$$

is a subgroup of $S_n$, which is isomorphic to the (finite) direct product

$$S_{\lambda_1}\times S_{\lambda_2}\times\cdots.$$

$S_\lambda$ is called a *Young subgroup* corresponding to $\lambda\vDash n$ or, more explicitly, the Young subgroup corresponding to $\mathbf{n}^\lambda:=\{\mathbf{n}_1^\lambda,\mathbf{n}_2^\lambda,\ldots\}$. It is important, and clear, that all the subgroups $S_\lambda$ which correspond to $\lambda$ are conjugate to each other in $S_n$. Special cases are the Young subgroups $S_\alpha$ which correspond to a proper partition $\alpha\vdash n$. Obviously each $S_\lambda$, $\lambda\vDash n$, is of the form $S_\alpha$, $\alpha\vdash n$.

We shall be interested in representations of $S_n$ which are induced by certain one-dimensional representations of Young subgroups of $S_n$.

If $F$ denotes a field, then there are two trivial one-dimensional representations of $S_\lambda$, $\lambda \models n$, over $F$, at hand. The first one is the *identity representation* $IS_\lambda$ of $S_\lambda$. Its representation space is the one-dimensional vector space $F_1$ over $F$, where each element $\pi \in S_\lambda$ is mapped onto the identity element $\mathrm{id}_{F_1}$ of the general linear group $\mathrm{GL}(F_1)$ of $F_1$:

1.3.3
$$IS_\lambda : S_\lambda \to \mathrm{GL}(F_1) : \pi \mapsto \mathrm{id}_{F_1}.$$

The second trivial one-dimensional representation of $S_\lambda$, $\lambda \models n$, over $F$ is the *alternating representation* $AS_\lambda$ of $S_\lambda$, where $\pi$ is mapped onto $\pm \mathrm{id}_{F_1}$, depending on the sign of $\pi$:

1.3.4
$$AS_\lambda : S_\lambda \to \mathrm{GL}(F_1) : \pi \mapsto \mathrm{sgn}\,\pi \cdot \mathrm{id}_{F_1}.$$

If $\mu$ is another improper partition of $n$, then we can form $IS_\lambda$, $IS_\mu$, and $AS_\mu$ and induce these representations into $S_n$, obtaining the representations

$$IS_\lambda \uparrow S_n, \quad IS_\mu \uparrow S_n, \quad AS_\mu \uparrow S_n$$

of $S_n$.

We would like to evaluate the intertwining numbers

$$i(IS_\lambda \uparrow S_n, IS_\mu \uparrow S_n) \quad \text{and} \quad i(IS_\lambda \uparrow S_n, AS_\mu \uparrow S_n),$$

in order to get an idea about common irreducible constituents of these induced representations of $S_n$.

Denoting by $\downarrow$ the restriction of representations to subgroups, Mackey's intertwining number theorem (cf. Curtis and Reiner [1962, (44.5)]) tells us the following:

**1.3.5**
$$\begin{aligned} i(IS_\lambda \uparrow S_n, IS_\mu \uparrow S_n) &= \sum_{S_\lambda \pi S_\mu} i\left(IS_\lambda \downarrow S_\lambda \cap \pi S_\mu \pi^{-1}, IS_\mu^{(\pi)} \downarrow S_\lambda \cap \pi S_\mu \pi^{-1}\right) \\ &= \sum_{S_\lambda \pi S_\mu} i\left(I\left(S_\lambda \cap \pi S_\mu \pi^{-1}\right), I\left(S_\lambda \cap \pi S_\mu \pi^{-1}\right)\right) \\ &= \sum_{S_\lambda \pi S_\mu} 1. \end{aligned}$$

The sum has to be taken over the double cosets $S_\lambda \pi S_\mu$ of $S_\lambda$ and $S_\mu$ in $S_n$ and $IS_\mu^{(\pi)}(\pi\rho\pi^{-1}) := IS_\mu(\rho)$, $\rho \in S_\mu$. This shows that the intertwining number of $IS_\lambda \uparrow S_n$ and $IS_\mu \uparrow S_n$ is equal to the number of double cosets $S_\lambda \pi S_\mu$ of $S_\lambda$ and $S_\mu$ in $S_n$.

For the second intertwining number we obtain analogously

**1.3.6**
$$i(IS_\lambda \uparrow S_n, AS_\mu \uparrow S_n) = \sum_{S_\lambda \pi S_\mu} i\left(I\left(S_\lambda \cap \pi S_\mu \pi^{-1}\right), A\left(S_\lambda \cap \pi S_\mu \pi^{-1}\right)\right).$$

Now both $I(S_\lambda \cap \pi S_\mu \pi^{-1})$ and $A(S_\lambda \cap \pi S_\mu \pi^{-1})$, the identity representation and the alternating representation of the intersection $S_\lambda \cap \pi S_\mu \pi^{-1}$, are irreducible. Since the intersection is a direct product of symmetric groups, the two irreducible representations are equal if and only if $S_\lambda \cap \pi S_\mu \pi^{-1}$ is trivial or the characteristic char $F$ is equal to 2. Therefore

$$\textbf{1.3.7} \qquad i(IS_\lambda \uparrow S_n, AS_\mu \uparrow S_n) = \begin{cases} \sum\limits_{S_\lambda \pi S_\mu} 1 & \text{if } \operatorname{char} F = 2, \\ \sum\limits_{\substack{S_\lambda \pi S_\mu \\ S_\lambda \cap \pi S_\mu \pi^{-1} = \{1\}}} 1 & \text{otherwise.} \end{cases}$$

1.3.5 and 1.3.7 show that the desired intertwining numbers can be expressed as numbers of specific double cosets of Young subgroups. This suggests that we should examine more closely the double cosets $S_\lambda \pi S_\mu$. The following lemma will turn out to be crucial in this context:

**1.3.8** LEMMA. $\rho \in S_\lambda \pi S_\mu \Leftrightarrow \forall i, k\ (|\mathbf{n}_i^\lambda \cap \pi[\mathbf{n}_k^\mu]| = |\mathbf{n}_i^\lambda \cap \rho[\mathbf{n}_k^\mu]|)$.

*Proof.*

(i) $\Rightarrow$: If $\rho \in S_\lambda \pi S_\mu$, say $\rho = \sigma\pi\tau$, $\sigma \in S_\lambda$, $\tau \in S_\mu$, then for each $k$,

$$\rho[\mathbf{n}_k^\mu] = \sigma\pi\tau[\mathbf{n}_k^\mu] = \sigma\pi[\mathbf{n}_k^\mu],$$

so that for each $i$ and $k$ we have:

$$\mathbf{n}_i^\lambda \cap \rho[\mathbf{n}_k^\mu] = \mathbf{n}_i^\lambda \cap \sigma\pi[\mathbf{n}_k^\mu] = \sigma[\mathbf{n}_i^\lambda \cap \pi[\mathbf{n}_k^\mu]].$$

This yields the statement, since $\sigma$ is a bijection.

(ii) $\Leftarrow$: The assumption

$$\forall i, k \quad (|\mathbf{n}_i^\lambda \cap \pi[\mathbf{n}_k^\mu]| = |\mathbf{n}_i^\lambda \cap \rho[\mathbf{n}_k^\mu]|)$$

implies that for a fixed $i$ the subsets $\mathbf{n}_i^\lambda \cap \pi[\mathbf{n}_k^\mu]$ and the subsets $\mathbf{n}_i^\lambda \cap \rho[\mathbf{n}_k^\mu]$ form two dissections of $\mathbf{n}_i^\lambda$ into subsets which can be collected into pairs

$$(\mathbf{n}_i^\lambda \cap \pi[\mathbf{n}_k^\mu], \mathbf{n}_i^\lambda \cap \rho[\mathbf{n}_k^\mu]), \qquad 1 \leqslant k,$$

of subsets of equal order. Hence for each $i$ there exist $\sigma_i \in S_i^\lambda$, which satisfy

$$\forall k \quad (\sigma_i[\mathbf{n}_i^\lambda \cap \pi[\mathbf{n}_k^\mu]] = \mathbf{n}_i^\lambda \cap \rho[\mathbf{n}_k^\mu]).$$

The product $\sigma := \sigma_1\sigma_2 \cdots \in S_\lambda$ of such permutations $\sigma_i$ then satisfies the equations

$$\forall k \quad (\sigma\pi[\mathbf{n}_k^\mu] = \rho[\mathbf{n}_k^\mu]).$$

Thus there exist $\tau \in S_\mu$ such that $\rho = \sigma\pi\tau$, as stated. ∎

This shows that the double coset $S_\lambda \pi S_\mu$ is characterized by the numbers

$$z_{ik} := |\mathbf{n}_i^\lambda \cap \pi[\mathbf{n}_k^\mu]|, \qquad 1 \leq i, k.$$

Now assume that $\lambda_i = \mu_i = 0$ if $i > n$. Then we have an injective mapping

$$f: S_\lambda \pi S_\mu \mapsto (z_{ik})$$

from the set of double cosets $S_\lambda \pi S_\mu$ into the set of all the $n \times n$ matrices over $\mathbb{N}_0$. It is clear that the image of $f$ is exactly the set of $n \times n$ matrices $(z_{ik})$ which satisfy

1.3.9

$$\text{(i)}\ \forall 1 \leq i, k \leq n \quad (z_{ik} \in \mathbb{N}_0),$$

$$\text{(ii)}\ \sum_{i=1}^{n} z_{ik} = \mu_k \quad \text{and} \quad \sum_{k=1}^{n} z_{ik} = \lambda_i.$$

Summarizing, we have proved the following important result:

**1.3.10** THEOREM. *If $\lambda = (\lambda_1, \ldots, \lambda_n)$ and $\mu = (\mu_1, \ldots, \mu_n)$ are improper partitions of $n$ with corresponding Young subgroups $S_\lambda$ and $S_\mu$, then the mapping*

$$f: S_\lambda \pi S_\mu \mapsto (z_{ik} := |\mathbf{n}_i^\lambda \cap \pi[\mathbf{n}_k^\mu]|)$$

*establishes a bijection between the set of double cosets of $S_\lambda$ and $S_\mu$ in $S_n$ and the set of $n \times n$ matrices $(z_{ik})$ over $\mathbb{N}_0$ which satisfy*

$$\sum_{i=1}^{n} z_{ik} = \mu_k \quad \textit{and} \quad \sum_{k=1}^{n} z_{ik} = \lambda_i.$$

For the number of these double cosets we obtain in particular:

**1.3.11** COROLLARY. *The number of double cosets $S_\lambda \pi S_\mu$ is equal to the number of $n \times n$ matrices over $\mathbb{N}_0$ with prescribed vector $\lambda = (\lambda_1, \ldots, \lambda_n)$ as vector of row sums and prescribed vector $\mu = (\mu_1, \ldots, \mu_n)$ as vector of column sums.*

If we restrict our attention to double cosets $S_\lambda \pi S_\mu$ with the *trivial intersection property*

1.3.12 $$S_\lambda \cap \pi S_\mu \pi^{-1} = \{1\},$$

and restrict $f$ to this subset, we obtain as the image of this restriction the set of $n \times n$ 0-1 matrices with $\lambda$ as row-sum vector and $\mu$ as column-sum vector:

**1.3.13** COROLLARY. *The number of double cosets $S_\lambda \pi S_\mu$ with the trivial intersection property* 1.3.12 *is equal to the number of $n \times n$* 0-1 *matrices with row sums $\lambda_i$ and column sums $\mu_k$.*

For numerical purposes it is useful to notice that the number of $n \times n$ matrices over $\mathbb{N}_0$ with row sums $\lambda_i$ and column sums $\mu_k$ is equal to the coefficient of

$$x^\lambda y^\mu := x_1^{\lambda_1} \cdots x_n^{\lambda_n} y_1^{\mu_1} \cdots y_n^{\mu_n}$$

in the formal power series

$$\prod_{i,k \in \mathbb{N}} (1 - x_i y_k)^{-1} := \prod_{i,k} \left( \sum_{\nu=0}^{\infty} x_i^\nu y_k^\nu \right),$$

while the number of $n \times n$ 0-1 matrices with row sums $\lambda_i$ and column sums $\mu_k$ is equal to the coefficient of $x^\lambda y^\mu$ in the formal power series

$$\prod_{i,k} (1 + x_i y_k).$$

This yields

**1.3.14** COROLLARY. *The number of double cosets $S_\lambda \pi S_\mu$ is equal to the coefficient of $x^\lambda y^\mu$ in*

$$\prod_{i,k} (1 - x_i y_k)^{-1},$$

*while the number of such double cosets with the trivial intersection property is equal to the coefficient of $x^\lambda y^\mu$ in*

$$\prod_{i,k} (1 + x_i y_k).$$

If we take for example $\lambda := (2,1,0)$ and $\mu := (1,1,1)$, then we have to consider

$$(1-x_1y_1)^{-1}(1-x_1y_2)^{-1}(1-x_1y_3)^{-1}(1-x_2y_1)^{-1}(1-x_2y_2)^{-1}\cdots$$
$$\cdots(1-x_3y_3)^{-1}=(1+x_1y_1+\cdots)(1+x_1y_2+\cdots)\cdots(1+x_2y_3+\cdots)\cdots$$
$$=1+3x_1^2x_2y_1y_2y_3+\cdots.$$

Hence there exist just 3 double-cosets $S_{(2,1,0)}\pi S_{(1,1,1)}$ in $S_3$ (which is of course trivial from $S_{(1,1,1)}=\{1\}$, but we wanted to demonstrate the numerical method of examining a generating function). An application of 1.3.5 yields

$$i(IS_{(2,1,0)}\uparrow S_3, IS_{(1,1,1)}\uparrow S_3)=3.$$

This shows how the formal power series

$$\prod_{i,k}(1-x_iy_k)^{-1} \quad \text{and} \quad \prod_{i,k}(1+x_iy_k)$$

enable us to evaluate the intertwining numbers

$$i(IS_\lambda\uparrow S_n, IS_\mu\uparrow S_n) \quad \text{and} \quad i(IS_\lambda\uparrow S_n, AS_\mu\uparrow S_n)$$

of representations of $S_n$ which are induced from one-dimensional representations of Young-subgroups. Since intertwining numbers can often be interpreted in terms of the multiplicities of irreducible constituents common to the representations involved, such a method for calculating intertwining numbers is very useful.

## 1.4 The Diagram Lattice

In the last section we considered pairs of representations $IS_\lambda\uparrow S_n$ and $AS_\mu\uparrow S_n$ of $S_n$ and expressed their intertwining numbers $i(IS_\lambda\uparrow S_n, AS_\mu\uparrow S_n)$ in terms of numbers of double cosets and of numbers of 0-1 matrices.

In order to apply these results we shall now take for $\lambda$ and $\mu$ specific pairs of partitions $\alpha$ and $\beta$ of $n$. It is our aim to show that certain pairs $(\alpha,\beta)$ of partitions of $n$ have the property

1.4.1 $$i(IS_\alpha\uparrow S_n, AS_\beta\uparrow S_n)=1.$$

In the case when the characteristic of the groundfield $F$ does not divide $n!$, 1.4.1 means that these two induced representations have a uniquely determined irreducible constituent in common, and that this constituent is contained in both of the induced representations with multiplicity 1. It will in fact turn out that we can obtain in this way a complete system of ordinary irreducible representations of $S_n$.

In order to do this we keep the partition

$$\alpha=(\alpha_1,\ldots,\alpha_h)\vdash n$$

fixed. This partition $\alpha$ can be illustrated by the corresponding *Young diagram* $[\alpha]$, which consists of $n$ *nodes* $\times$ placed in rows. The $i$th row of $[\alpha]$ consists of $\alpha_i$ nodes, $1\leqslant i$, and all the rows start in the same column:

1.4.2
$$[\alpha]:=\begin{array}{llll} \times\ \times\ \ldots\ldots\ldots\ \times & \alpha_1 \text{ nodes} \\ \times\ \times\ \ldots\ldots\ \times & \alpha_2 \text{ nodes} \\ \quad\ \vdots & \quad\vdots \\ \times\ \times\ \ldots\ \times & \alpha_h \text{ nodes} \end{array}$$

The partition $\alpha:=(3,2,1^2)$ for example can be visualized by

$$[3,2,1^2]=\begin{array}{lll} \times & \times & \times \\ \times & \times & \\ \times & & \\ \times & & \end{array}$$

(we write $[3,2,1^2]$ instead of $[(3,2,1^2)]$).

Recalling that $\alpha_i\geqslant\alpha_{i+1}$, $1\leqslant i$, we see that the lengths $\alpha_i'$ of the columns of $[\alpha]$ form another partition $\alpha'$ of $n$:

1.4.3
$$\alpha':=(\alpha_1',\alpha_2',\ldots),\qquad \text{where}\quad \alpha_i':=\sum_{\substack{j\\ \alpha_j\geqslant i}} 1.$$

This partition $\alpha'$ is called the *partition associated with* $\alpha$. Correspondingly $[\alpha']$ is called the *Young diagram associated with* $[\alpha]$. $[\alpha']$ arises from $[\alpha]$ by simply interchanging rows and columns, i.e. by reflecting $[\alpha]$ in its main diagonal: e.g.

$$[3,2,1^2]=\begin{array}{lll} \times & \times & \times \\ \times & \times & \\ \times & & \\ \times & & \end{array} \quad \text{yields}\quad \big[(3,2,1^2)'\big]=\begin{array}{llll} \times & \times & \times & \times \\ \times & \times & & \\ \times & & & \end{array}=[4,2,1].$$

Partitions $\alpha$ and Young diagrams $[\alpha]$ where $\alpha=\alpha'$ are called *self-associated*.

We aim to characterize the partitions $\beta\vdash n$ which satisfy for a given $\alpha\vdash n$ the inequality

$$i\big(IS_\alpha\uparrow S_n,\ AS_{\beta'}\uparrow S_n\big)\neq 0.$$

This can be done in terms of a certain partial order on the set

1.4.4 $$P(n) := \{\gamma \mid \gamma \vdash n\}$$

of all the partitions of $n$. The partial order will not be the natural *lexicographic* order $\leqslant$, which is defined as follows:

1.4.5 $$\alpha < \beta: \quad \Leftrightarrow \quad \exists i \; (\alpha_1 = \beta_1, \ldots, \alpha_{i-1} = \beta_{i-1}, \alpha_i < \beta_i).$$

It is clear that $\leqslant$ is a total order, so that the order diagram is always linear. The partial order $\trianglelefteq$ which we have in mind is defined in terms of the partial sums

$$\sum_1^i \gamma_\nu$$

of the parts $\gamma_\nu$ of the partitions in question:

1.4.6 $$\alpha \trianglelefteq \beta: \quad \Leftrightarrow \quad \forall i \left( \sum_1^i \alpha_\nu \leqslant \sum_1^i \beta_\nu \right).$$

In the case when this holds we say that $\beta$ *dominates* $\alpha$ and call $\trianglelefteq$ the *dominance order*. (When $\lambda, \mu \models n$, $\lambda \trianglelefteq \mu$ is defined similarly.)

It is easy to see that the dominance order differs from the lexicographic order on $P(n)$ if and only if $n \geqslant 6$. The order diagram of $(P(6), \trianglelefteq)$ is

1.4.7

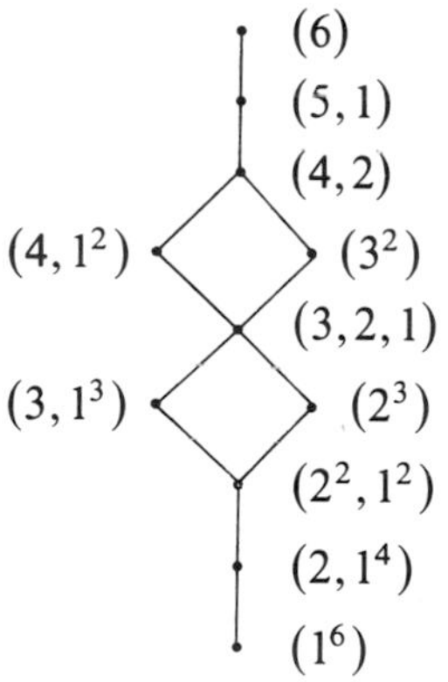


It is obvious that the following is true for all $\alpha, \beta \vdash n$:

**1.4.8** $$a \trianglelefteq \beta \quad \Rightarrow \quad \alpha \leqslant \beta.$$

It will be useful to have a characterization of partitions $\alpha$ and $\beta$ of $n$ which are *neighbors* with respect to $\trianglelefteq$, a situation which we denote by $\alpha \lessdot \beta$:

1.4.9 $$\alpha \lessdot \beta \;:\Leftrightarrow\; \left[\alpha \lhd \beta \text{ and } \nexists \gamma \vdash n \; (\alpha \lhd \gamma \lhd \beta)\right].$$

We shall prove that this can be expressed very nicely in terms of the corresponding Young diagrams $[\alpha]$ and $[\beta]$. In fact we shall show that $\alpha \lessdot \beta$ if and only if $[\beta]$ arises from $[\alpha]$ by moving one node from the end of a certain row of $[\alpha]$ upwards to the end of another row, say from the $j$th row to the $i$th row:

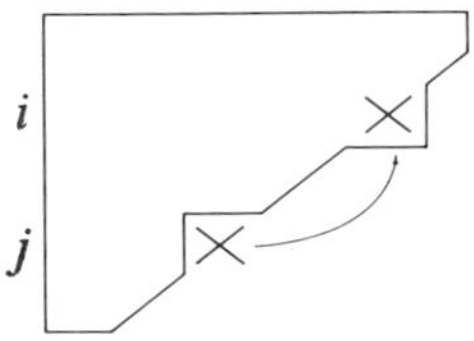


Moreover this step has to be as small a step as possible, which means either $i=j-1$ or $i<j-1$ and $\alpha_i=\alpha_{i+1}=\cdots=\alpha_j$. In the latter case the rim of $[\alpha]$ has the form

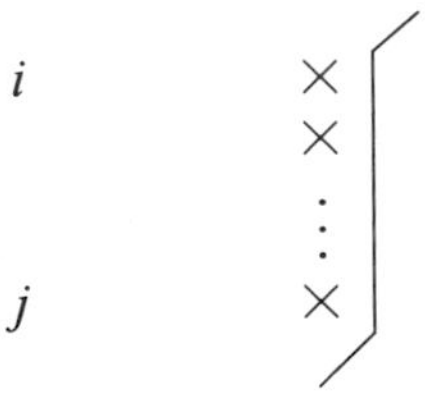


More formally this reads as follows:

**1.4.10** THEOREM. *If* $\alpha=(\alpha_1,\alpha_2,\ldots)$ *and* $\beta=(\beta_1,\beta_2,\ldots)$ *are partitions of* $n$, *then* $\alpha \lessdot \beta$ *if and only if there exist* $i$ *and* $j$ *such that*

(i) $1 \leqslant i < j$,
(ii) $\beta_j=\alpha_j-1$ *and* $\beta_i=\alpha_i+1$, *while for* $\nu \neq i, j$ *we have* $\alpha_\nu=\beta_\nu$,
(iii) $i=j-1$, *or* $\alpha_i=\alpha_j$.

*Proof.*

(1) Assuming $\alpha \lessdot \beta$, we put

$$i := \min\{\nu \mid \alpha_\nu \neq \beta_\nu\}$$

$$j := \min\left\{t \mid \sum_1^t \alpha_\nu = \sum_1^t \beta_\nu,\; i<t\right\},$$

so that obviously

$$1 \leqslant i < j \leqslant n,$$

and

$$\alpha_j > \beta_j \geqslant 0, \qquad \beta_{j+1} \geqslant \alpha_{j+1}.$$

If $i > 1$, then $\alpha_i < \beta_i \leqslant \beta_{i-1} = \alpha_{i-1}$; so that in this case

$$\alpha_i + 1 \leqslant \beta_i \leqslant \beta_{i-1} = \alpha_{i-1}.$$

Also $\alpha_j > \beta_j \geqslant \beta_{j+1} \geqslant \alpha_{j+1}$. Therefore

$$\alpha_j - 1 \geqslant \alpha_{j+1}.$$

Hence in any case, whether $i > 1$ or not, we obtain

$$\alpha \lhd \gamma := (\alpha_1, \ldots, \alpha_{i-1}, \alpha_i + 1, \alpha_{i+1}, \ldots, \alpha_{j-1}, \alpha_j - 1, \alpha_{j+1}, \ldots) \unlhd \beta,$$

so that $\gamma = \beta$ by assumption, and $[\beta]$ arises from $[\alpha]$ by moving exactly one node upwards from the end of the $j$th row of $[\alpha]$ to the end of the $i$th row of $[\alpha]$.

If $i \neq j-1$ and $\alpha_i \neq \alpha_j$, then $i < j-1$ and $\alpha_i > \alpha_j$. We can therefore put

$$t := 1 + \min\{\nu \mid \alpha_\nu > \alpha_{\nu+1},\ i \leqslant \nu < j\}.$$

We notice that $i < t \leqslant j$.

If $t = j$, then $\alpha_i = \cdots = \alpha_{j-1} > \alpha_j > 0$, and thus

$$\alpha \lhd (\alpha_1, \ldots, \alpha_{i-1}, \alpha_i + 1, \alpha_{i+1}, \ldots, \alpha_{j-2}, \alpha_{j-1} - 1, \alpha_j, \ldots) \lhd \beta,$$

which contradicts the assumption $\alpha \lhd\!\cdot\, \beta$. If $t < j$, then $\alpha_i = \cdots = \alpha_{t-1} > \alpha_t \geqslant \cdots \geqslant \alpha_j > 0$, so that in this case we have

$$\alpha \lhd (\alpha_1, \ldots, \alpha_{t-1}, \alpha_t + 1, \ldots, \alpha_{j-1}, \alpha_j - 1, \alpha_{j+1}, \ldots) \lhd \beta,$$

which also contradicts $\alpha \lhd\!\cdot\, \beta$.

(2) Let $\alpha$ and $\beta$ satisfy (i)–(iii), and let $\gamma$ be a partition of $n$ with the property $\alpha \lhd \gamma \unlhd \beta$. Then its parts $\gamma_\nu$ satisfy:

(a) $\forall \nu < i\ (\alpha_\nu = \gamma_\nu = \beta_\nu)$,
(b) $\forall \nu > j\ (\alpha_\nu = \gamma_\nu = \beta_\nu)$.

Hence if $i = j-1$, $\alpha \lhd \gamma$ implies that $\gamma_i = \alpha_i + 1$ and $\gamma_{i+1} = \alpha_i - 1$, and then $\gamma = \beta$.

In the case when $i \neq j-1$, we obtain from (iii) that

$$\alpha_i = \cdots = \alpha_j.$$

By $\alpha \lhd \gamma \unlhd \beta$, (a), (b), and part (1) of this proof, there exist $k$ and $l$ such that $i \leq k < l \leq j$ and

$$\gamma_k = \alpha_k + 1, \qquad \gamma_l = \alpha_l - 1, \qquad \gamma_\nu = \alpha_\nu \quad \text{for } \nu \neq k, l.$$

But this can hold only if $k = i$ and $l = j$; for otherwise

$$\alpha_k = \gamma_k - 1 \leq \gamma_{k-1} - 1 = \alpha_{k-1} - 1 < \alpha_{k-1} = \alpha_i$$

or

$$\alpha_l = \gamma_l + 1 \geq \gamma_{l+1} + 1 = \alpha_{l+1} + 1 > \alpha_{l+1} = \alpha_j$$

would contradict $\alpha_i = \cdots = \alpha_j$. Therefore $\gamma = \beta$ in this case too. ■

This can be used in the proof of

**1.4.11** LEMMA. $\alpha \unlhd \beta \Leftrightarrow \beta' \unlhd \alpha'$.

*Proof.* If $\alpha \unlhd \beta$, then there exists a chain of partitions $\alpha^\nu \vdash n$, $0 \leq \nu \leq r$, which satisfies

$$\alpha = \alpha^0 \lhd \alpha^1 \lhd \cdots \lhd \alpha^r = \beta.$$

1.4.10 yields that $\alpha^i \lhd \alpha^{i+1}$ implies

$$(\alpha^{i+1})' \lhd (\alpha^i)'.$$

Thus we obtain the chain

$$\beta' = (\alpha^r)' \lhd \cdots \lhd (\alpha^0)' = \alpha',$$

which shows $\beta' \unlhd \alpha'$. The converse is true by symmetry. ■

1.4.11 means that the mapping

1.4.12 $$-': P(n) \rightarrowtail\!\!\!\!\rightarrow P(n) : \alpha \mapsto \alpha'$$

is an order antiisomorphism.

The dominance order is described in terms of the partial sums

1.4.13 $$\sigma_i^\alpha := \sum_1^i \alpha_\nu$$

of the parts of $\alpha$. Using these expressions we can define for any $\alpha, \beta \vdash n$

1.4.14 $$\alpha \wedge \beta := \gamma, \qquad \text{where} \quad \sigma_i^\gamma := \min\{\sigma_i^\alpha, \sigma_i^\beta\}, \quad 1 \leqslant i,$$

and

1.4.15 $$\alpha \vee \beta := (\alpha' \wedge \beta')'.$$

It is not difficult to see that $\alpha \vee \beta$ is the supremum and $\alpha \wedge \beta$ the infimum of $\alpha$ and $\beta$ with respect to $\trianglelefteq$. Thus:

**1.4.16.** *The dominance order* $\trianglelefteq$ *induces a lattice structure of* $P(n)$.

We shall not stress this fact, but it should be mentioned that this lattice structure is the combinatorial background of a large part of the ordinary representation theory of $S_n$, as will soon become apparent.

We call this lattice $(P(n), \wedge, \vee)$ the *diagram lattice* of order $n$, since the name "partition lattice" might be misleading: it is already the standard name for a different lattice structure.

In order to apply the structure of the diagram lattice $P(n)$, we need to characterize the dominance $\alpha \trianglelefteq \beta$ of partitions $\alpha$ and $\beta$ of $n$ in terms related to representations of symmetric groups. And there is in fact a classical characterization of $\alpha \trianglelefteq \beta$ in terms of the existence of certain 0-1 matrices which we can easily apply to representations of symmetric groups by way of 1.3.13. This characterization of $\alpha \trianglelefteq \beta$ which we have in mind is the Gale-Ryser theorem, one of the most important existence theorems in combinatorics, where it is used mainly to prove the existence of incidence structures, by way of demonstrating the existence of corresponding incidence matrices. It reads as follows:

**1.4.17** THEOREM OF GALE AND RYSER. *If* $\alpha = (\alpha_1, \alpha_2, \ldots)$ *and* $\beta = (\beta_1, \beta_2, \ldots)$ *are partitions of* $n$, *then there exist* 0-1 *matrices with row sums* $\alpha_1, \alpha_2, \ldots$ *and column sums* $\beta_1', \beta_2', \ldots$ *if and only if* $\alpha \trianglelefteq \beta$.

We do not intend to give a proof of this theorem here, since the standard one is purely combinatorial. (That the existence of such a 0-1 matrix implies $\alpha \trianglelefteq \beta$ is trivial; the other direction is proved by exhibiting an algorithm which describes the construction of such a 0-1 matrix. The interested reader is referred to the book of H. J. Ryser [1963].)

Applying 1.3.7 and 1.3.13, we obtain a translation of this characterization of $\alpha \trianglelefteq \beta$ into representation-theoretical terms:

**1.4.18** THEOREM OF RUCH AND SCHÖNHOFER. *If* $\alpha$ *and* $\beta$ *are partitions of* $n$, *and* $S_\alpha$ *and* $S_{\beta'}$ *are Young subgroups of* $S_n$ *which correspond to* $\alpha$ *and* $\beta'$, *and if the groundfield* $F$ *has* $\operatorname{char} F \neq 2$, *then* $i(IS_\alpha \uparrow S_n, AS_{\beta'} \uparrow S_n) \neq 0$ *if and only if* $\alpha \trianglelefteq \beta$.

The proof of this theorem given by Ruch and Schönhofer does not use the Gale-Ryser theorem but another characterization of $\alpha \trianglelefteq \beta$ in terms of so-called Young tableaux. A *Young-tableau* $t^\alpha$ with Young diagram $[\alpha]$ (sometimes called an *$\alpha$-tableau* for short) arises from $[\alpha]$ by replacing the nodes $\times$ of $[\alpha]$ by the points $i$ of $\mathbf{n}=\{1,\ldots,n\}$. For example, here are two of the 7! Young tableaux with diagram $[3^2,1]$:

$$\begin{array}{ccc} 1 & 4 & 6 \\ 2 & 5 & 7 \\ 3 & & \end{array} \qquad \begin{array}{ccc} 5 & 1 & 6 \\ 2 & 4 & 3 \\ 7 & & \end{array}$$

The tableau obtained by replacing the nodes by 1,2,... in order down successive columns is denoted by $t_1^\alpha$:

1.4.19
$$t_1^\alpha := \begin{array}{ccc} 1 & \alpha_1'+1 & \cdots\cdots\cdots \\ 2 & \alpha_1'+2 & \cdots\cdots \\ \vdots & \vdots & \\ \alpha_1' & \alpha_1'+\alpha_2' & \cdots \end{array}$$

**1.4.20** LEMMA. *If $\alpha, \beta \vdash n$ and $t^\alpha$ is an $\alpha$-tableau, then $\alpha \trianglelefteq \beta$ if and only if there exists a $\beta$-tableau $t^\beta$ such that any two points which occur in $t^\alpha$ in the same row occur in $t^\beta$ in different columns.*

*Proof.* Assume the existence of $t^\alpha$, $t^\beta$ with the given property. Then $\alpha_1 \leqslant \beta_1$, since the points in the first row of $t^\alpha$ occur in different columns of $t^\beta$. As this also holds for the points in the second row of $t^\alpha$, there must be $\alpha_2$ places in different columns of $t^\beta$ which have not yet been accounted for. Thus $\alpha_2 \leqslant \beta_1+\beta_2-\alpha_1$. Continuing in this way, we get $\alpha \trianglelefteq \beta$.

Conversely, if $\alpha \trianglelefteq \beta$ then $t^\beta$ can be constructed algorithmically from an arbitrary $t^\alpha$ (see Dress [1979], Thürlings [1977]). One can proceed recursively by taking the last point of the last row (row $h$, say) of $t^\alpha$, and putting it at the end of row $j$ of $t^\beta$, where $j := \max\{i \mid \beta_i \geqslant \alpha_h\}$. ■

The last lemma and the results of the preceding section give the following characterizations of the dominance order in terms of group theory, representation theory, and combinatorics:

**1.4.21** THEOREM. *If $\alpha$ and $\beta$ are partitions of $n$, then the following properties of $\alpha$ and $\beta$ are equivalent when* char $F \neq 2$:

(i) $\alpha \trianglelefteq \beta$,

(ii) *for any two Young subgroups $S_\alpha$ and $S_{\beta'}$ corresponding to $\alpha$ and $\beta'$, there exist double cosets $S_\alpha \pi S_{\beta'}$ with the trivial intersection property $S_\alpha \cap \pi S_{\beta'} \pi^{-1} = \{1\}$,*

(iii) *for any two Young subgroups* $S_\alpha$ *and* $S_{\beta'}$ *corresponding to* $\alpha$ *and* $\beta'$ *and their representations* $IS_\alpha$ *and* $AS_{\beta'}$ *over* $F$, *we have* $i(IS_\alpha \uparrow S_n, AS_{\beta'} \uparrow S_n) \neq 0$,

(iv) *there exist* 0-1 *matrices with* $\alpha$ *as vector of row sums and* $\beta'$ *as vector of column sums*,

(v) *for each* $\alpha$-*tableau* $t^\alpha$ *there exists a* $\beta$-*tableau* $t^\beta$ *where any two points which occur in* $t^\alpha$ *in the same row occur in* $t^\beta$ *in different columns*,

(vi) *the coefficient of* $x^\alpha y^{\beta'}$ *in* $\prod_{i,k}(1+x_i y_k)$ *is* $\neq 0$.

We have seen already that (ii), (iii), (iv), and (vi) are equivalent and that (i) and (v) are equivalent. The rest of the statement is obtained from the Gale-Ryser theorem or from the Ruch-Schönhofer theorem, neither of which we have proved yet. Since we are going to apply some of these results, it should be pointed out that we shall use only the trivial part of the Gale-Ryser theorem, which gives $\alpha \trianglelefteq \beta$ from the existence of the relevant 0-1 matrix. A proof of the Ruch-Schönhofer theorem will be given later so that in fact the story will be complete.

## 1.5 Young Subgroups as Horizontal and Vertical Groups of Young Tableaux

In the last section we introduced the notion of a Young tableau $t^\alpha$ with Young diagram $[\alpha]$. If we are given such a Young tableau $t^\alpha$, then its rows and its columns define two dissections of **n**: the first one is a dissection of **n** into pairwise disjoint subsets of orders $\alpha_i$, the second one is a dissection of **n** into pairwise disjoint subsets of orders $\alpha'_j$. For example

$$\begin{matrix} 1 & 2 & 3 \\ 4 & 5 & \end{matrix}$$

defines by its rows the dissection of **5** into subsets $\{1,2,3\}$ and $\{4,5\}$, while it defines by its columns the dissection of **5** into subsets $\{1,4\}$, $\{2,5\}$, and $\{3\}$.

Correspondingly we obtain from the rows of $t^\alpha$ a Young subgroup $S_\alpha$, which is called the *horizontal group* of $t^\alpha$, and which will be denoted by

$$H^\alpha,$$

or, if necessary, by $H(t^\alpha)$. The columns of $t^\alpha$ define a Young subgroup $S_{\alpha'}$, which is called the *vertical group* of $t^\alpha$, and which will be denoted by

$$V^\alpha,$$

or, if necessary, by $V(t^\alpha)$. These specific Young subgroups $H^\alpha$ and $V^\alpha$ obviously have the property

**1.5.1** $$H^\alpha \cap V^\alpha = \{1\}.$$

Therefore a Young tableau $t^\alpha$ *displays in a certain sense the double coset* $S_\alpha \pi S_{\alpha'}$ *with the trivial intersection property* $S_\alpha \cap \pi S_{\alpha'} \pi^{-1} = \{1\}$. For if we are given $S_\alpha$ and $S_{\alpha'}$, then we need only consider a tableau $t^\alpha$, which has $S_\alpha$ as its horizontal group, and take a $\pi$ such that $\pi S_{\alpha'} \pi^{-1}$ is the vertical group of $t^\alpha$.

It is the aim of this section to place several properties of horizontal and vertical groups of tableaux at the reader's disposal.

For a while we shall keep the diagram $[\alpha]$ fixed, so that we need write only $t$ for the tableau in question and $H$ and $V$ for its horizontal and vertical groups. If now $\pi \in S_n$, then we define $\pi t$ to be the tableau which arises from $t$ by an application of $\pi$ to the points in the tableau $t$:

1.5.2 $$\text{If } t = \begin{matrix} \cdot\;\cdot\;\cdot\;\cdot\;\cdot\;\cdot\;\cdot\;\cdot\;\cdot\;\cdot \\ \cdot\;\cdot\;\cdot\; i \;\cdot\;\cdot\;\cdot \\ \cdot\;\cdot\;\cdot\;\cdot\;\cdot\;\cdot \end{matrix}\;, \quad \text{then} \quad \pi t := \begin{matrix} \cdot\;\cdot\;\cdot\;\cdot\;\cdot\;\cdot\;\cdot\;\cdot\;\cdot\;\cdot \\ \cdot\;\cdot\;\cdot\; \pi(i) \;\cdot\;\cdot\;\cdot \\ \cdot\;\cdot\;\cdot\;\cdot\;\cdot\;\cdot \end{matrix}$$

It is clear that the following holds:

**1.5.3** $$H(\pi t) = \pi H(t) \pi^{-1}, \quad \textit{and} \quad V(\pi t) = \pi V(t) \pi^{-1}.$$

In Chapter 3 it will emerge that elements of the following form are essentially (i.e. up to a scalar factor) idempotent elements of the group algebra $FS_n$ of $S_n$ over the groundfield $F$, char $F \nmid n!$, which generate minimal left ideals:

1.5.4 $$e := \sum_{\pi \in V} \sum_{\rho \in H} \operatorname{sgn} \pi \cdot \pi \rho .$$

Thus it will be useful to have a few results to hand on permutations of the form $\pi\rho$, $\pi \in V$, $\rho \in H$.

A trivial but useful remark is that for every $\pi \in V$, $\rho \in H$,

**1.5.5** $$t' := \pi \rho t = \rho' \pi t, \qquad \textit{where} \quad \rho' := \pi \rho \pi^{-1} \in H(\pi t).$$

It means that instead of first carrying out $\rho$, which leaves the points of $t$ in their rows, and then $\pi$, which usually does not leave the points of $\rho t$ in their columns, we may first apply $\pi$, which leaves the points of $t$ in their columns, and then apply $\rho'$, which leaves the points of $\pi t$ in their rows.

We use this remark in order to prove that for $\rho \in H(t)$, $\pi \in V(t)$, we have:

**1.5.6** LEMMA. *Any two points which occur in the same column of* $t$ *occur in* $t' := \pi \rho t$ *in different rows.*

*Proof.* This follows immediately from $t' = \rho' \pi t$. For two points in the same column of $t$ are in the same column of $\pi t$ and hence in different rows, and they do not leave their row when we shift from $\pi t$ to $\rho' \pi t$, since $\rho' \in H(\pi t)$. ■

But the converse is also true:

**1.5.7** LEMMA. *If every two points which occur in the same column of $t$, occur in different rows of $t' := \sigma t$, then $\sigma$ is of the form $\pi\rho$ for suitable $\pi \in V(t)$, $\rho \in H(t)$.*

*Proof.* By assumption, all the points of the first row of $t'$ occur in different columns of $t$, so that a vertical permutation applied to $t$ moves them into the first row. Leaving these points fixed, another element of $V(t)$ moves the points of the second row of $t'$ into the second row, and so on. Hence there exist $\pi \in V(t)$ such that $\pi t$ contains each point in the same row as does $t'$. A final application of a suitable $\rho' \in H(\pi t) = \pi H(t)\pi^{-1}$ yields $t' = \rho'\pi t = \pi\rho t$, for $\rho := \pi^{-1}\rho'\pi$. ■

**1.5.8** LEMMA. *If $\sigma \notin VH$, then there exist transpositions $\pi \in V$, $\rho \in H$ such that $\pi\sigma\rho = \sigma$.*

*Proof.* Since $\sigma \notin VH$, we obtain from 1.5.7 that there exist two points, say $i$ and $k$, which occur in $t$ in the same column, and which appear in $t' := \sigma t$ in the same row. Let $\tau := (ik)$ denote their transposition. Then $\tau \in V(t)$ and $\tau \in H(t') = \sigma H(t)\sigma^{-1}$, so that $\sigma^{-1}\tau\sigma \in H(t)$. Therefore $\pi := \tau$ and $\rho := \sigma^{-1}\tau\sigma$ fulfill the statement. ■

With these results in mind, we study the following elements of the group algebra $FS_n$ (where $F$ is an *arbitrary* field):

1.5.9 $$\mathcal{V} := \sum_{\pi \in V} \operatorname{sgn}\pi \cdot \pi \quad \text{and} \quad \mathcal{H} := \sum_{\rho \in H} \rho,$$

together with their product

1.5.10 $$e := \mathcal{V}\mathcal{H} = \sum_{\pi \in V} \sum_{\rho \in H} \operatorname{sgn}\pi \cdot \pi\rho.$$

We notice that it suffices to take the sum over all those $\sigma \in S_n$ which are of the form $\pi\rho$, $\pi \in V$, $\rho \in H$:

**1.5.11** $$e = \sum_{\pi\rho} \operatorname{sgn}\pi \cdot \pi\rho,$$

for if $\sigma = \pi_1\rho_1 = \pi_2\rho_2$, then $\pi_2^{-1}\pi_1 = \rho_2\rho_1^{-1} \in H \cap V = \{1\}$, so that both $\pi_1 = \pi_2$ and $\rho_1 = \rho_2$.

A very important property of $e$ is

**1.5.12** $$\forall \pi \in V, \rho \in H \quad (\pi e \rho = \operatorname{sgn}\pi \cdot e).$$

Besides this, we need to know which products $ee'$ are zero. Since $ee' = \mathcal{V}\mathcal{H}\mathcal{V}'\mathcal{H}'$, it will be useful to consider expressions of the form $\mathcal{H}\mathcal{V}'$.

**1.5.13** LEMMA. *Suppose that $t$ and $t'$ are tableaux with the same diagram and $\mathcal{H} := \mathcal{H}(t)$, $\mathcal{V}' := \mathcal{V}(t')$. Then $\mathcal{H}\mathcal{V}' = 0$ if and only if there exist two points which occur in $t$ in the same row and in $t'$ in the same column.*

*Proof.*

(i) If there exist two such points, say $i$ and $k$, then there exist subgroups $\overline{H}$ and $\overline{V}'$ in $H$ and $V'$ such that

$$H = \overline{H} \,\dot\cup\, (ik)\overline{H}, \qquad V' = \overline{V}' \,\dot\cup\, (ik)\overline{V}'.$$

Hence

$$\mathcal{H}\mathcal{V}' = \Big(\sum_{\rho\in\overline{H}} \rho\Big)(1+(ik))(1-(ik))\Big(\sum_{\pi\in\overline{V}'} \operatorname{sgn}\pi\cdot\pi\Big) = 0.$$

(ii) Conversely, if no such pair of points exists, then by 1.5.7

$$t = \pi'\rho' t', \qquad \pi' \in V', \quad \rho' \in H'.$$

This yields

$$\begin{aligned}
\mathcal{H}\mathcal{V}' &= \mathcal{H}(t)\mathcal{V}(t')\\
&= \mathcal{H}(\pi'\rho' t')\mathcal{V}(t')\\
&= \pi'\rho'\mathcal{H}(t')\rho'^{-1}\pi'^{-1}\mathcal{V}(t')\\
&= \pi'\mathcal{H}(t')(\operatorname{sgn}\pi'^{-1})\mathcal{V}(t')\\
&= \operatorname{sgn}\pi'\cdot\pi'\mathcal{H}'\mathcal{V}'\\
&\neq 0.
\end{aligned}$$

■

The second part of the proof can be rephrased as follows. Suppose that $t$ and $t'$ have the property that every pair of points which occur in $t$ in the same row occur in different columns of $t'$. Then we may construct a new tableau $t''$ by following the mechanical procedure sketched below:

1.5.14

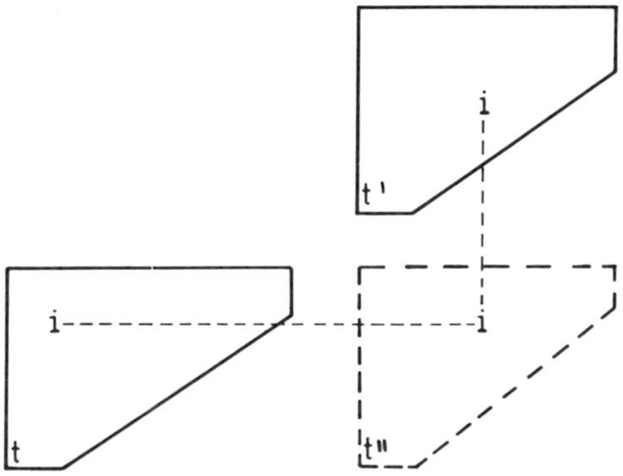

For the new tableau $t''$ we have

**1.5.15.** $\mathcal{H}''=\mathcal{H}$, $\mathcal{V}''=\mathcal{V}'$, *and therefore* $\mathcal{H}\mathcal{V}'=\mathcal{H}''\mathcal{V}''\neq 0$.

If we write $t''=\sigma't'=\sigma t$, then $\sigma'\in V'$ and $\sigma\in H$, so that $\sigma'^{-1}\sigma\in V'H$. This proves:

**1.5.16** COROLLARY. *Suppose that $t$ and $t'$ are tableaux with the same diagram, and $t'=\tau t$. Then $\mathcal{H}\mathcal{V}'\neq 0$ if and only if $\tau=\pi'\rho$ for some $\pi'\in V'$, $\rho\in H$.*

The argument used in the first part of the proof of 1.5.13 does not use the fact that $t$ and $t'$ are of the same diagram. Hence our next remark is

**1.5.17** LEMMA. *If $\alpha\not\trianglelefteq\beta$ and $t^\alpha$, $t^\beta$ are tableaux with diagrams $[\alpha]$,$[\beta]$, then there exist two points which occur in $t^\alpha$ in the same row while they occur in $t^\beta$ in the same column. Also $\mathcal{H}^\alpha\mathcal{V}^\beta=0$, and hence $e^\alpha e^\beta=0$.*

*Proof.* The existence of two such points is immediate from 1.4.20, and now part (i) of the proof of 1.5.13 completes the proof. ■

In Chapter 3 we shall see that the elements $e^\alpha$ are in fact essentially primitive idempotents, so that the results we have obtained can be interpreted in terms of irreducible representations of $S_n$.

## Exercises

1.1 Prove that $A_5$ is simple in the following two ways:

(a) by considering the orders of the conjugacy classes of $A_5$,

(b) by showing that the assumption $\{1\}\lhd N\lhd A_5$ yields $|N|=5$, or $|N|=20$ (consider $N\cap A_4$), so that $N$ and hence also $A_5$ would have a normal Sylow 5-subgroup, which yields a contradiction.

1.2 Prove (by induction on $n$ and arguments similar to those used in Exercise 1.1(b)) that $A_n$ is simple for $n>4$.

1.3 (a) Show that, for $n\neq 6$, each automorphism $\varphi$ of $S_n$ is inner, i.e. of the form $\pi\mapsto\rho\pi\rho^{-1}$, for a suitable $\rho\in S_n$. (*Hint:* What happens to a class of elements of order 2 if we apply $\varphi$?)

(b) Prove that the group of inner automorphisms of $S_6$ is of index 2 in the automorphism group of $S_6$. (*Hint:* There exists a subgroup $U\simeq S_5$ of $S_6$ which is not a stabilizer of a point.)

1.4 Prove that, for $n\neq 4$, $S_n$ does not contain a subgroup of index $k$ where $2<k<n$ (while in $S_4$ the Sylow 2-subgroups are of index 3).

CHAPTER 2

# *Ordinary Irreducible Representations and Characters of Symmetric and Alternating Groups*

The present chapter contains a characterization of the ordinary irreducible representations of symmetric groups as common constituents of monomial representations induced from Young subgroups. Besides this, the ordinary irreducible characters are given as $\mathbb{Z}$-linear combinations of permutation characters induced from Young subgroups, and an important recursion formula is derived. As an application, the ordinary irreducible characters and representations of $A_n$ are given. It is shown furthermore, that $S_n$ is characterized by its character table. An examination of the values of the irreducible characters leads to the notions of cores and quotients of partitions which later on are of great importance for modular purposes. Finally, the Littlewood-Richardson rule, which is one of the most useful results of the whole theory, is derived.

## 2.1 The Ordinary Irreducible Representations of $S_n$

By an *ordinary* representation of a group $G$ we mean a finite-dimensional representation of $G$ over the field $\mathbb{C}$ of complex numbers. It will turn out that each ordinary representation of $S_n$ can be realized over the rational field, so that $\mathbb{Q}$ (and thus every field) is a splitting field for $S_n$. Hence it does not make much difference whether we assume that the groundfield $F$ is of a characteristic which does not divide $n!$ or we just consider the ordinary representations of $S_n$.

ENCYCLOPEDIA OF MATHEMATICS and Its Applications, Gian-Carlo Rota (ed.). Vol. 16: G. D. James and A. Kerber, The Representation Theory of the Symmetric Group
ISBN 0-201-13515-9

We know from general representation theory of finite groups that $S_n$ has the same number of ordinary irreducible representations as conjugacy classes. 1.2.6 has shown that a conjugacy class of $S_n$ is characterized by the cycle partition $\gamma \vdash n$ of its elements. For each such partition $\gamma$ of $n$ we constructed Young subgroups $S_\gamma$ and considered their trivial one-dimensional representations $IS_\gamma$, the identity representation of $S_\gamma$, and $AS_\gamma$, the alternating representation of $S_\gamma$. In Section 1.3 we expressed the intertwining number

$$i(IS_\alpha \uparrow S_n, AS_\beta \uparrow S_n),$$

$\alpha, \beta \vdash n$, of the induced representations in terms of numbers of double cosets and in terms of numbers of 0-1 matrices with prescribed row and column sums. The particular intertwining number

$$i(IS_\alpha \uparrow S_n, AS_{\alpha'} \uparrow S_n),$$

for example, is equal to the number of 0-1 matrices with row sums $\alpha_i$ and column sums $\alpha'_j$. But it is not difficult to see that there is exactly one 0-1 matrix which satisfies these conditions on its row and column sums, namely the matrix

2.1.1
$$\begin{pmatrix} 1 & \cdots & \cdots & 1 & 0 & \cdots & \cdots & 0 \\ \vdots & & & & & & & \vdots \\ 1 & \cdots & 1 & 0 & \cdots & \cdots & \cdots & 0 \end{pmatrix}$$

where the 1's are placed in each row as far to the left as possible. This yields

**2.1.2** $$i(IS_\alpha \uparrow S_n, AS_{\alpha'} \uparrow S_n) = 1.$$

A careful interpretation of these equations yields several important results.

First of all we remember that these equations hold whenever the groundfield $F$ is of characteristic $\neq 2$ (cf. 1.3.7). Therefore let us interpret 2.1.2 first as a result on representations of $S_n$ over the field $\mathbb{Q}$.

Since the intertwining number $i(\cdot, \cdot)$ is "bilinear" and has all its values in $\mathbb{N}_0$, 2.1.2 implies that there is exactly one irreducible $\mathbb{Q}$-representation of $S_n$, say $D$, which is contained in both $IS_\alpha \uparrow S_n$ and $AS_{\alpha'} \uparrow S_n$. Furthermore $i(D, D) = 1$, and $D$ is contained in both $IS_\alpha \uparrow S_n$ and $AS_{\alpha'} \uparrow S_n$ with multiplicity 1. Since $i(D, D) = 1$, $D$ must be absolutely irreducible, so that it is also an irreducible $\mathbb{C}$-representation of $S_n$. Altogether this yields the following basic result:

**2.1.3** THEOREM. *If $\alpha$ is a partition of n and if $S_\alpha$ and $S_{\alpha'}$ are Young subgroups of $S_n$, which correspond to $\alpha$ and $\alpha'$, then the induced representations $IS_\alpha \uparrow S_n$ and $AS_{\alpha'} \uparrow S_n$ have exactly one ordinary irreducible constituent in common.*

*Furthermore this irreducible constituent can be realized over* $\mathbb{Q}$, *and it is contained with multiplicity* 1 *in both* $IS_\alpha \uparrow S_n$ *and* $AS_{\alpha'} \uparrow S_n$.

As any two Young subgroups of $S_n$ which correspond to the same partition $\alpha' \vdash n$ are conjugate subgroups, the characters of $IS_\alpha \uparrow S_n$ and $AS_{\alpha'} \uparrow S_n$ depend only on the partition $\alpha$ of the integer $n$ and *not* on the dissections of the set **n** which yield $S_\alpha$ and $S_{\alpha'}$. Hence the same must be true for the common irreducible constituent mentioned in Theorem 2.1.3; it depends only on $\alpha$. We therefore denote this uniquely determined constituent, and also its equivalence class of representations, by $[\alpha]$, since there is no danger of confusing this with the corresponding Young diagram. Abusing the symbol $\cap$, we have

2.1.4 $$[\alpha] := IS_\alpha \uparrow S_n \cap AS_{\alpha'} \uparrow S_n.$$

Two special cases are easy to identify, namely the identity representation $IS_n$ of $S_n$ and the alternating representation $AS_n$ of $S_n$. We denoted by $(1^n)$ the partition $(1, \ldots, 1)$ of $n$. Since the only corresponding Young subgroup is $S_{(1^n)} = \{1\}$, both $IS_{(1^n)} \uparrow S_n$ and $AS_{(1^n)} \uparrow S_n$ are equal to the regular representation $RS_n$ of $S_n$:

**2.1.5** $$IS_{(1^n)} \uparrow S_n = AS_{(1^n)} \uparrow S_n = RS_n.$$

Furthermore from $S_{(n)} = S_n$ it follows that

**2.1.6** $$IS_{(n)} \uparrow S_n = IS_n \quad \textit{and} \quad AS_{(n)} \uparrow S_n = AS_n.$$

Thus 2.1.4 implies

**2.1.7** $$[n] = IS_n, \quad \textit{and} \quad [1^n] = AS_n.$$

It is important to perceive the close connection between $[\alpha]$ and $[\alpha']$. Since

$$AS_n \otimes (IS_\alpha \uparrow S_n) = AS_\alpha \uparrow S_n$$

and

$$AS_n \otimes (AS_{\alpha'} \uparrow S_n) = IS_{\alpha'} \uparrow S_n$$

(use the standard argument for characters $\mu$ of $G$ and $\nu$ of $H \leqslant G$ that $\mu \otimes \nu \uparrow G = (\mu \downarrow H \otimes \nu) \uparrow G$), we have for each $\alpha \vdash n$:

**2.1.8** $$[\alpha'] = [\alpha] \otimes [1^n].$$

This means that the ordinary irreducible representation $[\alpha']$ differs from $[\alpha]$ only on the odd permutations, and there just by the sign.

The set

2.1.9 $$\{[\alpha] \mid \alpha \vdash n\}$$

is a subset of the set of equivalence classes of ordinary irreducible representations of $S_n$. We would like to show that 2.1.9 is already the complete set of these equivalence classes. A preliminary consideration shows

**2.1.10** LEMMA. *The multiplicity* $i(IS_\alpha \uparrow S_n, [\beta])$ *of* $[\beta]$ *in* $IS_\alpha \uparrow S_n$ *is nonzero only if* $\alpha \trianglelefteq \beta$.

*Proof.* 2.1.4 shows that $i(AS_{\beta'} \uparrow S_n, [\beta]) = 1$. The assumption that $i(IS_\alpha \uparrow S_n, [\beta]) \neq 0$ yields therefore $i(IS_\alpha \uparrow S_n, AS_{\beta'} \uparrow S_n) \neq 0$. Hence there exist 0-1 matrices with row sums $\alpha_i$ and column sums $\beta'_j$, which implies $\alpha \trianglelefteq \beta$. ■

Notice that in this proof we did only use the trivial part of the Gale-Ryser theorem, as we promised at the end of Section 1.4.

We are now in a position to prove the main theorem of this section.

**2.1.11** THEOREM. $\{[\alpha] \mid \alpha \vdash n\}$ *is the complete set of equivalence classes of ordinary irreducible representations of* $S_n$.

*Proof.* We need only show that $[\alpha] = [\beta]$ implies $\alpha = \beta$, for then the cardinality of $\{[\alpha] \mid \alpha \vdash n\}$ is equal to the number of conjugacy classes of $S_n$, so that this system must be complete. But if $[\alpha] = [\beta]$, we can argue in the following way:

$$(IS_\alpha \uparrow S_n, [\beta]) = (IS_\alpha \uparrow S_n, [\alpha]) = 1 = (IS_\beta \uparrow S_n, [\beta]) = (IS_\beta \uparrow S_n, [\alpha]).$$

Thus by 2.1.10 we have both $\alpha \trianglelefteq \beta$ and $\beta \trianglelefteq \alpha$, which imply $\alpha = \beta$, and we are done. ■

Hence each ordinary irreducible representation of $S_n$ is realizable over $\mathbb{Q}$, so that $\mathbb{Q}$ is a splitting field for $S_n$. This yields (apply (83.7) in Curtis and Reiner [1962]):

**2.1.12** THEOREM. *Each field is a splitting field for* $S_n$.

This is not true for alternating groups; $A_3$ provides a counterexample.

Let us conclude this section with a numerical example. By 2.1.11, we know that $[3]$, $[2,1]$, and $[1^3]$ are all the equivalence classes of ordinary irreducible representations of $S_3$. 2.1.7 yields

$$[3] = IS_3 \quad \text{and} \quad [1^3] = AS_3.$$

Since the squares of the dimensions add up to 6, [2, 1] has dimension 2. Since $\alpha := (2,1)$ is self-associated, 2.1.8 yields that the character of [2, 1] vanishes on the class of transpositions. This gives the character table of $S_3$ up to one remaining entry:

$$\begin{array}{c} \quad (1^3) \quad (2,1) \quad (3) \\ \begin{pmatrix} 1 & -1 & 1 \\ 2 & 0 & * \\ 1 & 1 & 1 \end{pmatrix} \begin{array}{l} [1^3] \\ [2,1] \\ [3] \end{array} \end{array}$$

The remaining entry turns out to be $-1$ by the orthogonality relations, say. Hence the complete character table of $S_3$ is

$$\begin{array}{c} \quad (1^3) \quad (2,1) \quad (3) \\ \begin{pmatrix} 1 & -1 & 1 \\ 2 & 0 & -1 \\ 1 & 1 & 1 \end{pmatrix} \begin{array}{l} [1^3] \\ [2,1] \\ [3] \end{array} \end{array}$$

But we are far from being able to evaluate the character table for every $n$, so that we need to examine more closely the permutation characters induced by Young subgroups.

Before we start doing this we should not forget to mention a result which follows from 1.1.26, 2.1.7, and the well known form of the composition series of symmetric groups:

**2.1.13** THEOREM. *All the ordinary irreducible representations of* $S_n$ *are faithful except* $[n]$ (*for* $n \geq 2$), $[1^n]$ (*for* $n \geq 3$), *and* $[2^2]$ (*for* $n=4$).

## 2.2 The Permutation Characters Induced by Young Subgroups

In the preceding section we constructed with the aid of the representations $IS_\alpha \uparrow S_n$ a complete system of ordinary irreducible representations of $S_n$. The former are transitive permutation representations, and we would like to examine them more thoroughly.

In order to do this, we put the $p(n) := |P(n)|$ proper partitions $\alpha^i$ of $n$ in lexicographic order:

**2.2.1** $$\alpha^1 = (1^n) < \alpha^2 < \cdots < \alpha^{p(n)} = (n)$$

and define a matrix

2.2.2 $$M_n := (m_{ik}), \qquad 1 \leq i, k \leq p(n),$$

with the following multiplicities as entries:

2.2.3 $$m_{ik} := \left(IS_{\alpha^i} \uparrow S_n, [\alpha^k]\right).$$

2.1.3 together with 1.4.8 and 2.1.10 shows that $M_n$ is an upper triangular matrix with 1's along the main diagonal:

**2.2.4** $$M_n = \begin{pmatrix} 1 & & & * \\ & \cdot & & \\ & & \cdot & \\ 0 & & & \cdot \\ & & & & 1 \end{pmatrix}.$$

This has several important consequences. We denote by

2.2.5 $$\zeta_\beta^\alpha$$

the value of the character $\zeta^\alpha$ of $[\alpha]$ on the conjugacy class of elements with cycle partition $\beta$, and by

2.2.6 $$\xi_\beta^\alpha$$

we denote the value of the character $\xi^\alpha$ of $IS_\alpha \uparrow S_n$ on this same class. Then 2.2.3 yields

2.2.7 $$\xi^{\alpha^i} = \sum_k m_{ik} \zeta^{\alpha^k}.$$

If we put the values $\zeta_\beta^\alpha$ and $\xi_\beta^\alpha$ into matrices, say

2.2.8 $$Z_n := \left(\zeta_{\beta^k}^{\alpha^i}\right) \text{ and } \Xi_n := \left(\xi_{\beta^k}^{\alpha^i}\right), \qquad 1 \leqslant i, k \leqslant p(n),$$

then $Z_n$ is the character table of $S_n$, and $\Xi_n$ satisfies (cf. 2.2.7)

$$\Xi_n = M_n Z_n,$$

so that we obtain from det $M_n \neq 0$ (cf. 2.2.4):

**2.2.9** $$Z_n = M_n^{-1} \Xi_n.$$

$M_n$ is a matrix over $\mathbb{N}_0$ and has determinant 1; hence $M_n^{-1}$ is a matrix over $\mathbb{Z}$, so that by 2.2.9 each ordinary irreducible character $\zeta^\alpha$ of $S_n$ is a $\mathbb{Z}$-linear combination of the permutation characters $\xi^\beta$. As there are exactly $p(n)$ such permutation characters, this $\mathbb{Z}$-linear combination is uniquely determined. Thus the following holds:

**2.2.10** THEOREM. *The ring* char$(S_n) = \oplus_{i=1}^{p(n)} \mathbb{Z}\zeta^{\alpha^i}$ *of generalized ordinary characters of* $S_n$ *possesses besides its* $\mathbb{Z}$*-basis* $\{\zeta^\alpha \mid \alpha \vdash n\}$ *the* $\mathbb{Z}$*-basis* $\{\xi^\alpha \mid \alpha \vdash n\}$, *which consists of characters of transitive permutation representations of* $S_n$.

If we want to express $\zeta^\alpha$ in terms of this basis of permutation characters, we need to know $M_n^{-1}$ (or $M_n$). It should be mentioned that $M_n$ can be evaluated using enumerations of double cosets. For $M_n$ is triangular and has 1's along its main diagonal, while the scalar product of the $i$th and the $j$th row of $M_n$ satisfies (apply 1.3.5)

2.2.11
$$\sum_k m_{ik}m_{jk} = i(IS_{\alpha^i}\uparrow \mathrm{S}_n, IS_{\alpha^j}\uparrow \mathrm{S}_n)$$
$$= \sum_{S_{\alpha^i}\pi S_{\alpha^j}} 1.$$

This number of double cosets can be found as a certain coefficient in $\prod(1-x_iy_j)^{-1}$ (cf. 1.3.14). A numerical example illustrates this: $M_3$ is of the form (notice $(1^3)<(2,1)<(3)$)

$$M_3 = \begin{pmatrix} 1 & * & * \\ 0 & 1 & * \\ 0 & 0 & 1 \end{pmatrix},$$

and we know

$$(IS_{(3)}\uparrow \mathrm{S}_3, IS_{(2,1)}\uparrow \mathrm{S}_3) = \sum_{\mathrm{S}_3\pi S_{(2,1)}} 1 = 1,$$

$$(IS_{(3)}\uparrow \mathrm{S}_3, IS_{(1^3)}\uparrow \mathrm{S}_3) = \sum_{\mathrm{S}_3\pi S_{(1^3)}} 1 = 1,$$

so that

$$M_3 = \begin{pmatrix} 1 & * & 1 \\ 0 & 1 & 1 \\ 0 & 0 & 1 \end{pmatrix}.$$

Furthermore,

$$(IS_{(2,1)}\uparrow \mathrm{S}_3, IS_{(1^3)}\uparrow \mathrm{S}_3) = \sum_{S_{(2,1)}\pi S_{(1^3)}} 1 = |\mathrm{S}_3 : S_{(2,1)}| = 3,$$

so that finally

$$M_3 = \begin{pmatrix} 1 & 2 & 1 \\ 0 & 1 & 1 \\ 0 & 0 & 1 \end{pmatrix}.$$

We have obtained from 2.2.11 and 1.3.14 the following result:

**2.2.12** THEOREM. *The matrix* $M_n$ *can be evaluated from the coefficients of the generating function* $\prod_{i,k}(1-x_i y_k)^{-1}$.

Since $IS_{(1^n)}\uparrow S_n = RS_n$ (cf. 2.1.5), $M_n$ contains in its first row the dimensions

2.2.13 $$f^\alpha := \zeta^\alpha(1)$$

of the ordinary irreducible representations $[\alpha]$ of $S_n$. And as $i(IS_{(n)}\uparrow S_n, IS_\alpha\uparrow S_n) = i(IS_n, IS_\alpha\uparrow S_n) = 1$, it contains in its last column only 1's, so that we have

2.2.14 $$M_n = \begin{pmatrix} 1=f^{\alpha^1} & \cdots & f^{\alpha^k} & \cdots & f^{\alpha^{p(n)}}=1 \\ & \cdot & & & \cdot \\ & & \cdot & * & \cdot \\ & 0 & & \cdot & \cdot \\ & & & & 1 \end{pmatrix}$$

Furthermore the scalar product of the first row of $M_n$ with its $i$th row is just the index $|S_n : S_{\alpha^i}|$, $1 \leqslant i \leqslant p(n)$.

But the knowledge of $M_n$ alone does not suffice to evaluate $Z_n$. We also need $\Xi_n$, the matrix of the permutation characters. Since $IS_\alpha\uparrow S_n$ is the permutation representation of $S_n$ on the left cosets of $S_\alpha$ it is in principle possible to evaluate $\xi^\alpha(\pi)$ by checking which left cosets of $S_\alpha$ remain fixed under left multiplication by $\pi$. If we want to do this, it makes life easier to visualize the left cosets by so-called *tabloids*, which may be introduced as equivalence classes of tableaux. We call two tableaux $t$ and $t'$ (with the same diagram) *row equivalent* if and only if $t'$ arises from $t$ by a horizontal permutation:

2.2.15 $$t \sim t' \quad :\Leftrightarrow \quad \exists \pi \in H(t) \ (t' = \pi t).$$

The equivalence class of $t$ is denoted by

2.2.16 $$\{t\}.$$

In other words: $\{t\}$ arises from $t$ by neglecting the order of points in the rows. We shall indicate this by drawing lines between the rows of $t$. It is obvious that the tabloids of shape $[\alpha]$ are in one-to-one correspondence with the left cosets of $S_\alpha$ and admit an action of $S_n$ which is equivalent to the left multiplication of $S_n$ on the cosets of $S_\alpha$. For example the (2, 1)-tableaux are

$$\begin{matrix}1&2\\3&\end{matrix},\quad \begin{matrix}1&3\\2&\end{matrix},\quad \begin{matrix}2&1\\3&\end{matrix},\quad \begin{matrix}2&3\\1&\end{matrix},\quad \begin{matrix}3&1\\2&\end{matrix},\quad \begin{matrix}3&2\\1&\end{matrix},$$

while the tabloids with diagram [2, 1] are

$$\frac{\overline{1\ 2}}{\underline{3}},\quad \frac{\overline{1\ 3}}{\underline{2}},\quad \frac{\overline{2\ 3}}{\underline{1}}.$$

$1_{S_3}$ leaves each one of them fixed, (12) leaves just the first one fixed, and (123) has no fixed tabloid. This enables us to evaluate $\xi^{(2,1)}$: $\xi^{(2,1)}(1)=3$, $\xi^{(2,1)}((12))=1$, $\xi^{(2,1)}((123))=0$. Thus we obtain for $\Xi_3$, since $IS_{(1^3)}\uparrow S_3=RS_3$ and $IS_{(3)}\uparrow S_3=IS_3$,

$$\Xi_3=\begin{pmatrix}6&0&0\\3&1&0\\1&1&1\end{pmatrix}.$$

We now obtain $Z_3$ in the following way:

$$Z_3=M_3^{-1}\Xi_3=\begin{pmatrix}1&-2&1\\0&1&-1\\0&0&1\end{pmatrix}\begin{pmatrix}6&0&0\\3&1&0\\1&1&1\end{pmatrix}=\begin{pmatrix}1&-1&1\\2&0&-1\\1&1&1\end{pmatrix}.$$

A further result on $M_n$ can be derived from 2.1.10. For this lemma says that $m_{ik}\neq 0$ only if $\alpha^i\trianglelefteq\alpha^k$. This means for the characters

**2.2.17**
$$\xi^\alpha=\zeta^\alpha+\sum_{\beta\rhd\alpha}(\xi^\alpha,[\beta])\zeta^\beta$$

and yields a lot of zeros in the matrix $M_n$, $n\geqslant 6$.

We would like to derive a result, which in a sense reverses 2.2.17, and which will turn out to allow a representation theoretical proof of the Ruch-Schönhofer theorem. We have already seen that $i(IS_\alpha\uparrow S_n, IS_\beta\uparrow S_n)$ is equal to the number of matrices over $\mathbb{N}_0$ with $\alpha$ as vector of row sums and $\beta$ as vector of column sums. We can apply this if we happen to know the number of such matrices. The following lemma gives the number of matrices in a particular case:

**2.2.18** Lemma. *If $r_1$, $r_2$, $c_1$, and $c_2$ are nonnegative integers with the property $r_1+r_2=c_1+c_2$, then the number of $2\times 2$ matrices over $\mathbb{N}_0$ with row sums $r_1$, $r_2$ and column sums $c_1$, $c_2$ is equal to $1+\min\{r_1,r_2,c_1,c_2\}$.*

*Proof.* If for example $r_1=\min\{r_1, r_2, c_1, c_2\}$, then we have the following $1+r_1$ choices for the entries of the first row of such a $2\times 2$ matrix:

$$\begin{pmatrix} r & r_1-r \\ * & * \end{pmatrix}, \qquad 0\leqslant r\leqslant r_1.$$

Each of these choices yields exactly one $2\times 2$ matrix with row sums $r_i$ and column sums $c_i$, since $r_1$ was assumed to be the minimum, so that a suitable second row can be found and is uniquely determined. In the case when $r_2$, $c_1$, or $c_2$ is the minimum, an analogous argument yields the statement. ∎

This helps in the proof of

**2.2.19** LEMMA. *If* $\alpha=(\alpha_1, \alpha_2)\vdash n$, $\alpha_2>0$, *and* $\alpha^*:=(\alpha_1+1, \alpha_2-1)$, *then*

(i) $[\alpha]+IS_{\alpha^*}\uparrow S_n=IS_\alpha\uparrow S_n$,

(ii) *the dimension* $f^\alpha$ *of* $[\alpha]$ *satisfies*

$$f^\alpha=\binom{n}{\alpha_2}-\binom{n}{\alpha_2-1},$$

(iii) $IS_\alpha\uparrow S_n=\sum_{\nu=0}^{\alpha_2}[n-\nu, \nu]$.

*Proof.* (i): 2.2.18 yields for the inner product of the generalized character $\xi^\alpha-\xi^{\alpha^*}$ with itself:

$$i(IS_\alpha\uparrow S_n, IS_\alpha\uparrow S_n)+i(IS_{\alpha^*}\uparrow S_n, IS_{\alpha^*}\uparrow S_n)-2i(IS_\alpha\uparrow S_n, IS_{\alpha^*}\uparrow S_n)$$
$$=(\alpha_2+1)+(\alpha_2-1+1)-2(\alpha_2-1+1)=1.$$

Hence $\xi^\alpha-\xi^{\alpha^*}$ is $\pm 1$ times an irreducible character of $S_n$. Since $(IS_\alpha\uparrow S_n,[\alpha])=1$, but $(IS_{\alpha^*}\uparrow S_n,[\alpha])=0$ (by 2.1.10), we obtain $\xi^\alpha-\xi^{\alpha^*}=\zeta^\alpha$, as stated.

(ii): The statement follows from the dimensions

$$\dim(IS_\alpha\uparrow S_n)=\frac{n!}{\alpha_1!\alpha_2!}=\binom{n}{\alpha_2},$$

$$\dim(IS_{\alpha^*}\uparrow S_n)=\frac{n!}{(\alpha_1+1)!(\alpha_2-1)!}=\binom{n}{\alpha_2-1}$$

by an application of (i).

(iii): This part follows immediately from part (i), by induction on $\alpha_2$. ∎

**2.2.20** THEOREM. *If* $\alpha, \beta \vdash n$, *then there is a representation* $D_{\beta\alpha}$ *of* $S_n$ *with* $IS_\beta \uparrow S_n + D_{\beta\alpha} = IS_\alpha \uparrow S_n$ *if and only if* $\alpha \lhd \beta$.

*Proof.*

(i) If there exists $D_{\beta\alpha}$ with $IS_\beta \uparrow S_n + D_{\beta\alpha} = IS_\alpha \uparrow S_n$, then $(IS_\beta \uparrow S_n, [\beta]) = 1$ shows that $(IS_\alpha \uparrow S_n, [\beta]) \geqslant 1$; thus $\alpha \trianglelefteq \beta$ by 2.1.10. Also $\alpha \neq \beta$.

(ii) If $\alpha \lhd \beta$, then there exists a chain of partitions $\alpha^i$ such that

$$\alpha = \alpha^0 \lhd \alpha^1 \lhd \cdots \lhd \alpha^r = \beta.$$

This shows that it suffices to prove the existence of a $D_{\beta\alpha}$ of the required form only for the case $\alpha \lhd \beta$. Assuming that $\alpha \lhd \beta$, $\beta$ is of the form

$$\beta = (\alpha_1, \ldots, \alpha_{i-1}, \alpha_i + 1, \alpha_{i+1}, \ldots, \alpha_{j-1}, \alpha_j - 1, \alpha_{j+1}, \ldots)$$

for suitable $i$ and $j$ (cf. 1.4.10). If we denote by # the outer tensor-product multiplication of representations, then

$$\begin{aligned} IS_\alpha \uparrow S_n &= ([\alpha_1] \# \cdots \# [\alpha_n]) \uparrow S_n \\ &= \Big( [\alpha_i] \# [\alpha_j] \# \Big( \mathop{\#}_{\nu \neq i, j} [\alpha_\nu] \Big) \Big) \uparrow S_n \end{aligned}$$

which by 2.2.19(i) equals

$$\begin{aligned} &\Big( \big( [\alpha_i, \alpha_j] + ([\alpha_i + 1] \# [\alpha_j - 1]) \uparrow S_{\alpha_i + \alpha_j} \big) \# \Big( \mathop{\#}_{\nu \neq i, j} [\alpha_\nu] \Big) \Big) \uparrow S_n \\ &\qquad = \Big( [\alpha_i, \alpha_j] \# \Big( \mathop{\#}_{\nu \neq i, j} [\alpha_\nu] \Big) \Big) \uparrow S_n + IS_\beta \uparrow S_n \\ &\qquad = D_{\beta\alpha} + IS_\beta \uparrow S_n, \end{aligned}$$

where

$$D_{\beta\alpha} := \Big( [\alpha_i, \alpha_j] \# \Big( \mathop{\#}_{\nu \neq i, j} [\alpha_\nu] \Big) \Big) \uparrow S_n. \qquad \blacksquare$$

Expressed in terms of the matrix $M_n$ this reads as follows:

**2.2.21** COROLLARY. *For given i and j, each entry* $m_{ik}$ *of the ith row of* $M_n$ *is greater than or equal to the corresponding entry* $m_{jk}$ *of the jth row of* $M_n$ *if and only if* $\alpha^i \trianglelefteq \alpha^j$.

Furthermore we obtain from 2.2.20 and 2.1.10:

**2.2.22** COROLLARY. $(IS_\alpha \uparrow S_n, [\beta]) > 0 \Leftrightarrow \alpha \trianglelefteq \beta$.

Finally we obtain the Ruch-Schönhofer theorem:

**2.2.23** COROLLARY. $(IS_{\alpha}\uparrow S_n, AS_{\beta'}\uparrow S_n)\neq 0 \Leftrightarrow \alpha \trianglelefteq \beta$.

For 2.2.22 yields the $\Leftarrow$ part, while the other direction of the statement follows from the trivial part of the Gale-Ryser theorem. This completes the story and gives at the same time a representation theoretical proof of both the Gale-Ryser theorem and the Ruch-Schönhofer theorem.

## 2.3 The Ordinary Irreducible Characters as $\mathbb{Z}$-linear Combinations of Permutation Characters

An important result of the last section was Theorem 2.2.10, which says that each ordinary irreducible character $\zeta^{\alpha}$ of $S_n$ is a uniquely determined $\mathbb{Z}$-linear combination of the permutation characters $\xi^{\beta}$. It is the aim of the present section to derive a direct method for determining these linear combinations.

The expression for $\zeta^{\alpha}$ in terms of the $\xi^{\beta}$ is of a determinantal form, so we first introduce a multiplication of representations $[\alpha]$ of symmetric groups. If $m$ and $n$ are natural numbers, then $S_m \times S_n$ can be embedded into $S_{m+n}$ in a natural way:

2.3.1 $$S_m \times S_n \hookrightarrow S_{m+n},$$

since $S_{\mathbf{n}} \hat{=} S_{\mathbf{m+n}\setminus\mathbf{m}}$ (where $\mathbf{m+n}:=\{1,\ldots,m+n\}$).

Hence if $\alpha \vdash m$ and $\beta \vdash n$, then the (ordinary irreducible) representation $[\alpha]\#[\beta]$ of $S_m \times S_n$ defines a representation of a subgroup of $S_{m+n}$ which is isomorphic to $S_m \times S_n$, and we can induce into $S_{m+n}$. The resulting representation is denoted by $[\alpha]\,[\beta]$ and called the *outer product* of $[\alpha]$ and $[\beta]$:

2.3.2 $$[\alpha]\,[\beta]:=[\alpha]\#[\beta]\uparrow S_{m+n}.$$

It is easy to see that this multiplication is associative (apply (43.3) in Curtis and Reiner [1962]), and it is commutative. Hence for example

**2.3.3** $$IS_{\alpha}\uparrow S_n = [\alpha_1]\,[\alpha_2]\cdots[\alpha_h].$$

With the aid of this multiplication we now define a determinant which corresponds to $\alpha=(\alpha_1,\alpha_2,\ldots)\vdash n$ as follows:

2.3.4 $|[\alpha_i+j-i]|=$

$$\begin{vmatrix} [\alpha_1] & [\alpha_1+1] & [\alpha_1+2] & \ldots & [\alpha_1+h-1] \\ [\alpha_2-1] & [\alpha_2] & [\alpha_2+1] & \ldots & [\alpha_2+h-2] \\ \cdots & \cdots & \cdots & \cdots & \cdots \\ [\alpha_h-h+1] & [\alpha_h-h+2] & [\alpha_h-h+3] & \ldots & [\alpha_h] \end{vmatrix}$$

when $\alpha_{h+1}=0$, subject to the definitions

2.3.5 $$[r]:=\begin{cases}1 & \text{if } r=0 \text{ (which is consistent with 2.3.3)},\\ 0 & \text{if } r<0.\end{cases}$$

(We put $1\cdot[r]:=[r]$ and $0\cdot[r]:=0$.)

Note that 2.3.5 ensures that the determinant's definition is independent of the choice of $h$ (subject only to $\alpha_{h+1}=0$). For example if $\alpha:=(3,1^2)\vdash 5$, we get

$$\begin{vmatrix}[3] & [4] & [5]\\ 1 & [1] & [2]\\ 0 & 1 & [1]\end{vmatrix}=[3][1][1]-[3][2]-[4][1]+[5]$$

$$=\begin{vmatrix}[3] & [4] & [5] & [6]\\ 1 & [1] & [2] & [3]\\ 0 & 1 & [1] & [2]\\ 0 & 0 & 0 & 1\end{vmatrix}.$$

Since minus signs appear, the determinant 2.3.4 has to be interpreted as a generalized "representation". The character of our example is

$$\xi^{(3,1^2)}-\xi^{(3,2)}-\xi^{(4,1)}+\xi^{(5)}.$$

If we had $M_5$ at hand, then we would see that this character is equal to $\zeta^{(3,1^2)}$, so that the above determinant which corresponds to $\alpha=(3,1^2)$ yields the desired expression for $\zeta^\alpha$ as a $\mathbb{Z}$-linear combination of the permutation characters $\xi^\beta$. And it is in fact true that for any partition $\alpha$, the determinant 2.3.4 gives this expression, for short:

$$[\alpha]=|[\alpha_i+j-i]|.$$

It is our aim to provide a proof of this.

In order to do this, we first generalize our definition of permutation characters $\xi^\alpha$ slightly. We defined an improper partition $\lambda$ of $n$ to be a sequence of nonnegative integers whose sum is $n$. Thus, $\lambda$ may be regarded as an element of $\mathbb{Z}^{\mathbb{N}}$, the set of mappings from $\mathbb{N}$ into $\mathbb{Z}$. Now assume only that $\lambda\in\mathbb{Z}^{\mathbb{N}}$ and $\sum_{i=1}^{\infty}\lambda_i=n$ (so that $\lambda$ still has finite support, but we no longer assume that its parts are nonnegative). Then $\lambda$ will be called a *composition* of $n$. (The reader will be encouraged to hear that we do not intend to introduce any more terms to describe elements of $\mathbb{Z}^{\mathbb{N}}$.) He who likes his definitions to be as general as possible may want occasionally to drop the assumption that $\sum_{i=1}^{\infty}\lambda_i=n$ (e.g. in 2.3.6).

Sometimes, when using compositions of $n$, we shall impose the restriction that $\lambda_i=0$ for $i>n$ (this works automatically for proper partitions)—a restriction which could, in fact, be avoided in all places where we make it, but this would force the introduction of some really unwieldy notation.

To continue with the story, let $\lambda$ be a composition of $n$, and put

2.3.6 $$\xi^\lambda := \begin{cases} \text{the character of } IS_\lambda \uparrow S_n & \text{if } \lambda \models n, \\ \text{the function mapping } S_n \text{ to } 0 & \text{otherwise.} \end{cases}$$

Although it is unusual to consider "the zero character", as in the case where $\lambda$ is not an improper partition of $n$, these zero characters behave in the obvious way. The properties we need are these:

(i) (the zero character of $S_m$) # (a character of $S_k$) = the zero character of $S_{m+k}$.

(ii) Inducing or restricting a zero character gives a zero character.

These results follow by inspecting the processes involved; alternatively, they may be taken as the definitions of the restriction of a zero character, etc.

Now regard $S_n$ as the subgroup of $S_{\mathbb{N}}$ consisting of the permutations fixing each point $i>n$. Then, by definition, every $\pi$ which belongs to $S_{\mathbb{N}}$ belongs to some $S_n$. (Remember that we chose to define a permutation to be a bijection fixing all but finitely many points.) Hence we may extend the function sgn to $S_{\mathbb{N}}$ in a well-defined manner. For every composition $\lambda$ of $n$, and every $\pi \in S_{\mathbb{N}}$, $\lambda \circ \pi$ is a composition of $n$, and

$$\xi^{\lambda\circ\pi} = \xi^\lambda.$$

We define addition on $\mathbb{Z}^{\mathbb{N}}$ pointwise. Then

2.3.7 $$\lambda - \mathrm{id} + \pi = (\lambda_1 - 1 + \pi(1), \lambda_2 - 2 + \pi(2), \ldots)$$

is a composition of $n$.

Using this notation, and having 2.3.6 in mind, we put for compositions $\lambda$ of $n$

2.3.8 $$\chi^\lambda := \sum_{\pi\in S_{\mathbb{N}}} \operatorname{sgn}\pi \cdot \xi^{\lambda-\mathrm{id}+\pi}.$$

Since $\sum_{i=1}^\infty \lambda_i = n$, there exists $h \in \mathbb{N}_0$ such that $\lambda_i = 0$ for all $i>h$; then for all $\pi \in S_{\mathbb{N}} \setminus S_h$, $\lambda - \mathrm{id} + \pi$ has a negative part. Therefore, the sum in 2.3.8 is finite and

$$\forall i \geq h, \quad \chi^\lambda = \sum_{\pi\in S_i} \operatorname{sgn}\pi \cdot \xi^{\lambda-\mathrm{id}+\pi}.$$

Thus $\chi^\lambda$ is a generalized character of $S_n$. In particular, if $\lambda := \alpha \vdash n$, then $\chi^\alpha$ is clearly the character of the determinant $\|[\alpha_i + j - i]\|$, so that it is our aim to prove $\chi^\alpha = \zeta^\alpha$.

We divide the proof of this basic result into several steps. A preliminary lemma gives an important property of $\chi^\lambda$:

**2.3.9 Lemma.** *If $\lambda$ is a composition of $n$, and*

$$\mu := (\lambda_1, \ldots, \lambda_{i-1}, \lambda_{i+1} - 1, \lambda_i + 1, \lambda_{i+2}, \ldots),$$

*then* $\chi^\mu = -\chi^\lambda$.

*Proof.* We put $\tau := (i, i+1) \in S_{\mathbb{N}}$. Then we have

$$(\mu - \mathrm{id} + \pi\tau)(j) = \begin{cases} (\lambda - \mathrm{id} + \pi)(j), & \text{if } j \neq i, i+1, \\ (\lambda - \mathrm{id} + \pi)(i+1), & \text{if } j = i, \\ (\lambda - \mathrm{id} + \pi)(i) & \text{if } j = i+1. \end{cases}$$

This implies $\xi^{\mu - \mathrm{id} + \pi\tau} = \xi^{\lambda - \mathrm{id} + \pi}$, so that

$$\chi^\lambda = \sum_\pi \operatorname{sgn} \pi \cdot \xi^{\lambda - \mathrm{id} + \pi} = \sum_\pi (-\operatorname{sgn} \pi\tau) \xi^{\mu - \mathrm{id} + \pi\tau} = -\chi^\mu. \qquad \blacksquare$$

**2.3.10 Lemma.** *Suppose that $\lambda$ is a composition of $n = m + k$. Then*

(i) $\xi^\lambda \downarrow S_m \times S_k = \sum_{\mu \models k} \xi^{\lambda - \mu} \# \xi^\mu$,
(ii) $\chi^\lambda \downarrow S_m \times S_k = \sum_{\mu \models k} \chi^{\lambda - \mu} \# \xi^\mu$.

*Proof.* (i): Both sides of the equation equal zero unless $\lambda$ is an improper partition of $m + k$. Assume, therefore, that $\lambda \models m + k$. (The sum over $\mu$ has only finitely many nonzero terms, since $\xi^{\lambda - \mu} \# \xi^\mu = 0$ unless $\lambda - \mu \models m$.) Mackey's subgroup theorem then yields

$$\xi^\lambda \downarrow \mathrm{S}_m \times \mathrm{S}_k = \sum_{\mathrm{S}_m \times \mathrm{S}_k \pi S_\lambda} I\left((\mathrm{S}_m \times \mathrm{S}_k) \cap \pi S_\lambda \pi^{-1}\right) \uparrow \mathrm{S}_m \times \mathrm{S}_k.$$

By 1.3.10 the double-coset $\mathrm{S}_m \times \mathrm{S}_k \pi S_\lambda$ is characterized by the $2 \times n$ matrix

$$\begin{pmatrix} \cdots |\mathbf{m} \cap \pi[\mathbf{n}_j^\lambda]| \cdots \\ \cdots |(\mathbf{n} \setminus \mathbf{m}) \cap \pi[\mathbf{n}_j^\lambda]| \cdots \end{pmatrix}, 1 \leq j \leq n.$$

It is therefore uniquely determined by

$$\mu := \left(\ldots, |(\mathbf{n} \setminus \mathbf{m}) \cap \pi[\mathbf{n}_j^\lambda]|, \ldots\right) \models k.$$

Furthermore, as $(S_m \times S_k) \cap \pi S_\lambda \pi^{-1} \hat{=} S_{\lambda-\mu} \times S_\mu$, we have

$$I\big((S_m \times S_k) \cap \pi S_\lambda \pi^{-1}\big) \uparrow S_m \times S_k = IS_{\lambda-\mu} \uparrow S_m \# IS_\mu \uparrow S_k.$$

The character of this representation is

$$\xi^{\lambda-\mu} \# \xi^\mu,$$

and so (i) is proved.

(ii): The definition 2.3.8 of $\chi^\lambda$ yields

$$\chi^\lambda \downarrow S_m \times S_k = \sum_\pi \operatorname{sgn} \pi \cdot \xi^{\lambda - \mathrm{id} + \pi} \downarrow S_m \times S_k,$$

and by (i), this equals

$$\sum_{\mu \vDash k} \sum_\pi \operatorname{sgn} \pi \cdot \xi^{(\lambda-\mu)-\mathrm{id}+\pi} \# \xi^\mu = \sum_{\mu \vDash k} \chi^{\lambda-\mu} \# \xi^\mu. \qquad \blacksquare$$

If $\lambda$ is a composition of $n$, and $\mu$ is a composition of $k$, we define

2.3.11 $$\chi^{\lambda/\mu} := \sum_{\pi \in S_{\mathbb{N}}} \operatorname{sgn} \pi \cdot \xi^{\lambda - \mathrm{id} - (\mu - \mathrm{id}) \circ \pi}.$$

We claim that the following is true:

**2.3.12** Lemma. *If $\lambda$ is a composition of $m+k$, then $\chi^\lambda \downarrow S_m \times S_k = \sum_{\beta \vDash k} \chi^{\lambda/\beta} \# \chi^\beta$.*

*Proof.* The proof of 2.3.10(ii) has already shown that

$$\chi^\lambda \downarrow S_m \times S_k = \sum_{\pi \in S_{\mathbb{N}}} \operatorname{sgn} \pi \sum_{\mu \vDash k} \xi^{\lambda - \mathrm{id} + \pi - \mu} \# \xi^\mu.$$

Replacing $\mu$ by $\mu \circ \pi$ we get (as $\xi^{\mu \circ \pi} = \xi^\mu$)

$$\chi^\lambda \downarrow S_m \times S_k = \sum_\pi \operatorname{sgn} \pi \sum_{\mu \vDash k} \xi^{\lambda - \mathrm{id} - (\mu - \mathrm{id}) \circ \pi} \# \xi^\mu.$$

Now, for each $\mu$, $\mu_i$ is eventually zero, and thus there exist uniquely determined $\beta \vDash k$ and $\sigma \in S_{\mathbb{N}}$ such that

$$(\beta - \mathrm{id}) \circ \sigma = \mu - \mathrm{id}$$

and

$$\beta_1 - 1 \geqslant \beta_2 - 2 \geqslant \cdots.$$

This enables us to proceed as follows (recall that all the sums are finite, so there is no problem about rearranging):

$$\begin{aligned}
\chi^{\lambda}\downarrow S_m\times S_k &= \sum_{\pi\in S_N}\operatorname{sgn}\pi\sum_{\sigma\in S_N}\sum_{\beta}\xi^{\lambda-\mathrm{id}-(\beta-\mathrm{id})\circ\sigma\circ\pi}\#\xi^{(\beta-\mathrm{id})\circ\sigma+\mathrm{id}}\\
&= \sum_{\pi,\sigma}\operatorname{sgn}(\sigma\pi)\operatorname{sgn}\sigma\sum_{\beta}\xi^{\lambda-\mathrm{id}-(\beta-\mathrm{id})\circ\sigma\pi}\#\xi^{\beta-\mathrm{id}+\sigma^{-1}}\\
&= \sum_{\beta}\sum_{\rho\in S_N}\operatorname{sgn}\rho\,\xi^{\lambda-\mathrm{id}-(\beta-\mathrm{id})\circ\rho}\#\sum_{\sigma\in S_N}\operatorname{sgn}\sigma\,\xi^{\beta-\mathrm{id}+\sigma^{-1}}\\
&= \sum_{\beta}\chi^{\lambda/\beta}\#\chi^{\beta}\\
&= \sum_{\beta\vdash k}\chi^{\lambda/\beta}\#\chi^{\beta}.
\end{aligned}$$

The last equation holds because $\chi^{\beta}=0$ if $\beta_i-i=\beta_{i+1}-(i+1)$ (by 2.3.9). ■

The lemma shows the importance of the generalized characters $\chi^{\lambda/\beta}$ of $S_{n-k}$.

**2.3.13** THEOREM. *If $\alpha\vdash n$, $k\leqslant n$, and $\beta\vdash k$, then*

(i) $\chi^{\alpha/\beta}\neq 0$ *only if $\alpha_i\geqslant\beta_i$ for all $i$,*

(ii) $(\chi^{\alpha/\beta},\xi^{(n-k)})=\begin{cases}1 & \text{if } \alpha_1\geqslant\beta_1\geqslant\alpha_2\geqslant\beta_2\geqslant\alpha_3\geqslant\cdots,\\ 0 & \text{otherwise}.\end{cases}$

*Proof.* (i): We consider the determinant by which $\chi^{\alpha/\beta}$ is defined (cf. 2.3.11). It is (cf. 2.3.4)

$$|[\alpha_i-i-(\beta_j-j)]|.$$

As the sequences $\alpha-\mathrm{id}$ and $\beta-\mathrm{id}$ are strictly decreasing, an entry $[\alpha_i-i-(\beta_j-j)]=0$, i.e. $\alpha_i-i-(\beta_j-j)<0$, implies that all the other entries to the left and below this one are also 0. Therefore the determinant vanishes when a diagonal entry is zero, i.e. when $0>\alpha_i-i-(\beta_i-i)=\alpha_i-\beta_i$. Hence $\chi^{\alpha/\beta}\neq 0$ implies $\beta_i\leqslant\alpha_i$.

(ii): For each composition $\lambda$ of $n-k$ we have (cf. 2.2.14)

$$(\xi^{\lambda},\xi^{(n-k)})=\begin{cases}1 & \text{if } \lambda\vDash n-k,\\ 0 & \text{otherwise.}\end{cases}$$

Thus, if $\delta$ denotes the Kronecker symbol,

$$\begin{aligned}
(\chi^{\alpha/\beta},\xi^{(n-k)}) &= \sum_{\pi}\operatorname{sgn}\pi\prod_i\delta_{\alpha_i-i-(\beta_{\pi(i)}-\pi(i)),\geqslant 0}\\
&= |\delta_{\alpha_i-i-(\beta_j-j),\geqslant 0}|.
\end{aligned}$$

If $\alpha_1 \geqslant \beta_1 \geqslant \alpha_2 \geqslant \beta_2 \geqslant \alpha_3 \geqslant \cdots$, then this determinant has 1's along its main diagonal and 0's below, and hence it is equal to 1. Otherwise it is not of this triangular form, so that by the monotonicity of the sequences $\alpha - \mathrm{id}$ and $\beta - \mathrm{id}$ two columns are equal, and hence the determinant must vanish in this case. ■

This result turns out to be crucial in the proof of

**2.3.14** YOUNG'S RULE (FIRST VERSION). *For each $\lambda \models n$ such that $\lambda_i = 0$ when $i > n$ and every partition $\alpha$ of $n$, we have that the inner product $(\chi^\alpha, \xi^\lambda)$ is equal to the number of $(n-1)$-tuples $(\beta^{(1)}, \ldots, \beta^{(n-1)})$ such that*

(i) $\forall 1 \leqslant i \leqslant n-1 \ (\beta^{(i)} \vdash \sum_{j=1}^{i} \lambda_j)$,
(ii) $\forall 1 \leqslant j \leqslant n \ (\beta_j^{(1)} \leqslant \beta_j^{(2)} \leqslant \cdots \leqslant \beta_j^{(n-1)} \leqslant \alpha_j)$,
(iii) $\forall j > 1,\ i \geqslant 1 \ (\beta_j^{(i)} \leqslant \beta_{j-1}^{(i-1)})$, *if we set* $\beta^{(0)} := (0)$ *and* $\beta^{(n)} := \alpha$.

*Proof.* Applying 2.3.12, we have

$$(\chi^\alpha, \xi^\lambda) = \left( \chi^\alpha \downarrow S_\lambda,\ \mathop{\#}_{i=1}^{n} \xi^{(\lambda_i)} \right)$$

$$= \sum_{\beta^{(n-1)} \vdash \sum_1^{n-1} \lambda_j} \left( \chi^{\alpha/\beta^{(n-1)}} \# \chi^{\beta^{(n-1)}},\ \xi^{(\lambda_n)} \# \mathop{\#}_{1}^{n-1} \xi^{(\lambda_j)} \right)$$

$$= \sum_{\beta^{(n-1)}} \left( \chi^{\alpha/\beta^{(n-1)}},\ \xi^{(\lambda_n)} \right) \left( \chi^{\beta^{(n-1)}},\ \mathop{\#}_{i=1}^{n-1} \xi^{(\lambda_j)} \right),$$

which is by 2.3.13

$$= \sum_{\beta^{(n-1)}} \left( \chi^{\beta^{(n-1)}},\ \mathop{\#}_{i=1}^{n-1} \xi^{(\lambda_j)} \right),$$

if the sum is taken over all the $\beta^{(n-1)}$ which satisfy

$$\beta^{(n-1)} \vdash \sum_{i=1}^{n-1} \lambda_j \quad \text{and} \quad \alpha_1 \geqslant \beta_1^{(n-1)} \geqslant \alpha_2 \geqslant \beta_2^{(n-1)} \geqslant \cdots.$$

Another application of 2.3.12 and 2.3.13 yields

$$(\chi^\alpha, \xi^\lambda) = \sum_{\beta^{(n-1)}} \sum_{\beta^{(n-2)}} \left( \chi^{\beta^{(n-2)}},\ \mathop{\#}_{i=1}^{n-2} \xi^{(\lambda_j)} \right),$$

where the sums have to be taken over $\beta^{(n-1)}$ and $\beta^{(n-2)}$ subject to the following conditions:

(i) $\beta^{(n-1)} \vdash \sum_1^{n-1} \lambda_j$, $\beta^{(n-2)} \vdash \sum_1^{n-2} \lambda_j$,
(ii) $\alpha_1 \geqslant \beta_1^{(n-1)} \geqslant \alpha_2 \geqslant \beta_2^{(n-1)} \geqslant \cdots$,
(iii) $\beta_1^{(n-1)} \geqslant \beta_1^{(n-2)} \geqslant \beta_2^{(n-1)} \geqslant \beta_2^{(n-2)} \geqslant \cdots$.

Further iterations yield

$$\begin{aligned}(\chi^\alpha, \xi^\lambda) &= \sum_{\beta^{(n-1)}} \sum_{\beta^{(n-2)}} \cdots \sum_{\beta^{(1)}} \left(\chi^{\beta^{(1)}}, \xi^{(\lambda_1)}\right) \\ &= \sum_{\beta^{(n-1)}} \cdots \sum_{\beta^{(1)}} 1 \\ &= \sum_{(\beta^{(1)},\ldots,\beta^{(n-1)})} 1,\end{aligned}$$

where the sum is taken over all $(n-1)$-tuples $(\beta^{(1)},\ldots,\beta^{(n-1)})$ subject to the conditions described in the statement. ■

We are now in a position to prove the main result of this section:

**2.3.15** THEOREM. *For each $\alpha \vdash n$ we have that*

$$\chi^\alpha = \zeta^\alpha,$$

*the character of the ordinary irreducible representation $[\alpha]$ of $S_n$. Thus, in particular, the set*

$$\{\chi^\alpha \mid \alpha \vdash n\}$$

*is the complete set of ordinary irreducible characters of $S_n$. Each $\zeta^\alpha$ can be expressed in the determinantal form*

$$\zeta^\alpha = \sum_{\pi \in S_n} \operatorname{sgn} \pi \cdot \xi^{\alpha - \mathrm{id} + \pi}$$

*as a linear combination of permutation characters $\xi^\lambda$, $\lambda \vDash n$, with coefficients $0, \pm 1$. We can therefore express the representation $[\alpha]$ itself as a generalized "representation" in the following* determinantal form:

$$[\alpha] = |[\alpha_i - i + j]|$$

*subject to the conventions $[r] := 1$ if $r = 0$, and $[r] := 0$ if $r < 0$.*

*Proof.*

(i) We prove first that for any $\lambda \models n$ (with $\lambda_i = 0$ for $i > n$), $(\chi^\alpha, \xi^\lambda) \neq 0$ implies $\alpha \trianglerighteq \lambda$: This follows from Young's rule, for condition 2.3.14(iii) yields $\beta_2^{(1)} = \beta_3^{(2)} = \cdots = \beta_n^{(n-1)} = 0$. Hence $\beta^{(1)} \vdash \lambda_1$ together with $\alpha_1 \geqslant \beta_1^{(1)}$ gives $\alpha_1 \geqslant \lambda_1$ (by $\beta_2^{(1)} = 0$); $\beta^{(2)} \vdash \lambda_1 + \lambda_2$ together with $\alpha_1 \geqslant \beta_1^{(2)}$, $\alpha_2 \geqslant \beta_2^{(2)}$ yields $\alpha_1 + \alpha_2 \geqslant \beta_1^{(2)} + \beta_2^{(2)} \geqslant \lambda_1 + \lambda_2$ (by $\beta_3^{(2)} = 0$), and so on.

(ii) $(\chi^\alpha, \xi^\alpha) = 1$: By Young's rule $(\chi^\alpha, \xi^\alpha)$ is equal to the number of $(n-1)$-tuples $(\beta^{(1)}, \ldots, \beta^{(n-1)})$ which satisfy (see (i))

$$\alpha_1 \geqslant \beta_1^{(1)} \geqslant \alpha_1; \qquad \alpha_1 \geqslant \beta_1^{(2)}, \quad \alpha_2 \geqslant \beta_2^{(2)}, \quad \alpha_1 + \alpha_2 \geqslant \beta_1^{(2)} + \beta_2^{(2)} \geqslant \alpha_1 + \alpha_2; \ldots$$

Thus $((\alpha_1), (\alpha_1, \alpha_2), \ldots, (\alpha_1, \ldots, \alpha_{n-1}))$ is the only possible $(n-1)$-tuple.

(iii) $(\chi^\alpha, \chi^\alpha) = 1$: If $(\chi^\alpha, \xi^{\alpha - \mathrm{id} + \pi}) \neq 0$, then $\alpha - \mathrm{id} + \pi \models n$ and $\alpha \trianglerighteq \alpha - \mathrm{id} + \pi$, by (i). But $\alpha - \mathrm{id} + \pi \trianglerighteq \alpha$, as it is very easy to see, and so $\alpha = \alpha - \mathrm{id} + \pi$ and $\pi = 1$. Together with part (ii), this shows that $\xi^\alpha$ is the unique summand $\xi^\lambda$ of $\chi^\alpha$ for which $(\chi^\alpha, \xi^\lambda) \neq 0$. Therefore, $(\chi^\alpha, \chi^\alpha) = (\chi^\alpha, \xi^\alpha) = 1$.

(iv) In order finally to identify $\chi^\alpha$ with $\zeta^\alpha$, we consider the determinantal form $|[\alpha_i - i + j]|$ which shows that $[\alpha]$ is contained (use 2.2.22) in the main diagonal term only. Thus $(\chi^\alpha, \zeta^\alpha) = 1$. But (iii) shows that $\chi^\alpha$ is $\pm$ an irreducible character, and so $\chi^\alpha = \zeta^\alpha$. ■

A numerical example is

$$[2,1] = \begin{vmatrix} [2] & [3] \\ 1 & [1] \end{vmatrix} = [2][1] - [3].$$

But

$$[2][1] = IS_{(2,1)} \uparrow S_3 = IS_2 \uparrow S_3$$

has the character

$$\xi^{(2,1)}(\pi) = a_1(\pi),$$

for $\pi \in S_3$ induces on the left cosets of $S_2$ in $S_3$ the same permutation as on the tabloids

$$\frac{\overline{1\ 2}}{\underline{3}}, \quad \frac{\overline{1\ 3}}{\underline{2}}, \quad \frac{\overline{2\ 3}}{\underline{1}}.$$

Thus

$$\zeta^{(2,1)}(\pi) = a_1(\pi) - 1.$$

This yields again the character table of $S_3$:

$$Z_3=\begin{pmatrix} 1 & -1 & 1 \\ 2 & 0 & -1 \\ 1 & 1 & 1 \end{pmatrix}.$$

More generally we have by 2.3.15 for each $n \geqslant 2$:

**2.3.16**
$$\text{(i) } [n-1,1]=[n-1][1]-[n],$$
$$\text{(ii) } \zeta^{(n-1,1)}(\pi)=a_1(\pi)-1.$$

(This of course also follows from 2.2.19(iii).) The representation

$$[n-1][1],$$

which has the character

$$\xi^{(n-1,1)}(\pi)=a_1(\pi),$$

is sometimes called the *natural representation* of $S_n$.

Another useful corollary of the determinantal formula is the following result on the values of the ordinary irreducible characters of $S_n$ on the class of elements which consist of a single $n$-cycle:

**2.3.17** $$\zeta^\alpha((1\cdots n))=\begin{cases} (-1)^r & \text{if } \alpha=(n-r,1^r),\quad 0\leqslant r\leqslant n-1, \\ 0 & \text{otherwise}. \end{cases}$$

*Proof.* The only Young subgroup which contains $n$-cycles is the Young subgroup $S_\beta := S_n$. Thus $\xi^\beta((1\cdots n))\neq 0$ only if $\beta=(n)$. The determinantal form therefore shows $\zeta^\alpha((1\cdots n))\neq 0$ only if a summand $\pm\xi^{(n)}$ occurs in

$$\zeta^\alpha=\sum_\pi \operatorname{sgn}\pi\cdot\xi^{\alpha-\mathrm{id}+\pi}.$$

But this is the case only if $\|[\alpha_i+j-i]\|$ is of the form

$$\|[\alpha_i+j-i]\|=\begin{vmatrix} [\alpha_1] & \cdots & [n] \\ & & \vdots \\ & & [\alpha_h] \end{vmatrix}$$

($h$ minimal such that $\alpha_{h+1}=0$), so that we must have $\alpha_1+h-1=n$ (cf. 2.3.4), i.e. $h=n-\alpha_1+1$, or equivalently $\alpha$ is of the form $\alpha=(n-r,1^r)$ for a suitable $r\leqslant n-1$.

If on the other hand $\alpha=(n-r,1^r)$, then

$$\zeta^\alpha((1\cdots n))=(-1)^{h-1}\xi^{(n)}((1\cdots n))=(-1)^{h-1}$$
$$=(-1)^r,$$

as stated. ■

This shows that $\zeta^\alpha((1\cdots n))\neq 0$ if and only if $[\alpha]$ is a Γ-shaped diagram:

$$[n-r,1^r]=\begin{array}{lllll}\times & \times & \cdots & \times & \times\\ \times &&&&\\ \vdots &&&&\\ \times &&&&\\ \times &&&&\end{array}$$

Such diagrams are called *hooks* and play an important part in the representation theory of $S_n$. We would like to conclude this section with a proof of a formula which expresses the dimension $f^\alpha$ of $[\alpha]$ in terms of hooks.

It was T. Nakayama who first introduced the concept of hooks, by which we mean specific Γ-shaped subsets of diagrams $[\alpha]$. In order to make this precise, we call the node of $[\alpha]$ which lies in the $i$th row and $j$th column the $(i, j)$-*node* of $[\alpha]$. Then we denote by

2.3.18 $$H_{ij}^\alpha$$

the $(i, j)$-*hook* of $[\alpha]$, which consists of the $(i, j)$-node, called the *corner* of the hook, and all the nodes to the right of it in the same row together with all the nodes lower down and in the same column as the corner:

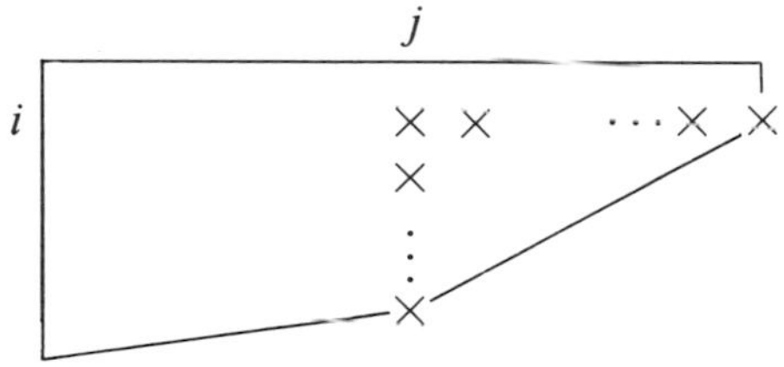


The $(i, k)$-nodes, $k>j$, form the *arm* of the hook, while the $(l, j)$-nodes, $l>i$, form the *leg* of the hook. The $(i, \alpha_i)$-node is called the *hand* of the hook, while the $(\alpha'_j, j)$-node is called the *foot*:

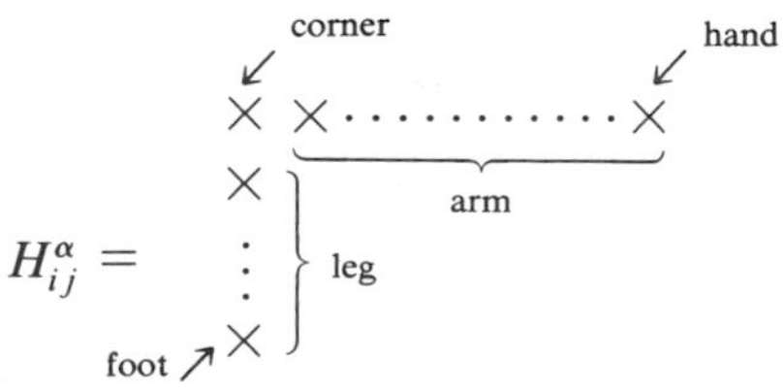

The number $h_{ij}^{\alpha}$ of nodes of $H_{ij}^{\alpha}$, i.e.

2.3.19
$$h_{ij}^{\alpha} := \alpha_i - j + \alpha'_j - i + 1$$

is called the *length* of $H_{ij}^{\alpha}$. The number $l_{ij}^{\alpha}$ of nodes in the leg of $H_{ij}^{\alpha}$, i.e.

2.3.20
$$l_{ij}^{\alpha} := \alpha'_j - i$$

is called the *leg length* of $H_{ij}^{\alpha}$.

To $H_{ij}^{\alpha}$ there corresponds a part of the *rim* of $[\alpha]$ which is of the same length. It consists of the nodes on the rim between the hand and the foot of $H_{ij}^{\alpha}$, including the hand and the foot node. To $H_{1,1}^{(3,2,1^2)}$ for example there correspond the encircled nodes of $[3,2,1^2]$ as follows:

$$\begin{array}{lll} \times & \otimes & \otimes \\ \otimes & \otimes & \\ \otimes & & \\ \otimes & & \end{array}$$

This *associated part of the rim* will be denoted by

$$R_{ij}^{\alpha}.$$

It is important to notice that the result

$$[\alpha] \backslash R_{ij}^{\alpha}$$

of removing $R_{ij}^{\alpha}$ from $[\alpha]$ is again a Young diagram: e.g.

$$[3,2,1^2] \backslash R_{1,1}^{(3,2,1^2)} = \begin{array}{lll} \times & \not\times & \not\times \\ \not\times & \not\times & \\ \not\times & & \\ \times & & \end{array} = \times = [1].$$

We intend to show that the dimension $f^{\alpha}$ of $[\alpha]$ is just $n!$ divided by the product of all the hook lengths:

**2.3.21** THEOREM. $f^{\alpha} = n!/\prod_{i,j} h_{ij}^{\alpha}$.

*Proof.* Notice that for each composition $\lambda = (\lambda_1, \lambda_2, \ldots)$ of $n$ we have

$$\xi^{\lambda}(1) = \frac{n!}{\lambda_1! \lambda_2! \cdots},$$

if we put (cf. 2.3.5) $1/r! = 0$ when $r < 0$. Thus (assuming that $\alpha$ has precisely

$h$ nonzero parts) the determinantal formula gives

$$f^\alpha = \chi^\alpha(1) = n! \left| \frac{1}{(\alpha_i + j - i)!} \right|, \qquad \text{an } h \times h \text{ determinant.}$$

Now

$$\left| \frac{1}{(\alpha_i + j - i)!} \right| = \frac{1}{\prod_i h_{i1}^\alpha !} \left| \frac{h_{i1}^\alpha !}{(\alpha_i + j - i)!} \right| = \frac{1}{\prod_i h_{i1}^\alpha !} \left| \prod_{r=1}^{h-j} (h_{i1}^\alpha + r + j - h) \right|.$$

Elementary transformations of the determinant on the right hand side of this equation yield the Vandermonde determinant

$$|(h_{i1}^\alpha)^{h-j}| = \prod_{i<j} (h_{i1}^\alpha - h_{j1}^\alpha).$$

Therefore

2.3.22
$$f^\alpha = n! \frac{\prod_{i<j} (h_{i1}^\alpha - h_{j1}^\alpha)}{\prod_i h_{i1}^\alpha !}.$$

This equation shows that $f^\alpha$ can be expressed in terms of the hooks $H_{i1}^\alpha$ which have their corner in the first column.

To complete the proof we show that for $1 \leqslant i \leqslant h$ we have

$$\prod_{j=i+1}^h (h_{i1}^\alpha - h_{j1}^\alpha) \prod_{\nu=1}^{\alpha_i} h_{i\nu}^\alpha = h_{i1}^\alpha !. \tag{1}$$

On each side of (1) there are $h_{i1}^\alpha$ factors. The factors $h_{i1}^\alpha - h_{j1}^\alpha$ are strictly increasing as $j$ increases, while the $h_{i\nu}^\alpha$ are strictly decreasing as $\nu$ increases. Furthermore the factors on the left-hand side are all $\leqslant h_{i1}^\alpha$. It therefore suffices to prove that the factors on the left-hand side are pairwise different. This can be done by a verification of

$$h_{i1}^\alpha - h_{j1}^\alpha < h_{i\nu}^\alpha < h_{i1}^\alpha - h_{j+1,1}^\alpha \tag{2}$$

for a suitable $j$, which depends on $\nu$. We put $j := \alpha'_\nu$, so that $\alpha_j \geqslant \nu$, while $\alpha_{j+1} < \nu$. Thus

$$h_{i1}^\alpha - h_{j1}^\alpha = \alpha_i + h - i - (\alpha_j + h - j) \leqslant \alpha_i - i - \nu + j < \alpha_i - i - \nu + j + 1 = h_{i\nu}^\alpha,$$

and

$$h_{i1}^\alpha - h_{j+1,1}^\alpha = \alpha_i + h - i - (\alpha_{j+1} + h - j - 1) > \alpha_i - i - \nu + j + 1 = h_{i\nu}^\alpha.$$

This proves (2), which implies (1), which then implies the statement. ∎

This formula relates the prime divisors of $f^\alpha$, $n!$, and the hook lengths $h_{ij}^\alpha$, so it is helpful when we come to consider modular representations.

## 2.4 A Recursion Formula for the Irreducible Characters

In the preceding section we saw how $\zeta^\alpha$ can be expressed as a $\mathbb{Z}$-linear combination of permutation characters $\xi^\beta$, so that we are now able to calculate the character table $Z_n$ of $S_n$ from the table $\Xi_n$ of the characters $\xi^\beta$ by suitable applications of the determinantal form of $[\alpha]$. But this is still unsatisfactory; we would prefer to evaluate $Z_n$ either directly or recursively. To obtain a recursion formula for $\zeta^\alpha$ we need to know what happens when $[\alpha]$, $\alpha$ a partition of $n$, is restricted to subgroups $S_m$, $m<n$. In order to get an idea how we may proceed in the general case, we discuss first the restriction down to $S_{n-1}$, the stabilizer of the point $n$. That is, we would like to know the irreducible constituents of the restriction $[\alpha]\downarrow S_{n-1}$. $S_{n-1}$ is a Young subgroup corresponding to the partition $(n-1,1)$, and we know the determinantal formula of $[\alpha]$, which expresses $\zeta^\alpha$ in terms of the $\xi^\beta$. This raises the question, how the restriction $\xi^\beta\downarrow S_{n-1}$ can be expressed in terms of the $\xi^\gamma$, $\gamma\vdash n-1$.

If $\lambda$ is a composition of $n$, we put

2.4.1
$$\lambda^{i-}:=(\lambda_1,\ldots,\lambda_{i-1},\lambda_i-1,\lambda_{i+1},\ldots),$$
$$\lambda^{i+}:=(\lambda_1,\ldots,\lambda_{i-1},\lambda_i+1,\lambda_{i+1},\ldots).$$

Then 2.3.10 yields (since $\xi^{\lambda^{i-}}=0$ if $\lambda^{i-}$ is not an improper partition of $n-1$; in particular, $\xi^{\lambda^{i-}}=0$ for all sufficiently large $i$):

**2.4.2** LEMMA.

$$\xi^\lambda\downarrow S_{n-1}=\sum_{i=1}^{\infty}\xi^{\lambda^{i-}}.$$

Applying this to the determinantal form of $\zeta^\alpha$, we obtain

$$\begin{aligned}\zeta^\alpha\downarrow S_{n-1}&=\sum_{\pi\in S_n}\operatorname{sgn}\pi\left(\xi^{\alpha-\mathrm{id}+\pi}\downarrow S_{n-1}\right)\\&=\sum_\pi\operatorname{sgn}\pi\sum_i\xi^{(\alpha-\mathrm{id}+\pi)^{i-}}\\&=\sum_i\sum_\pi\operatorname{sgn}\pi\cdot\xi^{(\alpha-\mathrm{id}+\pi)^{i-}}=\sum_i\chi^{\alpha^{i-}}.\end{aligned}$$

By 2.3.9 we have that $\chi^{\alpha^{i-}}\neq 0$ only if $\alpha_i-1\geqslant\alpha_{i+1}$. Therefore only summands for which $\alpha_i > \alpha_{i+1}$, and hence $\alpha^{i-} \vdash n-1$ contribute to $\zeta^\alpha \downarrow S_{n-1}$. This proves the first half of the following theorem, the second half of which follows by Frobenius's reciprocity law:

**2.4.3** THE BRANCHING THEOREM. *If $\alpha = (\alpha_1, \alpha_2, \ldots)$ is a partition of $n$, then we have for the restriction of $[\alpha]$ to the stabilizer $S_{n-1}$ of the point $n$*

$$[\alpha]\downarrow S_{n-1} = \sum_{\substack{i \\ \alpha_i > \alpha_{i+1}}} [\alpha^{i-}].$$

*On the other hand, if $S_n$ denotes the stabilizer of the point $n+1$ in $S_{n+1}$, then we have for the induced representation*

$$[\alpha]\uparrow S_{n+1} = \sum_{\substack{i \\ \alpha_i < \alpha_{i-1}}} [\alpha^{i+}]$$

(*put* $\alpha_0 := \infty$).

For example

$$[3,2,1^2]\downarrow S_6 = [2^2,1^2] + [3,1^3] + [3,2,1],$$

while

$$[3,2,1^2]\uparrow S_8 = [4,2,1^2] + [3^2,1^2] + [3,2^2,1] + [3,2,1^3].$$

This branching theorem shows how to evaluate $\zeta^\alpha(\pi)$ if $\pi$ contains 1-cycles and if the character table of $S_{n-1}$ is known. In the case where $\pi$ does not contain 1-cycles, we know $\zeta^\alpha(\pi)$ only if $\pi$ is an $n$-cycle (cf. 2.3.17). Hence if $\pi$ contains a $k$-cycle, say, then a formula would be welcome which expresses $\zeta^\alpha(\pi)$ in terms of certain $\zeta^\beta(\rho)$, $\beta \vdash n-k$, $\rho$ arising from $\pi$ by cancelling this $k$-cycle. We shall derive such a formula, and it should be clear from the foregoing that we need first to restrict from $S_n$ down to $S_{n-k} \times S_k$, the subgroup of elements which leave the subsets $\mathbf{n-k}$ and $\mathbf{n \backslash n-k}$ invariant. An immediate corollary of 2.3.10(ii) is

**2.4.4** LEMMA. *Suppose that $\lambda$ is a composition of $n = m+k$. If $\pi \in S_n$ contains a $k$-cycle, while $\rho \in S_m$ has the cycle type*

$$(a_1(\pi), \ldots, a_{k-1}(\pi), a_k(\pi)-1, a_{k+1}(\pi), \ldots, a_m(\pi)),$$

*then*

$$\chi^\lambda(\pi) = \sum_{\mu \vDash k} \chi^{\lambda-\mu}(\rho)\xi^\mu_{(k)}.$$

In order to apply this we remember that $\xi^{\mu}_{(k)} \neq 0$ if and only if $\mu$ has the form

$$\mu = (0,\ldots,0,k,0,\ldots),$$

in which case $\xi^{\mu}_{(k)} = 1$. Therefore,

**2.4.5**
$$\chi^{\lambda}(\pi) = \sum_{i=1}^{\infty} \chi^{(\lambda_1,\ldots,\lambda_{i-1},\lambda_i - k,\lambda_{i+1},\ldots)}(\rho).$$

Now we consider the summands in 2.4.5 for the case when $\lambda = \alpha$ is a partition of $n$. An $m$-fold application of 2.3.9 yields

2.4.6
$$\chi^{(\alpha_1,\ldots,\alpha_{i-1},\alpha_i - k,\alpha_{i+1},\ldots)} = (-1)^m \chi^{(\alpha_1,\ldots,\alpha_{i-1},\alpha_{i+1}-1,\ldots,\alpha_{i+m}-1,\alpha_i - k + m,\alpha_{i+m+1},\ldots)}.$$

It is not difficult to see that the existence of an $m \geqslant 0$ such that the composition on the right-hand side of 2.4.6 is a proper partition of $n-k$ is equivalent to the fact that the corresponding diagram is obtained from $[\alpha]$ by cancelling the rim part $R^{\alpha}_{ij}$ of a (uniquely determined) hook $H^{\alpha}_{ij}$ of length $k$ of $[\alpha]$. In this case $m = l^{\alpha}_{ij}$, and we obtain

$$\chi^{(\alpha_1,\ldots,\alpha_{i-1},\alpha_i - k,\alpha_{i+1},\ldots)} = (-1)^{l^{\alpha}_{ij}} \chi^{[\alpha]\backslash R^{\alpha}_{ij}}.$$

If no such $m$ exists, it follows from the considerations above and 2.3.8 that the character in 2.4.6 vanishes. This yields the desired recursion formula for $\zeta^{\alpha}(\pi)$:

**2.4.7** The Murnaghan-Nakayama formula. *If $\alpha$ is a partition of $n$, if $k \in \mathbf{n}$, and $\pi \in S_n$ contains a $k$-cycle, while $\rho \in S_{n-k}$ is of cycle type $(a_1(\pi),\ldots, a_{k-1}(\pi), a_k(\pi)-1, a_{k+1}(\pi),\ldots, a_{n-k}(\pi))$, then we have the following recursion formula for $\zeta^{\alpha}(\pi)$:*

$$\zeta^{\alpha}(\pi) = \sum_{\substack{i,j \\ h^{\alpha}_{ij} = k}} (-1)^{l^{\alpha}_{ij}} \zeta^{[\alpha]\backslash R^{\alpha}_{ij}}(\rho)$$

*(recall that $\zeta^{[0]} = 1$).*

For example

$$\begin{aligned}\zeta^{(3^3)}((1234)(56)(789)) &= \zeta^{(2^3)}((1234)(56)) - \zeta^{(3,2,1)}((1234)(56)) \\ &\quad + \zeta^{(3^2)}((1234)(56)) \\ &= -\zeta^{(2,1^2)}((1234)) + \zeta^{(2^2)}((1234)) \\ &\quad - \zeta^{(2^2)}((1234)) + \zeta^{(3,1)}((1234)) \\ &= -1-1 = -2.\end{aligned}$$

It is clear that a diagram $[\alpha]$ has exactly one hook of length $h_{11}^{\alpha}$. Thus

$$\zeta^{\alpha}_{(h_{11}^{\alpha},\ldots,h_{kk}^{\alpha})}=(-1)^{\alpha_1'-1}\zeta^{[\alpha]\setminus R_{11}^{\alpha}}_{(h_{22}^{\alpha},\ldots,h_{kk}^{\alpha})},$$

and a repeated application of this argument yields

**2.4.8** COROLLARY. $\zeta^{\alpha}_{(h_{11}^{\alpha},\ldots,h_{kk}^{\alpha})}=(-1)^{\Sigma_1^k\alpha_i'-i}$.

It is also clear that $[\alpha]$ does not possess any hook of a length $>h_{11}^{\alpha}$, and that this implies

**2.4.9** COROLLARY. $\zeta^{\alpha}_{\beta}\neq 0\Rightarrow\beta\trianglelefteq(h_{11}^{\alpha},\ldots,h_{kk}^{\alpha})$.

This shows once again the importance of the diagram lattice.

Later on we shall need a few results on lowest dimensions of ordinary irreducible representations of symmetric groups. There are many results known, we would now like to derive a few of them by various applications of the branching theorem.

**2.4.10** THEOREM.

(i) *For each* $n$, $[n]$ *and* $[1^n]$ *are the only one-dimensional ordinary representations of* $S_n$.

(ii) *For* $2\leqslant n\neq 4$, *the lowest dimension* $\neq 1$ *of an ordinary irreducible representation of* $S_n$ *is* $n-1$ (*while for* $n:=4$, $[2^2]$ *has dimension* $2=n-2$). *For* $2\leqslant n\neq 6$, $[n-1,1]$ *and* $[2,1^{n-2}]$ *are the only ordinary irreducible representations of* $S_n$ *which are of dimension* $n-1$ (*while for* $n:=6, [3^2]$ *and* $[2^3]$ *are other irreducible representations of dimension* $5=n-1$).

(iii) *For* $2\leqslant n\neq 5$ *there is no ordinary irreducible representation* $[\alpha]$ *of* $S_n$ *which is of a dimension* $f^{\alpha}$ *such that*

$$n\leqslant f^{\alpha}\leqslant n+2$$

(*while for* $n:=5$ *we have* $[3,2]$ *and* $[2^2,1]$, *which are of dimension* $5=n$, *and* $[3,1^2]$ *of dimension* $6=n+1$).

*Proof.* (i): This part is an easy consequence of 1.1.25 and 1.1.26. If we prefer to avoid this argument, we may proceed by induction on $n$ as follows. If $[\alpha]$, $\alpha\vdash n$, is one-dimensional, the same holds for $[\alpha]\downarrow S_{n-1}$. Hence, by the induction hypothesis, $[\alpha]\downarrow S_{n-1}$ coincides with $[n-1]$ or $[1^{n-1}]$, and the branching theorem yields

$$[\alpha]\in\{[n],[n-1,1],[1^n],[2,1^{n-2}]\}.$$

But 2.3.21 has shown that both $[n-1,1]$ and $[2,1^{n-2}]$ are of degree $n-1$, which finishes the proof of part (i).

(ii): An inspection of the character table shows that the statement holds for $n \leqslant 8$. We can therefore assume

$$\alpha \vdash n \geqslant 9 \quad \text{and} \quad 1 < f^\alpha \leqslant n+2.$$

(a) If $[\alpha]\downarrow S_{n-1}$ contains one-dimensional constituents, we obtain from (i) and the branching theorem that

$$\alpha \in \{(n-1,1),(2,1^{n-2})\},$$

and 2.3.21 shows that both $[n-1,1]$ and $[2,1^{n-2}]$ are of dimension $n-1$.

(b) If $[\alpha]\downarrow S_{n-1}$ does not contain any one-dimensional constituent, then by induction it must be irreducible, for $S_{n-1}$ has no irreducible representation $[\beta]$ such that $1<f^\beta \leqslant 4$. Hence $\alpha$ must be rectangular, say $\alpha=(r^s)$. Since we intend to derive a contradiction to $f^\alpha \leqslant n+2$, we may consider either $\alpha=(r^s)$ or $\alpha'=(s^r)$, so that we can assume without restriction that $r>2$. But then

$$[\alpha]\downarrow S_{n-2} = \left[r^{s-2},(r-1)^2\right] + [r^{s-1}, r-2],$$

where none of the constituents can be one-dimensional, since $n-2>6$.

Hence the induction hypothesis yields the contradiction

$$f^\alpha \geqslant 2(n-3) \geqslant n+3.$$

(iii) follows immediately from the proof of (ii). ■

The proof of this theorem shows clearly that analogously one can derive further results concerning ordinary irreducible representations of low dimension.

We next undertake to extend the Murnaghan-Nakayama formula so that it can be applied to the generalized characters $\chi^{\lambda/\mu}$ introduced in 2.3.11. To do this we need the following result:

**2.4.11 Lemma.** *If $\lambda$ is a composition of $n+r$, $\nu$ a composition of $r$, and $m+k=n$, then*

(i) $\chi^{\lambda/\nu}\downarrow S_m \times S_k = \sum_{\mu \models k} \chi^{(\lambda-\mu)/\nu} \# \xi^\mu$,

(ii) *and if $\pi \in S_n$ contains a $k$-cycle, while $\rho \in S_m$ is of cycle type $(a_1(\pi),\ldots,a_{k-1}(\pi),a_k(\pi)-1,a_{k+1}(\pi),\ldots,a_m(\pi))$, then*

$$\chi^{\lambda/\nu}(\pi) = \sum_{\mu \models k} \chi^{(\lambda-\mu)/\nu}(\rho)\xi^\mu_{(k)}.$$

*Proof.* (i): We obtain from 2.3.11 that

$$\chi^{\lambda/\nu}\downarrow \mathrm{S}_m\times\mathrm{S}_k=\sum_{\pi}\operatorname{sgn}\pi\cdot\xi^{\lambda-\mathrm{id}-(\nu-\mathrm{id})\circ\pi}\downarrow \mathrm{S}_m\times\mathrm{S}_k,$$

so that an application of 2.3.10 yields

$$=\sum_{\pi}\operatorname{sgn}\pi\sum_{\mu\vDash k}\xi^{\lambda-\mu-\mathrm{id}-(\nu-\mathrm{id})\circ\pi}\#\xi^{\mu}$$

$$=\sum_{\mu\vDash k}\left(\sum_{\pi}\operatorname{sgn}\pi\cdot\xi^{\lambda-\mu-\mathrm{id}-(\nu-\mathrm{id})\circ\pi}\right)\#\xi^{\mu}$$

$$=\sum_{\mu\vDash k}\chi^{(\lambda-\mu)/\nu}\#\xi^{\mu},$$

the last equation coming from 2.3.11.

(ii): As

$$\chi^{\lambda/\nu}(\pi)=\chi^{\lambda/\nu}\downarrow \mathrm{S}_m\times\mathrm{S}_k(\pi),$$

we obtain from (i)

$$=\sum_{\mu\vDash k}\chi^{(\lambda-\mu)/\nu}(\rho)\xi^{\mu}_{(k)}.\qquad\blacksquare$$

Again we remember that $\xi^{\mu}_{(k)}\neq 0$ if and only if $\mu$ is of the form $\mu=(0,\ldots,0,k,0,\ldots)$, in which case $\xi^{\mu}_{(k)}=1$. Therefore, the following must be true for $\lambda$, $\nu$, $\pi$, and $\rho$ as in 2.4.11:

**2.4.12** $$\chi^{\lambda/\nu}(\pi)=\sum_{i=1}^{\infty}\chi^{(\lambda_1,\ldots,\lambda_{i-1},\lambda_i-k,\lambda_{i+1},\ldots)/\nu}(\rho).$$

In order to derive from this the desired generalization of the Murnaghan-Nakayama formula 2.4.7, we associate with $\chi^{\alpha/\beta}$, $\alpha\vdash m+k$, $\beta\vdash k$, $\beta_i\leqslant\alpha_i$, $1\leqslant i\leqslant k$, a diagram which will be denoted by

2.4.13 $$[\alpha/\beta]$$

and called a *skew diagram*. It arises from $[\alpha]$ by deleting the first $\beta_i$ nodes of its $i$th row, $1\leqslant i\leqslant k$, e.g.

2.4.14

$$[8,7^2,5/4,3^2]=[9,8^2,6/5,4^2,1]=\begin{array}{cccccccc} & & & & \times & \times & \times & \times\\ & & & \times & \times & \times & \times & \\ & & & \times & \times & \times & \times & \\ \times & \times & \times & \times & \times & & & \end{array}.$$

As we can delete from a skew diagram $[\alpha/\beta]$ in an obvious manner all the parts $R_{ij}^{\alpha/\beta}$ of its rim which correspond to hooks $H_{ij}^{\alpha}$ such that $R_{ij}^{\alpha}$ does not contain any node of the subdiagram $[\beta]$ of $[\alpha]$, we can generalize 2.4.7 as follows ($\alpha, \beta, \pi, \rho$ as above, and cf. the derivation of 2.4.7 from 2.4.5):

**2.4.15** THE MURNAGHAN-NAKAYAMA FORMULA FOR SKEW DIAGRAMS.

$$\chi^{\alpha/\beta}(\pi)=\sum_{\substack{i,j\\ h_{ij}^{\alpha}=k}}(-1)^{l_{ij}^{\alpha}}\chi^{[\alpha/\beta]\setminus R_{ij}^{\alpha/\beta}}(\rho).$$

The generalized character $\chi^{\alpha/\beta}$ is in fact a character. Therefore the diagram $[\alpha/\beta]$ corresponds to a representation, which will be called a *skew representation*. It will also be denoted by $[\alpha/\beta]$. More precisely, we shall prove next the following result ($\alpha \vdash m+k$, $\beta \vdash k$, as above):

**2.4.16** $$[\alpha/\beta]=\sum_{\gamma\vdash m}([\alpha],[\beta][\gamma])[\gamma].$$

*Proof.* The statement is equivalent to

$$(\chi^{\alpha/\beta},\zeta^{\gamma})=(\zeta^{\alpha}\downarrow S_m\times S_k,\zeta^{\beta}\#\zeta^{\gamma}),$$

i.e., we have to verify that

2.4.17 $$\frac{1}{m!}\sum_{\pi\in S_m}\chi^{\alpha/\beta}(\pi)\zeta^{\gamma}(\pi)=\frac{1}{m!k!}\sum_{(\rho,\sigma)\in S_m\times S_k}\zeta^{\alpha}(\rho\sigma)\zeta^{\gamma}(\rho)\zeta^{\beta}(\sigma).$$

But the right-hand side of this equation is by 2.3.12

$$=\frac{1}{m!k!}\sum_{(\rho,\sigma)}\sum_{\delta\vdash k}\chi^{\alpha/\delta}(\rho)\zeta^{\delta}(\sigma)\zeta^{\beta}(\sigma)\zeta^{\gamma}(\rho)$$

$$=\frac{1}{m!}\sum_{\rho}\sum_{\delta\vdash k}\chi^{\alpha/\delta}(\rho)\zeta^{\gamma}(\rho)\frac{1}{k!}\sum_{\sigma\in S_k}\zeta^{\delta}(\sigma)\zeta^{\beta}(\sigma),$$

which by the orthogonality relations is equal to the left-hand side of 2.4.17, as required. ■

Later on the Littlewood-Richardson rule will show how the multiplicities $([\alpha],[\beta][\gamma])$ in 2.4.16 can be evaluated.

## 2.5 Ordinary Irreducible Representations and Characters of $A_n$

As soon as a result on representation of $S_n$ is derived, it is usual to ask what follows for the alternating group $A_n$. The standard method to derive the corresponding result for $A_n$ from that of $S_n$ is to apply Clifford's theory of representations of groups with normal subgroups. This theory is easy to use, since the factor group $S_n/A_n$ is cyclic, so that we need not deal with projective representations. Moreover the factor group is of prime order (if $n>1$), so that the inertia group of any ordinary irreducible representation of $A_n$ is either $A_n$ itself or $S_n$.

Let us first deduce from the system

$$\{[\alpha] \mid \alpha \vdash n\}$$

of ordinary irreducible representations of $S_n$ a complete system of equivalence classes of ordinary irreducible representations of $A_n$. A preliminary remark follows immediately from 2.1.8. For each $\alpha \vdash n$ we have:

**2.5.1** $$[\alpha]\downarrow A_n = [\alpha']\downarrow A_n.$$

Thus $[\alpha]$ and $[\alpha']$ are *associated representations* with respect to $A_n$ in the sense of Clifford's theory: their restrictions to $A_n$ have irreducible constituents in common. (This is the reason for calling $\alpha'$ the partition "associated" with $\alpha$.) Clifford's theory tells us (cf. Curtis and Reiner [1962, § 49]) that the irreducible constituents of their restrictions are conjugates of each other. Moreover the restriction is just equal to a certain multiple of a complete class of conjugate irreducible representations of $A_n$.

Let us consider an irreducible constituent $D$ of $[\alpha]\downarrow A_n$:

2.5.2 $$[\alpha]\downarrow A_n = [\alpha']\downarrow A_n = D + \cdots.$$

$D$ is either self-conjugate with respect to $S_n$—i.e. (assuming $n>1$)

2.5.3 $$D = D^{(12)},$$

where for each $\pi \in A_n$

2.5.4 $$D^{(12)}((12)\pi(12)) := D(\pi)$$

—or $D$ is not self-conjugate, which means that $D$ and $D^{(12)}$ are inequivalent irreducible representations of $A_n$. $D$ is self-conjugate if and only if its inertia group is $S_n$. Then $D$ can be extended to a representation $\tilde{D}$ of $S_n$ and gives rise to the following two irreducible representations of $S_n$:

2.5.5 $$D_1 := \tilde{D} \otimes IS_n, \text{ and } D_2 := \tilde{D} \otimes AS_n.$$

$D$ is not self-conjugate if and only if its inertia group is $A_n$, in which case it yields by induction an irreducible representation of $S_n$:

2.5.6 $$D_3 := D \uparrow S_n.$$

Furthermore each irreducible representation of $S_n$ is of one of these forms $D_i$, $i = 1, 2, 3$, and both $[\alpha]$ and $[\alpha']$ arise from $D$ in the way described.

Thus $[\alpha]$ is of the form $D_1$ or $D_2$ if and only if $[\alpha] \neq [\alpha']$, i.e. if and only if $\alpha \neq \alpha'$. (It is this fact that irreducible representations of $S_n$ which are associated with respect to $S_n$ form easily recognizable pairs, which does not hold if the characteristic of the field is 2, and in this case life is not as easy.) We therefore have

**2.5.7** THEOREM. *Suppose that $\alpha$ is a partition of $n > 1$.*

(i) *If $\alpha \neq \alpha'$, then $[\alpha]\downarrow A_n = [\alpha']\downarrow A_n$ is irreducible, while*

(ii) *if $\alpha = \alpha'$, then $[\alpha]\downarrow A_n = [\alpha']\downarrow A_n$ splits into two irreducible and conjugate representations $[\alpha]^{\pm}$ of $A_n$, i.e., $[\alpha]^{+(12)}$, defined by*

$$[\alpha]^{+(12)}((12)\pi(12)) := [\alpha]^{+}(\pi), \qquad \pi \in A_n,$$

*is equivalent to $[\alpha]^{-}$.*

*A complete system of equivalence classes of ordinary irreducible representations of $A_n$ is therefore*

$$\{[\alpha]\downarrow A_n \mid \alpha \neq \alpha'\} \cup \{[\alpha]^{\pm} \mid \alpha = \alpha' \vdash n\}.$$

If for example $n := 3$, then we obtain the system

$$\{[3]\downarrow A_3 = [1^3]\downarrow A_3, [2,1]^{+}, [2,1]^{-}\}.$$

This shows that the self-associated partitions of $n$ form an important subset $SA(n)$ of $P(n)$:

2.5.8 $$SA(n) := \{\alpha \mid \alpha' = \alpha \vdash n\}.$$

The next question which arises concerns the character table. We ask how the table of $A_n$ can be obtained from that of $S_n$. We remark first that 2.5.7 implies that the only problem is the evaluation of the characters of $A_n$ which are of the form $\zeta^{\alpha\pm}$.

If $\alpha$ is a self-associated partition of $n$, then the diagram $[\alpha]$ is symmetric with respect to its main diagonal. This implies that the hooks $H^\alpha_{ii}$, $1 \leqslant i \leqslant k :=$ (length of the main diagonal), of $[\alpha]$ have their arms and legs of the same length. The partition

2.5.9 $$h(\alpha) := (h^\alpha_{11}, \ldots, h^\alpha_{kk}),$$

formed by the lengths of these *main hooks* $H^\alpha_{ii}$ of $[\alpha]$, characterizes a conjugacy class of $S_n$, which splits over $A_n$ (c.f. 1.2.10); we therefore call it a *split partition* and denote the set of all the split partitions by $\mathrm{SP}(n)$:

2.5.10 $$\mathrm{SP}(n) := \left\{ \alpha \mid \alpha \vdash n \begin{array}{l} \textit{and the nonzero parts of } \alpha \\ \textit{are pairwise different and odd} \end{array} \right\}.$$

It is easy to see that $h(\cdot)$ provides a bijection from $\mathrm{SA}(n)$ onto $\mathrm{SP}(n)$:

**2.5.11** $$h\colon \mathrm{SA}(n) \rightarrowtail\!\!\!\!\!\!\twoheadrightarrow \mathrm{SP}(n)\colon \alpha \to (h^\alpha_{11}, \ldots, h^\alpha_{kk}).$$

Hence $h(\alpha)$ falls under suspicion; it turns out that in fact the values of $\zeta^{\alpha\pm}$ on the classes $C^{h(\alpha)\pm}$ are the only values which require a special description, while on any other class $\zeta^{\alpha\pm}$ has half the value of $\zeta^\alpha$. The value of $\zeta^\alpha_{h(\alpha)}$ is given by 2.4.8:

**2.5.12** Lemma. *If $\alpha$ is a self-associated partition of $n$, and $k$ denotes the length of the main diagonal of $[\alpha]$, then*

$$\zeta^\alpha_{h(\alpha)} = (-1)^{(n-k)/2}.$$

**2.5.13** Theorem. *If $\alpha = \alpha'$ is a self-associated partition of $n > 1$, then the values of $\zeta^{\alpha\pm}$ are*

$$\zeta^{\alpha\pm}_{h(\alpha)+} = \tfrac{1}{2}\left( \zeta^\alpha_{h(\alpha)} \pm \sqrt{\zeta^\alpha_{h(\alpha)} \Pi_i h^\alpha_{ii}} \right),$$

$$\zeta^{\alpha\pm}_{h(\alpha)-} = \tfrac{1}{2}\left( \zeta^\alpha_{h(\alpha)} \mp \sqrt{\zeta^\alpha_{h(\alpha)} \Pi_i h^\alpha_{ii}} \right),$$

*for a suitable numbering of the constituents of $[\alpha]\downarrow A_n$, while on all the other classes with cycle partitions $\gamma \neq h(\alpha)$ we have*

$$\zeta^{\alpha\pm}_\gamma = \zeta^\alpha_\gamma / 2,$$

*and $\zeta^\alpha_\gamma$ is an even integer.*

*Proof.* We consider the following generalized character of $A_n$:

$$\varphi^\alpha := (\zeta^{\alpha+} - \zeta^{\alpha-})(\overline{\zeta^{\alpha+}} - \overline{\zeta^{\alpha-}}),$$

where

$$\overline{\zeta^{\alpha\pm}}(\pi):=\zeta^{\alpha\pm}(\pi^{-1})=\overline{\zeta^{\alpha\pm}(\pi)},$$

i.e., $\overline{\zeta^{\alpha\pm}}$ denotes the character contragredient to $\zeta^{\alpha\pm}$. We would like to show that $\varphi^\alpha$ is the restriction to $A_n$ of a class function $\tilde{\varphi}^\alpha$ of $S_n$, which is in fact a scalar multiple of the indicator function of the class $C^{h(\alpha)}$ of $S_n$.

(i) *The function* $\tilde{\varphi}^\alpha$. The definitions of $[\alpha]^+$ and $[\alpha]^-$ yield that

$$\zeta^{\alpha+}_{\gamma+}=\zeta^{\alpha-}_{\gamma-} \text{ and } \zeta^{\alpha+}_{\gamma-}=\zeta^{\alpha-}_{\gamma+} \qquad \text{for} \quad \gamma\in SP(n)$$

and

$$\zeta^{\alpha+}_{\gamma}=\zeta^{\alpha-}_{\gamma} \qquad \text{for} \quad \gamma\in P(n)\backslash SP(n).$$

Therefore

$$\forall\gamma\in SP(n) \quad (\varphi^\alpha_{\gamma+}=\varphi^\alpha_{\gamma-}),$$

and $\varphi^\alpha$ is the restriction to $A_n$ of a class function $\tilde{\varphi}^\alpha$ of $S_n$, where $\tilde{\varphi}^\alpha:=\frac{1}{2}\varphi^\alpha\uparrow S_n$. $\tilde{\varphi}^\alpha$ is given by

$$\tilde{\varphi}^\alpha_\gamma=\begin{cases}|\zeta^{\alpha+}_{\gamma+}-\zeta^{\alpha-}_{\gamma+}|^2=|\zeta^{\alpha+}_{\gamma-}-\zeta^{\alpha-}_{\gamma-}|^2 & \text{if} \quad \gamma\in SP(n),\\ 0 & \text{if} \quad \gamma\in P(n)\backslash SP(n).\end{cases}$$

Since for each $\beta\in SA(n)$

$$(\varphi^\alpha,\zeta^{\beta+})=(\varphi^\alpha,\zeta^{\beta-}),$$

we have

$$(\tilde{\varphi}^\alpha,\zeta^\beta)\in\mathbb{Z} \qquad \text{for each} \quad \beta\in SA(n).$$

(ii) $\tilde{\varphi}^\alpha$ *as indicator function*. We know that $\tilde{\varphi}^\alpha$ is nonzero only on splitting classes. Let $\varepsilon$ denote an element of $SP(n)$ which is minimal with respect to the dominance order satisfying

$$\tilde{\varphi}^\alpha_\varepsilon\neq 0.$$

Such an $\varepsilon$ exists, since $\tilde{\varphi}^\alpha\neq 0$ by $\zeta^{\alpha+}\neq\zeta^{\alpha-}$. We have

$$\left(\tilde{\varphi}^\alpha,\zeta^{h^{-1}(\varepsilon)}\right)=\frac{1}{n!}\sum_{\beta\vdash n}|C^\beta|\tilde{\varphi}^\alpha_\beta\zeta^{h^{-1}(\varepsilon)}_\beta.$$

By 2.4.9 and the minimality of $\varepsilon$ we obtain

$$\mathbb{Z} \ni \left(\tilde{\varphi}^\alpha, \zeta^{h^{-1}(\varepsilon)}\right) = \frac{|C^\varepsilon|}{n!}\tilde{\varphi}^\alpha_\varepsilon \zeta^{h^{-1}(\varepsilon)}_\varepsilon = \frac{|C^\varepsilon|}{n!}\tilde{\varphi}^\alpha_\varepsilon (-1)^{(n-j)/2},$$

$j$ being the length of the main diagonal in $[h^{-1}(\varepsilon)]$ (cf. 2.5.12). This proves

$$\frac{|C^\varepsilon|}{n!}\tilde{\varphi}^\alpha_\varepsilon \in \mathbb{Z}\backslash\{0\}. \tag{1}$$

On the other hand, since $\tilde{\varphi}^\alpha$ vanishes outside $A_n$,

$$\begin{aligned}\frac{1}{n!}\sum_{\beta \vdash n}|C^\beta|\tilde{\varphi}^\alpha_\beta &= \frac{1}{2}\cdot\frac{2}{n!}\sum_{\beta \text{ even}}|C^\beta|\varphi^\alpha_\beta \\ &= \frac{1}{2}(\zeta^{\alpha+}-\zeta^{\alpha-}, \zeta^{\alpha+}-\zeta^{\alpha-}) \\ &= \frac{1}{2}\cdot 2 = 1. \end{aligned}\tag{2}$$

Now, the definition of $\tilde{\varphi}^\alpha$ gives

$$\tilde{\varphi}^\alpha_\beta \geqslant 0 \qquad (\beta \vdash n). \tag{3}$$

We obtain from (1), (2), and (3)

$$\frac{|C^\beta|}{n!}\tilde{\varphi}^\alpha_\beta = \begin{cases} 1 & \text{if } \beta = \varepsilon, \\ 0 & \text{otherwise.}\end{cases} \tag{4}$$

This shows that $\tilde{\varphi}^\alpha$ is a multiple of the indicator function of $C^\varepsilon$.

(iii) *$\zeta^{\alpha\pm}$ on classes other than $C^\varepsilon$.* If $\gamma \neq \varepsilon$, then for classes of $A_n$ with cycle partition $\gamma$, (4) gives

$$0 = \zeta^{\alpha+}_\gamma - \zeta^{\alpha-}_\gamma.$$

Since

$$[\alpha]\downarrow A_n = [\alpha]^+ + [\alpha]^-,$$
$$\zeta^{\alpha+}_\gamma = \zeta^{\alpha-}_\gamma = \tfrac{1}{2}\zeta^\alpha_\gamma,$$

as has been stated. (We show below that $\varepsilon = h(\alpha)$.)

(iv) *$\zeta^{\alpha\pm}$ on $C^\varepsilon$.* Since

$$\tilde{\varphi}^\alpha_\varepsilon = \frac{n!}{|C^\varepsilon|} = |\zeta^{\alpha+}_{\varepsilon\pm} - \zeta^{\alpha-}_{\varepsilon\pm}|^2,$$

we have

$$|\zeta_{\varepsilon\pm}^{\alpha+}-\zeta_{\varepsilon\pm}^{\alpha-}|=\sqrt{\frac{n!}{|C^{\varepsilon}|}}=\sqrt{\Pi_i\varepsilon_i}$$

(for the last equation cf. 1.2.15; we use $\varepsilon\in\mathrm{SP}(n)$, and note 1.2.10). Applying this to

$$\zeta_{\varepsilon}^{\alpha}=\zeta_{\varepsilon+}^{\alpha+}+\zeta_{\varepsilon+}^{\alpha-}=\zeta_{\varepsilon-}^{\alpha+}+\zeta_{\varepsilon-}^{\alpha-},$$

we find

$$\zeta_{\varepsilon+}^{\alpha\pm}=\zeta_{\varepsilon-}^{\alpha\mp}=\tfrac{1}{2}\left(\zeta_{\varepsilon}^{\alpha}\pm\sqrt{\Pi\varepsilon_i e^{ix}}\right)$$

for a suitable $x\in[0,2\pi)$.

(v) $\varepsilon=h(\alpha)$. Supposing $\varepsilon\neq h(\alpha)$, we obtain from (iii)

$$\zeta_{h(\alpha)+}^{\alpha+}=\tfrac{1}{2}\zeta_{h(\alpha)}^{\alpha}=\tfrac{1}{2}(-1)^{(n-k)/2}\in\mathbb{Q}\backslash\mathbb{Z},$$

which cannot happen, since a character value which lies in $\mathbb{Q}$ is an integer.

(vi) $e^{ix}=\zeta_{h(\alpha)}^{\alpha}$. If $C^{h(\alpha)\pm}$ is ambivalent, then $\zeta_{h(\alpha)\pm}^{\alpha\pm}\in\mathbb{R}$, so that we have $x=0$, i.e.

$$e^{ix}=1=\zeta_{h(\alpha)}^{\alpha},$$

for if $\delta:=h(\alpha)$, then by 2.5.12,

$$\begin{aligned}\zeta_{h(\alpha)}^{\alpha}&=(-1)^{\Sigma_i((h_{ii}^{\alpha}-1)/2)}\\&=(-1)^{\Sigma_i((\delta_i-1)/2)}\\&=1,\end{aligned}$$

where the last equation follows from 1.2.13. If $C^{h(\alpha)\pm}$ is not ambivalent, then

$$\zeta_{h(\alpha)+}^{\alpha+}=\overline{\zeta_{h(\alpha)-}^{\alpha+}},$$

so that for $\delta:=h(\alpha)$ and by 1.2.13 we get

$$\pm\sqrt{e^{ix}}=i=\left((-1)\Sigma_j{}^{((\delta_j-1)/2)}\right)^{1/2}=\sqrt{\zeta_{h(\alpha)}^{\alpha}}\,.$$

This completes the proof. ∎

The character table of $A_3$ therefore is

2.5.14

| | $(3)$ | $(2,1)^+$ | $(2,1)^-$ |
|---|---|---|---|
| $[3]$ | 1 | 1 | 1 |
| $[2,1]^+$ | 1 | $\frac{1}{2}(-1+\sqrt{3}\,i)$ | $\frac{1}{2}(-1-\sqrt{3}\,i)$ |
| $[2,1]^-$ | 1 | $\frac{1}{2}(-1-\sqrt{3}\,i)$ | $\frac{1}{2}(-1+\sqrt{3}\,i)$ |

Also for the alternating group we need a few results concerning the ordinary irreducible representations of low dimension.

**2.5.15** THEOREM.

(i) *For* $n\neq 3,4$ *the representation* $[n]\downarrow A_n=[1^n]\downarrow A_n$ *is the only one-dimensional ordinary representation of* $A_n$ *(while* $[2,1]^{\pm}$, $[2^2]^{\pm}$ *are other one-dimensional representations of* $A_3$, $A_4$*).*

(ii) *For* $n\neq 5$, *the lowest dimension* $\neq 1$ *of an ordinary irreducible representation of* $A_n$ *is* $n-1$ *(while for* $n:=5$ *we have the representations* $[3,1^2]^{\pm}$ *of dimension* 3*). For* $2\leqslant n\neq 3,6$, $[n-1,1]\downarrow A_n=[2,1^{n-2}]\downarrow A_n$ *is the only ordinary irreducible representation of* $A_n$ *which is of dimension* $n-1$ *(while for* $n:=3$ *we have no such representation, and for* $n:=6$ *we have both* $[5,1]\downarrow A_6=[2,1^4]\downarrow A_6$ *and* $[3^2]\downarrow A_6=[2^3]\downarrow A_6$ *of dimension* 5*).*

*Proof.* (i): The statement is seen to be true for $n\leqslant 4$ by inspecting the character tables of $S_n$, $n\leqslant 4$, and recalling 2.5.7. From 2.4.10 we get that for $n\geqslant 5$ there is no $\alpha\vdash n$ such that $f^{\alpha}=2$; in particular no such $\alpha$ exists with $\alpha=\alpha'$. The statement therefore follows from 2.5.7.

(ii): Since 2.4.10 is true, we need only show that for $2\leqslant n\neq 3,6$ there are no partitions $\alpha\vdash n$ with the properties

$$\alpha=\alpha' \quad\text{and}\quad f^{\alpha}=2n-2.$$

That this holds for $n\leqslant 8$ we can see by looking at the character tables.

Let us therefore assume $\alpha\vdash n\geqslant 9$, $\alpha=\alpha'$, and $f^{\alpha}=2n-2$.

(a) If $[\alpha]\downarrow S_{n-1}$ is reducible, then it is of the form

$$[\alpha]\downarrow S_{n-1}=[\beta]+[\gamma],$$

where $f^{\beta}=f^{\gamma}=n-1$. This follows from 2.4.10, since $\alpha=\alpha'$ implies that no one-dimensional constituent can occur. But 2.4.10 also gives

$$\beta,\gamma\in\{(n-2,1),(2,1^{n-3})\},$$

so that by the branching theorem

$$\alpha\in\{(n-1,1),(n-2,2),(n-2,1^2),(3,1^{n-3})(2^2,1^{n-4}),(2,1^{n-2})\},$$

which contradicts $\alpha=\alpha'$, for we assumed $n\geqslant 6$.

(b) If $[\alpha]\downarrow S_{n-1}$ is irreducible, then $\alpha$ is rectangular, say $\alpha=(r^s)$. Since $\alpha=\alpha'$, we have $r=s\geqslant 3$, and therefore

$$[\alpha]\downarrow S_{n-2}=[r^{s-2},(r-1)^2]+[r^{s-1},r-2].$$

Both these constituents are of dimension $\geqslant(n-2)+3$ (apply 2.4.10(iii)), contradicting $f^\alpha=2n-2$. ■

## 2.6 $S_n$ is Characterized by its Character Table

We denoted the character table of $S_n$ by $Z_n$:

$$Z_n=\left(\zeta^{\alpha^i}_{\alpha^k}\right),$$

where $i$ is the row index, $k$ the column index, and $i<j$ if and only if $\alpha^i<\alpha^j$ with respect to the lexicographic order.

If now $G$ denotes a finite group, which has for a certain numbering of its ordinary irreducible representations and conjugacy classes this same matrix as character table, then we say that $G$ has $Z_n$ as character table. We would like to prove the following theorem:

**2.6.1** THEOREM. *If a finite group $G$ has $Z_n$ as character table, then $G$ is isomorphic to $S_n$.*

*Proof.* The assumption that $G$ has the same character table $Z_n$ over $\mathbb{C}$ as has $S_n$ means that there exist two bijections $\Phi$ and $\varphi$, with $\Phi$ between the sets of irreducible representations of $S_n$ and $G$ over $\mathbb{C}$, and $\varphi$ between the sets of conjugacy classes of $S_n$ and $G$, which have the following property: Putting

$$[\alpha]^{\hat{}}:=\Phi([\alpha]),\quad\text{and}\quad C^{\check{\beta}}:=\varphi(C^\beta),$$

then for each $\alpha$, $\beta\vdash n$

2.6.2 $$\zeta^{\hat{\alpha}}_{\check{\beta}}=\zeta^\alpha_\beta,$$

$\zeta^{\hat{\alpha}}_{\check{\beta}}$ denoting the value of the character of $[\alpha]^{\hat{}}$ on $C^{\check{\beta}}$. Our method of proof will be to show that the representation $[n-1,1]^{\hat{}}$ maps $G$ isomorphically onto a group of reflections which by its order must be isomorphic to $S_n$. (The theorem is trivial if $n=0$ or 1, so that we may assume $(n-1,1)\vdash n$.)

(i) $[n-1,1]\hat{}$ *is a real representation.* 2.2.19(iii) implies

$$[n-1,1]+[n]=IS_{(n-1,1)}\uparrow S_n.$$

Hence for each $\pi\in S_n$ we have:

$$\zeta^{(n-1,1)}(\pi)=a_1(\pi)-1\geqslant -1.$$

Thus by 2.6.2, for each $g\in G$

$$\zeta^{(n-1,1)\hat{}}(g)\geqslant -1,$$

so that

2.6.3 $$\frac{1}{|G|}\sum_{g\in G}\zeta^{(n-1,1)\hat{}}(g^2)>-1,$$

for we can write $>-1$ instead of $\geqslant -1$, since

$$\zeta^{(n-1,1)\hat{}}_{(1^n)\check{}}=\zeta^{(n-1,1)}_{(1^n)}=n-1>-1.$$

It is well known that for any ordinary irreducible character $\hat{\zeta}$ of $G$ the expression

2.6.4 $$\frac{1}{|G|}\sum_{g\in G}\hat{\zeta}(g^2)$$

can take the values $-1,0$, and 1 only, depending on whether a representation with character $\hat{\zeta}$ is not equivalent to a real representation, but has real character, or has a nonreal character, or is equivalent to a real representation. Since $\zeta^{(n-1,1)}$ has real values only, this holds for $\zeta^{(n-1,1)\hat{}}$ as well (apply 2.6.2), so that we obtain from 2.6.3 that $[n-1,1]\hat{}$ is equivalent to a real representation.

(ii) $C^{(2,1^{n-2})\check{}}$ *consists of involutions* (*if* $n>4$). Let us denote by $\mathcal{C}^{\alpha}$, $\mathcal{C}^{\check{\alpha}}$ the class sums of $C^{\alpha}, C^{\check{\alpha}}$, respectively. It is well known that the character table of $G$ yields the coefficients $a_{\check{\alpha}\check{\beta}\check{\gamma}}$ in

$$\mathcal{C}^{\check{\alpha}}\mathcal{C}^{\check{\beta}}=\sum_{\gamma\vdash n}a_{\check{\alpha}\check{\beta}\check{\gamma}}\mathcal{C}^{\check{\gamma}}.$$

This shows that the following equation (which holds for $S_n$, $n \geq 4$):

$$\mathcal{C}^{(2,1^{n-2})}\mathcal{C}^{(2,1^{n-2})} = \frac{n(n-1)}{2}\mathcal{C}^{(1^n)} + 2\mathcal{C}^{(2^2,1^{n-4})} + 3\mathcal{C}^{(3,1^{n-3})}$$

gives

2.6.5 $$\mathcal{C}^{(2,1^{n-2})^\vee}\mathcal{C}^{(2,1^{n-2})^\vee} = \frac{n(n-1)}{2}\mathcal{C}^{(1^n)^\vee} + 2\mathcal{C}^{(2^2,1^{n-4})^\vee} + 3\mathcal{C}^{(3,1^{n-3})^\vee}.$$

The character table determines the orders of the conjugacy classes (apply the orthogonality relations), so that for $n \geq 4$:

2.6.6
$$\begin{aligned}
&\text{(i) } |C^{(1^n)^\vee}| = |C^{(1^n)}| = 1,\\
&\text{(ii) } |C^{(2^2,1^{n-4})^\vee}| = |C^{(2^2,1^{n-4})}| = \frac{n(n-1)(n-2)(n-3)}{8},\\
&\text{(iii) } |C^{(3,1^{n-3})^\vee}| = |C^{(3,1^{n-3})}| = \frac{n(n-1)(n-2)}{3},\\
&\text{(iv) } |C^{(2,1^{n-2})^\vee}| = |C^{(2,1^{n-2})}| = \frac{n(n-1)}{2}.
\end{aligned}$$

Let $\overset{*}{C}$ denote the conjugacy class of $G$, which consists of the squares of the elements in the class $\check{C}$ of $G$:

$$\overset{*}{C} := \{g^2 \mid g \in \check{C}\}.$$

Since we have for their orders and the orders of the centralizers of $g \in \check{C}$ and $g^2 \in \overset{*}{C}$

$$|\check{C}| = |\overset{*}{C}|\frac{|C_G(g^2)|}{|C_G(g)|},$$

$|\overset{*}{C}|$ divides $|\check{C}|$.

2.6.5 shows that the class of squares of elements $g \in C^{(2,1^{n-2})^\vee}$ must be either $C^{(1^n)^\vee}$, $C^{(2^2,1^{n-4})^\vee}$, or $C^{(3,1^{n-3})^\vee}$. But 2.6.6 shows that for $n > 4$, $C^{(1^n)^\vee}$ is the only one whose order divides

$$|C^{(2,1^{n-2})^\vee}| = \frac{n(n-1)}{2}.$$

Together with the corresponding $[\alpha]$, some of the $[\alpha]^\wedge$ are faithful, so that we have

$$C^{(1^n)^\vee} = \{1_G\},$$

and hence we can conclude

$$\left\{g^2 \mid g \in C^{(2,1^{n-2})^{\vee}}\right\} = \{1_G\},$$

so that $C^{(2,1^{n-2})^{\vee}}$ in fact consists of involutions.

(iii) $[n-1,1]^{\wedge}(g)$, $g \in C^{(2,1^{n-2})^{\vee}}$, *is a reflection* (*if* $n>4$). In (i) we saw that $[n-1,1]^{\wedge}$ is a real representation of $G$, so we may assume that for each $g \in G$ the mapping $[n-1,1]^{\wedge}(g)$ is a real orthogonal one. The eigenvalues of such matrices are of absolute value 1. Since by (ii) $C^{(2,1^{n-2})^{\vee}}$ consists of involutions, $[n-1,1]^{\wedge}(g)$, $g \in C^{(2,1^{n-2})^{\vee}}$, has eigenvalues $\pm 1$ (since $[n-1,1]^{\wedge}(g)^2 = \mathrm{id}$) only. But the sum of the eigenvalues of $[n-1,1]^{\wedge}(g)$ is its trace $\zeta^{(n-1,1)^{\wedge}}(g)$, for which we have by 2.6.2

$$\forall g \in C^{(2,1^{n-2})^{\vee}} \quad \left(\zeta^{(n-1,1)^{\wedge}}(g) = \zeta^{(n-1,1)}((12)) = a_1((12)) - 1 = n-3\right).$$

Hence $[n-1,1]^{\wedge}(g)$, $g \in C^{(2,1^{n-2})^{\vee}}$, has exactly $n-2$ eigenvalues $+1$ and one eigenvalue $-1$. Thus each such mapping is a reflection.

(iv) $G \simeq S_n$ *if* $n>4$. The character table $Z_n$ of $G$ yields the coefficients $a_{\check{\alpha}\check{\beta}\check{\gamma}}$, from which we can see that $C^{(2,1^{n-2})^{\vee}}$ generates $G$. Hence by (iii) for $n>4$ the image of $[n-1,1]^{\wedge}$ is a faithful representation of $G$ as an irreducible group of reflections. These groups are known (see e.g. Benson and Grove, [1971, Theorem 5.3.1]), so that we obtain from $|G| = |S_n| = n!$ that

$$\forall n>4 \quad (G \simeq S_n).$$

(v) $G \simeq S_n$ *if* $n \leq 4$. This follows by inspecting the groups of orders 1, 2, 6, 24 and their character tables. ■

## 2.7 Cores and Quotients of Partitions

Let us return to an examination of the values $\zeta^{\alpha}(\pi)$ of an ordinary irreducible character of $S_n$. The Murnaghan-Nakayama formula 2.4.7 allows us to evaluate $\zeta^{\alpha}(\pi)$ recursively by removing parts $R_{ij}^{\alpha}$ from the rim of $[\alpha]$ and then from the rim of $[\alpha] \backslash R_{ij}^{\alpha}$, and so on. The lengths of these parts which have to be removed correspond to lengths $k$, $l$, and so on of cyclic factors of $\pi$, so that the evaluation of $\zeta^{\alpha}(\pi)$ comes down to the evaluation of character values of $S_{n-k}$, $S_{n-k-l}$, and so on.

We ask what can be said if we restrict attention to a particular cycle length, say $q$.

While a hook $H_{ij}^{\alpha}$ of length $q$ will be called a *q-hook*, the corresponding part $R_{ij}^{\alpha}$ of the rim of $[\alpha]$ will be called a *rim q-hook*. It is very natural to ask whether the process of removing rim $q$-hooks from $[\alpha]$ always ends up with the same diagram which does not contain any further $q$-hook.

Diagrams $[\alpha]$ which do not contain any $q$-hook are called $q$-*cores*, a name which we shall also use for the partition $\alpha$. For example the zero diagram $[0]$ is the only 1-core, while the 2-cores are just the triangular diagrams: $[0]$ and the diagrams

2.7.1
$$\times, \quad \begin{array}{c} \times \ \times \\ \times \end{array}, \quad \begin{array}{c} \times \ \times \ \times \\ \times \ \times \\ \times \end{array}, \quad \begin{array}{c} \times \ \times \ \times \ \times \\ \times \ \times \ \times \\ \times \ \times \\ \times \end{array}, \quad \dots .$$

For $q \geqslant 3$, the complete set of $q$-cores is rather complicated to describe.

We ask whether we obtain a uniquely determined $q$-core

2.7.2
$$[\tilde{\alpha}]$$

from $[\alpha]$ by successive removals of rim $q$-hooks. We shall shortly see that this is indeed the case. For example the 3-core of $[7,5,4,3,2]$ is $[4,2]$, although the pictures below indicate at least two different ways of reaching it:

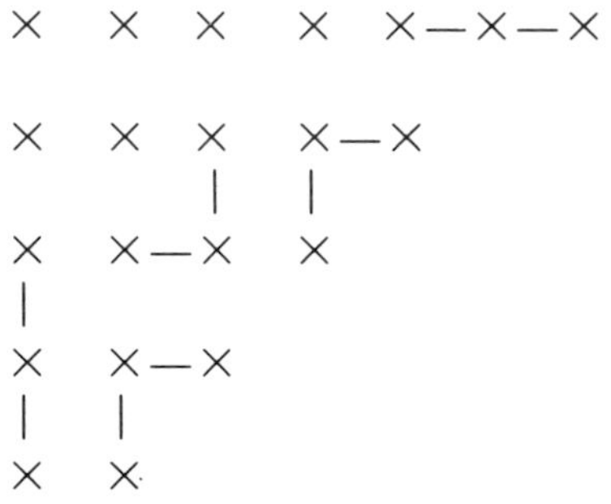

$$\begin{array}{cccccc} \times & \times & \times & \times & \times-\times-\times \\ \\ \times & \times & \times & \times-\times \\ & & & | & | \\ \times & \times-\times & \times \\ | \\ \times & \times-\times \\ | & | \\ \times & \times \end{array} \qquad \text{and} \qquad \begin{array}{cccc} \times & \times & \times & \times & \times-\times-\times \\ \\ \times & \times & \times-\times-\times \\ \\ \times-\times & \times-\times \\ | & | \\ \times & \times & \times \\ | \\ \times-\times \end{array}$$

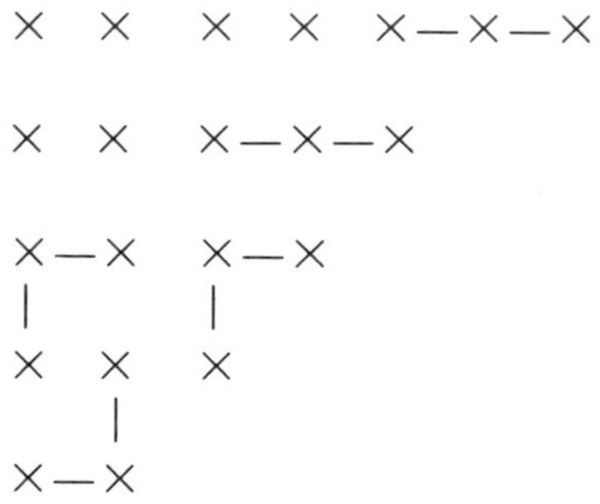

In order to prove that we always finish up with the same $q$-core, we shall give an algorithm which yields this core, and first of all we need an algorithmic description of the removal of a rim $q$-hook.

It is clear that a diagram $[\alpha]=[\alpha_1, \dots, \alpha_h]$ $(\alpha_h \neq 0)$ or its partition $\alpha$ is uniquely determined by its *first-column hook lengths*

2.7.3
$$h_i^\alpha := h_{i1}^\alpha = \alpha_i - i + h, \qquad 1 \leqslant i \leqslant h.$$

For example, if we are told that $h_4 := 2$, $h_3 := 4$, $h_2 := 5$, and $h_1 := 7$ are all the first-column hook lengths, then we deduce that $\alpha = (4, 3^2, 2)$. The method of reconstructing $\alpha$ is simply to use the equations

2.7.4
$$\alpha_h = h_h, \qquad \alpha_{h-1} = h_{h-1} - 1, \dots, \qquad \alpha_1 = h_1 - h + 1.$$

Notice that in this way we get a partition $\alpha = (\alpha_1, \dots, \alpha_r)$ from each sequence of strictly decreasing nonnegative integers

2.7.5 $$\beta_1 > \beta_2 > \cdots > \beta_r$$

by putting

2.7.6 $$\alpha_i := \beta_i + i - r, \qquad 1 \leqslant i \leqslant r.$$

If $\beta_r = 0$, we get a sequence having some zero parts $\alpha_i$ at the end, which does not matter. Thus from $9,7,6,4,1,0$ we obtain $\alpha = (4,3^2,2)$, a partition which we also obtain both from $8,6,5,3,0$ and from $7,5,4,2$. We now formally define this generalization of first-column hook lengths.

Let $\lambda$ denote a composition of some integer. In accordance with 2.3.8 we may define a generalized representation of a symmetric group by putting

2.7.7 $$[\lambda] := |[\lambda_i - i + j]|,$$

subject to the conventions 2.3.5. This equals the determinant of a finite matrix, since the subdeterminant

2.7.8 $$|[\lambda_i - i + j]|, \qquad 1 \leqslant i, j \leqslant r,$$

does not depend on $r$ provided that $r \geqslant \max\{i \mid \lambda_i \neq 0\}$. It is clear that for each such $r$, 2.7.8 is uniquely determined by its last column, i.e. by the $r$-tuple

2.7.9 $$(\lambda_1 - 1 + r, \lambda_2 - 2 + r, \ldots, \lambda_r - r + r).$$

Each such sequence is called *a sequence of $\beta$-numbers* for $\lambda$. 2.7.3 shows that this concept generalizes the concept of first-column hook lengths of partitions.

Now, if $(\beta_1, \beta_2, \ldots, \beta_r)$ is a sequence of $\beta$-numbers for $\lambda$ and

$$\mu := (\lambda_1, \ldots, \lambda_{i-1}, \lambda_{i+1} - 1, \lambda_i + 1, \lambda_{i+2}, \ldots),$$

then $[\mu] = -[\lambda]$. Also,

$$(\beta_1, \ldots, \beta_{i-1}, \beta_{i+1}, \beta_i, \beta_{i+2}, \ldots, \beta_r)$$

is a sequence of $\beta$-numbers for $\mu$. (All we have done is exchange the $i$th and $(i+1)$th rows of the determinant; cf. 2.3.9.) This means that $[\lambda]$ vanishes if two $\beta$-numbers for $\lambda$ are equal. The same happens if some $\beta$-number is negative. In all other cases, there is a unique permutation $\pi$ such that $\beta_{\pi(1)} > \beta_{\pi(2)} > \cdots > \beta_{\pi(r)} \geqslant 0$. If $(\beta_{\pi(1)}, \ldots, \beta_{\pi(r)})$ is a sequence of $\beta$-numbers for $\alpha$, then

2.7.10 $$\alpha \vdash \Sigma \lambda_i \quad \text{and} \quad [\lambda] = \operatorname{sgn} \pi \cdot [\alpha].$$

The sequences of $\beta$-numbers of partitions have the advantage that they are strictly decreasing:

2.7.11 $$\alpha_1 - 1 + r > \alpha_2 - 2 + r > \cdots > \alpha_r - r + r.$$

They can therefore be replaced by the *set* of these numbers, i.e. by

2.7.12 $$\{\alpha_1 - 1 + r, \ldots, \alpha_r - r + r\}.$$

Each such set will be called *a set of $\beta$-numbers for $\alpha$*. Thus each finite set of elements of $\mathbb{N}_0$, being a set of $\beta$-numbers, yields a sequence of $\beta$-numbers and in this way a partition.

Now the proof of 2.4.7 by an application of 2.4.5 shows clearly that the following is true:

**2.7.13** LEMMA. *Removing a rim q-hook $R_{ij}^{\alpha}$ from $[\alpha]$ means for each of its sequences $\beta_1, \ldots, \beta_r$ of $\beta$-numbers that a suitable $\beta_k$ is changed into $\beta_k - q$, and the resulting set $\{\beta_1, \ldots, \beta_{k-1}, \beta_k - q, \beta_{k+1}, \ldots, \beta_r\}$ is then a set of $\beta$-numbers for $[\alpha] \backslash R_{ij}^{\alpha}$. And conversely: if $\beta_1, \ldots, \beta_r$ is a sequence of $\beta$-numbers for $[\alpha]$ such that for a suitable k we have $0 \leqslant \beta_k - q \neq \beta_i$, for all $i \neq k$, then $\{\beta_1, \ldots, \beta_{k-1}, \beta_k - q, \beta_{k+1}, \ldots, \beta_r\}$ is a set of $\beta$-numbers for a diagram $[\gamma]$ which arises from $[\alpha]$ by removing a rim q-hook.*

$\beta$-numbers can be conveniently recorded on an abacus. Imagine that we have an abacus lying on a table with the runners going north-south and the abacus is viewed from the south:

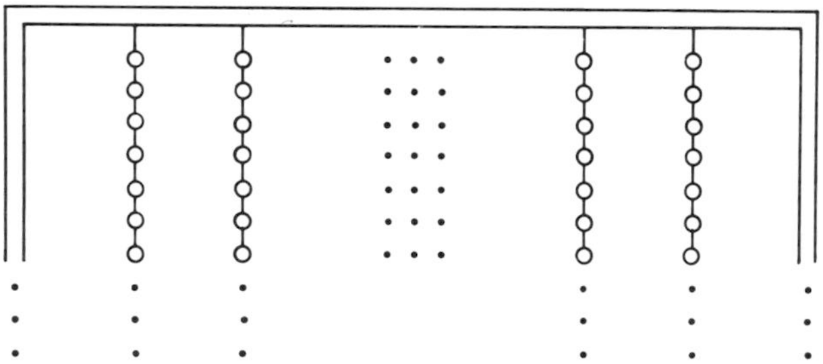

We assume that there are $q$ runners, called the 0th runner, 1st runner, ..., $(q-1)$th runner, from left to right and that the length of the runners and the supply of beads are both sufficiently large for the abacus frame not to interfere with our calculations. The possible bead positions are determined by assuming that all the beads are initially at the top and that we move beads only through one bead width at a time. Label the bead positions as below:

$$\begin{array}{ccccc} 0 & 1 & \cdots & q-2 & q-1 \\ q & q+1 & \cdots & 2q-2 & 2q-1 \\ \cdot & \cdot & & \cdot & \cdot \\ \cdot & \cdot & & \cdot & \cdot \\ \cdot & \cdot & & \cdot & \cdot \end{array}$$

For example, if $q{:}=3$, we have

$$\begin{array}{ccc} 0 & 1 & 2 \\ 3 & 4 & 5 \\ 6 & 7 & 8 \\ \cdot & \cdot & \cdot \\ \cdot & \cdot & \cdot \\ \cdot & \cdot & \cdot \end{array}$$

*A bead configuration is associated with a set of $\beta$-numbers (and hence a diagram or a partition) by letting the actual bead positions determine the $\beta$-numbers.* If again $q{:}=3$, then the bead configuration

2.7.14
$$\begin{array}{ccc} \bigcirc & \bigcirc & \bigcirc \\ \cdot & \cdot & \bigcirc \\ \bigcirc & \cdot & \cdot \\ \bigcirc & \cdot & \bigcirc \\ \cdot & \bigcirc & \cdot \\ \cdot & \cdot & \bigcirc \\ \cdot & \cdot & \cdot \end{array}$$

corresponds to the set of $\beta$-numbers $\{0,1,2,5,6,9,11,13,17\}$. These are $\beta$-numbers for the diagram whose first-column hook lengths are $2,3,6,8,10,14$, that is, for the diagram $[9,6,5,4,2^2]$. We notice the following:

**2.7.15.** *Given a bead configuration, the quickest way to find the first-column hook lengths of the corresponding diagram is to count the first gap as* 0, *and count on from there.*

We are now in a position to prove the desired theorem, which says that for any $q$ and every $\alpha \vdash n$ there exists a uniquely determined $q$-core $[\tilde{\alpha}]$ which is obtained from $[\alpha]$ by successive removals of rim $q$-hooks as far as we can do this, no matter which of the removable rim $q$-hooks of $[\alpha]$ and the resulting diagrams we remove first:

**2.7.16** THEOREM. *Each diagram has a uniquely determined q-core.*

*Proof.* By 2.7.13, removing a rim $q$-hook is equivalent to sliding a bead one space up. The bead configuration corresponding to a $q$-core of the diagram occurs when all the beads are as high as they will go. This configuration is clearly independent of the order in which we slide the beads.    ■

The 3-core of $[9,6,5,4,2^2]$ is $[3,1]$, and the bead configuration of the 3-core is (cf. 2.7.14)

$$
\begin{array}{ccc}
\bigcirc & \bigcirc & \bigcirc \\
\bigcirc & \bigcirc & \bigcirc \\
\bigcirc & \cdot & \bigcirc \\
\cdot & \cdot & \bigcirc \\
\cdot & \cdot & \cdot
\end{array}
$$

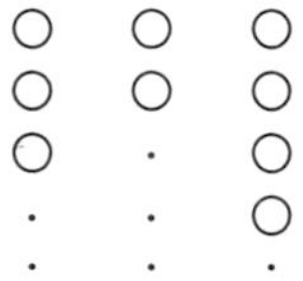

The partition $\tilde{\alpha}$ of the $q$-core $[\tilde{\alpha}]$ of $[\alpha]$ will be called the *$q$-core of* $\alpha$. The number of rim $q$-hooks which must be removed from $[\alpha]$ in order to obtain $[\tilde{\alpha}]$ is called the *$q$-weight* both of $[\alpha]$ and $\alpha$. An immediate corollary of the proof of 2.7.16 is

**2.7.17** THEOREM. *The number of partitions of $n$ with $q$-core $\tilde{\alpha}$ depends only on the $q$-weight $w$ of $\alpha$ and is equal to*

$$
b(w) := \sum_{(w_0, \ldots, w_{q-1})} p(w_0) \cdots p(w_{q-1}),
$$

*where the sum is taken over all the $q$-tuples $(w_0, \ldots w_{q-1})$ of nonnegative integers such that $\sum_{i=0}^{q-1} w_i = w$, and where again $p(w_i)$ denotes the number of proper partitions of $w_i$.*

Let us return to formula 2.4.5 and see how we can use this concept of a $q$-core in the evaluation of character values. For this purpose we introduce *numbered bead configurations* by giving each of the beads one of the numbers $1, \ldots, r$, if the bead configuration in question consists of $r$ beads.

When the beads are numbered in one way or the other, the bead configuration corresponds to an injective $r$-tuple $(l_1, \ldots, l_r)$ of nonnegative integers, $l_i$ denoting the place number of the $i$th bead. For example, if we number 2.7.14 as follows:

2.7.18

$$
\begin{array}{ccc}
\bigcirc_3 & \bigcirc_2 & \bigcirc_1 \\
\cdot & \cdot & \bigcirc_5 \\
\bigcirc_4 & \cdot & \cdot \\
\bigcirc_6 & \cdot & \bigcirc_8 \\
\cdot & \bigcirc_7 & \cdot \\
\cdot & \cdot & \bigcirc_9 \\
\cdot & \cdot & \cdot
\end{array}
$$

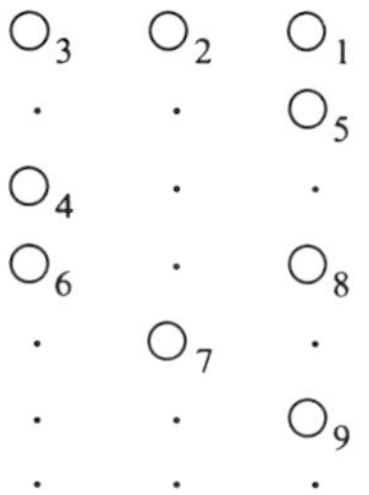

then the corresponding $r$-tuple is $(2, 1, 0, 6, 5, 9, 13, 11, 17)$.

Given such an $r$-tuple $(l_1, \ldots, l_r)$, we define the $r$-tuple $\lambda$ by $\lambda_i := l_{r-i+1} - r + i$ $(1 \leqslant i \leqslant r)$. Then $[\lambda] \neq 0$. For example, 2.7.18 gives $\lambda = (9, 4, 7, 4, 1, 3, -2, 0, 2)$. Forgetting the numbering of the beads, we obtain the set $\{l_1, \ldots, l_r\}$ and hence the corresponding partition $\alpha$ which satisfies $[\alpha] = \pm[\lambda]$. The sign stems from the necessary exchanges, and it is equal to the sign of the permutation which maps the numbering of the beads which led to $\lambda$ onto the natural numbering of the beads in the configuration of $[\alpha]$; e.g., the *natural numbering* of 2.7.14 is

2.7.19
$$\begin{array}{ccc} \bigcirc_1 & \bigcirc_2 & \bigcirc_3 \\ \cdot & \cdot & \bigcirc_4 \\ \bigcirc_5 & \cdot & \cdot \\ \bigcirc_6 & \cdot & \bigcirc_7, \\ \cdot & \bigcirc_8 & \cdot \\ \cdot & \cdot & \bigcirc_9 \\ \cdot & \cdot & \cdot \end{array}$$

so that the sign in question is

$$\operatorname{sgn}\begin{pmatrix} 1 & 2 & 3 & 4 & 5 & 6 & 7 & 8 & 9 \\ 3 & 2 & 1 & 5 & 4 & 6 & 8 & 7 & 9 \end{pmatrix} = -1,$$

and in fact (see 2.3.9):

$$[9,4,7,4,1,3,-2,0,2] = -[9,6,5,4,2^2].$$

Bearing this in mind, we consider an element $\rho\pi \in S_n$, $\rho := (i_1, \ldots, i_q)$ a $q$-cycle, $\pi \in S_n$, leaving $i_1, \ldots, i_q$ fixed. If $\lambda$ is a composition of $n$, then by 2.4.5

2.7.20
$$\chi^\lambda(\rho\pi) = \sum_\mu \chi^\mu(\pi),$$

where the sum is taken over all the compositions $\mu$ of $n-q$ arising from numbered bead configurations obtained by moving exactly one bead one step upwards in the numbered bead configuration of $\lambda$.

If we start this procedure with the naturally numbered bead configuration for a partition $\alpha$ of $q$-weight $w$, and apply it $w$ times, then we always end up with the same numbered bead configuration, no matter in which order we move the beads. This final bead configuration is still a numbered configuration if we keep the numbers when moving the beads, and it arises from a permutation $\sigma$ of the numbers in a naturally numbered bead configuration of $\tilde{\alpha}$ which contains the same number of beads. Let us illustrate this by an example before we go on.

The configuration 2.7.19 yields

2.7.21
$$\begin{array}{ccc} \bigcirc_1 & \bigcirc_2 & \bigcirc_3 \\ \bigcirc_5 & \bigcirc_8 & \bigcirc_4 \\ \bigcirc_6 & \cdot & \bigcirc_7 \\ \cdot & \cdot & \bigcirc_9 \end{array},$$

while the naturally numbered bead configuration of the 3-core

$$[9,6,\tilde{5},4,2^2]=[3,1]$$

is

2.7.22
$$\begin{array}{ccc} \bigcirc_1 & \bigcirc_2 & \bigcirc_3 \\ \bigcirc_4 & \bigcirc_5 & \bigcirc_6 \\ \bigcirc_7 & \cdot & \bigcirc_8 \\ \cdot & \cdot & \bigcirc_9 \end{array}$$

This yields, by comparing 2.7.21, 22,

2.7.23
$$\begin{aligned} \sigma &= \begin{pmatrix} 1 & 2 & 3 & 4 & 5 & 6 & 7 & 8 & 9 \\ 1 & 2 & 3 & 5 & 8 & 4 & 6 & 7 & 9 \end{pmatrix} \\ &= (45876). \end{aligned}$$

Let us denote the number of ways of moving the beads upwards step by step as far as they will go by

2.7.24
$$f_q^\alpha.$$

Then the considerations above show that if $\rho$ consists of $w$ $q$-cycles while $\pi$ acts on the remaining $n-qw$ symbols,

**2.7.25**
$$\zeta^\alpha(\rho\pi)=\operatorname{sgn}\sigma\cdot f_q^\alpha\zeta^{\tilde{\alpha}}(\pi).$$

Now repeated applications of 2.4.15, the Murnaghan-Nakayama formula for skew diagrams, together with the foregoing arguments, yield

**2.7.26**
$$\chi^{\alpha/\tilde{\alpha}}(\rho)=\operatorname{sgn}\sigma\cdot f_q^\alpha,$$

so that we have proved the following:

**2.7.27** THEOREM. *If $\alpha\vdash n$ is of $q$-weight $w$, and $\rho\in S_n$ a product of $w$ $q$-cycles, while $\pi\in S_n$ acts on the remaining $n-qw$ points, then*

$$\zeta^\alpha(\rho\pi)=\chi^{\alpha/\tilde{\alpha}}(\rho)\zeta^{\tilde{\alpha}}(\pi),$$

*and*

$$\chi^{\alpha/\tilde{\alpha}}(\rho)=\operatorname{sgn}\sigma\cdot f_q^{\alpha}.$$

We shall show next how $f_q^\alpha$ can be evaluated. In order to do this we have to be a bit circumstantial, but we shall be repaid by meeting with another very interesting structure.

We remember that $f_q^\alpha$ is the number of different ways in which $\tilde{\alpha}$ can be obtained from $\alpha$ by successive removals of rim $q$-hooks, i.e. by successive moves of beads in the corresponding bead configuration. But the beads are distributed into $q$ runners. Examine each runner in turn. *Since all diagrams on $w$ nodes have* 1-*weight $w$, it is clear how a bead configuration on one runner corresponds to a diagram*. For example the three bead configurations

**2.7.28** 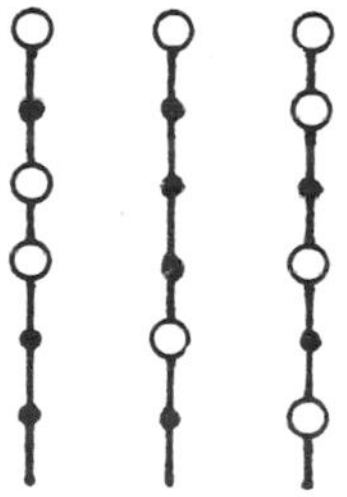 ,

correspond to the diagrams $[1^2], [3], [2,1]$. Therefore we may naturally associate a bead configuration on an abacus having $q$ runners with a sequence of $q$ diagrams, by examining each runner in turn.

Suppose, for the moment, that we impose the condition that the total number of beads on our abacus is a multiple of $q$. This ensures that each $q$-core is recorded on an abacus which is unique to within adding a constant number of beads to each runner. Suppose that to obtain the bead configuration corresponding to $[\alpha]$ from the bead configuration corresponding to its $q$-core $[\tilde{\alpha}]$, we make a series of moves on the $i$th runner corresponding to the diagram $[\alpha^{(i)}], 0\leqslant i\leqslant q-1$. Then the $q$-tuple of these diagrams,

2.7.29 $$[\alpha]_q:=[\alpha^{(0)}]_0[\alpha^{(1)}]_1\cdots[\alpha^{(q-1)}]_{q-1},$$

is called the *$q$-quotient* of $[\alpha]$.

For example 2.7.14 and 2.7.28 show that $[9,6,5,4,2^2]$ has 3-quotient $[1^2]_0[3]_1[2,1]_2$. The proof of 2.7.16 gives

**2.7.30** Theorem. *A diagram $[\alpha]$ is (for each $q\in\mathbb{N}$) uniquely determined by the pair $([\tilde{\alpha}];[\alpha]_q)$ consisting of its $q$-core $[\tilde{\alpha}]$ and its $q$-quotient $[\alpha]_q$.*

The desired number $f_q^\alpha$ of ways of getting $[\tilde{\alpha}]$ from $[\alpha]$ is closely related to the $q$-quotient $[\alpha]_q$. For we recall from the proof of 2.7.16 that (if $w_i$ denotes the number of nodes of $[\alpha^{(i)}]$), this number is equal to

$$\binom{w}{w_0, \ldots, w_{q-1}} = \frac{w!}{w_0! \cdots w_{q-1}!}$$

times the product over $i$ of the number of ways of moving the beads on the $i$th runner up as far as they will go. But by definition of the $q$-quotient $[\alpha]_q$ of $[\alpha]$, this number of ways of moving the beads on the $i$th runner up as far as they go is equal to the number of ways $[\alpha^{(i)}]$ can be brought to [0] by removing a rim 1-hook. On the other hand, the branching theorem yields that this number of ways of bringing $[\alpha^{(i)}]$ to [0] by removing 1-hooks is equal to the dimension,

2.7.31 $$f^{(i)},$$

of $[\alpha^{(i)}]$, so that we finally obtain

**2.7.32** $$f_q^\alpha = \binom{w}{w_0, \ldots, w_{q-1}} f^{(0)} \cdots f^{(q-1)}$$

**2.7.33** COROLLARY. *Let $\alpha \vdash n$ be of $q$-weight $w$, and $[\alpha]_q$ denote the $q$-quotient of $[\alpha]$, where $[\alpha^{(i)}]$ consists of $w_i$ nodes, $0 \leqslant i \leqslant q-1$. If $\rho$ consists of $b$ $q$-cycles, $b \geqslant w$, and $\pi$ acts on the points fixed by $\rho$, then*

$$\zeta^\alpha(\rho\pi) = \begin{cases} \operatorname{sgn}\sigma \binom{w}{w_0, \ldots, w_{q-1}} f^{(0)} \cdots f^{(q-1)} \zeta^{\tilde{\alpha}}(\pi) & \text{if } b = w, \\ 0 & \text{if } b > w. \end{cases}$$

We wish next to give a pictorial description of the $q$-quotient. In order to do this we introduce the *$q$-residue* of the $(i, j)$-node of $[\alpha]$, which we define to be the least nonnegative integer $r$ congruent to $j-i$ modulo $q$:

2.7.34 $$r \equiv j - i \pmod{q}, \qquad 0 \leqslant r < q.$$

For example, the 3-residues of $[9, 6, 5, 4, 2^2]$ are shown below:

| | | | | | | | | |
|---|---|---|---|---|---|---|---|---|
| 0 | 1 | 2 | 0 | 1 | 2 | 0 | 1 | 2 |
| 2 | 0 | 1 | 2 | 0 | 1 | | | |
| 1 | 2 | 0 | 1 | 2 | | | | |
| 0 | 1 | 2 | 0 | | | | | |
| 2 | 0 | | | | | | | |
| 1 | 2 | | | | | | | |

By construction, as we work round the rim of $[\alpha]$ from top to bottom, the $q$-residue decreases by 1 as we go from one node to the next. Thus:

**2.7.35.** *The length of a hook is divisible by $q$ iff the difference of the $q$-residues of the hand node and the foot node $\equiv -1$ modulo $q$.*

Given a diagram $[\alpha]$, draw lines through the rows whose last node has $q$-residue $i$ and through the columns whose last node has $q$-residue congruent to $i+1$, $0 \leqslant i \leqslant q-1$. Then by 2.7.35 the intersections of these lines mark just those nodes of $[\alpha]$ which are corners of hooks of length divisible by $q$ and which have hand nodes of residue class $i$. For example, if $q:=3$, we get for $i=0,1,2$ and $\alpha:=(9,6,5,4,2^2)$

2.7.36

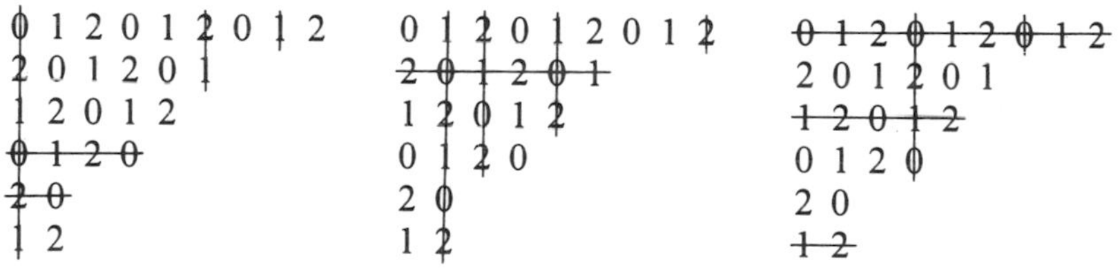

It is clear that the intersections of these lines give nodes forming a diagram shape. Call the $q$-tuple of these diagrams the *star $q$-diagram of* $[\alpha]$.

As 2.7.36 shows, the star 3-diagram of $[9,6,5,4,2^2]$ is equal to the 3-quotient of $[9,6,5,4,2^2]$, and in fact the following is true:

**2.7.37** Theorem. *The star $q$-diagram is the same as the $q$-quotient.*

Before proving this we need a couple of lemmas.

**2.7.38** Lemma. *Provided that our abacus has a multiple of $q$ beads (i.e our set of $\beta$-numbers has cardinality divisible by $q$), a bead on the $i$th runner corresponds to a row whose last node has $q$-residue $i$.*

*Proof.*

(i) Suppose that $\alpha$ has precisely $h$ nonzero parts, and the first gap in its abacus configuration occurs on the $k$th runner. Since the total number of beads is divisible by $q$, we have

$$k+h \equiv 0 \ (\mathrm{mod}\ q).$$

(ii) Suppose that $h_j^\alpha$ corresponds to a bead on the $i$th runner. Then $h_j^\alpha \equiv i-k \pmod q$ (see 2.7.15). Therefore $i \equiv h_j^\alpha - h \pmod q$, and

$$i \equiv h_j^\alpha - h = \alpha_j - j \pmod q,$$

as required. ■

**2.7.39** LEMMA. *The effect of removing the first column from* $[\alpha]$ *is recorded on the abacus by inserting a new bead in the first gap.*

This is clear from 2.7.15.

*Proof of Theorem 2.7.37.* Notice that if we count down a runner, the numbers we get being determined by the position of the beads on that runner, we obtain a set of $\beta$-numbers for the diagram corresponding to the configuration on that runner. (Counting down the first runner in the example 2.7.14 gives $0, 2, 3$ which is a set of $\beta$-numbers for $[1^2]$.)

Choose an abacus for $[\alpha]$ which contains a multiple of $q$ beads, and suppose that the first gap in the abacus configuration occurs on the $k$th runner. Adjoining a new bead in the first gap removes the first column from $[\alpha]$ (by 2.7.39) and from the $k$th part of the $q$-quotient (by the first paragraph of the proof). But removing the first column of $[\alpha]$ removes the first column of the $k$th part of the star $q$-diagram of $[\alpha]$ (by 2.7.38), and the result follows by induction. ■

Incidentally, Theorem 2.7.37 shows that

**2.7.40.** *The $q$-weight of a diagram equals the number of hook lengths divisible by $q$. In particular,* $[\alpha]$ *is a $q$-core iff no hook length is divisible by $q$.*

We conclude this section with a combinatorial result which will be useful when constructing decomposition matrices of symmetric groups.

Two diagrams, $[\alpha]$ and $[\beta]$, are said to have the same *$q$-content* if for each $r$, the number of nodes of $[\alpha]$ having $q$-residue $r$ (cf. 2.7.34) equals the number of nodes of $[\beta]$ having $q$-residue $r$. Thus, for example, if

| | | | | | |
|---|---|---|---|---|---|
| | 0 1 2 0 1 2 0 | | 0 1 2 0 1 | | 0 1 2 0 1 |
| | 2 0 1 2 0 | | 2 0 1 2 0 | | 2 0 1 2 0 |
| | 1 2 0 1 | | 1 2 0 | | 1 2 0 1 |
| $[\alpha]$: | 0 1 2 | $[\beta]$: | 0 1 2 | $[\gamma]$: | 0 1 2 |
| | 2 0 | | 2 0 | | 2 0 |
| | | | 1 2 | | 1 |
| | | | 0 | | 0 |

then $[\alpha]$ and $[\beta]$ have the same 3-content (each contains eight 0's, six 1's, and seven 2's), but $[\gamma]$ has a different 3-content ($[\gamma]$ contains eight 0's, seven 1's, and six 2's). The discrepancy occurs because $[\alpha]$ and $[\beta]$ have 3-core $[4,2]$, while the 3-core of $[\gamma]$ is $[2^2, 1^2]$.

**2.7.41** THEOREM. *Suppose that $\alpha$ and $\beta$ are partitions of the same integer. Then $[\alpha]$ and $[\beta]$ have the same $q$-core if and only if $[\alpha]$ and $[\beta]$ have the same $q$-content.*

*Proof.* A rim $q$-hook contains precisely one node of each $q$-residue. But the $q$-core of a diagram is obtained by removing rim $q$-hooks as often as possible. If $[\alpha]$ and $[\beta]$ have the same $q$-core, they therefore have the same $q$-content.

The other implication is not as easy. Suppose we have an abacus containing the same number of beads on each runner; in particular, our abacus has a multiple of $q$ beads. The configuration where all the beads are at the top corresponds to the empty diagram. Now build the diagram $[\alpha]$ by adding one node at a time. Adding a node whose $q$-residue is $i$ is recorded on the abacus by removing some bead from the $(i-1)$th runner and putting it the same distance south on the $i$th runner (with the obvious modification to this rule if $i=0$), by 2.7.38. If $[\alpha]$ contains $x$ nodes whose $q$-residue is $i$, and $y$ nodes whose $q$-residue is $i+1$, the number of beads on the $i$th runner has therefore increased by $x-y$, by the time we reach $[\alpha]$. On the assumption that $[\alpha]$ and $[\beta]$ have the same $q$-content, we deduce that the number of beads on each runner is the same in the configurations for $[\alpha]$ and $[\beta]$. But, as we have already remarked, the $q$-core of a diagram is determined by the number of beads on each runner. Therefore, $[\alpha]$ and $[\beta]$ have the same $q$-core. ■

## 2.8 Young's Rule and the Littlewood-Richardson Rule

The determinantal form 2.3.15 has shown us the way to write the irreducible character $\zeta^\alpha$ in terms of permutation characters $\xi^\lambda$. We now investigate how to reverse this process, and write $\xi^\lambda$ as a sum of irreducible characters. Having done this, we shall seek the irreducible constituents of (see 2.3.2)

2.8.1 $$[\alpha][\beta]=[\alpha]\#[\beta]\uparrow S_{m+n},$$

$\alpha \vdash m$, $\beta \vdash n$, $[\alpha]\#[\beta]$ the corresponding irreducible representation of $S_m \times S_n$. An algorithm, the so-called Littlewood-Richardson rule, will finally show us how the diagrams of the irreducible constituents $[\gamma]$ of 2.8.1 can be constructed from the diagrams $[\alpha]$ and $[\beta]$. First we consider particular cases.

**2.8.2** Young's rule (second version). *$[\alpha][n]=\sum[\gamma]$, where the sum has to be taken over all partitions $\gamma \vdash m+n$ such that for each $i$ we have $\alpha_i \leqslant \gamma_i \leqslant \alpha_{i-1}$ (if we put $\alpha_0 := \infty$).*

*Proof.* By 2.4.16

$$([\alpha][n],[\gamma]) = (\chi^{\gamma/\alpha}, \zeta^{(n)}) \\ = (\chi^{\gamma/\alpha}, \xi^{(n)}),$$

so that an application of 2.3.13 yields the statement. ∎

**2.8.3** Corollary. *The diagrams $[\gamma]$ of the irreducible constituents of $[\alpha][n]$ may be calculated by adding $n$ 2's to the diagram $[\alpha]$ in all possible ways such that no two 2's appear in the same column.*

For example if $\alpha := (3,2^2)$ and $n := 2$ we obtain:

| | | | | |
|---|---|---|---|---|
| × × × 2 2 | × × × 2 | × × × 2 | × × × | × × × |
| × × | × × 2 | × × | × × 2 | × × |
| × × | × × | × × | × × | × × |
| | | 2 | 2 | 2 2 |

Thus

$$[3,2^2][2] = [5,2^2]+[4,3,2]+[4,2^2,1]+[3^2,2,1]+[3,2^3].$$

We can apply this process repeatedly in order to get the constituents of

$$[\alpha_1][\alpha_2]\cdots[\alpha_h] = IS_\alpha \uparrow S_n. \tag{2.8.4}$$

For example, if we want to calculate [3][2][1], we first evaluate [3][2]:

| | | |
|---|---|---|
| 1 1 1 2 2 | 1 1 1 2 | 1 1 1 |
| | 2 | 2 2 |

This yields

$$[3][2] = [5]+[4,1]+[3,2],$$

so that the constituents of [3][2][1] are obtained as follows:

| | | | | |
|---|---|---|---|---|
| 1 1 1 2 2 3 | 1 1 1 2 2 | 1 1 1 2 3 | 1 1 1 2 | 1 1 1 2 |
| | 3 | 2 | 2 3 | 2 |
| | | | | 3 |
| | 1 1 1 3 | 1 1 1 | 1 1 1 | |
| | 2 2 | 2 2 3 | 2 2 | |
| | | | 3 | |

Therefore

$$[3][2][1]=[6]+2[5,1]+2[4,2]+[4,1^2]+[3^2]+[3,2,1].$$

This algorithm justifies the introduction of tableaux having repeated entries. A *generalized Young tableau* of *shape* $[\alpha]$, $\alpha\vdash n$, and *content* $\lambda$, $\lambda \models n$, arises from $[\alpha]$ by replacing the nodes of the diagram by positive integers in such a way that the integer $i$ occurs exactly $\lambda_i$ times; e.g.

$$\begin{matrix} 1 & 2 & 1 & 1 & 3 & 2 \\ 3 & 2 \end{matrix}$$

is a generalized Young tableau of shape $(6,2)$ and of content $(3^2,2)$. A Young tableau is therefore a generalized Young tableau of content $(1^n)$, for a certain $n$.

A generalized Young tableau is said to be *semistandard* if the numbers are nondecreasing along each row and strictly increasing down each column. For example

$$\begin{matrix} 1 & 1 & 1 & 2 & 2 & 2 \\ 3 & 3 \end{matrix} \qquad \begin{matrix} 1 & 1 & 1 & 2 & 2 & 3 \\ 2 & 3 \end{matrix} \qquad \begin{matrix} 1 & 1 & 1 & 2 & 3 & 3 \\ 2 & 2 \end{matrix}$$

are the only semistandard tableaux of shape $(6,2)$ and content $(3^2,2)$. In terms of this concept we can formulate

**2.8.5** Young's rule (third version). *The multiplicity* $([\beta_1]\cdots[\beta_k],[\alpha])$ *of* $[\alpha]$ *in* $IS_\beta\uparrow S_n$ *equals the number of semistandard tableaux of shape* $\alpha$ *and of content* $\beta$.

More generally the constituents $[\alpha]$, $\alpha\vdash m+n$, of the representation

2.8.6 $$[\beta][\delta_1]\cdots[\delta_l]=([\beta]\#(IS_\delta\uparrow S_n))\uparrow S_{m+n}, \qquad \beta\vdash m, \quad \delta\models n,$$

are obtained by first adding to the diagram $[\beta]$ $\delta_1$ nodes in all the admissible ways, then to each one of the resulting diagrams $\delta_2$ further nodes, and so on. This motivates the introduction of *skew tableaux* and of *generalized skew tableaux*. They arise from skew diagrams $[\alpha/\beta]$ if we replace the nodes by natural numbers; in the case of generalized skew tableaux repetitions may occur. When we are dealing with generalized semistandard skew tableaux we simply speak of semistandard skew tableaux.

Let

2.8.7 $$k_{\alpha/\beta,\lambda}$$

be the number of generalized semistandard skew tableaux of shape $\alpha/\beta$ and of content $\lambda$. For example, by Young's rule

**2.8.8** $$([\beta_1]\cdots[\beta_k],[\alpha])=k_{\alpha,\beta}.$$

If we are given a generalized skew tableau, say

2.8.9 $$\begin{matrix} & 1 & 1 \\ & 2 & \\ 2 & 3 & \end{matrix},$$

we can read its entries from *right to left* in each row and one row after the other, downwards, obtaining a sequence of positive integers. For example 2.8.9 yields

2.8.10 $$1\quad 1\quad 2\quad 3\quad 2.$$

Such a sequence is called a *lattice permutation* if and only if the following holds: For each place $j$ of the sequence the number of $i$'s which occur among the first $j$ elements is greater than or equal to the number of $(i+1)$'s, for each $i$. Hence 2.8.10 is a lattice permutation.

For the sake of simplicity we now allow the nodes of a diagram or a skew diagram to be replaced by elements of an arbitrary totally ordered set. We prove a combinatorial lemma which will turn out to be crucial:

**2.8.11** LEMMA. *Assume* $\alpha \vdash n_1+n_2+b$, $n_i, b\in\mathbb{N}$, $\beta\vdash n_2$, $\gamma\vdash n_1$. *Then the number of semistandard tableaux of shape* $\alpha/\gamma$ *which contain* $\beta_i$ *numbers* $i, 1\leq i\leq\beta_1'$, *together with* $b$ *symbols* $x$ *such that the points* $i$ *form a lattice permutation* (*if we read their entries along their rows from right to left and the rows downwards*), *is the same if the order is defined by*

$$x<1<2<\cdots<\beta_1',$$

*as if it is defined by*

$$1<2<\cdots<\beta_1'<x.$$

*Proof.* For $0\leq k\leq\beta_1'$ and the corresponding order $1<\cdots<k<x<k+1<\cdots<\beta_1'$ we denote by

$$M_k$$

the set of semistandard tableaux of shape $\alpha/\gamma$ which contain $\beta_i$ numbers $i$, $1\leq i\leq\beta_1'$, together with $b$ symbols $x$. The subset of $M_k$ which consists of such tableaux the points $i$ of which form a lattice permutation will be indicated by

$$A_k.$$

Our method of proof is to construct bijections

$$f_k: M_k \rightarrowtail\!\!\!\twoheadrightarrow M_{k-1}, \quad 1 \leqslant k \leqslant \beta_1',$$

which satisfy

$$f_1 \circ f_2 \circ \cdots \circ f_{\beta_1'}[A_{\beta_1'}] = A_0,$$

so that the statement immediately follows by definition of $A_{\beta_1'}$ and $A_0$ from the fact that the $f_k$ are bijections.

(i) *The bijections.* We define a mapping $f_k$ on $M_k$ as follows: If $T \in M_k$, then $f_k(T)$ arises from $T$ in the following way:

(a) In each column of $T$ which contains both $x$ and $k$, we interchange $x$ and $k$ (remember that $T$ is semistandard, so that each column of $T$ contains $x$ at most once and $k$ at most once, and $k$ occurs just above $x$ if the column contains both $x$ and $k$), and afterwards

(b) in each row containing both $x$'s and $k$'s left fixed under (a), we shift the $x$'s to the left of the $k$'s (remember that $T$ is semistandard, so that the $x$'s occur in $T$ just to the right of such $k$'s).

For example in the following situation:

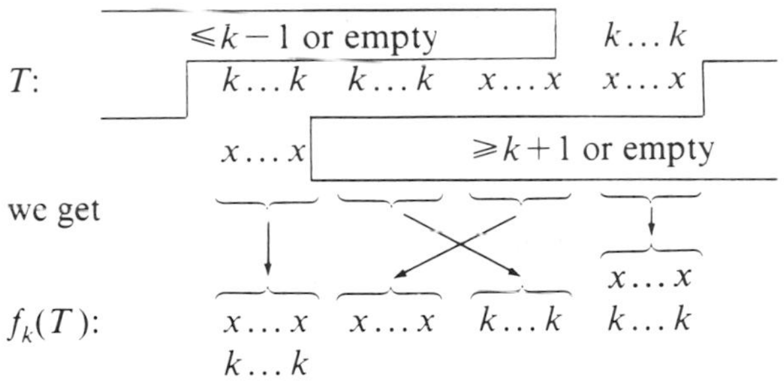


Obviously $f_k(T) \in M_{k-1}$, so that $f_k: M_k \to M_{k-1}$. $f_k$ is a bijection, since it is inverted by $g_k: M_{k-1} \to M_k$, which is defined by its operation on an element of $M_{k-1}$ as follows.

(a′) Interchange $x$ and $k$ when they occur in the same column, and then

(b′) in each row which contains both $x$'s and $k$'s fixed under (a′), shift the $x$'s to the right of the $k$'s.

(ii) $f_1 \circ \cdots \circ f_{\beta_1'}[A_{\beta_1'}] \subseteq A_0$. Take a $T_{\beta_1'} \in A_{\beta_1'}$. Suppose, inductively, that the $i$'s in each one of the following tableaux form a lattice permutation:

$$T_{\beta_1'}, f_{\beta_1'}(T_{\beta_1'}), \ldots, f_{k+1} \circ \cdots \circ f_{\beta_1'}(T_{\beta_1'}) =: T_k.$$

We have to show that this is also true for

$$T_{k-1} := f_k(T_k) = f_k \circ \cdots \circ f_{\beta_1'}(T_{\beta_1'}).$$

(Note that we do not prove $f_k[A_k] \subseteq A_{k-1}$; this is false.)

The definition of $f_k$ shows that the lattice property holds for all the numbers in $f_k(T_k)$, except, perhaps, between the numbers $k$ and $k+1$. In particular,

$$f_{\beta_1'}(T_{\beta_1'}) \in A_{\beta_1'-1}.$$

We may therefore assume that $k < \beta_1'$.

Suppose that there is a $k+1$ in the $i$th row of $T_k$. Then the required lattice property in $f_k(T_k)$ is true for this $k+1$ unless there is an $x$ to the left of it in the same row of $T_k$ and this $x$ lies below a $k$. We must examine the latter case more closely:

$j$th column
↓

$$T_k: \quad \begin{array}{ccccc} \leqslant k-1 & k\ldots k & k \;\ldots\; k & x \;\ldots\; x & x\ldots \\ \ldots x & \underbrace{x\ldots x}_{>0} & \underbrace{k+1\ldots k+1}_{a} & \underbrace{k+1\ldots k+1}_{b} & \geqslant k+2 \end{array} \quad \leftarrow i\text{th row}$$

Every $k+1$ in the $i$th row of $T_k$ has a $k$ or $x$ immediately above it, since $k<x<k+1$ and there is an $x$ in the $i$th row with a $k$ above it. Let $a(\geqslant 0) :=$ the number of $(k+1)$'s in the $i$th row of $T_k$ which lie below a $k$, and let $b(\geqslant 0) :=$ the number of $(k+1)$'s in the $i$th row of $T_k$ which lie below an $x$. Suppose that the last $k$ in the $(i-1)$th row of $T_k$ lies in the $j$th column. Since $T_k = f_{k+1}(T_{k+1})$,

$$\left|\{(i',j') \mid k \text{ is in the } (i',j') \text{ place of } T_k \text{ and } i'<i-1 \text{ or } i'=i-1 \text{ and } j'>j\}\right|$$
$$-\Big|\{(i',j') \mid k+1 \text{ is in the } (i',j') \text{ place of } T_k$$
$$\text{and } i'<i-1 \text{ or } i'=i-1 \text{ and } j'>j\}\Big| \geqslant b.$$

Therefore, (the number of $k$'s in the first $i-1$ rows of $f_k(T_k)$) $-$ (the number of $(k+1)$'s in the first $i-1$ rows of $f_k(T_k)$) $\geqslant b+a$. This shows that the lattice property holds for all the $(k+1)$'s in the $i$th row of $f_k(T_k)$, and completes the proof of (ii).

(iii) $g_{\beta'_1}\circ\cdots\circ g_1[A_0]\subseteq A_{\beta'_1}$ follows quite analogously. ■

**2.8.12** COROLLARY. *The multiplicity* $([\beta][\delta_1]\cdots[\delta_l],[\gamma])$, $\beta\vdash m$, $\delta\vdash n$, $\gamma\vdash m+n$, *is equal to the number of semistandard tableaux of shape* $\gamma$ *which contain* $\beta_i$ *symbols* $i$, $\delta_j$ *symbols* $\bar{j}$, $1\leqslant i\leqslant\beta'_1$, $1\leqslant j\leqslant l$, *subject to the ordering* $\bar{1}<\cdots<\bar{l}<1<\cdots<\beta'_1$ *and such that the* $i$*'s yield a lattice permutation when we read the rows from right to left and downwards.*

*Proof.* The considerations above have shown that

$$([\beta][\delta_1]\cdots[\delta_l],[\gamma])=k_{\gamma/\beta,\delta},$$

which is the number of semistandard tableaux of shape $\gamma/\beta$, of content $\delta$, i.e. containing $\delta_j$ symbols $j$, $1\leqslant j\leqslant l$. Now the generalized Young tableau

$$\begin{matrix} 1\cdots\cdots\cdots 1 \\ 2\cdots\cdots 2 \\ \vdots \\ \beta'_1\ldots\beta'_1 \end{matrix}$$

is the only semistandard one of shape and content $\beta$ where the $i$'s yield a lattice permutation. Thus $k_{\gamma/\beta,\delta}$ is equal to the number of semistandard tableaux of shape $\gamma$ which contain $\beta_i$ $i$'s and $\delta_j$ $\bar{j}$'s, subject to the order $1<\cdots<\beta'_1<\bar{1}<\cdots<\bar{l}$ and such that the $i$'s form a lattice permutation. Successive applications of 2.8.11 to $x:=\bar{1},\bar{2},\ldots,\bar{l}$ therefore yield the statement. ■

We are now in a position to prove the main theorem:

**2.8.13** THE LITTLEWOOD-RICHARDSON RULE. *The multiplicity* $([\alpha][\beta],[\gamma])$, $\alpha\vdash m$, $\beta\vdash n$, $\gamma\vdash m+n$, *is equal to the number of semistandard tableaux of shape* $\gamma/\alpha$ *and content* $\beta$ *which yield lattice permutations when we read their entries from the right to the left and downwards.*

*Proof.* Let $g_{\alpha\beta\gamma}$ denote this number of semistandard skew tableaux. 2.8.12 yields, if $\delta\vdash m$, $\beta\vdash n$, $\gamma\vdash m+n$,

$$([\delta_1]\cdots[\delta_l][\beta],[\gamma])=\sum_{\alpha\vdash m}k_{\alpha\delta}g_{\alpha\beta\gamma}.$$

But

$$[\delta_1]\cdots[\delta_l]=\sum_{\alpha\vdash m}k_{\alpha\delta}[\alpha],$$

so that for each $\delta \vdash m$:

$$\sum_{\alpha \vdash m} k_{\alpha\delta} g_{\alpha\beta\gamma} = ([\delta_1]\cdots[\delta_l][\beta],[\gamma])$$
$$= \sum_{\alpha \vdash m} k_{\alpha\delta}([\alpha][\beta],[\gamma]).$$

The statement now follows from the regularity of the matrix $(k_{\alpha\delta})$. ■

**2.8.14** Corollary. *The diagrams of the irreducible constituents* $[\gamma]$ *of* $[\alpha][\beta]$ *(counted with their multiplicities* $([\alpha][\beta],[\gamma])$*) are obtained as follows. In all possible ways, add to a fixed tableau of shape* $[\alpha]$ *semistandard tableaux of content* $\beta$ *which yield lattice permutations; the resulting tableaux will have the shapes* $[\gamma]$ *of the constituents of* $[\alpha][\beta]$.

*An alternative procedure is to take a symbol* $a_{ij}$ *for each i and j for which the* $(i, j)$*-node belongs to* $[\beta]$, *then*

(i) *Add to the diagram* $[\alpha]$ *all the symbols* $a_{1j}$. *After the additions no row of the compound tableau may contain more nodes than a preceding row.*

(ii) *Next add all the symbols* $a_{2j}$, *according to the same rule, and so on, until all the symbols have been added.*

(iii) *These additions must be such that for all i, if* $y<j$, $a_{iy}$ *goes in a later column than* $a_{ij}$, *and for all j, if* $x<i$, $a_{xj}$ *goes in an earlier row than* $a_{ij}$.

For example the decomposition of [3,2] [2,1] is obtained in the following way. Add the symbols of

$$[2,1]: \begin{array}{ll} a_{11} & a_{12} \\ a_{21} & \end{array},$$

to [3,2]. From the first row, keeping in mind 2.8.14(i) and (iii), we get

$$\begin{array}[t]{l} \times\times\times\, a_{12}\, a_{11} \\ \times\times \end{array} \qquad \begin{array}[t]{l} \times\times\times\, a_{11} \\ \times\times\, a_{12} \\ , \end{array} \qquad \begin{array}[t]{l} \times\times\times\, a_{11} \\ \times\times \\ a_{12} \end{array} \qquad \begin{array}[t]{l} \times\times\times \\ \times\times\, a_{11} \\ a_{12} \end{array} \qquad \begin{array}[t]{l} \times\times\times \\ \times\times \\ a_{12}\, a_{11} \end{array}$$

To these compound tableaux we add $a_{21}$ in all possible ways subject to 2.8.14 (ii) and (iii):

$$\begin{array}[t]{l} \times\times\times\, a_{12}\, a_{11} \\ \times\times\, a_{21} \end{array} \qquad \begin{array}[t]{l} \times\times\times\, a_{12}\, a_{11} \\ \times\times \\ a_{21} \end{array} \qquad \begin{array}[t]{l} \times\times\times\, a_{11} \\ \times\times\, a_{12}\, a_{21} \end{array} \qquad \begin{array}[t]{l} \times\times\times\, a_{11} \\ \times\times\, a_{12} \\ a_{21} \end{array}$$

$$\begin{array}[t]{l} \times\times\times\, a_{11} \\ \times\times\, a_{21} \\ a_{12} \end{array} \qquad \begin{array}[t]{l} \times\times\times\, a_{11} \\ \times\times \\ a_{12}\, a_{21} \end{array} \qquad \begin{array}[t]{l} \times\times\times\, a_{11} \\ \times\times \\ a_{12} \\ a_{21} \end{array} \qquad \begin{array}[t]{l} \times\times\times \\ \times\times\, a_{11} \\ a_{12}\, a_{21} \end{array}$$

$$\begin{array}[t]{l} \times\times\times \\ \times\times\, a_{11} \\ a_{12} \\ a_{21} \end{array} \qquad \begin{array}[t]{l} \times\times\times \\ \times\times \\ a_{12}\, a_{11} \\ a_{21} \end{array},$$

so that the required reduction is given by

$$[3,2][2,1]=[5,3]+[5,2,1]+[4^2]+2[4,3,1]+[4,2^2]+[4,2,1^2]$$
$$+[3^2,2]+[3^2,1^2]+[3,2^2,1].$$

## 2.9 Inner Tensor Products

The inner tensor product

2.9.1 $$[\alpha]\otimes[\beta]$$

of two ordinary irreducible representations $[\alpha]$ and $[\beta]$ of $S_n$ is an ordinary representation of $S_n$ which is in general reducible, so the question arises how to evaluate the decomposition of 2.9.1 into its irreducible constituents. No satisfactory answer to this question is known, but we can at least show how to tackle it by applications of the determinantal form and the Littlewood-Richardson rule. The determinantal form yields for the character $\zeta^\alpha$ of $[\alpha]$

2.9.2 $$\zeta^\alpha=\sum_\pi \operatorname{sgn}\pi\,\xi^{\alpha-\mathrm{id}+\pi},$$

so that the character $\zeta^\alpha\otimes\zeta^\beta$ of $[\alpha][\beta]$ satisfies

2.9.3 $$\zeta^\alpha\otimes\zeta^\beta=\sum_\pi \operatorname{sgn}\pi\,(\xi^{\alpha-\mathrm{id}+\pi}\otimes\zeta^\beta).$$

Now as $\xi^{\alpha-\mathrm{id}+\pi}$ is induced from $S_{\alpha-\mathrm{id}+\pi}$ (recall 2.3.3), a well-known formula on tensor products of induced characters (see e.g. Curtis and Reiner [1962, (38.5)]) yields

2.9.4 $$\xi^{\alpha-\mathrm{id}+\pi}\otimes\zeta^\beta=\zeta^\beta\downarrow S_{\alpha-\mathrm{id}+\pi}\uparrow S_n,$$

so that 2.9.3 implies

**2.9.5** $$\zeta^\alpha\otimes\zeta^\beta=\sum_\pi \operatorname{sgn}\pi\,(\zeta^\beta\downarrow S_{\alpha-\mathrm{id}+\pi}\uparrow S_n).$$

But the decomposition of $\zeta^\beta\downarrow S_{\alpha-\mathrm{id}+\pi}$ can be evaluated in principle by suitable applications of the Littlewood-Richardson rule, and induction from this decomposition into $S_n$ can be performed using the Littlewood-Richardson rule again. Hence the whole procedure can be carried out mechanically; we do not need the character table. An example illustrates this. We would like to evaluate the decomposition of

2.9.6 $$[3,2,1]\otimes[3,2,1].$$

First of all we remember from Section 2.3 that

2.9.7 $$[3,2,1]=\begin{vmatrix} [3] & [4] & [5] \\ [1] & [2] & [3] \\ 0 & 1 & [1] \end{vmatrix}$$
$$=[3][2][1]-[3][3]-[4][1][1]+[5][1],$$

so that we have to decompose the restrictions

2.9.8 $$\zeta^{(3,2,1)}\downarrow S_3\times S_2\times S_1, \quad \zeta^{(3,2,1)}\downarrow S_3\times S_3,$$

2.9.9 $$\zeta^{(3,2,1)}\downarrow S_4\times S_1\times S_1, \quad \zeta^{(3,2,1)}\downarrow S_5\times S_1.$$

Boring but easy applications of the Littlewood-Richardson rule together with Frobenius's reciprocity show the following:

2.9.10

(i) $[3,2,1]\downarrow S_3\times S_2\times S_1\uparrow S_6=[3][2][1]+[3][1^2][1]+3[2,1][2][1]$
$+3[2,1][1^2][1]+[1^3][2][1]+[1^3][1^2][1],$

(ii) $[3,2,1]\downarrow S_3\times S_3\uparrow S_6=2[3][2,1]+2[2,1][2,1]+2[2,1][1^3],$

(iii) $[3,2,1]\downarrow S_4\times S_1\times S_1\uparrow S_6=2[2,1^2][1][1]+2[2^2][1][1]+2[3,1][1][1],$

(iv) $[3,2,1]\downarrow S_5\times S_1\uparrow S_6=[2^2,1][1]+[3,1^2][1]+[3,2][1].$

Further applications of the Littlewood-Richardson rule therefore yield the desired result:

2.9.11

$$[3,2,1]\otimes[3,2,1]=[6]+2[5,1]+3[4,2]+4[4,1^2]+2[3^2]+5[3,2,1]$$
$$+4[3,1^3]+2[2^3]+3[2^2,1^2]+2[2,1^4]+[1^6].$$

Although we can evaluate the decomposition of $\zeta^\alpha\otimes\zeta^\beta$ in this way mechanically, the situation is still unsatisfactory, since we sometimes need more explicit information about the irreducible constituents of $[\alpha]\otimes[\beta]$. In order to get it, we go back to the permutation characters $\xi^\alpha$ and consider their inner tensor products $\xi^\alpha\otimes\xi^\beta$.

Now $\xi^\alpha$ is the character of the action of $S_n$ on the left cosets of $S_\alpha$, or, equivalently, on the set of tabloids of shape $[\alpha]$, i.e. on the set

2.9.12 $$\mathbf{n}_\alpha:=\{\mathbf{n}^\alpha \mid \mathbf{n}^\alpha \text{ a dissection of } \mathbf{n} \text{ into subsets of orders } \alpha_i\}$$

(cf. 1.3.2). Therefore $\xi^\alpha\otimes\xi^\beta$ is the character of the natural action of $S_n$ on the cartesian product

2.9.13 $$\mathbf{n}_\alpha \times \mathbf{n}_\beta$$

by

2.9.14 $$\pi(\mathbf{n}^\alpha,\mathbf{n}^\beta):=\left(\{\pi[\mathbf{n}_1^\alpha],\pi[\mathbf{n}_2^\alpha],\dots\},\{\pi[\mathbf{n}_1^\beta],\pi[\mathbf{n}_2^\beta],\dots\}\right).$$

The orbit of $(\mathbf{n}^\alpha,\mathbf{n}^\beta)$ is characterized by the orders

2.9.15 $$\nu_{ij}:=|\mathbf{n}_i^\alpha\cap\mathbf{n}_j^\beta|,$$

so that the stabilizer of this orbit is a Young subgroup $S_\nu$ corresponding to the dissection of $\mathbf{n}$ into the subsets $\mathbf{n}_i^\alpha\cap\mathbf{n}_j^\beta$. This yields

**2.9.16** LEMMA. *$\xi^\alpha\otimes\xi^\beta=\Sigma_\nu\xi^\nu$, if the sum is taken over all the $k\times k$ matrices $\nu=(\nu_{ij})$ such that $\nu_{ij}\in\mathbb{N}_0$ and*

$$\sum_i \nu_{ij}=\beta_j \quad \textit{and} \quad \sum_j \nu_{ij}=\alpha_i.$$

Hence at least the inner tensor product $\xi^\alpha\otimes\xi^\beta$ is explicitly given as a sum of permutation characters. For example

$$\xi^{(n-1,1)}\otimes\xi^{(n-1,1)}=\sum_\nu \xi^\nu,$$

where the sum is taken over all the matrices

$$\nu=\begin{pmatrix}\nu_{11} & \nu_{12}\\ \nu_{21} & \nu_{22}\end{pmatrix}$$

such that $\nu_{ij}\in\mathbb{N}_0$ and

$$\nu_{11}+\nu_{12}=\nu_{11}+\nu_{21}=n-1,$$
$$\nu_{12}+\nu_{22}=\nu_{21}+\nu_{22}=1.$$

The only matrices of this form are

$$\begin{pmatrix}n-1 & 0\\ 0 & 1\end{pmatrix} \quad\text{and}\quad \begin{pmatrix}n-2 & 1\\ 1 & 0\end{pmatrix},$$

so that we obtain

$$\xi^{(n-1,1)}\otimes\xi^{(n-1,1)}=\xi^{(n-1,1)}+\xi^{(n-2,1^2)}$$
$$=2\zeta^{(n)}+3\zeta^{(n-1,1)}+\zeta^{(n-2,2)}+\zeta^{(n-2,1^2)}.$$

It is sometimes useful to employ the following remark in place of Curtis and Reiner [1962, (38.5)]. Let $H \leqslant G$ be groups, and suppose that every element of $H$ is conjugate (in $H$) to its inverse, i.e., $H$ is ambivalent. If $\varphi$ and $\psi$ are characters of $G$, $\varepsilon$ the character of the identity representation of $H$, then

$$(\varepsilon \uparrow G, \varphi \otimes \psi) = (\varepsilon, (\varphi \otimes \psi) \downarrow H) = \frac{1}{|H|} \sum_{h \in H} \varphi(h) \psi(h)$$
$$= (\varphi \downarrow H, \psi \downarrow H) = \sum_{\chi} (\varphi \downarrow H, \chi)(\psi \downarrow H, \chi),$$

where the last sum is over the irreducible characters of $H$. This gives an alternative method for evaluating $(\zeta^\alpha \otimes \zeta^\beta, \xi^\gamma)$, and hence also $(\zeta^\alpha \otimes \zeta^\beta, \zeta^\gamma)$, for we may take $H$ to be the Young subgroup $S_\gamma$ when

**2.9.17**
$$(\zeta^\alpha \otimes \zeta^\beta, \xi^\gamma) = \sum_{\chi} (\zeta^\alpha \downarrow S_\gamma, \chi)(\zeta^\beta \downarrow S_\gamma, \chi).$$

The evaluation of $\zeta^\alpha \downarrow S_\gamma$ involves the Littlewood-Richardson rule again, and the calculation of $\zeta^\alpha \otimes \zeta^\beta$ is no quicker by this method than by the one already given. But viewing the problem this way is sometimes of assistance, for example in:

**2.9.18 Theorem.** *Suppose that* $\alpha, \beta, \gamma \vdash n$. *Then* $(\zeta^\alpha \otimes \zeta^\beta, \xi^\gamma) > 0$ *only if*

$$|\alpha - \beta| := \sum_{i=1}^{n} |\alpha_i - \beta_i| \leqslant 2(n - \gamma_1).$$

*Proof.* $(\zeta^\alpha \otimes \zeta^\beta, \xi^\gamma) > 0$ implies, by 2.9.17, that $\zeta^\alpha \downarrow S_\gamma$ and $\zeta^\beta \downarrow S_\gamma$ have a common constituent. Therefore, there are partitions $\delta^{(i)}$ of $\gamma_i$ such that

$$(\zeta^\alpha \downarrow S_\gamma, [\delta^{(1)}] \# [\delta^{(2)}] \# \cdots) > 0,$$

and

$$(\zeta^\beta \downarrow S_\gamma, [\delta^{(1)}] \# [\delta^{(2)}] \# \cdots) > 0.$$

By the Littlewood-Richardson rule, the diagram $[\delta^{(1)}]$ is a subdiagram of both the diagrams $[\alpha]$ and $[\beta]$. For this to be possible, we clearly require $|\alpha - \beta| \leqslant 2(n - \gamma_1)$. ∎

**2.9.19 Corollary.** $(\zeta^\alpha \otimes \zeta^\beta, \zeta^\gamma) > 0$ *holds only if* $|\alpha - \beta| \leqslant 2(n - \gamma_1)$.

*Proof.* $(\zeta^\alpha \otimes \zeta^\beta, \zeta^\gamma) > 0$ implies $(\zeta^\alpha \otimes \zeta^\beta, \xi^\gamma) > 0$, so that the statement follows from 2.9.18. ∎

**2.9.20** COROLLARY. $([\alpha]\otimes[\beta],[\gamma])>0$ *implies*

$$|\alpha_1-\beta_1|\leqslant n-\gamma_1\leqslant 2n-(\alpha_1+\beta_1).$$

*Proof.* 2.9.19 shows that the assumption implies

$$|\alpha-\beta|\leqslant 2(n-\gamma_1).$$

Now by the triangle inequality

$$\begin{aligned}|\alpha-\beta|&=|\alpha_1-\beta_1|+|\alpha_2-\beta_2|+\cdots+|\alpha_n-\beta_n|\\&\geqslant|\alpha_1-\beta_1|+|n-\alpha_1-(n-\beta_1)|\\&=2|\alpha_1-\beta_1|,\end{aligned}$$

so that we get

$$|\alpha_1-\beta_1|\leqslant n-\gamma_1.$$

As $([\alpha]\otimes[\beta],[\gamma])$ is symmetric in $\alpha$, $\beta$, and $\gamma$, we get from this last inequality that the following is also true:

$$\beta_1-\gamma_1\leqslant n-\alpha_1,$$

and hence also

$$n-\gamma_1\leqslant 2n-(\alpha_1+\beta_1),$$

which completes the proof. ■

This result shows how we can estimate the irreducible constituents $[\gamma]$ of $[\alpha]\otimes[\beta]$ in terms of the number $n-\gamma_1$ of nodes of $[\gamma]$ which do not belong to the first row. We call this number,

2.9.21 $$d_\gamma := n-\gamma_1,$$

the *depth* of $[\gamma]$ or $\gamma$. Using this notation, we can rephrase 2.9.20 as follows:

**2.9.22** THEOREM. *The depth* $d_\gamma$ *of each irreducible constituent* $[\gamma]$ *of* $[\alpha]\otimes[\beta]$ *satisfies the inequalities*

$$|d_\alpha-d_\beta|\leqslant d_\gamma\leqslant d_\alpha+d_\beta.$$

## Exercises

2.1 If $P \leqslant S_n$, then $P$ induces permutation groups $P_1$ and $P_2$ on the subsets of order $k$ and on the subsets of order $k-1$ of **n**. Using 2.2.19(iii), show that, provided $2 \leqslant k \leqslant n/2$, $P_1$ has at least as many orbits as has $P_2$.

2.2 Prove that the characters of the representations $[\alpha]^{\pm}$ of $A_n$ are real or non-real according as the number of nodes above or below the diagonal of the Young diagram $[\alpha]$ is even or odd.

CHAPTER 3

# *Ordinary Irreducible Matrix Representations of Symmetric Groups*

When investigating the representation theory of a given group, it is normal to follow the calculation of its ordinary irreducible characters by a consideration of its irreducible modules, and since we now know the characters of $S_n$ we should concern ourselves next with representations. We do this in the present chapter, by dealing with left ideals of the group algebra $\mathbb{Q}S_n$. At the same time, we investigate some of the matrix representations which arise.

In fact, the irreducible modules for $S_n$ are more easily described as Specht modules, and a shorter construction of the matrix representations is given in Chapter 7, but we present here Alfred Young's classical approach using the group algebra, since it provides an excellent example of the ring-theoretical tools from general theory being applied in a practical case.

## 3.1 A Decomposition of the Group Algebra $\mathbb{Q}S_n$ into Minimal Left Ideals

We remember from general representation theory that we can obtain the ordinary irreducible matrix representations of a given finite group $G$ at least theoretically as follows. The group algebra $\mathbb{C}G$ has a unique decomposition into simple two-sided ideals $I^\nu$, say:

3.1.1 $$\mathbb{C}G = \bigoplus_\nu I^\nu.$$

ENCYCLOPEDIA OF MATHEMATICS and Its Applications, Gian-Carlo Rota (ed.). Vol. 16: G. D. James and A. Kerber, The Representation Theory of the Symmetric Group
ISBN 0-201-13515-9

Each $I^\nu$ is, by Wedderburn's theorem, isomorphic to a full matrix ring over $\mathbb{C}$. Let $R^\nu$ denote one of these isomorphisms. Up to equivalence $I^\nu$ has exactly one irreducible matrix representation, and hence this matrix representation must be equivalent to $R^\nu$.

Each such system of irreducible matrix representations $R^\nu$ yields a complete system of pairwise inequivalent and irreducible ordinary matrix representations $D^\nu$ of $G$ as follows. If $g \in G$, then $1_{\mathbb{C}} \cdot g \in \mathbb{C}G$, so that we obtain a decomposition of this element with respect to the decomposition 3.1.1 of $\mathbb{C}G$, say

3.1.2 $$1_{\mathbb{C}} \cdot g = \sum_\nu \gamma_\nu .$$

Then we put

3.1.3 $$D^\nu(g) := R^\nu(\gamma_\nu).$$

This shows that the crucial step is an explicit description of the isomorphisms $R^\nu$ of $I^\nu$ onto a full matrix ring over $\mathbb{C}$, the existence of which is guaranteed by the theorem of Wedderburn.

The representation theory of $S_n$ is in fact one of the very rare cases in which such a description is available for a class of by no means trivial finite groups. Later on it will be shown that from the algorithm for symmetric groups we can derive algorithms for various other classes of finite groups, too.

In order to describe this algorithm we first decompose $\mathbb{Q}S_n$ into its simple two-sided ideals. Even more, we give a decomposition of $\mathbb{Q}S_n$ into minimal left ideals (by 2.1.12 it suffices to consider $\mathbb{Q}S_n$ instead of $\mathbb{C}S_n$). We shall do this by deriving a complete system of primitive idempotents, which generate minimal left ideals. For this purpose we recall the method which gave us the complete system of ordinary irreducible representations of $S_n$, in order to show how it fits into the general representation theory of finite groups and that it yields primitive idempotents.

We started off with a pair $S_\alpha, S_{\alpha'}$ of subgroups of $S_n$, say

$$S_\alpha := H_1^\alpha := H(t_1^\alpha) \quad \text{and} \quad S_{\alpha'} := V_1^\alpha := V(t_1^\alpha).$$

We took their one-dimensional representations

$$IH_1^\alpha \quad \text{and} \quad AV_1^\alpha,$$

and considered the induced representations

$$IH_1^\alpha \uparrow S_n \quad \text{and} \quad AV_1^\alpha \uparrow S_n .$$

These representations have the property that there exists exactly one double coset $V_1^\alpha \pi H_1^\alpha$ such that the restrictions of $IH_1^\alpha$ and $AV_1^\alpha$ coincide on the intersection $H_1^\alpha \cap \pi V_1^\alpha \pi^{-1}$; this particular double coset is $V_1^\alpha H_1^\alpha$.

Furthermore, for each $\sigma \in S_n \backslash V_1^\alpha H_1^\alpha$ there exist by 1.5.8 transpositions $\pi \in V_1^\alpha$ and $\rho \in H_1^\alpha$ such that $\pi\sigma = \sigma\rho$. Since these elements $\pi$ and $\rho$ are transpositions, we have

$$IH_1^\alpha(\pi) = \mathrm{id} \neq -\mathrm{id} = AV_1^\alpha(\rho).$$

This shows that $G := S_n$ and its subgroups $H := H_1^\alpha$ and $V := V_1^\alpha$, together with their representations $IH_1^\alpha$ and $AV_1^\alpha$ over $F := \mathbb{Q}$, satisfy the conditions of the following theorem:

**3.1.4** THEOREM. *Let $G$ denote a finite group and $F$ a field with* $\mathrm{char}\, F \nmid |G|$. *Furthermore let $H$ and $V$ be subgroups of $G$ with one-dimensional characters $\delta$ and $\varepsilon$ such that*

(i) $\delta \downarrow H \cap V = \varepsilon \downarrow H \cap V$, *and*

(ii) *for each $g \in G \backslash VH$ there exist $h \in H, v \in V$ such that $vg = gh$, while* $\delta(h) \neq \varepsilon(v)$.

*If we now put*

$$e := \sum_{v \in V} \sum_{h \in H} \varepsilon(v)\delta(h)vh,$$

*then this element of the group algebra $FG$ of $G$ over $F$ generates a minimal left ideal $FGe$, which affords an absolutely irreducible representation of $G$.*

*Proof.*

(1) We first prove that the following is true:

**3.1.5**
$$\left\{a \in FG \,|\, \forall h, v\left(vah = \varepsilon(v^{-1})\delta(h^{-1})a\right)\right\} = Fe.$$

(a) Every $\kappa e, \kappa \in F$, satisfies for each $v \in V, h \in H$

$$\begin{aligned} v\kappa eh &= \kappa \sum_{h',v'} \varepsilon(v')\delta(h')vv'h'h \\ &= \varepsilon(v^{-1})\delta(h^{-1})\kappa \sum_{h',v'} \varepsilon(vv')\delta(h'h)vv'h'h \\ &= \varepsilon(v^{-1})\delta(h^{-1})\kappa e. \end{aligned}$$

(b) Conversely let $a := \sum \kappa_g g \in FG$ satisfy for each $v \in V$, $h \in H$ the equation

3.1.6 $$\sum_g \kappa_g vgh = vah = \varepsilon(v^{-1})\delta(h^{-1})a = \varepsilon(v^{-1})\delta(h^{-1})\sum_g \kappa_g g.$$

Then we obtain $a = \kappa e$ by comparing coefficients as follows:

($\alpha$) In the case when $g$ is of the form $g = vh$, 3.1.6 gives

$$\kappa_g = \kappa_{vh} = \varepsilon(v)\delta(h)\kappa_1.$$

($\beta$) In the case when $g \in G \setminus VH$, we take $v \in V$, $h \in H$ such that $vg = gh^{-1}$ and $\delta(h^{-1}) \neq \varepsilon(v)$ (such $v$ and $h$ exist by assumption). We obtain from 3.1.6 that

$$\kappa_g = \kappa_{vgh} = \varepsilon(v)\delta(h)\kappa_g.$$

Since $\varepsilon(v)\delta(h) \neq 1$, this can be true only if $\kappa_g = 0$.

($\gamma$) As any $g \in VH$ can be written in $|H \cap V|$ ways as a product $g = vh$, we obtain from ($\alpha$) and ($\beta$) that

$$a = \frac{\kappa_1}{|H \cap V|} e,$$

as stated.

This completes the proof of 3.1.5.

(2) A particular $a \in FG$ which satisfies for each $v \in V$, $h \in H$

$$vah = \varepsilon(v^{-1})\delta(h^{-1})a$$

is the element $a := e^2$, so that we obtain from 3.1.5

$$e^2 = \kappa e$$

for a suitable $\kappa \in F$. Hence a proof of $\kappa \neq 0$ will show that $e$ is essentially idempotent. In order to verify this, we consider the linear mapping

$$\varphi : FG \to FG : a \mapsto ae,$$

i.e. right multiplication by $e$. We shall evaluate $\kappa$ by calculating the trace of $\varphi$ with respect to two different bases of $FG$. First we take a basis of $FG$, which is adapted to the subspace $FGe$, i.e. the first $\dim_F(FGe)$ elements of the basis span $FGe$. Then $\varphi$ is represented by a matrix $\Phi_1$ of the form

$$\Phi_1 = \begin{pmatrix} \kappa & & & & & * \\ & \ddots & & & & \\ & & \kappa & & & \\ & & & 0 & & \\ & 0 & & & \ddots & \\ & & & & & 0 \end{pmatrix}.$$

This gives our first equation for the trace:

$$\operatorname{trace} \varphi = \kappa \cdot \dim_F(FGe).$$

Secondly we take as basis of $FG$ the elements $1_F \cdot g$. Then $\varphi$ is represented by a matrix $\Phi_2$ of the form

$$\Phi_2 = \begin{pmatrix} \kappa_1 & & * \\ & \ddots & \\ * & & \kappa_1 \end{pmatrix},$$

$\kappa_1$ being the coefficient of $1_G$ in $e = \sum \kappa_g g$. Since $\kappa_1 = |H \cap V|$, we obtain as a second equation for the trace:

$$\operatorname{trace} \varphi = |G| \cdot |H \cap V|.$$

Comparing the two equations for the trace, we get

3.1.7 $$\kappa = \frac{|G| \cdot |H \cap V|}{\dim_F(FGe)}.$$

This is nonzero since $|G| \cdot |H \cap V| \neq 0$ by $\operatorname{char} F \nmid |G|$.

(3) In (2) we proved that

$$\hat{e} := \frac{1}{\kappa} e$$

is an idempotent of $FG$. In order to show that $FGe = FG\hat{e}$ is minimal, we need only prove that $\hat{e}FG\hat{e}$ is a division ring. But again by 3.1.5 we obtain immediately that

$$\hat{e}FG\hat{e} = eFGe = Fe \simeq F.$$

(4) Since by (3) $eFGe$ is not only a division ring but also isomorphic to $F$, $FGe$ is even absolutely irreducible. ■

This theorem elucidates the representation-theoretical background of our method of characterizing the ordinary irreducible representations of $S_n$ as "intersections" (cf. 2.1.4) of monomial representations induced from Young subgroups. It shows furthermore that the elements

3.1.8
$$e^\alpha := \mathcal{V}^\alpha \mathcal{H}^\alpha := \sum_{\pi \in V^\alpha} \sum_{\rho \in H^\alpha} \operatorname{sgn}\pi \cdot \pi\rho \in \mathbb{Q}S_n,$$

which we have defined already in 1.5.10, are in fact essentially idempotent and generate minimal left ideals in $\mathbb{Q}S_n$. Putting (cf. 3.1.7)

3.1.9
$$\kappa^\alpha := \frac{n!}{f^\alpha},$$

we obtain

**3.1.10** THEOREM. *For each tableau $t^\alpha$ the element*

$$\hat{e}^\alpha := \frac{1}{\kappa^\alpha} e^\alpha = \frac{f^\alpha}{n!} \sum_{\pi \in V^\alpha} \sum_{\rho \in H^\alpha} \operatorname{sgn}\pi \cdot \pi\rho$$

*is a primitive idempotent in the group algebra $\mathbb{Q}S_n$ of $S_n$ over $\mathbb{Q}$. The minimal left ideal*

$$L^\alpha := \mathbb{Q}S_n \hat{e}^\alpha$$

*affords the absolutely irreducible representation $[\alpha]$ of $S_n$.*

If we let $t^\alpha$ run just through the set of all the tableaux $t_1^\alpha$, $\alpha \vdash n$, (see 1.4.19) we obtain a set of irreducible left $S_n$-modules $L_1^\alpha$, which by the results of Chapter 2 form a complete set, i.e.

**3.1.11** COROLLARY. *The minimal left ideals $L_1^\alpha := \mathbb{Q}S_n \hat{e}_1^\alpha$, $\alpha \vdash n$, constitute a complete set of irreducible and pairwise inequivalent left $\mathbb{Q}S_n$-modules.*

For a fixed $\alpha \vdash n$ there exist $n!$ tableaux $t^\alpha$ and therefore $n!$ primitive idempotents $\hat{e}^\alpha$, so we have obtained $|P(n)|n!$ minimal left ideals $L^\alpha$ in $\mathbb{Q}S_n$, while the desired direct decomposition of $\mathbb{Q}S_n$ into minimal left ideals consists of only $\sum_{\alpha \vdash n} f^\alpha$ minimal left ideals. We therefore have to pick a suitable subset out of the set of all the tableaux. For a given $\alpha \vdash n$ we have to take exactly $f^\alpha$ tableaux $t_i^\alpha$ with diagram $[\alpha]$. The question is which subset (if any) will do what we want.

We get an idea which subset might be a suitable one by considering a combinatorial interpretation of $f^\alpha$. For the branching theorem 2.4.3 tells us that

$$f^\alpha = \zeta^\alpha(1_{S_n})$$

is equal to the number of ways we can reach [0] from the diagram $[\alpha]$ by successively cancelling just one node and subject to the condition that after each cancellation a diagram (or finally [0]) remains. Or vice versa: $f^\alpha$ is equal to the number of ways in which the nodes of $[\alpha]$ can be replaced by the points $i \in \mathbf{n}$ in such a way that the points of the resulting tableau in each row and in each column occur in their natural order (reading the rows from left to right and the columns downwards). These tableaux are called *standard* Young-tableaux. The standard (3,2)-tableaux for example are

3.1.12 $$\begin{matrix} 1 & 3 & 5 \\ 2 & 4 & \end{matrix}, \quad \begin{matrix} 1 & 2 & 5 \\ 3 & 4 & \end{matrix}, \quad \begin{matrix} 1 & 3 & 4 \\ 2 & 5 & \end{matrix}, \quad \begin{matrix} 1 & 2 & 4 \\ 3 & 5 & \end{matrix}, \quad \begin{matrix} 1 & 2 & 3 \\ 4 & 5 & \end{matrix}.$$

This implies the following

**3.1.13** COROLLARY. *$f^\alpha$ is equal to the number of standard α-tableaux.*

Let us denote the set of all the $n!$ α-tableaux by $t[\alpha]$:

3.1.14 $$t[\alpha] := \{t^\alpha \mid t^\alpha \text{ is an } \alpha\text{-tableau}\},$$

and put

3.1.15 $$\mathrm{st}[\alpha] := \{t^\alpha \mid t^\alpha \in t[\alpha] \text{ standard}\}.$$

We would like now to number the elements of $\mathrm{st}[\alpha]$ with respect to the location of the point $n$, the point $n-1$, and so on. This ordering is a suitable one for later applications.

We first notice that if $t^\alpha \in \mathrm{st}[\alpha]$, then the point $n$ occurs in $t^\alpha$ both at the end of a row and at the end of a column. Hence the tableau $t^{\alpha*}$, which arises from $t^\alpha$ by removing the point $n$ is a standard tableau again. This allows us to define *the last letter sequence* $t_1^\alpha, \ldots, t_{f^\alpha}^\alpha$ of the elements of $\mathrm{st}[\alpha]$ recursively as follows:

**3.1.16** DEFINITION. The numbers $i$ and $k$ of $t_i^\alpha, t_k^\alpha \in \mathrm{st}[\alpha]$, satisfy $i<k$ if and only if either

(i) the point $n$ occurs in $t_i^\alpha$ in a higher row than in $t_k^\alpha$, or
(ii) $t_i^{\alpha*} = t_j^\beta$ and $t_k^{\alpha*} = t_h^\beta$, $\beta \vdash n-1$, where $j<h$.

For simplicity, we define

3.1.17 $$t_i^\alpha < t_k^\alpha \quad :\Leftrightarrow \quad i<k.$$

For example

3.1.18
$$\begin{matrix}1&3&5\\2&4\end{matrix} < \begin{matrix}1&2&5\\3&4\end{matrix} < \begin{matrix}1&3&4\\2&5\end{matrix} < \begin{matrix}1&2&4\\3&5\end{matrix} < \begin{matrix}1&2&3\\4&5\end{matrix}.$$

It is clear from 3.1.16 that the following is true:

**3.1.19** LEMMA. *For $t_i^\alpha$, $t_k^\alpha \in \mathrm{st}[\alpha]$ we denote by $[\alpha^{(i,j)}]$ and $[\alpha^{(k,j)}]$ the diagrams of $t_i^{\alpha*\cdots*}$, $t_k^{\alpha*\cdots*}$ ($j$ asterisks), $0 \leqslant j \leqslant n$. Then $t_i^\alpha < t_k^\alpha$ if and only if there exists an $h \in \mathbb{N}$ such that*

$$\alpha^{(i,j)} = \alpha^{(k,j)}, \qquad 0 \leqslant j \leqslant h-1,$$

*while*

$$\alpha^{(i,h)} \lhd \alpha^{(k,h)}.$$

This characterization of the last letter sequence of the elements of $\mathrm{st}[\alpha]$ will be used in the proof of

**3.1.20** LEMMA. *If $t_i^\alpha < t_k^\alpha$, then $\hat{e}_k^\alpha \hat{e}_i^\alpha = 0$.*

*Proof.* Let $p_1, \ldots, p_r, n$ be the points in the row of $n$ in $t_k^\alpha$ and $q_1, \ldots, q_s, n$ be the points in the column of $n$ in $t_i^\alpha$, so that

3.1.21
$$\mathcal{H}_k^\alpha = (1 + (p_1 n) + \cdots + (p_r n))\mathcal{H}_k^{\alpha*},$$

where $\mathcal{H}_k^{\alpha*}$ has the obvious meaning, and analogously

3.1.22
$$\mathcal{V}_i^\alpha = \mathcal{V}_i^{\alpha*}(1 - (q_1 n) - \cdots - (q_s n)).$$

(i) We shall first prove by induction on $n$ that $\mathcal{H}_k^\alpha \mathcal{V}_i^\alpha = 0$.
(a) If $[\alpha^{(i,1)}] = [\alpha^{(k,1)}]$, then $t_i^\alpha < t_k^\alpha$ yields $t_i^{\alpha*} < t_k^{\alpha*}$, so that $\mathcal{H}_k^{\alpha*}\mathcal{V}_i^{\alpha*} = 0$ by the induction hypothesis. This together with 3.1.21 and 3.1.22 yields $\mathcal{H}_k^\alpha \mathcal{V}_i^\alpha = 0$.
(b) If $[\alpha^{(i,1)}] \neq [\alpha^{(k,1)}]$, then by 3.1.19, $\alpha^{(i,1)} \lhd \alpha^{(k,1)}$. By 1.5.17, again $\mathcal{H}_k^{\alpha*}\mathcal{V}_i^{\alpha*} = 0$ and therefore $\mathcal{H}_k^\alpha \mathcal{V}_i^\alpha = 0$.
(ii) The result of (i) yields immediately that

$$e_k^\alpha e_i^\alpha = \mathcal{V}_k^\alpha \mathcal{H}_k^\alpha \mathcal{V}_i^\alpha \mathcal{H}_i^\alpha = 0,$$

and hence

$$\hat{e}_k^\alpha \hat{e}_i^\alpha = \frac{1}{(\kappa^\alpha)^2} e_k^\alpha e_i^\alpha = 0. \qquad \blacksquare$$

This shows that the $\hat{e}_i^\alpha, 1 \leqslant i \leqslant f^\alpha$, are pairwise orthogonal. Thus

**3.1.23** $$\forall \alpha, i, k \quad \left(i \neq k \ \Rightarrow \ \mathbb{Q}S_n\hat{e}_i^\alpha \cap \mathbb{Q}S_n\hat{e}_k^\alpha = \{0\}\right),$$

and hence the following holds:

**3.1.24** THEOREM. *The minimal left ideals*

$$L_i^\alpha := \mathbb{Q}S_n\hat{e}_i^\alpha, \qquad \alpha \vdash n, \quad 1 \leqslant i \leqslant f^\alpha,$$

*yield the following decomposition of $\mathbb{Q}S_n$ into simple two-sided ideals $I^\alpha$:*

$$\mathbb{Q}S_n = \bigoplus_{\alpha \vdash n} I^\alpha,$$

*where*

$$I^\alpha = \bigoplus_{i=1}^{f^\alpha} L_i^\alpha.$$

*The generating elements $\hat{e}_i^\alpha$ of the minimal left ideals $L_i^\alpha$ are pairwise orthogonal primitive idempotents, so that in particular*

$$\hat{e}_i^\alpha \hat{e}_k^\beta = \delta_{\alpha\beta}\delta_{ik}\hat{e}_i^\alpha.$$

## 3.2 The Seminormal Basis of $\mathbb{Q}S_n$

It is our aim now to construct a $\mathbb{Q}$-basis

3.2.1 $$\{\varepsilon_{ik}^\alpha \mid \alpha \vdash n, \ 1 \leqslant i, k \leqslant f^\alpha\}$$

of $\mathbb{Q}S_n$, which is adapted to the decomposition of $\mathbb{Q}S_n$ into simple two-sided ideals given in the preceding section, i.e. 3.2.1 should satisfy

3.2.2 $$I^\alpha = \bigoplus_{i=1}^{f^\alpha} \mathbb{Q}S_n\hat{e}_i^\alpha = \langle\langle \varepsilon_{ik}^\alpha \mid 1 \leqslant i, k \leqslant f^\alpha \rangle\rangle_{\mathbb{Q}}.$$

The elements of this basis will also satisfy the following orthogonality relations:

3.2.3 $$\varepsilon_{ij}^\alpha \varepsilon_{kl}^\beta = \delta_{\alpha\beta}\delta_{jk}\varepsilon_{il}^\alpha.$$

In order to find such a basis, we start as in the usual proof of Wedderburn's theorem by considering $(f^\alpha)^2$ isomorphisms between the direct summands of 3.2.2. We therefore define permutations $\pi_{ik}^\alpha \in S_n$ by

3.2.4 $$t_i^\alpha = \pi_{ik}^\alpha t_k^\alpha, \qquad 1 \leqslant i, k \leqslant f^\alpha.$$

It is clear that

**3.2.5** $$(\pi_{ik}^\alpha)^{-1} = \pi_{ki}^\alpha \quad \text{and} \quad \pi_{ij}^\alpha \pi_{jk}^\alpha = \pi_{ik}^\alpha.$$

With the aid of these permutations we put

3.2.6 $$e_{ik}^\alpha := e_i^\alpha \pi_{ik}^\alpha, \qquad \text{in particular} \quad e_{ii}^\alpha = e_i^\alpha.$$

Since obviously

3.2.7 $$\pi_{ik}^\alpha H_k^\alpha \pi_{ki}^\alpha = H_i^\alpha \quad \text{and} \quad \pi_{ik}^\alpha V_k^\alpha \pi_{ki}^\alpha = V_i^\alpha,$$

we have

**3.2.8** $$\text{(i) } e_{ik}^\alpha = e_i^\alpha \pi_{ik}^\alpha = \pi_{ik}^\alpha e_k^\alpha = \frac{1}{\kappa^\alpha} e_i^\alpha \pi_{ik}^\alpha e_k^\alpha,$$
$$\text{(ii) } e_i^\alpha e_{ik}^\alpha = \kappa^\alpha e_{ik}^\alpha = e_{ij}^\alpha e_{jk}^\alpha.$$

This yields

3.2.9 $$\mathbb{Q}S_n e_i^\alpha e_{ik}^\alpha = \mathbb{Q}S_n e_i^\alpha \pi_{ik}^\alpha = \mathbb{Q}S_n \pi_{ik}^\alpha e_k^\alpha = \mathbb{Q}S_n e_k^\alpha.$$

Hence the elements $e_{ik}^\alpha$, $1 \leqslant i, k \leqslant f^\alpha$, provide $(f^\alpha)^2$ isomorphisms between the minimal left ideals $\mathbb{Q}S_n e_i^\alpha$, $1 \leqslant i \leqslant f^\alpha$. But unfortunately they do not satisfy the orthogonality relations 3.2.3. In order to force 3.2.3 we carry out a recursive orthogonalization procedure. For this we again denote by $t_i^{\alpha *}$ the standard tableau which arises from $t_i^\alpha$ by deleting the point $n$, by $t_i^{\alpha **}$ the one which arises from $t_i^{\alpha *}$ by deleting $n-1$, and so on, until after $n$ steps we arrive at the empty tableau

3.2.10 $$t^\varnothing = \varnothing$$

Let $e_i^{\alpha *}, e_i^{\alpha **}, \ldots, \hat{e}_i^{\alpha *}, \hat{e}_i^{\alpha **}, \ldots$ be the corresponding elements of the subalgebras $\mathbb{Q}S_{n-1}, \mathbb{Q}S_{n-2}, \ldots$ of $\mathbb{Q}S_n$.

We are now in a position to define the *seminormal units* $\varepsilon_i^\alpha$ recursively as follows (we put $\kappa_i^{\alpha *} := (n-1)!/f_i^{\alpha *}$, $\kappa_i^{\alpha **} := (n-2)!/f_i^{\alpha **}, \ldots$):

3.2.11 $$\varepsilon_i^\alpha := \frac{1}{\kappa_i^\alpha} \varepsilon_i^{\alpha *} e_i^\alpha \varepsilon_i^{\alpha *},$$
$$\varepsilon_i^{\alpha *} := \frac{1}{\kappa_i^{\alpha *}} \varepsilon_i^{\alpha **} e_i^{\alpha *} \varepsilon_i^{\alpha **},$$
$$\cdots\cdots\cdots\cdots\cdots\cdots$$
$$\varepsilon_i^\varnothing := 1_{\mathbb{Q}S_n}.$$

**3.2.12 LEMMA.**

(i) *For each* $a \in \mathbb{Q}S_n$ *we have* $e_{ij}^{\alpha} a e_{hk}^{\alpha} = \lambda \cdot \kappa^{\alpha} e_{ik}^{\alpha}$, *where* $\lambda := [$*coeff. of* $1_{S_n}$ *in* $e_{hj}^{\alpha} a]$.

(ii) $e_i^{\alpha} \varepsilon_i^{\alpha *} e_i^{\alpha} = \kappa^{\alpha} \cdot e_i^{\alpha}$.

*Proof.* (i): From 3.1.5, 3.2.5, and 3.2.6 we get

$$\mathbb{Q}e_i^{\alpha} = e_i^{\alpha} \mathbb{Q}S_n e_i^{\alpha} = e_{ij}^{\alpha} \pi_{ji}^{\alpha} \mathbb{Q}S_n \pi_{ih}^{\alpha} e_{hi}^{\alpha} = e_{ij}^{\alpha} \mathbb{Q}S_n e_{hi}^{\alpha}.$$

This implies that for a suitable $\mu \in \mathbb{Q}$

$$e_{ij}^{\alpha} a e_{hi}^{\alpha} = \mu e_i^{\alpha}.$$

This yields by comparing the coefficients of $1_{S_n}$ on both sides that

$$\mu = \left[\text{coeff. of } 1_{S_n} \text{ in } e_{ij}^{\alpha} a e_{hi}^{\alpha}\right].$$

But for all $b, c \in \mathbb{Q}S_n$ the following is true:

**3.2.13** $$\left[\text{coeff. of } 1_{S_n} \textit{ in } b \cdot c\right] = \left[\text{coeff. of } 1_{S_n} \text{ in } c \cdot b\right].$$

Hence

$$\begin{aligned} \mu &= \left[\text{coeff. of } 1_{S_n} \text{ in } e_{hi}^{\alpha} e_{ij}^{\alpha} a\right] \\ &= \left[\text{coeff. of } 1_{S_n} \text{ in } \kappa^{\alpha} e_{hj}^{\alpha} a\right], \end{aligned}$$

from which (i) follows.

(ii): We obtain from (i) that

$$e_i^{\alpha} \varepsilon_i^{\alpha *} e_i^{\alpha} = \lambda \kappa^{\alpha} e_i^{\alpha},$$

where

$$\lambda = \left[\text{coeff. of } 1_{S_n} \text{ in } e_i^{\alpha} \varepsilon_i^{\alpha *}\right].$$

It remains to show that $\lambda = 1$. We prove this by recursion. Another application of (i) gives

$$e_i^{\alpha *} \varepsilon_i^{\alpha *} e_i^{\alpha *} = \lambda^* \kappa_i^{\alpha *} e_i^{\alpha *},$$

where

$$\lambda^* = \left[\text{coeff. of } 1_{S_n} \text{ in } e_i^{\alpha *} \varepsilon_i^{\alpha *}\right].$$

Since $\varepsilon_i^{\alpha *} \in \mathbb{Q}S_{n-1}$, the coefficient of $1_{S_n}$ is the same in both $e_i^{\alpha} \varepsilon_i^{\alpha *}$ and $e_i^{\alpha *} \varepsilon_i^{\alpha *}$, i.e. $\lambda = \lambda^*$, and we can try to prove (ii) by induction on $n$.

I ($n:=2$). $\varepsilon_i^{\alpha*}=e_i^{\alpha*}=1_{S_2}$, so that $1=\lambda^*=\lambda$.
II ($n>2$). The induction hypothesis is

$$(1)\qquad \forall \beta\vdash n-1,\quad 1\leqslant j\leqslant f^\beta\left(e_j^\beta\varepsilon_j^{\beta*}e_j^\beta=\kappa^\beta e_j^\beta\right).$$

This implies (use 3.2.11):

$$e_i^{\alpha*}\varepsilon_i^{\alpha*}e_i^{\alpha*}=\frac{1}{\kappa_i^{\alpha*}}e_i^{\alpha*}\varepsilon_i^{\alpha**}e_i^{\alpha*}\varepsilon_i^{\alpha**}e_i^{\alpha*}$$
$$\underset{(1)}{=}e_i^{\alpha*}\varepsilon_i^{\alpha**}e_i^{\alpha*}$$
$$\underset{(1)}{=}\kappa_i^{\alpha*}e_i^{\alpha*}.$$

This proves $1=\lambda^*=\lambda$, and therefore (ii) is true. ■

We are now in a position to prove the main theorem of this section:

**3.2.14** THEOREM. *The elements*

$$\varepsilon_{ik}^\alpha:=\frac{1}{\kappa^\alpha}\varepsilon_i^{\alpha*}e_{ik}^\alpha\varepsilon_k^{\alpha*}$$

*satisfy the orthogonality relations*

$$\varepsilon_{ij}^\alpha\varepsilon_{hk}^\beta=\delta_{\alpha\beta}\delta_{jh}\varepsilon_{ik}^\alpha.$$

*They form a* $\mathbb{Q}$*-basis of* $\mathbb{Q}S_n$, *called the* seminormal basis *of* $\mathbb{Q}S_n$.

*Proof.*

(i) Since $e_{ik}^\alpha=e_i^\alpha\pi_{ik}$ and $e_i^\alpha\in I^\alpha$, we have for each $\alpha$, $i$, and $k$

$$e_{ik}^\alpha\in I^\alpha=\bigoplus_i \mathbb{Q}S_n\hat{e}_i^\alpha,$$

so by the definition of $\varepsilon_{ik}^\alpha$

$$\varepsilon_{ik}^\alpha\in I^\alpha.$$

Elements of $\mathbb{Q}S_n$ which are contained in different simple two-sided ideals of $\mathbb{Q}S_n$ annihilate each other, so that we obtain

$$\forall i,j,h,k\quad\left(\alpha\neq\beta\Rightarrow\varepsilon_{ij}^\alpha\varepsilon_{hk}^\beta=0\right).$$

(ii) We shall now use induction on $n$ to prove

$$\varepsilon_{ij}^{\alpha}\varepsilon_{hk}^{\alpha}=\delta_{jh}\varepsilon_{ik}^{\alpha}.$$

I ($n:=1$): There is the standard tableau 1 only, so that $i=j=h=k=1$ and then

$$\varepsilon_{11}^{\alpha}\varepsilon_{11}^{\alpha}=\varepsilon_{11}^{(1)}\varepsilon_{11}^{(1)}=1_{\mathbb{Q}S_n}\cdot 1_{\mathbb{Q}S_n}=1_{\mathbb{Q}S_n}=\varepsilon_{11}^{(1)}.$$

II ($n>1$): If $0\neq\varepsilon_{ij}^{\alpha}\varepsilon_{hk}^{\alpha}$, then by definition of $\varepsilon_{ij}^{\alpha}$ we have

$$0\neq\varepsilon_{ij}^{\alpha}\varepsilon_{hk}^{\alpha}=\left(\frac{1}{\kappa^{\alpha}}\right)^{2}\varepsilon_{i}^{\alpha *}e_{ij}^{\alpha}\varepsilon_{j}^{\alpha *}\varepsilon_{h}^{\alpha *}e_{hk}^{\alpha}\varepsilon_{k}^{\alpha *}, \tag{1}$$

so that necessarily

$$\varepsilon_{j}^{\alpha *}\varepsilon_{h}^{\alpha *}\neq 0.$$

Hence by the induction hypothesis $j=h$. And in this case, from (1) (use $(\varepsilon_{j}^{\alpha *})^2=\varepsilon_{j}^{\alpha *}$),

$$\begin{aligned}\varepsilon_{ij}^{\alpha}\varepsilon_{hk}^{\alpha}&=\left(\frac{1}{\kappa^{\alpha}}\right)^{2}\varepsilon_{i}^{\alpha *}e_{ij}^{\alpha}\varepsilon_{j}^{\alpha *}e_{jk}^{\alpha}\varepsilon_{k}^{\alpha *}\\&=\left(\frac{1}{\kappa^{\alpha}}\right)^{2}\varepsilon_{i}^{\alpha *}\pi_{ij}^{\alpha}e_{j}^{\alpha}\varepsilon_{j}^{\alpha *}e_{j}^{\alpha}\pi_{jk}^{\alpha}\varepsilon_{k}^{\alpha *},\end{aligned}$$

and by 3.2.12

$$\begin{aligned}&=\frac{1}{\kappa^{\alpha}}\varepsilon_{i}^{\alpha *}\pi_{ij}^{\alpha}e_{j}^{\alpha}\pi_{jk}^{\alpha}\varepsilon_{k}^{\alpha *}\\&=\varepsilon_{ik}^{\alpha}.\end{aligned}$$

This proves the orthogonality relations.

(iii) The orthogonality relations imply that the $\varepsilon_{ik}^{\alpha}$ are linearly independent. Since there are $n!=\sum_{\alpha}(f^{\alpha})^2$ such elements $\varepsilon_{ik}^{\alpha}$, they form a $\mathbb{Q}$-basis of $\mathbb{Q}S_n$. ∎

Another formula, which will be of use later on, is

**3.2.15** $$\varepsilon_{i}^{\alpha}e_{ik}^{\alpha}\varepsilon_{k}^{\alpha}=\kappa^{\alpha}\varepsilon_{ik}^{\alpha}.$$

*Proof.* From 3.2.11, $\varepsilon_{j}^{\alpha}=\varepsilon_{jj}^{\alpha}$, and $(\varepsilon_{j}^{\alpha *})^2=\varepsilon_{j}^{\alpha *}$, we get

$$\varepsilon_{j}^{\alpha}\varepsilon_{j}^{\alpha *}=\varepsilon_{j}^{\alpha *}\varepsilon_{j}^{\alpha}=\varepsilon_{j}^{\alpha}.$$

This yields

$$\begin{aligned}\varepsilon_i^\alpha e_{ik}^\alpha \varepsilon_k^\alpha &= \varepsilon_i^\alpha \varepsilon_i^{\alpha *} e_{ik}^\alpha \varepsilon_k^{\alpha *} \varepsilon_k^\alpha \\ &= \kappa^\alpha \varepsilon_i^\alpha \varepsilon_{ik}^\alpha \varepsilon_k^\alpha \\ &= \kappa^\alpha \varepsilon_{ik}^\alpha. \end{aligned}$$

■

Let us now recall from general representation theory how we can obtain from such a basis $\{\varepsilon_{ik}^\alpha\}$ the corresponding matrices. The elements $\varepsilon_{ik}^\alpha$, $1 \leqslant i, k \leqslant f^\alpha$, form a $\mathbb{Q}$-basis of the simple two-sided ideal

**3.2.16** $$I^\alpha = \langle\langle \varepsilon_{ik}^\alpha \mid 1 \leqslant i, k \leqslant f^\alpha \rangle\rangle_{\mathbb{Q}}.$$

For each $a \in \mathbb{Q}S_n$ we have uniquely determined $a^\alpha \in I^\alpha$, $\alpha \vdash n$, such that

$$a = \sum_{\alpha \vdash n} a^\alpha.$$

Hence by 3.2.16 there are uniquely determined $\sigma_{ik}^\alpha(a)$ such that

3.2.17 $$a^\alpha = \sum_{i,k=1}^{f^\alpha} \sigma_{ik}^\alpha(a) \varepsilon_{ik}^\alpha,$$

while

**3.2.18** $$a = \sum_{\alpha \vdash n} \sum_{i,k=1}^{f^\alpha} \sigma_{ik}^\alpha(a) \varepsilon_{ik}^\alpha.$$

The mapping

3.2.19 $$\sigma^\alpha : a^\alpha \mapsto \left(\sigma_{ik}^\alpha(a)\right)$$

is a representation of $I^\alpha$; this follows easily from 3.2.14. It is irreducible and faithful, since by Wedderburn's theorem $\sigma^\alpha[I^\alpha]$ is a full matrix ring. This irreducible representation is uniquely determined up to equivalence. Restricting it to the group elements, we get the desired matrix representation corresponding to $[\alpha]$. We call it the *seminormal form* of the representation $[\alpha]$, since it is obtained from the seminormal basis. It should be mentioned that the basis consisting of the $\varepsilon_{ik}^\alpha$ also yields a decomposition of $\mathbb{Q}S_n$ into minimal left ideals:

3.2.20 $$\mathbb{Q}S_n \varepsilon_{ik}^\alpha = \langle\langle \varepsilon_{1k}^\alpha, \ldots, \varepsilon_{f^\alpha k}^\alpha \rangle\rangle_{\mathbb{Q}},$$

so that in particular the following holds:

**3.2.21**
$$\text{(i) } I^\alpha = \bigoplus_{k=1}^{f^\alpha} \mathbb{Q}S_n \varepsilon_{1k}^\alpha,$$
$$\text{(ii) } \mathbb{Q}S_n \varepsilon_{1k}^\alpha = \langle\langle \varepsilon_{1k}^\alpha, \ldots, \varepsilon_{f^\alpha k}^\alpha \rangle\rangle_{\mathbb{Q}}.$$

## 3.3 The Representing Matrices

We are faced with the problem of evaluating the coefficients

$$\sigma_{ik}^{\alpha}(\pi), \qquad \pi \in S_n.$$

The equation

3.3.1
$$a = \sum_{\alpha \vdash n} \sum_{i,k} \sigma_{ik}^{\alpha}(a) \varepsilon_{ik}^{\alpha}$$

yields (use 3.2.14)

3.3.2
$$\varepsilon_i^{\alpha} a \varepsilon_k^{\alpha} = \sigma_{ik}^{\alpha}(a) \varepsilon_{ik}^{\alpha}.$$

**3.3.3** LEMMA. $\sigma_{ik}^{\alpha}(\pi) = \kappa^{\alpha}$[*coeff. of* $\pi^{-1}$ *in* $\varepsilon_{ki}^{\alpha}$].

*Proof.* From 3.3.2 we obtain

$$\varepsilon_i^{\alpha} \pi \varepsilon_{ki}^{\alpha} = \varepsilon_i^{\alpha} \pi \varepsilon_k^{\alpha} \varepsilon_{ki}^{\alpha} = \sigma_{ik}^{\alpha}(\pi) \varepsilon_{ik}^{\alpha} \varepsilon_{ki}^{\alpha} = \sigma_{ik}^{\alpha}(\pi) \varepsilon_i^{\alpha}. \tag{1}$$

But

$$\begin{aligned}[\text{coeff. of } \pi^{-1} \text{ in } \varepsilon_{ki}^{\alpha}] &= [\text{coeff. of } 1 \text{ in } \pi \varepsilon_{ki}^{\alpha}] \\ &= [\text{coeff. of } 1 \text{ in } \pi \varepsilon_{ki}^{\alpha} \varepsilon_i^{\alpha}],\end{aligned}$$

which is by 3.2.13

$$= [\text{coeff. of } 1 \text{ in } \varepsilon_i^{\alpha} \pi \varepsilon_{ki}^{\alpha}],$$

which is by (1)

$$\begin{aligned} &= [\text{coeff. of } 1 \text{ in } \sigma_{ik}^{\alpha}(\pi) \varepsilon_i^{\alpha}] \\ &= \sigma_{ik}^{\alpha}(\pi)[\text{coeff. of } 1 \text{ in } \varepsilon_i^{\alpha}].\end{aligned}$$

But the coefficient of $1_{S_n}$ in $\varepsilon_i^{\alpha}$ is $(\kappa^{\alpha})^{-1}$, for $\dim(\mathbb{Q}S_n \varepsilon_i^{\alpha}) = f^{\alpha}$ is the trace of the right multiplication of $\mathbb{Q}S_n$ by $\varepsilon_i^{\alpha}$, and this trace is also equal to $n!$ times the coefficient of $1_{S_n}$ in $\varepsilon_i^{\alpha}$ (cf. the proof of 3.1.4). ∎

It is trivial that $\varepsilon_{ik}^{\alpha}$ is represented under $\sigma^{\alpha}$ by an $f^{\alpha}$-rowed square matrix with a 1 at the intersection of the $i$th row and $k$th column and zeros elsewhere:

$$\text{3.3.4} \qquad \sigma^\alpha(\varepsilon_{ik}^\alpha)=\begin{pmatrix} & & & 0 & & & \\ & & & \vdots & & & \\ & & & 0 & & & \\ 0 & \cdots & 0 & 1 & 0 & \cdots & 0 \\ & & & 0 & & & \\ & & & \vdots & & & \\ & & & 0 & & & \end{pmatrix} \begin{matrix} \\ \\ \\ i, \\ \\ \\ \\ \end{matrix}$$

$$\qquad\qquad\qquad\qquad k$$

while for $\beta\neq\alpha$, $\sigma^\alpha(\varepsilon_{ik}^\beta)$ is the zero matrix. Putting

$$\text{3.3.5} \qquad \varepsilon^\alpha := \sum_{i=1}^{f^\alpha} \varepsilon_{ii}^\alpha,$$

we obtain an element of $\mathbf{Q}S_n$ which is represented by the $f^\alpha$-rowed identity matrix $I_{f^\alpha}$:

$$\textbf{3.3.6} \qquad \sigma^\alpha(\varepsilon^\alpha)=I_{f^\alpha}.$$

This shows for example that $\varepsilon^\alpha$ is the identity element of $I^\alpha$. Hence furthermore

$$\textbf{3.3.7} \qquad 1_{\mathbf{Q}S_n}=\sum_{\alpha\vdash n}\varepsilon^\alpha.$$

**3.3.8** LEMMA. *For each $\alpha\vdash n$ and $1\leqslant i\leqslant f^\alpha$ we have*

$$\varepsilon_i^{\alpha *}=\sum_{(\beta,j)}\varepsilon_j^\beta$$

*if the sum is taken over all pairs $(\beta, j)$, $\beta\vdash n, 1\leqslant j\leqslant f^\beta$, such that $t_j^{\beta *}=t_i^{\alpha *}$, i.e. over all the standard tableaux which arise from $t_i^{\alpha *}$ by adding an n-node.*

*Proof.* 3.2.11 yields

$$\varepsilon_j^\beta\varepsilon_i^{\alpha *}\varepsilon_k^\beta=\left(\frac{1}{\kappa^\beta}\right)^2\varepsilon_j^{\beta *}e_j^\beta\varepsilon_j^{\beta *}\varepsilon_i^{\alpha *}\varepsilon_k^{\beta *}e_k^\beta\varepsilon_k^{\beta *}.$$

If this is nonzero, then $t_j^{\beta *}=t_i^{\alpha *}=t_k^{\beta *}$ by 3.2.14, so that $j=k$ and $t_j^\beta$ arises from $t_i^{\alpha *}$ by adding an $n$-node. In this case we can proceed as follows:

$$\varepsilon_j^\beta\varepsilon_i^{\alpha *}\varepsilon_k^\beta=\left(\frac{1}{\kappa^\beta}\right)^2\varepsilon_j^{\beta *}e_j^\beta\varepsilon_j^{\beta *}e_j^\beta\varepsilon_j^{\beta *},$$

so that by 3.2.12

$$\varepsilon_j^\beta \varepsilon_i^{\alpha *} \varepsilon_k^\beta = \frac{1}{\kappa^\beta} \varepsilon_j^{\beta *} e_j^\beta \varepsilon_j^{\beta *} = \varepsilon_j^\beta .$$

Hence we obtain from 3.3.4 and 3.3.2 that $\varepsilon_i^{\alpha *}$ is represented under $\sigma^\beta$ by the same matrix as $\varepsilon_j^\beta$. Since the representation $\sigma^\beta$ of $I^\beta$ is faithful, we are done. ■

It should be remarked that 3.3.8 is another version of the branching theorem 2.4.3. In order to make this clear, we keep the diagram $[\alpha]$ fixed for the rest of this section. For simplicity we skip the index $\alpha$ and denote by $\sigma^i$ the representation corresponding to the diagram $[\alpha^{i-}]$ (cf. 2.4.1) if $\alpha^{i-} \vdash n-1$.

**3.3.9 Lemma.** *If $S_{n-1} \leqslant S_n$ denotes the stabilizer of the point $n$, then we have for each $\pi \in S_{n-1}$ the following decomposition of the matrix $\sigma(\pi) := \sigma^\alpha(\pi)$ into a direct sum of matrices:*

$$\sigma(\pi) = \sigma^1(\pi) \dotplus \sigma^2(\pi) \dotplus \cdots \dotplus \sigma^h(\pi),$$

*where the summand $\sigma^i(\pi)$ is to be neglected when $\alpha^{i-}$ is not a proper partition of $n-1$.*

*Proof.* For $\pi \in S_{n-1}$ we have both

$$\pi = \sum_{\alpha \vdash n} \sum_{i,k} \sigma_{ik}^\alpha(\pi) \varepsilon_{ik}^\alpha \quad \text{and} \quad \pi = \sum_{\beta \vdash n-1} \sum_{j,l} \sigma_{jl}^\beta(\pi) \varepsilon_{jl}^\beta .$$

By 3.3.2, 3.2.11, and the second equation,

$$\sigma_{ik}^\alpha(\pi) \varepsilon_{ik}^\alpha = \varepsilon_i^\alpha \pi \varepsilon_k^\alpha = \left(\frac{1}{\kappa^\alpha}\right)^2 \sum_{\beta \vdash n-1} \sum_{j,l} \sigma_{jl}^\beta(\pi) \varepsilon_i^{\alpha *} e_i^\alpha \varepsilon_i^{\alpha *} \varepsilon_{jl}^\beta \varepsilon_k^{\alpha *} e_k^\alpha \varepsilon_k^{\alpha *} .$$

This together with 3.2.14 shows that $\sigma_{ik}^\alpha(\pi)$ is 0 except (possibly) if there exist $\beta \vdash n-1$, $j$, and $l$ which satisfy

$$t_i^{\alpha *} = t_j^\beta \quad \text{and} \quad t_k^{\alpha *} = t_l^\beta . \tag{1}$$

Hence $\sigma_{ik}^\alpha(\pi) \neq 0$ only if both $t_i^{\alpha *}$ and $t_k^{\alpha *}$ have the same diagram $[\beta]$.

If on the other hand $t_i^{\alpha *}$ and $t_k^{\alpha *}$ have the same diagram $[\beta]$, i.e. if $t_i^\alpha$ and $t_k^\alpha$ have the point $n$ in the same position, then we can use 3.3.3:

$$\sigma_{ik}^\alpha(\pi) = \kappa^\alpha \left[\text{coeff. of } \pi^{-1} \text{ in } \varepsilon_{ki}^\alpha\right],$$

which is by 3.2.14 and 3.2.6

$$= \left[\text{coeff. of } \pi^{-1} \text{ in } \varepsilon_k^{\alpha *} e_k^\alpha \pi_{ki}^\alpha \varepsilon_i^{\alpha *}\right]. \tag{2}$$

From (1) we get that $\pi_{ki}^{\alpha}=\pi_{lj}^{\beta}\in S_{n-1}$, so that in (2) only such summands of $e_k^{\alpha}$ which leave $n$ fixed contribute to the coefficient of $\pi^{-1}$. Thus

$$\sigma_{ik}^{\alpha}(\pi)=\left[\text{coeff. of } \pi^{-1} \text{ in } \varepsilon_l^{\beta} e_l^{\beta} \pi_{lj}^{\beta} \varepsilon_j^{\beta}\right],$$

which is by 3.2.15

$$=\left[\text{coeff. of } \pi^{-1} \text{ in } \kappa^{\beta} \varepsilon_{lj}^{\beta}\right]$$

$$=\sigma_{jl}^{\beta}(\pi),$$

by 3.3.3. Done. ■

This shows that our ordering of the seminormal basis of $I^{\alpha}$ with respect to the last letter sequence of the standard tableaux $t_i^{\alpha}$ is adapted to the decomposition of the restriction of the seminormal form $\sigma^{\alpha}$ of $[\alpha]$ to the subgroup $S_{n-1}$:

**3.3.10** THEOREM. *The irreducible matrix representation $\sigma^{\alpha}$ of $S_n$ decomposes when it is restricted to $S_{n-1}$ as follows:*

$$\sigma^{\alpha}\downarrow S_{n-1}=\sigma^{\alpha^{1-}}\dotplus\cdots\dotplus\sigma^{\alpha^{h-}},$$

*where the summand $\sigma^{\alpha^{i-}}$ is to be neglected when $\alpha^{i-}$ is not a proper partition of $n-1$.*

This shows clearly why we used the last letter sequence of the standard tableaux. Later on other orderings will be used, too.

**3.3.11** LEMMA. *The matrix of $e_h^{\alpha}$ is of the form*

$$\sigma^{\alpha}(e_h^{\alpha})=\begin{array}{c} \\ \\ \\ \\ \\ \end{array}\left(\begin{array}{c|cccc} & h & & & \\ & * & & & \\ & \vdots & & * & \\ & * & & & \\ & \kappa^{\alpha} & * & \cdots & * \\ \hline 0 & & & & \end{array}\right)\begin{array}{l} \\ \\ \\ \\ h \\ \\ \end{array}.$$

*Proof.*

(i) From 3.2.15 we get that

$$\varepsilon_h^{\alpha} e_h^{\alpha} \varepsilon_h^{\alpha}=\kappa^{\alpha}\varepsilon_h^{\alpha};$$

this yields the entry $\sigma_{hh}^{\alpha}(e_h^{\alpha})=\kappa^{\alpha}$, as stated.

(ii) If $k<i$ and $n$ occurs in $t_i^\alpha$ in a later row than in $t_k^\alpha$, then the diagrams $[\gamma]$ of $t_i^{\alpha *}$ and $[\delta]$ of $t_k^{\alpha *}$ satisfy $\delta \lhd \gamma$, so that by 1.5.17 there exist two points, which occur in the same row of $t_i^{\alpha *}$ (and hence also of $t_i^\alpha$) as well as in the same column of $t_k^{\alpha *}$ (and hence also of $t_k^\alpha$). Thus $\mathcal{H}_i^\alpha \mathcal{V}_k^\alpha = 0$ and also for each $\pi \in S_{n-1}$

$$\pi \mathcal{H}_i^\alpha \pi^{-1} \mathcal{V}_k^\alpha = 0,$$

and therefore

$$\forall a \in \mathbb{Q}S_{n-1} \quad (e_i^\alpha a e_k^\alpha = 0).$$

This implies that in particular

$$\varepsilon_i^\alpha e_h^\alpha \varepsilon_k^\alpha = \left(\frac{1}{\kappa^\alpha}\right)^2 \varepsilon_i^{\alpha *} e_i^\alpha \varepsilon_i^{\alpha *} e_h^\alpha \varepsilon_k^{\alpha *} e_k^\alpha \varepsilon_k^{\alpha *} = 0,$$

both if $h<i$ while $n$ occurs in $t_i^\alpha$ in a later row than in $t_h^\alpha$, and also if $k<h$ while $n$ occurs in $t_h^\alpha$ in a later row than in $t_k^\alpha$. This shows that comparing the subdivision of the matrices into boxes corresponding to 3.3.10, we have zeros in the boxes to the left of or lower down than the box containing the entry $\sigma_{hh}^\alpha(e_h^\alpha)$:

$\sigma^\alpha(e_h^\alpha) =$ 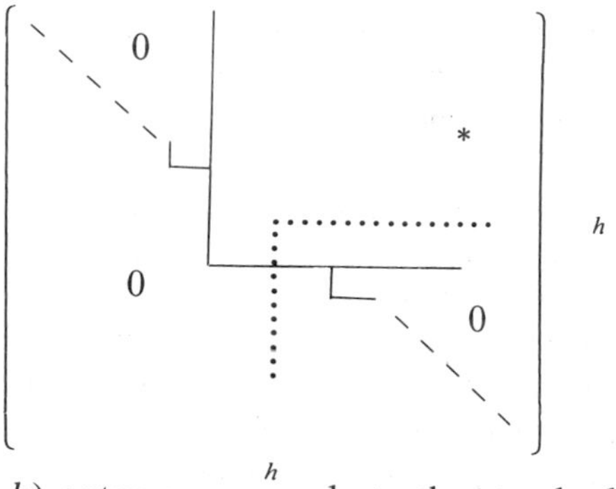


The box with the $(h, h)$ entry corresponds to the standard tableaux with diagram $[\alpha]$ which have $n$ in the same position as has $t_h^\alpha$.

We next refine this argument by considering only those tableaux which have both $n$ and $n-1$ in the same position as has $t_h^\alpha$. To do this we notice the following fact. If a standard tableau $t$ has besides $n$ the points $p_1, \ldots, p_r$ (respectively $q_1, \ldots, q_s$) in one of its rows (columns), i.e. if

$t =$ 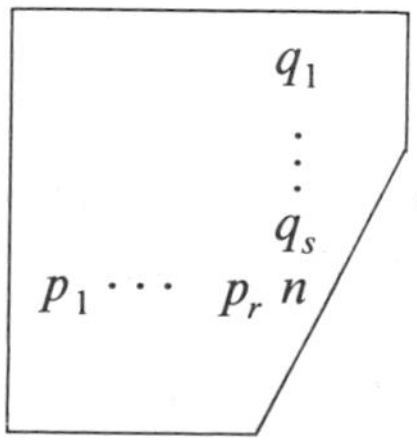, then

$$\mathcal{H}=(1+(p_1,n)+\cdots+(p_r,n))\mathcal{H}^*,$$

and

$$\mathcal{V}=\mathcal{V}^*(1-(q_1,n)-\cdots-(q_s,n)),$$

so that

$$e=\mathcal{V}\mathcal{H}=\mathcal{V}^*bc\mathcal{H}^*$$

for suitable $b,c\in\mathbf{Q}S_n$. Hence for a suitable $a\in\mathbf{Q}S_n$ we have

$$\begin{aligned}\varepsilon_i^\alpha e_h^\alpha \varepsilon_k^\alpha &= \left(\frac{1}{\kappa^\alpha}\right)^2 \varepsilon_i^{\alpha*} e_i^\alpha \varepsilon_i^{\alpha*} e_h^\alpha \varepsilon_k^{\alpha*} e_k^\alpha \varepsilon_k^{\alpha*}\\ &= \left(\frac{1}{\kappa^\alpha}\right)^2 \varepsilon_i^{\alpha*} e_i^\alpha \varepsilon_i^{\alpha**} e_i^{\alpha*} \varepsilon_i^{\alpha**} \mathcal{V}_h^{\alpha*} a \mathcal{H}_h^{\alpha*} \varepsilon_k^{\alpha**} e_k^{\alpha*} \varepsilon_k^{\alpha**} e_k^\alpha \varepsilon_k^{\alpha*}\\ &=0\end{aligned}$$

when $t_i^\alpha$ and $t_h^\alpha$ have $n$ in the same position but $n-1$ occurs in $t_i^\alpha$ in a later row than in $t_h^\alpha$ or in $t_h^\alpha$ later than in $t_k^\alpha$. An iterative application of this argument proves the statement. ■

We are now in a position to evaluate $\sigma^\alpha(\pi)$. The transpositions of successive points generate $S_n$ (cf. 1.1.17), so we need only evaluate the matrices which represent transpositions of successive points.

We start with $\sigma^\alpha((n-1,n))$, for which we use the fact that $\tau:=(n-1,n)$ lies in the centralizer of $S_{n-2}$, the subgroup of $S_n$ which stabilizes the points $n$ and $n-1$. The method of proof is of course induction on $n$, so that we can apply the induction hypothesis to $\sigma^\alpha(\pi), \pi\in S_{n-2}$, using the branching as described in 3.3.10.

The last letter sequence of the standard tableaux yields a subdivision of $\sigma^\alpha(\pi)$ which corresponds to tableaux having both $n$ and $n-1$ in the same position, say in the $i$th and $k$th row, respectively, so that this box belongs to the diagram $[\alpha^{(i-)k-}]$: to $[\alpha^{i,k}]$ for short. Hence

3.3.12 $\quad \forall \pi\in S_{n-2} \quad \left(\sigma(\pi)=\sigma^{1,1}(\pi)\dot{+}\sigma^{1,2}(\pi)\dot{+}\cdots\dot{+}\sigma^{h,h}(\pi)\right).$

($\sigma^{i,k}(\pi)$ has to be neglected when $\alpha^{(i-)k-}$ is not a proper partition of $n-2$.) We notice that in this decomposition of $\sigma^\alpha(\pi)$, $\sigma^{k,i}(\pi)$ occurs together with $\sigma^{i,k}(\pi)$ except for the case when either $i=k$ or $k=i-1$ and $\alpha_{i-1}=\alpha_i>\alpha_{i+1}$. All the other boxes occurring in 3.3.12 belong to pairwise inequivalent irreducible representations of $S_{n-2}$, since they correspond to pairwise different diagrams.

We now subdivide the matrix $\sigma^\alpha(\tau)$, $\tau:=(n-1,n)$, correspondingly into boxes $\sigma^{i,j;k,l}(\tau)$:

3.3.13 $$\sigma^\alpha(\tau)=\begin{pmatrix} & \vdots & \\ \cdots & \sigma^{i,j;k,l}(\tau) & \cdots \\ & \vdots & \end{pmatrix} \begin{matrix} \\ (i,j). \\ \\ \end{matrix}$$
$$(k,l)$$

Since

3.3.14 $$\forall \pi\in S_{n-2}\quad (\sigma^\alpha(\pi)\sigma^\alpha(\tau)=\sigma^\alpha(\tau)\sigma^\alpha(\pi)),$$

we obtain for the boxes of 3.3.13

3.3.15 $$\sigma^{i,j}(\pi)\sigma^{i,j;k,l}(\tau)=\sigma^{i,j;k,l}(\tau)\sigma^{k,l}(\pi).$$

Hence by Schur's lemma and $\sigma^{i,j}(\pi)=\sigma^{j,i}(\pi)$ (if they exist):

**3.3.16** $$\sigma^{i,j;k,l}(\tau)=\begin{cases}\gamma_{k,l}^{i,j}\cdot I & \textit{if}\quad \{i,j\}=\{k,l\},\\ 0\textit{-matrix} & \textit{otherwise}.\end{cases}$$

This shows that we are left with the problem of evaluating the scalar factors $\gamma$ in

**3.3.17**
$$\text{(i) } \sigma^{i,i;i,i}(\tau)=\gamma_{i,i}^{i,i}\cdot I,$$
$$\text{(ii) } \sigma^{i,i-1;i,i-1}(\tau)=\gamma_{i,i-1}^{i,i-1}\cdot I \qquad \textit{if}\quad \alpha_i=\alpha_{i-1},$$
$$\text{(iii) } \begin{pmatrix}\sigma^{i,j;i,j}(\tau) & \sigma^{i,j;j,i}(\tau)\\ \sigma^{j,i;i,j}(\tau) & \sigma^{j,i;j,i}(\tau)\end{pmatrix}=\begin{pmatrix}\gamma_{i,j}^{i,j}\cdot I & \gamma_{j,i}^{i,j}\cdot I\\ \gamma_{i,j}^{j,i}\cdot I & \gamma_{j,i}^{j,i}\cdot I\end{pmatrix}.$$

The first result concerning the entries is obtained from $\tau^2=1$, a fact which yields

**3.3.18**
$$\text{(i) } \left(\gamma_{i,i}^{i,i}\right)^2=\left(\gamma_{i,i-1}^{i,i-1}\right)^2=1,$$
$$\text{(ii) } \left(\gamma_{i,j}^{i,j}\right)^2+\gamma_{j,i}^{i,j}\cdot\gamma_{i,j}^{j,i}=\left(\gamma_{j,i}^{j,i}\right)^2+\gamma_{j,i}^{i,j}\cdot\gamma_{i,j}^{j,i}=1,$$
$$\text{(iii) } \gamma_{i,j}^{i,j}\cdot\gamma_{j,i}^{i,j}+\gamma_{j,i}^{j,i}\cdot\gamma_{j,i}^{i,j}=\gamma_{j,i}^{j,i}\cdot\gamma_{i,j}^{j,i}+\gamma_{i,j}^{i,j}\cdot\gamma_{i,j}^{j,i}=0.$$

In order to evaluate $\gamma_{i,i}^{i,i}$, we consider a standard tableau $t_j^\alpha$ which contains both $n$ and $n-1$ in its $i$th row. (Recall that if no such tableau exists, $\sigma^{i,i}(\pi)$ does not occur in 3.3.12.) Then $\tau\in H_j^\alpha$ and hence

3.3.19 $$e_j^\alpha=e_j^\alpha\tau.$$

We use this in

3.3.20 $$\kappa^\alpha \varepsilon_j^\alpha = \varepsilon_j^{\alpha *} e_j^\alpha \varepsilon_j^{\alpha *} = \varepsilon_j^{\alpha *} e_j^\alpha \tau \varepsilon_j^{\alpha *}.$$

We shall obtain $\gamma_{i,i}^{i,i}$ by equating suitable coefficients of the matrices which represent the elements on each side of Equation 3.3.20. The matrix $\sigma^\alpha(\kappa^\alpha \varepsilon_j^\alpha)$ which represents the left-hand side of 3.3.20 has $\kappa^\alpha$ in the $(j, j)$ place and zeros elsewhere (cf. 3.3.4). Let us therefore evaluate the corresponding coefficient of the matrix representing the right-hand side of 3.3.20. $\sigma^\alpha(\varepsilon_j^{\alpha *})$ has a 1 at $(j, j)$ and zeros elsewhere in the $j$th row and the $j$th column, by 3.3.8. $\sigma^\alpha(e_j^\alpha)$ has $\kappa^\alpha$ at $(j, j)$ by 3.3.11. Thus $\sigma^\alpha(\varepsilon_j^{\alpha *} e_j^\alpha)$ has the same $j$th row as $\sigma^\alpha(e_j^\alpha)$, while $\sigma^\alpha(\tau \varepsilon_j^{\alpha *})$ has at $(j, j)$ the coefficient $\gamma_{i,i}^{i,i}$, and zeros elsewhere in the $j$th column. Hence

$$\kappa^\alpha = \sigma_{jj}^\alpha(\kappa^\alpha \varepsilon_j^\alpha) = \sigma_{jj}^\alpha(\varepsilon_j^{\alpha *} e_j^\alpha \tau \varepsilon_j^{\alpha *}) = \kappa^\alpha \gamma_{i,i}^{i,i}.$$

This proves

**3.3.21** $$\gamma_{i,i}^{i,i} = 1.$$

$\gamma_{i,i-1}^{i,i-1}$ is obtained quite analogously. For in this case $\tau \in V_j^\alpha$, for a suitable $j$, so that $\tau e_j^\alpha = -e_j^\alpha$, and hence

**3.3.22** $$\gamma_{i,i-1}^{i,i-1} = -1.$$

Let us now evaluate $\gamma_{i,j}^{i,j}$, $\gamma_{j,i}^{i,j}$, $\gamma_{i,j}^{j,i}$, and $\gamma_{j,i}^{j,i}$. To do this we choose $k$ and $l$ such that $t_k^\alpha$ contains $n-1$ in its $i$th row and $n$ in its $j$th row, and where

$$\tau t_l^\alpha = t_k^\alpha.$$

Without loss of generality we can assume $i<j$ and hence $k>l$. Since $\tau = \pi_{kl}^\alpha$, we get

3.3.23 $$\kappa^\alpha \varepsilon_{kl}^\alpha = \varepsilon_k^{\alpha *} e_k^\alpha \tau \varepsilon_l^{\alpha *}.$$

We again compare suitable coefficients of the matrices which represent the elements on each side of this equation. $\sigma^\alpha(\tau)$ contains the submatrix

$$\begin{pmatrix} \gamma_{i,j}^{i,j} & \gamma_{j,i}^{i,j} \\ \gamma_{i,j}^{j,i} & \gamma_{j,i}^{j,i} \end{pmatrix}$$

at the intersections of the $l$th and $k$th rows and columns, and zeros elsewhere in these rows and columns. Comparing the coefficients $(k, l)$ of the matrices representing the elements on each side of 3.3.23, we obtain

**3.3.24** $$\gamma_{i,j}^{j,i}=1 \quad \textit{when} \quad i<j$$

This together with 3.3.18 yields for the remaining coefficients:

**3.3.25**
$$\text{(i) } \gamma_{i,j}^{i,j}=-\gamma_{j,i}^{j,i},$$
$$\text{(ii) } \gamma_{j,i}^{i,j}=1-\left(\gamma_{j,i}^{j,i}\right)^2=1-\left(\gamma_{i,j}^{i,j}\right)^2.$$

This shows that we have reduced our problems to the evaluation of $\gamma_{i,j}^{i,j}$.

To complete the story, we introduce the concept of the axial distance between two points in a tableau in terms of which $\gamma_{i,j}^{i,j}$ will be expressed eventually. Let $t$ denote a tableau, and let the points $r$ and $s$ occur in $t$ in the positions $(i_r, j_r)$ and $(i_s, j_s)$, respectively. Then the absolute values

$$|i_r-j_r| \quad \text{and} \quad |i_s-j_s|$$

measure the distances of $r$ and $s$ from the main diagonal:

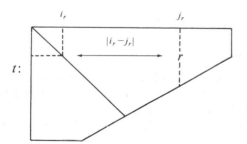

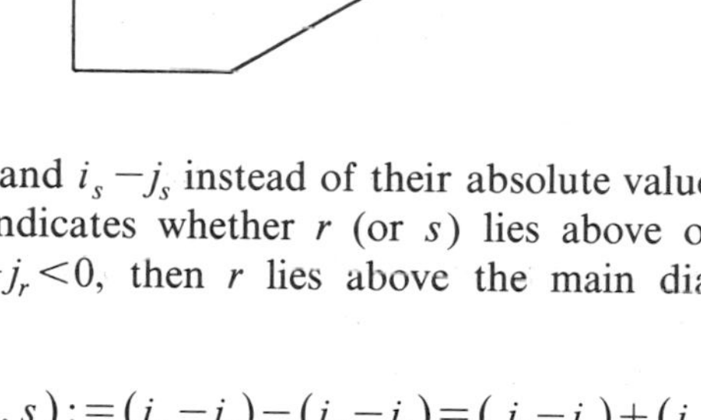


If we take $i_r-j_r$ and $i_s-j_s$ instead of their absolute values, then the sign of this difference indicates whether $r$ (or $s$) lies above or below the main diagonal. If $i_r-j_r<0$, then $r$ lies above the main diagonal. Hence the difference

3.3.26 $$d(r,s):=(i_s-j_s)-(i_r-j_r)=(j_r-j_s)+(i_s-i_r)$$

counts the number of steps we need to go from $r$ to $s$, if steps to the left and downwards contribute $+1$, while steps to the right and upwards contribute $-1$. The resulting integer $d(r,s)$ is independent of the path we choose; we call it the *axial distance* between $r$ and $s$ in $t$. The axial distance between $r$ and $s$ in $t_j^\alpha$ will be denoted by

3.3.27 $$d_j^\alpha(r,s).$$

It satisfies

**3.3.28** $$\forall q, r, s \in \mathbf{n} \quad \left(d_j^\alpha(r,q)+d_j^\alpha(q,s)=d_j^\alpha(r,s)\right).$$

Using this notation, the final result will be:

**3.3.29** THE SEMINORMAL FORM OF $[\alpha]$ (cf. 7.3.13). *If $t_1^\alpha,\ldots,t_{f^\alpha}^\alpha$ is the last letter sequence of the standard $\alpha$-tableaux, then the matrix*

$$\sigma^\alpha((m-1,m))=\left(\sigma_{ik}^\alpha((m-1,m))\right)$$

*which represents the transposition $(m-1,m)$, $1<m\leqslant n$, under $\sigma^\alpha$ reads as follows:*

(i) $\sigma_{ii}^\alpha((m-1,m))=\pm 1$ *if $t_i^\alpha$ contains $m-1$ and $m$ in the same* $\begin{smallmatrix}\textit{row}\\ \textit{column}\end{smallmatrix}$,

(ii) *if $i<k$ and $t_i^\alpha=(m-1,m)t_k^\alpha$, then we have the following submatrix:*

$$\begin{pmatrix}\sigma_{ii}^\alpha((m-1,m)) & \sigma_{ik}^\alpha((m-1,m))\\ \sigma_{ki}^\alpha((m-1,m)) & \sigma_{kk}^\alpha((m-1,m))\end{pmatrix}=\begin{pmatrix}-d_i^\alpha(m-1,m)^{-1} & 1-d_i^\alpha(m-1,m)^{-2}\\ 1 & d_i^\alpha(m-1,m)^{-1}\end{pmatrix},$$

(iii) $\sigma_{ik}^\alpha((m-1,m))=0$ *everywhere else.*

*Proof.* By induction on $n$.

(1) For $n=2$ the statement holds, as is very easy to see.

(2) Now let $n$ be greater than 2, and assume the statement to be true for $S_{n-1}$, so that it holds for $\tau:=(m-1,m)$, $m<n$, by 3.3.10. It remains to prove the statement for $\tau:=(n-1,n)$. By 3.3.25 this amounts to a verification of

3.3.30 $$\gamma_{r,s}^{r,s}=-d_i^\alpha(n-1,n)^{-1} \qquad \text{when} \quad r<s,$$

where $t_i^\alpha$ denotes a standard tableau with $n$ in its $r$th row and $n-1$ in its $s$th row. Let $j$ be defined by $t_i^\alpha=(n-1,n)t_j^\alpha$.

Together with $t_i^\alpha$ and $t_j^\alpha$, we consider

$$t_k^\alpha:=(n-2,n-1)t_i^\alpha, \qquad t_l^\alpha:=(n-2,n-1)t_j^\alpha$$

(if they exist) and use

$$\begin{aligned}(n-2,n)&=(n-1,n)(n-2,n-1)(n-1,n)\\ &=(n-2,n-1)(n-1,n)(n-2,n-1).\end{aligned}$$

We shall equate the $(j,i)$ coefficient of the matrices representing these two expressions for $(n-2,n)$. The induction hypothesis tells us that $\sigma^\alpha((n-2,n-1))$ has the following submatrix at the intersections of the $i$th and $j$th rows with the $i$th and $j$th columns:

$$\begin{pmatrix} -d_i^\alpha(n-2,n-1)^{-1} & 0 \\ 0 & -d_j^\alpha(n-2,n-1)^{-1} \end{pmatrix}.$$

(If $t_k^\alpha$ or $t_l^\alpha$ exist, these are not the only nonzero entries of these rows and columns, but the others do not matter in the ensuing argument. If $t_k^\alpha$ or $t_l^\alpha$ do not exist, then this is still the correct submatrix.) Hence $\sigma^\alpha((n-1,n))\sigma^\alpha((n-2,n-1))$ has at these places the submatrix (put $\gamma := \gamma_{r,s}^{r,s}$)

$$\begin{pmatrix} +\gamma & 1-\gamma^2 \\ 1 & -\gamma \end{pmatrix}\begin{pmatrix} -d_i^\alpha(n-2,n-1)^{-1} & 0 \\ 0 & -d_j^\alpha(n-2,n-1)^{-1} \end{pmatrix}$$
$$= \begin{pmatrix} -\gamma d_i^\alpha(n-2,n-1)^{-1} & -(1-\gamma^2)d_j^\alpha(n-2,n-1)^{-1} \\ -d_i^\alpha(n-2,n-1)^{-1} & +\gamma d_j^\alpha(n-2,n-1)^{-1} \end{pmatrix}.$$

Thus

$$\sigma_{ji}^\alpha((n-1,n)(n-2,n-1)(n-1,n))$$
$$= -\gamma d_i^\alpha(n-2,n-1)^{-1} + \gamma d_j^\alpha(n-2,n-1)^{-1},$$

while

$$\sigma_{ji}^\alpha((n-2,n-1)(n-1,n)(n-2,n-1))$$
$$= d_i^\alpha(n-2,n-1)^{-1} d_j^\alpha(n-2,n-1)^{-1}.$$

This implies

$$\begin{aligned} \frac{1}{\gamma} &= -d_j^\alpha(n-2,n-1) + d_i^\alpha(n-2,n-1) \\ &= -d_i^\alpha(n-2,n) + d_i^\alpha(n-2,n-1) \\ &= -d_i^\alpha(n-1,n), \end{aligned}$$

the last equation by 3.3.28. This proves 3.3.30, and we are done. ■

For an example we take $\alpha:=(3,2)$. The last letter sequence of the standard tableaux is

3.3.31 $$\begin{array}{ccc} 1 & 3 & 5 \\ 2 & 4 & \end{array}, \quad \begin{array}{ccc} 1 & 2 & 5 \\ 3 & 4 & \end{array}, \quad \begin{array}{ccc} 1 & 3 & 4 \\ 2 & 5 & \end{array}, \quad \begin{array}{ccc} 1 & 2 & 4 \\ 3 & 5 & \end{array}, \quad \begin{array}{ccc} 1 & 2 & 3 \\ 4 & 5 & \end{array},$$

so that we obtain

3.3.32 $$\sigma^{(3,2)}((12))=\begin{pmatrix} -1 & & & & \\ & 1 & & & \\ & & -1 & & \\ & & & 1 & \\ & & & & 1 \end{pmatrix},$$

while

3.3.33 $$\sigma^{(3,2)}((23))=\begin{pmatrix} \frac{1}{2} & \frac{3}{4} & & & \\ 1 & -\frac{1}{2} & & & \\ & & \frac{1}{2} & \frac{3}{4} & \\ & & 1 & -\frac{1}{2} & \\ & & & & 1 \end{pmatrix},$$

3.3.34 $$\sigma^{(3,2)}((34))=\begin{pmatrix} -1 & & & & \\ & 1 & & & \\ & & 1 & & \\ & & & \frac{1}{3} & \frac{8}{9} \\ & & & 1 & -\frac{1}{3} \end{pmatrix},$$

and

3.3.35 $$\sigma^{(3,2)}((45))=\begin{pmatrix} \frac{1}{2} & & \frac{3}{4} & & \\ & \frac{1}{2} & & \frac{3}{4} & \\ 1 & & -\frac{1}{2} & & \\ & 1 & & -\frac{1}{2} & \\ & & & & 1 \end{pmatrix}.$$

This shows how easy it is to write down $\sigma^{\alpha}((m-1,m))$.

**3.3.36** COROLLARY. *For each* $\pi\in S_n$, $\sigma^{\alpha}(\pi)$ *is a matrix over* $\mathbb{Q}$.

## 3.4 The Orthogonal and the Natural Form of $[\alpha]$

For certain purposes (like applications in science) it is more suitable to have orthogonal matrices as representing matrices. A. Young gave such a form, too:

**3.4.1** THE ORTHOGONAL FORM OF $[\alpha]$. *We obtain a representation* $\omega^\alpha$ *which is equivalent to* $\sigma^\alpha$ *but which has orthogonal representing matrices if we replace the submatrix* 3.3.29(ii) *by*

$$\begin{pmatrix} -d_i^\alpha(m-1,m)^{-1} & \sqrt{1-d_i^\alpha(m-1,m)^{-2}} \\ \sqrt{1-d_i^\alpha(m-1,m)^{-2}} & d_i^\alpha(m-1,m)^{-1} \end{pmatrix}.$$

*Proof.* We construct a matrix $H^\alpha$ which satisfies

$$\forall \pi \in S_n \quad \left((H^\alpha)^{-1}\sigma^\alpha(\pi)H^\alpha = \omega^\alpha(\pi)\right).$$

In order to do this, for a given standard tableau $t$ we denote the last point of its $i$th row by $a_i$ and assuming that $\alpha=(\alpha_1,\ldots,\alpha_h)$ with $\alpha_h>0$, we put

3.4.2 $$\varphi_t(n) := \begin{cases} 1 & \text{if} \quad n=a_h, \\ \prod_{i=j+1}^{h} \left(1+d(a_i,n)^{-1}\right) & \text{if} \quad n=a_j,\ j<h. \end{cases}$$

Furthermore we define

3.4.3 $$\varphi_t(n-1) := \varphi_{t^*}(n-1),$$

and so on, so that we can put

3.4.4 $$\psi_i := \varphi_{t_i^\alpha}(n)\varphi_{t_i^\alpha}(n-1)\cdots\varphi_{t_i^\alpha}(1), \qquad 1 \leqslant i \leqslant f^\alpha.$$

If $i<k$ and $t_i^\alpha=(m-1,m)t_k^\alpha$, then for each $j<m-1$ and for each $j>m$ we have

$$\varphi_{t_i^\alpha}(j) = \varphi_{t_k^\alpha}(j).$$

Moreover

$$\varphi_{t_i^\alpha}(m-1) = \varphi_{t_k^\alpha}(m),$$

so that altogether (check how $\varphi_{t_i^\alpha}(m)$ and $\varphi_{t_k^\alpha}(m-1)$ differ)

3.4.5 $$\frac{\psi_i}{\psi_k} = 1 - d_i^\alpha(m-1,m)^{-2}.$$

Putting

3.4.6
$$H^\alpha := \begin{pmatrix} \sqrt{\psi_1} & & 0 \\ & \ddots & \\ 0 & & \sqrt{\psi_{f^\alpha}} \end{pmatrix},$$

we find that

$$\omega^\alpha((m-1,m)) := (H^\alpha)^{-1}\sigma^\alpha((m-1,m))H^\alpha$$

differs from $\sigma^\alpha((m-1,m))$ just in the submatrix 3.3.29(ii) and has the stated form. It is clear that this matrix is orthogonal. ∎

For example

3.4.7
$$\omega^{(3,2)}((12)) = \begin{pmatrix} -1 & & & & \\ & 1 & & & \\ & & -1 & & \\ & & & 1 & \\ & & & & 1 \end{pmatrix},$$

$$\omega^{(3,2)}((2,3)) = \begin{pmatrix} \frac{1}{2} & \frac{1}{2}\sqrt{3} & & & \\ \frac{1}{2}\sqrt{3} & -\frac{1}{2} & & & \\ & & \frac{1}{2} & \frac{1}{2}\sqrt{3} & \\ & & \frac{1}{2}\sqrt{3} & -\frac{1}{2} & \\ & & & & 1 \end{pmatrix},$$

$$\omega^{(3,2)}((3,4)) = \begin{pmatrix} -1 & & & & \\ & 1 & & & \\ & & 1 & & \\ & & & \frac{1}{3} & \frac{2}{3}\sqrt{2} \\ & & & \frac{2}{3}\sqrt{2} & -\frac{1}{3} \end{pmatrix},$$

$$\omega^{(3,2)}((4,5)) = \begin{pmatrix} \frac{1}{2} & & \frac{1}{2}\sqrt{3} & & \\ & \frac{1}{2} & & \frac{1}{2}\sqrt{3} & \\ \frac{1}{2}\sqrt{3} & & -\frac{1}{2} & & \\ & \frac{1}{2}\sqrt{3} & & -\frac{1}{2} & \\ & & & & 1 \end{pmatrix}.$$

**3.4.8** COROLLARY. *$\omega^\alpha$ provides an irreducible matrix representation of $S_n$ onto real orthogonal matrices. It corresponds to the basis consisting of the* orthogonal units

$$\sigma_{ik}^{\alpha} := \sqrt{\frac{\psi_i}{\psi_k}}\, \varepsilon_{ik}^{\alpha}.$$

*This basis* $\{\sigma_{ik}^{\alpha} \mid \alpha \vdash n,\ 1 \leqslant i, k \leqslant f^{\alpha}\}$ *will be called the* orthogonal basis *of* $\mathbb{R}S_n$.

Later on in this book we shall consider modular representations associated with the ordinary irreducible representations $[\alpha]$ of $S_n$. Since we shall be interested mainly in their irreducible constituents, it does not matter which associated modular representation we take. Therefore it will be easiest to start with a matrix form of $[\alpha]$ such that its matrices are matrices over $\mathbb{Z}$, so that we need only reduce the coefficients modulo the prime characteristic $p$ in order to get an associated $p$-modular representation. In practice, we shall do this directly, but it is interesting to see how to perform the calculation inside $\mathbb{Q}S_n$. For this reason we shall now derive Young's natural form $\nu^{\alpha}$ of $[\alpha]$, which will turn out to be a matrix representation over $\mathbb{Z}$.

To do this we recall from 3.2.12 and 3.1.24 that the elements

$$e_{ik}^{\alpha} := e_i^{\alpha}\pi_{ik}^{\alpha} = \pi_{ik}^{\alpha} e_k^{\alpha} = \frac{1}{\kappa^{\alpha}} e_i^{\alpha}\pi_{ik}^{\alpha} e_k^{\alpha} = \kappa^{\alpha}\hat{e}_i^{\alpha}\pi_{ik}^{\alpha}\hat{e}_k^{\alpha}$$

satisfy

**3.4.9** $\quad e_{ij}^{\alpha} e_{kl}^{\beta} = \delta_{\alpha\beta}\kappa^{\alpha}\mu_{kj}^{\alpha} e_{il}^{\alpha}, \qquad$ *where* $\quad \mu_{kj}^{\alpha} := \left[\textit{coeff. of } 1 \textit{ in } e_{kj}^{\alpha}\right].$

We put these coefficients into a matrix $M^{\alpha}$:

3.4.10 $$M^{\alpha} := (\mu_{rs}^{\alpha})$$

and prove

**3.4.11** Lemma. *$M^{\alpha}$ is a lower triangular matrix, i.e., $\mu_{rs}^{\alpha} = 0$ if $r < s$. The diagonal entries $\mu_{ii}^{\alpha}$ are* 1, *so that in particular* $\det M^{\alpha} = 1$.

*Proof.* By induction on $n$. If we partition $M^{\alpha}$ into boxes according to the last letter sequence, the boxes along the main diagonal are lower triangular and have 1's along their main diagonal by the induction hypothesis. When the index pair $(r, s)$, $r < s$, does not belong to a box on the main diagonal, then $t_r^{\alpha *}$ and $t_s^{\alpha *}$ have diagrams $[\gamma]$ and $[\delta]$ such that $\gamma \lhd \delta$, and hence by 1.5.17 both $\mathcal{H}_s^{\alpha *}\mathcal{V}_r^{\alpha *} = 0 = \mathcal{H}_s^{\alpha}\mathcal{V}_r^{\alpha}$, which implies $e_s^{\alpha} e_r^{\alpha} = 0$, so that $\mu_{rs}^{\alpha} = 0$ (cf. 3.1.20). ∎

This shows in particular that $M^\alpha$ is invertible, say

3.4.12 $$L^\alpha := (\lambda^\alpha_{jk}) := (M^\alpha)^{-1}.$$

We notice that $L^\alpha$ is a matrix over $\mathbb{Z}$ by 3.4.11.
Using the entries of $L^\alpha$ we define *natural units* by

3.4.13 $$\gamma^\alpha_{ik} := \frac{1}{\kappa^\alpha} \sum_{j=1}^{f^\alpha} e^\alpha_{ij} \lambda^\alpha_{jk},$$

and prove that they also satisfy the orthogonality relations:

**3.4.14** LEMMA. $\gamma^\alpha_{ij}\gamma^\beta_{kl} = \delta_{\alpha\beta}\delta_{jk}\gamma^\alpha_{il}$.

*Proof.*

$$\gamma^\alpha_{ij}\gamma^\beta_{kl} = \frac{1}{\kappa^\alpha\kappa^\beta} \sum_{r,s} e^\alpha_{ir}\lambda^\alpha_{rj} e^\beta_{ks}\lambda^\beta_{sl}$$
$$= \frac{1}{\kappa^\alpha\kappa^\beta} \sum_{r,s} \lambda^\alpha_{rj}\lambda^\beta_{sl} e^\alpha_{ir} e^\beta_{ks},$$

which is by 3.4.9

$$= \delta_{\alpha\beta}\left(\frac{1}{\kappa^\alpha}\right)^2 \sum_{r,s} \lambda^\alpha_{rj}\lambda^\alpha_{sl}\kappa^\alpha\mu^\alpha_{kr} e^\alpha_{is}$$
$$= \delta_{\alpha\beta}\frac{1}{\kappa^\alpha} \sum_s \lambda^\alpha_{sl}\left(\sum_r \mu^\alpha_{kr}\lambda^\alpha_{rj}\right) e^\alpha_{is}$$
$$= \delta_{\alpha\beta}\delta_{jk}\gamma^\alpha_{il}.$$ ∎

**3.4.15** LEMMA. $\forall \alpha, i, k\,(\gamma^\alpha_{ik} \neq 0)$.

*Proof.* 3.4.9 shows that $\gamma^\alpha_{ik} e^\alpha_{ki} = e^\alpha_{ii} \neq 0$. ∎

This together with 3.4.14 yields:

**3.4.16** THEOREM. *The natural units form a basis of* $\mathbb{Q}S_n$, *which is called the* natural basis.

The corresponding matrix representation $\nu^\alpha$ of $S_n$,

3.4.17 $$\nu^\alpha : \pi \mapsto (\nu^\alpha_{ik}(\pi)), \qquad \nu^\alpha_{ik}(\pi) := [\text{coeff. of } \pi \text{ in } \gamma^\alpha_{ik}],$$

is called the *natural form* of $[\alpha]$. It remains to show that it is integral.

**3.4.18** LEMMA. $\forall \alpha, i, k, \pi\ (\nu_{ik}^{\alpha}(\pi) \in \mathbb{Z})$.

*Proof.* From 3.4.14 we obtain (cf. the proof of 3.3.3) that

$$\begin{aligned}\nu_{ik}^{\alpha}(\pi) &= \kappa^{\alpha}\left[\text{coeff. of } \pi^{-1} \text{ in } \gamma_{ki}^{\alpha}\right],\\ &= \left[\text{coeff. of } \pi^{-1} \text{ in } \sum_j e_{kj}^{\alpha} \lambda_{ji}^{\alpha}\right],\end{aligned}$$

but $\sum e_{kj}^{\alpha} \lambda_{ji}^{\alpha} \in \mathbb{Z}S_n \subseteq \mathbb{Q}S_n$. ∎

## Exercises

3.1 Read the papers of Malzan [1969], Puttaswamaiah [1963], Puttaswamaiah and Robinson [1964], Specht [1938], and Thrall [1941] in order to see how the irreducible matrix representations of the alternating group can be obtained and what their main properties are.

3.2 Derive an algorithm which directly yields a rational integral form of the representing matrices, or study the corresponding sections in the first edition of Boerner's book [1955].

CHAPTER 4

# Representations of Wreath Products

Since the ordinary irreducible characters and representations of $S_n$ are now at our disposal, we should try to apply them to the representation theory of a wider class of groups. Apart from the alternating groups, which we have already discussed, the most natural groups to consider are the complete monomial groups $G \operatorname{wr} S_n$, $G$ being a given finite group. These groups are wreath products, so we introduce the concept of wreath products and derive their representations over algebraically closed fields.

## 4.1 Wreath Products

Let $G$ be a group and $H$ a subgroup of $S_n$. Denoting by $G^{\mathbf{n}}$ the set of all mappings from $\mathbf{n}=\{1,\ldots,n\}$ into $G$:

4.1.1 $$G^{\mathbf{n}}:=\{f \mid f:\mathbf{n}\to G\},$$

we put

4.1.2 $$G \operatorname{wr} H := G^{\mathbf{n}} \times H = \{(f;\pi) \mid f:\mathbf{n}\to G \text{ and } \pi\in H\}.$$

For $f\in G^{\mathbf{n}}$ and $\pi\in H$ we define $f_\pi \in G^{\mathbf{n}}$ by

4.1.3 $$f_\pi := f\circ\pi^{-1}.$$

ENCYCLOPEDIA OF MATHEMATICS and Its Applications, Gian-Carlo Rota (ed.). Vol. 16: G. D. James and A. Kerber, The Representation Theory of the Symmetric Group
ISBN 0-201-13515-9

It is easy to check that for $\pi' \in H$,

**4.1.4** $$(f_\pi)_{\pi'} = f_{\pi'\pi}.$$

Furthermore we introduce a multiplication on $G^{\mathbf{n}}$ by

4.1.5 $$(ff')(i) := f(i) \cdot f'(i), i \in \mathbf{n}.$$

Using this, one easily verifies that $G \operatorname{wr} H$ together with the composition defined by

4.1.6 $$(f;\pi)(f';\pi') := (ff'_\pi; \pi\pi')$$

forms a group, the *wreath product* of $G$ by $H$. Its order (if $G$ is finite) is

**4.1.7** $$|G \operatorname{wr} H| = |G|^n |H|.$$

If we define $e \in G^{\mathbf{n}}$ by

4.1.8 $$e(i) := 1_G, \qquad i \in \mathbf{n}$$

and for $f \in G^{\mathbf{n}}$ the mapping $f^{-1} \in G^{\mathbf{n}}$ by

4.1.9 $$f^{-1}(i) := f(i)^{-1}, \qquad i \in \mathbf{n},$$

then we obtain for the identity element in $G \operatorname{wr} H$ and for the inverse of $(f;\pi) \in G \operatorname{wr} H$:

**4.1.10** $$1_{G \operatorname{wr} H} = (e; 1_H) \quad \text{and} \quad (f;\pi)^{-1} = (f_{\pi^{-1}}^{-1}; \pi^{-1}),$$

where

4.1.11 $$f_{\pi^{-1}}^{-1} := (f_{\pi^{-1}})^{-1} = (f^{-1})_{\pi^{-1}}.$$

We shall sometimes describe $(f;\pi) \in G \operatorname{wr} H$ more explicitly by displaying the values of $f$, i.e., we write

$$(f(1), \ldots, f(n); \pi)$$

instead of $(f;\pi)$ if it seems advisable.

The following normal subgroup of $G \operatorname{wr} H$ is of particular importance:

4.1.12 $$G^* := \{(f; 1_H) \mid f \in G^{\mathbf{n}}\}.$$

$G^*$ is called the *base group* of $G \operatorname{wr} H$. It is the direct product of $n$ copies $G_i$

of $G$, where

4.1.13 $$G_i := \{(f;1_H) \mid \forall j \neq i\ (f(j) = 1_G)\} \simeq G.$$

The subgroup

4.1.14 $$H' := \{(e;\pi) \mid \pi \in H\} \simeq H$$

is a complement of $G^*$, so we have

**4.1.15** $$\begin{aligned} &(\text{i})\ G \operatorname{wr} H = G^* \cdot H', \\ &(\text{ii})\ G^* = \prod_i G_i \trianglelefteq G \operatorname{wr} H, \\ &(\text{iii})\ G^* \cap H' = \{1_{G \operatorname{wr} H}\} = \{(e;1_H)\}. \end{aligned}$$

Another interesting subgroup is the diagonal subgroup of $G^*$:

4.1.16 $$\begin{aligned} \operatorname{diag} G^* &:= \{(f;1_H) \mid f \text{ constant}\} \\ &= \{(g,\ldots,g;1_H) \mid g \in G\} \simeq G, \end{aligned}$$

so that

4.1.17 $$\begin{aligned} \operatorname{diag} G^* \cdot H' &= \{(f;\pi) \mid f \text{ constant}\} \\ &= \{(g,\ldots,g;\pi) \mid g \in G, \pi \in H\} \\ &\simeq G \times H. \end{aligned}$$

This shows that we have nice embeddings of $G$, $H$, and $G \times H$ in $G \operatorname{wr} H$.

If $G$ is a permutation group of finite degree, say $G \leqslant S_m$, then we obtain a permutation representation $\psi$ of $G \operatorname{wr} H$ as follows:

4.1.18

$$\psi : G \operatorname{wr} H \to S_{mn} : (f,\pi) \mapsto \begin{pmatrix} (j-1)m+i \\ (\pi(j)-1)m + f(\pi(j))(i) \end{pmatrix}_{1 \leqslant i \leqslant m,\ 1 \leqslant j \leqslant n}.$$

This faithful representation of $G \operatorname{wr} H$, where $G \leqslant S_m$ and $H \leqslant S_n$, is in a sense a natural one, since $\psi[G_1]$ acts on $\{1,\ldots,m\} \subseteq \{1,\ldots,mn\}$ in the same way as $G$ acts on $\{1,\ldots,m\}$, and the restriction of $\psi[G_1]$ to $\{1,\ldots,m\}$ is just $G$. Furthermore $\psi[G_2]$ acts on $\{m+1,\ldots,2m\}$ and $G$ acts on $\mathbf{m}$ in similar ways, etc. Also $\psi[H']$ permutes the subsets $\{1,\ldots,m\}$, $\{m+1,\ldots,2m\},\ldots,$ $\{(n-1)m+1,\ldots,mn\}$ in the same way as $H$ acts on $\{1,\ldots,n\}$; e.g., the element

$$(f;\pi) := ((12),(123),1;(23)) \in S_3 \operatorname{wr} S_3$$

is mapped under $\psi$ onto

$$\underbrace{(12)(456)}_{\psi((f;1_H))}\ \underbrace{(47)(58)(69)}_{\psi((e;\pi))}=(12)(475869)\in S_9.$$

As an application of this permutation representation we obtain a nice description of the centralizers of elements in $S_n$. First of all we notice that if $C_m:=\langle(1,\dots,m)\rangle\leqslant S_m$, then $\psi[C_m \operatorname{wr} S_n]$ is just the centralizer of the permutation

$$(1,\dots,m)(m+1,\dots,2m)\cdots((n-1)m+1,\dots,nm)\in S_{nm}$$

(this follows from 1.2.1 and 1.2.15). In order to derive therefrom an expression for the centralizer of a general element $\pi\in S_n$ we need some more notation. If $P\leqslant S_\Omega$, $P'\leqslant S_{\Omega'}$, where $\Omega\cap\Omega'=\varnothing$, then we understand by the *direct sum* $P\oplus P'$ of these two permutation groups the following group acting on $\Omega\cup\Omega'$: The underlying set is the cartesian product $P\times P'$, and the action is defined by

$$(\pi,\pi')(\omega''):=\begin{cases}\pi(\omega'') & \text{if } \omega''\in\Omega,\\ \pi'(\omega'') & \text{if } \omega''\in\Omega'.\end{cases}$$

Hence for example $S_\alpha=\oplus_i S_{\alpha_i}$, where $S_{\alpha_i}$ acts on suitable subsets. Using this notation we obtain for the centralizer of $\pi\in S_n$

**4.1.19** $$C_{S_n}(\pi)\mathrel{\hat{=}}\bigoplus_{\substack{i\\ a_i(\pi)>0}}\psi\left[C_i \operatorname{wr} S_{a_i(\pi)}\right].$$

Another important application of the concept of wreath products is the construction of Sylow subgroups of $S_n$. In fact this construction was mentioned by A. Cauchy in Vol. III of his *Traité d'Analyse* which was published in 1844, although Sylow did not prove his theorem until 1872. If $p$ denotes a prime number and $P^i$ a $p$-Sylow-subgroup of $S_{p^i}$, then for $C_p:=\langle(1,\dots,p)\rangle$ we have

$$|P^i \operatorname{wr} C_p|=|P^i|^p\cdot p,$$

and it is easy to see that this is the order of the $p$-Sylow-subgroups of $S_{p^{i+1}}$. Hence $\psi[P^i \operatorname{wr} C_p]$ is a $p$-Sylow-subgroup of $S_{p^{i+1}}$. Thus by induction we get

**4.1.20** $\psi[\cdots\psi[\psi\underbrace{[C_p \operatorname{wr} C_p] \operatorname{wr} C_p]\dots \operatorname{wr} C_p}_{i \text{ factors } C_p}]$ is a $p$-Sylow-subgroup of $S_{p^i}$.

If now

$$n=\sum_{i=0}^{t} b_i p^i, \qquad 0\leqslant b_i<p,$$

is the $p$-adic decomposition of $n$, then the exponent $\nu_p(n!)$ of the maximal power of $p$ dividing $n!$ satisfies

4.1.21
$$\begin{aligned}\nu_p(n!)&=b_1\nu_p(p!)+b_2\nu_p(p^2!)+\cdots+b_t\nu_p(p^t!)\\ &=b_1+b_2(1+p)+\cdots+b_t(1+p+\cdots+p^{t-1}).\end{aligned}$$

This shows that the following is true:

**4.1.22.** *If $P_n$ denotes a p-Sylow-subgroup of $S_n$ and $n=\sum_{i=0}^{t} b_i p^i$, $0\leqslant b_i<p$, then*

$$P_n \triangleq \bigoplus_{i=1}^{t}\left(\bigoplus^{b_i} P^i\right),$$

*where $\oplus^{b_i}P^i:=P^i\oplus\cdots\oplus P^i$ ($b_i$ direct summands).*

An easy check verifies that the wreath product is associative in the following sense:

**4.1.23**

$$G\leqslant S_l,\ H\leqslant S_m,\ I\leqslant S_n \quad\Rightarrow\quad \psi[\psi[G\,\mathrm{wr}\,H]\,\mathrm{wr}\,I]\triangleq\psi[G\,\mathrm{wr}\,\psi[H\,\mathrm{wr}\,I]].$$

Hence we can abbreviate $\psi[\psi[G\,\mathrm{wr}\,H]\,\mathrm{wr}\,I]$ by

$$\psi[G\,\mathrm{wr}\,H\,\mathrm{wr}\,I].$$

Applying this to 4.1.20, we notice that

**4.1.24**
$$P^i=\psi[C_p\,\mathrm{wr}\cdots\mathrm{wr}\,C_p]=:\psi[\mathrm{wr}\,{}^iC_p].$$

This together with 4.1.22 shows that each $p$-Sylow-subgroup of $S_n$ is a direct sum of wreath products of the form $\psi[\mathrm{wr}\,{}^iC_p]$. cf.

A third kind of subgroup of symmetric groups which will occur later on is the normalizer of a subgroup of the form

$$\bigoplus^{n} S_m:=\underbrace{S_m\oplus\cdots\oplus S_m}_{n\text{ summands}}\leqslant S_{mn}.$$

The number of subgroups of $S_{mn}$ conjugate to $\oplus^n S_m$ (which are exactly the subgroups of $S_{mn}$ similar to $\oplus^n S_m$) is

$$\frac{1}{n!}\binom{mn}{m}\binom{mn-m}{m}\cdots\binom{m}{m}=\frac{(mn)!}{m!^n n!}=|S_{mn}:\psi[S_m \operatorname{wr} S_n]|.$$

Hence $|N_{S_{mn}}(\oplus^n S_m)|=m!^n n!=|\psi[S_m \operatorname{wr} S_n]|$. Since $\psi[S_m \operatorname{wr} S_n]$ is similar to a subgroup contained in this normalizer, we have

4.1.25 $$N_{S_{mn}}\left(\overset{n}{\oplus} S_m\right)\triangleq\psi[S_m \operatorname{wr} S_n].$$

Thus the permutation representation $\psi$ of $G \operatorname{wr} H$, where $G \leqslant S_m$ and $H \leqslant S_n$, allows a detailed description of centralizers of elements, $p$-Sylow-subgroups, and normalizers of subgroups of form $\oplus^n S_m$ in symmetric groups.

Another permutation representation $\theta$ of $G \operatorname{wr} H$ is important in the theory of enumeration under group action in combinatorics. $\theta$ maps $G \operatorname{wr} H$ into the symmetric group on

4.1.26 $$\mathbf{m}^{\mathbf{n}}:=\{\varphi \mid \varphi:\mathbf{n}\to\mathbf{m}\}$$

as follows:

4.1.27 $$\theta: G \operatorname{wr} H\to S_{\mathbf{m}^{\mathbf{n}}}:(f;\pi)\mapsto\begin{pmatrix}\varphi\\ \varphi'\end{pmatrix},$$

where $\varphi'\in\mathbf{m}^{\mathbf{n}}$ is defined by

4.1.28 $$\varphi'(i):=f(i)\left(\varphi\left(\pi^{-1}(i)\right)\right),\qquad i\in\mathbf{n}.$$

The image of $\theta$ is denoted by

4.1.29 $$[G]^H:=\theta[G \operatorname{wr} H]$$

and called the *exponentiation* of $G$ by $H$. The subgroup

4.1.30 $$G^H:=\theta[\operatorname{diag} G^*\cdot H']$$

is called the *power* of $G$ by $H$. Another important subgroup which has no particular name is

4.1.31 $$E^H:=\theta[H'].$$

We notice

**4.1.32** $$E^H \leqslant G^H \leqslant [G]^H$$

and remark that $\theta$, as well as $\psi$, is a faithful representation.

Finally it should be mentioned that the series of Weyl groups (but not the exceptional Weyl groups) are special cases of the groups defined above. The *Weyl groups* are defined to be certain permutation groups acting on certain subsets ("root systems") of euclidean spaces. It is not difficult to show that the following holds:

**4.1.33.**

(i) *The Weyl group of type* $A_n$ $(n \geqslant 1)$ *is similar to* $S_{n+1}$.

(ii) *The Weyl group of type* $B_n$ $(n \geqslant 2)$ *is similar to* $\psi[S_2 \operatorname{wr} S_n]$, *as is the Weyl group of type* $C_n$ $(n \geqslant 3)$.

(iii) *The Weyl group of type* $D_n$ $(n \geqslant 4)$ *is similar to* $\psi[S_2 \operatorname{wr} S_n] \cap A_{2n}$.

Therefore the results which follow apply in particular to the series of Weyl groups.

## 4.2 The Conjugacy Classes of $G \operatorname{wr} S_n$

The wreath products of the form $G \operatorname{wr} S_n$, $n \in \mathbb{N}$, are of particular importance, and they are called the *complete monomial groups* over $G$. This name points to the fact that $G \operatorname{wr} S_n$ can be considered as a group of monomial matrices which arise from the $n!$ permutation matrices of dimension $n$ by replacing the 1's of the matrices by the elements $g$ of $G$ in all possible ways. The multiplication of these matrices is defined with the aid of the multiplication of $G$.

We shall describe the conjugacy classes of these groups, and to do this we consider an element $(f;\pi) \in G \operatorname{wr} S_n$, assuming that $G$ contains countably many conjugacy classes $C^1, C^2, \ldots$. If

$$\pi = \prod_{\nu=1}^{c(\pi)} \left(j_\nu \pi(j_\nu) \cdots \pi^{l_\nu - 1}(j_\nu)\right)$$

in cycle notation (so that 1.1.15 holds), then we can associate with its $\nu$th cyclic factor $(j_\nu \cdots \pi^{l_\nu-1}(j_\nu))$ the uniquely determined element

4.2.1 $$g_\nu(f;\pi) := f(j_\nu) f\left(\pi^{-1}(j_\nu)\right) \cdots f\left(\pi^{-l_\nu+1}(j_\nu)\right)$$
$$= f f_\pi \cdots f_{\pi^{l(\nu)-1}}(j_\nu)$$

of $G$, which we call the $\nu$th *cycle product* of $(f;\pi)$ or the cycle product associated with $(j_\nu \cdots \pi^{l_\nu-1}(j_\nu))$ with respect to $f$. (Here and below, for typographical convenience we sometimes write $l(\nu)$ instead of $l_\nu$, etc.)

We have thus $c(\pi)$ cycle products which are associated with the cyclic factors of $\pi$ with respect to $f$. $a_k(\pi)$ of them arise from the $a_k(\pi)$ cyclic factors of $\pi$ of length $k$. Now let $a_{ik}(f;\pi)$ be the number of cycle products which are associated with a $k$-cycle of $\pi$ and which belong to the conjugacy class $C^i$ of $G$. We put these nonnegative integers $a_{ik}(f;\pi)$ together into a matrix

4.2.2 $$a(f;\pi):=(a_{ik}(f;\pi)).$$

The matrix has $n$ columns and as many rows as there are conjugacy classes in $G$. Its entries satisfy the following conditions:

**4.2.3**
$$\text{(i) } a_{ik}(f;\pi)\in\mathbb{N}_0,$$
$$\text{(ii) } \sum_i a_{ik}(f;\pi)=a_k(\pi),$$
$$\text{(iii) } \sum_{i,k} k a_{ik}(f;\pi)=n.$$

We call this matrix $a(f;\pi)$ the *type* of $(f;\pi)$, since it generalizes the cycle type of the elements of $S_n$ which was introduced in Section 1.2.

One notices that for each matrix $(b_{ik})$ of the stated size the coefficients of which satisfy

4.2.4
$$\text{(i) } b_{ik}\in\mathbb{N}_0,$$
$$\text{(ii) } \sum_{i,k} k\cdot b_{ik}=n,$$

there are elements of $G \operatorname{wr} S_n$ which are of type $(b_{ik})$. It is our aim now to show that two elements of $G \operatorname{wr} S_n$ are conjugate if and only if they are of the same type.

The first remark concerns the definition 4.2.1 of the cycle product. Since we are merely interested in the type of $(f;\pi)$, we need only determine a cycle product up to conjugation within $G$. Therefore we show first that the convention that $j_\nu$ in 4.2.1 denotes the least symbol of the cycle in question is unnecessary here, i.e. we show

**4.2.5** $$\forall \nu, s \quad \left(f\cdots f_{\pi^{l(\nu)-1}}(j_\nu)\sim f\cdots f_{\pi^{l(\nu)-1}}(\pi^s(j_\nu))\right).$$

*Proof.* We can assume $0\leqslant s\leqslant l_\nu-1$. Then

$$f\cdots f_{\pi^{l(\nu)-1}}(\pi^s(j_\nu))=f(\pi^s(j_\nu))\cdots f(\pi(j_\nu))f(j_\nu)\cdots f(\pi^{s+1}(j_\nu)).$$

But in a group $G$ each product $ab$ of elements $a, b \in G$ is conjugate to $ba$, so that we can proceed to deduce

$$\sim f(j_\nu) f(\pi^{-1}(j_\nu)) \cdots f(\pi^{s+1}(j_\nu)) f(\pi^s(j_\nu)) \cdots f(\pi(j_\nu)). \qquad \blacksquare$$

Another useful remark is

**4.2.6** $a(f;\pi) = a((e;\pi')(f;\pi)(e;\pi')^{-1}) = a((f';1)(f;\pi)(f';1)^{-1})$.

*Proof.* Since

$$(e;\pi')(f;\pi)(e;\pi')^{-1} = (f_{\pi'};\pi'\pi\pi'^{-1}),$$
$$(f';1)(f;\pi)(f';1)^{-1} = (f'ff_\pi'^{-1};\pi),$$

it suffices to check that the elements on the right-hand sides of these equations are both of type $a(f;\pi)$.

(i) $\pi'\pi\pi'^{-1}$ contains the cyclic factor $(\pi'(j_\nu)\cdots\pi'\pi^{l_\nu-1}(j_\nu))$. Its cycle product with respect to $f_{\pi'}$ is by 4.2.5 a conjugate of

$$f_{\pi'}\cdots(f_{\pi'})_{(\pi'\pi\pi'^{-1})^{l(\nu)-1}}(\pi'(j_\nu)) = f_{\pi'}f_{\pi'\pi}\cdots f_{\pi'\pi^{l(\nu)-1}}(\pi'(j_\nu)) = g_\nu(f;\pi).$$

(ii)

$$\begin{aligned} g_\nu(f'ff_\pi'^{-1};\pi) &= (f'ff_\pi'^{-1})(f'ff_\pi'^{-1})_\pi\cdots(f'ff_\pi'^{-1})_{\pi^{l(\nu)-1}}(j_\nu)\\ &= f'f\cdots f_{\pi^{l(\nu)-1}}f'^{-1}(j_\nu)\\ &= f'(j_\nu)g_\nu(f;\pi)f'(j_\nu)^{-1} \sim g_\nu(f;\pi). \qquad \blacksquare \end{aligned}$$

A final lemma before we prove that the type characterizes a conjugacy class:

**4.2.7** Lemma. *If $a(f;\pi) = a(f';\pi')$, then there is a $\pi'' \in S_n$ satisfying $\pi = \pi''\pi'\pi''^{-1}$ and with the property that for each $\nu$, $g_\nu(f;\pi) \sim g_\nu(f'_{\pi''};\pi)$.*

*Proof.* $a(f;\pi) = a(f';\pi')$ implies (cf. 4.2.3 (ii)) that $\pi \sim \pi'$, say

$$\pi = \tilde{\pi}\pi'\tilde{\pi}^{-1}.$$

The set of all these $\tilde{\pi}$ forms a right coset of the centralizer of $\pi$. Since

$$(e;\tilde{\pi})(f';\pi')(e;\tilde{\pi})^{-1} = (f'_{\tilde{\pi}};\pi),$$

4.2.6 yields

(1) $$a(f;\pi)=a(f';\pi')=a(f'_{\tilde{\pi}};\pi).$$

But

$$g_\nu(f;\pi)=f\cdots f_{\pi^{l(\nu)-1}}(j_\nu),$$

while

$$g_\nu(f'_{\tilde{\pi}};\pi)=f'_{\tilde{\pi}}\cdots f'_{\pi^{l(\nu)-1}\tilde{\pi}}(j_\nu).$$

Since (1) is valid, there is a $\pi^*$ in the centralizer of $\pi$, so that for each $\nu$ we have $g_\nu(f;\pi)\sim g_\nu(f'_{\pi^*\tilde{\pi}};\pi)$, i.e. (cf. 1.2.1)

$$f'_{\pi^*\tilde{\pi}}\cdots f'_{\pi^{l(\nu)-1}\pi^*\tilde{\pi}}(j_\nu)\sim f\cdots f_{\pi^{l(\nu)-1}}(j_\nu).$$

Thus $\pi'':=\pi^*\tilde{\pi}$ fulfills the statement. ∎

We are now in the position to characterize the conjugacy classes:

**4.2.8** THEOREM. *Two elements $(f;\pi)$ and $(f';\pi')$ of $G \operatorname{wr} S_n$ are conjugate if and only if $a(f;\pi)=a(f';\pi')$.*

*Proof.*

(i) If $(f;\pi)\sim(f';\pi')$, then there are $f''$ and $\pi''$ such that

$$\begin{aligned}(f';\pi')&=(f'';\pi'')(f;\pi)(f'';\pi'')^{-1}\\&=(f'';1)(e;\pi'')(f;\pi)(e;\pi'')^{-1}(f'';1)^{-1},\end{aligned}$$

and hence by 4.2.6 we have $a(f';\pi')=a(f;\pi)$.

(ii) If on the other hand $a(f;\pi)=a(f';\pi')$, then by 4.2.7 there is a $\pi''\in S_n$ satisfying $\pi=\pi''\pi'\pi''^{-1}$ and $g_\nu(f;\pi)\sim g_\nu(f'_{\pi''};\pi)$. Now $(f'_{\pi''};\pi)=(e;\pi'')(f';\pi)(e;\pi'')^{-1}$. Hence it suffices to prove

(1) $$(f'_{\pi''};\pi)\sim(f;\pi).$$

Since $g_\nu(f;\pi)\sim g_\nu(f'_{\pi''};\pi)$, there exist elements $g_{j_\nu}\in G$ such that

$$\forall\nu\quad\left(f\cdots f_{\pi^{l(\nu)-1}}(j_\nu)=g_{j_\nu}(f'_{\pi''}\cdots f'_{\pi^{l(\nu)-1}\pi''}(j_\nu))g_{j_\nu}^{-1}\right).$$

Keeping such $g_{j_\nu}$, $1\leqslant\nu\leqslant c(\pi)$, fixed and starting with $r:=0$, we obtain from the equations

(2) $$f(\pi^{-r}(j_\nu))=g_{\pi^{-r}(j_\nu)}f'_{\pi''}(\pi^{-r}(j_\nu))g^{-1}_{\pi^{-r-1}(j_\nu)}$$

elements $g_{\pi^{-r-1}(j_\nu)}$ which are uniquely determined $(1 \leqslant r \leqslant l_\nu - 1)$. Having done this for every cyclic factor of $\pi$, we define the mapping $f^*: \mathbf{n} \to G$ by

$$f^*(i) := g_i, \qquad i \in \mathbf{n}.$$

It is easy to check that this mapping satisfies

$$(f^*;1)(f'_{\pi''};\pi)(f^*;1)^{-1} = (f^* f'_{\pi''} f^{*-1}_{\pi};\pi) = (f;\pi)$$

(cf. (2)), which proves (1). ■

This characterization of the conjugacy classes of $G \operatorname{wr} S_n$ by types gives the number of conjugacy classes of $G \operatorname{wr} S_n$. There are as many conjugacy classes as there are types, i.e. matrices $(b_{ik})$ which consist of $n$ columns and as many rows as there are conjugacy classes in $G$, and whose elements satisfy 4.2.4. We derive a formula for this number.

**4.2.9** Lemma. *If $s \in \mathbb{N}$ is equal to the number of conjugacy classes of $G$ and if $p(m)$ denotes the number of partitions of $m$, then the number of conjugacy classes of $G \operatorname{wr} S_n$ is equal to*

$$\sum_{(n)} p(n_1) \cdots p(n_s)$$

*if the sum is taken over all the $s$-tuples $(n_1 \ldots, n_s)$ over $\mathbb{N}_0$ such that $\sum n_i = n$.*

*Proof.* If $(a_{ik})$ is a type of $G \operatorname{wr} S_n$, then the integers

$$n_i := \sum_k k \cdot a_{ik}$$

form an $s$-tuple $(n_1, \ldots, n_s)$ with the properties

$$n_i \in \mathbb{N}_0 \quad \text{and} \quad \sum_i n_i = n.$$

And if $(a_{ik})$ runs through all the types, then each admissible $s$-tuple arises in this way.

The $n_i$ defined above is the sum of the elements of the $i$th row of $(a_{ik})$ weighted by their column number. Therefore if the other rows are kept fixed, there are exactly $p(n_i)$ possibilities for the $i$th row to be the row of a type of $G \operatorname{wr} S_n$. This proves the assertion. ■

If we wish to derive the order of the conjugacy class of $G \operatorname{wr} S_n$ which is characterized by the type $(a_{ik})$, $G$ being a finite group with $s$ conjugacy classes, then we put

$$a_k := \sum_i a_{ik}, \qquad 1 \leqslant k \leqslant n.$$

There are $n!/\prod_k k^{a_k} a_k!$ elements of type $(a_1, \ldots, a_n)$ contained in $S_n$ (cf. 1.2.15). We fix one of them; call it $\pi$. The $a_k$ cyclic factors of $\pi$ which are of length $k$ can be distributed into the $s$ conjugacy classes of $G$ in the following number of ways in accordance with the given type $(a_{ik})$:

$$\binom{a_k}{a_{1k}}\binom{a_k - a_{1k}}{a_{2k}} \cdots \binom{a_k - a_{1k} - \cdots - a_{s-1,k}}{a_{sk}} = \frac{a_k!}{a_{1k}! \cdots a_{sk}!}.$$

Now let $f: \mathbf{n} \to G$ be a mapping which yields such a distribution of the cycle products. It remains to show how much freedom of choice is left for the values of $f$. In order to ensure that

$$g_\nu(f; \pi) = f \cdots f_{\pi^{l(\nu)-1}}(j_\nu) \in C^i,$$

we may choose the values $f(j_\nu)$, $f(\pi^{-1}(j_\nu)), \ldots, f(\pi^{-l_\nu+2}(j_\nu))$ at will, since we can always find a suitable $f(\pi^{-l_\nu+1}(j_\nu)) \in G$ such that the complete product $g_\nu(f; \pi)$ is an element of $C^i \subseteq G$. Hence there exist

$$\prod_{k=1}^{n} \frac{a_k!}{a_{1k}! \cdots a_{sk}!} \prod_{i=1}^{s} \left(|G|^{k-1}|C^i|\right)^{a_{ik}}$$

mappings $f: \mathbf{n} \to G$ such that $a(f; \pi) = (a_{ik})$ for our fixed $\pi \in S_n$. Multiplying this number of mappings by the number $n!/\prod_k k^{a_k} a_k!$ of admissible $\pi \in S_n$, we obtain the desired result:

**4.2.10** Lemma. *If $G$ is a finite group consisting of $s$ conjugacy classes, then the class of elements of type $(a_{ik})$ in $G\,\mathrm{wr}\,S_n$ is of order*

$$\frac{|G\,\mathrm{wr}\,S_n|}{\prod_{i,k} a_{ik}!\left(k|G|/|C^i|\right)^{a_{ik}}}$$

If we want to evaluate the order of a single element $(f; \pi) \in G\,\mathrm{wr}\,S_n$, we consider the equation

**4.2.11** $$\forall r \in \mathbb{N} \quad \left((f, \pi)^r = (f \cdots f_{\pi^{r-1}}; \pi^r)\right).$$

It implies that the order of $(f; \pi)$ is a multiple of the order of $\pi$. The order of $\pi$ is the least common multiple of the lengths of the cyclic factors of $\pi$ (cf.

1.2.14). Now let the exponent $r$ in 4.2.11 be a multiple of the length of each cyclic factor of $\pi$. If $i\in\mathbf{n}$ and $j_\nu$ is the least point in the cyclic factor of $\pi$ which contains $i$, say $\pi^\mu(j_\nu)=i$, then

$$f\cdots f_{\pi^{r-1}}(i)=f(\pi^\mu(j_\nu))\cdots f(\pi(j_\nu))g_\nu(f;\pi)^w f(\pi(j_\nu))^{-1}\cdots f(\pi^\mu(j_\nu))^{-1}$$

if $r=w\cdot l_\nu$. This shows that the following holds (cf. 1.2.14):

**4.2.12.** $|\langle(f;\pi)\rangle|=\operatorname{lcm}\{kw_i|a_{ik}(f;\pi)>0\}$, *if* $w_i$ *denotes the order of the elements in* $C^i\subseteq G$.

**4.2.13** LEMMA. *If* $G$ *is ambivalent, then* $G\operatorname{wr}S_n$ *is ambivalent.*

*Proof.* If

$$\pi=\prod_\nu\left(j_\nu\cdots\pi^{l_\nu-1}(j_\nu)\right),$$

in cycle notation, then

$$\pi^{-1}=\prod_\nu\left(j_\nu\cdots\pi^{-l_\nu+1}(j_\nu)\right)$$

in cycle notation. Hence

$$g_\nu\left((f;\pi)^{-1}\right)=g_\nu\left(f_{\pi^{-1}}^{-1};\pi^{-1}\right)=f_{\pi^{-1}}^{-1}\cdots f_{\pi^{-l(\nu)}}^{-1}(j_\nu)=\left(g_\nu(f;\pi)\right)^{-1}.$$

If $G$ is ambivalent, then $(g_\nu(f;\pi))^{-1}$ is a conjugate of $g_\nu(f;\pi)$, so that in this case we have $a(f;\pi)=a((f;\pi)^{-1})$. ∎

This yields the first result on the character table of a complete monomial group:

**4.2.14** COROLLARY. *If* $G$ *has a real character table, then* $G\operatorname{wr}S_n$ *also has its character table over* $\mathbb{R}$.

A special case is

**4.2.15** COROLLARY. *For each* $m$ *and* $n$ *the wreath product* $S_m\operatorname{wr}S_n$ *is ambivalent.*

For general wreath products $G\operatorname{wr}H$ we notice the following:

**4.2.16** LEMMA. *Ambivalency of* $G\operatorname{wr}H$ *implies the ambivalency of both* $G$ *and* $H$.

*Proof.* If $G\,\mathrm{wr}\,H$ is ambivalent, then for every $(f;\pi)\in G\,\mathrm{wr}\,H$ there exist $(f';\pi')\in G\,\mathrm{wr}\,H$ which satisfy

$$(f';\pi')(f;\pi)(f';\pi')^{-1}=(f'f_{\pi'}f'^{-1}_{\pi'\pi\pi'^{-1}};\pi'\pi\pi'^{-1})=(f_{\pi^{-1}}^{-1};\pi^{-1}).$$

This implies that $\pi$ is conjugate to $\pi^{-1}$, i.e. $H$ is ambivalent. Choosing a constant $f\in G^{\mathbf{n}}$, say $f(i):=g$, $1\leq i\leq n$, then for $\pi:=1$ we obtain from the equation above that there exist $(f';\pi')\in G\,\mathrm{wr}\,H$ which satisfy $f'ff'^{-1}=f'f_{\pi'}f'^{-1}=f^{-1}$. But this implies that $g$ and $g^{-1}$ are conjugates. Hence $G$ has to be ambivalent as well. ■

A numerical example is provided by $S_3\,\mathrm{wr}\,S_2$:

$$\psi[S_3\,\mathrm{wr}\,S_2]=(\{1,(12),(13),(23),(123),(132)\}$$
$$\oplus\{1,(45),(46),(56),(456),(465)\})\cdot\{1,(14)(25)(36)\}\leq S_6.$$

Numbering the conjugacy classes of $S_3$ in the following way:

$$C^1:=\{1_{S_3}\},\qquad C^2:=\{(12),(13),(23)\},\qquad C^3:=\{(123),(132)\},$$

then the types of $S_3\,\mathrm{wr}\,S_2$ are

$$\begin{pmatrix}2&0\\0&0\\0&0\end{pmatrix},\ \begin{pmatrix}0&0\\2&0\\0&0\end{pmatrix},\ \begin{pmatrix}0&0\\0&0\\2&0\end{pmatrix},\ \begin{pmatrix}1&0\\1&0\\0&0\end{pmatrix},\ \begin{pmatrix}1&0\\0&0\\1&0\end{pmatrix},$$
$$\begin{pmatrix}0&0\\1&0\\1&0\end{pmatrix},\ \begin{pmatrix}0&1\\0&0\\0&0\end{pmatrix},\ \begin{pmatrix}0&0\\0&1\\0&0\end{pmatrix},\ \begin{pmatrix}0&0\\0&0\\0&1\end{pmatrix}.$$

Images $\psi((f;\pi))$ of representatives $(f;\pi)$ of these classes are the elements

$$1_{S_6},\quad (12)(45),\quad (123)(456),\quad (12),\quad (123),$$
$$(12)(456),\quad (14)(25)(36),\quad (15)(2634),\quad (153426).$$

The orders of the classes are 1, 9, 4, 6, 4, 12, 6, 18, 12.

For later purposes it is useful to notice the following fact concerning the cycle partition of the image $\psi((f;(1\cdots n)))$, $(f;(1\cdots n))\in S_m\,\mathrm{wr}\,S_n$ in $S_{mn}$:

**4.2.17.** *If $\alpha=(\alpha_1,\ldots,\alpha_h)\vdash m$ denotes the cycle partition of $g_1(f;(1\cdots m))$, then*

$$\alpha(\psi((f;(1\cdots n))))=(n\alpha_1,\ldots,n\alpha_h)=:n\alpha.$$

This is already suggested by the preceding examples and is easy to check. 4.2.17 shows clearly that the $\nu$th cyclic factor $(j_\nu \cdots \pi^{l_\nu - 1}(j_\nu))$ of $\nu$ contributes to the cycle partition $\alpha(\psi((f;\pi)))$, $(f;\pi) \in S_m \operatorname{wr} S_n$, just the summand

4.2.18 $$l_\nu \cdot \alpha(g_\nu(f;\pi)),$$

so we obtain the cycle partition of $\psi((f;\pi))$ as the following sum of cycle partitions of the cycle products:

**4.2.19** $$\alpha(\psi((f;\pi))) = \sum_{\nu=1}^{c(\pi)} l_\nu \cdot \alpha(g_\nu(f;\pi)).$$

## 4.3 Representations of Wreath Products over Algebraically Closed Fields

Let $G$ denote a group, $F$ a field, and $D$ a representation of $G$ over $F$ with representation space $V$. We show first how $D$ yields in a natural way a representation of $G \operatorname{wr} H$, $H$ a given subgroup of $S_n$.

In order to do this we form the $n$-fold tensor power $\otimes^n V$ of the vector space $V$ with itself:

4.3.1 $$\overset{n}{\otimes} V := V \otimes_F \cdots \otimes_F V, \qquad n \text{ factors}.$$

We recall from multilinear algebra that the additive group of this vector space is generated by the tensors $v_1 \otimes \cdots \otimes v_n$, i.e.

**4.3.2** $$\overset{n}{\otimes} V = \langle v_1 \otimes \cdots \otimes v_n \mid v_i \in V \rangle.$$

Let $\{b_1, \ldots, b_m\}$ be an $F$-basis of $V$: for short,

$$V = \langle\langle b_1, \ldots, b_m \rangle\rangle_F.$$

For each mapping $\varphi: \mathbf{n} \to \mathbf{m}$, i.e. $\varphi \in \mathbf{m}^{\mathbf{n}}$, we put

4.3.3 $$b_\varphi := b_{\varphi(1)} \otimes \cdots \otimes b_{\varphi(n)},$$

and obtain in this way a basis of $\otimes^n V$:

**4.3.4** $$\overset{n}{\otimes} V = \langle\langle b_\varphi \mid \varphi \in \mathbf{m}^{\mathbf{n}} \rangle\rangle_F.$$

The representation $D$ turns $V$ into a left $G$-module:

4.3.5 $$\forall g \in G, v \in V \quad gv := D(g)v,$$

so that $\otimes^n V$ turns into a left $G \operatorname{wr} H$-module if we define the action of $(f;\pi) \in G \operatorname{wr} H$ on an element $b_\varphi$ of the basis by

4.3.6 $$(f;\pi)b_\varphi := f(1)b_{\varphi(\pi^{-1}(1))} \otimes \cdots \otimes f(n)b_{\varphi(\pi^{-1}(n))}.$$

This yields for the action of $(f;\pi)$ on a generating element:

**4.3.7** $$(f;\pi)v_1 \otimes \cdots \otimes v_n = f(1)v_{\pi^{-1}(1)} \otimes \cdots \otimes f(n)v_{\pi^{-1}(n)}.$$

This last equation shows clearly the irrelevance of the chosen basis of $V$.

4.3.6 defines a representation of $G \operatorname{wr} H$ with representation space $\otimes^n V$. Its restriction to the base group $G^*$ is essentially the outer tensor power $\#^n D$ of $D$ with itself. This suggests denoting the representation defined in 4.3.6 by

4.3.8 $$\left(\overset{n}{\#} D\right)^{\sim}$$

A numerical example illustrates this. We would like to evaluate the matrix representing $((12),1,(123);(132))$ under the representation $(\#^3[2,1])^{\sim}$ of $S_3 \operatorname{wr} S_3$. We notice first that

$$\left(\overset{3}{\#}[2,1]\right)^{\sim}((12),1,(123);(132))$$
$$= \left(\overset{3}{\#}[2,1]\right)^{\sim}((12),1,(123);1_{S_3})\left(\overset{3}{\#}[2,1]\right)^{\sim}(e;(132)).$$

The seminormal form of $[2,1]$ yields (cf. 3.3.29)

$$[2,1]((12)) = \begin{pmatrix} -1 & 0 \\ 0 & 1 \end{pmatrix}, \qquad [2,1]((23)) = \begin{pmatrix} \frac{1}{2} & \frac{3}{4} \\ 1 & -\frac{1}{2} \end{pmatrix},$$

so that

$$[2,1]((123)) = \begin{pmatrix} -\frac{1}{2} & -\frac{3}{4} \\ 1 & -\frac{1}{2} \end{pmatrix}.$$

Let $\{b_1, b_2\}$ be the corresponding ordered basis of the representation space $V$ of $[2,1]$. Then $(e;(132))$ acts on the 8 elements $b_\varphi$ of the basis of

$\otimes^3 V$ by *place permutation* as follows:

$$\begin{array}{lcl}
b_{\varphi_1} := b_1 \otimes b_1 \otimes b_1 & \mapsto & b_1 \otimes b_1 \otimes b_1 = b_{\varphi_1}, \\
b_{\varphi_2} := b_1 \otimes b_1 \otimes b_2 & \mapsto & b_1 \otimes b_2 \otimes b_1 = b_{\varphi_3}, \\
b_{\varphi_3} := b_1 \otimes b_2 \otimes b_1 & \mapsto & b_2 \otimes b_1 \otimes b_1 = b_{\varphi_5}, \\
b_{\varphi_4} := b_1 \otimes b_2 \otimes b_2 & \mapsto & b_2 \otimes b_2 \otimes b_1 = b_{\varphi_7}, \\
b_{\varphi_5} := b_2 \otimes b_1 \otimes b_1 & \mapsto & b_1 \otimes b_1 \otimes b_2 = b_{\varphi_2}, \\
b_{\varphi_6} := b_2 \otimes b_1 \otimes b_2 & \mapsto & b_1 \otimes b_2 \otimes b_2 = b_{\varphi_4}, \\
b_{\varphi_7} := b_2 \otimes b_2 \otimes b_1 & \mapsto & b_2 \otimes b_1 \otimes b_2 = b_{\varphi_6}, \\
b_{\varphi_8} := b_2 \otimes b_2 \otimes b_2 & \mapsto & b_2 \otimes b_2 \otimes b_2 = b_{\varphi_8}.
\end{array}$$

Hence with respect to this ordered basis $\{b_{\varphi_1}, \ldots, b_{\varphi_8}\}$ of $\otimes^3 V$ we obtain

$$\left(\overset{3}{\#}[2,1]\right)^{\sim}(e;(132)) = \begin{pmatrix}
1 & 0 & 0 & 0 & 0 & 0 & 0 & 0 \\
0 & 0 & 0 & 0 & 1 & 0 & 0 & 0 \\
0 & 1 & 0 & 0 & 0 & 0 & 0 & 0 \\
0 & 0 & 0 & 0 & 0 & 1 & 0 & 0 \\
0 & 0 & 1 & 0 & 0 & 0 & 0 & 0 \\
0 & 0 & 0 & 0 & 0 & 0 & 1 & 0 \\
0 & 0 & 0 & 1 & 0 & 0 & 0 & 0 \\
0 & 0 & 0 & 0 & 0 & 0 & 0 & 1
\end{pmatrix}.$$

Furthermore we have

$$\left(\overset{3}{\#}[2,1]\right)^{\sim}((12),1,(123);1) = \begin{pmatrix} -1 & 0 \\ 0 & 1 \end{pmatrix} \otimes \begin{pmatrix} 1 & 0 \\ 0 & 1 \end{pmatrix} \otimes \begin{pmatrix} -\frac{1}{2} & -\frac{3}{4} \\ 1 & -\frac{1}{2} \end{pmatrix}$$

$$= \begin{pmatrix}
\frac{1}{2} & \frac{3}{4} & 0 & 0 & 0 & 0 & 0 & 0 \\
-1 & \frac{1}{2} & 0 & 0 & 0 & 0 & 0 & 0 \\
0 & 0 & \frac{1}{2} & \frac{3}{4} & 0 & 0 & 0 & 0 \\
0 & 0 & -1 & \frac{1}{2} & 0 & 0 & 0 & 0 \\
0 & 0 & 0 & 0 & -\frac{1}{2} & -\frac{3}{4} & 0 & 0 \\
0 & 0 & 0 & 0 & 1 & -\frac{1}{2} & 0 & 0 \\
0 & 0 & 0 & 0 & 0 & 0 & -\frac{1}{2} & -\frac{3}{4} \\
0 & 0 & 0 & 0 & 0 & 0 & 1 & -\frac{1}{2}
\end{pmatrix},$$

so that altogether

$$\left(\overset{3}{\#}[2,1]\right)^{\sim}((12),1,(123);(132))$$

$$=\begin{pmatrix} \frac{1}{2} & 0 & 0 & 0 & \frac{3}{4} & 0 & 0 & 0 \\ -1 & 0 & 0 & 0 & \frac{1}{2} & 0 & 0 & 0 \\ 0 & \frac{1}{2} & 0 & 0 & 0 & \frac{3}{4} & 0 & 0 \\ 0 & -1 & 0 & 0 & 0 & \frac{1}{2} & 0 & 0 \\ 0 & 0 & -\frac{1}{2} & 0 & 0 & 0 & -\frac{3}{4} & 0 \\ 0 & 0 & 1 & 0 & 0 & 0 & -\frac{1}{2} & 0 \\ 0 & 0 & 0 & -\frac{1}{2} & 0 & 0 & 0 & -\frac{3}{4} \\ 0 & 0 & 0 & 1 & 0 & 0 & 0 & -\frac{1}{2} \end{pmatrix}.$$

We would like to know the character of the representation $(\#^n D)^{\sim}$. If again $g_\nu(f;\pi)$, $1\leq\nu\leq c(\pi)$, are the cycle products of $\pi$ with respect to $f$ (see 4.2.1), we obtain the following result, which is very useful:

**4.3.9 Lemma.** $\forall (f;\pi)\in G \quad \chi^{(\#^n D)^{\sim}}(f;\pi)=\prod_{\nu=1}^{c(\pi)}\chi^D(g_\nu(f;\pi))$.

*Proof.* By 4.3.4 the trace of $(f;\pi)$ is just

$$\chi^{(\#^n D)^{\sim}}(f;\pi)=\sum_{\varphi}\left[\text{coeff. of } b_\varphi \text{ in } (f;\pi)b_\varphi\right].$$

Let $\mathbb{D}$ denote the matrix representation corresponding to $D$ with respect to the ordered basis $\{b_1,\ldots,b_m\}$ of $V$, and put

$$\forall g\in G \quad ((d_{ik}(g)):=\mathbb{D}(g)).$$

Then by 4.3.6

$$(f;\pi)b_\varphi=\left(\sum_{i_1}d_{i_1,\varphi(\pi^{-1}(1))}(f(1))b_{i_1}\right)\otimes\cdots\otimes\left(\sum_{i_n}d_{i_n,\varphi(\pi^{-1}(n))}(f(n))b_{i_n}\right)$$

$$=\sum_{1\leq i_1,\ldots,i_n\leq n}d_{i_1,\varphi(\pi^{-1}(1))}(f(1))\cdots d_{i_n,\varphi(\pi^{-1}(n))}(f(n))b_{i_1}\otimes\cdots\otimes b_{i_n}$$

$$=\sum_{\psi}\left(\prod_j d_{\psi(j),\varphi(\pi^{-1}(j))}(f(j))\right)b_\psi.$$

In particular the coefficient of $b_\varphi$ in this sum is equal to

$$\prod_{j=1}^{n} d_{\varphi(j),\varphi(\pi^{-1}(j))}(f(j)),$$

the factors of which can be rearranged in order to obtain

$$\prod_{\nu=1}^{c(\pi)} d_{\varphi(j_\nu),\varphi(\pi^{-1}(j_\nu))}(f(j_\nu))d_{\varphi(\pi^{-1}(j_\nu)),\varphi(\pi^{-2}(j_\nu))}(f(\pi^{-1}(j_\nu)))\cdots.$$

Taking the sum of these expressions over all $\varphi\in\mathbf{m^n}$, we obtain for the desired character

$$\begin{aligned}\chi^{(\#^n D)^\sim}(f;\pi)&=\prod_\nu \operatorname{trace}\mathbb{D}(g_\nu(f;\pi))\\&=\prod_\nu \chi^D(g_\nu(f;\pi)),\end{aligned}$$

as was stated. ∎

Let us illustrate 4.3.9 by a few examples:

**4.3.10**

$$\text{(i) } f^{(\#^n D)^\sim}:=\chi^{(\#^n D)^\sim}(e;1)=m^n,$$

$$\text{(ii) } \chi^{(\#^n D)^\sim}(e;\pi)=m^{c(\pi)},$$

$$\text{(iii) } \chi^{(\#^n D)^\sim}(f;1)=\prod_{i=1}^{n}\chi^D(f(i)),$$

$$\text{(iv) } \chi^{(\#^n D)^\sim}(g,\ldots,g;1)=\chi^D(g)^n,$$

$$\text{(v) } \chi^{(\#^n D)^\sim}(g,\ldots,g;(1\cdots n))=\chi^D(g^n),$$

$$\text{(vi) } \chi^{(\#^n D)^\sim}(g,\ldots,g;\pi)=\prod_{k=1}^{n}\chi^D(g^k)^{a_k(\pi)}.$$

The restriction of $(\#^n D)^\sim$ to the complement $H'$ of the base group $G$ yields a representation of $H$ which depends only on the dimension $m$ of $V$. We therefore denote it by $P_n^m$:

4.3.11 $$\forall \pi\in H\left(P_n^m(\pi):=\left(\overset{n}{\#}D\right)^\sim(e;\pi)\right).$$

This is in fact a representation of $S_n$. 4.3.10(ii) shows that we have for its character:

**4.3.12** $$\forall\pi\in S_n\quad\left(\chi^{P_n^m}(\pi)=m^{c(\pi)}\right).$$

Besides this restriction of $(\#^n D)^{\sim}$ to $H'$ or $S'_n$, we have the restriction of $(\#^n D)^{\sim}$ to the diagonal subgroup $\operatorname{diag} G^*$ of the base group (cf. 4.1.16) which is, as a representation of $G$, the $n$-fold inner tensor power $\otimes^n D$ of $D$ with itself:

4.3.13 $$\forall g \in G \left( \overset{n}{\otimes} D(g) := \left( \overset{n}{\#} D \right)^{\sim} (g, \ldots, g; 1_{S_n}) \right).$$

This shows that we have:

**4.3.14** $\forall g \in G, \pi \in S_n$

$$\left( \left( \overset{n}{\#} D \right)^{\sim} (g, \ldots, g; \pi) = \overset{n}{\otimes} D(g) \cdot P_n^m(\pi) = P_n^m(\pi) \cdot \overset{n}{\otimes} D(g) \right).$$

Later on we shall return to this equation.

In the earlier part of this section we saw how a representation $D$ of $G$ yields in a very natural way a representation $(\#^n D)^{\sim}$ of $G \operatorname{wr} H$. It is even easier to see that a representation $D''$ of $H$ gives a representation $D'$ of $G \operatorname{wr} H$, for we need only put for $(f; \pi) \in G \operatorname{wr} H$:

4.3.15 $$D'(f; \pi) := D''(\pi).$$

Hence given representations $D$ of $G$ and $D''$ of $H$ yield in a very natural way the inner tensor product of certain representations of $G \operatorname{wr} H$:

4.3.16 $$(D; D'') := \left( \overset{n}{\#} D \right)^{\sim} \otimes D'.$$

If again $I$ denotes the identity representation of the group in question, then we have in particular

**4.3.17** $$\left( \overset{n}{\#} D \right)^{\sim} = (D; I) \quad \textit{and} \quad D' = (I; D'').$$

It is our aim now to show that each irreducible representation of $G \operatorname{wr} H$ over an algebraically closed field can be induced from a suitable product of such representations $(D; D'')$. *Let us therefore assume for the rest of this section that the ground field $F$ is algebraically closed and that $G$ is finite.*

Let $D^1, \ldots, D^r$ denote a complete set of pairwise inequivalent and irreducible representations of $G$ over $F$ with corresponding representation modules $V^1, \ldots, V^r$ and corresponding matrix representations $\mathbb{D}^1, \ldots, \mathbb{D}^r$. Then each irreducible $F$-representation of $G^*$ is of the form

4.3.18 $$D^* := D_1 \# \cdots \# D_n =: \underset{i}{\#} D_i,$$

where $D_i \in \{D^1,\ldots, D^r\}$, with representation module

4.3.19 $$V^* := V_1 \# \cdots \# V_n =: \mathop{\#}_i V_i,$$

where $V_i := V^j$, if $D_i = D^j$. The underlying vector space is

4.3.20 $$V_1 \otimes_F \cdots \otimes_F V_n =: \bigotimes_i V_i.$$

If $n_j$ denotes the number of factors $D_i$ of $D^*$ which are equal to $D^j$, $1 \leqslant j \leqslant r$, then

4.3.21 $$(n) := (n_1,\ldots, n_r)$$

is called the *type* of $D^*$.

We denote by $S_{n_j}$ the subgroup of $S_n$ $(\geqslant H)$ consisting of just those permutations which move at most the indices $i$ of factors $D_i$ equal to $D^j$, if $n_j > 0$, or $S_{n_j} := \{1_{S_n}\}$ if $n_j = 0$.

Now the *inertia group* of $D^*$ is defined as

4.3.22 $$G\,\mathrm{wr}\, H_{D^*} := \{(f;\pi) \mid D^{*(f;\pi)} \sim D^*\},$$

where $\sim$ denotes equivalence of representations and $D^{*(f;\pi)}$ is as usual the representation conjugate to $D^*$ as follows:

4.3.23 $$D^{*(f;\pi)}\big((f;\pi)(f';1)(f;\pi)^{-1}\big) := D^*((f';1)).$$

Since $G^* \leqslant G\,\mathrm{wr}\, H_{D^*}$, the inertia group is obviously a product

4.3.24 $$G\,\mathrm{wr}\, H_{D^*} = G^* \cdot H'_{D^*}$$

of $G^*$ with a subgroup $H'_{D^*}$ of the complement $H'$ of $G^*$. $H'_{D^*}$ will be called the *inertia factor* of $D^*$:

4.3.25 $$H'_{D^*} := \{(e;\pi) \mid D^{*(e;\pi)} \sim D^*\}.$$

We notice that

**4.3.26** $$D^{*(e;\pi)}(f';1) = D^*(f'_{\pi^{-1}};1),$$

and would like to show how $H'_{D^*}$ can be expressed in terms of the subgroups $S_{n_j}$ introduced above.

**4.3.27** LEMMA. *If we put* $S'_{n_j} := \{(e;\pi) \mid \pi \in S_{n_j}\}$, $1 \leqslant j \leqslant r$, *and*

$$S'_{(n)} := \prod_j S'_{n_j} \leqslant S'_n,$$

*then we have for the inertia factor $H'_{D^*}$ of $D^*$*

$$H'_{D^*} = H' \cap S'_{(n)},$$

*so that for the inertia group of $D^*$ the following holds:*

$$G \operatorname{wr} H_{D^*} = G^* \cdot \left(H \cap S_{(n)}\right)' = G \operatorname{wr} \left(H \cap S_{(n)}\right).$$

*Proof.* We have

$$\begin{aligned} \mathbb{D}^{*(e;\pi)}((f;1)) &= \mathbb{D}^*\left((e;\pi)^{-1}(f;1)(e;\pi)\right) \\ &= \mathbb{D}^*((f_{\pi^{-1}};1)) \\ &= \mathbb{D}_1(f(\pi(1))) \times \cdots \times \mathbb{D}_n(f(\pi(n))). \end{aligned}$$

The question is for which $\pi \in H$ this representation is equivalent to $\mathbb{D}^*$. Since both $\mathbb{D}^*$ and $\mathbb{D}^{*(e;\pi)}$ are irreducible representations, we can use a character-theoretical argument.

(i) If $\pi \in H \cap S_{(n)}$, then the trace of $\mathbb{D}_i(f(\pi(i)))$ is equal to the trace of $\mathbb{D}_{\pi(i)}(f(\pi(i)))$, so that in this case for each $f$:

$$\operatorname{trace} \mathbb{D}^*((f;1)) = \operatorname{trace} \mathbb{D}^{*(e;\pi)}((f;1))$$

and hence $(H \cap S_{(n)})' \leq H'_{D^*}$.

(ii) On the other hand if $(e;\pi)$ belongs to the inertia factor, we have for each $f$

$$\operatorname{trace} \mathbb{D}^*((f;1)) = \operatorname{trace} \mathbb{D}^{*(e;\pi)}((f;1)) = \operatorname{trace} \mathbb{D}^*((f_{\pi^{-1}};1)).$$

If we choose $(f;1) \in G_i$, i.e. $f(j)=1$ for each $j \neq i$, then we obtain the equation (put $f_j := \dim(D_j)$):

$$\left(\prod_{j \neq i} f_j\right) \operatorname{trace} \mathbb{D}_i(f(i)) = \left(\prod_{j \neq \pi^{-1}(i)} f_j\right) \operatorname{trace} \mathbb{D}_{\pi^{-1}(i)}(f(i)).$$

This holds for each $f(i) \in G$. Let us consider the associated Brauer characters (if $\operatorname{char} F = p$) or the traces (if $\operatorname{char} F = 0$), $\varphi_j$. Then we can simplify the equation to obtain

$$\varphi_i(f(i)) f_{\pi^{-1}(i)} = \varphi_{\pi^{-1}(i)}(f(i)) f_i, \qquad f(i) \in G.$$

Since characters and Brauer characters of irreducible modular representations are linearly independent, we obtain $\varphi_i = \varphi_{\pi^{-1}(i)}$, so that $D_i \sim D_{\pi(i)}$ and hence $\pi \in H \cap S_{(n)}$. ∎

According to Clifford's theory of representations of groups with nontrivial normal subgroups, we now have to extend $D^*$ to a representation of its inertia group. It is not always possible to extend a given linear representation of a normal subgroup to a linear representation of its inertia group. But in our case here, where $G^*$ is the normal subgroup, we are fortunately able to extend $D^*$ to a linear representation of $G\operatorname{wr} H_{D^*}$, as we already saw at the beginning of this section when we extended $\#^n D$ of $G^*$ to $(\#^n D)^\sim$ of its inertia group $G \operatorname{wr} S_n$ in $G \operatorname{wr} S_n$. The corresponding extension procedure yields from the left $G^*$-module $\#_i V_i$, which affords $D^*$ a left $G\operatorname{wr} H_{D^*}$-module $(\#_i V_i)^\sim$, where the action of $(f;\pi) \in G\operatorname{wr} H_{D^*}$ is defined as in 4.3.7:

$$(f;\pi)v_1 \otimes \cdots \otimes v_n := f(1)v_{\pi^{-1}(1)} \otimes \cdots \otimes f(n)v_{\pi^{-1}(n)}.$$

We denote this extension of $D^*$ by

$$(D^*)^\sim.$$

It is important to notice what this means for the corresponding matrix representation $(\mathbb{D}^*)^\sim$. If

4.3.28 $$\mathbb{D}^*((f;1)) = \mathbb{D}_1(f(1)) \times \cdots \times \mathbb{D}_n(f(n)) = \left(d^1_{i_1 j_1}(f(1)) \cdots d^n_{i_n j_n}(f(n))\right)$$

then by 4.3.7 for each $(f;\pi) \in G\operatorname{wr} H_{D^*}$

**4.3.29** $$(\mathbb{D}^*)^\sim((f;\pi)) = \left(d^1_{i_1 j(\pi^{-1}(1))}(f(1)) \cdots d^n_{i_n j(\pi^{-1}(n))}(f(n))\right),$$

which means that

**4.3.30.** $(\mathbb{D}^*)^\sim((f;\pi))$ *arises from* $\mathbb{D}^*(f;1)$ *by a suitable column permutation.*

Now let $D''$ be an irreducible $F$-representation of $H \cap S_{(n)}$, and define $D'$ by

4.3.31 $$D'((f;\pi)) := D''(\pi),$$

for each $(f;\pi) \in G\operatorname{wr} H_{D^*}$. Multiplying these two representations yields the inner tensor product

$$(D^*)^\sim \otimes D'$$

with the representing matrices

**4.3.32** $$(\mathbb{D}^*)^\sim \otimes \mathbb{D}'((f;\pi)) = (\mathbb{D}^*)^\sim((f;\pi)) \times \mathbb{D}'((f;\pi)).$$

Clifford's theory says that $(D^*)^{\sim}\otimes D'$ is an irreducible $F$-representation of $G \operatorname{wr} H_{D^*}$, and that the induced representation

$$(D^*)^{\sim}\otimes D'\uparrow G\operatorname{wr} H$$

is an irreducible $F$-representation of $G\operatorname{wr}H$, but even more (cf. Clifford's original paper [1937]):

**4.3.33.** *$D:=(D^*)^{\sim}\otimes D'\uparrow G\operatorname{wr}H$ is irreducible, and every irreducible $F$-representation of $G\operatorname{wr}H$ is of this form.*

It remains to investigate which representations $D^*$ and $D'$ (or $D''$) have to run through in order that $D$ shall run through exactly a complete system of pairwise inequivalent and irreducible $F$-representations of $G\operatorname{wr}H$. Using Clifford's notation, we call two irreducible representations of $G\operatorname{wr}H$ *associated* with respect to $G^*$ if their restrictions to $G^*$ have an irreducible constituent in common. From Clifford's theory we know that there is a 1-1 correspondence between the classes of associated representations of $G\operatorname{wr}H$ and the classes of representations of $G^*$ which are conjugate with respect to $G\operatorname{wr}H$. (Two representations $D^*$ and $D^{**}$ are *conjugates* with respect to $G\operatorname{wr}H$ if there exist $(f;\pi)\in G\operatorname{wr}H$ such that $D^{*(f;\pi)}\sim D^{**}$.) And this correspondence is as follows. The restriction to $G^*$ of every element out of a class of associated representations is (up to a multiplicity) just the corresponding class of conjugate representations.

Hence it suffices that in $D$ the representation $D^*$ runs through a complete system of representatives of the classes of conjugate representations of $G^*$. Moreover Clifford's theory yields that associated representations differ only in the factor $D'$. This yields the following main theorem:

**4.3.34 Theorem.** *The irreducible $F$-representation $D=(D^*)^{\sim}\otimes D'\uparrow G\operatorname{wr}H$ runs just through a complete system of pairwise inequivalent and irreducible $F$-representations of $G\operatorname{wr}H$ if $D^*$ defined by* 4.3.18 *runs through a complete system of pairwise not conjugate but irreducible $F$-representations of $G^*$, and, while $D^*$ remains fixed, $D''$ runs through a complete system of pairwise inequivalent and irreducible $F$-representations of $H\cap S_{(n)}$.*

## 4.4 Special Cases and Properties of Representations of Wreath Products

In the case when $H=S_n$, two representations of $G^*$ are conjugates if and only if they are of the same type. Hence $D$ runs through a complete system of irreducible $F$-representations of the complete monomial group if $D^*$ runs through a complete system of irreducible $F$-representations with pairwise different types and $D''$, while $D^*$ is kept fixed, runs through a complete system of pairwise inequivalent and irreducible $F$-representations of $S_{(n)}$. Furthermore each irreducible $F$-representation $D''$ of $S_{(n)}$, where $(n)=(n_1,\ldots,n_s)$, is equivalent to a representation of the form

$$D''=D_1''\#\cdots\#D_s'',$$

where $D_i''$ is an irreducible $F$-representation of $S_{n_i}$, $1\leqslant i\leqslant s$. Since

4.4.1
$$G\,\mathrm{wr}\,S_{(n)}\simeq \bigtimes_i G\,\mathrm{wr}\,S_{n_i},$$

we have in this case (cf. 4.3.16):

4.4.2
$$(D^*)^{\sim}\otimes D'=(D_1;D_1'')\#\cdots\#(D_s;D_s'')=:\mathop{\#}_i(D_i;D_i''),$$

so that we obtain

**4.4.3** THEOREM. *The representation*

$$D:=\prod_{i=1}^{s}(D_i;D_i''):=\left(\mathop{\#}_{i=1}^{s}(D_i;D_i'')\right)\uparrow G\,\mathrm{wr}\,S_n$$

*runs just through a complete system of pairwise inequivalent and irreducible $F$-representations of $G\,\mathrm{wr}\,S_n$ if $D^*=\#_iD_i$ runs through a complete system of representations of $G^*$ which are of different type $(n)=(n_1,\ldots,n_s)$, while if $D^*$ of type $(n)$ is kept fixed, $D''=\#_iD_i''$ runs through a complete system of irreducible $F$-representations of $S_{(n)}$.*

In accordance with 4.2.9 we obtain the following corollary on the number of ordinary irreducible representations:

**4.4.4** COROLLARY. *The number of ordinary irreducible representations of $G\,\mathrm{wr}\,S_n$ is equal to*

$$\sum_{(n)}p(n_1)\cdots p(n_s)$$

*if $s$ denotes the number of conjugacy classes of $G$, $p(m)$ the number of partitions of $m$, $p(0):=1$, and the sum is taken over all the types $(n)=(n_1,\ldots,n_s)$.*

As an example we derive the ordinary irreducible representations of $S_3\,\mathrm{wr}\,S_2$.

(i) *The representations of the base group* $S_3^*$, *their types, inertia groups and inertia factors.* $S_3^* \simeq S_3 \times S_3$ has the following ordinary irreducible representations:

$$[3]\#[3],\ [3]\#[2,1],\ [3]\#[1^3],\ [2,1]\#[3],\ [2,1]\#[2,1],$$
$$[2,1]\#[1^3],\ [1^3]\#[3],\ [1^3]\#[2,1],\ [1^3]\#[1^3].$$

With respect to the ordering $D^1:=[3]$, $D^2:=[2,1]$, $D^3:=[1^3]$ of the ordinary irreducible representations of $S_3$, the types of these representations are

$$(2,0,0),\ (1,1,0),\ (1,0,1),\ (1,1,0),\ (0,2,0),$$
$$(0,1,1),\ (1,0,1)\ (0,1,1),\ (0,0,2).$$

Hence a complete system of irreducible ordinary representations of $S_3^*$ with pairwise different types is

$$\{[3]\#[3],\ [3]\#[2,1],\ [3]\#[1^3],\ [2,1]\#[2,1],\ [2,1]\#[1^3],\ [1^3]\#[1^3]\}.$$

The corresponding inertia groups are

$$S_3 \operatorname{wr} S_2,\ S_3^*,\ S_3^*,\ S_3 \operatorname{wr} S_2,\ S_3^*,\ S_3 \operatorname{wr} S_2.$$

The inertia factors are

$$S_2',\ S_1',\ S_1',\ S_2',\ S_1',\ S_2'.$$

(ii) *The representations of* $S_3 \operatorname{wr} S_2$: The ordinary irreducible representations of $S_2$ are $[2]$ and $[1^2]$, the only one of $S_1$ is $[1]$. Thus we get for the ordinary irreducible representations of $S_3 \operatorname{wr} S_2$

$$([3];[2])=([3]\#[3])^{\sim}\otimes[2]'=([3]\#[3])^{\sim},$$
$$([3];[1^2])=([3]\#[3])^{\sim}\otimes[1^2]',$$
$$([3];[1])([2,1];[1])=(([3]\#[2,1])^{\sim}\otimes([1]\#[1])')\uparrow S_3 \operatorname{wr} S_2$$
$$=[3]\#[2,1]\uparrow S_3 \operatorname{wr} S_2,$$
$$([3];[1])([1^3];[1])=(([3]\#[1^3])^{\sim}\otimes([1]\#[1])')\uparrow S_3 \operatorname{wr} S_2$$
$$=[3]\#[1^3]\uparrow S_3 \operatorname{wr} S_2,$$
$$([2,1];[2])=([2,1]\#[2,1])^{\sim}\otimes[2]'=([2,1]\#[2,1])^{\sim},$$
$$([2,1];[1^2])=([2,1]\#[2,1])^{\sim}\otimes[1^2]',$$
$$([2,1];[1])([1^3];[1])=(([2,1]\#[1^3])^{\sim}\otimes([1]\#[1])')\uparrow S_3 \operatorname{wr} S_2$$
$$=[2,1]\#[1^3]\uparrow S_3 \operatorname{wr} S_2,$$
$$([1^3];[2])=([1^3]\#[1^3])^{\sim}\otimes[2]'=([1^3]\#[1^3])^{\sim},$$
$$([1^3];[1^2])=([1^3]\#[1^3])^{\sim}\otimes[1^2]'.$$

Their dimensions are 1, 1, 4, 2, 4, 4, 4, 1, 1 in agreement with

$$1^2+1^2+4^2+2^2+4^2+4^2+4^2+1^2+1^2=72=|S_3 \,\mathrm{wr}\, S_2|.$$

(iii) *Representing matrices.* As a numerical example we shall evaluate the matrix of

$$([2,1];[1^2])((e;(12))).$$

First of all we notice that

$$[2,1]\#[2,1]((e;1))=\begin{pmatrix}1&0\\0&1\end{pmatrix}\begin{pmatrix}1&0\\0&1\end{pmatrix}$$

$$=\begin{pmatrix}1&0&0&0\\0&1&0&0\\0&0&1&0\\0&0&0&1\end{pmatrix}$$

$$=\left(d^1_{i_1j_1}(1)d^2_{i_2j_2}(1)\right),$$

so that (use 4.3.29)

$$([2,1]\#[2,1])^{\sim}((e;(12)))=\left(d^1_{i_1j_2}(1)d^2_{i_2j_1}(1)\right)=\begin{pmatrix}1&0&0&0\\0&0&1&0\\0&1&0&0\\0&0&0&1\end{pmatrix}$$

(we have to transpose the second and the third columns). From $[1^2]((12))=(-1)$, we obtain now

$$([2,1]\#[2,1])^{\sim}\otimes[1^2]'((e;(12)))=\begin{pmatrix}1&0&0&0\\0&0&1&0\\0&1&0&0\\0&0&0&1\end{pmatrix}\times(-1)$$

$$=\begin{pmatrix}-1&0&0&0\\0&0&-1&0\\0&-1&0&0\\0&0&0&-1\end{pmatrix}.$$

Another case of particular interest is the wreath product $G \,\mathrm{wr}\, H$, $G$ being abelian, e.g. the wreath products $C_m \,\mathrm{wr}\, S_n$, $C_m$ a cyclic group of order $m$, the so-called *generalized symmetric groups*. In this case the procedure for getting $D$ from $D^*$ becomes much simpler, since $D^*$ is one-dimensional, so that

$$(D^*)^{\sim}(f;\pi)=D^*(f;1)=\prod_i D_i(f(i)).$$

If $\pi_1,\ldots,\pi_{|H:H\cap S_{(n)}|}$ is a complete system of representatives of the left cosets of the inertia factor $H\cap S_{(n)}$ of $D^*$ in $H$, and $D''$ is again an irreducible representation of $H\cap S_{(n)}$, we use the notation

$$(\mathbb{D}^*)^{\sim}\dot{\otimes}\,\mathbb{D}'\left(\left(f_{\pi_i^{-1}};\pi_i^{-1}\pi\pi_k\right)\right)$$
$$:=\begin{cases}(\mathbb{D}^*)^{\sim}\times\mathbb{D}'\left(\left(f_{\pi_i^{-1}};\pi_i^{-1}\pi\pi_k\right)\right) & \text{if } \pi_i^{-1}\pi\pi_k\in S_{(n)}\cap H,\\ \text{0-matrix} & \text{otherwise.}\end{cases}$$

With this notation in mind we obtain for the representing matrices:

4.4.5
$$\begin{aligned}\mathbb{D}((f;\pi))&=\left((\mathbb{D}^*)^{\sim}\dot{\otimes}\,\mathbb{D}'\left(\left(f_{\pi_i^{-1}};\pi_i^{-1}\pi\pi_k\right)\right)\right)\\&=\left((\mathbb{D}^*)^{\sim}\left(\left(f_{\pi_i^{-1}};1\right)\right)\cdot\dot{\mathbb{D}}''\left(\pi_i^{-1}\pi\pi_k\right)\right),\end{aligned}$$

where

$$\dot{\mathbb{D}}''\left(\pi_i^{-1}\pi\pi_k\right):=\begin{cases}\mathbb{D}''\left(\pi_i^{-1}\pi\pi_k\right) & \text{if } \pi_i^{-1}\pi\pi_k\in H\cap S_{(n)},\\ \text{0-matrix,} & \text{otherwise.}\end{cases}$$

Since on the other hand

$$(\mathbb{D}''\uparrow H)(\pi)=\left(\dot{\mathbb{D}}''\left(\pi_i^{-1}\pi\pi_k\right)\right),$$

we have

**4.4.6** LEMMA. *If $D^*$ is one-dimensional, then we have for the representing matrices of $D=(D^*)^{\sim}\otimes D'\uparrow G\,\mathrm{wr}\,H$:*

$$\mathbb{D}((f;\pi))=\left(\mathbb{D}^*\left(\left(f_{\pi_i^{-1}};1\right)\right)\cdot\dot{\mathbb{D}}''\left(\pi_i^{-1}\pi\pi_k\right)\right),$$

*i.e., the $|H:H\cap S_{(n)}|^2$ submatrices of this matrix $\mathbb{D}((f;\pi))$ are, up to their numerical factors $\mathbb{D}^*((f_{\pi_i^{-1}};1))$, just the submatrices of the induced representation $(\mathbb{D}''\uparrow H)(\pi)$.*

Another conclusion which can be drawn from the fact that $(\mathbb{D}^*)^{\sim}((f;\pi))$ arises from $\mathbb{D}^*((f;1))$ by a suitable column permutation concerns the question whether $G \operatorname{wr} H$ is an $M$-group. A finite group is called an *M-group*, if every irreducible representation of $G$ over an algebraically closed field $F$ with char $F \nmid |G|$ is induced by a one-dimensional representation of a suitable subgroup.

**4.4.7** THEOREM. *If $G$ and the subgroups $H \cap \mathrm{S}_{(n)}$ are $M$-groups then $G \operatorname{wr} H$ is an $M$-group, too.*

*Proof.* If $F$ is algebraically closed and char $F \nmid |G|$, then every irreducible representation $\mathbb{D}^i$ of $G$ (being an $M$-group), and hence also of $G^*$, is equivalent to a representation whose matrices contain in every row and column exactly one nonzero entry . This holds also for $\mathbb{D}''$ of $H \cap \mathrm{S}_{(n)}$, and hence both $\mathbb{D}^*$ and $\mathbb{D}'$ and therefore $(\mathbb{D}^*)^{\sim} \otimes \mathbb{D}'$ and even $\mathbb{D} = (\mathbb{D}^*)^{\sim} \otimes \mathbb{D}' \uparrow G \operatorname{wr} H$ can be assumed to be of this monomial form. Since $\mathbb{D}$ is irreducible, we obtain from a well-known theorem (cf. Huppert I, [1967] V, 18.9) that this implies that $\mathbb{D}$ is induced from a one-dimensional representation of a suitable subgroup. ∎

We notice from this proof that properties carry over from $G$ and the groups $H \cap \mathrm{S}_{(n)}$ to $G \operatorname{wr} H$ if they remain valid under forming outer tensor products and inner tensor products of representations as well as under column permutations and induction. For example:

**4.4.8** THEOREM. *If $K \subset F$ is a splitting field for $G$ and all the subgroups $H \cap \mathrm{S}_{(n)}$, then $K$ is a splitting field for $G \operatorname{wr} H$, too, i.e. every irreducible $F$-representation of $G \operatorname{wr} H$ is realizable in $K$.*

**4.4.9** COROLLARY. *Each field is a splitting field for $\mathrm{S}_m \operatorname{wr} \mathrm{S}_n$.*

Further properties show up as soon as we consider the characters. In the same way as we got 4.3.9, we obtain for the character $\chi^{(D^*)^{\sim} \otimes D'}$ of $(D^*)^{\sim} \otimes D'$

**4.4.10**
$$\chi^{(D^*)^{\sim} \otimes D'}((f;\pi)) = \chi^{D'}(\pi) \prod_{\nu=1}^{c(\pi)} \chi^{D_{j(\nu)}}(g_\nu(f;\pi)).$$

Since by 4.3.34 each irreducible character of $G \operatorname{wr} H$ arises from such a character by induction, we obtain various corollaries on the characters of $G \operatorname{wr} H$, a typical one being (cf. also 4.2.16 and 4.4.8):

**4.4.11** COROLLARY. *If the ordinary irreducible characters of $G$ as well as of the subgroups $H \cap \mathrm{S}_{(n)}$ of $H$ are rational integral (real), then all the ordinary irreducible characters of $G \operatorname{wr} H$ are rational integral (real) as well.*

## Exercises

4.1 Describe the centralizer $C_{G\,\mathrm{wr}\,\mathrm{S}_n}((f;\pi))$ of $(f;\pi)$ in $G\,\mathrm{wr}\,\mathrm{S}_n$ in order to determine which conjugacy classes of $G\,\mathrm{wr}\,\mathrm{S}_n$ split over $G\,\mathrm{wr}\,\mathrm{A}_n$ and which split over

$$(G\,\mathrm{wr}\,\mathrm{S}_n)_N := \left\{(f;\pi)\in G\,\mathrm{wr}\,\mathrm{S}_n \,\middle|\, \prod_1^n f(i)\in N\right\},$$

for a given subgroup $N$ of index 2 in $G$.

4.2 Deduce from Exercise 4.1 a characterization of the conjugacy classes of the Weyl groups of type $D_n$ (cf. 4.1.33).

4.3 Which of the ordinary irreducible representations of $\mathrm{S}_2\,\mathrm{wr}\,\mathrm{S}_n$ split over $(\mathrm{S}_2\,\mathrm{wr}\,\mathrm{S}_n)_{\mathrm{A}_2}$, and which of them form pairs of associated representations?

4.4 Prove along the lines of Exercise 4.3 that the character tables of Weyl groups of type $D_n$ are rational integral.

CHAPTER 5

# Applications to Combinatorics and Representation Theory

In this chapter we consider some applications of the preceding results on the representation theory of symmetric groups and of wreath products to combinatorics and to the representation theory of general finite groups.

The applications to combinatorics cover the so-called Pólya enumeration theory which is an important part of combinatorics and which has at present its main applications in the theory of enumeration of isomorphism types of discrete structures such as graphs and boolean functions. This theory is covered because a quite general problem in the theory is concerned with the action of the wreath product $[G]^H = \theta[G \operatorname{wr} H]$ on the set $\mathbf{m}^{\mathbf{n}}$ of all the mappings from $\mathbf{n}$ into $\mathbf{m}$.

The applications to the representation theory of a general finite group $G$ arise from the fact that the direct product $G \times S_n$ is nicely embedded in $G \operatorname{wr} S_n$. Hence for any ordinary representation $D$ of $G$ we have that $\otimes^n D$ breaks up into representations $D \boxdot [\alpha]$, $\alpha \vdash n$, of $G$, the so-called symmetrized products of $D$ with the ordinary irreducible representations $[\alpha]$ of $S_n$. Conversely each irreducible constituent $D_i$ of $\otimes^n D$ yields an ordinary representation $D \triangle_n D_i$ of $S_n$, the so-called $n$th permutrized product of $D$ and $D_i$. The values of the characters of these representations $D \triangle_n D_i$ appear quite often in the representation theory of $G$, so that we can derive certain results on the representation theory of $G$ by simply recognizing that these numbers are values of characters of symmetric groups. We shall return to the theory of symmetrized products in Chapter 8, and more concrete examples of the concept are given there.

Another application is provided by the notion of the plethysm of representations of $G \leqslant S_m$ and $H \leqslant S_n$ which arises from the fact that $G \operatorname{wr} H$ is a

ENCYCLOPEDIA OF MATHEMATICS and Its Applications, Gian-Carlo Rota (ed.).
Vol. 16: G. D. James and A. Kerber, The Representation Theory of the Symmetric Group
ISBN 0-201-13515-9

subgroup of $S_{mn}$ in a natural way. Furthermore we can apply the representation theory of $S_n$ to the representation theory of multiply transitive subgroups of $S_n$ by showing that certain representations remain irreducible when we restrict. We illustrate this in the case of the Mathieu groups. Also, characterizations of multiple transitivity are given in terms of the cycle structure of the elements of the group in question.

## 5.1 The Pólya Theory of Enumeration

This theory is concerned with the set

$$\mathbf{m}^{\mathbf{n}} := \{\varphi \mid \varphi : \mathbf{n} \to \mathbf{m}\}$$

and certain equivalence relations defined on $\mathbf{m}^{\mathbf{n}}$ with the aid of two given permutation groups $G \leqslant S_m$ and $H \leqslant S_n$. The problem is to get as much information as possible on the equivalence classes.

Before we formulate a typical example which is taken from graph theory, where the main applications of enumeration theory lie, we introduce three equivalence relations $\sim_i$, $i=1,2,3$, on $\mathbf{m}^{\mathbf{n}}$ which cover most of the problems in this area:

5.1.1

$$\text{(i)}\ \varphi \sim_1 \psi \quad :\Leftrightarrow \quad \exists \pi \in H \quad (\varphi = \psi \circ \pi^{-1}),$$

$$\text{(ii)}\ \varphi \sim_2 \psi \quad :\Leftrightarrow \quad \exists \pi \in H, g \in G \quad (\varphi = g \circ \psi \circ \pi^{-1}),$$

$$\text{(iii)}\ \varphi \sim_3 \psi \quad :\Leftrightarrow \quad \exists (f;\pi) \in G \operatorname{wr} H \quad \forall i \in \mathbf{n} \quad \big(\varphi(i) = f(i)\big(\psi(\pi^{-1}(i))\big)\big).$$

The definitions 4.1.26 up to 4.1.32 of the groups $E^H$, $G^H$, and $[G]^H$ show immediately that the classes of $\sim_1$, $\sim_2$, and $\sim_3$ are the orbits of these permutation groups in their action on $\mathbf{m}^{\mathbf{n}}$.

First of all we want to know the total number of orbits; later on we shall ask for details like the number of orbits with specific properties or for a complete system of representatives and so on.

Before we try to answer such questions we provide a typical example. We compute the number of nonisomorphic graphs with $p$ points. We consider a graph with $p$ points as a mapping

$$\varphi \in \mathbf{2}^{\mathbf{p}^{[2]}},$$

where

$$\mathbf{p}^{[2]} := \left\{1, \ldots, \binom{p}{2}\right\}$$

is taken to be the set of the $\binom{p}{2}$ pairs of points of the graph, and $\varphi(i) := 1$

if and only if the $i$th pair of points is disconnected, while $\varphi(i):=2$ if and only if the $i$th pair of points is connected. For $H$ we take the group

$$\mathrm{S}_p^{[2]}$$

which is induced by $\mathrm{S}_p$ (acting on the $p$ points) on the set $\mathbf{p}^{[2]}$ of the $\binom{p}{2}$ pairs of points.

It is clear that an orbit of the group

$$E^{(\mathrm{S}_p^{[2]})}$$

consists of a complete class of isomorphic graphs $\varphi$. Hence the number of nonisomorphic graphs with $p$ points is equal to the number of orbits of this group, which can be evaluated by an application of the Cauchy-Frobenius lemma (which is usually called "Burnside's lemma"; we follow Neumann [1979]) as follows: If $\pi \in \mathrm{S}_p$ induces $\pi^{[2]}$ on $\mathbf{p}^{[2]}$, then $\pi^{[2]}$ leaves exactly

$$2^{c(\pi^{[2]})}$$

mappings $\varphi$ fixed, for it is easy to see that $\varphi$ is left fixed if and only if $\varphi$ is constant on each cyclic factor of $\pi^{[2]}$. Hence the Cauchy-Frobenius lemma yields the following answer to our problem:

**5.1.2.** *The number of nonisomorphic graphs with p points is equal to*

$$\frac{1}{p!} \sum_{\pi \in \mathrm{S}_p} 2^{c(\pi^{[2]})}.$$

If $p:=4$, a complete system of representatives of the isomorphism classes is

5.1.3

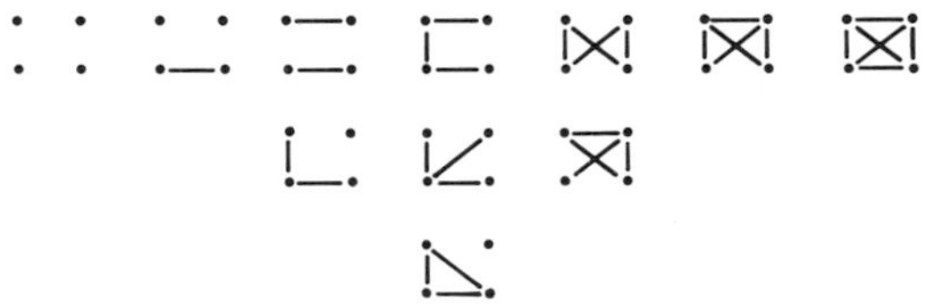

This is a typical problem where the acting permutation group is of the form $E^H$. We call it a *Pólya-type* problem. An obvious generalization of 5.1.2 is

**5.1.4** Pólya's theorem, constant form. *The number of orbits of $E^H$ on $\mathbf{m}^{\mathbf{n}}$ is equal to*

$$\frac{1}{|H|} \sum_{\pi \in H} m^{c(\pi)}.$$

Problems where the group is of the form $G^H$ are called *de Bruijn–type* problems. A typical problem of this kind arises from the foregoing one by asking for the number of classes of graphs with $p$ points which are isomorphic up to complementation. (The graph $\varphi$ is said to be the one complementary to $\psi$ if $\varphi(i)=1$ if and only if $\psi(i)=2$.)

Representatives of these classes for $p:=4$ are (cf. 5.1.3)

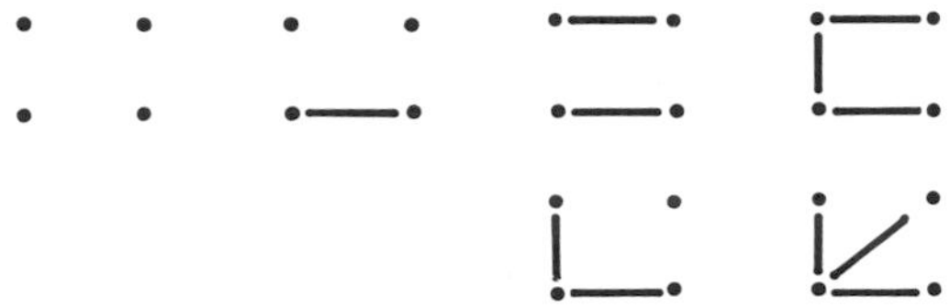

In order to solve this problem we need only take $H:=\mathrm{S}_p^{[2]}$ as before and put $G:=\mathrm{S}_2$. It is obvious that the orbits of

$$\mathrm{S}_2^{(\mathrm{S}_p^{[2]})}$$

are just the desired equivalence classes of graphs.

Since $G^H=\theta[\operatorname{diag} G^*\cdot H']$, the representation of $G^H$ on $\mathbf{m}^{\mathbf{n}}$ is just the restriction

5.1.5 $$\left(\overset{n}{\#} N\right)^{\sim} \downarrow \operatorname{diag} G^*\cdot H',$$

if $N$ denotes the natural representation of $G$ which has the character

5.1.6 $$\forall g\in G \quad \chi^N(g)=a_1(g).$$

Thus our character formulae 4.3.9 and 4.3.10(vi) together with the Cauchy-Frobenius lemma yield the following two theorems:

**5.1.7** POWER-GROUP ENUMERATION THEOREM, CONSTANT FORM. *The number of orbits of the power group $G^H$ on $\mathbf{m}^{\mathbf{n}}$ is equal to*

$$\frac{1}{|G||H|}\sum_{(g,\pi)\in G\times H}\prod_{k=1}^{n} a_1(g^k)^{a_k(\pi)}.$$

**5.1.8** EXPONENTIATION-GROUP ENUMERATION THEOREM, CONSTANT FORM. *The number of orbits of the exponentiation group $[G]^H$ on $\mathbf{m}^{\mathbf{n}}$ is equal to*

$$\frac{1}{|G|^n|H|}\sum_{(f;\pi)\in G\mathrm{wr}H}\prod_{\nu=1}^{c(\pi)} a_1(g_\nu(f;\pi)).$$

An immediate corollary of 5.1.7 is the following solution of the graph enumeration problem introduced above:

**5.1.9.** *The number of classes of graphs on p points which are isomorphic up to complementation is equal to*

$$\frac{1}{2\cdot p!}\left(\sum_{\substack{\pi\in S_p\\ 2\nmid i\ \Rightarrow\ a_i(\pi^{[2]})=0}} 2^{c(\pi^{[2]})}+\sum_{\pi\in S_p} 2^{c(\pi^{[2]})}\right).$$

For

$$\prod_k a_1(g^k)^{a_k(\pi^{[2]})}=\begin{cases}2^{c(\pi^{[2]})} & \text{if } g=(12) \text{ and each } a_{2i+1}(\pi^{[2]})=0, \text{ or } g=1,\\ 0 & \text{otherwise.}\end{cases}$$

A combination of results on de Bruijn–type problems with results on Pólya-type problems yields further theorems of interest, as the following example shows. It is easy to see that twice the number of classes mentioned in 5.1.9 minus the number of classes in 5.1.2 is equal to the number of isomorphism classes which remain fixed under complementation. Thus

**5.1.10.** *The number of self-complementary isomorphism classes of graphs on p points is equal to*

$$\frac{1}{p!}\sum_{\substack{\pi\in S_p\\ 2\nmid k\ \Rightarrow\ a_k(\pi^{[2]})=0}} 2^{c(\pi^{[2]})}.$$

Enumeration problems where the exponentiation group $[G]^H$ is involved are very rare. Furthermore, each such problem can be reduced to a Pólya-type problem, since it is not $G$ which really matters but the orbits of $G$ (see the exercises). However, problems often occur which are Pólya-type but where we have an exponentiation group in the exponent: $H=[P]^Q$. A typical example is the enumeration of classes of boolean functions.

Up to now we have used only the Cauchy-Frobenius lemma together with our results on characters of wreath products. But there is of course much more theory involved in enumeration under group action.

Having evaluated the total number of orbits, one asks next for the number of orbits of a given content, say for the number of isomorphism classes of graphs with $p$ points and $e$ edges. The graph $\varphi\in\mathbf{2}^{\mathbf{p}^{[2]}}$ has $e$ edges if and only if $\varphi$ takes the value 2 $e$ times, i.e.

$$e=|\varphi^{-1}[\{2\}]|.$$

Hence more generally we say the $\varphi\in\mathbf{m}^{\mathbf{n}}$ is of *content* $\lambda:=(\lambda_1,\ldots,\lambda_m)$ if and only if for each $i \in \mathbf{m}$

$$\lambda_i = |\varphi^{-1}[\{i\}]|.$$

Asking for the number of classes of functions $\varphi$ of content $\lambda$ is therefore the same as asking for the generating function

$$\gamma \in \mathbb{Q}[z_1, \ldots, z_m],$$

where the monomial

$$z^\lambda := z_1^{\lambda_1} \cdots z_m^{\lambda_m}$$

has as its coefficient the number of classes of functions of content $\lambda$. For example, by 5.1.3 this generating function for the enumeration of graphs of 4 points by content is just the polynomial

5.1.11 $$z_1^6 + z_1^5 z_2 + 2z_1^4 z_2^2 + 3z_1^3 z_2^3 + 2z_1^2 z_2^4 + z_1 z_2^5 + z_2^6.$$

We would like to describe a systematic approach to this enumeration of classes by content.

One starts with a *weight function*

5.1.12 $$w : \mathbf{m} \to \mathbb{Q}[z_1, \ldots, z_m],$$

which yields the mapping

5.1.13 $$w^* : \mathbf{m}^\mathbf{n} \to \mathbb{Q}[z_1, \ldots, z_m] : \varphi \mapsto \prod_{i \in \mathbf{n}} w(\varphi(i)).$$

We call the functions $w^*$ which arise in this way *multiplicative weights* on $\mathbf{m}^\mathbf{n}$. When $w$ is constant on the orbits of $G$ the resulting multiplicative weight $w^*$ is constant on each orbit of $[G]^H$, i.e. $w^*$ is $[G]^H$*-compatible*, and hence also $G^H$- and $E^H$-compatible. (In fact $w^*$ is $E^H$-compatible for each weight function $w$; we need not assume that $w$ is $G$-compatible in the case of a Pólya-type problem.)

Using such a compatible weight $w^*$ on $\mathbf{m}^\mathbf{n}$, we can define a weight $W$ on the set $\Omega$ of orbits $\omega$ by putting

$$W : \Omega \to \mathbb{Q}[z_1, \ldots, z_m] : \omega \mapsto w^*(\varphi) \qquad \text{for any} \quad \varphi \in \omega.$$

It is our aim to choose $w$ in such a way that the polynomial

5.1.14 $$\sum_{\omega \in \Omega} W(\omega)$$

is the generating function for the enumeration problem in question. Putting $w=1$ for example, 5.1.14 yields just the number of orbits, while the generating function 5.1.11 for our graph-theoretic problem introduced above is obtained by putting $w(1):=z_1$, $w(2):=z_2$.

The orbits of functions are sometimes called *patterns*, while the generating function 5.1.14 is called the *pattern inventory*. We need a systematic approach to the evaluation of the pattern inventory.

We remember that Pólya's theorem, the power-group enumeration theorem, and the exponentiation-group enumeration theorem were obtained by an application of our results on characters of wreath products and the Cauchy-Frobenius lemma to the exponentiation group $[G]^H=\theta[G\operatorname{wr}H]$; we would like to show that the same character together with a variation of the Cauchy-Frobenius lemma yields the corresponding theorems in weighted form.

The variation of the Cauchy-Frobenius lemma is:

**5.1.15** CAUCHY-FROBENIUS LEMMA, WEIGHTED FORM. *Let $\delta: A\to S_M$ denote a permutation representation of the finite group A on the finite set M. Let furthermore $w^*: M\to\mathbb{Q}[x_1,\dots,x_r]$ denote a mapping which is constant on the orbits $\omega$ of $\delta[A]$, $W(\omega)$ denoting this value. Then*

$$\sum_{\omega} W(\omega)=\frac{1}{|\delta[A]|}\sum_{\pi\in\delta[A]}\sum_{\substack{m\in M\\ \pi(m)=m}} w^*(m)$$

$$=\frac{1}{|A|}\sum_{a\in A}\sum_{\substack{m\in M\\ \delta(a)(m)=m}} w^*(m).$$

The proof of this lemma is essentially the same as the usual proof of the Cauchy-Frobenius lemma in its constant form via a double count of the entries of a matrix $(b_{ma})$, where

$$b_{ma}:=\begin{cases} w^*(m) & \text{if } \delta(a)(m)=m,\\ 0 & \text{otherwise.}\end{cases}$$

We leave the proof to the reader.

From the equations of 5.1.15 we see that in order to apply this weighted form we have to determine the fixed points of $\delta(a)$, $a\in A$, and evaluate their weights. The first part of this has usually already been done if we have evaluated the permutation character of $\delta$. We consider the example

$$\theta: G\operatorname{wr}H\to S_{\mathbf{m}^{\mathbf{n}}}.$$

$\varphi \in \mathbf{m}^{\mathbf{n}}$ is left fixed by $\theta((f;\pi))$, $(f;\pi)\in G \operatorname{wr} H$, if and only if the following is true (see 4.1.28):

**5.1.16.**

(i) *For each* $\nu$, $\varphi(j_\nu)$ *is a fixed point of* $g_\nu(f;\pi)$, *and*
(ii) *the further values of* $\varphi$ *on this cycle of* $\pi$ *are obtained from* $\varphi(j_\nu)$ *as follows*:

$$\varphi(j_\nu)=f(j_\nu)\big(\varphi\big(\pi^{-1}(j_\nu)\big)\big)=f(j_\nu)f\big(\pi^{-1}(j_\nu)\big)\big(\varphi\big(\pi^{-2}(j_\nu)\big)\big)=\cdots.$$

In order to evaluate the weight of such a $\varphi$ we recall that $w$ is assumed to be constant on the orbits of $G$, so that

$$w(\varphi(j_\nu))=w\big(\varphi\big(\pi^{-1}(j_\nu)\big)\big)=\cdots=w\big(\varphi\big(\pi^{-l_\nu+1}(j_\nu)\big)\big).$$

Hence such a $\varphi$, which is left fixed by $\theta((f;\pi))$, has the weight

$$w^*(\varphi)=\prod_\nu w(\varphi(j_\nu))^{l_\nu}.$$

This gives (as we can freely choose the values $\varphi(j_\nu)$; cf. 5.1.16(i))

5.1.17
$$\sum_{\substack{\varphi\\ \theta((f;\pi))(\varphi)=\varphi}} w^*(f)=\prod_{\nu=1}^{c(\pi)}\left(\sum_{\substack{j\\ g_\nu(f;\pi)(j)=j}} w(j)^{l_\nu}\right).$$

This together with the weighted form of the Cauchy-Frobenius lemma yields:

**5.1.18** Exponentiation-group enumeration theorem, weighted form. *If $w$ is a $G$-compatible weight function on* $\mathbf{m}$, *then the pattern inventory of the exponentiation group* $[G]^H$ *acting on* $\mathbf{m}^{\mathbf{n}}$ *is equal to*

$$\frac{1}{|G|^n|H|}\sum_{(f;\pi)\in G\operatorname{wr}H}\prod_{\nu=1}^{c(\pi)}\sum_{\substack{j\\ g_\nu(f;\pi)(j)=j}} w(j)^{l_\nu}.$$

Restricting our attention to the subgroup $\operatorname{diag} G^*\cdot H'$ yields the corresponding theorem for the action of $G^H$:

**5.1.19** Power-group enumeration theorem, weighted form. *If $w$ is a $G$-compatible weight function on* $\mathbf{m}$, *then the pattern inventory of the power* *group $G^H$ acting on* $\mathbf{m}^{\mathbf{n}}$ *is equal to*

$$\frac{1}{|G||H|}\sum_{(g,\pi)\in G\times H}\prod_{k=1}^{n}\left[\sum_{\substack{j\\ g^k(j)=j}} w(j)^k\right]^{a_k(\pi)}.$$

A further restriction down to $E^H$ yields:

**5.1.20** PÓLYA'S THEOREM, WEIGHTED FORM. *For any weight function w on* $\mathbf{m}$, *we obtain as the pattern inventory of $E^H$ acting on* $\mathbf{m}^{\mathbf{n}}$:

$$\frac{1}{|H|}\sum_{\pi\in H}\prod_{k=1}^{n}\left(\sum_{j=1}^{m} w(j)^k\right)^{a_k(\pi)}.$$

The next definition enables us to systematize our approach. If $H\leqslant S_n$, then the following element of the polynomial ring $\mathbb{Q}[x_1,\dots,x_n]$ is called the *cycle index* of $H$:

5.1.21 $$\mathrm{Cyc}(H):=\frac{1}{|H|}\sum_{\pi\in H}\prod_{k=1}^{n} x_k^{a_k(\pi)}.$$

We shall sometimes display the indeterminates by writing

$$\mathrm{Cyc}(H;x_1,\dots,x_n)$$

instead of $\mathrm{Cyc}(H)$.

Using this last notation we define what we understand by *Pólya insertion* of a polynomial $p=p(z_1,\dots,z_m)\in\mathbb{Q}[z_1,\dots,z_m]$ in $\mathrm{Cyc}(H)$:

5.1.22 $$\mathrm{Cyc}(H|p):=\mathrm{Cyc}(H;p(z_1,\dots,z_m),\dots,p(z_1^n,\dots,z_m^n))$$
$$:=\frac{1}{|H|}\sum_{\pi\in H}\prod_{k=1}^{n} p(z_1^k,\dots,z_m^k)^{a_k(\pi)}.$$

This yields a simple formulation for 5.1.20.

**5.1.23.** *The pattern inventory of $E^H$ is* $\mathrm{Cyc}(H|\sum_{j=1}^m w(j))$.

For our graph-theoretic enumeration problem we get:

**5.1.24.**

(i) *The number of isomorphism classes of graphs with p points is equal to*

$$\mathrm{Cyc}\left(S_p^{[2]}|2\right)=\mathrm{Cyc}\left(S_p^{[2]};2,2,\dots\right).$$

(ii) *The number of self-complementary isomorphism classes of graphs with p points is equal to*

$$\mathrm{Cyc}\left(\mathrm{S}_p^{[2]};0,2,0,2,\ldots\right).$$

We list the cycle indices of some particular groups:

**5.1.25.**

(i) $\mathrm{Cyc}(\{1_{\mathrm{S}_n}\})=x_1^n$,

(ii) $\mathrm{Cyc}(C_n:=\langle(1,\ldots,n)\rangle)=\frac{1}{n}\Sigma_{i|n}\Phi(i)x_i^{n/i}$, *where* $\Phi$ *denotes the Euler function, i.e.* $\Phi(i):=|\{k\in\mathbb{N}\,|\,k\leqslant i$ *and* $(k,i)=1\}|$,

(iii) $\mathrm{Cyc}(\mathrm{S}_n)=\Sigma_a\Pi_{k=1}^n\frac{1}{a_k!}\left(\frac{x_k}{k}\right)^{a_k}$,

(iv) $\mathrm{Cyc}(\mathrm{A}_n)=\Sigma_a(1+(-1)^{a_2+a_4+\cdots})\Pi_{k=1}^n\frac{1}{a_k!}\left(\frac{x_k}{k}\right)^{a_k}$,

(v) $\mathrm{Cyc}(D_n)=\frac{1}{2}\mathrm{Cyc}(C_n)+\begin{cases}\frac{1}{2}x_1x_2^{(n-1)/2} & \text{if } n \text{ is odd},\\ \frac{1}{4}\left(x_2^{n/2}+x_1^2x_2^{(n-2)/2}\right) & \text{if } n \text{ is even}.\end{cases}$

An important case is the weight function

$$w:\mathbf{m}\to\mathbb{Q}[z_1,\ldots,z_m]:i\mapsto z_i,$$

where we obtain the so-called *group reduction function*

5.1.26
$$\mathrm{Grf}(H):=\mathrm{Cyc}\left(H\,|\,\sum_i z_i\right)$$
$$=\frac{1}{|H|}\sum_{\pi\in H}\prod_{k=1}^{n}\left(z_1^k+\cdots+z_m^k\right)^{a_k(\pi)}.$$

Returning to representation theory, we define the *generalized cycle index* of $H$ with respect to an ordinary character $\chi^D$ of $H$ by

5.1.27
$$\mathrm{Cyc}(H,D):=\frac{1}{|H|}\sum_{\pi\in H}\chi^D(\pi^{-1})\prod_{k=1}^{n}x_k^{a_k(\pi)}.$$
$= \mathrm{ch}(\mathrm{ind}_H^{S_n}\chi^D)$

It is very easy to see that the following is true for $H\leqslant\mathrm{S}_n$:

**5.1.28**
$$\mathrm{Cyc}(H)=\mathrm{Cyc}(\mathrm{S}_n,IH\uparrow\mathrm{S}_n)$$
$$=\sum_{\alpha\vdash n}(IH\uparrow\mathrm{S}_n,[\alpha])\,\mathrm{Cyc}(\mathrm{S}_n,[\alpha]).$$

so that in particular

**5.1.29**
$$\mathrm{Grf}(H)=\sum_{\alpha\vdash n}(IH\uparrow\mathrm{S}_n,[\alpha])\,\mathrm{Grf}(\mathrm{S}_n,[\alpha]).$$

The polynomials

5.1.30 $$\{\alpha\}:=\mathrm{Grf}(S_n,[\alpha])=\frac{1}{n!}\sum_{\pi\in S_n}\zeta^{\alpha}(\pi^{-1})\prod_{k=1}^{n}(x_1^k+\cdots+x_n^k)^{a_k(\pi)}$$

are called *Schur-functions* or *S-functions* for short. They are very important. There is in fact a way of deriving most of the results of the ordinary representation theory of $S_n$ by considering these symmetric polynomials, i.e. a polynomial approach to the representation theory of $S_n$.

A few remarks concerning the evaluation of the cycle-index polynomial $\mathrm{Cyc}(H)$ are in order. It is known as soon as we know the cycle type for each conjugacy class of $H$, and the latter can be calculated using the permutation character $a_1(\cdot)$. Obviously for each $\pi\in H$ we have

**5.1.31** $$\forall k\in\mathbb{N}\quad\left(a_1(\pi^k)=\sum_{i|k}i\cdot a_i(\pi)\right),$$

so that by an application of Moebius inversion we obtain

**5.1.32** $$a_i(\pi)=\frac{1}{i}\sum_{k|i}\mu\left(\frac{i}{k}\right)a_1(\pi^k).$$

Thus $a_i(\pi)$ can be obtained from the permutation character of $H$ as soon as we know to which conjugacy class $\pi^k$ belongs, for each $k\in\mathbb{N}$.

Another useful remark is that we obtain the cycle index of a direct sum of permutation groups just by multiplication:

**5.1.33** $$\mathrm{Cyc}(H\oplus K)=\mathrm{Cyc}(H)\cdot\mathrm{Cyc}(K).$$

Furthermore if $H_1\leqslant S_{n_1}$, $H_2\leqslant S_{n_2}$, then they define a group $H_1\otimes H_2$ acting on $\mathbf{n}_1\times\mathbf{n}_2$ as follows. The underlying set of $H_1\otimes H_2$ is $H_1\times H_2$, and the action is defined by

$$(\pi_1,\pi_2)(i_1,i_2):=(\pi_1(i_1),\pi_2(i_2)).$$

Then we have, as it is not difficult to check,

**5.1.34** $$\mathrm{Cyc}(H_1\otimes H_2)=\frac{1}{|H_1||H_2|}\sum_{(\pi_1,\pi_2)}\prod_{i,k=1}^{n_1,n_2}x_{\mathrm{lcm}(i,k)}^{\gcd(i,k)a_i(\pi_1)a_k(\pi_2)}.$$

The cycle index of the *composition*

5.1.35 $$G[H]:=\psi[H\,\mathrm{wr}\,G]\leqslant S_{mn}$$

of $G \leqslant S_m$ and $H \leqslant S_n$ (cf. 4.1.18) follows directly from 4.2.19. It is the following "composition" of the cycle indices of $G$ and $H$:

**5.1.36** $$\operatorname{Cyc}(G)[\operatorname{Cyc}(H)] := \frac{1}{|G|} \sum_{\pi \in G} \prod_{k=1}^{m} \left( \frac{1}{|H|} \sum_{\rho \in H} \prod_{i=1}^{n} x_{ik}^{a_i(\rho)} \right)^{a_k(\pi)}.$$

Our description of the basic ideas of combinatorial enumeration theory would be incomplete without mentioning J. H. Redfield, who published in 1927 (ten years before Pólya's famous paper appeared) an article which was overlooked for many years, although it anticipated the most important results of the whole theory. The trouble with Redfield's results was that they were formulated in terms of his "cup" and "cap" product of polynomials, which at first glance seem to be quite obscure. (But with the aid of representation theory they can be elucidated without much effort.) This may also be the reason why Redfield's results play their own role and usually they either need special consideration or are left out of the discussion. But since they are very useful, we would like to include them by giving the outline of an approach which unifies both Redfield's results and the preceding way of describing Pólya's and de Bruijn's theorems as well as the exponentiation-group case. Besides this, the more general problem which will be formulated and solved may turn out to be helpful for further enumerative purposes.

Redfield's paper deals with the so-called *superposition* of isomorphism classes of graphs. We don't need a precise definition of this kind of problem, since an example gives quite a good idea of what is meant.

The different superpositions of the two isomorphism classes

5.1.37 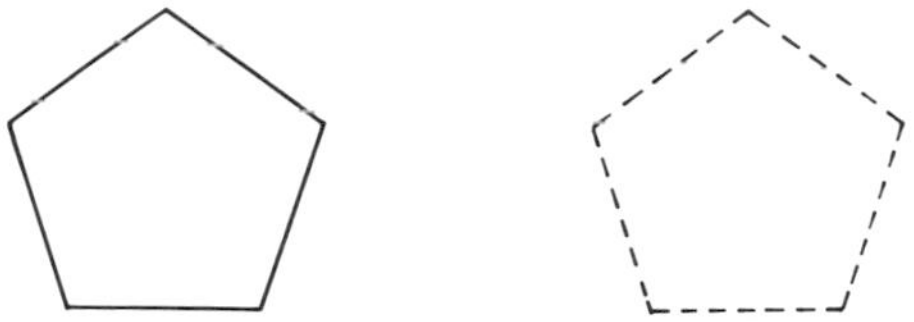

are

5.1.38 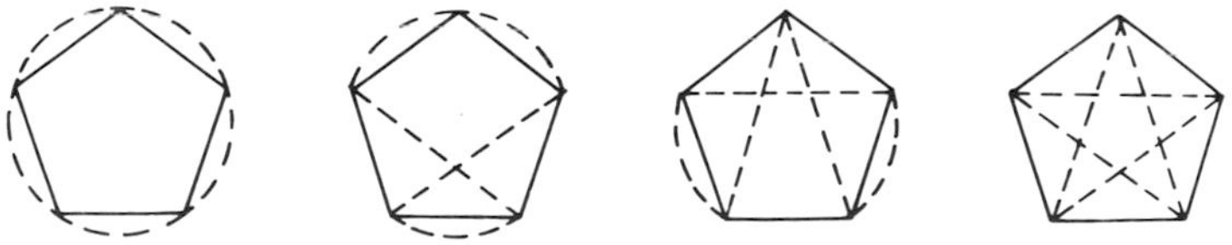

We would like to formulate an enumeration problem of which the question of the number of superpositions of isomorphism classes of graphs can be shown to be a special case. For given $m, n \in \mathbb{N}$ and an $s$ where $1 \leqslant s \leqslant m$, we consider the set

5.1.39 $$M_{\langle s \rangle}$$

of all the $n\times s$ matrices

$$A=(a_{ik}),$$

the rows of which are injective $s$-tuples over $\mathbf{m}$, i.e. each row consists of $s$ different elements of $\mathbf{m}$. We notice that

**5.1.40** $$|M_{\langle s\rangle}|=\left(\binom{m}{s}s!\right)^n.$$

In order to define a suitable group which will act on $M_{\langle s\rangle}$, we introduce the following generalization of the wreath product:

**5.1.41** Definition. If $H\leqslant S_n$ has the orbits $\omega_1,\ldots,\omega_r\subseteq\mathbf{n}$, and if correspondingly $G_1,\ldots,G_r$ are subgroups of $S_m$, then we call the following group *the generalized wreath product* $(G_1,\ldots,G_r)\operatorname{wr}H$ of $G_1,\ldots,G_r$ with $H$. The underlying set (which of course heavily depends on the given numbering of the orbits $\omega_i$) is defined to be

$$\left\{(f;\pi)\,|\,f:\mathbf{n}\to\bigcup_k G_k,\ \forall i\in\mathbf{n}\ (i\in\omega_j\Rightarrow f(i)\in G_j),\ \pi\in H\right\}.$$

The multiplication law remains as it was:

$$(f;\pi)(f';\pi'):=(ff'_\pi;\pi\pi').$$

A permutation representation $\theta_{\langle s\rangle}$ of this group on $M_{\langle s\rangle}$ is defined as follows:

5.1.42 $$\theta_{\langle s\rangle}:(G_1,\ldots,G_r)\operatorname{wr}H\to S_{M_{\langle s\rangle}}:(f;\pi)\mapsto\begin{pmatrix}(a_{ik})\\(f(i)(a_{\pi^{-1}(i),k}))\end{pmatrix}.$$

It is clear that for $s:=1$ and $G_1:=\cdots:=G_r:=G$ we obtain the representation $\theta$ of $G\operatorname{wr}H$ which was mentioned above.

Besides $\theta_{\langle s\rangle}$, we now consider a permutation representation $\eta$ of $S_s$ on $M_{\langle s\rangle}$:

5.1.43 $$\eta:S_s\to S_{M_{\langle s\rangle}}:\sigma\mapsto\begin{pmatrix}(a_{ik})\\(a_{i,\sigma^{-1}(k)})\end{pmatrix}.$$

Two elements in the same orbit of $\eta[S_s]$ are called *column-equivalent matrices*, for obvious reasons.

The crucial fact is that for each $A\in M_{\langle s\rangle}$ and every $(f;\pi)\in(G_1,\ldots,G_r)\operatorname{wr}H$, $\sigma\in S_s$, we have

**5.1.44** $$\theta_{\langle s\rangle}((f;\pi))\eta(\sigma)(A)=\eta(\sigma)\theta_{\langle s\rangle}((f;\pi))(A).$$

That this holds is obvious, and it will allow us later on to apply the following result:

**5.1.45** De Bruijn's lemma. *If $\mu: P \to S_N$ and $\nu: Q \to S_N$ denote permutation representations of finite groups $P$ and $Q$ on a finite set $N$ such that $\mu[P]$ is contained in the normalizer of $\nu[Q]$, i.e.*

$$\forall p \in P, q \in Q, i \in N \quad \exists q' \in Q \quad \mu(p)\nu(q)(i) = \nu(q')\mu(p)(i),$$

*then we can define a permutation representation $\bar{\mu}$ of $P$ on the set $\bar{N}$ of orbits $\bar{i}$, $i \in N$, of $\nu[Q]$ on $N$ as follows:*

$$\bar{\mu}: P \to S_{\bar{N}}: p \mapsto \left( \frac{\bar{i}}{\overline{\mu(p)(i)}} \right).$$

*The character of this permutation representation satisfies*

$$\chi^{\bar{\mu}}(p) = a_1(\bar{\mu}(p)) = \frac{1}{|Q|} \sum_{q \in Q} \left| \{ i \mid i \in N \text{ and } \mu(p)(i) = \nu(q)(i) \} \right|.$$
$$= \frac{1}{|Q|} \sum_q a_1\left(\mu(p^{-1})\nu(q)\right).$$

*Proof.* The check that $\bar{\mu}$ defines a permutation representation is very easy. To derive the character we notice that

$$a_1(\bar{\mu}(p)) = \sum_{\substack{\bar{i} \\ \overline{\mu(p)(i)} = \bar{i}}} 1$$
$$= \sum_{\substack{i \\ \overline{\mu(p)(i)} = \bar{i}}} \frac{1}{|\bar{i}|}.$$

And if $\nu[Q]_i$ denotes the stabilizer of $i$ in $\nu[Q]$, then we have

$$= \sum_{\substack{i \\ \overline{\mu(p)(i)} = \bar{i}}} \frac{|\nu[Q]_i|}{|\nu[Q]|}$$
$$= \frac{1}{|Q|} \sum_{\substack{i \\ \overline{\mu(p)(i)} = \bar{i}}} \sum_{\substack{q \\ \nu(q)(i) = i}} 1$$
$$= \frac{1}{|Q|} \sum_{\substack{i \\ \overline{\mu(p)(i)} = \bar{i}}} \sum_{\substack{q \\ \nu(q)(i) = \mu(p)(i)}} 1$$
$$= \frac{1}{|Q|} \sum_q \sum_{\substack{i \\ \mu(p)(i) = \nu(q)(i)}} 1. \qquad \blacksquare$$

5.1.44 shows that $\mu:=\theta_{\langle s\rangle}$ and $\nu:=\eta$ satisfy the conditions of de Bruijn's lemma, so that we can use it in the proof of

**5.1.46 THEOREM.** *$\theta_{\langle s\rangle}$ and $\eta$ yield a representation $\bar{\theta}_{\langle s\rangle}$ of $(G_1,\ldots,G_r)\operatorname{wr}H$ on the set $\overline{M}_{\langle s\rangle}$ of classes of column-equivalent matrices $A\in M_{\langle s\rangle}$. Its character satisfies*

$$a_1\big(\bar{\theta}_{\langle s\rangle}((f;\pi))\big)=\frac{1}{s!}\sum_{\sigma\in S_s}\prod_{\nu=1}^{c(\pi)}\prod_{k=1}^{s}\binom{a_k(g_\nu(f;\pi))}{a_k(\sigma^{l_\nu})}a_k(\sigma^{l_\nu})!k^{a_k(\sigma^{l(\nu)})}.$$

It remains to prove the stated identity for the character. By 5.1.45 we have to evaluate the number of matrices $A=(a_{ik})\in M_{\langle s\rangle}$, the coefficients $a_{ik}$ of which satisfy

$$(1)\qquad a_{i,\sigma^{-1}(k)}=f(i)a_{\pi^{-1}(i),k}.$$

In order to do this we discuss the degree of freedom we have in choosing such $a_{ik}$. If (in cycle notation)

$$\pi=\prod_{\nu=1}^{c(\pi)}\big(j_\nu\pi(j_\nu)\cdots\pi^{l_\nu-1}(j_\nu)\big),$$

then by iteration of (1) we get

$$(2)\qquad \forall t\in\mathbb{N}\quad a_{i,\sigma^{-t}(k)}=f(i)\cdots f\big(\pi^{-t+1}(i)\big)\big(a_{\pi^{-t}(i),k}\big),$$

so that in particular the following must hold:

$$(3)\qquad \forall\nu,k\quad a_{j_\nu,\sigma^{-l(\nu)}(k)}=g_\nu(f;\pi)\big((a_{j_\nu,k})\big).$$

Let us now denote by $s_{\nu\lambda}$, $1\leqslant\lambda\leqslant c(\sigma^{-l_\nu})$, the least points in the cyclic factors of $\sigma^{-l_\nu}$. Equations (2) and (3) show clearly that the coefficients

$$(4)\qquad a_{j_\nu,s_{\nu\lambda}}$$

determine the matrix $A$ completely. Thus we can freely choose at most these entries (4). But still we have to take (3) into account, which says that $a_{j_\nu,s_{\nu\lambda}}$ must occur in $g_\nu(f;\pi)$ and in $\sigma^{-l_\nu}$ in cycles of the same length (recall that the rows of $A$ are injective). Finally (2) shows that, for $\lambda\neq\mu$, $a_{j_\nu,s_{\nu\lambda}}$ and $a_{j_\nu,s_{\nu\mu}}$ must occur in different cyclic factors of $g_\nu(f;\pi)$.

Thus for a fixed $\nu$ the entries $a_{j_\nu,s_{\nu\lambda}}$ in the row with number $j_\nu$ are obtained as images $\psi(s_{\nu\lambda})$ of the $s_{\nu\lambda}$ under a mapping $\psi$ from $\mathbf{s}$ into $\mathbf{m}$ which must be injective and which has to map cyclic factors of $\sigma^{-l_\nu}$ onto cyclic factors of $g_\nu(f;\pi)$. Conversely each such injection defines a matrix which satisfies (1), and different injections yield different matrices $A$.

As the entries (4) determine $A$ uniquely, the number of such $A$ is equal to the product (over $\nu$) of the number of injections as they are described above, i.e. it is equal to

$$\prod_{\nu=1}^{c(\pi)} \prod_{k=1}^{s} \binom{a_k(g_\nu(f;\pi))}{a_k(\sigma^{-l_\nu})} a_k(\sigma^{-l_\nu})! k^{a_k(\sigma^{-l(\nu)})}$$

This yields the statement. ■

In order to derive from this the most important results of Redfield we need only to put $s := m$, in which case we get

**5.1.47** $$a_1\left(\bar{\theta}_{\langle m\rangle}((f;1))\right) = \frac{1}{m!} \sum_{\sigma \in S_m} \prod_{\nu=1}^{n} \prod_{k=1}^{m} \binom{a_k(f(\nu))}{a_k(\sigma)} a_k(\sigma)! k^{a_k(\sigma)}.$$

This can be simplified considerably, for a summand on the right-hand side of 5.1.47 is $\neq 0$ only if all the $f(\nu)$ have the same cycle type as has $\sigma$, in which case the summand in question is equal to

$$\left(\prod_k a_k(\sigma)! k^{a_k(\sigma)}\right)^n.$$

Applying this to 5.1.47, we obtain

**5.1.48**

$$a_1\left(\bar{\theta}_{\langle m\rangle}((f;1))\right) = \begin{cases} \left(\prod_k a_k! k^{a_k}\right)^{n-1} & \text{if } \forall \nu \ (a(f(\nu)) = (a_1, \ldots, a_m)) \\ 0 & \text{otherwise.} \end{cases}$$

This yields Redfield's method of calculating the number of orbits of $\bar{\theta}_{\langle m\rangle}[(G_1, \ldots, G_n) \operatorname{wr} \{1_{S_n}\}]$ (which is, by definition of superposition, the number of superpositions of $n$ isomorphism classes of graphs on $m$ points and with automorphism groups $G_1, \ldots, G_n$).

Redfield introduced the *cap operation* $\cap$ on the set $\mathbb{Q}[x_1, \ldots, x_m]$ as the linear extension of the following operation on $n \geq 2$ monomials:

5.1.49 $$x_1^{c_1} \cdots x_m^{c_m} \cap x_1^{d_1} \cdots x_m^{d_m} \cap \cdots \cap x_1^{e_1} \cdots x_m^{e_m}$$

$$:= \begin{cases} \left(\prod_k a_k! k^{a_k}\right)^{n-1} & \text{if } c_i = \cdots = e_i = a_i, \\ 0 & \text{otherwise.} \end{cases}$$

$p_\lambda \cap p_\mu = \langle p_\lambda, p_\mu \rangle$

$f \cap g \cap h = \langle f, g * h \rangle$

etc.

In terms of this notation an immediate consequence of 5.1.48 is

**5.1.50** REDFIELD'S ENUMERATION THEOREM. *The number of orbits of* $\bar{\theta}_{\langle m\rangle}[(G_1,\ldots,G_n)\operatorname{wr}\{1_{S_n}\}]$ *is equal to the cap product*

$$\bigcap_i \operatorname{cyc}(G_i).$$

For an example we take $G_1:=G_2:=D_5$, the dihedral group of order 10, which is the automorphism group of the following isomorphism class of graphs:

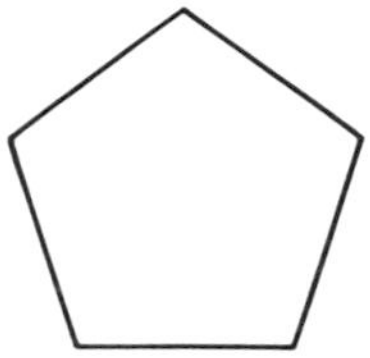

The cycle index is

$$\operatorname{Cyc}(D_5)=\tfrac{1}{10}\left(x_1^5+4x_5+5x_1x_2^2\right),$$

so that by 5.1.50 the number of superpositions of two isomorphism classes of graphs of this kind is

$$\begin{aligned}\operatorname{Cyc}(D_5)\cap\operatorname{Cyc}(D_5)&=\tfrac{1}{100}\left(x_1^5\cap x_1^5+16(x_5\cap x_5)+25\left(x_1x_2^2\cap x_1x_2^2\right)\right)\\&=\tfrac{1}{100}(5!+16\times 5+25\times 8)\\&=4,\end{aligned}$$

in accordance with 5.1.38.

Redfield's enumeration theorem provides the solution of various enumerative problems, e.g.

**5.1.51.** *The number of orbits of the power group* $G^H$, $G\leqslant S_n$, $H\leqslant S_n$ *on the set of the bijective mappings* $\varphi\in\mathbf{n}^{\mathbf{n}}$ *is equal to the cap product*

$$\operatorname{Cyc}(G)\cap\operatorname{Cyc}(H).$$

*Proof.* The set of bijections $\varphi\in\mathbf{n}^{\mathbf{n}}$ is in a natural way bijective to the set $\overline{M}_{\langle n\rangle}$ of classes of column equivalent $2\times n$ matrices with injective $n$-tuples over $\mathbf{n}$ as rows. This natural bijection maps an equivalence class of mappings onto an orbit of $\bar{\theta}_{\langle n\rangle}[(G,H)\operatorname{wr}\{1_{S_2}\}]=G^H$ (restricted to the set of bijections). Thus 5.1.50 yields the statement. ■

But bijections $\varphi\in\mathbf{n}^{\mathbf{n}}$ are just the permutations of $\mathbf{n}$, i.e. the elements $\sigma\in S_n$. Thus 5.1.51 can be rephrased in the following way:

**5.1.52.** $\mathrm{Cyc}(G)\cap\mathrm{Cyc}(H)$, $G\leqslant S_n\geqslant H$, *is equal to the number of double cosets* $G\pi H$ *in* $S_n$.

This yields the desired representation-theoretical interpretation of the cap product of cycle indices:

**5.1.53 Corollary.** *If* $G, H\leqslant S_n$, *then*

$$\mathrm{Cyc}(G)\cap\mathrm{Cyc}(H)=(IG\uparrow S_n, IH\uparrow S_n).$$

In order to deduce from Redfield's theorem the number of isomorphism classes of graphs with $p$ points and $e$ edges, we take

$$n=\binom{p}{2},\qquad G=S_e\oplus S_{n-e},\quad\text{and}\quad H=S_p^{[2]}$$

in 5.1.51. Given a bijection $\varphi\in\mathbf{n}^{\mathbf{n}}$, we associate with it a graph on $p$ points and $e$ edges by saying that the $i$th pair of points is connected iff $\varphi(i)\leqslant e$. Then, under the action of $G^H$, two bijections are in the same orbit of $G$ iff they are associated with the same graph, and they are in the same orbit of $G^H$ iff the corresponding graphs are isomorphic. Thus, 5.1.51 and 5.1.53 give:

**5.1.54.** *The number of isomorphism classes of graphs with p points and e edges is equal to*

$$\mathrm{Cyc}\left(S_p^{[2]}\right)\cap\mathrm{Cyc}\left(S_e\oplus S_{\binom{p}{2}}\right)=\left(IS_p^{[2]}\uparrow S_{\binom{p}{2}}, I\left(S_e\oplus S_{\binom{p}{2}-e}\right)\uparrow S_{\binom{p}{2}}\right).$$

More generally we obtain in the same way

**5.1.55 Theorem.** *The number of classes (under* $\sim_1$*) of mappings* $\varphi\in\mathbf{m}^{\mathbf{n}}$ *which consist of elements of content* $\lambda=(\lambda_1,\ldots,\lambda_m)\vDash n$ *is equal to* $\mathrm{Cyc}(H)\cap\mathrm{Cyc}(S_\lambda)=(IH\uparrow S_n, IS_\lambda\uparrow S_n)$.

This, together with our previous knowledge of the representations of $S_n$ induced from Young subgroups, yields various results on the monotonicity of such numbers, e.g.

**5.1.56 Corollary.** *If* $\mu\trianglelefteq\lambda$, *then the number of classes (under* $\sim_1$*) of mappings of content* $\lambda$ *is less than or equal to the number of classes of content* $\mu$.

Thus for example the number of isomorphism classes of graphs on $p$ points and with $e$ edges is nondecreasing for

$$0 \leqslant e \leqslant \frac{1}{2}\binom{p}{2}$$

as $e$ increases.

Besides the cap operation $\cap$ Redfield introduced a *cup operation* $\cup$ as the linear extension of the following operation on $\geqslant 2$ monomials:

5.1.57

$$x_1^{c_1}\cdots x_m^{c_m} \cup \cdots \cup x_1^{d_1}\cdots x_m^{d_m} := \left[x_1^{c_1}\cdots x_m^{c_m} \cap \cdots \cap x_1^{d_1}\cdots x_m^{d_m}\right] x_1^{c_1}\cdots x_m^{c_m}.$$

Obviously the following holds for $G, H \leqslant S_n$:

**5.1.58.** $\mathrm{Cyc}(G) \cap \mathrm{Cyc}(H)$ *is equal to the sums of the coefficients in* $\mathrm{Cyc}(G) \cup \mathrm{Cyc}(H)$.

This indicates that an examination of the cup product of cycle indices might turn out to be useful. In fact one can write each cup product of cycle indices as a sum of cycle indices of stabilizers. This will be proved with the aid of the following lemma. (There is also an easy representation-theoretical proof of it, and the cup product has a natural representation-theoretical interpretation as well; this is left to the exercises.)

**5.1.59** REDFIELD'S LEMMA. *Let* $\delta: G \to S_M$ *denote a permutation representation of a finite group* $G$ *on a finite set* $M$; $\chi: G \to \mathbb{Q}[x_1, \ldots, x_n]$, $n \in \mathbb{N}$, *a class function*; $\omega_1, \ldots, \omega_r$ *the orbits of* $\delta[G]$ *on* $M$; $m_i \in \omega_i$; *and*

$$G_i := \{g \mid g \in G \text{ and } \delta(g)(m_i) = m_i\}.$$

*Then we have*

$$\frac{1}{|G|}\sum_{g \in G} \chi(g) a_1(\delta(g)) = \sum_{i=1}^{r} \frac{1}{|G_i|} \sum_{g \in G_i} \chi(g).$$

*Proof.* We consider the $|M| \times |G|$ matrix $(a_{mg})$, where

$$a_{mg} := \begin{cases} \chi(g) & \text{if } \delta(g)(m) = m, \\ 0 & \text{otherwise.} \end{cases}$$

The sum of all its entries is equal to

(1) $$\sum_{g} a_1(\delta(g))\chi(g),$$

expressed as the sum of column sums. On the other hand the sum of its row sums is

$$(2) \qquad \sum_{m} \sum_{\substack{g \\ \delta(g)(m)=m}} \chi(g) = \sum_{i} |\omega_i| \sum_{g \in G_i} \chi(g)$$

(the last equation holds because elements in the same orbit have conjugate stabilizers and since $\chi$ is a class function). Comparing (1) and (2), we obtain the statement. ■

**5.1.60** Redfield's decomposition theorem. *If* $G_1, \ldots, G_n \leqslant S_m$, *and* $\omega_1, \ldots, \omega_r$ *denote the orbits of* $\bar{\theta}_{\langle m\rangle}[(G_1, \ldots, G_n)\operatorname{wr}\{1_{S_n}\}]$, $A_i \in \omega_i$, *while*

$$P_i := \left\{ (f;1) | \bar{\theta}_{\langle m\rangle}((f;1))(A_i) = A_i \right\},$$

*then*

$$\operatorname{Cyc}(G_1) \cup \cdots \cup \operatorname{Cyc}(G_n) = \sum_{i=1}^{r} \operatorname{Cyc}(P_i).$$

*Proof.* We apply 5.1.59 to $\bar{\theta}_{\langle m\rangle}$ and the class function $\chi$: $(G_1, \ldots, G_n)\operatorname{wr}\{1_{S_n}\} \simeq G_1 \times \cdots \times G_n \to \mathbb{Q}_m[x_1, \ldots, x_m]$, defined by

$$\chi(\pi_1, \ldots, \pi_n) := \begin{cases} \prod_k x_k^{a_k} & \text{if } \ a(\pi_i) = (a_1, \ldots, a_m), \ \ 1 \leqslant i \leqslant n, \\ 0 & \text{otherwise.} \end{cases}$$

Since, by 5.1.48,

$$a_1\left(\bar{\theta}_{\langle m\rangle}((\pi_1, \ldots, \pi_n; 1))\right) = \begin{cases} \left( \prod_k k^{a_k} a_k! \right)^{n-1} & \text{if } \ a(\pi_i) = (a_1, \ldots, a_m), \\ 0 & \text{otherwise,} \end{cases}$$

the statement follows from Redfield's lemma. ■

The representation-theoretical interpretation of this result is clear from the fact that a transitive permutation representation is equivalent to the representation induced by the identity representation of a stabilizer.

This may suffice concerning enumeration and enumeration by content of classes of mappings $\varphi \in \mathbf{m}^{\mathbf{n}}$ with respect to symmetry groups $E^H$, $G^H$, $[G]^H$. But some remarks about the construction of systems of representatives of these classes are well in order, since this is the most difficult part of the theory and not very much is known about this kind of problem. It can be solved for Pólya-type problems, i.e. in the case when $E^H$ is the symmetry group, and therefore also for the case $[G]^H$, since such problems can be transformed into Pólya-type problems.

In order to show how a complete system of representatives of the classes of mappings $\varphi \in \mathbf{m}^{\mathbf{n}}$ under $\sim_1$ (i.e. of the orbits of $E^H$) can be constructed, we remember 5.1.53, from which we read off that the number of such classes which consist of elements of content $\lambda=(\lambda_1,\ldots,\lambda_m)\models n$ is equal to the number of double cosets

5.1.61 $$(S_{\lambda_1}\oplus\cdots\oplus S_{\lambda_m})\pi H.$$

Therefore it is not surprising that we can get from a system of representatives of such double cosets the desired system of representatives of classes of mappings $\varphi \in \mathbf{m}^{\mathbf{n}}$. How we can do this will be described next. We recall what it means that a $\varphi \in \mathbf{m}^{\mathbf{n}}$ is of content $\lambda$, namely that

5.1.62 $$\forall i \in \mathbf{m} \quad \lambda_i = |\varphi^{-1}[\{i\}]|.$$

Correspondingly we form a dissection $\mathbf{n}^\lambda$ of the set $\mathbf{n}$ into subsets $\mathbf{n}_i^\lambda$, say

5.1.63 $$\mathbf{n}_i^\lambda := \left\{ j \,\middle|\, \sum_{\nu=1}^{i-1} \lambda_\nu < j \leqslant \sum_{\nu=1}^{i} \lambda_\nu \right\}.$$

Then

5.1.64 $$\mathbf{n} = \dot{\bigcup}\ \mathbf{n}_i^\lambda \quad \text{and} \quad \forall i \in \mathbf{m}\ |\mathbf{n}_i^\lambda| = \lambda_i.$$

Then with $\pi \in \mathrm{S}_n$ we associate a $\varphi_\pi \in \mathbf{m}^{\mathbf{n}}$ by putting

5.1.65 $$\varphi_\pi(i) := j \qquad \text{if} \quad \pi(i) \in \mathbf{n}_j^\lambda.$$

It is clear that $\varphi_\pi$ runs through all the mappings of content $\lambda$, if $\pi$ runs through all the elements of $\mathrm{S}_n$.

Furthermore, the following holds:

**5.1.66** $$\varphi_\pi = \varphi_\tau \quad \Leftrightarrow \quad \exists \sigma \in S_{\mathbf{n}_1^\lambda} \oplus \cdots \oplus S_{\mathbf{n}_m^\lambda}\ (\tau = \sigma\pi).$$

This yields

**5.1.67.** *The inverse image of $\varphi_\pi$ under $\pi \mapsto \varphi_\pi$ is just the right coset*

$$(S_{\mathbf{n}_1^\lambda} \oplus \cdots \oplus S_{\mathbf{n}_m^\lambda})\pi \subseteq \mathrm{S}_n.$$

*An immediate consequence is*

**5.1.68** $$\varphi_\pi \sim_1 \varphi_\tau \quad \Leftrightarrow \quad \tau \in (S_{\mathbf{n}_1^\lambda} \oplus \cdots \oplus S_{\mathbf{n}_m^\lambda})\pi H.$$

We have therefore proved the following:

**5.1.69** THEOREM. *A complete system of representatives of the double cosets $S_\lambda \pi H \subseteq S_n$ is mapped under $\pi \mapsto \varphi_\pi$ (see 5.1.65) bijectively onto a complete system of representatives of the orbits of mappings $\varphi \in \mathbf{m}^{\mathbf{n}}$ of content $\lambda$ under $E^H$.*

This describes in particular a redundancy-free algorithm for the construction of such representatives. But since the evaluation of representatives of double cosets is very cumbersome, one does not get very far. The graphs can be evaluated for up to 8 points (there are 12346 classes of them) using a big computer.

To conclude with an example, we consider $\lambda := (3,2)$ and $H := D_5 = \langle (12345), (13)(45) \rangle$. We have to evaluate a system of representatives of the double cosets

$$\left(S_{\{1,2,3\}} \oplus S_{\{4,5\}}\right) \pi D_5 .$$

Each such double coset is a union of right cosets $S_\lambda \rho$, and we remember that these can be represented by tabloids, in our case by the tabloids

$$\begin{array}{l}\hline 1\ 2\ 3\\ \hline 4\ 5\\ \hline\end{array},\ \begin{array}{l}\hline 1\ 2\ 4\\ \hline 3\ 5\\ \hline\end{array},\ \begin{array}{l}\hline 1\ 2\ 5\\ \hline 3\ 4\\ \hline\end{array},\ \begin{array}{l}\hline 1\ 3\ 4\\ \hline 2\ 5\\ \hline\end{array},\ \begin{array}{l}\hline 1\ 3\ 5\\ \hline 2\ 4\\ \hline\end{array},$$

$$\begin{array}{l}\hline 1\ 4\ 5\\ \hline 2\ 3\\ \hline\end{array},\ \begin{array}{l}\hline 2\ 3\ 4\\ \hline 1\ 5\\ \hline\end{array},\ \begin{array}{l}\hline 2\ 3\ 5\\ \hline 1\ 4\\ \hline\end{array},\ \begin{array}{l}\hline 2\ 4\ 5\\ \hline 1\ 3\\ \hline\end{array},\ \begin{array}{l}\hline 3\ 4\ 5\\ \hline 1\ 2\\ \hline\end{array}.$$

As the present $\lambda = (3,2)$ is 2-rowed, we need only take notice of the last row, i.e. we need only consider

$$\begin{array}{l}\hline 4\ 5\\ \hline\end{array},\ \begin{array}{l}\hline 3\ 5\\ \hline\end{array},\ \begin{array}{l}\hline 3\ 4\\ \hline\end{array},\ \begin{array}{l}\hline 2\ 5\\ \hline\end{array},\ \begin{array}{l}\hline 2\ 4\\ \hline\end{array},$$

$$\begin{array}{l}\hline 2\ 3\\ \hline\end{array},\ \begin{array}{l}\hline 1\ 5\\ \hline\end{array},\ \begin{array}{l}\hline 1\ 4\\ \hline\end{array},\ \begin{array}{l}\hline 1\ 3\\ \hline\end{array},\ \begin{array}{l}\hline 1\ 2\\ \hline\end{array}.$$

Applying $D_5 = \{((13)(45))^a (12345)^b \mid 0 \leqslant a \leqslant 1,\ 0 \leqslant b \leqslant 4\}$, we see that from $\underline{4\quad 5}$ we obtain

$$\begin{array}{l}\hline 4\ 5\\ \hline\end{array},\ \begin{array}{l}\hline 1\ 5\\ \hline\end{array},\ \begin{array}{l}\hline 1\ 2\\ \hline\end{array},\ \begin{array}{l}\hline 2\ 3\\ \hline\end{array},\ \begin{array}{l}\hline 3\ 4\\ \hline\end{array},$$

while from $\underline{3\quad 5}$ we obtain

$$\begin{array}{l}\hline 3\ 5\\ \hline\end{array},\ \begin{array}{l}\hline 1\ 4\\ \hline\end{array},\ \begin{array}{l}\hline 2\ 5\\ \hline\end{array},\ \begin{array}{l}\hline 1\ 3\\ \hline\end{array},\ \begin{array}{l}\hline 2\ 4\\ \hline\end{array}.$$

Hence there are exactly 2 double cosets; they are represented by

$$\frac{\overline{1\ \ 2\ \ 3}}{\underline{4\ \ 5}} \quad \text{and} \quad \frac{\overline{1\ \ 2\ \ 4}}{\underline{3\ \ 5}},$$

i.e. by

$$1_{S_5} \quad \text{and} \quad (34).$$

Applying $\pi \mapsto \varphi_\pi$, we get

$$\varphi_1 : \begin{matrix} 1 \mapsto 1 \\ 2 \mapsto 1 \\ 3 \mapsto 1 \\ 4 \mapsto 2 \\ 5 \mapsto 2 \end{matrix} \quad \text{and} \quad \varphi_2 : \begin{matrix} 1 \mapsto 1 \\ 2 \mapsto 1 \\ 3 \mapsto 2 \\ 4 \mapsto 1 \\ 5 \mapsto 2. \end{matrix}$$

These mappings represent the two classes of mappings $\varphi \in \mathbf{2}^{\mathbf{5}}$ under $E^{D_5}$ of content (3,2) which can be interpreted as colorings of a regular pentagon or "necklaces" with 5 beads in 2 colors, and of content (3,2):

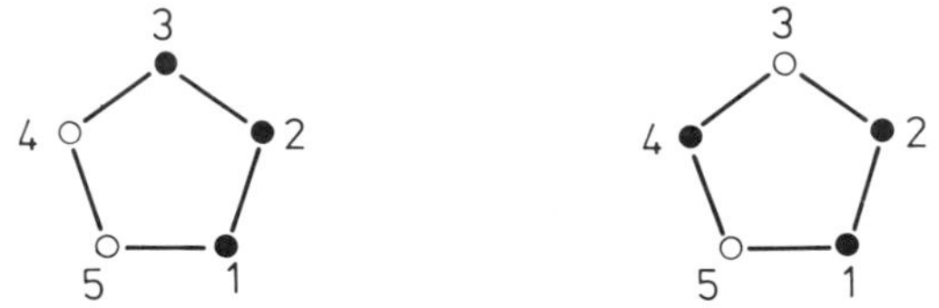


This shows how tabloids can be used for orbit calculations. (It is clear that we could have avoided the double cosets, as a tabloid already can be considered as a function of a certain content.)

## 5.2 Symmetrization of Representations

In order to discuss, in this and the following sections of the present chapter, a few applications to general representation theory of finite groups, we return to 4.3.14, where we noticed that the representations $\otimes^n D$ of $G$ and $P_n^m$ of $S_n$ afforded by $\otimes^n V$ centralize each other:

**5.2.1** $$\forall g \in G, \pi \in S_n \quad \overset{n}{\otimes} D(g) \circ P_n^m(\pi) = P_n^m(\pi) \circ \overset{n}{\otimes} D(g).$$

The reason for this very useful fact is that the subgroup $\operatorname{diag} G^* \cdot S_n'$ (cf. 4.1.17) provides an embedding of the direct product $G \times S_n$ into $G \operatorname{wr} S_n$:

**5.2.2** $$G \times S_n \simeq \operatorname{diag} G^* \cdot S_n' \hookrightarrow G \operatorname{wr} S_n.$$

Since 5.2.1 holds, certain decompositions of $\otimes^n D$ and $P_n^m$ are closely related as it will be shown next. The argument which we shall use holds under suitable assumptions on the groundfield whenever we have centralizing representations of two groups acting on the same vector space. In order to show this, we assume that we are given two groups $G_1$ and $G_2$ with matrix representations $\mathbb{D}^1$ and $\mathbb{D}^2$ on a vector space $V$ such that

5.2.3 $$\forall g_i \in G_i \quad \mathbb{D}^1(g_1)\circ\mathbb{D}^2(g_2)=\mathbb{D}^2(g_2)\circ\mathbb{D}^1(g_1).$$

Furthermore we assume that $\mathbb{D}^1$ is completely reducible into absolutely irreducible representations $\mathbb{D}_j^1$ of $G_1$.

Without loss of generality we can assume that this reduction of $\mathbb{D}^1$ has already been carried out and that an adapted basis has been chosen, so that $\mathbb{D}^1$ looks like

$$\mathbb{D}^1(g_1)=\left(\begin{array}{ccccccc}
\mathbb{D}_1^1(g_1) & & & & & & \\
 & \ddots & & & 0 & & \\
 & & \mathbb{D}_1^1(g_1) & & & & \\
 & & & \ddots & & & \\
 & & & & \mathbb{D}_r^1(g_1) & & \\
 & 0 & & & & \ddots & \\
 & & & & & & \mathbb{D}_r^1(g_1)
\end{array}\right)
\begin{array}{l}
\left.\begin{array}{l} \\ \\ \\ \end{array}\right\} n_1 \text{ times} \\
\vdots \\
\left.\begin{array}{l} \\ \\ \\ \end{array}\right\} n_r \text{ times}
\end{array}$$

The direct summand of $V$ which affords $n_j$ times the irreducible constituent $\mathbb{D}_j^1$ is called the *homogeneous component of type* $\mathbb{D}_j^1$ of $V$ (as a left $G_1$-module). It is a uniquely determined subspace, and if $G_1$ is finite and the characteristic of the groundfield does not divide $|G_1|$, it is equal to

5.2.4 $$\left(\sum_{g\in G_1}\left(\textit{trace of } \mathbb{D}_j^1(g^{-1})\right)g\right)V.$$

This subdivision of $\mathbb{D}^1(g_1)$ into boxes along the main diagonal yields a subdivision of the whole matrix $\mathbb{D}^1(g_1)$ into boxes. We subdivide $\mathbb{D}^2(g_2)$ correspondingly into boxes $\mathbb{D}_{kl}^{ij}(g_2)$ of the same sizes:

5.2.5 $\mathbb{D}^2(g_2)=$

$$\left(\begin{array}{cccccccc}
\mathbb{D}_{11}^{11}(g_2) & \cdots & \mathbb{D}_{1n_1}^{11}(g_2) & \mathbb{D}_{11}^{12}(g_2) & \cdots & \mathbb{D}_{1n_2}^{12}(g_2) & \cdots \\
\vdots & & \vdots & \vdots & & \vdots & \\
\mathbb{D}_{n_11}^{11}(g_2) & \cdots & \mathbb{D}_{n_1n_1}^{11}(g_2) & \mathbb{D}_{n_11}^{12}(g_2) & \cdots & \mathbb{D}_{n_1n_2}^{12}(g_2) & \cdots \\
\mathbb{D}_{11}^{21}(g_2) & \cdots & \mathbb{D}_{1n_1}^{21}(g_2) & \mathbb{D}_{11}^{22}(g_2) & \cdots & \mathbb{D}_{1n_2}^{22}(g_2) & \cdots \\
\vdots & & \vdots & \vdots & & \vdots & \\
\mathbb{D}_{n_21}^{21}(g_2) & \cdots & \mathbb{D}_{n_2n_1}^{21}(g_2) & \mathbb{D}_{n_21}^{22}(g_2) & \cdots & \mathbb{D}_{n_2n_2}^{22}(g_2) & \cdots \\
\vdots & & \vdots & \vdots & & \vdots & 
\end{array}\right).$$

Using this notation, we obtain from 5.2.3 that for a fixed $g_2 \in G_2$ we have for each $g_1 \in G_1$, $1 \leq k \leq n_i$, $1 \leq l \leq n_j$

5.2.6 $$\mathbb{D}^{ij}_{kl}(g_2)\mathbb{D}^1_j(g_1) = \mathbb{D}^1_i(g_1)\mathbb{D}^{ij}_{kl}(g_2).$$

Since $\mathbb{D}^1_j$ is assumed to be absolutely irreducible, we obtain from Schur's lemma

**5.2.7** $$\mathbb{D}^{ij}_{kl}(g_2) = \begin{cases} 0\text{-}matrix & \text{if } i \neq j, \\ d^i_{kl}(g_2) \cdot I & \text{if } i = j, \end{cases}$$

where $I$ denotes the identity matrix whose number of rows equals the dimension of $\mathbb{D}^1_i$ and where $d^i_{kl}(g_2)$ is an element of the groundfield $F$.

Using the constants $d^i_{kl}(g_2)$, we define

5.2.8 $$\mathbb{D}^2_i(g_2) := \left(d^i_{kl}(g_2)\right), \qquad 1 \leq k, l \leq n_i, \quad 1 \leq i \leq r.$$

From 5.2.5 and 5.2.7 we obtain that there exists a permutation matrix $T$ such that for each $g_2 \in G_2$ we have

$$T\mathbb{D}^2(g_2)T^{-1} = \left(\begin{array}{ccccccc} \mathbb{D}^2_1(g_2) & & & & & & \\ & \ddots & & & 0 & & \\ & & \mathbb{D}^2_1(g_2) & & & & \\ & & & \ddots & & & \\ & & & & \mathbb{D}^2_r(g_2) & & \\ & 0 & & & & \ddots & \\ & & & & & & \mathbb{D}^2_r(g_2) \end{array}\right) \begin{array}{l} \left.\vphantom{\begin{array}{c}a\\a\\a\end{array}}\right\} f^1 \text{ times} \\ \vdots \\ \left.\vphantom{\begin{array}{c}a\\a\\a\end{array}}\right\} f^r \text{ times} \end{array}$$

where $f^i$ denotes the dimension of $\mathbb{D}^1_i$. We notice in particular that the multiplicity $n_j$ of $\mathbb{D}^1_j$ in $\mathbb{D}^1$ is equal to the dimension of $\mathbb{D}^2_j$, while the dimension $f^j$ of $\mathbb{D}^1_j$ is equal to the multiplicity of $\mathbb{D}^2_j$ in $\mathbb{D}^2$. An application of this quite general argument to 5.2.1 instead of 5.2.3 yields

**5.2.9** THEOREM. *Let $G$ denote a group and $D$ an ordinary representation of $G$ on the vector space $V$, while $n \in \mathbb{N}$. Then the homogeneous component of type $[\alpha]$, $\alpha \vdash n$, of $\otimes^n V$ (as left $S_n$-module) affords, if it is nonzero, $f^\alpha$ times a representation of $G$. We denote this representation by $D \boxdot [\alpha]$ and call it* symmetrization *of $D$ by $[\alpha]$. Its dimension is equal to the multiplicity $(P^m_n, [\alpha])$*

*of* $[\alpha]$ *in* $P_n^m$, *and it satisfies, for each* $g \in G$, $\pi \in S_n$, *and with respect to a suitable basis of* $\otimes^n V$,

$$\overset{n}{\otimes}\mathbb{D}(g)\circ\mathbb{P}_n^m(\pi)=\underset{\alpha\vdash n}{\dot{+}}(\mathbb{D}\boxdot[\![\alpha]\!](g)\times[\![\alpha]\!](\pi)).$$

*(We allow* $\mathbb{D}\boxdot[\![\alpha]\!]$ *to denote the zero "representation" of* $G$ *if* $[\alpha]$ *is not an irreducible constituent of* $P_n^m$, *so that the corresponding homogeneous component of type* $[\alpha]$ *is the zero subspace. Thus* $\mathbb{D}\boxdot[\![\alpha]\!]$ *is defined for each* $\alpha \vdash n$, *but the corresponding summand in the equation above may have to be neglected.)*

An immediate corollary is that the representation $P_n^m$ of $S_n$ on $\otimes^n V$ induces a direct decomposition of $\otimes^n D$ into representations $D\boxdot[\alpha]$ (which are in general reducible) as follows (put $\pi := 1_{S_n}$ in the equation of 5.2.9):

**5.2.10**
$$\overset{n}{\otimes} D=\sum_{\alpha\vdash n} f^{\alpha}\cdot(D\boxdot[\alpha]).$$

It is clear that the above argument can be reversed in order to get from $\otimes^n D$ a decomposition of $P_n^m$; this will be discussed in the following section. 5.2.10 implies for example that

**5.2.11**
$$\begin{aligned}&\text{(i) } D=D\boxdot[1],\\ &\text{(ii) } D\otimes D=D\boxdot[2]+D\boxdot[1^2],\\ &\text{(iii) } D\otimes D\otimes D=D\boxdot[3]+2(D\boxdot[2,1])+D\boxdot[1^3].\end{aligned}$$

and so on.

If again $\zeta^\alpha$ denotes the character of $[\alpha]$, then the equation in 5.2.9 together with 4.3.10(vi) yields

**5.2.12**
$$\forall g\in G, \pi\in S_n \quad \prod_k \chi^D(g^k)^{a_k(\pi)}=\sum_{\alpha\vdash n}\chi^{D\boxdot[\alpha]}(g)\cdot\zeta^\alpha(\pi).$$

An application of the orthogonality relations yields for the character of $D\boxdot[\alpha]$:

**5.2.13**
$$\chi^{D\boxdot[\alpha]}(g)=\frac{1}{n!}\sum_{\pi\in S_n}\zeta^\alpha(\pi^{-1})\prod_{k=1}^{n}\chi^D(g^k)^{a_k(\pi)}.$$

We would like to discuss a few basic results on the dimension and irreducibility of $D\boxdot[\alpha]$.

**5.2.14** THEOREM. *The dimension $f^{D\boxdot[\alpha]}$ of $D\boxdot[\alpha]$ equals the number of semistandard $\alpha$-tableaux with entries $\leqslant f^D$. A more explicit formula is*

$$f^{D\boxdot[\alpha]}=\frac{f^\alpha}{n!}\prod_{\substack{i,j\\1\leqslant i\leqslant\alpha_1'\\1\leqslant j\leqslant\alpha_i}}(f^D-i+j)=\prod_{i,j}\frac{f^D-i+j}{h_{ij}^\alpha}.$$

*Proof.*

(i) 5.2.9 shows that

$$f^{D\boxdot[\alpha]}=\left(P_n^{f^D},[\alpha]\right)=\sum_{\beta=(\beta_1,\ldots,\beta_{f^D})\vDash n}(IS_\beta\uparrow S_n,[\alpha]),$$

so that Young's rule immediately yields the first part of the statement.

(ii) By definition of symmetrization, $f^{D\boxdot[\alpha]}$ is just the multiplicity of $[\alpha]$ in $P_n^m$ (recall that $m=f^D$), which is $(f^\alpha)^{-1}$ times the dimension of the homogeneous component of type $[\alpha]$ of $\otimes^n V$ as a left $S_n$-module, $V$ being the representation space of $D$. On the other hand, the dimension of this homogeneous component is just the trace of the corresponding projection operator

(1) $$E_\alpha:=\frac{f^\alpha}{n!}\sum_\pi\zeta^\alpha(\pi^{-1})\pi,$$

which projects $\otimes^n V$ onto this homogeneous component. Furthermore this centrally primitive idempotent $E_\alpha$ is the unit of the simple component $I^\alpha$ of the group algebra $\mathbb{C}S_n$ and hence it is the sum of the minimal idempotents $e_i^\alpha$:

(2) $$E_\alpha=\sum_{i=1}^{f^\alpha}e_i^\alpha=\sum_{i=1}^{f^\alpha}\frac{f^\alpha}{n!}\mathcal{V}_i^\alpha\mathcal{H}_i^\alpha.$$

This yields

(3) $$f^{D\boxdot[\alpha]}=\frac{1}{f^\alpha}\operatorname{trace}E_\alpha=\frac{1}{n!}\sum_{i=1}^{f^\alpha}\operatorname{trace}\mathcal{V}_i^\alpha\mathcal{H}_i^\alpha,$$

so that we shall be done once we have shown that the following holds:

(4) $$\forall k\quad \operatorname{trace}\mathcal{V}_k^\alpha\mathcal{H}_k^\alpha=\prod_{i,j}(f^D-i+j).$$

In order to prove this we first notice that by 4.3.10(ii)

(5) $$\operatorname{trace}\mathcal{V}_k^\alpha\mathcal{H}_k^\alpha=\sum_{\substack{\pi\in H_k^\alpha\\\rho\in V_k^\alpha}}\operatorname{sgn}\rho\cdot(f^D)^{c(\rho\pi)}.$$

Now we proceed by induction on $n$.

(a) If $n:=1$, then $\alpha:=(1)$ is the only partition, hence $f^\alpha=1$, and in fact, since $\mathcal{V}_k^\alpha\mathcal{H}_k^\alpha=1_{S_1}$, we have

$$\operatorname{trace}\mathcal{V}_k^\alpha\mathcal{H}_k^\alpha=f^D.$$

(b) If $n>1$ and $\alpha=(\alpha_1,\ldots,\alpha_h)$, we put

$$\hat\alpha:=(\alpha_1,\ldots,\alpha_{h-1},\alpha_h-1)\vdash n-1.$$

The tableau which arises from $t_k^\alpha$ by cancelling the last symbol of the last row will be denoted by $\hat t_k^\alpha$; its horizontal and vertical groups, by $\hat H_k^\alpha, \hat V_k^\alpha$. We shall prove that

$$(6)\qquad (f^D-h+\alpha_h)\sum_{\substack{\pi\in\hat H_k^\alpha\\ \rho\in\hat V_k^\alpha}}\operatorname{sgn}\rho\cdot(f^D)^{c(\rho\pi)}=\sum_{\substack{\pi\in H_k^\alpha\\ \rho\in V_k^\alpha}}\operatorname{sgn}\rho\cdot(f^D)^{c(\rho\pi)}.$$

This shows that (4) is true by the induction hypothesis (apply (5)). To prove (6), we embed $S_{n-1}$ into $S_n$ as follows. If $z$ is the cancelled symbol, i.e. if

$t_k^\alpha=$ 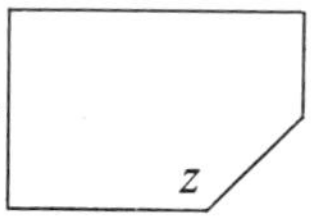


then $S_{n-1}$ is embedded into $S_n$ as the stabilizer of $\{z\}$. The subgroups corresponding to $\hat H_k^\alpha, \hat V_k^\alpha$ under this embedding are denoted by $\tilde H_k^\alpha, \tilde V_k^\alpha$.

We now decompose $H_k^\alpha$ and $V_k^\alpha$ into cosets with respect to $\tilde H_k^\alpha$ and $\tilde V_k^\alpha$ using the symbols in the row and in the column of $z$:

$t_k^\alpha=$ 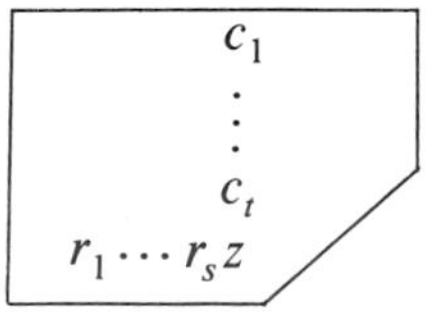


We have

$$\mathcal{V}_k^\alpha=(1-(zc_1)-\cdots-(zc_t))\tilde{\mathcal{V}}_k^\alpha,$$

and

$$\mathcal{H}_k^\alpha = \tilde{\mathcal{H}}_k^\alpha (1 + (zr_1) + \cdots + (zr_s)),$$

so that

(7)

$$\mathcal{V}_k^\alpha \mathcal{H}_k^\alpha = \tilde{\mathcal{V}}_k^\alpha \tilde{\mathcal{H}}_k^\alpha - \sum_i (zc_i) \tilde{\mathcal{V}}_k^\alpha \tilde{\mathcal{H}}_k^\alpha + \sum_j \tilde{\mathcal{V}}_k^\alpha \tilde{\mathcal{H}}_k^\alpha (zr_j) - \sum_{i,j} (zc_i) \tilde{\mathcal{V}}_k^\alpha \tilde{\mathcal{H}}_k^\alpha (zr_j).$$

In order to apply the induction hypothesis to (7) we notice that

$$\text{trace}\left(\tilde{\mathcal{V}}_k^\alpha \tilde{\mathcal{H}}_k^\alpha\right) = f^D \cdot \text{trace}\left(\hat{\mathcal{V}}_k^\alpha \hat{\mathcal{H}}_k^\alpha\right) \tag{8}$$

($\hat{\mathcal{V}}_k^\alpha \hat{\mathcal{H}}_k^\alpha$ acts on $\otimes^{n-1} V$) and that

$$\text{trace}\left((zc_i)\tilde{\mathcal{V}}_k^\alpha \tilde{\mathcal{H}}_k^\alpha\right) = \text{trace}\left(\hat{\mathcal{V}}_k^\alpha \hat{\mathcal{H}}_k^\alpha\right) = \text{trace}\left(\tilde{\mathcal{V}}_k^\alpha \tilde{\mathcal{H}}_k^\alpha (zr_j)\right). \tag{9}$$

(Multiplication of a summand $\pm\pi$ of $\tilde{\mathcal{V}}_k^\alpha \tilde{\mathcal{H}}_k^\alpha$ by $(zc_i)$ or $(zr_j)$ gives $c((zc_i)\pi) = c(\pi(zr_j)) = c(\pi) - 1$, since $\pi$ leaves $z$ fixed. Now use (8) to get (9).)

Later on we shall see that furthermore the following holds:

$$\text{trace}\left((zc_i)\tilde{\mathcal{V}}_k^\alpha \tilde{\mathcal{H}}_k^\alpha (zr_j)\right) = 0, \tag{10}$$

so that we obtain from (7) by an application of (8) and (9)

$$\text{trace}\left(\mathcal{V}_k^\alpha \mathcal{H}_k^\alpha\right) = \left(f^D - (h-1) + (\alpha_h - 1)\right) \text{trace}\left(\hat{\mathcal{V}}_k^\alpha \hat{\mathcal{H}}_k^\alpha\right),$$

an equation which implies (6).

It remains to prove (10). To do this we bring the element $a$ of $t_k^\alpha$ in the $i$th row and $j$th column into the game:

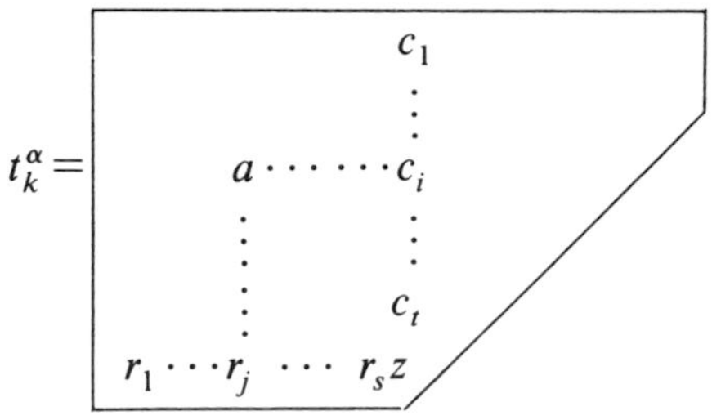


It satisfies

$$\tilde{\mathcal{V}}_k^\alpha = (1 - (ar_j))X \quad \text{and} \quad \tilde{\mathcal{H}}_k^\alpha = Y(1 + (ac_i))$$

for suitable $X$ and $Y$ in the subalgebra $\mathbb{C}S_{n-1}$. This gives

$$\begin{aligned}\operatorname{trace}\big((zc_i)\mathcal{V}_k^{\alpha}\mathcal{H}_k^{\alpha}(zr_j)\big) &= \operatorname{trace}((zc_i)(1-(ar_j))XY(1+(ac_i))(zr_j)) \\ &= \operatorname{trace}\big([(1+(ac_i))(zr_j)(zc_i)(1-(ar_j))]\,XY\big) \\ &= \operatorname{trace}\big([((zc_i)(c_ir_j)-(za)(c_ir_j))(1-(ar_j))]\,XY\big) \\ &= \operatorname{trace}(((zc_i)-(za))W).\end{aligned}$$

Note that $W:=(c_ir_j)(1-(ar_j))XY$ is in $\mathbb{C}S_{n-1}$; hence it is obvious that

$$\operatorname{trace}((zc_i)W)=\operatorname{trace}((za)W),$$

so that we are done. ■

5.2.14 shows that $f^{D\boxdot[\alpha]}$ is obtained by first substituting for the $(i,j)$-node of $[\alpha]$ the number $f^D-i+j$, i.e. by forming

$$\begin{array}{lllll} f^D & f^D+1 & \cdots & \cdots & f^D+\alpha_1-1 \\ f^D-1 & f^D & \cdots & f^D+\alpha_2-2 & \\ \cdots & \cdots & \cdots & \cdots & \\ f^D-h+1 & \cdots & f^D-h+\alpha_h & & \end{array}$$

and then dividing the product of all these numbers by the product of all the hook lengths in $[\alpha]$. The resulting quotient is 0 if and only if $f^D<h$. Hence we obtain the following criterion for the existence of the symmetrized product $D\boxdot[\alpha]$ as a corollary of 5.2.14:

**5.2.15** THEOREM. *The symmetrized product* $D\boxdot[\alpha]$ *is nonzero if and only if* $\alpha_1'\leqslant f^D$.

This shows e.g. that $D\boxdot[n]$ always exists (i.e. it is nonzero) and that $D\boxdot[1^n]$ exists if and only if $n\leqslant f^D$. For the dimensions of these particular products we obtain from 5.2.14:

**5.2.16**

$$\text{(i) } f^{D\boxdot[n]}=\binom{f^D+n-1}{n},$$

$$\text{(ii) } f^{D\boxdot[1^n]}=\binom{f^D}{n}.$$

Both these symmetrized products are of particular importance. $D\boxdot[n]$ is called the *symmetric part* of $\otimes^n D$, while $D\boxdot[1^n]$ is called the *antisymmetric part* of $\otimes^n D$. Their representation spaces are

5.2.17 $$E_{(n)} \overset{n}{\otimes} V = \left( \frac{1}{n!} \sum_{\pi \in S_n} \pi \right) \overset{n}{\otimes} V$$

and

5.2.18 $$E_{(1^n)} \overset{n}{\otimes} V = \left( \frac{1}{n!} \sum_{\pi} \operatorname{sgn} \pi \cdot \pi \right) \overset{n}{\otimes} V,$$

respectively. The elements of

5.2.19 $$B := \langle P_n^m[S_n] \rangle = \left\{ \sum_{\pi \in S_n} a_\pi P_n^m(\pi) \mid a_\pi \in \mathbb{C} \right\} \subseteq \operatorname{End}_{\mathbb{C}} \left( \overset{n}{\otimes} V \right),$$

like $E_{(n)}$ and $E_{(1^n)}$, are called *symmetry operators*. The image of such a symmetry operator is called a *symmetry class* in $\otimes^n V$.

By 5.2.9 and 5.2.14, the decomposition of $P_n^m$ into its irreducible constituents is

**5.2.20** $$\forall m, n \in N \quad P_n^m = \sum_{\alpha \vdash n} \left( \prod_{i,j} \frac{m-i+j}{h_{ij}^\alpha} \right) [\alpha].$$

We now investigate the case where $V$ is an $m$-dimensional vector space over $\mathbb{C}$ and $G$ is the general linear group,

$$G := \mathrm{GL}(V).$$

It turns out that if we take $D$ to be the natural representation of $G$, then all the symmetrized products of $D$ are either zero or irreducible. Here we mean by the natural representation the identity mapping $\mathrm{id}_{\mathrm{GL}(V)}$, i.e.

5.2.21 $$\mathrm{id}_{\mathrm{GL}(V)} : \mathrm{GL}(V) \to \mathrm{GL}(V) : g \mapsto g.$$

The corresponding symmetrized inner product is denoted by $\langle \alpha \rangle$:

5.2.22 $$\langle \alpha \rangle := \mathrm{id}_{\mathrm{GL}(V)} \boxdot [\alpha].$$

In order to prove that $\langle \alpha \rangle$ is either zero or irreducible, we consider besides the subalgebra $B$ (cf. 5.2.19) of $\operatorname{End}_{\mathbb{C}}(\otimes^n V)$ the subalgebra $C$ generated by

5.2.23 $$I_m^n[\mathrm{GL}(V)] := \overset{n}{\otimes} \mathrm{id}_{\mathrm{GL}(V)}[\mathrm{GL}(V)],$$

i.e., we consider

5.2.24
$$C := \langle I_m^n[\mathrm{GL}(V)] \rangle .$$

In the approach we are adopting here the following theorem is crucial:

**5.2.25** THEOREM.

(i) $C = \mathrm{End}_{\mathbb{C}S_n}(\otimes^n V)$.
(ii) $\otimes^n V$ *is a completely reducible* $\mathrm{GL}(V)$*-module.*

*Proof.* We consider representing matrices for the elements of the subalgebras $B$ and $C$ of $\mathrm{End}_{\mathbb{C}}(\otimes^n V)$ which correspond to the basis given in 4.3.4:

$$\overset{n}{\otimes} V = \langle\langle b_\varphi \mid \varphi \in \mathbf{m}^\mathbf{n} \rangle\rangle_{\mathbb{C}}.$$

(a) Denote the matrix representing $\pi \in S_n$ under $P_n^m$ by $\mathbb{P}_n^m(\pi) = (a_{\varphi\psi}(\pi))$. We obtain from 4.3.6 for its coefficients

(1)
$$a_{\varphi\psi}(\pi) = \delta_{\varphi, \psi\circ\pi^{-1}},$$

if $\delta$ denotes the usual Kronecker symbol.

(b) An element $\gamma \in \mathrm{End}_{\mathbb{C}}(\otimes^n V)$ is contained in $\mathrm{End}_{\mathbb{C}S_n}(\otimes^n V)$ if and only if for its representing matrix

$$M(\gamma) = (a_{\varphi\psi}(\gamma))$$

the following holds:

$$\forall \pi \in S_n \quad M(\gamma)\mathbb{P}_n^m(\pi) = \mathbb{P}_n^m(\pi) M(\gamma),$$

i.e.

$$\forall \pi, \varphi, \psi \quad \sum_\xi a_{\varphi\xi}(\gamma) a_{\xi\psi}(\pi) = \sum_\xi a_{\varphi\xi}(\pi) a_{\xi\psi}(\gamma).$$

By (1) this is equivalent to

$$\forall \pi, \varphi, \psi \quad a_{\varphi\psi}(\gamma) = a_{\varphi\circ\pi^{-1}, \psi\circ\pi^{-1}}(\gamma).$$

This shows that $\gamma$ is contained in $\mathrm{End}_{\mathbb{C}S_n}(\otimes^n V)$ if and only if the coefficient function $(\varphi, \psi) \mapsto a_{\varphi\psi}(\gamma)$ is constant on each orbit of $S_n$ on $\mathbf{m}^\mathbf{n} \times \mathbf{m}^\mathbf{n}$ under the action

$$\pi(\varphi, \psi) := (\varphi\circ\pi^{-1}, \psi\circ\pi^{-1}).$$

We notice that this permutation representation of $S_n$ on $\mathbf{m^n} \times \mathbf{m^n}$ is just $\otimes^2 P_n^m$, so that the number of orbits is equal to the multiplicity $(\otimes^2 P_n^m, [n])$, which is

$$\frac{1}{n!}\sum_{\pi} m^{2c(\pi)} = \frac{1}{n!}\sum_{\pi} (m^2)^{c(\pi)} = \binom{m^2+n-1}{n}.$$

The last equation is obtained from 5.2.16(i). This yields for the $\mathbb{C}$-dimension of $\mathrm{End}_{\mathbb{C}S_n}(\otimes^n V)$

$$\dim_{\mathbb{C}}\left(\mathrm{End}_{\mathbb{C}S_n}\left(\overset{n}{\otimes} V\right)\right) = \binom{m^2+n-1}{n}.$$

If we put

$$E_{\varphi\psi} := \begin{pmatrix} & 0 & & \begin{matrix}0\\ \vdots\\ 0\end{matrix} & & 0 & \\ 0 & \cdots & 0 & 1 & 0 & \cdots & 0 \\ & 0 & & \begin{matrix}0\\ \vdots\\ 0\end{matrix} & & 0 & \end{pmatrix} \begin{matrix} \\ \varphi \\ \\ \end{matrix}$$
$$\qquad\qquad\qquad\qquad\qquad\psi$$

and (denoting the orbit of $(\varphi, \psi)$ by $[\varphi, \psi]$)

$$E_{[\varphi,\psi]} := \sum_{(\xi,\zeta)\in[\varphi,\psi]} E_{\xi\zeta},$$

then if $\mathfrak{T}$ is a transversal of the orbits, we have a $\mathbb{C}$-basis of $\mathrm{End}_{\mathbb{C}S_n}(\otimes^n V)$:

$$\mathrm{End}_{\mathbb{C}S_n}\left(\overset{n}{\otimes} V\right) = \langle\langle E_{[\varphi,\psi]} | (\varphi,\psi) \in \mathfrak{T} \rangle\rangle_{\mathbb{C}}.$$

(c) In order to prove statement (i), we notice that by 5.2.1 $C \subseteq \mathrm{End}_{\mathbb{C}S_n}(\otimes^n V)$. If $\mathrm{End}_{\mathbb{C}S_n}(\otimes^n V)^*$ denotes the vector space dual to $\mathrm{End}_{\mathbb{C}S_n}(\otimes^n V)$, then $C \neq \mathrm{End}_{\mathbb{C}S_n}(\otimes^n V)$ if and only if there exists a nonzero linear functional $f \in \mathrm{End}_{\mathbb{C}S_n}(\otimes^n V)^*$ which restricts to the zero functional on $C$. Denoting by $E_{[\varphi,\psi]^*}$ the elements of the dual basis, then $f \in \mathrm{End}_{\mathbb{C}S_n}(\otimes^n V)^*$ is of the form

$$f = \sum_{(\varphi,\psi)\in\mathfrak{T}} b_{(\varphi,\psi)} E_{[\varphi,\psi]^*}.$$

An application of $f$ to $g\otimes\cdots\otimes g\in I_m^n[\mathrm{GL}(V)]$ yields (writing $g:=(d_{ik}(g))$ in matrix form):

$$0=f(g\otimes\cdots\otimes g)=\sum_{(\varphi,\psi)\in\mathfrak{T}} b_{(\varphi,\psi)}E_{[\varphi,\psi]^*}\left(\sum_{\xi,\zeta}\prod_j d_{\xi(j),\zeta(j)}(g)E_{\xi\zeta}\right)$$

$$=\sum_{(\varphi,\psi)\in\mathfrak{T}} b_{(\varphi,\psi)}\prod_j d_{\varphi(j),\psi(j)}(g).$$

This shows that the corresponding polynomial

$$P_f:=\sum_{(\varphi,\psi)\in\mathfrak{T}} b_{(\varphi,\psi)}\prod_j x_{\varphi(j),\psi(j)}\in\mathbb{C}[x_{11},x_{12},\ldots,x_{mm}]$$

has the property that for each non-singular $m\times m$ matrix $(a_{ik})$, the result of making the substitutions $x_{ik}\mapsto a_{ik}$ in $P_f$ is zero. Therefore, $P_f(a_{ik})\cdot\det(a_{ik})=0$ for all $m\times m$ matrices $(a_{ik})$. Since $\mathbb{C}$ is infinite, $P_f$ must be the zero polynomial. But for distinct $(\varphi,\psi)\in\mathfrak{T}$, the monomials $\prod_j x_{\varphi(j),\psi(j)}$ are distinct, so $b_{(\varphi,\psi)}=0$ for all $(\varphi,\psi)$. Thus there is no nonzero linear functional $f$ in $\mathrm{End}_{\mathbb{C}S_n}(\otimes^n V)$ which restricts to the zero functional on $C$.

This proves $C=\mathrm{End}_{\mathbb{C}S_n}(\otimes^n V)$. Assertion (ii) follows from (i), 5.2.24, and the semisimplicity of $C$. ∎

**5.2.26 Theorem.** *If $\alpha=(\alpha_1,\ldots,\alpha_h)\vdash n$, and $m$ is a natural number with $\alpha_1'\leqslant m$, then $\langle\alpha\rangle$ is an irreducible representation of* $\mathrm{GL}(m,\mathbb{C})$ *over* $\mathbb{C}$.

*Proof.* Recall that $B$ is defined to be $\langle P_n^m[\mathrm{S}_n]\rangle$. Now, $\langle\alpha\rangle$ is afforded by $e\otimes^n V$, where $e$ is a primitive idempotent in $E_\alpha B$. (We may therefore assume that $\langle\alpha\rangle$ is afforded by $\mathcal{V}^\alpha\mathcal{H}^\alpha\otimes^n V$.) If $\langle\alpha\rangle$ were reducible, then 5.2.25(ii) would yield a nontrivial $\mathrm{GL}(V)$-decomposition

$$e\overset{n}{\otimes}V=V_1\oplus V_2$$

Let $W$ be a $\mathrm{GL}(V)$-complement to $e\otimes^n V$ in $\otimes^n V$, and let $e_i$ be the projection of $\otimes^n V$ onto $V_i$ with kernel $W\oplus V_j$, $i\neq j$. Then $e_1$ and $e_2$ belong to the centralizer $\mathrm{End}_C(\otimes^n V)$ of $C$. Since $B$ has the double-centralizer property, we get $e=e_1+e_2$, contradicting the primitivity of $e$. ∎

Theorem 5.2.26 gives many irreducible representations of $\mathrm{GL}(V)$. But every group with a finite-dimensional complex representation has a homomorphic image inside some general linear group, so the concept of symmetrization enables us to produce many new representations from a given one.

A well-known theorem of Burnside states that if $D$ is a faithful representation of a finite group $G$, then $\otimes^0 D := I, \otimes^1 D, \ldots, \otimes^s D$ ($s$ being the number of different values $\chi^D$ takes) between them contain all the irreducible representations of $G$ as composition factors. An interactive computer program which does symmetrization and subtracts known irreducible characters from new ones can therefore be used to great effect on the problem of determining the ordinary character table of a given finite group, once a faithful character has been found.

In Chapter 8 we shall investigate the representation theory of general linear groups more fully, proving that $\mathrm{id}_{GL(V)} \boxdot [\alpha]$ is irreducible whenever $\mathrm{char}\, F = 0$, and showing how to find an irreducible composition factor when $\mathrm{char}\, F = p$.

The applications of symmetrization $D \boxdot [\alpha]$ of representations $D$ of a group $G$ by representations $[\alpha]$ of symmetric groups give rise to the question whether it might be possible and even useful occasionally to symmetrize $D$ by an ordinary irreducible representation $D^i$ of a subgroup $P$ of $S_n$. That such a symmetrization

5.2.27 $$D \boxdot D^i$$

is well defined is trivial from the definition of $D \boxdot [\alpha]$, since we merely applied Schur's lemma in order to define $D \boxdot [\alpha]$, but we never used the fact that the permutation group acting on $\otimes^n V$ was the full symmetric group. 5.2.27 is nonzero if and only if the centrally primitive idempotent $E_i$, which corresponds to $D^i$, i.e.

5.2.28 $$E_i := \frac{1}{|P|} \sum_{\pi \in P} \zeta^i(\pi^{-1})\pi,$$

satisfies

5.2.29 $$E_i \overset{n}{\otimes} V \neq \{0\},$$

which holds if and only if $D^i$ is contained in the restriction of $P_n^m$:

5.2.30 $$\left(P_n^m \downarrow P, D^i\right) > 0.$$

Or, equivalently, it holds if and only if there exists an $\alpha \vdash n$ such that

5.2.31 $$\left([\alpha] \downarrow P, D^i\right) > 0 \quad \text{and} \quad \alpha_1' \leq m.$$

A closer examination of the homogeneous component of type $D^i$ is

5.2.32
$$E_i \overset{n}{\otimes} V = E_i\left(\bigoplus_{\alpha\vdash n} E_\alpha \overset{n}{\otimes} V\right)$$
$$= \bigoplus_{\alpha\vdash n}\left(E_i E_\alpha \overset{n}{\otimes} V\right)$$
$$= \bigoplus_{\alpha\vdash n} f^{D^i}\!\left([\alpha]\downarrow P, D^i\right)\hat{e}_1^{\alpha} \overset{n}{\otimes} V,$$

where again (see Chapter 3) $\hat{e}_1^{\alpha}$ denotes the minimal idempotent corresponding to the first standard tableau of shape $[\alpha]$. 5.2.32 can be rephrased in the following way:

**5.2.33**
$$D\boxdot D^i = \sum_{\alpha\vdash n} f^{D^i}\!\left([\alpha]\downarrow P, D^i\right) D\boxdot[\alpha].$$

This equation shows in particular that the decomposition $\otimes^n D = \sum_\alpha f^\alpha(D\boxdot[\alpha])$ of $\otimes^n D$ into symmetrized products $D\boxdot[\alpha]$ with ordinary irreducible representations of $S_n$ is at least as fine as the decomposition into symmetrized products $D\boxdot D^i$ with ordinary irreducible representations of a subgroup $P$ of $S_n$. Nevertheless it is occasionally useful to symmetrize by representations of subgroups $P \leqslant S_n$. A surprising example is a proof of the Amitsur-Levitzki theorem on the standard polynomial identity:

**5.2.34** Amitsur-Levitzki theorem. *If $M_1,\ldots,M_{2n}$ are $n\times n$-matrices over* $\mathbb{C}$, *then the following* standard polynomial identity *holds*:

$$[M_1,\ldots,M_{2n}] := \sum_{\pi\in S_{2n}} \operatorname{sgn}\pi\, M_{\pi(1)}\cdot\cdots\cdot M_{\pi(2n)} = 0.$$

*Proof.*

(i) We first prove that the mapping

$$\sum_{\pi\in S_{2n+1}} \operatorname{sgn}\pi\left(e;\pi(1\cdots 2n+1)\pi^{-1}\right):\ \overset{2n+1}{\otimes}\mathbb{C}^n \to \overset{2n+1}{\otimes}\mathbb{C}^n$$

which maps the tensor $v_1\otimes\cdots\otimes v_{2n+1} = \cdots\otimes v_i\otimes\cdots$ onto

$$\sum_{\pi\in S_{2n+1}} \operatorname{sgn}\pi\left(\cdots\otimes v_{\pi(2n+1\cdots 1)\pi^{-1}(i)}\otimes\cdots\right)$$

is the zero mapping. In order to prove this, we notice first that

$$\langle n+1, 1^n\rangle = \mathrm{id}_{GL(n,\mathbb{C})}\boxdot[n+1, 1^n]$$

is the zero "representation" by 5.2.15. Hence 5.2.33 together with

$$[n+1,1^n]\downarrow A_{2n+1}=[n+1,1^n]^+ +[n+1,1^n]^-$$

(see 2.5.7) yields that both

$$\mathrm{id}_{GL(n,\mathbb{C})}\boxdot[n+1,1^n]^\pm$$

are zero. Expressed in terms of the corresponding centrally primitive idempotents $E^\pm_{(n+1,1^n)}$, this reads as follows:

$$E^\pm_{(n+1,1^n)} \overset{2n+1}{\otimes} \mathbb{C}^n=\{0\}.$$

Hence also

$$\left(E^+_{(n+1,1^n)}-E^-_{(n+1,1^n)}\right) \overset{2n+1}{\otimes} \mathbb{C}^n=\{0\}.$$

But the corresponding irreducible characters $\zeta^{(n+1,1^n)^\pm}$ of $A_{2n+1}$ differ on the two classes $(2n+1)^\pm$ of the $(2n+1)$-cycles only (recall 2.5.12, 13), so that $E^+_{(n+1,1^n)}-E^-_{(n+1,1^n)}$ is a multiple of

$$\sum_{\rho\in(2n+1)^+}\rho-\sum_{\tau\in(2n+1)^-}\tau=\text{a multiple of}\sum_{\pi\in S_{2n+1}}\operatorname{sgn}\pi\cdot\pi(1\cdots2n+1)\pi^{-1}.$$

This yields our first statement.

(ii) We prove now that

$$\operatorname{trace}[M_1,\ldots,M_{2n+1}]=0$$

for arbitrary $n\times n$ matrices $M_i$ over $\mathbb{C}$.

$$\operatorname{trace}[M_1,\ldots,M_{2n+1}]=\operatorname{trace}\sum_\pi \operatorname{sgn}\pi\cdot M_{\pi(1)}\cdots M_{\pi(2n+1)}$$

$$=\sum_\pi \operatorname{sgn}\pi \operatorname{trace} M_{\pi(1)}\cdots M_{\pi(2n+1)}$$

$$=\sum_\pi \operatorname{sgn}\pi\operatorname{trace}\left(\overset{2n+1}{\#}\mathrm{id}_{\mathrm{End}(V)}\right)^\sim\left(M_1,\ldots,M_{2n+1};\pi(1\cdots2n+1)\pi^{-1}\right)$$

(cf. the proof of 4.3.9). Hence

$$\operatorname{trace}[M_1,\ldots,M_{2n+1}]=\operatorname{trace}\Big(\mathop{\#}^{2n+1}\operatorname{id}_{\operatorname{End}(V)}\Big)^{\sim}$$
$$\times\Big(\big(M_1,\ldots,M_{2n+1};1\big)\circ\sum_{\pi}\operatorname{sgn}\pi\big(e;\pi(1\cdots 2n+1)\pi^{-1}\big)\Big),$$

and the right-hand side is 0 by (i).

(iii) We are now in a position to prove the statement. In order to do this we first remark that

$$S_{2n+1}=S_{2n}\cdot\langle(1\cdots 2n+1)\rangle,$$

if $S_{2n}$ denotes the stabilizer of $\{2n+1\}$ in $S_{2n+1}$. This together with (ii) gives us

$$\begin{aligned}0&=\operatorname{trace}[M_1,\ldots,M_{2n+1}]\\&=\sum_{\pi\in S_{2n+1}}\operatorname{sgn}\pi\cdot\operatorname{trace}M_{\pi(1)}\cdots M_{\pi(2n+1)}\\&=\sum_{\sigma\in S_{2n}}\operatorname{sgn}\sigma\sum_{\tau\in\langle(1\cdots 2n+1)\rangle}\operatorname{sgn}\tau\cdot\operatorname{trace}M_{\sigma\tau(1)}\cdots M_{\sigma\tau(2n+1)}\\&=(2n+1)\operatorname{trace}M_{2n+1}\cdot\sum_{\sigma\in S_{2n}}\operatorname{sgn}\sigma M_{\sigma(1)}\cdots M_{\sigma(2n)}\\&=(2n+1)\operatorname{trace}M_{2n+1}\cdot[M_1,\ldots,M_{2n}].\end{aligned}$$

Since this holds for arbitrary $n\times n$ matrices $M_{2n+1}$, the statement $[M_1,\ldots,M_{2n}]=0$ must be true. ■

Besides 5.2.33 there are many other useful equations—for example the following one: If $D$ is an ordinary representation of a finite group $G$, then for any $\alpha\vdash m$, $\beta\vdash n$

**5.2.35** $$(D\boxdot[\alpha])\otimes(D\boxdot[\beta])=D\boxdot([\alpha][\beta]).$$

(This sometimes holds for ordinary representations $D$ of infinite groups also, e.g. if $G:=\mathrm{GL}(k,\mathbb{C})$, in which case in the proof the characters have to be replaced by $S$-functions or a different approach adopted.)

*Proof.* For the character of the right-hand side on $g\in G$ we have

5.2.36 $$\chi^{D\boxdot([\alpha][\beta])}(g)=\frac{1}{(m+n)!}\sum_{\pi\in S_{m+n}}\chi^{[\alpha][\beta]}(\pi)\prod_{k=1}^{m+n}\chi^{D}(g^k)^{a_k(\pi)}.$$

But the product

$$f_g(\pi):=\prod_k \chi^D(g^k)^{a_k(\pi)}, \qquad \pi\in S_{m+n},$$

is the value of a class function $f_g$ on $S_{m+n}$, so that the right-hand side of 5.2.36 is an inner product of class functions, to which we can apply Frobenius's reciprocity theorem, obtaining:

$$\begin{aligned}\chi^{D\boxdot([\alpha][\beta])}(g)&=\left(\chi^{[\alpha][\beta]},f_g\right)\\&=\left(\chi^{[\alpha]\#[\beta]},f_g\downarrow S_m\oplus S_n\right)\\&=\frac{1}{m!n!}\sum_{\pi=\rho\sigma}\zeta^\alpha(\rho^{-1})\zeta^\beta(\sigma^{-1})\prod_{j,k=1}^{m,n}\chi^D(g^j)^{a_j(\rho)}\chi^D(g^k)^{a_k(\sigma)}\\&=\chi^{D\boxdot[\alpha]}(g)\chi^{D\boxdot[\beta]}(g)\\&=\chi^{(D\boxdot[\alpha])\otimes(D\boxdot[\beta])}(g).\end{aligned}$$

■

An immediate corollary is the following determinantal form:

**5.2.37** $$D\boxdot[\alpha]=|D\boxdot[\alpha_i+j-i]|^{\otimes}.$$

It reduces the decomposition of symmetrized products $D\boxdot[\alpha]$ to the decomposition of "completely symmetric" representations $D\boxdot[r]$ together with decompositions of inner tensor products.

For inner tensor products $D\otimes D'$ of ordinary representations of a finite group $G$ we find (if $\gamma\vdash n$)

**5.2.38** $$(D\otimes D')\boxdot[\gamma]=\sum_{\alpha,\beta\vdash n}([\alpha]\otimes[\beta],[\gamma])(D\boxdot[\alpha])\otimes(D'\boxdot[\beta]).$$

*Proof.* If $g\in G$ and $(\rho,\sigma)\in S_n\oplus S_n$, we write

$$f_g(\rho,\sigma):=\prod_{k=1}^{n}\chi^D(g^k)^{a_k(\rho)}\chi^{D'}(g^k)^{a_k(\sigma)}.$$

Then the character of the right-hand side of 5.2.38 on $g\in G$ is equal to

$$\begin{aligned}&\sum_{\alpha,\beta}([\alpha]\otimes[\beta],[\gamma])\frac{1}{n!^2}\sum_{(\rho,\sigma)\in S_n\oplus S_n}\zeta^\alpha(\rho^{-1})\zeta^\beta(\sigma^{-1})f_g(\rho,\sigma)\\&=\frac{1}{n!^2}\sum_{(\rho,\sigma)}\left(\sum_{\alpha,\beta}([\alpha]\otimes[\beta],[\gamma])\chi^{[\alpha]\#[\beta]}(\rho,\sigma)\right)\prod_k\chi^D(g^k)^{a_k(\rho)}\chi^{D'}(g^k)^{a_k(\sigma)}.\end{aligned}$$

Now $[\gamma]$ of $S_n$ can be considered as a representation of diag$(S_n\oplus S_n)$, in which case we have

$$[\alpha]\#[\beta]\downarrow\text{diag}(S_n\oplus S_n)=[\alpha]\otimes[\beta]=\sum_\gamma([\alpha]\otimes[\beta],[\gamma])[\gamma],$$

or, equivalently,

$$[\gamma]\uparrow S_n \oplus S_n = \sum_{\alpha,\beta} ([\alpha]\otimes[\beta],[\gamma])[\alpha]\#[\beta].$$

In addition, the product $f_g(\rho,\sigma)$, defined above is, for each $g\in G$, the value of a class function $f_g$ on $S_n\oplus S_n$, so that the character of the right-hand side of 5.2.38, evaluated at $g$, is equal to

$$\begin{aligned}([\gamma]\uparrow S_n\oplus S_n, f_g) &= ([\gamma], f_g\downarrow S_n\oplus S_n)\\ &= \frac{1}{n!}\sum_{\pi\in S_n} \zeta^{\gamma}(\pi^{-1})\prod_k \chi^{D}(g^k)^{a_k(\pi)}\chi^{D'}(g^k)^{a_k(\pi)}\\ &= \chi^{(D\otimes D')\boxdot[\gamma]}(g).\end{aligned}$$

■

Analogously one obtains

**5.2.39** $$(D\#D')\boxdot[\gamma] = \sum_{\alpha,\beta}([\alpha]\otimes[\beta],[\gamma])(D\boxdot[\alpha])\#(D'\boxdot[\beta]).$$

If we restrict attention to symmetric groups, we find the following:

**5.2.40** Theorem of associates (in case of symmetrization). *If* $\alpha\vdash n$, $\beta\vdash n$, *then*

$$[\alpha']\boxdot[\beta] = \begin{cases}([\alpha]\boxdot[\beta])\otimes[1^n] & \text{if } n \text{ is odd}\\ [\alpha]\boxdot[\beta] & \text{if } n \text{ is even.}\end{cases}$$

*Proof.*

$$\begin{aligned}\chi^{[\alpha']\boxdot[\beta]}(\pi) &= \frac{1}{n!}\sum_{\rho\in S_n}\zeta^{\beta}(\rho^{-1})\prod_{k=1}^{n}\zeta^{\alpha'}(\pi^k)^{a_k(\rho)}\\ &= \frac{1}{n!}\sum_{\rho\in S_n}\zeta^{\beta}(\rho^{-1})\prod_k \zeta^{\alpha}(\pi^k)^{a_k(\rho)}(\operatorname{sgn}\pi)^{ka_k(\rho)}\\ &= (\operatorname{sgn}\pi)^n\chi^{[\alpha]\boxdot[\beta]}(\pi).\end{aligned}$$

■

## 5.3 Permutrization of Representations

The preceding section was devoted to the symmetrizations $D\boxdot[\alpha]$. These representations yield the decomposition

$$\overset{n}{\otimes} D = \sum_{\alpha \vdash n} f^{\alpha}(D\boxdot[\alpha])$$

of $\otimes^n D$. The basic equation for the proof of this was

$$\overset{n}{\otimes} D(g) \circ P_n^m(\pi) = P_n^m(\pi) \circ \overset{n}{\otimes} D(g),$$

so that Schur's lemma could be applied to the matrices $\otimes^n \mathbb{D}(g)$ under the assumption that $P_n^m$ is completely reducible into absolutely irreducible representations. Hence an application of Schur's lemma to $\mathbb{P}_n^m(\pi)$ yields analogously (cf. 5.2.9):

**5.3.1** THEOREM. *Let $G$ denote a finite group, and $D$ an ordinary representation of $G$ on the vector space $V$, while $n \in \mathbb{N}$. If $D_i$ denotes an ordinary irreducible representation of dimension $f^i$ of $G$, then the homogeneous component of type $D_i$ of $\otimes^n V$ (as a left $G$-module) affords, if it is nonzero, $f^i$ times a representation of $S_n$. We denote this representation by $D \triangle_n D_i$ and call it the* $n$th permutrization *of $D$ by $D_i$. Its dimension is equal to the multiplicity $(\otimes^n D, D_i)$ of $D_i$ in $\otimes^n D$, and it satisfies, for each $g \in G$, $\pi \in S_n$, and with respect to a suitable basis of $\otimes^n V$,*

$$\overset{n}{\otimes} \mathbb{D}(g) \cdot \mathbb{P}_n^m(\pi) = \dot{+}_{i} (\mathbb{D}_i(g) \times \mathbb{D} \triangle_n \mathbb{D}_i(\pi)).$$

*(We allow $D \triangle_n D_i$ to denote the zero "representation" of $S_n$ if $D_i$ is not an irreducible constituent of $\otimes^n D$, so that the corresponding homogeneous component of type $D_i$ is the zero subspace. Thus $D \triangle_n D_i$ is defined for each $i$, but the corresponding summand in the above equation may have to be neglected.)*

To 5.2.10 there corresponds now the decomposition

**5.3.2**
$$P_n^m = \sum_i f^i(D \triangle_n D_i).$$

Notice that this holds for each $m$-dimensional representation $D$ of any finite group $G$.

From the equation in 5.3.1 we get that the character of $D \triangle_n D_i$ satisfies

**5.3.3**
$$\forall g \in G, \pi \in S_n \quad \prod_{k=1}^{n} \chi^D(g^k)^{a_k(\pi)} = \sum_i \zeta^i(g) \chi^{D \triangle_n D_i}(\pi),$$

where $\zeta^i$ is the character of $D_i$, so that we obtain by an application of the orthogonality relations

**5.3.4** $$\chi^{D \triangle_n D_i}(\pi) = \frac{1}{|G|} \sum_{g \in G} \zeta^i(g^{-1}) \prod_{k=1}^{n} \chi^D(g^k)^{a_k(\pi)}. \quad = \langle \psi^{\tau(\pi)}(\chi^D), \zeta^i \rangle_G$$

Comparing this with 5.2.13, we get the following very close connection between symmetrization and permutrization which is in a sense a duality:

**5.3.5** DUALITY THEOREM. *For each ordinary representation $D$ of a finite group $G$, for each $n \in \mathbb{N}$, each $\alpha \vdash n$ and each ordinary irreducible representation $D_i$ of $G$ we have the following equality of multiplicities:*

$$(D \boxdot [\alpha], D_i) = (D \triangle_n D_i, [\alpha]).$$

*This multiplicity is just the multiplicity of $D_i \# [\alpha]$ in the representation of $G \times S_n$ afforded by $\otimes^n V$.*

This shows that we can define for $D$ and each $n \in \mathbb{N}$ a matrix $S(D, n)$, the rows of which correspond to the partitions $\alpha$ of $n$ (in lexicographic order, say) and the columns of which correspond to the ordinary irreducible representations $D_k$ of $G$ as follows:

5.3.6 $$S(D, n) := (s_{\alpha k} := (D \boxdot [\alpha], D_k)).$$

The duality theorem says that $S(D, n)$ contains in its $k$th column the multiplicities

**5.3.7** $$s_{\alpha k} = (D \triangle_n D_k, [\alpha]),$$

so that its columns yield the decompositions of the permutrizations $D \triangle_n D_k$ while its rows yield the decompositions of the symmetrizations $D \boxdot [\alpha]$. A numerical example is

$$S([2,1^2],3) = \begin{array}{c} \begin{array}{ccccc} [1^4] & [2,1^2] & [2^2] & [3,1] & [4] \end{array} \\ \begin{pmatrix} 0 & 0 & 0 & 0 & 1 \\ 0 & 1 & 1 & 1 & 0 \\ 1 & 2 & 0 & 1 & 0 \end{pmatrix} \end{array} \begin{array}{c} \\ [1^3] \\ [2,1] \\ [3] \end{array} .$$

The multiplicities $s_{\alpha k}$ allow the following symmetric version of the equations in 5.3.1:

**5.3.8** $$\overset{n}{\otimes} \mathbb{D}(g) \cdot \mathbb{P}_n^m(\pi) = \dot{+}_{\alpha} \dot{+}_{k} s_{\alpha k} (\mathbb{D}_k(g) \times [\![\alpha]\!](\pi)).$$

Let us return to formula 5.3.4, which yields the character of $D \triangle_n D_i$. We would like to point to a few consequences of the fact that by 5.3.4 each expression of the form

$$\frac{1}{|G|} \sum_{g \in G} \zeta^i(g^{-1}) \prod_{k=1}^{n} \chi^D(g^k)^{a_k(\pi)}$$

is the value of a character of a symmetric group.

Since the character table of $S_n$ is a matrix over $\mathbb{Z}$, we obtain as the first result

**5.3.9** THEOREM. *Let $G$ denote a finite group with ordinary representations $D$ and $D'$. Then for any $a_k \in \mathbb{N}_0$, $1 \leqslant k \leqslant n$, we have*

$$\frac{1}{|G|} \sum_{g \in G} \chi^{D'}(g^{-1}) \prod_{k=1}^{n} \chi^D(g^k)^{a_k} \in \mathbb{Z}.$$

A particular case is (put $a_n := 1$, $a_1 := a_2 := \cdots := a_{n-1} := 0$)

**5.3.10**
$$\frac{1}{|G|} \sum_g \zeta^i(g^{-1}) \chi^D(g^n) \in \mathbb{Z}.$$

But this number is the multiplicity of $\zeta^i$ in the class function $\chi_n^D$ defined by

5.3.11
$$\chi_n^D(g) := \chi^D(g^n),$$

so that we obtain

**5.3.12** COROLLARY. *For each ordinary representation $D$ of a finite group $G$,* 5.3.11 *defines a generalized character, for the multiplicity of $\zeta^i$ in $\chi_n^D$ is equal to*

$$\chi^{D \triangle_n D_i}((1 \cdots n)).$$

Let us denote by $D_1$ the identity representation of $G$ and compare the particular multiplicity

**5.3.13**
$$\chi^{D_i \triangle_n D_1}((1 \cdots n)) = \frac{1}{|G|} \sum_g \zeta^i(g^n)$$

with the dimension of $D_i \triangle_n D_1$:

**5.3.14**
$$\chi^{D_i \triangle_n D_1}(1) = \frac{1}{|G|} \sum_g \zeta^i(g)^n.$$

This leads us to a characterization of ordinary irreducible characters of $G$ where the averages on the right-hand sides of 5.3.13, 14 coincide:

**5.3.15** THEOREM. *If $\zeta^i$ is an ordinary irreducible character of a finite group $G$ and $n \in \mathbb{N}$, then*

$$\frac{1}{|G|}\sum_g \zeta^i(g^n) = \frac{1}{|G|}\sum_g \zeta^i(g)^n$$

*if and only if $D_i \triangle_n D_1$ is of the form*

$$D_i \triangle_n D_1 = b_1^i[n] + b_2^i[1^n],$$

*where the multiplicity $b_2^i$ vanishes if $n$ is even. If on the other hand $D_i \triangle_n D_1$ is of this particular form $b_1^i[n]+b_2^i[1^n]$, then still (i.e. whether $n$ is even or odd, no matter if $b_2^i \neq 0$)*

$$\frac{1}{|G|}\sum_g \zeta^i(g^n) \equiv \frac{1}{|G|}\sum \zeta^i(g)^n \pmod 2.$$

*Proof.*

(i) A comparison of 5.3.13 and 5.3.14 shows that the averages coincide if and only if $(1\cdots n)$ is contained in the kernel of $D_i \triangle_n D_1$. But we know from 2.1.13 that $[n]$ and $[1^n]$ (for each $n \in \mathbb{N}$) together with the representation $[2^2]$ of $S_4$ are the only ordinary irreducible representations of symmetric groups which are not faithful. Since $[2^2]$ does not contain (1234) in its kernel (for $\zeta^{(2^2)}((1234))=0$), the first part of the statement is true.

(ii) If $D_i \triangle_n D_1 = b_1^i[n]+b_2^i[1^n]$, then by 5.3.14

$$\frac{1}{|G|}\sum \zeta^i(g)^n = \chi^{D_i \triangle_n D_1}(1) = b_1^i + b_2^i,$$

while by 5.3.13 and 2.4.8,

$$\frac{1}{|G|}\sum \zeta^i(g^n) = \chi^{D_i \triangle_n D_1}((1\cdots n)) = b_1^i + (-1)^{n-1} b_2^i \equiv b_1^i + b_2^i \pmod 2.$$

■

This is the first case where we meet a congruence modulo a prime number. We would like to show now that our result, that each expression of the form

$$\frac{1}{|G|}\sum_g \chi^{D'}(g^{-1}) \prod_k \chi^D(g^k)^{a_k}$$

is the value of a character of a symmetric group, opens an easy way to a systematic use of congruences. (This can of course be done in other ways, too, since characters are sums of roots of unity and the binomial theorem and number-theoretic arguments may be applied, but such arguments need case-by-case proofs.)

Let $p$ denote a prime number. If $\pi \in S_n$, then we know that this element is of the form $\pi = \rho\sigma$, where $\rho$ denotes the *p-regular component* of $\pi$, i.e. $p$ does not divide the order of $\rho$, while the order of $\sigma$ is a power of $p$ and $\rho\sigma = \sigma\rho$. We know from general representation theory that for each character $\chi$ of $S_n$ we have

**5.3.16** $$\chi(\pi) \equiv \chi(\rho) \pmod{p}.$$

Hence for example

**5.3.17** $$\chi^{D \triangle_p D_i}((1\cdots p)) \equiv \chi^{D \triangle_p D_i}(1) \pmod{p}.$$

A particular case is the following generalization of a part of 5.3.15:

**5.3.18** $$\frac{1}{|G|}\sum_g \zeta^i(g^p) \equiv \frac{1}{|G|}\sum_g \zeta^i(g)^p \pmod{p}.$$

In terms of multiplicities 5.3.17 reads as follows:

**5.3.19** $$(\chi_p^D, \zeta^i) \equiv ((\chi^D)^p, \zeta^i) \pmod{p},$$

where $(\chi^D)^p$ denotes the character of $\otimes^p D$, and $\chi_p^D$ is the class function defined in 5.3.11. Let us abbreviate this by

**5.3.20** $$\chi_p^D \equiv (\chi^D)^p \pmod{p}.$$

Since $\chi^{D \triangle_p D_i}((1\cdots p))$ is linear in $D$, and $\varphi^p$ is linear in $\varphi$ mod $p$, we obtain more generally

**5.3.21** THEOREM. *For each generalized character $\varphi$ of $G$ we have*

$$\varphi_p \equiv \varphi^p \pmod{p}.$$

We shall apply this in the proof of

**5.3.22** THEOREM. *Suppose that $\chi^R$ is the character of the regular representation $R$ of $G$, and that $n$ divides $|G|$. Then*

$$\chi_n^R \equiv 0 \pmod{n},$$

*i.e., each multiplicity in the generalized character* $\chi_n^R$ *of G is divisible by n, i.e.,* $(1/n)\chi_n^R$ *is a generalized character.*

*Proof.*

(i) We first show that we need only prove the statement when $n$ is a power of a prime. This can be seen as follows: Assume $n=s\cdot t$, where $s$ and $t$ are relatively prime, and assume furthermore that the statement holds for $s$ and $t$, i.e.

$$\chi_s^R \equiv 0 \pmod{s} \quad \text{and} \quad \chi_t^R \equiv 0 \pmod{t}.$$

Then since

$$\chi_n^R = (\chi_s^R)_t = (\chi_t^R)_s,$$

we have $\chi_n^R \equiv 0$ both modulo $s$ and modulo $t$, so that each multiplicity is divisible by both $s$ and $t$ and hence also by $n$, since we assumed $(s,t)=1$.

(ii) Let $n$ be a power of a prime, say $n=p^r$. We use induction on $r$.

(I) $r:=0$. This case is trivial.

(II) $r>0$. The induction hypothesis yields the existence of a generalized character $\varphi$ which satisfies

$$\chi_{p^{r-1}}^R = p^{r-1}\varphi. \tag{1}$$

We have to prove that $\chi_{p^r}^R \equiv 0 \pmod{p^r}$. As

$$\chi_{p^r}^R = (\chi_{p^{r-1}}^R)_p = p^{r-1}\varphi_p,$$

it suffices to prove that $\varphi_p \equiv 0 \pmod{p}$ or (apply 5.3.21) that

$$\varphi^p \equiv 0 \pmod{p}. \tag{2}$$

From the definition (1) of $\varphi$ we get that for each $g \in G$

$$\varphi^p(g) = \begin{cases} \left(\dfrac{|G|}{p^{r-1}}\right)^p & \text{if } g^{p^{r-1}}=1, \\ 0 & \text{otherwise} \end{cases}$$

$$= \left(\frac{|G|}{p^{r-1}}\right)^{p-1} \varphi(g),$$

and hence

$$\varphi^p = \left(\frac{|G|}{p^{r-1}}\right)^{p-1} \varphi. \tag{3}$$

Since $p^r$ divides $|G|$, $p$ divides $|G|/p^{r-1}$ and therefore (3) shows that (2) is true. ■

This proof indicates that there is a close connection between divisibility questions and the count of roots of unity in groups. We would like to give a more detailed description of these connections. For $g \in G$ and $n \in \mathbb{N}$ we put

5.3.23 $$R_n(g) := \{x \mid x \in G \text{ and } g = x^n\};$$

thus $R_n(g)$ consists of the *n th roots* of $g$. The order of $R_n(g)$ will be denoted by

5.3.24 $$r_n(g) := |R_n(g)|.$$

The next lemma shows that $r_n(g)$ can be expressed nicely in terms of permutrized characters:

**5.3.25** LEMMA. $r_n(g) = \sum_i \chi^{D_i \triangle_n D_1}((1 \cdots n))\zeta^i(g)$.

*Proof.* The following equation is obvious:

$$\sum_{x \in G} \zeta^i(x^n) = \sum_{g \in G} r_n(g)\zeta^i(g).$$

An application of an orthogonality relation yields therefrom the statement. ■

As a corollary we obtain for the number $r_n(1)$ of $n$th roots of unity:

**5.3.26** $$r_n(1) = \sum_i \chi^{D_i \triangle_n D_1}((1 \cdots n))f^i = \chi^{R \triangle_n D_1}((1, \ldots, n)).$$

Since hook representations $[\alpha] := [n-r, 1^r]$, $0 \leq r < n$, are the only ones such that $\zeta^\alpha((1 \cdots n)) \neq 0$, and since

$$\zeta^{(n-r,1^r)}((1 \cdots n)) = (-1)^r,$$

we get from 5.3.26 $r_n(1)$ as an alternating sum:

**5.3.27** $$r_n(1) = \sum_{r=0}^{n-1} (R \triangle_n D_1, [n-r, 1^r])(-1)^r.$$

Since

$$r_n(1)=\chi^{R\triangle_n D_1}((1\cdots n))=(\chi_n^R,\zeta^1),$$

we obtain from 5.3.26 and 5.3.22:

**5.3.28** THEOREM. *For each divisor $n$ of $G$ we have*

$$r_n(1)\equiv 0 \pmod{n}.$$

There is an old conjecture of Frobenius which says that $r_n(1)=n$ implies that $R_n(1)$ is a subgroup of $G$. That this is true under certain additional conditions is shown in the next theorem:

**5.3.29** THEOREM. *If $n$ divides the order of a finite group $G$ but is relatively prime to the dimension of every ordinary irreducible representation of $G$, while $r_n(1)=n$, then $R_n(1)$ is an abelian normal subgroup of $G$.*

*Proof.*

(i) In order to prove that $R_n(1)$ is a normal subgroup, we show that $\chi_n^R$ is a character in this case, so that $R_n(1)$ is the kernel of a representation and hence normal. To do this we notice that for each $i$

$$(\chi_n^R,\zeta^i)=\sum_{g\in R_n(1)}\zeta^i(g). \tag{1}$$

But $R_n(1)$ is a union of conjugacy classes $C_j$, while the class multipliers

$$\omega_j^i:=\frac{|C_j|\zeta_j^i}{f^i} \qquad \left(\text{where}\quad f^i:=\zeta^i(1)\right)$$

are algebraic integers. Hence $(\chi_n^R,\zeta^i)$ is a sum of algebraic integers with common factor $f^i$, so that

$$(\chi_n^R,\zeta^i)\equiv 0 \pmod{f^i}.$$

But 5.3.22 has shown that in addition

$$(\chi_n^R,\zeta^i)\equiv 0 \pmod{n}.$$

The assumption $(f^i,n)=1$ therefore gives

$$(\chi_n^R,\zeta^i)\equiv 0 \pmod{nf^i}.$$

This together with $r_n(1)=n$ (so that the sum in (1) consists of $n$ summands, each of which is of absolute value $\leqslant f^i$, and at least one of which is $f^i$) gives

$$(\chi_n^R, \zeta^i) \in \{0, nf^i\},$$

and hence in particular $(\chi_n^R, \zeta^i) \geqslant 0$. Thus $\chi_n^R$ is a character.

(ii) We consider an arbitrary irreducible character $\hat{\zeta}$ of $R_n(1)$ together with a $\zeta^i$ which contains $\hat{\zeta}$ when it is restricted:

$$0 \neq (\zeta^i \downarrow R_n(1), \hat{\zeta}).$$

$R_n(1)$ is a normal subgroup, and hence by Clifford's theory $D_i \downarrow R_n(1)$ consists of $\hat{\zeta}(1)$-dimensional irreducibles only. Hence $\hat{\zeta}(1)$ divides both $n = |R_n(1)|$ and $f^i$, so that the assumption $(f^i, n) = 1$ yields $\hat{\zeta}(1) = 1$. ■

The special case $n := 2$ is of particular interest. 5.3.26 gives for the number $r_2(1)$ of involutions in $G$:

**5.3.30** $$r_2(1) = \sum_i \chi^{D_i \triangle_2 D_1}((12)) \cdot f^i = \chi^{R \triangle_2 D_1}((12)).$$

The coefficients

5.3.31 $$c_i := \chi^{D_i \triangle_2 D_1}((12)) = \frac{1}{|G|} \sum_g \zeta^i(g^2)$$

are well known in the representation theory of finite groups, for $c_i$ determines the *kind* of $D_i$. the ordinary irreducible representation $D_i$ of $G$ is said to be of the *first* kind if it is equivalent to a real representation. It is said to be of the *second* or *third* kind if it is not equivalent to a real representation, according as $\zeta^i$, its character, is real-valued or not. Thus this well known invariant $c_i$ has nicely turned out to be a character value of $S_2$.

We know from general representation theory that the following holds:

**5.3.32** $$c_i = \begin{cases} 1 & \text{if } D_i \text{ is of the first kind,} \\ -1 & \text{if } D_i \text{ is of the second kind,} \\ 0 & \text{if } D_i \text{ is of the third kind.} \end{cases}$$

Part of these results is immediate from the definition of symmetrization: $c_i \in \{0, 1, -1\}$ follows from

$$f^{D_i \triangle_2 D_1} \leqslant 1$$

and the character table of $S_2$. That $c_i = 0$ if and only if $\zeta^i$ is not real is not difficult to show. We notice that 5.3.32 implies:

**5.3.33.** *$D_i$ is of the first (second, third) kind if and only if $D_i \triangle_2 D_1$ is equal to* [2] ([$1^2$], *zero*).

For the count of square roots of $g \in G$ we obtain from 5.3.25:

**5.3.34**
$$r_2(g) = \sum_i c_i \zeta^i(g) = \sum_{\substack{i \\ c_i = 1}} \zeta^i(g) - \sum_{\substack{j \\ c_j = -1}} \zeta^j(g);$$

in particular

**5.3.35**
$$r_2(1) = \sum_{\substack{i \\ c_i = 1}} f^i - \sum_{\substack{j \\ c_j = -1}} f^j.$$

The equations 5.3.34, 35 show clearly the close connection between the kinds of the ordinary irreducible representations of $G$ and the existence of square roots. A typical example is

**5.3.36.** *If there exists a $g \in G$ such that $r_2(g) > r_2(1)$, i.e. $g$ has more square roots than 1, then $G$ has ordinary irreducible representations of the second kind, and hence in particular $\mathbb{R}$ is not a splitting field for $G$.*

*Proof.* We have by 5.3.34, 35

$$r_2(g) - r_2(1) = \sum_{\substack{i \\ c_i = 1}} \left(\zeta^i(g) - f^i\right) - \sum_{\substack{j \\ c_j = -1}} \left(\zeta^j(g) - f^j\right).$$

If $G$ had no representation of the second kind, then

$$r_2(g) - r_2(1) = \sum_{\substack{i \\ c_i = 1}} \left(\zeta^i(g) - f^i\right) \leqslant 0.$$ ∎

An example is provided by $G := Q_8 := \{\pm 1, \pm j, \pm k, \pm l\}$, the quaternion group of order 8, where $-1$ has the six square roots $\pm j$, $\pm k$ and $\pm l$, while 1 has only two square roots, namely $\pm 1$. Hence $Q_8$ has ordinary representations which cannot be written over $\mathbb{R}$.

There are various other results which can be obtained similarly. In certain applications of representation theory in physics it is useful to know whether an ordinary irreducible representation $D_i$ is a so-called *simple-phase repre-*

*sentation*, i.e. its character satisfies

5.3.37
$$\frac{1}{|G|}\sum_g \zeta^i(g^3)=\frac{1}{|G|}\sum_g \zeta^i(g)^3.$$

$G$ is called a *simple-phase group* if and only if each $D_i$ is simple-phase. From our earlier results we have:

**5.3.38** THEOREM. *The following properties of an ordinary irreducible representation $D_i$ with character $\zeta^i$ are equivalent*:

(i) $D_i$ *is simple-phase*,

(ii) $(1/|G|)\Sigma_g \zeta^i(g^3)=(1/|G|)\Sigma_g \zeta^i(g)^3$,

(iii) $\chi^{D_i \triangle_3 D_1}((123))=\chi^{D_i\triangle_3 D_1}(1)=(\otimes^3 D_i, D_1)$,

(iv) $(D_i \triangle_3 D_1, [2,1])=0$,

(v) $D_i \triangle_3 D_1$ *has one-dimensional constituents at most*,

(vi) $(D_i \boxdot [2,1], D_1)=0$.

More generally one might call an ordinary irreducible representation $D_i$ of $G$ an *n-simple-phase representation* if and only if $D_i \triangle_n D_1$ has at most one-dimensional constituents. Numerical calculations have shown that $[\alpha]$, $\alpha \vdash n \leqslant 5$, is 3-simple-phase, while $[3,2,1]$ is not, so that $S_6$ is not a simple-phase group. If we call a finite group $G$ an *n-simple-phase group* if and only if each $D_i$ is $n$-simple-phase, then we have

**5.3.39** THEOREM. *If $n$ is odd and if there exists a $g \in G$ such that $r_n(g)>r_n(1)$, then $G$ is not $n$-simple-phase.*

*Proof.* If $G$ were $n$-simple-phase and $n$ odd, then (see 5.3.25)

$$r_n(1)=\sum_i \left(\overset{n}{\otimes} D_i, D_1\right) f^i \geqslant \sum_i \left(\overset{n}{\otimes} D_i, D_1\right)\zeta^i(g)=r_n(g). \qquad \blacksquare$$

The Duality Theorem 5.3.5 yields the following characterization of $n$-simple-phase representations in terms of symmetrization:

**5.3.40** THEOREM. *$D_i$ is $n$-simple-phase if and only if $D_1$ occurs at most in the symmetric part $D_i \boxdot [n]$ or in the antisymmetric part $D_i \boxdot [1^n]$ of $\otimes^n D_i$, i.e., $(D_i \boxdot [\alpha], D_1)=0$ when $(n)\neq\alpha\neq(1^n)$.*

We would like to provide a few numerical results concerning ordinary irreducible representations of symmetric groups being $n$-simple phase or not. Computer calculations yielded the list in 5.3.41, the rows of which correspond to the partitions $\alpha \vdash m$, $m \leqslant 7$. The columns are numbered by $n$, $1 \leqslant n \leqslant 6$, and the entries $*, +, -, 0$ mean that the corresponding $[\alpha]\triangle_n[m]$

either does not only contain one-dimensional constituents, has $[n]$ as its only irreducible constituent, has $[1^n]$ as its only irreducible constituent, or does not exist.

**5.3.41**

| | 1 | 2 | 3 | 4 | 5 | 6 |
|---|---|---|---|---|---|---|
| [1] | + | + | + | + | + | + |
| [2] | + | + | + | + | + | + |
| $[1^2]$ | − | + | 0 | + | 0 | + |
| [3] | + | + | + | + | + | + |
| [2,1] | * | + | + | * | * | * |
| $[1^3]$ | − | + | 0 | + | 0 | + |
| [4] | + | + | + | + | + | + |
| [3,1] | * | + | + | * | * | * |
| $[2^2]$ | * | + | + | * | * | * |
| $[2,1^2]$ | * | + | − | * | * | * |
| $[1^4]$ | − | + | 0 | + | 0 | + |
| [5] | + | + | + | + | + | + |
| [4,1] | * | + | + | * | * | * |
| [3,2] | * | + | + | * | * | * |
| $[3,1^2]$ | * | + | − | * | * | * |
| $[2^2,1]$ | * | + | + | * | * | * |
| $[2,1^3]$ | * | + | 0 | * | * | * |
| $[1^5]$ | − | + | 0 | + | 0 | + |
| [6] | + | + | + | + | + | + |
| [5,1] | * | + | + | * | * | * |
| [4,2] | * | + | + | * | * | * |
| $[4,1^2]$ | * | + | − | * | * | * |
| $[3^2]$ | * | + | 0 | * | − | * |
| [3,2,1] | * | + | * | * | * | * |
| $[3,1^3]$ | * | + | − | * | * | * |
| $[2^3]$ | * | + | + | * | * | * |
| $[2^2,1^2]$ | * | + | 0 | * | * | * |
| $[2,1^4]$ | * | + | 0 | * | − | * |
| $[1^6]$ | − | + | 0 | + | 0 | + |
| [7] | + | + | + | + | + | + |
| [6,1] | * | + | + | * | * | * |
| [5,2] | * | + | + | * | * | * |
| $[5,1^2]$ | * | + | − | * | * | * |
| [4,3] | * | + | + | * | * | * |
| [4,2,1] | * | + | * | * | * | * |
| $[4,1^3]$ | * | + | − | * | * | * |
| $[3^2,1]$ | * | + | − | * | * | * |
| $[3,2^2]$ | * | + | + | * | * | * |
| $[3,2,1^2]$ | * | + | * | * | * | * |
| $[3,1^4]$ | * | + | + | * | * | * |
| $[2^3,1]$ | * | + | + | * | * | * |
| $[2^2,1^3]$ | * | + | 0 | * | * | * |
| $[2,1^6]$ | * | + | 0 | * | 0 | * |
| $[1^7]$ | − | + | 0 | + | 0 | + |

An easy check of characters yields the following

**5.3.42** THEOREM OF ASSOCIATES (IN CASE OF PERMUTRIZATION). *If* $\alpha, \beta \vdash m$, $\gamma \vdash n$, *then we have the following equations of multiplicities*:

$$([\alpha'] \triangle_n [\beta], [\gamma]) = \begin{cases} ([\alpha] \triangle_n [\beta], [\gamma]) & \text{if } n \text{ is even}, \\ ([\alpha] \triangle_n [\beta'], [\gamma]) & \text{if } n \text{ is odd}. \end{cases}$$

An immediate consequence is

**5.3.43** COROLLARY.

(i) *If* $n \in \mathbb{N}$ *is even, then* $[\alpha]$ *is n-simple-phase if and only if* $[\alpha']$ *is n-simple-phase.*

(ii) *If* $n \in \mathbb{N}$ *is odd and* $([\alpha] \triangle_n [1^n], [\gamma]) = 0$ *for each* $[\gamma]$ *of dimension* $f^\gamma > 1$, *then* $[\alpha']$ *is also n-simple-phase.*

Returning to 5.3.26, the count of $n$th roots of the unit element of $G$, we can find further interesting connections between numbers of solutions of equations in $G$ and the values of specific permutrization characters. Let $\mathbb{D}_i$ denote an ordinary irreducible matrix representation of $G$. Then it is easy to verify that for each $n_1 \in \mathbb{N}$ the matrix sum

$$\sum_{g_1 \in G} \mathbb{D}_i(g_1^{n_1})$$

commutes with each of the matrices $\mathbb{D}_i(h)$, $h \in G$, so that by Schur's lemma it is a multiple of the identity matrix $\mathbb{I}$. The numerical factor follows from 5.3.13, which yields:

5.3.44 $$\sum_{g_1} \mathbb{D}_i(g_1^{n_1}) = \frac{|G|}{f^i} \chi^{D_i \triangle_{n(1)} D_1}((1 \cdots n_1)) \mathbb{I}\,.$$

(Here and below we write $\triangle_{n_j}$ as $\triangle_{n(j)}$ for typographical convenience.) From the product of $k$ such equations we derive

**5.3.45**

$$\frac{1}{|G|} \sum_{g_1, \ldots, g_k} f^i \mathbb{D}_i(g_1^{n_1} \cdots g_k^{n_k}) = \left(\frac{|G|}{f^i}\right)^{k-1} \prod_{j=1}^{k} \chi^{D_i \triangle_{n(j)} D_1}((1 \cdots n_j)) \mathbb{I}\,.$$

Now let $C$ denote a conjugacy class of $G$, and consider an element $g \in C$. 5.3.45 leads to the equation

$$\frac{1}{|G|}\sum_{g_1,\ldots,g_k} f^i \mathbb{D}_i(g_1^{n_1}\cdots g_k^{n_k}g^{-1})$$

$$=\left(\frac{|G|}{f^i}\right)^{k-1}\cdot\left(\prod_j \chi^{D_i \triangle_{n(j)} D_1}((1\cdots n_j))\right)\cdot\mathbb{D}_i(g^{-1}).$$

Summing over $i$ and taking traces, we find

$$\frac{1}{|G|}\sum_{g_1,\ldots,g_k} \chi^R(g_1^{n_1}\cdots g_k^{n_k}g^{-1})$$

$$=|G|\sum_i\left(\frac{|G|}{f^i}\right)^{k-2}\cdot\left(\prod_j \chi^{D_i \triangle_{n(j)} D_1}((1\cdots n_j))\right)\cdot\frac{1}{|C|}\cdot\frac{|C|\zeta^i(g^{-1})}{f^i}.$$

The left-hand side is equal to the number of solutions $(g_1,\ldots,g_k)\in G^k$ of the equation

$$g_1^{n_1}\cdots g_k^{n_k}=g,$$

so that the number of solutions of

$$g_1^{n_1}\cdots g_k^{n_k}\in C$$

must be equal to $|C|$ times the right-hand side. But in terms of central characters $\omega^i$ the right-hand side times $|C|$ is equal to

$$|G|\sum_i\left(\frac{|G|}{f^i}\right)^{k-2}\cdot\left(\prod_j \chi^{D_i \triangle_{n(j)} D_1}((1\cdots n_j))\right)\cdot\omega^i(g^{-1}).$$

Thus, for $k\geqslant 2$, $|G|^{-1}\{\gcd_i(|G|/f^i)\}^{-k+2}$ times the number of solutions of $g_1^{n_1}\cdots g_k^{n_k}\in C$ is both a rational number and an algebraic integer, and therefore it must be a rational integer. Thus we have proved

**5.3.46** THEOREM. *For given $n_j\in\mathbb{N}$, $1\leqslant j\leqslant k$, and a conjugacy class $C\subseteq G$, the number of solutions $(g_1,\ldots,g_k)\in G^k$ of the equation*

$$g_1^{n_1}\cdots g_k^{n_k}\in C$$

*is equal to*

$$|G|\sum_i\left(\frac{|G|}{f^i}\right)^{k-2}\cdot\left(\prod_j \chi^{D_i \triangle_{n(j)} D_1}((1\cdots n_j))\right)\cdot\omega^i(g^{-1})$$

*for any $g \in C$. For $k \geqslant 2$ this number is divisible by $|G|$ times the greatest common divisor of the natural numbers $(|G|/f^i)^{k-2}$.*

Applying this to the case $k=2$ and $n_1=n_2=2$, we obtain that the number of pairs $(g_1, g_2) \in G \times G$ such that $g_1^2 g_2^2 = 1$ is equal to

5.3.47
$$|G| \sum_i c_i^2,$$

which, by 5.3.32, is equal to $|G|$ times the number of ordinary irreducible characters of $G$ which are $\mathbb{R}$-valued. Let us compare this with the number of ambivalent classes of $G$.

We denote by $C_j$ a conjugacy class of $G$, by $C_{j*}$ the class consisting of the inverses of the elements of $C_j$. Then in terms of Kronecker symbols, $\Sigma_j \delta_{j,j*}$ is the number of ambivalent classes of $G$, and the orthogonality relations imply that

5.3.48
$$\begin{aligned}\sum_j \delta_{j,j*} &= \sum_j \frac{1}{|G|} \sum_i |C_j| \zeta_j^i \zeta_j^i \\ &= \sum_i \frac{1}{|G|} \sum_j |C_j| \zeta_j^i \zeta_j^i.\end{aligned}$$

Hence the number of ambivalent classes is equal to the number of ordinary irreducible characters $\zeta^i$ which are $\mathbb{R}$-valued (so that $\zeta_j^i = \zeta_{j*}^i$ for each $j$).

Furthermore 5.3.25 shows that

5.3.49
$$\begin{aligned}\frac{1}{|G|} \sum_g r_2(g)^2 &= \frac{1}{|G|} \sum_g \Big( \sum_i c_i \zeta^i(g) \Big)^2 \\ &= \sum_{i,k} c_i c_k \frac{1}{|G|} \sum_g \zeta^i(g) \zeta^k(g^{-1}) \\ &= \sum_i c_i^2.\end{aligned}$$

Thus a comparison of 5.3.47, 5.3.48, and 5.3.49 yields:

**5.3.50** COROLLARY. *The number of solutions $(g_1, g_2) \in G \times G$ which satisfy $g_1^2 g_2^2 = 1$ is equal to the sum $\Sigma_g r_2(g)^2$ of the squares of the numbers of square roots as well as to the number of ambivalent classes of $G$ and the number of ordinary irreducible characters of $G$ which are $\mathbb{R}$-valued.*

These considerations show clearly the importance of the numbers $\chi^{D_i \triangle_n D_1}((1 \cdots n))$ for the group $G$ and its representation theory. Since $\chi^{D_i \triangle_2 D_1}((12))$ is usually denoted by $c_i$, we shall abbreviate these numbers as follows: For each $n \in \mathbb{N}$ we put

5.3.51 $$c_{i,n} := \chi^{D_i \triangle_n D_1}((1 \cdots n)) \quad (\in \mathbb{Z}),$$

so that in particular

**5.3.52** $$c_{i,2} = c_i \quad \textit{and} \quad c_{i,1} = \delta_{i,1}.$$

In terms of this notation the number of solutions of the equation

$$g_1^{n_1} \cdots g_k^{n_k} = 1$$

is equal to

**5.3.53** $$|G| \sum_i \left( \frac{|G|}{f^i} \right)^{k-2} \prod_j c_{i,n_j}.$$

Furthermore we have

**5.3.54**
$$\text{(i)} \quad \chi^{R \triangle_n D_1}((1 \cdots n)) = \sum_i f^i c_{i,n}.$$
$$\text{(ii)} \quad r_n(g) = \sum_i \zeta^i(g) c_{i,n}.$$

But we should not forget that permutrization also deals with much more general expressions like

$$\chi^{R \triangle_n D_1}(\pi) = \frac{1}{|G|} \sum_{g \in G} \prod_{k=1}^{n} \chi^R(g^k)^{a_k(\pi)}.$$

In order to interprete this expression as a cardinal number we recall that $R \triangle_n D_1$ is a representation of $S_n$ which is afforded by the homogeneous component of type $D_1$ of $\otimes^n R$ of $G$, or, equivalently, by the set of orbits of the following action of $G$ on the set of $n$-tuples:

$$g(g_1, \ldots, g_n) := (gg_1, \ldots, gg_n).$$

As the action of $\pi$ is by place permutation, $\chi^{R \triangle_n D_1}(\pi)$ denotes the number of sets of such $n$-tuples which are left invariant both under the action of each $g \in G$ and also under place permutation by $\pi$; i.e., we have proved

**5.3.55 Theorem.** $\chi^{R \triangle_n D_1}(\pi) = \frac{1}{|G|} \Sigma_g \Pi_k \chi^R(g^k)^{a_k(\pi)}$ *is equal to the number of sets of n-tuples* $(g_1, \ldots, g_n) \in G^n$ *which are invariant under both of the*

*following actions*:

$$\forall g \in G \quad g(g_1,\ldots,g_n):=(gg_1,\ldots,gg_n),$$

*and*

$$\pi(g_1,\ldots,g_n):=(g_{\pi^{-1}(1)},\ldots,g_{\pi^{-1}(n)}).$$

Analogous interpretations can be given for any permutation representation $P$ of $G$, say:

**5.3.56** THEOREM. *If $PG$ is a permutation representation of $G$ on a finite set $\Omega$, then*

$$\chi^{P\triangle_n D_1}(\pi)=\frac{1}{|G|}\sum_g \prod_{k=1}^{n} \chi^P(g^k)^{a_k(\pi)}$$

*is equal to the number of orbits of the natural action of $PG$ on $\Omega^n$ which remain invariant under place permutation by $\pi$, i.e. under*

$$(\omega_1,\ldots,\omega_n)\mapsto(\omega_{\pi^{-1}(1)},\ldots,\omega_{\pi^{-1}(n)}).$$

The case $n=2$ is quite well known: If $G$ acts on $\Omega$, then

$$\frac{1}{|G|}\sum_g a_1(g^2)$$

is the number of orbits of $G$ on $\Omega^2$ which are "symmetric", i.e. which together with $(\omega_1,\omega_2)$ also contain $(\omega_2,\omega_1)$.

## 5.4 Plethysms of Representations

An opportunity to apply the representation theory of wreath products to the representation theory of symmetric groups arises from the fact that the normalizer of a direct sum

5.4.1 $$\bigoplus^n S_m := S_m \oplus \cdots \oplus S_m, \qquad n \text{ summands},$$

in $S_{mn}$ is equal to $\psi[S_m \operatorname{wr} S_n]$ when the first summand in 5.4.1 is taken to be the symmetric group on the subset $\{1,\ldots,m\}$ of $\{1,\ldots,mn\}$, the second summand on $\{m+1,\ldots,2m\}$, and so on—i.e., we have

**5.4.2** $$N_{S_{mn}}\left(\bigoplus^n S_m\right)=\psi[S_m \operatorname{wr} S_n].$$

As obviously

**5.4.3** $$\overset{n}{\oplus} S_m \leq \psi[S_m \operatorname{wr} S_n] \leq S_{mn},$$

we can carry out each induction from a representation of $\oplus^n S_m$ to $S_{mn}$ in two steps, namely by inducing first up to the normalizer, and then up to $S_{mn}$. For example, if we take the $n$-fold outer power of $[\alpha]$, $\alpha \vdash m$:

5.4.4 $$\overset{n}{\#}[\alpha] := [\alpha] \# \cdots \# [\alpha], \qquad n \text{ factors},$$

as a representation of $\oplus^n S_m \simeq \times^n S_m$, then

**5.4.5** $$[\alpha]\cdots[\alpha] := \overset{n}{\#}[\alpha]\uparrow S_{mn} = \overset{n}{\#}[\alpha]\uparrow \psi[S_m \operatorname{wr} S_n]\uparrow S_{mn}.$$

Frobenius's reciprocity theorem together with our results on the representation theory of wreath products yields (cf. 4.3.16)

**5.4.6** $$\overset{n}{\#}[\alpha]\uparrow \psi[S_m \operatorname{wr} S_n] = \sum_{\beta \vdash n} f^\beta(\alpha; \beta)$$

if we abbreviate

5.4.7 $$(\alpha; \beta) := ([\alpha]; [\beta]).$$

Thus 5.4.5 gives

**5.4.8** $$[\alpha]\cdots[\alpha] = \sum_{\beta \vdash n} f^\beta((\alpha; \beta)\uparrow S_{mn}).$$

We therefore put

5.4.9 $$[\alpha] \odot [\beta] := (\alpha; \beta)\uparrow S_{mn},$$

and call this representation the "plethysm" of $[\alpha]$ and $[\beta]$. This allows us to rephrase 5.4.8 as follows:

**5.4.10** $$[\alpha]\cdots[\alpha] = \sum_{\beta \vdash n} f^\beta([\alpha] \odot [\beta]).$$

This equation shows that we are left with the problem of decomposing the plethysms $[\alpha]\odot[\beta]$ if we don't want to carry out the Littlewood-Richardson rule to calculate $[\alpha]\cdots[\alpha]$. But astonishingly this problem of decomposing $[\alpha]\odot[\beta]$ is still far from a satisfactory solution, and it seems to be difficult.

It is obvious how 5.4.9 can be generalized. If $G \leq S_m$, $H \leq S_n$, then $\psi[G \operatorname{wr} H] \leq S_{mn}$, so that we can define for any representations $D$ of $G$ and

$D''$ of $H$ (over the same field, of course) the representation

5.4.11 $$(D;D''):=\left(\overset{n}{\#}D\right)^{\sim}\otimes D'$$

of $\psi[G\operatorname{wr}H]$, and then define

5.4.12 $$D\odot D':=(D;D'')\uparrow S_{mn}.$$

This representation is called the *plethysm* of $D$ and $D''$.

But let us restrict our attention to plethysms $[\alpha]\odot[\beta]$ of ordinary irreducible representations of symmetric groups. As already mentioned, the reduction of $[\alpha]\odot[\beta]$ into its irreducible constituents is not known in general, but there are various results available which can be used in particular situations. For example the

**5.4.13** THEOREM OF ASSOCIATES. *If $\alpha\vdash m$, $\beta\vdash n$, $\gamma\vdash mn$, then we have for the corresponding multiplicities*:

$$([\alpha]\odot[\beta],[\gamma])=\begin{cases}([\alpha']\odot[\beta],[\gamma']) & \textit{if}\quad m \textit{ is even},\\ ([\alpha']\odot[\beta'],[\gamma']) & \textit{if}\quad m \textit{ is odd}.\end{cases}$$

*Proof.* Frobenius's reciprocity yields (if we write $S_m\operatorname{wr}S_n$ instead of $\psi[S_m\operatorname{wr}S_n]$, for simplicity)

$$([\alpha]\odot[\beta],[\gamma])=([\gamma]\downarrow S_m\operatorname{wr}S_n,(\alpha;\beta)),$$

and

$$\begin{aligned}([\alpha]\odot[\beta],[\gamma'])&=([\gamma']\downarrow S_m\operatorname{wr}S_n,(\alpha;\beta))\\ &=([\gamma]\downarrow S_m\operatorname{wr}S_n,(\alpha;\beta)\otimes([1^{mn}]\downarrow S_m\operatorname{wr}S_n)).\end{aligned}$$

Therefore, we are left with the problem of determining

$$(\alpha;\beta)\otimes([1^{mn}]\downarrow S_m\operatorname{wr}S_n).$$

This is the irreducible representation of $S_m\operatorname{wr}S_n$ (better: $\psi[S_m\operatorname{wr}S_n]$) which is associated to $(\alpha;\beta)$ with respect to the subgroup

$$\psi[S_m\operatorname{wr}S_n]\cap A_{mn}.$$

In order to solve our problem, we notice that $\psi((e;(12)))$ consists of $m$ 2-cycles together with $mn-2m$ 1-cycles, so that

$$\psi((e;(12)))\in\begin{cases}A_{mn} & \text{if}\quad m \text{ is even},\\ S_{mn}\backslash A_{mn} & \text{otherwise}.\end{cases}$$

Hence if $m$ is even, then $\psi[S_m \operatorname{wr} S_n] \cap A_{mn}$ is equal to

$$\left\{\psi((f;\pi)) \,\Big|\, \prod_i f(i) \text{ even}\right\},$$

while, if $m$ is odd, then $\psi[S_m \operatorname{wr} S_n] \cap A_{mn}$ is equal to

$$\left\{\psi((f;\pi)) \,\Big|\, \prod_i f(i) \text{ even iff } \pi \text{ is even}\right\}.$$

This yields for the character of $[1^{mn}] \downarrow S_m \operatorname{wr} S_n$

5.4.14 $$\zeta^{(1^{mn})}(\psi((f;\pi))) = \begin{cases} \operatorname{sgn} \prod_i f(i) & \text{if } m \text{ is even,} \\ \operatorname{sgn} \pi \cdot \operatorname{sgn} \prod_i f(i) & \text{if } m \text{ is odd.} \end{cases}$$

Now as

$$\zeta^{(\alpha;\beta)}((f;\pi)) = \zeta^{\beta}(\pi) \prod_\nu \zeta^{\alpha}(g_\nu(f;\pi)),$$

we have for the value of the character associated with $\zeta^{(\alpha;\beta)}$ with respect to $\psi[S_m \operatorname{wr} S_n] \cap A_{mn}$ taken on $\psi((f;\pi))$

$$\begin{cases} \zeta^{\beta}(\pi) \prod_\nu \zeta^{\alpha'}(g_\nu(f;\pi)) & \text{if } m \text{ is even,} \\ \zeta^{\beta'}(\pi) \prod_\nu \zeta^{\alpha'}(g_\nu(f;\pi)) & \text{if } m \text{ is odd.} \end{cases}$$

This yields the statement. ■

The proof of this theorem points to interesting problems concerning the relationship between irreducible representations of $S_m \operatorname{wr} S_n$ and its subgroups of index 2. We shall not go into details here; the interested reader is referred to the literature. But we should at least remark that there are further subgroups of index 2 in $S_m \operatorname{wr} S_n$ which can be treated in a similar way to that in which the alternating group is treated in connection with symmetric groups.

As we mentioned in 4.1.33, the Weyl groups of type $D_n$ are similar to the groups

5.4.15 $$\psi[S_2 \operatorname{wr} S_n] \cap A_{2n},$$

so that such considerations are in a certain sense generalizations of the representation theory of the (series of) Weyl groups.

Another theorem, perhaps the most important one, can also be obtained quite easily. With its help, the problem of decomposing $[\alpha]\odot[\beta]$ can be reduced to the decomposition of plethysms $[\alpha]\odot[r]$, $r\leqslant n$, since a determinantal form analogous to 2.3.15 will be derived. We need a preliminary lemma:

**5.4.16** LEMMA. $([\alpha]\odot[n_1])([\alpha]\odot[n_2])=[\alpha]\odot([n_1][n_2])$.

*Proof.* Using a well-known theorem on induced representations (see e.g. Curtis and Reiner [1962, (43.2)]) and the transitivity of induction, we obtain for the left-hand side, if $\alpha\vdash m$ and $n_1+n_2=n$,

$$([\alpha]\odot[n_1])([\alpha]\odot[n_2])$$
$$=((\alpha;(n_1))\#(\alpha;(n_2))\uparrow S_m \operatorname{wr}(S_{n_1}\oplus S_{n_2})\uparrow S_m \operatorname{wr} S_n \uparrow S_{mn}.$$

As $\psi[(S_m \operatorname{wr} S_{n_1})\oplus(S_m \operatorname{wr} S_{n_2})]=\psi[S_m \operatorname{wr}(S_{n_1}\oplus S_{n_2})]$, this is equal to

$$(\alpha;[n_1]\#[n_2])\uparrow S_m \operatorname{wr} S_n \uparrow S_{mn}=[\alpha]\odot([n_1][n_2]),$$

as stated. ■

Furthermore it is clear that plethysm is distributive on the right hand side: If $D'=\sum_i b_i D_i'$, $D_i'$ irreducible, then

**5.4.17**
$$D\odot D'=\sum_i b_i(D\odot D_i').$$

Combining this, 5.4.16, and the determinantal form 2.3.15, we obtain:

**5.4.18** THEOREM. $[\alpha]\odot[\beta]=\|[\alpha]\odot[\beta_i+j-i]\|$.

For example

$$[\alpha]\odot[2,1]=\begin{vmatrix}[\alpha]\odot[2] & [\alpha]\odot[3]\\ 1 & [\alpha]\odot[1]\end{vmatrix}$$
$$=([\alpha]\odot[2])([\alpha]\odot[1])-[\alpha]\odot[3]$$
$$=([\alpha]\odot[2])\cdot[\alpha]-[\alpha]\odot[3].$$

The plethysm $[\alpha]\odot[\beta_i+j-i]$ is of the form $[\alpha]\odot[r]$, $r\leqslant n$. If we know the decomposition of these plethysms, then we can get the decomposition of $[\alpha]\odot[\beta]$ by 5.4.18 and suitable applications of the Littlewood-Richardson rule.

An obvious generalization of 5.4.16 is

**5.4.19** $$([\alpha]\odot[\beta])([\alpha]\odot[\gamma])=[\alpha]\odot([\gamma][\beta]).$$

The associativity of the wreath product yields

**5.4.20** $$([\alpha]\odot[\beta])\odot[\gamma]=[\alpha]\odot([\beta]\odot[\gamma]).$$

It is interesting to see that the plethysm of representations also occurs in connection with symmetrization $\boxdot$, for this kind of product of representations is *not* associative:

**5.4.21** $$(F\boxdot[\alpha])\boxdot[\beta]=F\boxdot([\alpha]\odot[\beta]).$$

*Proof.*

$$\begin{aligned}
\chi^{F\boxdot([\alpha]\odot[\beta])}(g)&=\frac{1}{(mn)!}\sum_{\sigma\in S_{mn}}\chi^{[\alpha]\odot[\beta]}(\sigma^{-1})\prod_{i=1}^{mn}\chi^F(g^i)^{a_i(\sigma)}\\
&=\frac{1}{m!^n n!}\sum_{(f;\pi)\in S_m \operatorname{wr} S_n}\zeta^{(\alpha;\beta)}(f;\pi)\prod_{i=1}^{mn}\chi^F(g^i)^{a_i(\psi(f;\pi))}\\
&=\frac{1}{m!^n n!}\sum_{(f;\pi)}\zeta^{\beta}(\pi^{-1})\prod_{\nu}\zeta^{\alpha}(g_\nu(f;\pi))\prod_{k=1}^{m}\chi^F(g^{k\cdot l_\nu})^{a_k(g_\nu(f;\pi))}\\
&=\frac{1}{m!^n n!}\sum_{\pi}\zeta^{\beta}(\pi^{-1})\sum_{f}\prod_{\nu}\zeta^{\alpha}(g_\nu(f;\pi))\prod_{k}\chi^F(g^{k\cdot l_\nu})^{a_k(g_\nu(f;\pi))}\\
&=\frac{1}{n!}\sum_{\pi}\zeta^{\beta}(\pi^{-1})\prod_{j=1}^{n}\left(\frac{1}{m!}\sum_{\rho\in S_m}\zeta^{\alpha}(\rho^{-1})\prod_{k=1}^{m}\chi^F(g^{j\cdot k})^{a_k(\rho)}\right)^{a_j(\pi)}\\
&=\frac{1}{n!}\sum_{\pi\in S_n}\zeta^{\beta}(\pi^{-1})\prod_{j=1}^{n}\chi^{F\boxdot[\alpha]}(g^j)^{a_j(\pi)}\\
&=\chi^{(F\boxdot[\alpha])\boxdot[\beta]}(g).
\end{aligned}$$

■

In order to describe some of the very few results known about decompositions of plethysms $[\alpha]\odot[\beta]$, we start with the following remark:

**5.4.22** $$[2]\odot[n]\downarrow S_{2n-1}=[2]\odot[n-1]\uparrow S_{2n-1}.$$

*Proof.* By definition of plethysm

$$[2]\odot[n]\downarrow S_{2n-1}=IS_2 \operatorname{wr} S_n\uparrow S_{2n}\downarrow S_{2n-1},$$

which by Mackey's subgroup theorem (see e.g. Curtis and Reiner [1962, (44.2)]) equals

$$\sum_{S_2 \operatorname{wr} S_n \pi S_{2n-1}} I(S_2 \operatorname{wr} S_n \cap \pi S_{2n-1} \pi^{-1}) \uparrow S_{2n-1}$$

Now as $S_2 \operatorname{wr} S_n$ is transitive, there is exactly one such double coset, namely $(S_2 \operatorname{wr} S_n)S_{2n-1}$, and furthermore it is clear that $S_2 \operatorname{wr} S_n \cap S_{2n-1} = S_2 \operatorname{wr} S_{n-1}$, so that we can proceed as follows:

$$\begin{aligned}[2]\odot[n]\downarrow S_{2n-1} &= IS_2 \operatorname{wr} S_{n-1} \uparrow S_{2n-1}\\ &= [2]\odot[n-1]\uparrow S_{2n-1}.\end{aligned}$$ ■

This will help us to prove (cf. 4.2.17, 4.2.18 for the notation):

**5.4.23** THEOREM. $[2]\odot[n] = \sum_{\alpha \vdash n}[2\alpha]$.

*Proof.* By induction on $n$.

(i) The case $n := 1$ is obvious: $[2]\odot[1] = [2]$.

(ii) If $n > 1$, the induction hypothesis together with 5.4.22 and the branching rule yields that $[2]\odot[n]\downarrow S_{2n-1}$ is equal to $\sum_\beta[\beta]$, where $\beta$ runs through the partitions of $2n-1$ which contain exactly one odd part. Another application of the branching rule shows that for these $\beta$ the following is true:

$$\sum_\beta [\beta] = \sum_{\alpha \vdash n} [2\alpha]\downarrow S_{2n-1}.$$

It therefore suffices to prove that $[2]\odot[n]$ cannot contain any constituent $[\gamma]$ with odd parts $\gamma_i$. Since for each such $\gamma$ every constituent of $[\gamma]\downarrow S_{2n-1}$ has exactly one odd part, $\gamma$ must be of the form $\gamma = (\gamma_1, \gamma_2)$, $\gamma_i$ odd. Hence it remains to show that no $\gamma = (\gamma_1, \gamma_2)$, $\gamma_i$ odd, can be a constituent of $[2]\odot[n]$.

In order to prove this we notice that by 5.4.22 together with the induction hypothesis, $[2]\odot[n]$ must be multiplicity-free. Now $[2n]$ is a constituent of $[2]\odot[n]$ (which is a permutation representation). We would like to check that $[2n-1,1]$ cannot occur. As $[2n-1,1]\downarrow S_{2n-1} = [2n-2,1] + [2n-1]$, the occurrence of both $[2n]$ and $[2n-1,1]$ in $[2]\odot[n]$ would imply that $[2n-1]$ occurs twice in $[2]\odot[n-1]\uparrow S_{2n-1}$, which is in fact multiplicity-free. Thus $[2n]$ is a constituent of $[2]\odot[n]$, while $[2n-1,1]$ is not.

In order to prove that $[2n-2,2]$ is a constituent of $[2]\odot[n]$, we consider $[2n-2,1]$, a constituent of $[2]\odot[n-1]\uparrow S_{2n-1}$. It satisfies

$$[2n-2,1]\uparrow S_{2n} = [2n-1,1] + [2n-2,2] + [2n-2,1^2].$$

Hence $[2n-2,1]$ arises from the restriction of $[2n-2,2]$, for we have seen already that neither $[2n-1,1]$ nor $[2n-2,1^2]$ occur in $[2]\odot[n]$.

This gives us an idea how to proceed inductively. Let us assume that we have shown that $[2n-k, k]$, $k$ even, is a constituent of $[2]\odot[n]$. We have to check that $[2n-k-1, k+1]$ does not occur, while $[2n-k-2, k+2]$ does. As

$$[2n-k-1, k+1]\downarrow S_{2n-1} = [2n-k-2, k+1] + [2n-k-1, k],$$

$$[2n-k, k]\downarrow S_{2n-1} = [2n-k-1, k] + [2n-k, k-1],$$

$[2n-k-1, k+1]$ cannot occur, for otherwise $[2]\odot[n-1]\uparrow S_{2n-1}$ would not be multiplicity-free. On the other hand $[2n-k-2, k+1]$ is a constituent of $[2]\odot[n-1]\uparrow S_{2n-1}$, and

$$[2n-k-2, k+1]\uparrow S_{2n} = [2n-k-1, k+1] + [2n-k-2, k+2] + [2n-k-2, k+1, 1],$$

so that $[2n-k-2, k+1]$ stems from the restriction of $[2n-k-2, k+2]$, which therefore must occur in $[2]\odot[n]$. This completes the proof. ■

Hence for example

5.4.24 (i) $[2]\odot[3] = [6] + [4,2] + [2^3]$,

(ii) $[2]\odot[4] = [8] + [6,2] + [4^2] + [4,2^2] + [2^4]$.

This can be used in

5.4.25
$$[2]\odot[3,1] = \begin{vmatrix} [2]\odot[3] & [2]\odot[4] \\ 1 & [2]\odot[1] \end{vmatrix}$$
$$= ([2]\odot[3])[2] - [2]\odot[4]$$
$$= [7,1] + [6,2] + [5,3] + [5,2,1] + [4,3,1] + [4,2^2] + [3,2^2,1].$$

An application of 5.4.13 to the last equation yields

$$[1^2]\odot[3,1] = [2,1^6] + [2^2,1^4] + [2^3,1^2] + [3,2,1^3] + [3,2^2,1] + [3^2,1^2] + [4,3,1].$$

An application of plethysm is provided by a proof of certain results on the so-called *Gaussian coefficient*

5.4.26 $$p(k; m, n)$$

which, for $k, m, n \in \mathbb{N}$, is defined to be the number of partitions of $k$ into $\leqslant m$ parts, each one being $\leqslant n$. The corresponding polynomial

5.4.27
$$G_{m,n}(x) := \sum_{k=0}^{mn} p(k; m, n) x^k$$

is called a *Gaussian polynomial*. It is known that the sequence of coefficients of this polynomial is *unimodal*, i.e. (if $[i]$ denotes the largest integer $\leqslant i$)

5.4.28
$$p(0; m, n) \leqslant p(1; m, n) \leqslant \cdots \leqslant p([mn/2]; m, n)$$
$$= p([(mn+1)/2]; m, n) \geqslant \cdots \geqslant p(mn; m, n).$$

In order to prove this and other results, we recall that $S_m[S_n] \simeq S_n \operatorname{wr} S_m$ acts on $\mathbf{m} \times \mathbf{n}$ (see 4.1.18, 5.1.35) and therefore correspondingly on the set

5.4.29
$$\{0,1\}^{\mathbf{m} \times \mathbf{n}}$$

(cf. Section 5.1), which can be considered as the set of $m \times n$ matrices $(a_{ik})$ over $\{0,1\}$. By 4.1.38 an element $(f; \pi) \in S_n \operatorname{wr} S_m$ induces the mapping

$$(a_{ik}) \mapsto (a_{\pi(i), f(\pi(j))(i)}).$$

Thus each orbit of $S_m[S_n]$ on 5.4.29 is characterized by a 0-1 matrix of the form

5.4.30
$$\begin{pmatrix} 1 & 1 & \cdots & \cdots & \cdots & 1 & 1 & \\ \cdots & \cdots & \cdots & \cdots & \cdots & \cdots & & \\ 1 & 1 & \cdots & 1 & 1 & & & \\ & & & & & & & 0 \end{pmatrix},$$

i.e. by a partition of $k := \sum a_{ik}$ into $\leqslant m$ parts of length $\leqslant n$. The results of Section 5.1 together with the definition of plethysm therefore give

**5.4.31** THEOREM.

(i) $p(k; m, n) = ([mn-k][k], [n] \odot [m]) = (\sum_{j=0}^{k} [mn-j, j], [n] \odot [m])$ *if* $k \leqslant mn/2$.

(ii) $p(k; m, n) \leqslant p(k+1; m, n)$ *if* $k+1 \leqslant mn/2$.

(iii) $p(k; m, n) = p(mn-k; m, n)$.

(iv) 5.4.28 *holds*.

(v) $G_{m,n}(x) = \operatorname{Cyc}(S_m[S_n] \| 1+x)$.

*Proof.* (i) is an immediate consequence of 5.1.55, and it obviously implies (ii) and (iii). Thus also (iv) is true. Finally (v) is direct from 5.1.23. ∎

Besides this, we obtain from the definition of $p(k; m, n)$ by reflecting diagrams in their main diagonal that

**5.4.32.** *$p(k; m, n) = p(k; n, m)$; thus $G_{m,n}(x) = G_{n,m}(x)$.*

And since, by 5.4.31(i), the following holds:

$$\textbf{5.4.33} \qquad \begin{aligned}([mn-k, k],[n]\odot[m]) &= p(k; m, n) - p(k-1; m, n)\\ &= p(k; n, m) - p(k-1; n, m)\\ &= ([mn-k, k],[m]\odot[n]),\end{aligned}$$

our results on Gaussian coefficients yield the following result on plethysms:

**5.4.34** THEOREM. *For two-rowed diagrams $[\alpha]$, $\alpha \vdash mn$, we have the following equation for multiplicities:*

$$([n]\odot[m],[\alpha]) = ([m]\odot[n],[\alpha]).$$

Still open is the following generalization of 5.4.34:

**5.4.35** FOULKES'S CONJECTURE. *If $m \leqslant n$, then for each $\alpha \vdash mn$ we have*

$$([m]\odot[n],[\alpha]) \leqslant ([n]\odot[m],[\alpha]).$$

## 5.5 Multiply Transitive Groups

One of the very first applications of the representation theory of symmetric groups was by Frobenius, who showed that the character tables of the Mathieu groups can be derived from a result concerning the irreducibility of restrictions of ordinary irreducible representations of $S_n$ to multiply transitive subgroups. We shall derive this result and several related ones.

Let $P$ denote a subgroup of $S_n$ acting on $\mathbf{n}$. We derive first a few characterizations of multiple transitivity of $P$ in terms of the cycle structure of its elements.

For a natural number $k \leqslant n$ we denote by $\mathbf{n}^{[k]}$ the set of all the *k-subsets* of $\mathbf{n}$:

$$5.5.1 \qquad \mathbf{n}^{[k]} := \{M \mid M \subseteq \mathbf{n} \text{ and } |M| = k\}.$$

By $\mathbf{n}^{(k)}$ we denote the set of all the *k-tuples* over $\mathbf{n}$:

$$5.5.2 \qquad \mathbf{n}^{(k)} := \mathbf{n} \times \cdots \times \mathbf{n}, \ k \text{ factors},$$

while $\mathbf{n}^{\langle k\rangle}$ should denote the set of all the *injective k-tuples* over $\mathbf{n}$:

5.5.3 $$\mathbf{n}^{\langle k\rangle}:=\{(i_1,\dots,i_k)\mid i_\nu\in\mathbf{n},\ \forall \nu\neq\mu\ (i_\nu\neq i_\mu)\}.$$

It is clear that $P$ acts on each of these sets in a natural way. The corresponding permutation groups on $\mathbf{n}^{[k]}$, $\mathbf{n}^{(k)}$, and $\mathbf{n}^{\langle k\rangle}$ will be denoted by $P^{[k]}$, $P^{(k)}$, and $P^{\langle k\rangle}$, respectively, and will be called the *k-subsets group*, the *k-tuples group*, and the *injective-k-tuples group*. The permutations corresponding to $\pi\in P$ will be denoted by $\pi^{[k]}$, $\pi^{(k)}$, and $\pi^{\langle k\rangle}$, respectively.

In Section 5.1 we considered $S_p^{[2]}$, the permutation group induced by $S_p$ (on $\mathbf{p}$) on the set $\mathbf{p}^{[2]}$ of pairs of points. The permutation characters, i.e. the numbers of fixed points, are obviously

**5.5.4**
$$\text{(i)}\ a_1(\pi^{[k]})=\sum_{b\vdash\!\dashv k}\prod_{i=1}^{k}\binom{a_i(\pi)}{b_i},$$
$$\text{(ii)}\ a_1(\pi^{(k)})=a_1(\pi)^k,$$
$$\text{(iii)}\ a_1(\pi^{\langle k\rangle})=a_1(\pi)(a_1(\pi)-1)\dots(a_1(\pi)-k+1)=:(a_1(\pi))_k.$$

$P$ is by definition *k-fold transitive* if and only if $P^{\langle k\rangle}$ is transitive (so that in particular $k$-fold transitivity implies $(k-1)$-fold transitivity). By the Cauchy-Frobenius lemma and 5.5.4(iii) we obtain as the first characterization of $k$-fold transitivity:

**5.5.5** *$P$ is k-fold transitive if and only if*

$$|P|=\sum_{\pi\in P}(a_1(\pi))_k.$$

This is a characterization in terms of the character of $P^{\langle k\rangle}$. The second one will be in terms of the character of $P^{(k)}$, and it will be obtained by comparing the actions of $P^{(k)}$ and $S_n^{(k)}$.

$S_n$ is $k$-fold transitive, i.e. $S_n^{\langle k\rangle}$ is transitive. $S_n^{(k)}$ is intransitive if $k>1$ and $n>1$, and $(i_1,\dots,i_k),(j_1,\dots,j_k)\in\mathbf{n}^{(k)}$ are in the same orbit of $S_n^{(k)}$ if and only if the following holds:

5.5.6 $$\forall 1\leqslant\mu,\nu\leqslant k\quad i_\mu=i_\nu\ \Leftrightarrow\ j_\mu=j_\nu.$$

Hence the following is true:

**5.5.7** $S_n^{(k)}$ *possesses as many orbits as there are dissections of the set* $\mathbf{k}$.

But the number of dissections of a set of order $k$ is usually denoted by $B_k$ and called the *k th Bell number*. By 5.5.4(ii),

**5.5.8** $$\forall k\in\mathbb{N},\ n\geqslant k\quad\left(B_k=\frac{1}{n!}\sum_{\pi\in S_n}a_1(\pi)^k\right).$$

Bell numbers are tabulated; e.g.

5.5.9 $B_1=1,\quad B_2=2,\quad B_3=5,\quad B_4=15,\quad B_5=52,\quad B_6=203,\ldots.$

They are used in our second characterization of $k$-fold transitivity: As $P$ is $k$-fold transitive if and only if $P^{(k)}$ and $S_n^{(k)}$ have the same orbits on $\mathbf{n}^{(k)}$, we have:

**5.5.10** *$P$ is $k$-fold transitive if and only if*

$$B_k=\frac{1}{|P|}\sum_{\pi\in P}a_1(\pi)^k.$$

The third characterization uses the character of $P^{[k]}$:

**5.5.11** *$P$ is $k$-fold transitive if and only if for every choice of $b_1, b_2,\ldots, b_k\in\mathbb{N}_0$, we have*

$$\sum ib_i\leqslant k\quad\Rightarrow\quad\frac{1}{|P|}\sum_{\pi\in P}\binom{a_1(\pi)}{b_1}\cdots\binom{a_k(\pi)}{b_k}=\frac{1}{\prod_i i^{b_i}b_i!}.$$

*Proof.*

(i) If each such equation holds, then in particular

$$\frac{1}{|P|}\sum_{\pi\in P}\binom{a_1(\pi)}{k}=\frac{1}{k!},$$

so that $P$ is $k$-fold transitive by 5.5.5.

(ii) Now let $P$ be $k$-fold transitive and $b_1,\ldots, b_k\in\mathbb{N}_0$ such that $\sum ib_i=k$. The expression

$$\sum_{\pi\in P}\binom{a_1(\pi)}{b_1}\cdots\binom{a_k(\pi)}{b_k}$$

is equal to the number of ways of picking from the elements $\pi\in P$ just $b_1$ 1-cycles, $b_2$ 2-cycles,..., $b_k$ $k$-cycles. Each such choice $\{(i_1),\ldots,(i_{b_1})\}$, $\{(i_{b_1+1},i_{b_1+2}),\ldots,(i_{b_1+2b_2-1},i_{b_1+2b_2})\},\ldots$ yields a $k$-tuple $(i_1,\ldots,i_k)$, and the expression

$$(1)\qquad\left(\prod_{i=1}^{k}i^{b_i}b_i!\right)\left(\sum_{\pi\in P}\binom{a_1(\pi)}{b_1}\cdots\binom{a_k(\pi)}{b_k}\right)$$

is equal to the number of $k$-tuples which arise in this way, if each $k$-tuple is counted with its multiplicity. (Notice that, in order to form $(i_1,\ldots,i_k)$, we take first the chosen 1-cycles, then the chosen 2-cycles, and so on, respecting the order of the choices of 1-cycles, 2-cycles, etc., while from each chosen $i$-cycle we obtain $i$ different $i$-tuples by cyclically permuting the points.)

For a given $(i_1,\ldots,i_k)$ there always exists a permutation $\pi\in P$ from which it arises by a suitable choice, since $P$ is assumed to be $k$-fold transitive and hence there exists an element in $P$ which maps $(i_1,\ldots,i_k)$ onto

$$\bigl(i_1,\ldots,i_{b_1},i_{b_1+2},i_{b_1+1},\ldots,i_{b_1+2b_2},i_{b_1+2b_2-1},\\ i_{b_1+2b_2+2},i_{b_1+2b_2+3},i_{b_1+2b_2+1},\ldots\bigr).$$

If $(i_1,\ldots,i_k)$ arises from $\pi\in P$, then it arises exactly from the elements $\rho$ in the left coset

$$\pi P_{\{i_1\}\cdots\{i_k\}}$$

of the stabilizer of the points $i_1,\ldots,i_k$. Hence $(i_1,\ldots,i_k)$ occurs $|P_{\{i_1\}\cdots\{i_k\}}|$ times. But all these stabilizers are conjugate subgroups, since $P$ is $k$-fold transitive, so that each $k$-tuple arises with the same multiplicity $|P_{\{i_1\}\cdots\{i_k\}}|$. Furthermore, by the $k$-fold transitivity of $P$, there are exactly $|P:P_{\{i_1\}\cdots\{i_k\}}|$ pairwise different $k$-tuples, and hence if each one of them is counted with its multiplicity, there are $|P|$ of them. Thus (1) is equal to $|P|$, and this completes the proof, for $k$-fold transitivity implies $(k-1)$-fold transitivity. ∎

A few examples are (for (iii) put $b_2:=2$ and use (i) and (ii))

**5.5.12.**

(i) *If $P$ is 2-fold transitive, then*

$$\frac{1}{|P|}\sum_{\pi\in P}a_2(\pi)=\frac{1}{2}.$$

(ii) *If $P$ is 3-fold transitive, then*

$$\frac{1}{|P|}\sum_{\pi\in P}a_1(\pi)a_2(\pi)=\frac{1}{2}.$$

(iii) *If $P$ is 4-fold transitive, then both*

$$\frac{1}{|P|}\sum_{\pi\in P}a_2(\pi)^2=\frac{3}{4}\quad\text{and}\quad\frac{1}{|P|}\sum a_1(\pi)^2a_2(\pi)=1.$$

These examples show how we can get results on expressions of the form

5.5.13
$$\frac{1}{|P|}\sum_{\pi\in P} a_1(\pi)^{b_1}\cdots a_k(\pi)^{b_k}$$

recursively from 5.5.11 once $P$ is $(\sum ib_i)$-fold transitive. In order to provide a direct approach, we shall define a matrix in terms of whose coefficients we can formulate all the results of this form. We therefore introduce for each $i, k\in\mathbb{N}$ the number $t_{ik}$ defined by

5.5.14
$$t_{ik} := \frac{i^k}{(i\cdot k)!}\sum_{\pi\in S_{ik}} a_i(\pi)^k,$$

and form the matrix $T$ of all these numbers:

5.5.15
$$T := (t_{ik}).$$

It is a matrix with infinitely many rows and columns. Later on we shall prove that $P$ is $k$-fold transitive if and only if $\sum ib_i \leqslant k$ implies that 5.5.13 is equal to the following expression:

5.5.16
$$\prod_{i=1}^{k}\frac{t_{ib_i}}{i^{b_i}}$$

in terms of the coefficients of $T$.

But let us first show how $T$ can be evaluated and that it is a matrix over $\mathbb{N}$. In order to do this we introduce the *Stirling numbers $S(k, j)$ of the second kind*. They can defined as coefficients of the falling factorials

5.5.17
$$(x)_j := x(x-1)\cdots(x-j+1)$$

in the expression for $x^k$ in terms of falling factorials:

5.5.18
$$x^k = \sum_{j=0}^{k} S(k, j)(x)_j.$$

It is instructive to notice that this yields (put $x := a_1(\pi)$ and use 5.5.4)

**5.5.19**
$$a_1(\pi^{(k)}) = \sum_{j=0}^{k} S(k, j)a_1(\pi^{\langle j\rangle}),$$

which is also directly obtained from the fact that $S(k, j)$ is equal to the number of dissections of a set of $k$ elements into $j$ nonempty and disjoint subsets. Another immediate corollary of 5.5.18 is

**5.5.20**
$$\frac{1}{|P|}\sum_{\pi\in P} a_i(\pi)^k = \sum_{j=0}^{k} S(k,j)\frac{1}{|P|}\sum_{\pi\in P}(a_i(\pi))_j.$$

(5.5.19 and therefore also 5.5.20 can be inverted with the aid of *Stirling numbers of the first kind*, but we do not want to stress this fact.)

We derive now

**5.5.21** LEMMA. *For each $i, k\in\mathbb{N}$ we have*

$$\frac{1}{(i\cdot k)!}\sum_{\pi\in S_{ik}} a_i(\pi)^k = \sum_{j=0}^{k}\frac{S(k,j)}{i^j},$$

*so that in particular*

$$t_{ik} = \sum_{j=0}^{k} S(k,j)\cdot i^{k-j}\in\mathbb{N},$$

*and hence $T$ is a matrix over $\mathbb{N}$.*

*Proof.* For $j\leqslant k$ the symmetric group $S_{ik}$ is $(i\cdot j)$-fold transitive, so that by 5.5.11

$$\frac{1}{(i\cdot k)!}\sum_{\pi\in S_{ik}}\binom{a_i(\pi)}{j} = \frac{1}{i^j j!},$$

i.e.

$$\frac{1}{(i\cdot k)!}\sum_{\pi\in S_{ik}}(a_i(\pi))_j = \frac{1}{i^j}.$$

The statement now follows from 5.5.20. ■

This result shows that we can evaluate the coefficients of $T$ with the aid of one of the well-known explicit formulae for the Stirling numbers of the second kind and a programmable pocket calculator, say. The upper left-hand corner of $T$ is

5.5.22
$$T=\begin{pmatrix} 1 & 2 & 5 & 15 & 52 & 203 & \cdots \\ 1 & 3 & 11 & 49 & 257 & 1539 & \cdots \\ 1 & 4 & 19 & 109 & 742 & 5815 & \cdots \\ 1 & 5 & 29 & 201 & 1657 & 15821 & \cdots \\ 1 & 6 & 41 & 331 & 3176 & 35451 & \cdots \\ 1 & 7 & 55 & 505 & 5497 & 69823 & \cdots \\ \vdots & \vdots & \vdots & \vdots & \vdots & \vdots & \end{pmatrix}$$

We notice that the first row of $T$ contains the sequence of Bell numbers. Let us now prove the desired theorem.

**5.5.23** THEOREM. *$P \leqslant S_n$ is k-fold transitive if and only if for every choice of $b_1, b_2, \ldots, b_k \in \mathbb{N}_0$, the following holds*:

$$\sum ib_i \leqslant k \quad \Rightarrow \quad \frac{1}{|P|} \sum_{\pi \in P} a_1(\pi)^{b_1} \cdots a_k(\pi)^{b_k} = \prod_{\substack{i=1 \\ b_i > 0}}^{k} \frac{t_{ib_i}}{i^{b_i}}.$$

*Proof.*

(i) If $P$ is $k$-fold transitive and $\sum ib_i \leqslant k$, then from 5.5.18 we get

$$\frac{1}{|P|} \sum_{\pi \in P} \prod_{i=1}^{k} a_i(\pi)^{b_i} = \frac{1}{|P|} \sum_{\pi \in P} \prod_{i=1}^{k} \sum_{j_i=0}^{b_i} S(b_i, j_i)(a_i(\pi))_{j_i}$$

$$= \sum_{j_1, \ldots, j_k=0}^{b_1, \ldots, b_k} \left( \prod_{i=1}^{k} S(b_i, j_i) \right) \frac{1}{|P|} \sum_{\pi \in P} \prod_i (a_i(\pi))_{j_i},$$

which is by 5.5.11 (cf. the last equation in the proof of 5.5.21)

$$= \sum_{j_1, \ldots, j_k} \left( \prod_i S(b_i, j_i) \right) \prod_i i^{(-j_i)}$$

$$= \prod_{i=1}^{k} \sum_{j_i=0}^{b_i} \frac{S(b_i, j_i)}{i^{j_i}}$$

$$= \prod_i \frac{t_{ib_i}}{i^{b_i}}.$$

(ii) Conversely, suppose that $\sum ib_i \leqslant k$ implies that

$$\frac{1}{|P|} \sum_{\pi \in P} \prod_i a_i(\pi)^{b_i} = \prod_i \frac{t_{ib_i}}{i^{b_i}}.$$

Then in particular

$$\frac{1}{|P|} \sum a_1(\pi)^k = t_{1k} = B_k,$$

and hence $P$ is $k$-fold transitive by 5.5.10. ■

It is a reasonable guess that these results can be reformulated in terms of characters, and we would now like to emphasize the representation theoretical aspect. The permutation group $S_n^{\langle k \rangle}$ which is induced by $S_n$ on the set $\mathbf{n}^{\langle k \rangle}$ of injective $k$-tuples over $\mathbf{n}$ is a transitive permutation representation of $S_n$, and hence it is induced from the identity representation of the stabilizer of any injective $k$-tuple. The stabilizer of the particular $k$-tuple $(n-k+1, n-k+2,\ldots, n)$ for example is the subgroup

5.5.24 $$S_{(n-k,1^k)} := S_{n-k} \oplus S_1 \oplus \cdots \oplus S_1$$

of $S_n$. Hence $S_n^{\langle k \rangle}$ has the same character as

5.5.25 $$IS_{(n-k,1^k)} \uparrow S_n = [n-k][1]\cdots[1] =: [n-k][1]^k.$$

Thus if $P \leqslant S_n$, then $P^{\langle k \rangle}$ has the same character as has the following representation of $P$:

5.5.26 $$IS_{(n-k,1^k)} \uparrow S_n \downarrow P = [n-k][1]^k \downarrow P.$$

Since the number of orbits of a permutation representation equals the multiplicity of the identity representation in that permutation representation, we have the following equivalences:

**5.5.27** LEMMA. *$P \leqslant S_n$ is k-fold transitive if and only if*

$$1 = (IS_{(n-k,1^k)} \uparrow S_n \downarrow P, IP),$$

*i.e. if and only if*

$$1 = (IS_{(n-k,1^k)} \uparrow S_n, IP \uparrow S_n),$$

*or, equivalently, if and only if $IS_n = [n]$ is the only irreducible constituent which $IS_{(n-k,1^k)} \uparrow S_n$ and $IP \uparrow S_n$ have in common.*

The last of these characterizations leads us to the concept of the depth $d_\alpha$ of $[\alpha]$, which was defined in 2.9.21 as follows: If $\alpha = (\alpha_1, \ldots, \alpha_n) \vdash n$, then

5.5.28 $$d_\alpha := n - \alpha_1 = \sum_{i=2}^{n} \alpha_i.$$

It is clear from 2.2.22 that $[\alpha]$ is contained in $[n-k][1]^k$ if and only if $d_\alpha \leqslant k$, so that 5.5.27 reads in terms of depths as follows:

**5.5.29** LEMMA. *$P$ is k-fold transitive if and only if $[n]$ is the only constituent $[\alpha]$ of depth $d_\alpha \leqslant k$ which is contained in $IP \uparrow S_n$.*

As 2.2.4 holds, we have the permutation characters $\xi^\alpha$ of the representations $IS_\alpha \uparrow S_n$, $d_\alpha \leqslant k$, form a basis of the space $V_{n,k}$ generated by the irreducible characters $\zeta^{\beta}$, $d_{\beta}\leqslant k$:

**5.5.30**
$$V_{n,k}:=\langle\langle\zeta^{\beta}|d_{\beta}\leqslant k\rangle\rangle_{\mathbb{C}}=\langle\langle\xi^{\alpha}|d_{\alpha}\leqslant k\rangle\rangle_{\mathbb{C}}.$$

Thus by linearity we get

**5.5.31** LEMMA. *$P$ is $k$-fold transitive if and only if for each $\chi\in V_{n,k}$ we have*

$$(\chi^{IP\uparrow S_n},\chi)=(\xi^{(n)},\chi),$$

*or, equivalently, if and only if for each $\alpha\vdash n$ such that $d_{\alpha}\leqslant k$,*

$$(\chi^{IP\uparrow S_n},\xi^{\alpha})=(\xi^{(n)},\xi^{\alpha})=1.$$

The connection with the preceding results on the cycle structure of the elements of $P$ becomes clear if we express $\xi^{\alpha}$ as a polynomial function in the following way. If $\alpha=(\alpha_1,\ldots,\alpha_h)\vdash n$, then $IS_{\alpha}\uparrow S_n$ is the permutation representation of $S_n$ on the tabloids with diagram $[\alpha]$ (see Section 2.2). $\xi^{\alpha}(\pi)$ is therefore equal to the number of such tabloids which remain fixed under the action of $\pi$. But a tabloid remains fixed under $\pi$ if and only if the points contained in a cyclic factor of $\pi$ are collinear in the tabloid in question. Hence in terms of multinomial coefficients,

**5.5.32**
$$\xi^{\alpha}(\pi)=\sum_{a^{(1)},\ldots,a^{(h)}}\prod_i\binom{a_i(\pi)}{a_i^{(1)},\ldots,a_i^{(h)}},$$

where the sum has to be taken over all the types $a^{(1)},\ldots,a^{(h)}$ of $\alpha_1,\ldots,\alpha_h$ which have the property that their sum $\sum a^{(i)}$ is equal to the cycle type of $\pi$:

$$\sum_i a^{(i)}:=\left(\sum_i a_1^{(i)},\ldots,\sum_i a_n^{(i)}\right)=(a_1(\pi),\ldots,a_n(\pi)).$$

For example

**5.5.33**
$$\begin{aligned}
&\text{(i) } \xi^{(n)}(\pi)=1,\\
&\text{(ii) } \xi^{(n-1,1)}(\pi)=a_1(\pi),\\
&\text{(iii) } \xi^{(n-2,2)}(\pi)=\frac{a_1(\pi)(a_1(\pi)-1)}{2}+a_2(\pi),\\
&\text{(iv) } \xi^{(n-2,1^2)}(\pi)=a_1(\pi)(a_1(\pi)-1).
\end{aligned}$$

Since

$$\begin{aligned}
[n-1][1]&=[n]+[n-1,1],\\
[n-2][2]&=[n]+[n-1,1]+[n-2,2],\\
[n-2][1][1]&=[n]+2[n-1,1]+[n-2,2]+[n-2,1^2],
\end{aligned}$$

we obtain from 5.5.33 for the corresponding irreducible characters $\zeta^{\alpha}$:

**5.5.34**

$$\text{(i)}\quad \zeta^{(n)}(\pi)=1,$$

$$\text{(ii)}\quad \zeta^{(n-1,1)}(\pi)=a_1(\pi)-1,$$

$$\text{(iii)}\quad \zeta^{(n-2,2)}(\pi)=\frac{a_1(\pi)(a_1(\pi)-3)}{2}+a_2(\pi),$$

$$\text{(iv)}\quad \zeta^{(n-2,1^2)}(\pi)=\frac{(a_1(\pi)-1)(a_1(\pi)-2)}{2}-a_2(\pi).$$

More generally each $\xi^{\alpha}$ and hence also each $\zeta^{\alpha}$ is a polynomial function in the $a_i(\cdot)$, as we shall soon see. We would now like to derive a result on the weight of each such monomial function.

In order to do this we recall that

$$\binom{a_i(\pi)}{a_i^{(1)},\ldots,a_i^{(h)}}=\frac{a_i(\pi)!}{a_i^{(1)}!\cdots a_i^{(h)}!},$$

so that we can cancel the factor $a_i^{(1)}!$, obtaining

$$\binom{a_i(\pi)}{a_i^{(1)},\ldots,a_i^{(h)}}=\frac{(a_i(\pi))_{a_i^{(2)}+\cdots+a_i^{(h)}}}{a_i^{(2)}!\cdots a_i^{(h)}!}. \tag{5.5.35}$$

This together with 5.5.32 yields

**5.5.36** *$\xi^{\alpha}$ can be written as a polynomial function in the $a_i(\cdot)$, such that each of its monomial summands*

$$\kappa\prod_i a_i(\cdot)^{k_i},\qquad \kappa\in\mathbb{Q},$$

*is of weight $\Sigma_i ik_i\leqslant\alpha_2+\cdots+\alpha_h=d_{\alpha}$.*

Since $(\xi^{\alpha},\zeta^{\beta})>0$ only if $\beta\trianglerighteq\alpha$ (which implies $d_{\beta}\leqslant d_{\alpha}$), we obtain the following corollary of 5.5.36:

**5.5.37.** *$\zeta^{\alpha}$ can be written as a polynomial function in the $a_i(\cdot)$, such that each of its monomial summands is of weight $\leqslant d_{\alpha}$.*

**5.5.38** THEOREM. *If $P\leqslant S_n$ is $k$-fold transitive and $\alpha,\beta\vdash n$ are such that $d_{\alpha}+d_{\beta}\leqslant k$, then*

$$([\alpha]\downarrow P,[\beta]\downarrow P)=\delta_{\alpha\beta}.$$

*Proof.* Since $[\beta]$ is a real representation, we have

$$([\alpha]\downarrow P,[\beta]\downarrow P)=([\alpha]\otimes[\beta]\downarrow P, IP)$$
$$=([\alpha]\otimes[\beta], IP\uparrow S_n).$$

The assumption $d_\alpha+d_\beta\leqslant k$ together with 5.5.29 gives (cf. 2.9.22):

$$=([\alpha]\otimes[\beta],[n])$$
$$=([\alpha],[\beta])$$
$$=\delta_{\alpha\beta}.$$

■

**5.5.39** Corollary. *If $P\leqslant S_n$ is k-fold transitive, then $[\alpha]\downarrow P$ is irreducible when $d_\alpha\leqslant k/2$. Furthermore all these restrictions $[\alpha]\downarrow P$, $d_\alpha\leqslant k/2$, are pairwise inequivalent irreducible representations of P.*

Applying this to the Mathieu group $M_{11}\leqslant S_{11}$, which is 4-fold transitive, one gets:

**5.5.40** $[11]\downarrow M_{11}$, $[10,1]\downarrow M_{11}$, $[9,2]\downarrow M_{11}$, *and* $[9,1^2]\downarrow M_{11}$ *are pairwise inequivalent ordinary irreducible representations of* $M_{11}$.

This yields the following four rows of the character table of $M_{11}$ (the cycle structure of the conjugacy classes is indicated in the upper row (cf. 1.2.3); the order of the conjugacy class is given in the second row):

| | $(1^{11})$ | $(11)$ | $(11)$ | $(5^2,1)$ | $(4^2,1^3)$ | $(3^3,1^2)$ | $(6,3,2)$ | $(8,2,1)$ | $(8,2,1)$ | $(2^4,1^3)$ |
|---|---|---|---|---|---|---|---|---|---|---|
| | 1 | 720 | 720 | 1584 | 990 | 440 | 1320 | 990 | 990 | 165 |
| $\zeta^1$ | 1 | 1 | 1 | 1 | 1 | 1 | 1 | 1 | 1 | 1 |
| $\zeta^2$ | 10 | −1 | −1 | 0 | 2 | 1 | −1 | 0 | 0 | 2 |
| $\zeta^3$ | 44 | 0 | 0 | −1 | 0 | −1 | 1 | 0 | 0 | 4 |
| $\zeta^4$ | 45 | 1 | 1 | 0 | 1 | 0 | 0 | −1 | −1 | −3 |

## Exercises

5.1 Prove the following by suitable applications of 5.1.20.

(a) For each $z\in\mathbb{Z}$ and every prime number $p$,

$$z^p\equiv z \pmod p \qquad \text{(Fermat's theorem)}.$$

(b) For each $z\in\mathbb{Z}$, $0<n\in\mathbb{N}$, and every prime number $p$,

$$z^{(p^n)}\equiv z^{(p^{n-1})} \pmod{p^n}.$$

(c) For each $z \in \mathbb{Z}$ and $0 < n \in \mathbb{N}$ such that $(z, n) = 1$,

$$z^{\varphi(n)} \equiv 1 \pmod n \qquad \text{(Euler's theorem).}$$

(d) For each prime number $p$

$$(p-1)! \equiv (-1)^p \pmod p \qquad \text{(Wilson's theorem).}$$

5.2 (a) Show that

$$\mathrm{Grf}(S_n, D) = \sum_{\substack{\lambda \vdash n \\ \lambda_i = 0,\, i > m}} (D, IS_\lambda \uparrow S_n) z_1^{\lambda_1} \dots z_m^{\lambda_m}$$

for an arbitrary $n$, $D$ a representation of $S_n$.

(b) Derive from (a) and 5.2.20 a monotonicity condition for the coefficients of the monomials in $\mathrm{Grf}(S_n, D)$ and hence for the numbers counted by Pólya's theorem.

(c) Show that $\mathrm{Grf}(S_n, [1^n])$, $n \in \mathbb{N}$, are the elementary symmetric functions. How can $\mathrm{Grf}(S_n, AS_\lambda \uparrow S_n)$ be expressed in terms of elementary symmetric functions?

(d) Express $\mathrm{Grf}(S_n, AS_\lambda \uparrow S_n) \det(z_i^{m-j})_{i, j \in \mathbf{m}}$ as a linear combination of determinants $\det(z_i^{\alpha_j + m - j})_{i, j \in \mathbf{m}}$ with nonnegative integral coefficients.

(e) Show by comparing the solution of (d) with a rule "associated" to 2.8.5 that

$$\{\alpha\} = \mathrm{Grf}(S_n, [\alpha]) = \frac{\det\left(z_i^{\alpha_j + m - j}\right)}{\det\left(z_i^{m-j}\right)}.$$

5.3 Reduce exponentiation group enumeration problems to Pólya-type problems, and formulate the corresponding enumeration theorems in constant and in weighted form.

5.4 Give a representation-theoretic proof of Redfield's lemma.

5.5 Show that for groups $P, Q \leqslant S_n$ we have

$$\begin{aligned} \mathrm{Cyc}(P) \cup \mathrm{Cyc}(Q) &= \mathrm{Cyc}(S_n, IP \uparrow S_n \otimes IQ \uparrow S_n) \\ &= \sum_{\pi \in T} \mathrm{Cyc}(P \cap \pi Q \pi^{-1}), \end{aligned}$$

if $T$ denotes a transversal of the double cosets $P\pi Q \subseteq S_n$.

5.6 If $P \leqslant S_n$ has the ordinary irreducible character $\chi$, then we denote by $E_\chi$ the corresponding centrally primitive idempotent and call (for given natural number $m$) the subset $V_\chi^m(P) := E_\chi \otimes^n \mathbb{C}^m$ of $\otimes^n \mathbb{C}^m$ a *symmetry class* of tensors. Evaluate the quadruples $(m, n, P, \chi)$ for which $V_\chi^m(P)$ is irreducible and where $|S_n:P|\leqslant n$. (*Hint*: Use 5.2.14, 5.2.33, Exercise 1.4, and the fact that for $n\neq 6$ $S_n$ contains exactly one class of conjugate subgroups of index $n$, the stabilizers of a point.)

5.7 Show that the character value $\chi^{D\boxdot[\alpha]}(g)$ equals

$$\operatorname{Grf}(S_n,[\alpha];z_1,\ldots,z_{f^D})|_{z_i=\text{eigenvalues of } D(g)}.$$

5.8 Consider 5.3.8 in order to put symmetrization and permutrization into a more general frame as follows. If $G$ and $H$ are finite groups with ordinary irreducible representations $D_i$ of $G$ and $F_j$ of $H$, a representation $R:=\sum a_{ij}D_i\#F_j$ of $G\times H$ yields an operator $\rho$ which maps a representation $D:=\sum b_iD_i$ of $G$ onto

$$\rho(D):=\sum_{i,j} b_i a_{ij}F_j,$$

a representation of $H$.

(a) Rephrase symmetrization and permutrization in terms of such operators.

(b) Show that if $G\leqslant G'$, then $\rho'(D\uparrow G')=(\rho'\downarrow G\times H)(D)$.

(c) Use (b) for another proof of 5.2.33.

5.9 Prove that for partitions $\alpha,\beta$ of $m$ and $\gamma$ of $n$ the following is true:

$$([\alpha]+[\beta])\odot[\gamma]=\sum_{\substack{k,l\geqslant 0\\ k+l=n}}\ \sum_{\substack{\mu\vdash k\\ \nu\vdash l}}([\mu][\nu],[\gamma])([\alpha]\odot[\mu])([\beta]\odot[\nu]).$$

5.10 Verify 5.5.37 for the partitions $\alpha=(n)$, $(n-1,1)$, $(n-2,2)$, and $(n-2,1^2)$.

5.11 Prove by applications of 5.5.8 and 5.3.16 that for each prime number $p$ we have for the corresponding Bell number

$$B_p\equiv 2 \pmod p,$$

and

$$B_{p+1}\equiv 3 \pmod p.$$

5.12 Prove that for each finite group $G$ with an ordinary character $\chi$ and every polynomial $f\in\mathbb{Z}[x]$ we have

$$|G| \text{ divides } \sum_{g\in G} f(\chi(g)).$$

Show that if $\chi$ is integral-valued with different values $\chi_1=\chi(1),\chi_2,\ldots,\chi_t$ and if $k$ denotes the order of the kernel of $\chi$, then

$$|G| \text{ divides } k(\chi_1-\chi_2)\cdots(\chi_1-\chi_t).$$

CHAPTER 6

# *Modular Representations*

We turn now to the modular representation theory of symmetric groups; that is, we shall work over fields whose characteristic is a prime number $p$. There are two distinct ways of approaching the modular theory, either by working in terms of the characters or by way of a module-theoretic definition of the modular irreducibles. In this chapter we concentrate on the former, more classical approach. Here, the emphasis is on the $p$-block structure, and the central theorem, which is still called "Nakayama's conjecture" although it has long since been proved, provides a necessary and sufficient condition for two ordinary irreducible characters to lie in the same $p$-block. It is one of the most beautiful results in the entire theory of symmetric groups.

After presenting the theory of $p$-blocks, we shall look closely at the as yet unsolved problem of determining the decomposition matrices of $S_n$. It becomes clear in the end that some more powerful techniques are required, and we shall turn in the next chapter to one of the alternative ways of conducting an investigation of modular representations.

## 6.1 The $p$-block Structure of the Ordinary Irreducibles of $S_n$ and $A_n$; Generalized Decomposition Numbers

Recall from 2.1.12 that every field is a splitting field for $S_n$ and hence, in particular, $\mathbb{Z}_p$, the field of $p$ elements, is a splitting field. We may therefore restrict our attention to representations of $S_n$ over $\mathbb{Z}_p$, and call these the *p-modular representations of* $S_n$.

The ordinary irreducible representation $[\alpha]$ can be written over $\mathbb{Z}$, say in Young's natural form 3.4.18. Let $[\![\alpha]\!]$ denote this way of representing $S_n$ by

ENCYCLOPEDIA OF MATHEMATICS and Its Applications, Gian-Carlo Rota (ed.). Vol. 16: G. D. James and A. Kerber, The Representation Theory of the Symmetric Group
ISBN 0-201-13515-9

matrices with integer coefficients. A corresponding $p$-modular representation

$$\overline{[\![\alpha]\!]}$$

of $S_n$ is then obtained by simply reducing each entry of each representing matrix $[\![\alpha]\!](\pi)=(\nu_{ik}^{\alpha}(\pi))$ modulo $p$:

6.1.1 $$\overline{[\![\alpha]\!]}(\pi):=\left(\overline{\nu_{ik}^{\alpha}(\pi)}\right).$$

This is of course not the only way to get from $[\alpha]$ a corresponding $p$-modular representation $\overline{[\alpha]}$ of $S_n$, for there are other matrix representations of $S_n$ which are representations over $\mathbb{Z}$ and $\mathbb{Q}$-equivalent to $[\alpha]$. Although two such $\mathbb{Z}_p$-representations corresponding to $[\alpha]$ need not be $\mathbb{Z}_p$-equivalent, they all have the same irreducible constituents. We may therefore stick to 6.1.1, for we shall restrict attention to the irreducible constituents of the modular representations in question. This even allows us to use the notation

$$\overline{[\alpha]}$$

instead of $\overline{[\![\alpha]\!]}$.

As $\mathbb{Z}_p$ is a splitting field for $S_n$, there are just as many $p$-modular irreducibles of $S_n$ as there are conjugacy classes of $S_n$ which consist of $p$-regular elements. We call these classes the *$p$-regular classes* of $S_n$. It is clear from 1.2.14 that these classes correspond to partitions no part of which is divisible by $p$. Although this holds, we shall *not* call such partitions $p$-regular.

There is another set of partitions of $n$ which is in a natural 1-1 correspondence to the $p$-modular irreducibles of $S_n$. We shall call a partition $\alpha\vdash n$ or its diagram $[\alpha]$ *$p$-regular* if and only if it does not contain $p$ parts $\alpha_i\neq 0$ which are equal.

**6.1.2** LEMMA. *There are as many $p$-regular classes of* $S_n$ *as there are $p$-regular partitions $\alpha\vdash n$.*

There is an elegant proof of this by simplifying the ratio

6.1.3 $$\frac{(1-x^p)(1-x^{2p})\cdots}{(1-x)(1-x^2)\cdots}$$

in two ways. (Notice that $P(x):=(1-x)^{-1}(1-x^2)^{-1}(1-x^3)^{-1}\cdots=1+p(1)x+p(2)x^2+\cdots$ is the generating function for the enumeration of partitions of natural numbers.) Dividing $(1-x^m)$ in the denominator into

$(1-x^{mp})$ in the numerator yields the generating function for the number of partitions of $n$ in which no summand appears $p$ or more times. On the other hand by cancelling equal factors $(1-x^{mp})$ in the numerator and denominator we obtain the generating function for the number of partitions of $n$ containing no part of the form $mp$. We then get 6.1.2 by comparing coefficients.

There is also another proof, by induction on $n$. Although it is more complicated, it uses an argument which we shall apply again later on. We denote by $\delta_j$ the differences between the lengths of neighboring columns of $[\alpha]$. Thus, if the partition associated with $\alpha$ is $\alpha'=(\alpha_1',\alpha_2',\ldots)\vdash n$, then

6.1.4 $$\delta_j := \alpha_j' - \alpha_{j+1}', \quad 1 \leqslant j.$$

It is clear that $\alpha$ is completely defined by the nonnegative integers $\delta_j$. Furthermore $\alpha$ is $p$-regular if and only if each $\delta_j$ is less than $p$.

We now divide the $\delta_j$ by $p$, which leads us to the nonnegative integers $\delta_j^{(1)}$ and $\delta_j^{(2)}$ defined by

6.1.5 $$\delta_j = \delta_j^{(1)} + \delta_j^{(2)}p, \qquad 0 \leqslant \delta_j^{(1)} < p.$$

Then $\alpha$ is uniquely determined by the $\delta_j^{(1)}$ together with the $\delta_j^{(2)}$. We denote by

6.1.6 $$[\alpha_{(1)}] \quad \text{and} \quad [\alpha_{(2)}]$$

the diagrams determined by these numbers $\delta_j^{(1)}$ and $\delta_j^{(2)}$. As each $\delta_j^{(1)} < p$, we have

**6.1.7.** *$\alpha_{(1)}$ is p-regular, while $\alpha_{(2)} \neq (0)$ only if $\alpha$ is not p-regular.*

If $\alpha_{(2)} \vdash b$, then $\alpha_{(1)} \vdash n - bp$. Moreover

**6.1.8.** *$[\alpha_{(1)}]$ is obtained from $[\alpha]$ by removing b p-hooks of leg length $p-1$ successively from $[\alpha]$. Hence in particular for their p-cores we have*

$$\widetilde{\alpha_{(1)}} = \tilde{\alpha}.$$

Since $\alpha_{(1)}$ is uniquely determined by $\alpha$, we shall call $\alpha_{(1)}$ (or $[\alpha_{(1)}]$) the *p-regular partition* (or *diagram*) *corresponding to* $\alpha$ (or $[\alpha]$).

Now let $\beta \vdash m$ be $p$-regular and let $\gamma \vdash a$ be arbitrary. Then the pair $(\beta, \gamma)$ uniquely determines an $\alpha \vdash m + ap$, for we can put

$$\alpha_{(1)} := \beta \quad \text{and} \quad \alpha_{(2)} := \gamma.$$

Hence if we denote by $p'(n)$ the number of $p$-regular diagrams with $n$ nodes, then

**6.1.9**
$$p'(n)=p(n)-\sum_{a=1}^{t} p'(n-ap)p(a),$$

where $t$ is defined by $n=tp+r$, $0\leqslant r<p$.

In order to prove 6.1.2 by induction and with the aid of 6.1.9, we need only notice that an analogous equation holds for the number $p^*(n)$ of $p$-regular classes. To see this we compute $p(n)-p^*(n)$, the number of $p$-singular classes. If $\varepsilon=(\varepsilon_1,\ldots,\varepsilon_s)$ is the partition of a $p$-singular class, then at least one $\varepsilon_i$ is divisible by $p$. Let $\varepsilon_{i_1},\ldots,\varepsilon_{i_u}$ be the parts which are divisible by $p$, $i_1\leqslant\cdots\leqslant i_u$, say $\varepsilon_{i_\nu}=p\eta_\nu$. Then $(\varepsilon_{i_1},\ldots,\varepsilon_{i_u})$ uniquely determines the partition $(\eta_1,\ldots,\eta_u)$. Thus

**6.1.10**
$$p(n)-p^*(n)=\sum_{a=1}^{t} p^*(n-ap)p(a).$$

As $p^*(0)=p'(0)=1$, and $p^*(m)=p'(m)$ for each $m<n$ by the induction hypothesis, 6.1.2 follows by comparing 6.1.9 and 6.1.10. We may therefore denote by

6.1.11
$$p^*(n)$$

both the number of $p$-regular partitions of $n$ and the number of $p$-regular classes of $S_n$.

**6.1.12** COROLLARY. $S_n$ *possesses exactly* $p^*(n)$ *pairwise inequivalent and irreducible p-modular representations.*

This leads us to the notation

6.1.13
$$F_\beta$$

for the $p$-modular irreducibles of $S_n$, $\beta$ being a $p$-regular partition of $n$. Later on we shall see how $F_\beta$ can be defined in a natural way with the aid of the ordinary irreducible $[\beta]$.

Now let

6.1.14
$$d^1_{\alpha\beta}$$

denote the multiplicity of $F_\beta$ in $\overline{[\alpha]}$ (the meaning of the superscript 1 will become clear later on). Recall from the remarks at the beginning of this section that $d^1_{\alpha\beta}$ does not depend on the matrix form of $[\alpha]$ we choose.

Since we cannot assume that $\overline{[\alpha]}$ is completely reducible, $\overline{[\alpha]}$ need not be equivalent to $\sum_\beta d^1_{\alpha\beta} F_\beta$, but it has the same irreducible constituents. We shall indicate this by writing

6.1.15 $$\overline{[\alpha]} \approx \sum_\beta d^1_{\alpha\beta} F_\beta$$

(remember that the sum is taken over the set of $p$-regular partitions $\beta$ of $n$). These multiplicities $d^1_{\alpha\beta}$ are called the ($p$-modular) *decomposition numbers* of $[\alpha]$. They are put together into the matrix

6.1.16 $$D^1_{n,p} := \left(d^1_{\alpha\beta}\right),$$

which is called the ($p$-modular) *decomposition matrix* of $S_n$. It is a $p(n)\times p^*(n)$ matrix, and it reflects a good deal of the relationship between the ordinary and $p$-modular representation theory of $S_n$. Later on we shall consider a square matrix $D_{n,p}$, the first columns of which are formed by $D^1_{n,p}$; it will be called the generalized decomposition matrix of $S_n$.

In order to get information on $D^1_{n,p}$, we may first recall from the modular representation theory of finite groups that many of its entries are zero, since both the ordinary and the modular irreducibles of a given finite group fall into *p-blocks* in such a way that (for a suitable arrangement of rows and columns) the decomposition matrix breaks up into boxes which contain all its nonzero entries, say

6.1.17 $$D^1_{n,p} = \begin{pmatrix} \boxed{*} & & & 0 \\ & \boxed{*} & & \\ 0 & & \ddots & \\ & & & \boxed{*} \end{pmatrix}.$$

A precise definition of this structure, called the *p-block structure* of the ordinary and the $p$-modular irreducibles of $S_n$, will be given later on.

It is the main aim of this section to give a necessary and sufficient condition for two ordinary irreducibles of $S_n$ to fall into the same $p$-block. The first step towards this famous theorem will give us an idea which properties of the diagram $[\alpha]$ might be important. It uses the following theorem:

**6.1.18.** *An ordinary irreducible representation of a finite group $G$ is also p-modularly irreducible (i.e. each corresponding p-modular representation is irreducible) and forms a block if the power of p dividing its dimension is the same as the power of p dividing $|G|$.*

This theorem is easy to apply to our case $G := S_n$, for we have the dimension formula 2.3.18 at hand:

6.1.19
$$f^\alpha = \frac{n!}{\prod_{i,j} h_{ij}^\alpha}.$$

This equation together with 6.1.18 yields

**6.1.20** THEOREM. *When $\alpha \vdash n$ is a p-core, i.e. $\alpha = \tilde{\alpha}$, $\overline{[\alpha]}$ is irreducible and $[\alpha]$ forms a p-block of ordinary irreducibles, while $F_{\tilde{\alpha}} = F_\alpha := \overline{[\alpha]}$ forms a p-block of modular irreducibles. The corresponding entry of the decomposition matrix is $d_{\alpha\alpha}^1 = 1$.*

It is therefore a reasonable guess that the $p$-block of $S_n$ which contains $[\alpha]$ might be characterized by the $p$-core $[\tilde{\alpha}]$. T. Nakayama stated this as a conjecture in 1940:

**6.1.21** NAKAYAMA'S CONJECTURE. *Two ordinary irreducible representations $[\alpha]$ and $[\beta]$ of $S_n$ belong to the same p-block if and only if their p-cores are equal*: $[\tilde{\alpha}] = [\tilde{\beta}]$.

This conjecture was first proved in 1947 by Brauer and Robinson, and since then several different methods of proof have been found. To prepare a proof of it, we introduce generalized decomposition numbers, which are defined as follows. If $p$ is a prime number and $G$ a finite group with ordinary irreducible characters $\zeta^i$, irreducible Brauer characters $\varphi^k$ and decomposition numbers $d_{ik}^1$, then we have for any $p$-regular element $g \in G$

6.1.22
$$\zeta^i(g) = \sum_k d_{ik}^1 \varphi^k(g).$$

This can be generalized to arbitrary $g \in G$. For each $g \in G$ is a product of a uniquely determined $p$-element $x \in G$ and a uniquely determined $p$-regular $y \in G$ which commutes with $x$:

6.1.23
$$g = xy = yx, \qquad x \text{ a } p\text{-element}, \quad y \ p\text{-regular}.$$

Let us call $x$ the *p-component*, $y$ the *p-regular component* of $g$. The element $g$ runs through a complete system of representatives of the conjugacy classes of $G$ if in 6.1.23 $x$ runs through a complete system of the conjugacy classes of $G$ which consist of $p$-elements, the so-called *p-classes* of $G$, and $y$—while $x$ is kept fixed—runs through a complete system of representatives of the $p$-regular classes of the centralizer $C_G(x)$ of $x$ in $G$. As $C_G(1) = G$, the following result of Brauer generalizes 6.1.22:

**6.1.24.** *If $x \in G$ is a p-element of order $p^r$ and if $\hat{\varphi}^k$ are the irreducible Brauer characters of $C_G(x)$ with respect to p, then there exist algebraic integers $d_{ik}^x \in \mathbb{Q}(\varepsilon)$ ($\varepsilon$ a primitive $p^r$th root of unity) which depend only on x (but not on y) and which satisfy*

$$\zeta^i(xy) = \sum_k d_{ik}^x \hat{\varphi}^k(y),$$

*for each p-regular $y \in C_G(x)$.*

If $D^x$ indicates the matrix formed by these algebraic integers,

6.1.25 $$D^x := (d_{ik}^x),$$

then it can be shown that for any conjugate $x'$ of $x$, the matrix $D^{x'}$ arises from $D^x$ by a permutation of the columns. Thus for an investigation of these algebraic integers $d_{ik}^x$ we need only consider the matrices $D^{x_j}$ for a complete system $\{x_1 := 1, x_2, \ldots, x_t\}$ of representatives of the $p$-classes of $G$. If we have chosen such a system of representatives $x_j$, then we put

6.1.26 $$D^j := (d_{ik}^{x_j}), \qquad 1 \leqslant j \leqslant t.$$

The matrix $D$ consisting of these $D^j$ as submatrices,

6.1.27 $$D := (D^1, \ldots, D^t),$$

which is uniquely determined up to permutations of rows and columns, is called the *generalized decomposition matrix* of $G$ with respect to $p$. Its entries are called the *generalized decomposition numbers* of $G$ with respect to $p$. The first columns of $D$ contain $D^1$, the decomposition matrix of $G$. If $y_k$ are representatives of the $p$-regular classes of $C_G(x_j)$, then we put

6.1.28 $$Z^j := \left(\zeta^i(x_j y_k)\right), \qquad \Phi^j := \left(\hat{\varphi}^i(y_k)\right),$$

$\hat{\varphi}^i$ being the irreducible Brauer characters of $C_G(x_j)$. The matrices

6.1.29 $$\Phi := \dot{+}_{j=1}^{t} \Phi^j, \qquad Z := (Z^1, \ldots, Z^t),$$

satisfy the equation (by 6.1.24)

**6.1.30** $$Z = D\hat{\Phi}.$$

Later on we shall consider the evaluation of generalized decomposition matrices of $S_n$.

For our present purpose, the proof of Nakayama's conjecture, we need a very important theorem on $D$:

**6.1.31** BRAUER'S SECOND MAIN THEOREM (IN PART). *For each $p$-element $x \in G$ and each $p$-block $\hat{B}$ of $C_G(x)$, there exists exactly one $p$-block $B$ of $G$ such that the following holds: For any ordinary irreducible character $\zeta^i \notin B$ and every $\hat{\varphi}^j \in \hat{B}$ we have $d^x_{ij}=0$.*

Applying this to the symmetric group $S_n$, we obtain

**6.1.32** COROLLARY. *$[\alpha]$ and $[\beta]$ belong to the same $p$-block of $S_n$ as soon as there exists a $p$-element $\pi \in S_n \backslash \{1\}$ and $i, k \in \mathbb{N}$ such that both $d^{\pi}_{\alpha i}$ and $d^{\pi}_{\beta k}$ are $\neq 0$, while $\hat{\varphi}^i$ and $\hat{\varphi}^k$ belong to the same $p$-block $\hat{B}$ of $C_{S_n}(\pi)$.*

It is our aim to show that such a $\pi$ exists when $\alpha$ and $\beta$ have the same $p$-core. If $\pi \in S_n$ is a $p$-element and of type $(a_1(\pi),\ldots,a_n(\pi))$, then we have for its centralizer (see 4.1.19)

**6.1.33**
$$C_{S_n}(\pi) \triangleq \bigoplus_i \psi\left[C_{p^i} \operatorname{wr} S_{a_{p^i}}\right].$$

In order to apply 6.1.32 we need information on the blocks of such centralizers. By 6.1.33, $C_{S_n}(\pi)$ is a direct product of wreath products of the form

$$C_{p^i} \operatorname{wr} S_{a_{p^i}},$$

for which we know the following:

**6.1.34.** *If $G \neq \{1_G\}$ is a $p$-group, then $G \operatorname{wr} H$ possesses exactly one $p$-block.*

This follows from a lemma of Brauer, which says that a group possesses exactly one $p$-block when it contains a normal $p$-subgroup which includes its own centralizer (as the base group in such a $G \operatorname{wr} H$ does). We are now in a position to prove the first half of Nakayama's conjecture:

**6.1.35.** *If $\alpha, \beta \vdash n$ have the same $p$-core $\tilde{\alpha}=\tilde{\beta}$, then $[\alpha]$ and $[\beta]$ belong to the same $p$-block of $S_n$.*

*Proof.* Let $w$ denote the $p$-weight of $\alpha$ (and hence also of $\beta$ by $\tilde{\alpha}=\tilde{\beta}$) and put

$$\pi := (1 \cdots p) \cdots ((w-1)p+1,\ldots,wp) \in S_n.$$

For each $p$-regular $\rho \in C_{S_n}(\pi)$

$$\zeta^{\alpha}(\pi\rho) = \sum_k d^{\pi}_{\alpha k} \hat{\varphi}^k(\rho),$$

$\hat{\varphi}^k$ being the irreducible Brauer characters of

$$C_{S_n}(\pi) = \psi[C_p \operatorname{wr} S_w] \oplus S_{n-wp}.$$

This last equation yields that $\hat{\varphi}^k$ is of the form

$$\hat{\varphi}^k = \tilde{\varphi}^k_1 \tilde{\varphi}^k_2,$$

$\tilde{\varphi}^k_i$ being irreducible Brauer characters of the direct summands of $C_{S_n}(\pi)$. Thus if $\rho \in S_{n-wp}$,

6.1.36
$$\zeta^{\alpha}(\pi\rho) = \sum_k d^{\pi}_{\alpha k} \tilde{\varphi}^k_1(1) \tilde{\varphi}^k_2(\rho)$$

$$= z\zeta^{\tilde{\alpha}}(\rho),$$

for a certain integer $z \neq 0$ (use 2.7.26, 27). As $\zeta^{\tilde{\alpha}}$ yields an irreducible Brauer character of $S_{n-wp}$ by restriction to $p$-regular elements (see 6.1.20), which we also denote by $\zeta^{\tilde{\alpha}}$, and as the irreducible Brauer characters of $S_{n-wp}$ are linearly independent, 6.1.36 shows that there must be a $k$ such that

(1) $$\zeta^{\tilde{\alpha}} = \tilde{\varphi}^k_2 \quad \text{and} \quad d^{\pi}_{\alpha k} \neq 0.$$

Analogously we obtain that for a suitable $i$ we have

(2) $$\zeta^{\tilde{\beta}} = \tilde{\varphi}^i_2 \quad \text{and} \quad d^{\pi}_{\beta i} \neq 0.$$

Since by 6.1.34 $\psi[C_p \operatorname{wr} S_w]$ possesses exactly one $p$-block, the irreducible Brauer characters $\hat{\varphi}^k = \tilde{\varphi}^k_1 \zeta^{\tilde{\alpha}}$ and $\hat{\varphi}^i = \tilde{\varphi}^i_1 \zeta^{\tilde{\beta}} = \tilde{\varphi}^i_1 \zeta^{\tilde{\alpha}}$ belong to the same $p$-block of $C_{S_n}(\pi)$. Hence (1) and (2) yield the statement by an application of 6.1.32. ∎

To prove the converse we denote by $A$ the set of all the partitions of $n$ with $p$-core $\tilde{\alpha}$. A theorem of Osima says that this set,

6.1.37
$$A := \{[\beta] \mid \beta \vdash n \text{ and } \tilde{\beta} = \tilde{\alpha}\},$$

is proved to be a union of $p$-blocks as soon as we can show that for each $p$-regular $\tau$ and each $p$-irregular $\sigma$ we have

6.1.38 $$0=\sum_{\beta\in A}\zeta^{\beta}(\tau)\zeta^{\beta}(\sigma).$$

Here an element $\sigma\in S_n$ is called *p-irregular* if its order is divisible by $p$, i.e. it contains a cycle the length of which is divisible by $p$, say

$$\sigma=\rho\pi,\qquad \rho=(1\cdots cp),\quad \pi\in S_{n-cp}.$$

In order to check that 6.1.38 is true, we notice that by the orthogonality relations we have

$$0=\sum_{\alpha\vdash n}\zeta^{\alpha}(\tau)\zeta^{\alpha}(\rho\pi),$$

so that by 2.4.7 we get, if $M_\alpha$ denotes the set of all the partitions $\beta\vdash n-cp$ which can be obtained from $\alpha$ by removing a rim $cp$-hook of leg length $l_{\alpha\beta}$,

**6.1.39** $$\begin{aligned}0&=\sum_{\alpha\vdash n}\zeta^{\alpha}(\tau)\Big(\sum_{\beta\in M_\alpha}(-1)^{l_{\alpha\beta}}\zeta^{\beta}(\pi)\Big)\\&=\sum_{\beta\vdash n-cp}\zeta^{\beta}(\pi)\Big(\sum_{\alpha\in N_\beta}(-1)^{l_{\alpha\beta}}\zeta^{\alpha}(\tau)\Big),\end{aligned}$$

where $N_\beta$ denotes the set of partitions of $n$ which yield $\beta$ by removing a suitable rim $cp$-hook. The linear independence of irreducible characters therefore implies

**6.1.40** $$0=\sum_{\alpha\in N_\beta}(-1)^{l_{\alpha\beta}}\zeta^{\alpha}(\tau).$$

Let $C$ denote the set of partitions of $n-cp$ with $p$-core $\tilde{\alpha}$. Then

**6.1.41** $$\begin{aligned}&\text{(i)}\quad [\beta\in C \text{ and } \alpha\in N_\beta]\ \Rightarrow\ \alpha\in A,\\&\text{(ii)}\quad [\alpha\in A \text{ and } \beta\in M_\alpha]\ \Rightarrow\ \beta\in C.\end{aligned}$$

This is true because the removal of a rim $cp$-hook can be carried out by $c$ successive removals of rim $p$-hooks.

From 6.1.39 we derive with the aid of 6.1.40 and 6.1.41

$$\begin{aligned} 0 &= \sum_{\beta\in C} \zeta^\beta(\pi)\Big(\sum_{\alpha\in N_\beta} (-1)^{l_{\alpha\beta}}\zeta^\alpha(\tau)\Big)\\ &= \sum_{\alpha\in A} \zeta^\alpha(\tau)\Big(\sum_{\beta\in M_\alpha} (-1)^{l_{\alpha\beta}}\zeta^\beta(\pi)\Big)\\ &= \sum_{\alpha\in A} \zeta^\alpha(\tau)\zeta^\alpha(\rho\pi)\\ &= \sum_{\alpha\in A} \zeta^\alpha(\tau)\zeta^\alpha(\sigma). \end{aligned}$$

Thus 6.1.38 holds, and hence by Osima's theorem the set $\{[\beta] \| \beta\vdash n$ and $\tilde{\beta}=\tilde{\alpha}\}$ is a union of $p$-blocks. This has an immediate consequence

**6.1.42** COROLLARY. *If* $[\alpha]$ *and* $[\beta]$ *are in the same p-block of* $S_n$, *then we have for their p-cores* $\tilde{\alpha}=\tilde{\beta}$.

This completes the proof of Nakayama's conjecture. In order to provide a few examples we put $p:=2$. Let us see what can be said about the decomposition matrices. It is trivial that

$$D^1_{1,2}=(1) \quad \textit{and} \quad D^1_{2,2}=\begin{pmatrix}1\\1\end{pmatrix}.$$

For one-dimensional representations are always irreducible, and for $p=2$ we have $\overline{[n]}=\overline{[1^n]}$. This together with 6.1.20 yields furthermore

**6.1.43**
$$D^1_{3,2}=\begin{pmatrix}1&0\\1&0\\0&1\end{pmatrix}\begin{matrix}[3]\\ [1^3]\\ [2,1]\end{matrix}.$$

$S_4$ possesses exactly one 2-block, since

6.1.44 $$[(4)^{\sim}]=[(3,1)^{\sim}]=[(2^2)^{\sim}]=[(2,1^2)^{\sim}]=[(1^4)^{\sim}]=[0].$$

There are just two 2-regular classes, so that $D^1_{4,2}$ consists of 2 columns. But all what we can say at the moment about its entries is

**6.1.45**
$$D^1_{4,2}=\begin{pmatrix}1&0\\ *&*\\ *&*\\ *&*\\ 1&0\end{pmatrix}\begin{matrix}[4]\\ [3,1]\\ [2^2]\\ [2,1^2]\\ [1^4]\end{matrix}.$$

Thus further information is needed. In fact most of the forthcoming sections will be devoted to the evaluation of decomposition matrices.

As usual, we ask next for the $p$-block structure of $A_n$.

**6.1.46** THEOREM. *If $\alpha=\tilde{\alpha}$, then every irreducible constituent of $[\alpha]\downarrow A_n$ forms its own p-block and is modularly irreducible. If $\alpha\neq\tilde{\alpha}$, then to the p-block of an irreducible constituent of $[\alpha]\downarrow A_n$ there belong just the constituents of such restrictions $[\beta]\downarrow A_n$, where $\tilde{\beta}=\tilde{\alpha}$ or $\tilde{\beta}=\tilde{\alpha}'$.*

*Proof (see also Exercise 6.2):*

(i) Assume $\alpha=\tilde{\alpha}$.

(a) If $\alpha\neq\alpha'$, then by 2.5.7, $[\alpha]\downarrow A_n$ is irreducible. $\alpha=\tilde{\alpha}$ implies that $[\alpha]$ contains no $p$-hook, and hence $f^\alpha$ contains the same power of $p$ as do $|S_n|$ and $|A_n|$ (notice that $p\neq 2$ by $\alpha=\tilde{\alpha}$ and $\alpha\neq\alpha'$).

(b) If $\alpha=\alpha'$, then by 2.5.7 we have

$$[\alpha]\downarrow A_n=[\alpha]^+ + [\alpha]^-,$$

$[\alpha]^\pm$ being of dimension $f^{\alpha^\pm}=f^\alpha/2$. Thus, by $\alpha=\tilde{\alpha}$, $f^{\alpha^\pm}$ still contains the same power of $p$ as does $|A_n|$.

This shows that in both these cases the statement follows from 6.1.18.

(ii) Assume $\alpha\neq\tilde{\alpha}$, so that its $p$-weight $w$ is $>0$. By a theorem of Fong, each irreducible constituent of $[\beta]\downarrow A_n$, for each $\beta$ with $\tilde{\beta}=\tilde{\alpha}$, belongs to a conjugacy class of blocks of $A_n$, and each of these blocks contains the same number of constituents. But in our case here, this subset of the set of blocks of $A_n$ consists of a single block, since there do exist partitions $\beta\vdash n$ such that $\tilde{\beta}=\tilde{\alpha}$ while $\beta\neq\beta'$ (and hence $[\beta]\downarrow A_n$ is irreducible). Such a $\beta$ is for example

$$\beta:=(\tilde{\alpha}_1+wp,\tilde{\alpha}_2,\ldots).$$

Hence if $\alpha\neq\tilde{\alpha}$, there exists a $p$-block $\hat{B}$ of $A_n$ which contains the constituents of all the restrictions $[\beta]\downarrow A_n$ where $\tilde{\beta}=\tilde{\alpha}$. But this $p$-block $\hat{B}$ must also contain the constituents of all the restrictions $[\beta]\downarrow A_n$ where $\tilde{\beta}=\tilde{\alpha}'$, for by 2.5.7 we have $[\beta']\downarrow A_n=[\beta]\downarrow A_n$. We say that the two blocks $B_{\tilde{\alpha}}$ and $B_{\tilde{\alpha}'}$ of $S_n$ *cover* the block $\hat{B}$ of $A_n$.

Conversely these two blocks $B_{\tilde{\alpha}}$ and $B_{\tilde{\alpha}'}$ are the only blocks of $S_n$ which cover $\hat{B}$, for if $[\gamma]_A\in\hat{B}$ is an ordinary irreducible representation of $A_n$, then $B_{\tilde{\alpha}}$ contains constituents of $[\gamma]_A\uparrow S_n$. ($B$ covers $\hat{B}$ only if it contains constituents of a representation induced by an (ordinary) element of $\hat{B}$.) ■

It remains to provide a numerical example of a generalized decomposition matrix. We evaluate $D_{6,2}$. As each matrix of irreducible Brauer characters is nonsingular, we can use the equation (see 6.1.24)

6.1.47
$$D^\nu = Z^\nu(\Phi^\nu)^{-1}.$$

The 2-classes of $S_6$ are represented by

$$\pi_1 := 1, \qquad \pi_2 := (12), \qquad \pi_3 := (12)(34),$$
$$\pi_4 := (12)(34)(56), \qquad \pi_5 := (1234), \qquad \pi_6 := (1234)(56).$$

The centralizers of these elements are of the form

$$C(\pi_1) = S_6, \qquad C(\pi_2) = C_2 \oplus S_4, \qquad C(\pi_3) = C_2 \operatorname{wr} S_2 \oplus S_2,$$
$$C(\pi_4) = C_2 \operatorname{wr} S_3, \qquad C(\pi_5) = C_4 \oplus S_2, \qquad C(\pi_6) = C_4 \oplus C_2.$$

Hence we only need the matrices of Brauer characters of $S_1$, $S_2$, $S_3$, and $S_4$. These matrices are

$$\Phi_1 = \Phi_2 = (1) \quad \text{and} \quad \Phi_3 = \Phi_4 = \begin{pmatrix} 1 & 1 \\ 2 & -1 \end{pmatrix},$$

as we can see from the character tables and decomposition matrices of these groups (for $D^1_{4,2}$, see the following sections). Thus we obtain

$$\Phi^2 = \Phi_2 \times \Phi_4 = \begin{pmatrix} 1 & 1 \\ 2 & -1 \end{pmatrix}, \qquad \Phi^3 = \Phi_2 \times \Phi_2 = (1),$$
$$\Phi^4 = \Phi_3 = \begin{pmatrix} 1 & 1 \\ 2 & -1 \end{pmatrix}, \qquad \Phi^5 = \Phi_1 \times \Phi_2 = (1) = \Phi^6.$$

($\Phi^1$ has been omitted, since $D^1_{6,2}$ has to be evaluated separately; we take this matrix for granted. The point is to show that $D^i_{n,p}$, $i > 1$, can be evaluated from the decomposition matrices of $S_m$, $m < n$.) Because of

$$\begin{pmatrix} 1 & 1 \\ 2 & -1 \end{pmatrix}^{-1} = \begin{pmatrix} \frac{1}{3} & \frac{1}{3} \\ \frac{2}{3} & -\frac{1}{3} \end{pmatrix},$$

and

$$Z^2 = \begin{bmatrix} 1 & 1 \\ 3 & 0 \\ 3 & 0 \\ 2 & -1 \\ 1 & 1 \\ 0 & 0 \\ -1 & -1 \\ -2 & 1 \\ -3 & 0 \\ -3 & 0 \\ -1 & -1 \end{bmatrix} \quad \text{and} \quad Z^4 = \begin{bmatrix} 1 & 1 \\ -1 & -1 \\ 3 & 0 \\ -2 & 1 \\ -3 & 0 \\ 0 & 0 \\ 3 & 0 \\ 2 & -1 \\ -3 & 0 \\ 1 & 1 \\ -1 & -1 \end{bmatrix}$$

(see the character tables in Appendix I), we get for the desired generalized decomposition matrix of $S_6$ for $p := 2$

$$D_{6,2} = \left(\begin{array}{ccc|c||cc||c||cc||c||c}
1 & 0 & 0 & 0 & 1 & 0 & 1 & 1 & 0 & 1 & 1\\
1 & 1 & 0 & 0 & 1 & 1 & 1 & -1 & 0 & 1 & -1\\
1 & 1 & 1 & 0 & 1 & 1 & 1 & 1 & 1 & -1 & 1\\
2 & 1 & 1 & 0 & 0 & 1 & -2 & 0 & -1 & 0 & 0\\
1 & 0 & 1 & 0 & 1 & 0 & 1 & -1 & -1 & -1 & -1\\
1 & 0 & 1 & 0 & -1 & 0 & 1 & 1 & 1 & 1 & -1\\
2 & 1 & 1 & 0 & 0 & -1 & -2 & 0 & 1 & 0 & 0\\
1 & 1 & 1 & 0 & -1 & -1 & 1 & -1 & -1 & 1 & 1\\
1 & 1 & 0 & 0 & -1 & -1 & 1 & 1 & 0 & -1 & -1\\
1 & 0 & 0 & 0 & -1 & 0 & 1 & -1 & 0 & -1 & 1\\
\hline
0 & 0 & 0 & 1 & 0 & 0 & 0 & 0 & 0 & 0 & 0
\end{array}\right).$$

(Double bars separate the $D^\nu$; the columns of $D^3$, $D^5$, and $D^6$ agree (since $\Phi^\nu = (1)$) with the corresponding columns of the character table.)

If we multiply a generalized decomposition matrix $D_{n,p}$ by its transpose, then we obtain the *generalized Cartan matrix* $C_{n,p}$ of $S_n$:

6.1.48 $$D'_{n,p} D_{n,p} = C_{n,p} = \dotplus_{\nu=1}^{t} C^\nu_{n,p},$$

where $C^\nu_{n,p}$ denotes the *Cartan matrix* of the centralizer of the $p$-element $\pi_\nu$. For example

$$C_{6,2} = \begin{pmatrix} 16 & 8 & 8 \\ 8 & 6 & 4 \\ 8 & 4 & 6 \end{pmatrix} \dotplus (1) \dotplus \begin{pmatrix} 8 & 4 \\ 4 & 6 \end{pmatrix} \dotplus (16) \dotplus \begin{pmatrix} 8 & 4 \\ 4 & 6 \end{pmatrix} \dotplus (8) \dotplus (8).$$

The generalized decomposition numbers of $A_n$ can be evaluated similarly as long as the centralizers of the $p$-elements are direct sums of wreath products of the form $C_{p^i} \operatorname{wr} A_{a_{p^i}}$ or they have (1) as matrix of Brauer characters (so that the corresponding column agrees with a column of the character table). For example

$$D_{A_3,3} = \left(\begin{array}{c||c||c}
1 & 1 & 1 \\
1 & (-1+i\sqrt{3})/2 & (-1-i\sqrt{3})/2 \\
1 & (-1-i\sqrt{3})/2 & (-1+i\sqrt{3})/2
\end{array}\right).$$

This matrix, which is in fact equal to the character table of $A_3$, demonstrates that the generalized decomposition matrices of alternating groups are not in general rational integral. A result of Reynolds shows that if the values of the ordinary characters of a finite group $G$ on $p$-irregular elements are rational integral, then the generalized decompositions matrix of $G$ with respect to $p$ is rational integral. This together with our results on characters of $S_n$, $A_n$, and wreath products yields:

**6.1.49** THEOREM. *The generalized decomposition numbers of* $G \,\mathrm{wr}\, H$ *are rational integral if the character tables of* $G$ *as well as of the inertia factors* $H \cap S_{(n)}$ *are rational integral. In particular the generalized decomposition matrices of symmetric groups* $S_n$ *and of wreath products* $S_m \,\mathrm{wr}\, S_n$ *of symmetric groups are rational integral for each prime number* $p$. *In contrast to this, the generalized decomposition matrices of alternating groups* $A_n$ *are not in general rational integral, but for* $p:=2$ *they are.*

## 6.2 The Dimensions of a $p$-block; $u$-numbers; Defect Groups

Having determined the $p$-blocks of ordinary irreducibles of $S_n$, we ask for the number of ordinary and modular irreducibles in each block.

Nakayama's conjecture shows that the number of ordinary irreducibles in the block of $[\alpha]$ is equal to the number of partitions $\beta \vdash n$ with the property $\tilde{\beta}=\tilde{\alpha}$. By 2.7.30 a partition $\beta$ is uniquely determined by its $p$-core together with its $p$-quotient. Hence there are as many such $\beta \vdash n$ as there are $p$-quotients consisting of $w$ nodes, $w$ the $p$-weight of $\alpha$. We can therefore call *$w$ the $p$-weight of this block* and obtain

**6.2.1** THEOREM. *The number of ordinary irreducible representations in a* $p$*-block of weight* $w$ *of* $S_n$ *is equal to*

$$b(w):=\sum p(w_0)\cdots p(w_{p-1}),$$

*if the sum is taken over all the improper partitions* $(w_0,\ldots,w_{p-1})$ *of* $w$.

4.2.9 shows that $b(w)$ is equal to the number of ordinary irreducibles of $C_p \,\mathrm{wr}\, S_w$, a result which is unclear. A reason for it might be found by a closer examination of the Brauer correspondence used in the above proof of the Nakayama conjecture. The corresponding result for the number of modular irreducible representations is:

**6.2.2** THEOREM. *The number of modular irreducible representations in a* $p$*-block of weight* $w$ *of* $S_n$ *is equal to*

$$b^*(w):=\sum p(w_1)\cdots p(w_{p-1}),$$

*if the sum is taken over all the improper partitions* $(w_1,\ldots,w_{p-1})$ *of* $w$. *If* $\alpha$ *is any partition of* $n$ *which is of* $p$*-weight* $w$, *then* $b^*(w)$ *is equal to the number of* $p$*-regular diagrams with* $p$*-core* $\tilde{\alpha}$.

*Proof.* The proof of this theorem is by induction on $n$ and quite long-winded. We shall use another generalization of the decomposition numbers of $S_n$ which was introduced by M. Osima. It depends heavily on the fact that for each conjugacy class of $S_n$ the lengths of the cyclic factors of its elements which are divisible by $p$ are uniquely determined—i.e., a given $\pi \in S_n$ can be written in a unique way as a product

6.2.3
$$\pi = \tau\sigma,$$

where $\sigma$ either consists just of the cyclic factors of $\pi$ the lengths of which are divisible by $p$, or $\sigma := 1_{S_n}$ if no such cycles exist, and where $\tau$ acts on the remaining symbols. Hence $\tau$ is always $p$-regular.

We call $\sigma$ *the p-singular component* of $\pi$. If $\sigma \neq 1$, then it can be expressed as a product

6.2.4
$$\sigma = \sigma_1 \cdots \sigma_s$$

of $(b_i p)$-cycles $\sigma_i$, $1 \leqslant i \leqslant s$. The quantity $b := \sum b_i$ if $\sigma \neq 1$, and $b := 0$ if $\sigma = 1$, will be called the *p-weight of* $\sigma$.

If we put

6.2.5
$$r := \sum_{b=0}^{t} p(b),$$

$t$ being the largest integer $\leqslant n/p$, then we can choose a system of $r$ elements $\sigma^{(i)}$, say

6.2.6
$$\sigma^{(1)} := 1, \sigma^{(2)}, \ldots, \sigma^{(r)},$$

such that the $p$-singular component of any $\pi \in S_n$ is a conjugate of one of the $\sigma^{(i)}$. Without loss of generality we can assume that the $p$-weights of the $\sigma^{(i)}$ are increasing:

6.2.7
$$p\text{-weight of } \sigma^{(i)} \leqslant p\text{-weight of } \sigma^{(i+1)}, \qquad 1 \leqslant i < r.$$

Now every conjugacy class of $S_n$ contains elements of the form

6.2.8
$$\tau\sigma^{(i)},$$

where $i$ is uniquely determined by the class, and $\tau$ is a $p$-regular element of $S_{n-bp}$, $b$ being the $p$-weight of $\sigma^{(i)}$.

Thus if again $p^*(m)$ denotes the number of $p$-regular classes of $S_m$, then we have

**6.2.9**
$$p(n) = \sum_{b=0}^{t} p^*(n-bp)\,p(b).$$

The Murnaghan-Nakayama formula yields the existence of integers $h(\alpha,\beta)$ such that (if again $w$ is the $p$-weight of $\alpha$)

**6.2.10**
$$\zeta^{\alpha}(\tau\sigma^{(i)})=\begin{cases}\sum\limits_{\substack{\beta\vdash n-bp,\\ \tilde{\alpha}=\tilde{\beta}}} h(\alpha,\beta)\zeta^{\beta}(\tau) & \text{if } b\leqslant w,\\ 0 & \text{otherwise}.\end{cases}$$

As $\tau$ is $p$-regular, the value $\zeta^{\beta}(\tau)$ can be expressed in terms of irreducible Brauer characters $\hat{\varphi}^{\lambda}$ of $S_{n-bp}$ and its decomposition numbers $\hat{d}^{1}_{\beta\lambda}$, say

$$\zeta^{\beta}(\tau)=\sum_{p\text{-reg. }\lambda\vdash n-bp}\hat{d}^{1}_{\beta\lambda}\hat{\varphi}^{\lambda}(\tau).$$

Hence 6.2.10 yields the existence of integers $u^{i}_{\alpha\lambda}\in\mathbb{Z}$, for each $p$-regular $\lambda\vdash n-bp$, which satisfy

**6.2.11**
$$\zeta^{\alpha}(\tau\sigma^{(i)})=\sum_{\lambda}u^{i}_{\alpha\lambda}\hat{\varphi}^{\lambda}(\tau).$$

These integers are called the *u-numbers* of $S_n$ with respect to $p$.

In order to derive some of their properties, we notice first that

**6.2.12.** *If $\alpha\vdash n$ is of $p$-weight $w$ which is smaller than the $p$-weight $b$ of $\sigma^{(i)}\in S_n$, then for each $p$-regular $\lambda\vdash n-bp$ we have $u^{i}_{\alpha\lambda}=0$.*

Our next remark concerning the $u$-numbers uses the induction hypothesis (remember that we are in the middle of a proof of 6.2.2 by induction on $n$). It says that for $m<n$ the irreducible Brauer characters $\hat{\varphi}$ of $S_m$ in the $p$-block of $\gamma\vdash m$ can be parametrized by the $p$-regular diagrams $[\delta]$, where $\tilde{\delta}=\tilde{\gamma}$. Hence

**6.2.13.** *If $0<b\leqslant w$, then $u^{i}_{\alpha\lambda}=0$ for each $p$-regular $\lambda\vdash n-bp$ such that $\tilde{\lambda}\neq\tilde{\alpha}$.*

The case $b:=0$ is covered by

**6.2.14.** *For each $\alpha\vdash n$ and any $p$-regular $\lambda\vdash n$ we have $u^{1}_{\alpha\lambda}=d^{1}_{\alpha\lambda}$.*

This last remark shows that the $u$-numbers are in particular a generalization of the decomposition numbers of $S_n$. Later on an example will show that they differ from the generalized decomposition numbers, as should be expected from their definition. For each $1\leqslant i\leqslant r$ we arrange these numbers $u^{i}_{\alpha\lambda}$ into a matrix:

6.2.15
$$U^{i}:=\left(u^{i}_{\alpha\lambda}\right).$$

This is a $p(n) \times p^*(n-bp)$ matrix, $b$ the weight of $\sigma^{(i)}$. Let these $r$ matrices form the columns of another matrix $U$:

6.2.16 $$U=(U^1,\ldots,U^r).$$

6.2.9 shows that $U$ is a $p(n) \times p(n)$ matrix. And it is a matrix over $\mathbb{Z}$; this is clear from the definition of $u$-numbers.

Let $\Phi^{(b)} := (\hat{\varphi}^\lambda(\tau))$ denote the matrix of irreducible Brauer characters of $S_{n-bp}$, and put

6.2.17 $$A := \begin{pmatrix} \Phi^{(0)} & & & & & & 0 \\ & \ddots & & & & & \\ & & \Phi^{(b)} & & & & \\ & & & \ddots & & & \\ & & & & \Phi^{(b)} & & \\ & & & & & \ddots & \\ 0 & & & & & & \Phi^{(r)} \end{pmatrix} \Big\} \; p(b) \text{ times}.$$

6.2.11 shows that for a suitable arrangement of the rows and columns of the character table $Z$ of $S_n$ we have

**6.2.18** $$Z=UA.$$

As $Z$ is nonsingular, so is $U$ by 6.2.18:

**6.2.19** $$\det U \neq 0.$$

Now 6.2.11 together with 6.2.13 shows that the rows and columns of $U$ can be arranged in such a way that the resulting matrix $U'$ breaks up completely into (say) $q$ matrices $U_1',\ldots,U_q'$, each $U_i'$ corresponding to a $p$-block $B_i$ of $S_n$:

6.2.20 $$U' = \begin{pmatrix} U_1' & & 0 \\ & \ddots & \\ 0 & & U_q' \end{pmatrix}$$

Thus by 6.2.19 each $U_i'$ must be square and nonsingular. Hence each $U_i'$ consists of $b(w_i)$ rows and columns, if $w_i$ denotes the $p$-weight of the $p$-block $B_i$ of $S_n$.

Now let $\alpha \vdash n$ be of $p$-weight $w$, and let $b'$ be the number of modular irreducibles in its $p$-block $B_i$. The number of columns of the corresponding submatrix $U_i'$ of $U'$, by 6.2.11 and an application of the induction hypothesis, is equal to

6.2.21 $$b'+\sum_{v=0}^{w-1} b^*(v)p(w-v).$$

The number of rows of $U_i'$ is $b(w)$, so that 6.2.21 implies

6.2.22 $$b(w)=b'+\sum_{v=0}^{w-1} b^*(v)p(w-v),$$

an equation which yields

6.2.23 $$b'=b(w)-\sum_{v=0}^{w-1} b^*(v)p(w-v).$$

An application of the induction hypothesis

$$b^*(v)=\sum p(v_1)\cdots p(v_{p-1}).$$

to 6.2.23 gives

$$b'=\sum p(w_1)\cdots p(w_{p-1}),$$

as is claimed in 6.2.2.

It remains to show that $b^*(w)$ is equal to the number $\tilde{p}$ of $p$-regular diagrams with $p$-core $\tilde{\alpha}$. We would like to express $\tilde{p}$ in terms of $b(\cdot)$, $b^*(\cdot)$, and $p(\cdot)$.

There are $b(w)$ diagrams with $p$-core $\tilde{\alpha}$. If such a $[\beta]$ is not $p$-regular, then it has a nonzero $[\beta_{(2)}]$ (cf. 6.1.6). If $\beta_{(1)}\vdash n-ap$, then $a>0$. But there exist just $p^*(w-a)$ such $p$-regular $[\beta_{(1)}]$ (by the induction hypothesis) and $p(a)$ possibilities for $\beta_{(2)}\vdash a$. Thus

$$\tilde{p}=b(w)-\sum_{a=1}^{w} p^*(w-a)p(a)$$
$$=b(w)-\sum_{v=0}^{w-1} p^*(v)p(w-v),$$

so that we obtain from $b^*(0)=p^*(0)=1$ and a comparison with 6.2.23 the following equation by induction on $w$:

$$\tilde{p}=b^*(w)$$

This completes the proof of 6.2.2. ∎

In order to derive further properties of $u$-numbers, such as orthogonality relations, we restrict our attention to a fixed block $B$ of $S_n$. The ordinary orthogonality relations give

**6.2.24** $$\sum_{\alpha\vdash n} \zeta^\alpha(\tau\sigma^{(i)})\zeta^\alpha(\tau'\sigma^{(j)})=0 \quad \text{if} \quad i\neq j.$$

Applications of 6.2.11–6.2.13 to this equation yield

6.2.25 $$0=\sum_{\alpha\vdash n}\Big(\sum_{\lambda,\mu} u^i_{\alpha\lambda}u^j_{\alpha\mu}\Big)\hat{\varphi}^\lambda(\tau)\hat{\varphi}^\mu(\tau'),$$

if the inner sum is taken over all the $p$-regular $\lambda\vdash n-bp$, $\mu\vdash n-b'p$ such that $\tilde{\lambda}=\tilde{\mu}=\tilde{\alpha}$, $b$ the $p$-weight of $\sigma^{(i)}$, $b'$ the $p$-weight of $\sigma^{(j)}$. Thus

6.2.26 $$0=\sum_{w}\sum_{\lambda,\mu}\Big(\sum_{\alpha} u^i_{\alpha\lambda}u^j_{\alpha\mu}\Big)\hat{\varphi}^\lambda(\tau)\hat{\varphi}^\mu(\tau'),$$

where the inner sum is taken over all the $\alpha\vdash n$ which are of $p$-weight $w$ and which satisfy $\tilde{\alpha}=\tilde{\lambda}=\tilde{\mu}$, i.e. over all the $\alpha\vdash n$ which form a $p$-block $B$:

6.2.27 $$0=\sum_{\lambda,\mu}\Big(\sum_{\alpha\in B} u^i_{\alpha\lambda}u^j_{\alpha\mu}\Big)\hat{\varphi}^\lambda(\tau)\hat{\varphi}^\mu(\tau').$$

This holds for each $p$-regular $\tau\in S_{n-bp}$, $\tau'\in S_{n-b'p}$. Thus by the linear independence of irreducible Brauer characters we obtain the following orthogonality relations for $u$-numbers of $S_n$:

**6.2.28** $$\sum_{\alpha\in B} u^i_{\alpha\lambda}u^j_{\alpha\mu}=0 \quad \textit{if} \quad i\neq j.$$

Applying this to 6.2.25, we obtain

**6.2.29** $$\sum_{\alpha\in B} \zeta^\alpha(\tau\sigma^{(i)})\zeta^\alpha(\tau'\sigma^{(j)})=0 \quad \textit{if} \quad i\neq j.$$

If on the other hand $i=j$, then we have to replace 6.2.24 by

**6.2.30** $$\sum_{\alpha\vdash n} \zeta^\alpha(\tau\sigma^{(i)})\zeta^\alpha(\tau\sigma^{(i)})=\left|C_{S_n}(\tau\sigma^{(i)})\right|,$$

so that by 6.2.11

6.2.31 $$\sum_{\lambda}\Big(\sum_{\alpha\in B} u^i_{\alpha\lambda}\zeta^\alpha(\tau\sigma^{(i)})\Big)\hat{\varphi}^\lambda(\tau)=\left|C_{S_n}(\tau\sigma^{(i)})\right|.$$

Besides this we have the modular orthogonality relations at hand. If $\hat{\eta}^\lambda$ denotes the Brauer character of the principal indecomposable corresponding to $\hat{\varphi}^\lambda$, then

6.2.32 $$\sum_\lambda \hat{\varphi}^\lambda(\tau)\hat{\eta}^\lambda(\tau)=|C_{S_{n-bp}}(\tau)|.$$

As the matrix $(\hat{\varphi}^\lambda(\tau))$ is nonsingular, we can compare the coefficients of 6.2.31 and 6.2.32 to obtain

**6.2.33** $$\sum_{\alpha\in B} u^i_{\alpha\lambda}\zeta^\alpha(\tau\sigma^{(i)})=\frac{|C_{S_n}(\tau\sigma^{(i)})|}{|C_{S_{n-bp}}(\tau)|}\hat{\eta}^\lambda(\tau).$$

Another application of 6.2.11 now gives

$$\sum_{\alpha\in B} u^i_{\alpha\lambda}\sum_\mu u^i_{\alpha\mu}\hat{\varphi}^\mu(\tau)=\frac{|C_{S_n}(\tau\sigma^{(i)})|}{|C_{S_{n-bp}}(\tau)|}\hat{\eta}^\lambda(\tau)=|C_{S_{bp}}(\sigma^{(i)})|\hat{\eta}^\lambda(t).$$

Multiplying both sides of this by $\hat{\varphi}^\lambda(\tau)$ and summing over all the $p$-regular $\tau\in S_{n-bp}$, we get, if $(e_{\mu\lambda}):=\hat{C}^{-1}$, the inverse of the Cartan matrix $(\hat{c}_{\lambda\mu})$ of $S_{n-bp}$:

6.2.34 $$\sum_{\substack{\alpha\in B\\ \mu}} u^i_{\alpha\lambda}u^i_{\alpha\mu}e_{\mu\lambda}=|C_{S_{bp}}(\sigma^{(i)})|,$$

for we know from general theory that

$$\sum_\tau \hat{\eta}^\lambda(\tau)\hat{\varphi}^\lambda(\tau)=(n-bp)!.$$

Thus 6.2.34 finally yields

**6.2.35** $$\sum_{\alpha\in B} u^i_{\alpha\lambda}u^i_{\alpha\mu}=\hat{c}_{\lambda\mu}\prod_{k=1}^{s}(kp)^{b_k}b_k!,$$

if $(b_1,\ldots,b_s)$ is the $p$-type of $\sigma^{(i)}$. This corresponds to 6.1.48. It shows that $U'U$ is a direct sum of matrices which are closely related to Cartan matrices of symmetric groups.

As a numerical example we evaluate the matrix of $u$-numbers of $S_6$ with respect to $p:=2$. The largest integer $\leqslant n/p$ is $t:=3=6/2$ in this case. Hence

$$r:=\sum_{b=0}^{3}p(b)=1+1+2+3=7.$$

As corresponding elements of $p$-weight $0 \leqslant b \leqslant 3$ we choose

$$\sigma^{(1)} := 1, \quad \sigma^{(2)} := (12), \quad \sigma^{(3)} := (12)(34), \quad \sigma^{(4)} := (1234),$$
$$\sigma^{(5)} := (12)(34)(56), \quad \sigma^{(6)} := (12)(3456), \quad \sigma^{(7)} := (123456).$$

The matrices of irreducible Brauer characters of $S_{n-bp}$, $b>0$, are the following matrices $\Phi^{(b)}$:

$$\Phi^{(1)} = \begin{pmatrix} 1 & 1 \\ 2 & -1 \end{pmatrix}, \qquad \Phi^{(2)} = (1) = \Phi^{(3)}.$$

(As in the evaluation of $D_{6,2}$, we omit $\Phi^{(0)}$, the matrix of irreducible Brauer characters of $S_6$, in order to emphasize that the parts $U^i$, $i>1$, of $U$ can be evaluated with the aid of the character table of $S_6$ and a certain knowledge of the modular representation theory of the groups $S_m$, $m<n$, only.)

Since $\Phi^{(2)} = \Phi^{(3)} = (1)$, the columns of $U$ which correspond to $\sigma^{(3)}, \ldots, \sigma^{(6)}$ agree with the corresponding columns of the character table of $S_6$, while

$$U^2 = \left(\zeta^\alpha(\tau\sigma^{(2)})\right)\Phi^{(1)-1}, \qquad \tau \in \{1, (123)\}$$

$$= \begin{pmatrix} 1 & 1 \\ 3 & 0 \\ 3 & 0 \\ 2 & -1 \\ 1 & 1 \\ -1 & -1 \\ -2 & 1 \\ -3 & 0 \\ -3 & 0 \\ -1 & -1 \\ 0 & 0 \end{pmatrix} \begin{pmatrix} \frac{1}{3} & \frac{1}{3} \\ \frac{2}{3} & -\frac{1}{3} \end{pmatrix} = \begin{pmatrix} 1 & 0 \\ 1 & 1 \\ 1 & 1 \\ 0 & 1 \\ 1 & 0 \\ -1 & 0 \\ 0 & -1 \\ -1 & -1 \\ -1 & -1 \\ -1 & 0 \\ 0 & 0 \end{pmatrix}.$$

Altogether this yields (if we take $D^1_{6,2}$ for granted)

$$U = \left(\begin{array}{ccc|c||cc||c||c||c||c||c} 1 & 0 & 0 & 0 & 1 & 0 & 1 & 1 & 1 & 1 & 1 \\ 1 & 1 & 0 & 0 & 1 & 1 & 1 & 1 & -1 & -1 & -1 \\ 1 & 1 & 1 & 0 & 1 & 1 & 1 & -1 & 3 & 1 & 0 \\ 2 & 1 & 1 & 0 & 0 & 1 & -2 & 0 & -2 & 0 & 1 \\ 1 & 0 & 1 & 0 & 1 & 0 & 1 & -1 & -3 & -1 & 0 \\ 1 & 0 & 1 & 0 & -1 & 0 & 1 & 1 & 3 & -1 & 0 \\ 2 & 1 & 1 & 0 & 0 & -1 & -2 & 0 & 2 & 0 & -1 \\ 1 & 1 & 1 & 0 & -1 & -1 & 1 & 1 & -3 & 1 & 0 \\ 1 & 1 & 0 & 0 & -1 & -1 & 1 & -1 & 1 & -1 & 1 \\ 1 & 0 & 0 & 0 & -1 & 0 & 1 & -1 & -1 & 1 & -1 \\ \hline 0 & 0 & 0 & 1 & 0 & 0 & 0 & 0 & 0 & 0 & 0 \end{array}\right).$$

Comparing this with the matrix $D_{6,2}$ of generalized decomposition numbers of $S_6$ which we evaluated at the end of the last section, we see that the $u$-numbers of $S_n$ differ from the generalized decomposition numbers (there exist e.g. no 3's in $D_{6,2}$).

If $G$ is a finite group with a conjugacy class $C$, then a $p$-subgroup $D$ of $G$ is a *defect group of* $C$ if, for some $g\in C$, $D$ is a Sylow $p$-subgroup of the centralizer $C_G(g)$ of $g$. We abbreviate this by writing

6.2.36 $$D=(C_G(g))_p.$$

Hence a defect group of the conjugacy class of elements of type $(a_1,\ldots,a_n)$ of $S_n$ is for example the subgroup

**6.2.37** $$\bigoplus_i \left(\psi[C_i \operatorname{wr} S_{a_i}]\right)_p.$$

Since the Sylow $p$-subgroups of $C_G(g)$ are all conjugate, it follows that the set of defect groups of $C\subseteq G$ is a single class of conjugate $p$-subgroups in $G$. In particular, the order $p^d:=|D|$ is uniquely determined. $d$ is called the *defect of* $C$. If $B$ is a $p$-block of $G$ with $\mathbb{Q}$ as a splitting field, then a *defect group of* $B$ may be defined to be any subgroup $D\leqslant G$ which satisfies:

**6.2.38.**

(i) $|D|=p^\delta$, *where* $\delta$ *is the* defect of $B$, *i.e.*

$$\delta:=\nu_p(|G|)-\min_{Z_i\in B}\nu_p(f^i)$$

*($f^i$ the dimension of the ordinary irreducible $Z_i$, and $\nu_p$ (integer) indicates the exponent of p dividing the integer).*

(ii) *There exists a p-regular $g\in G$ such that*

$$D=(C_G(g))_p,$$

*and for the central character $\omega^i$ we have*

$$\omega^i(g):=\frac{|C^G(g)|\zeta^i(g)}{f^i}\not\equiv 0 \pmod p$$

*for each i such that $Z_i\in B$.*

Thus a defect group of a block is also a defect group of a class. It can be shown that the defect groups of $B$ also form a single class of conjugate $p$-subgroups of $G$. We would like to show first which subgroups of $S_n$ occur as defect groups of blocks. Afterwards we shall calculate the defect groups of the block of $S_n$ which contains $[\alpha]$.

**6.2.39 THEOREM.** *The defect groups of the p-blocks of* $S_n$ *are conjugates of subgroups of form*

$$D^b := \psi[C_p \,\mathrm{wr}\, (S_b)_p]$$

*where* $(S_b)_p$ *denotes a Sylow p-subgroup of* $S_b$.

*Proof.* We use a lemma of Brauer which says the following. If $g \in G$ is an element of the center of the defect group $D$ of $G$, then $D$ is a defect group of $C_G(g)$. Now assume that $D \neq \{1_{S_n}\}$ is a defect group of $S_n$. Then $D$ contains elements of order $p$ in its center. These elements consist of $p$-cycles and 1-cycles only. We choose one of them in such a way that it contains a maximal number of $p$-cycles. Call this element $\pi$, and denote by $b$ its number of cyclic factors of length $p$. Then (up to conjugation)

$$C_{S_n}(\pi) = \psi[C_p \,\mathrm{wr}\, S_b] \oplus S_{n-bp}.$$

Thus, by Brauer's lemma, $D$ is of the form

$$D = D_1 \oplus D_2,$$

where $D_1$ is a defect group of $\psi[C_p \,\mathrm{wr}\, S_b] \simeq C_p \,\mathrm{wr}\, S_b$, while $D_2$ is a defect group of $S_{n-bp}$.

As $C_p \,\mathrm{wr}\, S_b$ possesses exactly one block by 6.1.34, we have

$$D_1 = \psi[C_p \,\mathrm{wr}\, (S_b)_p].$$

Now $D_2 \neq \{1\}$ would imply the existence of a $\pi'$ in its center which is of order $p$, so that it would contain at least one $p$-cycle. But $\psi[C_p \,\mathrm{wr}\, (S_b)_p]$ contains in its center the element

$$\pi'' := \psi(((1 \cdots p), \ldots, (1 \cdots p); 1))$$

which consists of just $b$ $p$-cycles. Thus $D$ would contain in its center the element $\pi''\pi'$, which contains more than $b$ cycles of length $p$, in contradiction to the maximality of $b$. Hence $D = D_1 = \psi[C_p \,\mathrm{wr}\, (S_b)_p]$, as stated. ∎

This theorem suggests that $D^w = \psi[C_p \,\mathrm{wr}\, (S_w)_p]$, $w$ being the $p$-weight of $\alpha \vdash n$, might turn out to be a defect group of the block of $[\alpha]$. This is in fact true, and the above remarks on defect groups of blocks of finite groups show that in order to prove this we need only evaluate the defect $\delta_\alpha$ of the block $B$ of $[\alpha]$. By 6.2.38(i) we have

**6.2.40**
$$\delta_\alpha = \nu_p(n!) - \min_{[\beta] \in B} \nu_p(f^\beta).$$

This shows that we have to evaluate

6.2.41 $$\min_{[\beta]\in B} \nu_p(f^\beta).$$

In order to do this we remember that by 2.3.18

$$f^\beta = \frac{n!}{H^\beta},$$

where $H^\beta$ denotes the product of all the hook lengths in $[\beta]$. It is clear from the definition of the $p$-quotient $[\beta]_p$ of $[\beta]$ that if $a$ denotes the $p$-weight of $[\beta]$, then

**6.2.42** $$H^\beta = p^a H_p^\beta (H^\beta)',$$

where $H_p^\beta$ denotes the product of all the hook lengths of $[\beta]_p$, while $(H^\beta)'$ denotes the product of all the hook lengths of $[\beta]$ which are prime to $p$. This yields

$$\begin{aligned}\nu_p(f^\beta) &= \nu_p(n!) - \nu_p(H^\beta)\\ &= \nu_p(n!) - \left(a + \nu_p\left(H_p^\beta\right)\right)\\ &= \nu_p(n!) - \left(\nu_p((ap)!) - \nu_p(a!) + \nu_p\left(H_p^\beta\right)\right)\\ &= \nu_p(n!) - \nu_p((ap)!) + \nu_p\left(f_p^\beta\right).\end{aligned}$$

($f_p^\beta$ again denotes the dimension of $[\beta]_p$; see 2.7.32.) In order to get the minimal $\nu_p(f^\beta)$ from this, we only need to notice that there exists a $\beta \vdash n$ such that $\tilde{\beta} = \tilde{\alpha}$ and such that $f_p^\beta$ is prime to $p$, i.e. $\nu_p(f_p^\beta) = 0$ (e.g. if $[\beta]_p = [a]$). Thus

**6.2.43** $$\min_{[\beta]\in B} \nu_p(f^\beta) = \nu_p(n!) - \nu_p((wp)!),$$

and hence by 6.2.40 the defect $\delta_\alpha$ of the block of $[\alpha]$ satisfies

**6.2.44** $$\delta_\alpha = \nu_p((wp)!) = w + \nu_p(w!).$$

An immediate corollary is now

**6.2.45 Theorem.** *If $\alpha \vdash n$ is of p-weight w, then each defect group of the p-block*

*of* $S_n$ *which contains* $[\alpha]$ *is conjugate to the group*

$$D^w := \psi\left[C_p \operatorname{wr} (S_w)_p\right].$$

## 6.3 Techniques for Finding Decomposition Matrices

Armed with the information we have obtained so far, we turn now to the problem of evaluating the decomposition matrix of $S_n$ for a given integer $n$ and prime $p$. The most important tool we apply in this section will be the Nakayama conjecture, which tells us how to sort the ordinary irreducible representations into blocks.

Although there are many papers and several books which discuss modular representation theory, little has been written on the practical problem of finding the modular irreducible characters of a given group. Since the symmetric groups provide a nice concrete example, while still being sufficiently complicated, it is worthwhile to present here various techniques which can be tried in the context of an arbitrary finite group, illustrating the results in the case of $S_n$.

Most of the discussion will assume that the character table of $G$, the group under consideration, is known. Determining the modular irreducible characters is then equivalent to finding the decomposition matrix of $G$. It is also necessary to know something about the decomposition matrix of at least one "large" subgroup of $G$. These assumptions are pertinent in the case of the symmetric groups, since, in principle, the character table of $S_n$ is known, and a natural induction hypothesis is that the decomposition matrix of $S_{n-1}$ has been determined.

It must be made clear from the beginning that none of our methods are *guaranteed* to give the full decomposition matrix of $G$. Although they can be used to find the decomposition matrices of $S_n$ for small $n$, they eventually break down; the smallest $n$ for which the decomposition matrix of $S_n$ is unknown is $n=14$ and $p=2$.

We start by fixing some notation.

Let $x :=$ the number of $p$-regular classes of $G$.

Let $\phi_1, \phi_2, \ldots, \phi_x$ be the modular irreducible characters of $G$.

Let $\chi_1, \chi_2, \ldots, \chi_y$ be the ordinary irreducible characters of $G$, restricted to $p$-regular classes. For $S_n$, we shall abuse notation and write

$[\alpha] :=$ the ordinary character of $S_n$ corresponding to the diagram $[\alpha]$, restricted to $p$-regular classes.

We can write $\chi_i$ as $\sum_{j=1}^{x} d_{ij}\phi_j$. This defines the decomposition matrix $(d_{ij})$, whose rows are labeled by ordinary, and columns by modular, irreducible characters.

It will be proved below (6.3.50) that, with the row labels suitably ordered, the decomposition matrix of $S_n$ has the form

**6.3.1**
$$\begin{pmatrix} 1 & & & \\ & 1 & & 0 \\ & & \ddots & \\ & & & 1 \\ & * & & \end{pmatrix}.$$

The proof of this result is quite tricky, and a better proof is presented in Section 7.1. This latter proof has the advantage that it does not rely on Nakayama's conjecture, and shows clearly the structural reason why the result is true.

Sometimes it will be convenient to give explanations which assume that we know the decomposition matrix of our group has shape 6.3.1. It is not uncommon for a decomposition matrix to approximate to 6.3.1, and usually only slight changes in the arguments will be required to deal with decomposition matrices of other shapes.

It is known that for each $j$, $\sum_{i=1}^{y} d_{ij}\chi_i$ gives a projective indecomposable character of $G$. We shall identify a projective character $\sum_{i=1}^{y} c_i\chi_i$ with the column of integers $(c_i)$. For example:

**6.3.2.** *Let* $\mathbf{d}_j$ *be the y-long column vector* $(d_{ij})$. *Then* $\mathbf{d}_1, \mathbf{d}_2, \ldots, \mathbf{d}_x$ *are the projective indecomposable characters of* $G$.

The two results which are most helpful in finding decomposition matrices are the trivial:

**6.3.3.** *Any modular character of a group can be written as a linear combination, with nonnegative integer coefficients, of the modular irreducible characters.*

**6.3.4.** *Any projective character of a group can be written as a linear combination, with nonnegative integer coefficients, of the projective indecomposable characters.*

(We shall see later that one of these statements is redundant in the practical problem of determining the decomposition matrix.)

In particular, any modular character $\phi$ of $G$ can be written as $\phi = a_1\phi_1 + \cdots + a_x\phi_x$ with each $a_i \geqslant 0$.

**6.3.5.** *We shall say that the modular irreducible* $\phi_i$ *is a constituent of* $\phi$ *iff* $a_i \neq 0$. $a_i$ *is called the multiplicity of* $\phi_i$ *in* $\phi$.

**6.3.6.** *If* $\rho := b_1\phi_1 + \cdots + b_x\phi_x$ *is another modular character of* $G$, *we shall write* $\rho \subseteq \phi$ *iff* $\forall i\ b_i \leq a_i$.

The following results are useful, but not widely known:

**6.3.7 Theorem.** *For all primes* $q$, *the decomposition matrix of* $G$ *for the prime* $p$ *has full* $q$*-rank.*

*Proof.* The case when $q=p$ is proved in Curtis and Reiner [1962, 83.5], so suppose $q \neq p$. We must show that if $q$ divides all the entries in the column vector $a_1\mathbf{d}_1 + \cdots + a_x\mathbf{d}_x$, then $q$ divides all the integers $a_i$.

Let $D$ be the decomposition matrix (whose columns are $\mathbf{d}_1,\ldots,\mathbf{d}_x$), and $D^*$ be the matrix whose columns are $\mathbf{d}_1,\ldots,\mathbf{d}_{i-1}, \sum_{j=1}^{x} a_j\mathbf{d}_j, \mathbf{d}_{i+1},\ldots,\mathbf{d}_x$. It is easy to check that $\det(D^{*\prime}D^*) = a_i^2\det(D'D)$. But $q$ divides all the entries in the $i$th row of $D^{*\prime}D^*$, and so divides $a_i^2\det(D'D)$. Now, $D'D$ is the Cartan matrix, and it is known that the determinant of this is a power of $p$ (Curtis and Reiner [1962, 84.17]). Therefore, $q$ divides $a_i$, as required. ■

**6.3.8 Corollary.** *If* $\mathbf{c}$ *is a projective character, and all the entries in* $\mathbf{c}$ *are divisible by an integer* $q$, *then* $(1/q)\mathbf{c}$ *is a projective character.*

*Proof.* We know that $\mathbf{c} = a_1\mathbf{d}_1 + \cdots + a_x\mathbf{d}_x$, with each $a_i \geq 0$ (6.3.4). Every prime which divides all the entries in $\mathbf{c}$ must divide all the integers $a_i$, by 6.3.7, and the result follows. ■

Having presented our notation, we can now begin the task of explaining how to find the decomposition matrix.

First the ordinary characters of $G$ are sorted into blocks. This is simply a matter of reducing modulo $p$ (in general, modulo a prime ideal dividing $p$) the expression

$$\frac{|G|\chi_i(g)}{|C_G(g)|\chi_i(1_G)},$$

which gives, for every entry in the character table (for the moment, we still include $p$-singular classes), an algebraic integer. $\chi_i$ and $\chi_j$ are in the same block iff all the reduced entries in the row of the character table corresponding to $\chi_i$ are the same as the reduced entries in the row corresponding to $\chi_j$. When $G = S_n$, this partitioning of the ordinary characters into blocks is determined by Nakayama's conjecture. Having done this, we can concentrate on a particular block.

(It is sometimes easier, though, to give general results in terms of the whole decomposition matrix; for example, the decomposition matrix is "wedge-shaped" (6.3.1) iff every block is wedge-shaped—this being just a matter of rearranging the ordinary characters.)

General theory tells us that each block contains at least as many ordinary irreducible characters as modular irreducible characters. Thus, if a certain block contains only one ordinary character, it follows from 6.3.8 that the decomposition matrix for this block is (1). Compare 6.1.20.

In general, the largest $d$ such that $p^d$ divides $|G|/\chi(1)$ is called the *defect* of the ordinary character $\chi$. Brauer has proved that if a character has defect 1, then so do all the characters in the same block. He also developed a beautiful theory for such blocks, which shows that the decomposition matrix for the block can be described by a tree. In $S_p$, $[\alpha]$ has defect $1 \Leftrightarrow p|$ (product of the hook lengths in $[\alpha]$) $\Leftrightarrow [\alpha]$ is a hook diagram (recall that $[\alpha]$ is a hook diagram for $S_n \Leftrightarrow \exists x, y$ $(x+y=n$ and $[\alpha]=[x,1^y]$)). All the hook diagrams for $S_p$ belong to the same block (they all have empty cores), and the Brauer tree for this block is

**6.3.9**

$$\begin{array}{ccccccccccc} \times & — & \times & — & \times & — & \cdots & — & \times & — & \times \\ [p] & & [p-1,1] & & [p-2,1^2] & & & & [2,1^{p-2}] & & [1^p] \end{array}$$

Having sorted the characters into blocks, the next step is to find a set of $x$ column vectors which span the same space as $\mathbf{d}_1,\ldots,\mathbf{d}_x$. The explanation of how to do this is complicated by the fact that, in practice, the two methods about to be described should be combined. It will later be proved that method II is sufficient, by itself, for the symmetric groups.

**6.3.10** METHOD I *for finding linearly independent columns*. Because the character table for $G$ is nonsingular, all the ordinary characters, restricted to $p$-regular classes, can be written as a linear combination of some set of $x$ ordinary characters. By definition of "block", the relations can be divided into sets of relations, each set involving only characters from the same block. There is a lot of choice for the set $\chi_1,\ldots,\chi_x$ of ordinary characters to use, but when in doubt, use those with small degrees.

The relations $\chi_i=\sum_{j=1}^{x} m_{ij}\chi_j$ define a matrix of the form

**6.3.11**

$$\begin{array}{c} \\ \chi_1 \\ \chi_2 \\ \vdots \\ \chi_x \\ \vdots \end{array} \begin{array}{c} \begin{array}{cccc} \chi_1 & \chi_2 & \cdots & \chi_x \end{array} \\ \left(\begin{array}{cccc} 1 & & & \\ & 1 & 0 & \\ & 0 & \ddots & \\ & & & 1 \\ & & * & \end{array}\right) \end{array}$$

where, in general, some of the entries may be negative or not integers. It is easy to see that the columns of this matrix span the same space as $\mathbf{d}_1,\ldots,\mathbf{d}_x$.

**6.3.12.** Note that arranging the row labels by blocks simply gives several submatrices of the form 6.3.11.

**6.3.13** EXAMPLE. $p=2$ and $S_4$. $[1^4]=[4]$, $[2,1^2]=[3,1]$, and $[2^2]=[3,1]-[4]$ on 2-regular classes. Using [4] and [3, 1] as $\chi_1,\ldots,\chi_x$, we get

| | | |
|---|---|---|
| $[4]$ | 1 | |
| $[3,1]$ | | 1 |
| $[2^2]$ | $-1$ | 1 |
| $[2,1^2]$ | | 1 |
| $[1^4]$ | 1 | |

(by convention, omitted numbers are always zero). Clearly, this approximation to the decomposition matrix can be improved, because all the entries in the decomposition matrix have to be nonnegative. Since $[3,1]-[4]=[2^2]$ is a modular character, the identity modular character is a constituent of [3, 1]. Notice that this enables us to add the second *column* of the above matrix to the first, and obtain a better approximation to the decomposition matrix:

| | | |
|---|---|---|
| $[4]$ | 1 | |
| $[3,1]$ | 1 | 1 |
| $[2^2]$ | | 1 |
| $[2,1^2]$ | 1 | 1 |
| $[1^4]$ | 1 | |

Of course, this amounts to choosing [4] and $[2^2]$ as our $\chi_1,\ldots,\chi_x$.

This way of finding linearly independent columns is very boring, as it requires solving many linear equations. It is the only way of being sure to find $x$ linearly independent columns for a general group, but it should be used in conjunction with

**6.3.14** METHOD II *for finding linearly independent columns.* It is known that inducing a projective character from a subgroup $H$ of $G$ gives a projective character of $G$. Therefore, assuming that we know the decomposition matrix of $H$, we can get combinations of projective indecomposables of $G$—that is, combinations of the columns of the decomposition matrix of $G$. This is a way of constructing several linearly independent columns, although in general there is no guarantee that the right number (=the number of $p$-regular classes of $G$) will turn up. Indeed, it looks as though the maximum number of linearly independent columns we shall find is the number of $p$-regular classes of $H$. But this overlooks the fact that we may be able to break a column into different pieces, using block theory.

**6.3.15** EXAMPLE. For $p=2$, let us try to construct linearly independent columns for the decomposition matrix of $G:=S_5$. Take $H:=S_4$. The decomposition matrix for $S_4$ is

$$\begin{array}{lcc} [4] & 1 & \\ [3,1] & 1 & 1 \\ [2^2] & & 1 \\ [2,1^2] & 1 & 1 \\ [1^4] & 1 & \end{array}$$

so $[3,1]+[2^2]+[2,1^2]$ and $[4]+[3,1]+[2,1^2]+[1^4]$ are projective indecomposable characters of $S_4$. Inducing up to $S_5$,

$$\begin{aligned} [3,1]+[2^2]+[2,1^2]\uparrow S_5 &= [4,1]+[3,2]+[3,1^2]+[3,2] \\ &\quad +[2^2,1]+[3,1^2]+[2^2,1]+[2,1^3], \\ [4]+[3,1]+[2,1^2]+[1^4]\uparrow S_5 &= [5]+[4,1]+[4,1] \\ &\quad +[3,2]+[3,1^2]+[3,1^2] \\ &\quad +[2^2,1]+[2,1^3]+[2,1^3]+[1^5], \end{aligned}$$

and these give projective characters of $S_5$. Neither can be a projective indecomposable, because each involves characters from more than one block. Hence, splitting the above into blocks,

$$\begin{gathered} [4,1]+[2,1^3], \\ 2[3,2]+2[3,1^2]+2[2^2,1], \\ [5]+[3,2]+2[3,1^2]+[2^2,1]+[1^5], \end{gathered}$$

and

$$2[4,1]+2[2,1^3]$$

are all projective characters of $S_5$.

Using 6.3.8, the following columns represent projective characters of $S_5$:

**6.3.16**

$$\begin{array}{lccc} [5] & 1 & & \\ [3,2] & 1 & 1 & \\ [3,1^2] & 2 & 1 & \\ [2^2,1] & 1 & 1 & \\ [1^5] & 1 & & \\ [4,1] & & & 1 \\ [2,1^3] & & & 1 \end{array}$$

Since this matrix has the same number of columns as there are 2-regular classes of $S_5$, we have found columns which *span* the decomposition matrix.

The process of sorting into blocks the projective characters induced from $S_{n-1}$ to $S_n$ can be done in a completely mechanical way, known as *r-inducing*, and this is described in

**6.3.17.** *Suppose* $a_1[\alpha_1]+\cdots+a_s[\alpha_s]$ *is a projective indecomposable character of* $S_{n-1}$. *Choose a p-residue r. Add to each diagram* $[\alpha_i]$ *a node whose p-residue is r, in all possible ways, to yield a diagram for* $S_n$. *From* $a_1[\alpha_1]+\cdots+a_s[\alpha_s]$, *we thereby obtain* $b_1[\beta_1]+\cdots+b_t[\beta_t]$. *By Nakayama's conjecture and* 2.7.41, $b_1[\beta_1]+\cdots+b_t[\beta_t]$ *is a projective (not necessarily indecomposable) character of* $S_n$.

**6.3.18** Example. For $p=2$, $[3,1]+[2^2]+[2,1^2]$ is a projective indecomposable character of $S_4$. In terms of pictures, we have

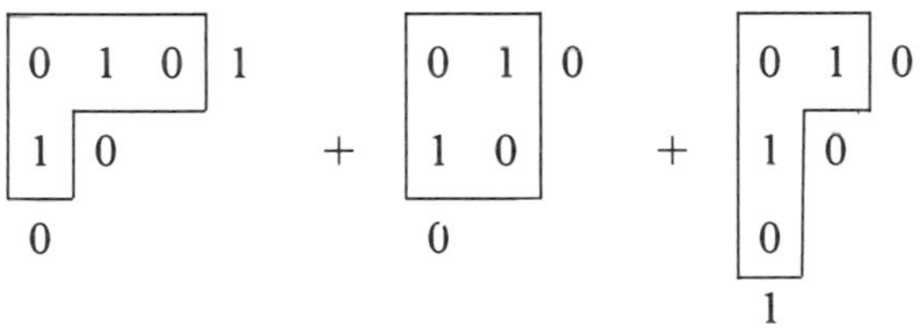

0-inducing gives

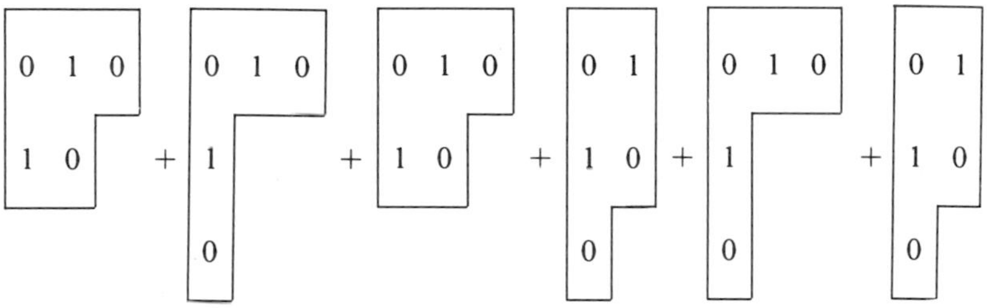

and 1-inducing yields

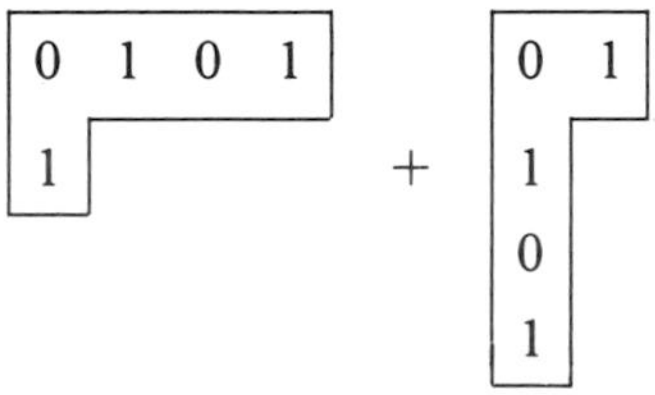

Therefore, $2[3,2]+2[2^2,1]+2[3,1^2]$ and $[4,1]+[2,1^3]$ are projective characters of $S_5$. The construction is precisely that used in Example 6.3.15.

If inducing projective indecomposables from $H$ to $G$ does not give $x$ linearly independent columns, a different subgroup can be examined. It is very remarkable that for $S_n$, $r$-inducing projective indecomposables from $S_{n-1}$ always gives the right number of linearly independent columns (cf. Example 6.3.15). When we come to prove this, we shall also show that, for $S_n$,

**6.3.19.** *Projective characters $\mathbf{c}_1, \mathbf{c}_2, \ldots, \mathbf{c}_x$ can be found which, with a suitable ordering of the ordinary characters, look like*

**6.3.20**
$$\begin{matrix} \chi_1 \\ \chi_2 \\ \vdots \\ \chi_x \\ \vdots \end{matrix} \begin{pmatrix} 1 & & & \\ & 1 & 0 & \\ & & \ddots & \\ & * & & 1 \\ \mathbf{c}_1 & \mathbf{c}_2 & \cdots & \mathbf{c}_x \end{pmatrix},$$

*where $x$ = the number of $p$-regular classes of $G$ (cf.* 6.3.16, *taking* [5], [3,2], *and* [4,1] *as the first three row labels).*

We now state a general result, which will later be shown to apply for $S_n$.

**6.3.21.** *Suppose we are fortunate enough to have found projective characters $\mathbf{c}_1, \ldots, \mathbf{c}_x$ of $G$ satisfying* 6.3.19. *Then the decomposition matrix of $G$ also has form* 6.3.20. *Further,*

**6.3.22**
$$\mathbf{d}_i = \mathbf{c}_i - a_{i+1}\mathbf{d}_{i+1} - \cdots - a_x\mathbf{d}_x$$

*for some nonnegative integers $a_i$. In particular $\mathbf{d}_x = \mathbf{c}_x$.*

*Proof.* This is trivial, because each $\mathbf{c}_i$ is a linear combination of $\mathbf{d}_1, \ldots, \mathbf{d}_x$ with nonnegative integer coefficients. ■

The form of 6.3.22 should be carefully noted. It is *not* true that the decomposition matrix is found from 6.3.20 by "subtracting columns to the right". It may happen that $\mathbf{d}_{x-1} = \mathbf{c}_{x-1} - \mathbf{c}_x$ and $\mathbf{d}_{x-2} = \mathbf{c}_{x-2} - \mathbf{d}_{x-1} = \mathbf{c}_{x-2} - \mathbf{c}_{x-1} + \mathbf{c}_x$. Thus, the nonassociativity of the minus operation complicates 6.3.22. What can be said (since the decomposition matrix has nonnegative entries) is

**6.3.23.** *The entries in our matrix of projective characters* (6.3.20) *are $\geqslant$ the corresponding entries in the decomposition matrix.*

**6.3.24** EXAMPLE. Assume the decomposition matrix for $p=3$ and $S_5$ given in Appendix I.E is correct. We shall find the decomposition matrix of $S_6$.

The first column of the decomposition matrix of $S_5$ gives the projective character $[5]+[2^2,1]$:

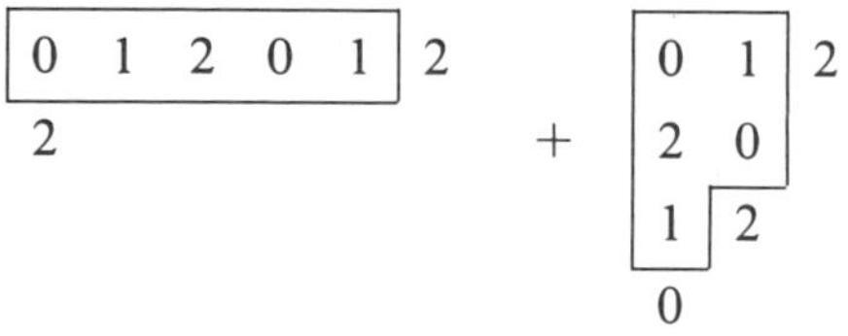


2-inducing gives the projective character

$$\mathbf{c}_1 := [6]+[5,1]+[3,2,1]+[2^3]$$

of $S_6$. In a similar fashion, we find the following projective characters in the principal block of $S_6$ ( $:=$ the block containing the identity character):

6.3.25

| | $\mathbf{c}_1$ | $\mathbf{c}_2$ | $\mathbf{c}_3$ | $\mathbf{c}_4$ | $\mathbf{c}_5$ |
|---|---|---|---|---|---|
| $[6]$ | 1 | | | | |
| $[5,1]$ | 1 | 1 | | | |
| $[4,1^2]$ | | 1 | 1 | | |
| $[3^2]$ | | 1 | | 1 | |
| $[3,2,1]$ | 1 | 1 | 1 | 1 | 1 |
| $[3,1^3]$ | | | 1 | | 1 |
| $[2^3]$ | 1 | | | | 1 |
| $[2,1^4]$ | | | | 1 | 1 |
| $[1^6]$ | | | | 1 | |

Now, $[4,2]$ and $[2^2,1^2]$ are in blocks of defect 0, and $S_6$ has 7 3-regular classes. Hence $\mathbf{c}_1,\ldots,\mathbf{c}_5$ span the principal block of $S_6$, and we can apply 6.3.21: $\mathbf{d}_5=\mathbf{c}_5$. $\mathbf{d}_5$ cannot be subtracted from $\mathbf{c}_4$ to leave nonnegative entries, so $\mathbf{d}_4=\mathbf{c}_4$. Neither $\mathbf{d}_5$ nor $\mathbf{d}_4$ can be subtracted from $\mathbf{c}_3$, so $\mathbf{d}_3=\mathbf{c}_3$. Continuing in this way, we deduce that 6.3.25 gives the principal block of the decomposition matrix.

All the decomposition matrices for $n\leqslant 10$ and $p$ odd can be found using the decomposition matrix of $S_{n-1}$ as in Example 6.3.24, with one notable exception, namely $n=8$ and $p=3$. Naturally, things do not always turn out so rosy. In 6.3.16, for example, the second column can be subtracted from the first. Having found column vectors which span the same space as the columns of the decomposition matrix, it may still be very difficult to determine the correct decomposition matrix. It must be noted that at this stage, no general results help. For instance, whatever integers $a_i$ are chosen in 6.3.22, det $D'D$ is the same. Therefore, we must go on and discuss *ad hoc* methods for determining the decomposition matrix.

Of the two results 6.3.3 and 6.3.4 which we claimed were helpful, we have used the second when applying the fact that

**6.3.26.** *If* $\mathbf{c}$ *is a projective character of* $H$, *then* $\mathbf{c}\uparrow G$ *is a projective character of* $G$.

To employ the first, notice that

**6.3.27.** *If* $\bar{\phi}$ *is a modular character of* $H$, *then* $\bar{\phi}\uparrow G$ *is a modular character of* $G$.

That the decomposition matrix of $G$ is "squeezed" between two possibilities by these facts is illustrated in

**6.3.28** EXAMPLE. Consider the nonprincipal 2-block of $S_7$. Assume that the decomposition matrix for $S_6$ given in Appendix I.E is correct. 1-induce the projective indecomposables

$$[5,1]+[4,2]+[4,1^2]+[3,1^3]+[2^2,1^2]+[2,1^4]$$

and

$$[4,2]+[4,1^2]+[3^2]+[2^3]+[3,1^3]+[2^2,1^2]$$

from $S_6$ to $S_7$. This gives the following projective characters for $S_7$ (using 6.3.8):

**6.3.29**

| | | |
|---|---|---|
| $[6,1]$ | 1 | |
| $[4,3]$ | 1 | 1 |
| $[4,1^3]$ | 2 | 1 |
| $[2^3,1]$ | 1 | 1 |
| $[2,1^5]$ | 1 | |

It can be checked from the character table that all the ordinary characters in this block can be written in terms of $[6,1]$ and $[4,3]$ on 2-regular classes (cf. method I above), so the given columns span the nonprincipal block. Thus 6.3.21 shows that either we have the decomposition matrix for the block, or the second column should be subtracted from the first. The problem, therefore, is to decide whether $z=0$ or 1 in the part of the decomposition matrix

**6.3.30**

| | | |
|---|---|---|
| $[6,1]$ | 1 | |
| $[4,3]$ | $z$ | 1 |

Now, the decomposition matrix of $S_6$ shows that the identity character is a constituent of [4, 2]. Hence

$$[4,2]-[6] \text{ is a modular character of } S_6,$$

and so $[5,2]+[4,3]+[4,2,1]-[7]-[6,1]$ is a modular character of $S_7$ (by using 6.3.27).

We already know that [6, 1] is a modular irreducible, but it is not a constituent of $[5,2]+[4,2,1]$, because it is in the wrong block. Hence [6, 1] is a constituent of [4, 3] and $z=1$.

**6.3.31.** Notice that inducing projective indecomposables showed that $z$ is *at most* 1, and inducing the modular irreducible proved that $z$ is *at least* 1.

This, in essence, is the best character-theory method for finding decomposition matrices. 6.3.26 is used to find a maximum value for an entry, and 6.3.27 determines a minimum value for the entry. If these values do not coincide, try a different subgroup.

Apparently, another way of obtaining information is to use

**6.3.32.** *If* **c** *is a projective character of* $G$, *then* $\mathbf{c}\downarrow H$ *is a projective character of* $H$, *and*

**6.3.33.** *If* $\phi$ *is a modular character of* $G$, *then* $\phi\downarrow H$ *is a modular character of* $H$.

To save spending futile effort working through the same set of answers twice, it is important to realize that

**6.3.34.** *The information gained from inducing projective characters from* $H$ (6.3.26) *is precisely the same as that found by restricting modular characters to* $H$ (6.3.33). *Similarly for* 6.3.27 *and* 6.3.32.

(This result is a consequence of the Frobenius reciprocity theorem, but it is not easy to write out a water-tight proof, and we shall not attempt to do so.)

**6.3.35** Example. We calculate again the value of $z$ in 6.3.30. $[6,1]\downarrow S_6=[6]+[5,1]$ and $[4,3]\downarrow S_6=[4,2]+[3^2]$. If $z>1$, then $[4,3]-2[6,1]$ would be a modular character of $S_7$. This would imply that $[4,2]+[3^2]-2[6]-2[5,1]$ is a modular character of $S_6$. Since this is false,

**6.3.36.** The technique of restricting modular characters has proved $z\leqslant 1$.

If $z<1$, then the second column in 6.3.29 must be subtracted from the first, giving a projective character $[6,1]+[4,1^3]+[2,1^5]$ of $S_7$. Restricting,

$[6]+[5,1]+[4,1^2]+[3,1^3]+[1^6]+[2,1^4]$ would be a projective character of $S_6$. This is false, so

**6.3.37.** The method of restricting projective characters has proved $z \geqslant 1$.

Compare 6.3.36 and 6.3.37 with 6.3.31, and see how 6.3.34 is illustrated. As is indicated in the above example, there is a process of *r*-restricting (6.3.32), corresponding to *r*-inducing (6.3.26), but in practice it is much quicker to combine *r*-inducing with 6.3.27. Which combination of techniques from 6.3.26, 6.3.27, 6.3.32, and 6.2.33 to use is a matter of taste. It is often more efficient to work entirely in terms of modular characters, especially when dealing with only part of the decomposition matrix.

When working only with modular characters, it is relevant to note that

**6.3.38.** *Once the right number of linearly independent rows in a block has been found, the whole block can be calculated.*

This is so because once the modular irreducible characters are known, all the ordinary characters in the block can be written in terms of them, if need be by looking at the *p*-regular classes in the character table. Often, a set of columns spanning the block will have been found before stage 6.3.38 is reached, and this provides a much simpler way of completing the full block.

We shall next work through a big example to show the combination of inducing projective and modular characters in action.

**6.3.39** EXAMPLE. *The decomposition matrix for* $S_9$ *when* $p=2$. The decomposition matrix of $S_8$ in Appendix I.E may be assumed correct. 1-inducing the 1st to 5th columns for $S_8$, we find the following projective characters in the nonprincipal block of $S_9$:

| | | | | | | | | | | |
|---|---|---|---|---|---|---|---|---|---|---|
| $[8,1]$ | 2 | 1 | | | | | 1 | 1 | | |
| $[6,3]$ | | 2 | 2 | 1 | | | | 2 | 1 | |
| $[6,1^3]$ | 2 | 3 | 2 | 1 | | | 1 | 3 | 1 | |
| $[4,3,2]$ | 6 | 1 | 2 | 1 | 3 | | 3 | 1 | 1 | 1 |
| $[4,3,1^2]$ | 6 | 3 | 4 | 2 | 3 | | 3 | 3 | 2 | 1 |
| $[4,2^2,1]$ | 6 | 3 | 4 | 2 | 3 | $\underset{6.3.8}{\Rightarrow}$ | 3 | 3 | 2 | 1 |
| $[4,1^5]$ | 2 | 3 | 2 | 1 | | | 1 | 3 | 1 | |
| $[3^2,2,1]$ | 6 | 1 | 2 | 1 | 3 | | 3 | 1 | 1 | 1 |
| $[2^3,1^3]$ | | 2 | 2 | 1 | | | | 2 | 1 | |
| $[2,1^7]$ | 2 | 1 | | | | | 1 | 1 | | |
| | | | | | | | $\mathbf{k}_1$ | $\mathbf{k}_2$ | $\mathbf{k}_3$ | $\mathbf{k}_4$ |

$S_9$ has 8 2-regular classes, and we shall soon find 5 linearly independent columns in the principal block. Therefore, the block under consideration contains 3 modular irreducible representations (cf. 6.2.2). Applying 6.3.21 to $\mathbf{k}_1, \mathbf{k}_3, \mathbf{k}_4$, we deduce that $\mathbf{k}_4$ is a projective indecomposable. Now, $\mathbf{k}_2 = \mathbf{k}_1 + 2\mathbf{k}_3 - 4\mathbf{k}_4$. Since $\mathbf{k}_1 - 4\mathbf{k}_4$ and $\mathbf{k}_3 - 2\mathbf{k}_4$ are not projective characters, we deduce that $\mathbf{k}_1 - 2\mathbf{k}_4$ and $\mathbf{k}_3 - \mathbf{k}_4$ are projective characters. We can now use 6.3.21 on $\mathbf{k}'_1 := \mathbf{k}_1 - 2\mathbf{k}_4$, $\mathbf{k}'_2 := \mathbf{k}_3 - \mathbf{k}_4$, and $\mathbf{k}'_3 := \mathbf{k}_4$.

Parts of the columns $\mathbf{k}'_1, \mathbf{k}'_2, \mathbf{k}'_3$ are recorded below:

**6.3.40**

| | | | |
|---|---|---|---|
| [8,1] | 1 | | |
| [6,3] | | 1 | |
| [4,3,2] | 1 | | 1 |
| $[4,3,1^2]$ | 1 | 1 | 1 |

The fourth row shows that

**6.3.41** $\quad [4,3,1^2] = [4,3,2] + [6,3] \quad$ on 2-regular classes.

(Note this trick, which avoids checking the character table.)

Remember that the numbers in 6.3.40 are *at most* the entries in the true decomposition matrix. We have to check, therefore, whether $[8,1] \subseteq [4,3,2]$.

Induce the modular character $[4,3,1] - [5,3] - [8]$ from $S_8$, and ignore the characters in the principal block of $S_9$. Hence $2[4,3,2] - [8,1]$ is a modular character of $S_9$ (using 6.3.41).

Having proved that $[8,1] \subseteq [4,3,2]$, we know that $\mathbf{k}'_3$ should not be subtracted from $\mathbf{k}'_1$, and 6.3.21 shows that $\mathbf{k}'_1$, $\mathbf{k}'_2$, and $\mathbf{k}'_3$ are columns of the decomposition matrix.

Now 0-induce the 1st, 2nd, 3rd, 6th, 4th, and 5th columns of $S_8$:

| | | | | | | | | | | | | |
|---|---|---|---|---|---|---|---|---|---|---|---|---|
| [9] | 1 | | | | | | | 1 | | | | |
| [7,2] | 1 | 2 | 1 | | | | | 1 | 1 | | | |
| $[7,1^2]$ | 2 | 2 | 1 | | | | | 2 | 1 | | | |
| [6,2,1] | 1 | 2 | 2 | 1 | | | | 1 | 1 | 1 | | |
| [5,4] | | 2 | 1 | | 2 | | | | 1 | | 1 | |
| [5,3,1] | 2 | 2 | 2 | 1 | 2 | 1 | | 2 | 1 | 1 | 1 | 1 |
| $[5,2^2]$ | 2 | | 1 | 1 | | 1 | | 2 | | 1 | | 1 |
| $[5,2,1^2]$ | 3 | 4 | 3 | 1 | 2 | 1 | | 3 | 2 | 1 | 1 | 1 |
| $[5,1^4]$ | 2 | 4 | 2 | | 2 | | | 2 | 2 | | 1 | |
| $[4^2,1]$ | 2 | 2 | 1 | | 2 | 1 | | 2 | 1 | | 1 | 1 |
| $[4,2,1^3]$ | 3 | 4 | 3 | 1 | 2 | 1 | $\underset{6.3.8}{\Rightarrow}$ | 3 | 2 | 1 | 1 | 1 |

| | | | | | | | $\mathbf{c}_1$ | $\mathbf{c}_2$ | $\mathbf{c}_3$ | $\mathbf{c}_4$ | $\mathbf{c}_5$ |
|---|---|---|---|---|---|---|---|---|---|---|---|
| $[3^3]$ | 2 | | | | | 1 | 2 | | | | 1 |
| $[3^2,1^3]$ | 2 | | 1 | 1 | | 1 | 2 | | 1 | | 1 |
| $[3,2^3]$ | 2 | 2 | 1 | | 2 | 1 | 2 | 1 | | 1 | 1 |
| $[3,2^2,1^2]$ | 2 | 2 | 2 | 1 | 2 | 1 | 2 | 1 | 1 | 1 | 1 |
| $[3,2,1^4]$ | 1 | 2 | 2 | 1 | | | 1 | 1 | 1 | | |
| $[3,1^6]$ | 2 | 2 | 1 | | | | 2 | 1 | | | |
| $[2^4,1]$ | | 2 | 1 | | 2 | | | 1 | | 1 | |
| $[2^2,1^5]$ | 1 | 2 | 1 | | | | 1 | 1 | | | |
| $[1^9]$ | 1 | | | | | | 1 | | | | |

The projective characters $\mathbf{c}_5$, $\mathbf{c}_4$, and $\mathbf{c}_3$ are columns of the decomposition matrix by 6.3.21. The first problem is whether we should take $\mathbf{c}_2$ as it stands, or subtract $\mathbf{c}_4$ from it.

We know that the decomposition matrix has columns starting opposite [9], [7,2], [6,2,1], [5,4], and [5,3,1]. Call the modular irreducible characters which correspond to these columns $\phi[9]$, $\phi[7,2]$, etc.

$[6,2]-[7,1]+[8]$ and $[5,3]-[6,2]$ are modular characters of $S_8$. Inducing to $S_9$, and ignoring the characters in the nonprincipal block, we find that the following are modular characters of $S_9$:

**6.3.42** $[6,2,1]-[7,1^2]+[9]=[6,2,1]-[7,2]$ (cf. 6.3.41),

**6.3.43** $[5,3,1]+[5,4]-[6,2,1]-[7,2]$.

The first of these shows that $\phi[7,2]\subseteq[6,2,1]$, and hence in the second $2\phi[7,2]\subseteq[5,3,1]+[5,4]$. Since $2\phi[7,2]\nsubseteq[5,3,1]$ (by 6.3.23), $\phi[7,2]\subseteq[5,4]$. This shows that $\mathbf{c}_4$ should not be subtracted from $\mathbf{c}_2$, which is therefore a column of the decomposition matrix.

Since $\mathbf{c}_2$ cannot be subtracted from $\mathbf{c}_1$, $\phi[9]\subseteq[7,2]$. From 6.3.42, $\phi[9]\subseteq[6,2,1]$. This proves that $\mathbf{c}_3$ should not be subtracted from $\mathbf{c}_1$. Since $2\phi[9]\subseteq[6,2,1]+[7,2]$ and $\phi[9]\nsubseteq[5,4]$, 6.3.43 shows that $2\phi[9]\subseteq[5,3,1]$. Therefore, $\mathbf{c}_5$ should not be taken from $\mathbf{c}_1$. We have now proved that $\mathbf{c}_1$ is a column of the decomposition matrix. The calculation is now complete, because it has been proved that $\mathbf{c}_1$, $\mathbf{c}_2$, $\mathbf{c}_3$, $\mathbf{c}_4$, and $\mathbf{c}_5$ are columns of the decomposition matrix.

This example does not instil a feeling of confidence that the methods used will always work. The complete answer can often be found this way for small symmetric groups, but there are several exceptions even for $n\leqslant 10$. These ambiguities, and how to resolve them, will be discussed later. For the moment, though, the plan is to prove that $r$-inducing always leads to the right number of linearly independent columns, and to show that the decomposition matrix of $S_n$ has the shape 6.3.20. As we warned at the beginning of the chapter, this proof is hard, and a nicer proof of the form of the decomposition matrix is given in Section 7.1.

Inspecting the decomposition matrices given in Appendix I, we notice that each new column begins opposite a $p$-regular partition. (Recall that a partition $\alpha$ is said to be $p$-regular if it does not contain $p$ parts $\alpha_i \neq 0$ which are equal; otherwise, $\alpha$ is $p$-singular. 6.1.2 shows that the number of inequivalent irreducible $p$-modular characters of $S_n$ equals the number of $p$-regular partitions.)

The awkward part of the $r$-inducing process is that it is by no means clear that (with row labels arranged in lexicographic order):

   (i) $r$-inducing always gives a column whose first nonzero entry is opposite a $p$-regular diagram.

   (ii) For each $p$-regular diagram $[\alpha]$, a column arises from $r$-inducing which starts opposite $[\alpha]$. Further, there is at least one such column all of whose entries are divisible by the entry opposite $[\alpha]$.

Let us examine some cases, and try to find a pattern.

**6.3.44** EXAMPLES.

   (i) When $p=2$, the column for $S_4$ starting opposite

$$\begin{array}{|ccc|} \hline 0 & 1 & 0 \\ \hline 1 \\ \hline \end{array}$$

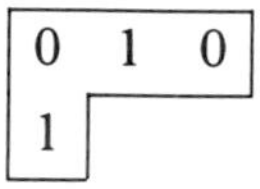

gives, when we 1-induce, a column starting with

$$\begin{array}{|cccc|} \hline 0 & 1 & 0 & 1 \\ \hline 1 \\ \hline \end{array}$$

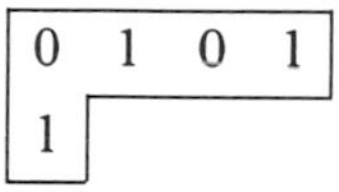

and on 0-inducing, a column beginning

$$2\times \begin{array}{|ccc|} \hline 0 & 1 & 0 \\ \hline 1 & 0 \\ \hline \end{array}$$

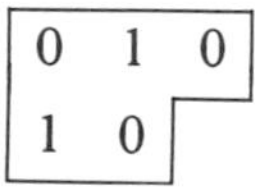

(cf. Example 6.3.18).

(ii) When $p=3$, consider the column for $S_7$ commencing with

$$\begin{array}{|lll|}\hline 0 & 1 & 2 \\ 2 & 0 & \\ 1 & & \\ 0 & & \\ \hline \end{array} .$$

0-inducing gives a column starting with

$$\begin{array}{|llll|}\hline 0 & 1 & 2 & 0 \\ 2 & 0 & & \\ 1 & & & \\ 0 & & & \\ \hline \end{array} .$$

1-inducing gives a column starting with

$$\begin{array}{|lll|}\hline 0 & 1 & 2 \\ 2 & 0 & 1 \\ 1 & & \\ 0 & & \\ \hline \end{array} .$$

2-inducing gives a column starting with

$$2\times \begin{array}{|lll|}\hline 0 & 1 & 2 \\ 2 & 0 & \\ 1 & 2 & \\ 0 & & \\ \hline \end{array} .$$

(This example can be checked by referring to the decomposition matrices in the appendix.)

(iii) When $p=2$, 1-inducing the column of $S_8$ starting with

$$\begin{array}{|llll|}\hline 0 & 1 & 0 & 1 \\ 1 & 0 & 1 & \\ 0 & & & \\ \hline \end{array}$$

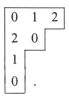

gives a column beginning with

$$3\times \begin{array}{|cccc} 0 & 1 & 0 & 1 \\ 1 & 0 & 1 & \\ 0 & 1 & & \end{array}$$

(cf. 6.3.39).

This suggests that to find a column starting opposite a $p$-regular diagram $[\alpha]$ of $S_n$, we look for the node which

(a) Can be removed from $[\alpha]$ to leave a $p$-regular diagram, and
(b) is the furthest right in $[\alpha]$ satisfying (a).

Suppose this node has $p$-residue $r$. It appears that we should $r$-induce from $S_{n-1}$ the column which starts opposite $[\alpha]$ with the node removed. Further, if there are $k-1$ higher nodes which could have been removed from $[\alpha]$, $r$-inducing seems to give the number $k$ opposite $[\alpha]$.

Thus, for example, in (ii) above, the 2 multiplying $[3,2^2,1]$ occurs because the line in the picture below intersects the diagram twice:

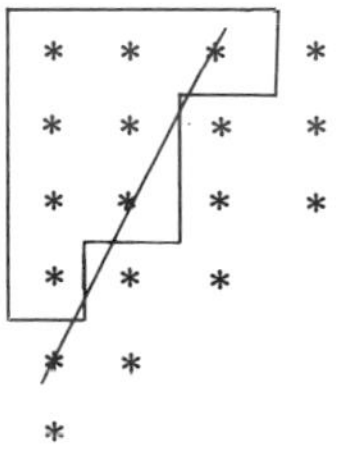

With this in mind, we define a *ladder* to be a straight line joining the vertex $(i,1)$ to the point $(1,(i-1)(p-1)^{-1}+1)$. By construction, every vertex lies in some ladder, and

**6.3.45.** *All the vertices in a ladder have the same p-residue.*

**6.3.46** Example. $p=3$

```
0 1 2 0
2 0 1
1 2 0
0 1
2 0
1
0
```

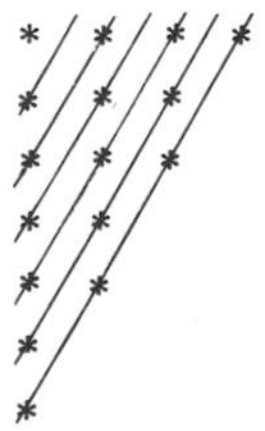

We notice that a diagram is $p$-regular iff every ladder which intersects the diagram in $i$ vertices has these $i$ vertices at the top of the ladder.

**6.3.47** EXAMPLE.

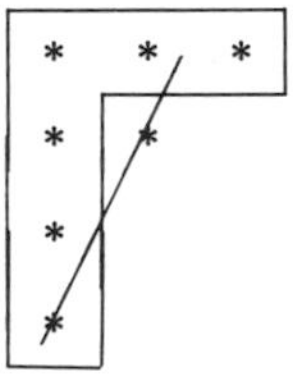

The ladder shown intersects $[3, 1^3]$ in one node, but this node is not at the top of the ladder. Therefore, $[3, 1^3]$ is 3-singular.

This interpretation of $p$-regularity alerts us to the fact that from any diagram $[\beta]$ we can construct a $p$-regular one, thus:

**6.3.48.** *For each ladder, if the ladder intersects* $[\beta]$ *in i nodes, replace these nodes by the top i vertices in the ladder. Call the new diagram* $[\beta]^R$.

It is easy to check that $[\beta]^R$ is, indeed, a diagram. The construction ensures that it is $p$-regular.

**6.3.49** EXAMPLE. $p=3$ Looking at

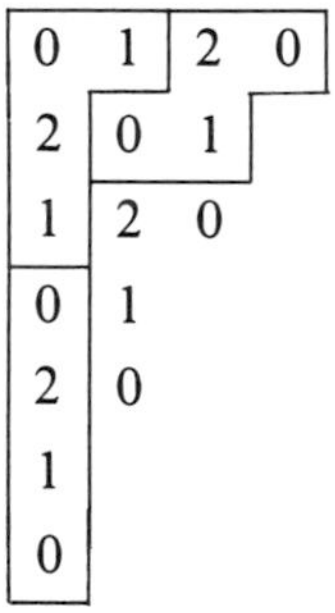


we see that $[2, 1^6]^R = [4, 3, 1]$.

Certainly $[\beta]^R \trianglerighteq [\beta]$, but the above example shows that there may be a $[\gamma]$ $(=[3^2, 1^2])$ such that $[\beta]^R \triangleright [\gamma] \triangleright [\beta]$. The diagram $[\beta]^R$ is obtained from $[\beta]$ simply by "squashing" $[\beta]$ from a particular angle.

Now we can state the main result of this section (cf. 7.3.6).

**6.3.50** THEOREM. *Arrange the ordinary characters of* $S_n$ *in lexicographic order. Then for each p-regular diagram* $[\alpha]$ *there is a column of the decomposi-*

*tion matrix whose first nonzero entry is opposite* $[\alpha]$. *This column has the following properties*:

**6.3.51.**

(i) *The entry opposite* $[\beta]$ *is nonzero only if* $[\alpha] \trianglerighteq [\beta]^R$.

(ii) *If* $[\alpha] = [\beta]^R$, *the entry opposite* $[\beta]$ *is* 1. *In particular, the entry opposite* $[\alpha]$ *is* 1.

(iii) *If the longest ladder intersecting* $[\alpha]$ *does so in k r-nodes, the entry opposite* $[\beta]$ *is zero unless k r-nodes can be removed from* $[\beta]$.

*Proof.* Assume the theorem is true for $n' < n$.

Let $[\alpha]$ be a $p$-regular diagram for $S_n$, and assume that the longest ladder, $L_\alpha$, intersecting $[\alpha]$ does so in $k$ $r$-nodes. (Note that we say "the longest ladder," not "the ladder with the longest intersection with $[\alpha]$".) Let $[\bar{\alpha}]$ be $[\alpha]$ with these $k$ nodes removed. Then $[\bar{\alpha}]$ is $p$-regular.

By the induction hypothesis, there is a projective indecomposable character $\xi$ of $S_{n-k}$ for which

**6.3.52.**

(i) the entry opposite $[\bar{\beta}]$ is nonzero only if $[\bar{\alpha}] \trianglerighteq [\bar{\beta}]^R$,

(ii) if $[\bar{\alpha}] = [\bar{\beta}]^R$, the entry opposite $[\bar{\beta}]$ is 1.

We wish to show that $\xi \uparrow S_n$ gives a column satisfying

**6.3.53.**

(i) all the entries are divisible by $k!$,

(ii) the entry opposite $[\beta]$ is nonzero only if $[\alpha] \trianglerighteq [\beta]^R$,

(iii) if $[\alpha] = [\beta]^R$, the entry opposite $[\beta]$ is $k!$,

(iv) if it is impossible to remove $k$ $r$-nodes from $[\beta]$, the entry opposite $[\beta]$ is zero.

Now, for any character $[\bar{\beta}]$ in the same block as $[\bar{\alpha}]$, and $[\beta]$ in the same block as $[\alpha]$, the branching theorem gives

$$([\bar{\beta}] \uparrow S_n, [\beta])_{S_n} = \begin{cases} k! & \text{if } \; k \text{ } r\text{-nodes can be added to } [\bar{\beta}] \text{ to give } [\beta], \\ 0 & \text{otherwise.} \end{cases}$$

The $k!$ arises because we can add the nodes in any order—notice that since they all have the same $p$-residue, no two added nodes are adjacent. This proves 6.3.53(i) and (iv).

6.3.53(ii) will follow from 6.3.52(i) once we have proved:

**6.3.54.** *If* $[\bar{\alpha}] \trianglerighteq [\bar{\beta}]^R$, *and* $[\beta]$ *is obtained from* $[\bar{\beta}]$ *by adding k r-nodes, then* $[\alpha] \trianglerighteq [\beta]^R$.

For 6.3.53(iii), we require:

**6.3.55.** *For each* $[\beta]$ *satisfying* $[\beta]^R = [\alpha]$ *there is a unique diagram* $[\bar{\beta}]$ *such that*

(i) *k r-nodes can be added to* $[\bar{\beta}]$ *to give* $[\beta]$,
(ii) $[\bar{\alpha}] \trianglerighteq [\bar{\beta}]^R$,

*and, in fact, this* $[\bar{\beta}]$ *satisfies* $[\bar{\alpha}] = [\bar{\beta}]^R$.

We shall come back later and prove these results. Assuming they are true, $\xi \uparrow S_n$ is a projective character of $S_n$ satisfying 6.3.53.

By 6.3.8, all the column entries can be divided by $k!$ to give a projective character for which 6.3.51 holds. This can be done for all $p$-regular diagrams of $S_n$. If $[\beta]^R = [\alpha]$, then the entry opposite $[\beta]$ in the $[\alpha]$ column is 1, and the later entries in the $[\beta]$ row are zero. Because the number of $p$-regular diagrams equals the number of $p$-regular classes, 6.3.21 can be applied. The theorem follows. ■

*Proof of* 6.3.54 *and* 6.3.55. Assume that $[\bar{\alpha}] \trianglerighteq [\bar{\beta}]^R$. Let the highest node of the ladder $L_\alpha$ be in the $l$th row. Using the obvious notation for row lengths, we have

$$\bar{\alpha}_1 = \bar{\alpha}_2 = \cdots = \bar{\alpha}_{l-1} = \bar{\alpha}_l + 1$$

and

$$\bar{\beta}_1^R \geqslant \bar{\beta}_2^R \geqslant \cdots \geqslant \bar{\beta}_{l-1}^R \geqslant \bar{\beta}_l^R.$$

Since $[\bar{\alpha}] \trianglerighteq [\bar{\beta}]^R$, $\bar{\beta}_l^R < \bar{\alpha}_l + 1$. This shows that $[\bar{\beta}]^R$ does not contain the highest node of $L_\alpha$, and because $[\bar{\beta}]^R$ is $p$-regular,

**6.3.56.** $[\bar{\beta}]^R$ does not contain any node in $L_\alpha$.

One need only note that $\bar{\beta}_1^R \leqslant \bar{\alpha}_1$ to deduce, also, that

**6.3.57.** $[\bar{\beta}]^R$ does not contain any node to the right of $L_\alpha$.

Now suppose that $k$ $r$-nodes can be added to $[\bar{\beta}]$ to give $[\beta]$. Since, by construction, $[\bar{\beta}]^R$ is a subset of $[\beta]^R$, $k$ $r$-nodes can be added to $[\bar{\beta}]^R$ to give $[\beta]^R$. By 6.3.56 and 6.3.57, none of the nodes added to $[\bar{\beta}]^R$ lie to the right of $L_\alpha$, and

$$\text{if} \quad [\beta]^R = [\alpha], \quad \text{then} \quad [\bar{\beta}]^R = [\bar{\alpha}].$$

More generally, the part of $[\beta]^R$ no lower than the $k$th node of $L_\alpha$ (which is in the $f$th row, say) is a subset of $[\alpha]$. Therefore,

$$\text{for} \quad j \leqslant f, \qquad \sum_{i=1}^{j} \alpha_i \geqslant \sum_{i=1}^{j} \beta_i^R,$$

and

$$\text{for} \quad j > f, \qquad \sum_{i=1}^{j} \alpha_i = k + \sum_{i-1}^{j} \bar{\alpha}_i \geqslant k + \sum_{i=1}^{j} \bar{\beta}_i^R \geqslant \sum_{i=1}^{j} \beta_i^R.$$

This proves 6.3.54.

To complete the proof of 6.3.55, we must show that, given $[\beta]$ with $[\beta]^R = [\alpha]$, a unique $[\bar{\beta}]$ exists. If $[\beta]^R = [\alpha]$, then $[\beta]$ intersects $L_\alpha$ in $k$ $r$-nodes. These nodes can be removed from $[\beta]$, since no two $r$-nodes are adjacent and $[\beta]$ does not contain any nodes to the right of $L_\alpha$. Removing these nodes is clearly the one and only way to obtain a $[\bar{\beta}]$ satisfying $[\bar{\beta}]^R = [\bar{\alpha}]$. ■

**6.3.58** COROLLARY. *The number of linearly independent projective characters found by r-inducing equals the number of columns of the decomposition matrix of* $S_n$.

*Proof.* We need note only that although the proof of Theorem 6.3.50 involved $r$-inducing from $S_{n-k}$, at least as many linearly independent projective characters will arise by $r$-inducing from $S_{n-1}$, because of the transitivity of the inducing process. ■

Interpreted in terms of the rows of the decomposition matrix, Theorem 6.3.50 proves:

**6.3.59.** *Let* $\phi[\alpha]$ *be the modular irreducible character which corresponds to the column of the decomposition matrix whose first nonzero entry is opposite the p-regular diagram* $[\alpha]$.

(i) $\{\phi[\alpha] \| \alpha$ *is a p-regular partition of n*$\}$ *forms a complete set of inequivalent, irreducible p-modular characters of* $S_n$.

(ii) *If* $\alpha$ *is p-regular,* $\phi[\alpha]$ *is a constituent of* $[\alpha]$ *with multiplicity* 1. *The other constituents of* $[\alpha]$ *have the form* $\phi[\gamma]$ *with* $\gamma$ *p-regular and* $[\gamma] \rhd [\alpha]$.

*More generally,*

(iii) *If* $\beta$ *is any partition of n, then* $[\beta]$ *contains* $\phi[\beta]^R$ *with multiplicity* 1, *and the other constituents of* $[\beta]$ *have the form* $\phi[\gamma]$ *with* $\gamma$ *p-regular and* $[\gamma] \trianglerighteq [\beta]^R$.

In terms of individual blocks, Theorem 6.3.50 says:

**6.3.60** THEOREM. *Within each block arrange the row labels so that all p-regular diagrams (in lexicographic order) come before all the p-singular diagrams (in lexicographic order). Then the block of the decomposition matrix looks like*

$$\begin{pmatrix} 1 & & & \\ & 1 & 0 & \\ & & \ddots & \\ & & & 1 \\ & * & & \end{pmatrix}.$$

*The columns will be labeled* $\phi[\alpha]$, *where* $\alpha$ *is p-regular, and automatically these labels will be in lexicographic order.*

*Further zeros can be inserted using the fact that if* $[\alpha]$ *is p-regular and* $[\alpha] \not\trianglerighteq [\beta]^R$, *then there is a zero at the intersection of the row labeled* $[\beta]$ *and the column labeled* $\phi[\alpha]$. *Theorem* 6.3.50(iii), *may also imply some zeros.*

**6.3.61** EXAMPLES. $p=2$, $n=13$. $\phi[7,5,1]$ is not a constituent of $[8,4,1]$, because $[7,5,1] \not\trianglerighteq [8,4,1]$; nor is it a constituent of $[7,1^6]$, because $[7,5,1] \not\trianglerighteq [7,1^6]^R = [7,6]$. By Theorem 6.3.50(iii), $\phi[7,6]$ is not a constituent of $[5,4^2]$, as is seen in the pictures below:

$$[7,6] = \begin{array}{ccccccc} 0 & 1 & 0 & 1 & 0 & 1 & \not{0} \\ 1 & 0 & 1 & 0 & 1 & \not{0} & \end{array}, \qquad [5,4^2] = \begin{array}{ccccc} 0 & 1 & 0 & 1 & 0 \\ 1 & 0 & 1 & 0 & \\ 0 & 1 & 0 & 1 & \end{array}.$$

It is clear that we have not got all the mileage we could out of theorem 6.3.50(iii), since it was not used in the induction hypothesis. An improvement of this result and other $r$-inducing information could be given, but it is hardly worthwhile formulating the statements, since they seem to be of no general value, and the results would become apparent in numerical calculations.

Several other corollaries of Theorem 6.3.50 can be listed.

**6.3.62** Corollary. *The first ordinary character in lexicographic order within each block is modularly irreducible.*

Because of 6.3.51(i), the decomposition matrix of $S_n$ will be "wedge-shaped" (6.3.1) if the *p*-regular ordinary characters are put at the beginning, in any order which agrees with the dominance order. It is clear that the first *p*-regular diagram we must take in each block is unique (add nodes to the first row of the core). Less clear is that there is always a *unique* *p*-regular diagram $[\alpha]$ satisfying:

**6.3.63.** $[\beta]$ is a *p*-regular diagram in the same block as $[\alpha] \Rightarrow [\alpha] \not\trianglerighteq [\beta]$.

But such a diagram obviously exists, and we can say of it:

**6.3.64** Corollary. *The column of the decomposition matrix corresponding to a p-regular diagram* $[\alpha]$ *satisfying* 6.3.63 *can be written down at once, since there is a* 1 *opposite* $[\beta]$ *if* $[\beta]^R = [\alpha]$, *and* 0 *otherwise.*

**6.3.65** Example. The 3-regular diagram in question for the principal 3-block of $S_6$ is $[3,2,1]$. The diagrams $[\beta]$ satisfying $[\beta]^R = [3,2,1]$ are $[3,2,1]$, $[3,1^3]$, $[2^3]$, and $[2,1^4]$. (See $\mathbf{c}_5$ in 6.3.25.)

A similar result is

**6.3.66** Corollary. *r-inducing provides an algorithm for writing down a complete set of linearly independent projective characters of* $S_n$.

*Proof.* It is not necessary to determine the decomposition matrix of $S_n$ before inducing further the projective characters found by *r*-inducing. ■

The matrices obtained by following Corollary 6.3.66 are presented, incorrectly, as decomposition matrices in Robinson's book.

Compare 6.3.11 with the next corollary of Theorem 6.3.50.

**6.3.67** Corollary. *Each p-singular character can be written as a linear combination, with integer coefficients, of the p-regular characters on p-regular classes.*

**6.3.68** Example. When $n \geq 5$, we find that $[n]+[n-2,2]=[n-2,1^2]$ on 2-regular classes. Such relations also follow from the Murnaghan-Nakayama formula (cf. Littlewood's paper [1951]).

It is clear that a similar result to Theorem 6.3.50 could be proved by interchanging the words "row" and "column" (referring to diagrams). In particular, the modular irreducible characters can be labeled not only by

$p$-regular diagrams, but also by diagrams which have no $p$ columns of the same length; that is, by diagrams $[\beta]$ whose associated diagram $[\beta']$ is $p$-regular.

We now return to the problem of how to proceed when inducing projective indecomposable and modular irreducible characters from $S_{n-1}$ to $S_n$ does not provide the whole decomposition matrix. The short answer is to try the same tricks on a different subgroup. When most of the decomposition matrix of a group is known, the plan of working entirely with modular irreducible characters (6.3.27 and 6.3.33) really comes into its own. As we have explained, there are equivalent methods using projective characters, but these do take longer.

This is a most frustrating stage of the game. Suppose we are trying to determine the multiplicity of a known modular irreducible $\phi$ in a character $\chi$, and we know that this multiplicity is 0 or 1. The two methods to try are:

**6.3.69.** Prove $(\chi-\phi)\downarrow H$ is not a modular character of $H$.

**6.3.70.** Find a modular character $\bar{\phi}$ of $H$ such that $\bar{\phi}\uparrow G$ cannot be a modular character of $G$ unless $\phi\subseteq\chi$.

These two methods are virtually independent, and there is a fifty-fifty chance that at any given time we are trying the wrong one. Two notes of warning:

**6.3.71.** If a subgroup $H$ fails to give the answer, it is no good trying 6.3.69 and 6.3.70 on a subgroup contained in $H$, because of the transitivity of the process involved.

**6.3.72.** The method of 6.3.70 will not work if $\bar{\phi}$ agrees with an ordinary character of $H$, because inducing an ordinary character gives an ordinary character.

We shall illustrate with $S_n$. For $n\leqslant 10$, there are only four problem cases not solved by inducing from $S_{n-1}$, namely

$$\text{The multiplicity of}\left\{\begin{array}{l}\phi[8] \text{ in } [5,3] \bmod 3\\ \phi[5] \text{ in } [3,2] \bmod 2\\ \phi[7] \text{ in } [5,2] \bmod 2\\ \phi[10] \text{ in } [6,3,1] \bmod 2\end{array}\right\}\text{is undecided.}$$

$$\text{It is}\left\{\begin{array}{l}0 \text{ or } 1\\ 0 \text{ or } 1\\ 0,\ 1, \text{ or } 2\\ 1,\ 2, \text{ or } 3.\end{array}\right\}.$$

**6.3.73** EXAMPLE. *The multiplicity of* $\phi[7]$ *in* $[5,2]$ *mod* 2 *is* 0. We must choose a subgroup of $S_7$ not contained in $S_6$. In passing, it must be noted that, in general, the decomposition matrix of $A_n$ is as hard to find as that of $S_n$, so we assume no knowledge about $A_7$. Since we do not want to become too involved in calculating the decomposition matrix of our subgroup, we choose one which has a simple decomposition matrix. A subgroup of odd order would be good, because

**6.3.74.** *If* $p\nmid|H|$, *the ordinary characters of H are modularly irreducible.*

The most obvious subgroup of $S_7$ to try is that generated by a 7-cycle. This does not work, because $([5,2]\downarrow H, 1)_H=2$. Try, therefore, a slightly bigger subgroup. The group $N$ generated by (1234567) and (235)(476) contains $1_{S_7}$, 6 7-cycles, and 14 elements whose cycle partition is $(3^2,1)$. Part of the character table of $S_7$ is

| | $(1^7)$ | $(3^2 1)$ | $(7)$ |
|---|---|---|---|
| $[5,2]$ | 14 | $-1$ | 0 |

Therefore $[5,2]\downarrow N$ does not contain the identity character. Since $([5,2]-[7])\downarrow N$ is not a modular character of $N$, we are finished.

**6.3.75** EXAMPLE: *The multiplicity of* $\phi[8]$ *in* $[5,3]$ *mod* 3 *is* 0. Bearing in mind the method of the last example, and looking at the character table of $S_8$, we soon come around to trying the group $H$ generated by (1234) and (15)(26)(37)(48) ($H\simeq C_4 \operatorname{wr} C_2$, and $3\nmid|H|=32$). Sure enough, $[5,3]\downarrow H$ does not contain the identity character, and the problem is resolved.

The above examples are just the start of many troubles. If $p^a>1$, and the power of $p$ dividing $n-p^a+1$ is exactly $a$, then there is difficulty in deciding the multiplicity of the identity character in $[n-p^a, p^a]$. This multiplicity is 0 (see 7.3.23), but apparently could be $>0$ if we use only $S_{n-1}$. The first problem like this for $p=5$ is in $[9,5]$, and no elementary character-theoretic method of finding the multiplicity of the identity character here is known.

**6.3.76** EXAMPLE: *The multiplicity of* $\phi[5]$ *in* $[3,2]$ mod 2 *is* 1. (Cf. Example 6.3.15.) This is an interesting problem. As the alternative multiplicity is 0, we should try 6.3.70 this time. But since $S_4$ fails, 6.3.71 and 6.3.72 imply that we have to use $A_5$. It is lucky that a trick works, because we certainly do not want to assume knowledge about $A_n$ in general. The idea is to see that $[3,1^2]$ splits when restricted to $A_5$. Since $\phi[3,2]$ is the only constituent of $[3,1^2]$ besides the trivial character, $\phi[3,2]$ must split into two modular characters of equal degrees. Thus $\phi[3,2]$ has even degree. But $[3,2]=a\phi[5]+\phi[3,2]$ with $a=0$ or 1, and $\deg[3,2]$ is odd. Therefore, $a=1$.

We shall use this last example to show that the most elementary properties of representation modules are sometimes more powerful than mere character-theoretic results (6.3.82 and 6.3.88).

It seems to be very hard to determine the multiplicity mod 2 of $\phi[10]$ in $[6,3,1]$, and it is likely that no character-theoretic argument, such as those discussed so far, solves the problem.

Since no example of the use of 6.3.70 when $H=S_{n-1}$ fails has yet been given, we illustrate with the instructive

**6.3.77** EXAMPLE. *When* $p=2$, $\phi[6,5]\subseteq[6,3,2]$. This does not follow by inducing modular irreducibles from $S_{10}$. We have just seen that $[5]\subseteq[3,2]$ mod 2. Therefore,

$$[3,2]\#[6]-[5]\#[6]$$

is a 2-modular character of $S_5\times S_6$. Induce up to $S_{11}$; using the Littlewood-Richardson rule,

$$[9,2]+[8,3]+[8,2,1]+[7,3,1]+[7,2^2]+[6,3,2] \\ -([11]+[10,1]+[9,2]+[8,3]+[7,4]+[6,5])$$

is a 2-modular character of $S_{11}$. Taking only those characters having 2-core $[2,1]$ gives $[6,3,2]-[10,1]-[6,5]$, and the result follows.

Before moving on to methods which involve more than just character theory, we mention two elementary methods which may be of use.

**6.3.78.** *Restricting projective indecomposables and modular irreducibles from a group containing G gives projective and modular characters of G.*

**6.3.79** EXAMPLE. $[3,2,1]$ is a projective indecomposable of $S_6$ modulo 2 (it is in a block of defect 0). Hence $[3,2]+[3,1^2]+[2^2,1]$ is a projective character of $S_5$.

**6.3.80.** *Tensoring two modular characters gives a modular character.*

This can be useful in one of two ways. We may have to subtract constituents from known modular characters in order to write the product as the sum of modular characters. Alternatively, it may be possible to prove $\phi_1\not\subseteq\phi_2$ by showing that $(\phi_2-\phi_1)\otimes\phi_3$ cannot be written as a linear combination of irreducible modular characters.

What can be done if character-theoretic methods fail to distinguish between the various possibilities for a decomposition matrix? Little is known in general, but in the following sections we show that much information about $S_n$ can be found by looking closely at the actual representation modules. It is encouraging to see that most of Theorem 6.3.50 (which has relied heavily on the Nakayama conjecture) can be proved quite simply by examining composition factors of the representation modules.

That some results can be found very easily by considering representation modules is illustrated by

**6.3.81** THEOREM. *Suppose $M$ is a permutation representation of $G$ on $n$ points, and $p$ divides $n$. Then the identity $p$-modular representation is a composition factor of $M$ at least twice.*

*Proof.* Let $v_1, v_2, \ldots, v_n$ be the $n$ points on which $G$ acts. Then we may regard $M$ as the vector space over a field of characteristic $p$ whose basis elements are $v_1, v_2, \ldots, v_n$. The following are submodules of $M$:

$$V_1 = \{a_1v_1 + \cdots + a_nv_n \mid a_1 + \cdots + a_n = 0\},$$
$$V_2 = \langle v_1 + v_2 + \cdots + v_n \rangle_F.$$

$M/V_1$ and $V_2$ are both the identity module. Since $p|n$, $V_1 \supseteq V_2$, and we are done. ■

Of course, the hypothesis of this theorem shows that there is a homomorphic image of $G$ in $S_n$, and the theorem is essentially about the group $S_n$.

**6.3.82** EXAMPLE. $[5]+[4,1]+[3,2]$ is the permutation character of $S_5$ on the subgroup $S_3 \times S_2$, of index 10. Therefore the identity $p$-modular character $\phi[5]$ is a constituent of $[4,1]+[3,2]$ if $p=2$ or 5. By Nakayama's conjecture, $\phi[5] \subseteq [3,2]$ if $p=2$, and $\phi[5] \subseteq [4,1]$ if $p=5$.

Very occasionally, a special piece of information is available which helps to find modular characters.

**6.3.83** EXAMPLE. $A_5 \simeq SL_2(4)$, the group of $2 \times 2$ matrices with determinant 1 over the field of order 4. Hence $A_5$ has a 2-dimensional representation over the field of order 4. Applying the field automorphism, an inequivalent representation is obtained. These are the two parts into which $\phi[3,2]$ splits (see Example 6.3.76).

In conclusion, we give two results which sometimes solve problems which appear not to yield to purely character-theoretic arguments. The first is an easy, and very weak, result analogous to the Frobenius reciprocity theorem. It is useful only when we have a representation which is "nearly" irreducible.

**6.3.84** THEOREM. *Suppose $M$ is an irreducible $G$-module, and $H$ is a subgroup of $G$. If $L$ is an $H$-submodule of $M$, then $M$ is a top composition factor of $L^G$.*

*Proof.* $L^G = L\otimes g_1 + \cdots + L\otimes g_n$, where the $g_i$'s are coset representatives of $H$ in $G$.

$$\theta : l\otimes g_i \to lg_i \qquad (l\in L)$$

is a nonzero $G$-module homomorphism from $L^G$ into $M$. Since $M$ is irreducible, we must have $L^G/\operatorname{Ker}\theta \simeq M$. ■

**6.3.85** EXAMPLE. For $p=2$, see the decomposition matrix of $S_{11}$. $[4^3]\downarrow S_{11} = [4^2,3] = 2\phi[11]+\phi[9,2]+\phi[5,4,2]$. Now, if $[4^3]$ is irreducible, either its representation module or the dual of this has an $S_{11}$ submodule admitting $\phi[11]$ or $\phi[9,2]$. But $\phi[11]\uparrow S_{12} \not\supseteq [4^3]$, $\phi[9,2]\uparrow S_{12} \not\supseteq [4^3]$, contradicting Theorem 6.3.84. Therefore, $[4^3]$ is reducible. The same argument works for any factor of $[4^3]$ which restricts to $S_{11}$ to contain $\phi[5,4,2]$ and another factor. Hence $[4^3]$ contains a factor (which must be irreducible) which restricts to $S_{11}$ to be $\phi[5,4,2]$.

**6.3.86** THEOREM. *If $\phi$ is any irreducible 2-modular character of a finite group $G$, then either $\phi$ is the identity character, or $\phi$ is not real, or the degree of $\phi$ is even.*

*Proof.* Suppose that $F$ is a field of characteristic 2, and $G$ is a finite absolutely irreducible subgroup of $GL(n, F)$ with $n>1$. Suppose too, that for every $g$ in $G$, $\operatorname{tr}(g)=\operatorname{tr}(g'^{-1})$, where $g'$ is the transpose of $g$. We must show that $n$ is even.

There is a nonsingular matrix $a$ such that for all $g\in G$ $g=ag'^{-1}a^{-1}$. By Schur's lemma, $a'=\mu a$ where $\mu\in F$. Therefore, $a=a''=\mu a'=\mu^2 a$. Thus $\mu=1$ and $a=a'$. Define a bilinear form $\Phi(\ ,\ )$ on $F^n$ by letting $\Phi(v,w)=vaw'$. By construction, this is a $G$-invariant symmetric bilinear form. $V:=\{v\,|\,\Phi(v,v)=0\}$ is an invariant subspace of $F^n$ of codimension at most 1. Therefore $V=F^n$, and the bilinear form is alternating. This shows that $n$ is even. ■

**6.3.87** COROLLARY. *All the irreducible 2-modular characters of $S_n$, except the identity one, have even degree. If $\phi$ is a 2-modular character of $S_n$, then the identity character has even multiplicity in $\phi$ if* $\deg\phi$ *is even, or odd multiplicity if* $\deg\phi$ *is odd.*

**6.3.88** EXAMPLE. $\deg[3,2]$ is odd, so it contains the identity 2-modular character.

## Exercises

6.1 Let $\tilde{\alpha}$ be a $p$-core. Using the technique employed for 6.1.9, prove that the number of $p$-regular partitions whose $p$-core is $\tilde{\alpha}$ equals the number of partitions whose associated partition is $p$-regular and whose $p$-core is $\tilde{\alpha}$.

6.2 Provide an alternative proof of the $p$-block structure of $A_n$ (Theorem 6.1.46), by considering the central character $\omega^i$, whose definition is

$$\omega^i(g) = \frac{|C^G(g)|\zeta^i(g)}{f^i}.$$

6.3 Prove that the Brauer tree for the principal $p$-block of $S_p$ is given by 6.3.9.

6.4 Write $n = x(p-1)+r$, where $0 \leqslant r < p-1$. Using 6.3.59, prove that $[1^n] = \phi[(x+1)^r, x^{p-1-r}]$.

6.5 Use the Littlewood-Richardson rule, and the fact that $[1^y] = [y]$ on 2-regular classes, to prove that

$$[x+1, 1^{y-1}] + [x, 1^y] = \sum_{i=0}^{y} [x+y-i, i]$$

on 2-regular classes. Deduce necessary and sufficient conditions for $[x, 1^y]$ to be 2-modularly irreducible.

CHAPTER 7

# Representation Theory of $S_n$ over an Arbitrary Field

Although much information about the $p$-modular representations of $S_n$ can be obtained in terms of characters, difficulties remain which apparently cannot be overcome by character-theory arguments alone. More results are found by examining the representation modules of $S_n$. It is useful to have a module for each ordinary irreducible representation. Such modules have already been constructed—the left ideals of the group algebra in Section 3.1—but it is easier to work in terms of Specht modules, which will be defined in the first section of this chapter. Each left ideal of Section 3.1 is isomorphic to some Specht module, and so every result could be interpreted in terms of the group algebra. Essentially, the advantage enjoyed by the method of examining Specht modules, modulo $p$, over that of looking at the $p$-modular components of the ordinary characters is that the order of the factors in a composition series can be noted.

## 7.1 Specht Modules

As the new approach to the representation theory of $S_n$ is to be characteristic-free, the first problem is to construct an $S_n$-module which has the same definition over every field. Certainly permutation modules have this property. It is reassuring to see that the ordinary irreducible representations of $S_n$ have been found in Chapter 2 by looking inside the permutation modules of $S_n$ on Young subgroups. It should be profitable, therefore, to study these permutation modules first, and try to find a submodule (which will be called a Specht module) which is a generalization of the ordinary irreducible representation arising from the partition $\beta$.

ENCYCLOPEDIA OF MATHEMATICS and Its Applications, Gian-Carlo Rota (ed.). Vol. 16: G. D. James and A. Kerber, The Representation Theory of the Symmetric Group
ISBN 0-201-13515-9

Let $F$ be an *arbitrary field*, and consider the $F$ vector space $M^\beta$ spanned by *β-tabloids*. Recall from 2.2.16 that a $\beta$-tabloid is a $\beta$-tableau with unordered row entries; formally, the tabloid $\{t\}$ containing $t$ is the equivalence class of $t$ under the equivalence relation

$$t_1 \sim t_2 \quad \Leftrightarrow \quad t_2 = \pi t_1 \quad \text{for some } \pi \in H^{t_1}.$$

As before, in examples we shall draw lines between the rows of $t$ to denote the tabloid $\{t\}$.

If we let $S_n$ act on the set of $\beta$-tabloids by

$$\pi\{t\} := \{\pi t\} \qquad (\pi \in S_n),$$

then this action is well defined, and extending the action to be $F$-linear turns $M^\beta$ into an $FS_n$-module. Clearly $S_n$ acts transitively on the set of $\beta$-tabloids, and the stabilizer of a given $\beta$-tabloid is a Young subgroup $S_\beta$. Therefore

**7.1.1.** *$M^\beta$ := the $FS_n$-module spanned by $\beta$-tabloids is isomorphic to the permutation module of $S_n$ on a Young subgroup $S_\beta$. $M^\beta$ is a cyclic module, generated by any one $\beta$-tabloid.*

A *β-polytabloid* is an element of $M^\beta$ of the form

$$e_t := \mathcal{V}^t\{t\},$$

where, as usual, $\mathcal{V}^t$ is the signed column sum for $t$.

**7.1.2 Example.** If

$$t = \begin{array}{ccc} 1 & 2 & 5 \\ 4 & 3 & \end{array},$$

then

$$\mathcal{V}^t = (1-(14))(1-(23))$$

and

$$e_t = \begin{array}{ccc} \hline 1 & 2 & 5 \\ \hline 4 & 3 & \\ \cline{1-2} \end{array} - \begin{array}{ccc} \hline 4 & 2 & 5 \\ \hline 1 & 3 & \\ \cline{1-2} \end{array} - \begin{array}{ccc} \hline 1 & 3 & 5 \\ \hline 4 & 2 & \\ \cline{1-2} \end{array} + \begin{array}{ccc} \hline 4 & 3 & 5 \\ \hline 1 & 2 & \\ \cline{1-2} \end{array}.$$

It is important to note that the polytabloid $e_t$ depends on the tableau $t$, not just the tabloid $\{t\}$.

The *Specht module* $S^\beta$ associated with the partition $\beta$ is defined to be the subspace of $M^\beta$ spanned by polytabloids. Now, $\pi\mathcal{V}^t=\mathcal{V}^{\pi t}\pi$, so $\pi e_t=e_{\pi t}$, and $S^\beta$ is an $FS_n$-submodule of $M^\beta$. Further,

**7.1.3.** *$S^\beta$ := the $FS_n$-module spanned by $\beta$-polytabloids is a cyclic module, generated by any one $\beta$-polytabloid.*

Notice that when $\beta=(n)$, $S^\beta=M^\beta$ = the identity $FS_n$-module. When $\beta=(1^n)$, $M^\beta$ is isomorphic to the regular representation module, and $S^\beta$ affords the alternating representation.

**7.1.4** LEMMA. *Each left ideal $FS_n\mathcal{V}^t\mathcal{H}^t$ ($t$ a $\beta$-tableau) of the group algebra defined in Section* 3.1 *is isomorphic to the Specht module $S^\beta$.*

*Proof.* The $FS_n$-homomorphism from $FS_n\mathcal{H}^t$ to $M^\beta$ given by $\mathcal{H}^t\to\{t\}$ is clearly a well-defined isomorphism, since $\pi$ fixes $\mathcal{H}^t\Leftrightarrow\pi\in H^t\Leftrightarrow\pi$ fixes $\{t\}$. Restricting this isomorphism to the subideal $FS_n\mathcal{V}^t\mathcal{H}^t$ proves the lemma. ■

This lemma illuminates the advantages of working with Specht modules, since the ideal $FS_n\mathcal{V}^t\mathcal{H}^t$ depends upon the tableau chosen, whereas $S^\beta$ depends only on the partition $\beta$. We have also replaced $\mathcal{H}^t$, which is a long sum of group elements, by the single object $\{t\}$.

An independent proof that Specht modules can be used to describe all the ordinary irreducible representations of $S_n$ will be given shortly. A few preliminary results are needed.

**7.1.5** LEMMA. *Suppose that $\alpha$, $\beta\vdash n$. Then*

(i) $\mathcal{V}^tM^\beta=0$ *if $t$ is an $\alpha$-tableau and $\alpha\not\trianglerighteq\beta$;*
(ii) $\mathcal{V}^tM^\beta=Fe_t$ *if $t$ is a $\beta$-tableau.*

*Proof.* Let $t^*$ be a $\beta$-tableau, and $a$ and $b$ be two numbers in the same row of $t^*$. Then

$$(1-(ab))\{t^*\}=\{t^*\}-\{(ab)t^*\}=0.$$

If $t$ is any other tableau, and $\mathcal{V}^t\{t^*\}\neq 0$, then $a$ and $b$ must lie in different columns of $t$; for, otherwise, we could select signed coset representatives $\sigma_1,\ldots,\sigma_k$ for the subgroup of $V^t$ consisting of 1 and $(ab)$, and obtain $\mathcal{V}^t\{t^*\}=(\sigma_1+\cdots+\sigma_k)(1-(ab))\{t^*\}=0$. If $t$ is an $\alpha$-tableau, $\mathcal{V}^t\{t^*\}\neq 0$ therefore implies $\alpha\trianglerighteq\beta$, by Lemma 1.4.20, and this proves the first result. If $t$ is a $\beta$-tableau, and $\mathcal{V}^t\{t^*\}\neq 0$, then the fact that the numbers from each row of $t^*$ must belong to different columns of $t$ shows that $\{t^*\}=\pi\{t\}$ for some $\pi\in V^t$. In this case, $\mathcal{V}^t\{t^*\}=\mathcal{V}^t\pi\{t\}=\pm\mathcal{V}^t\{t\}=\pm e_t$, and the second result follows. ■

Before going further, a bilinear form $\Phi(\ ,\ )$ is introduced on $M^\beta$.

**7.1.6.** Let

$$\Phi(\{t_1\},\{t_2\}) := \begin{cases} 1 & \text{if } \{t_1\}=\{t_2\}, \\ 0 & \text{if } \{t_1\}\neq\{t_2\}, \end{cases}$$

and extend to be bilinear on $M^\beta$.

This is a natural symmetric, $S_n$-invariant, bilinear form which is nonsingular on $M^\beta$.

We shall often use tricks like the following one: For $u, v \in M^\beta$,

$$\begin{aligned}\Phi(\mathcal{V}^t u, v) &= \sum_{\pi\in V^t} \Phi((\operatorname{sgn}\pi)\pi u, v) \\ &= \sum_{\pi\in V^t} \Phi(u, (\operatorname{sgn}\pi)\pi^{-1} v) \\ &= \sum_{\pi\in V^t} \Phi(u, (\operatorname{sgn}\pi)\pi v) \\ &= \Phi(u, \mathcal{V}^t v).\end{aligned}$$

With this preparation behind us, we give the crucial theorem:

**7.1.7 Theorem.** *If $A$ is an $FS_n$-submodule of $M^\beta$, then either $A \supseteq S^\beta$ or $A \subseteq S^{\beta\perp}$.*

*Proof.* Suppose that $a \in A$ and $t$ is a $\beta$-tableau. By 7.1.5(ii) $\mathcal{V}^t a$ is a multiple of $e_t$. If we can choose $a$ and $t$ so that this multiple is non-zero, then $e_t \in A$. Since $S^\beta$ is generated by $e_t$, we have $A \supseteq S^\beta$.

If for every $a$ and $t$, $\mathcal{V}^t a = 0$, then for all $a$ and $t$

$$0 = \Phi(\mathcal{V}^t a, \{t\}) = \Phi(a, \mathcal{V}^t\{t\}) = \Phi(a, e_t).$$

That is, $A \subseteq S^{\beta\perp}$. ∎

**7.1.8 Theorem.** *If $F$ is any field (in particular if $F = \mathbb{Q}$ or $\mathbb{Z}_p$), then $S^\beta/(S^\beta \cap S^{\beta\perp})$ is an absolutely irreducible $FS_n$-module, or zero.*

*Proof.* By Theorem 7.1.7, any submodule of $S^\beta$ is either $S^\beta$ itself or is contained in $S^\beta \cap S^{\beta\perp}$. Therefore, $S^\beta/(S^\beta \cap S^{\beta\perp})$ is irreducible or zero.

Choose a basis $e_1, \ldots, e_k$ for $S^\beta$ where each $e_i$ is a polytabloid. Then $\dim(S^\beta/(S^\beta \cap S^{\beta\perp}))$ is the rank of the *Gram matrix* (i.e. the matrix whose $(i, j)$th entry is $\Phi(e_i, e_j)$) with respect to this basis. But the entries in the

Gram matrix belong to the prime subfield of $F$, since polytabloids involve tabloids with coefficients $\pm 1$. Therefore, the rank of the Gram matrix is the same over $F$ as over the prime subfield, and so $S^\beta \cap S^{\beta\perp}$ does not increase in dimension if we extend $F$. Since $S^\beta/(S^\beta \cap S^{\beta\perp})$ is always irreducible when it is nonzero, it follows that it is absolutely irreducible. ■

When $F=\mathbb{Q}$, $S^\beta \cap S^{\beta\perp}=0$, and no two Specht modules for different partitions are isomorphic (this is proved in 7.1.14 below, but can readily be deduced in this special case using 7.1.5(i)). Since the number of partitions of $n$ equals the number of conjugacy classes of $S_n$, we have:

**7.1.9** THEOREM. *When $F=\mathbb{Q}$, the various Specht modules $S^\beta$ ($\beta \vdash n$) give all the ordinary irreducible representations of* $S_n$.

Returning to the case where $F$ is arbitrary, we have to decide when $S^\beta$ is contained in $S^{\beta\perp}$.

**7.1.10** LEMMA. *Suppose that the partition $\beta$ has exactly $z_j$ parts equal to $j$. Then*

(i) *for every pair of $\beta$-polytabloids $e_t$ and $e_{t^*}$, $\prod_{j=1}^{\infty} z_j!$ divides $\Phi(e_t, e_{t^*})$, and*

(ii) *for every $\beta$-polytabloid $e_t$, there is a $\beta$-polytabloid $e_{t^*}$ such that $\Phi(e_t, e_{t^*})=\prod_{j=1}^{\infty}(z_j!)^j$.*

*Proof.* Since $0!=1$, there is no problem about taking infinite products.

Define an equivalence relation $\sim$ on the set of $\beta$-tabloids by $\{t_1\}\sim\{t_2\}$ if and only if, for all $i$ and $j$, $i$ and $j$ belong to the same row of $\{t_1\}$ when $i$ and $j$ belong to the same row of $\{t_2\}$. Informally, this simply says that we can go from $\{t_1\}$ to $\{t_2\}$ by shuffling rows. The equivalence classes have size $\prod_{j=1}^{\infty} z_j!$.

Now, if $\{t_1\}$ is involved in $e_t$ and $\{t_1\}\sim\{t_2\}$, then the definition of a polytabloid shows that $\{t_2\}$ is involved in $e_t$, and whether the coefficients (which are $\pm 1$) are the same or have opposite signs depends only on $\{t_1\}$ and $\{t_2\}$. Therefore, any two $\beta$-polytabloids have a multiple of $\prod_{j=1}^{\infty} z_j!$ tabloids in common, and part (i) follows.

Next, let $t$ be any $\beta$-tableau, and obtain $t^*$ from $t$ by reversing the order of the numbers in each row of $t$. For example,

$$\text{if}\quad t=\begin{matrix}1&2&3&4\\5&6&7&\\8&9&10&\\11&&&\end{matrix}\qquad\text{then}\quad t^*=\begin{matrix}4&3&2&1\\7&6&5&\\10&9&8&\\11&&&\end{matrix}$$

Let $\pi \in V^t$ have the property that for every $i$, $i$ and $\pi i$ belong to rows of $t$

which have the same length. (In the example, $\pi$ can be any permutation in the group generated by (5,8), (6,9), (7,10).) Then $\{\pi t\}$ is involved in $e_t$ and in $e_{t*}$, with the same coefficient in each. It is easy to see that all tabloids common to $e_t$ and $e_{t*}$ have this form. Therefore, $\Phi(e_t, e_{t*})=\prod_{j=1}^{\infty}(z_j!)^j$, which proves part (ii). ■

**7.1.11** Corollary. *Suppose that $S^\beta$ is defined over a field of characteristic p. Then $S^\beta/(S^\beta \cap S^{\beta\perp})$ is nonzero if and only if $\beta$ is p-regular.*

*Proof.* By definition, $\beta$ is $p$-regular if and only if for every $j \geq 1$, $z_j < p$. If $\beta$ is $p$-regular, $\Phi(e_t, e_{t*}) \neq 0$ for the $\beta$-polytabloids $e_t$ and $e_{t*}$ in part (ii) of the lemma. Therefore, $S^\beta \not\subseteq S^{\beta\perp}$.

If $\beta$ is $p$-singular, $\Phi(e_t, e_{t*})=0$ for every pair of $\beta$-polytabloids, by part (i) of the lemma. Therefore, $S^\beta \subseteq S^{\beta\perp}$. ■

**7.1.12** Corollary. *Suppose that $S^\beta$ is defined over a field of characteristic p, and $\beta$ is p-regular. Then, for every $\beta$-tableau t, there is a $\pi \in S_n$ such that $\mathcal{V}^t\pi e_t$ is a nonzero multiple of $e_t$.*

*Proof.* By part (ii) of the lemma, given $t$, we can choose $\pi$ such that $\Phi(e_t, \pi e_t)=\prod_{j=1}^{\infty}(z_j!)^j$. But $\mathcal{V}^t\pi e_t$ is a multiple of $e_t$, by 7.1.5(ii); say $\mathcal{V}^t\pi e_t = h e_t$. Now,

$$h = h\Phi(\{t\}, e_t) = \Phi(\{t\}, he_t) = \Phi(\{t\}, \mathcal{V}^t\pi e_t)$$

$$= \Phi(\mathcal{V}^t\{t\}, \pi e_t) = \Phi(e_t, \pi e_t) = \prod_{j=1}^{\infty}(z_j!)^j.$$

Since $\beta$ is $p$-regular, $h \neq 0$. ■

We now give $S^\beta/(S^\beta \cap S^{\beta\perp})$ a name:

**7.1.13.** Assume that char $F=p$ (prime or $=\infty$) and that $\beta$ is $p$-regular. Let $D^\beta := S^\beta/(S^\beta \cap S^{\beta\perp})$.

The main theorem of this section is:

**7.1.14** Theorem. *Suppose that F is an arbitrary field, and* char $F=p$ *(prime or $=\infty$). Then*

(i) *As $\beta$ varies over p-regular partitions of n, $D^\beta$ varies over a complete set of inequivalent irreducible $FS_n$-modules.*

(ii) *$D^\beta$ is self-dual and absolutely irreducible.*

(iii) *All the composition factors of $S^\beta \cap S^{\beta\perp}$ and of $M^\beta/S^\beta$ are isomorphic to modules $D^\alpha$ with $\alpha \rhd \beta$.*

*Proof.* Suppose that some composition factor of $M^\beta$ is isomorphic to $D^\alpha$. $D^\alpha$ is a homomorphic image of $S^\alpha$, so there is a nonzero $FS_n$-homomorphism $\theta$ from $S^\alpha$ into a quotient of $M^\beta$. By Corollary 7.1.12, there exist $\alpha$-tableaux $t$ and $t^*$ such that $\mathcal{T}^t e_{t^*} = he_t$, where $h \neq 0$. But

$$\text{7.1.15} \qquad \theta(e_t) = h^{-1}\theta(he_t) = h^{-1}\theta(\mathcal{T}^t e_{t^*}) = \mathcal{T}^t h^{-1}\theta(e_{t^*}).$$

Since $\theta \neq 0$, 7.1.5(i) gives $\alpha \trianglerighteq \beta$.

Now, if $D^\alpha \simeq D^\beta$, then $M^\beta$ has a composition factor isomorphic to $D^\alpha$. Therefore, $\alpha \trianglerighteq \beta$. Similarly, $\beta \trianglerighteq \alpha$, so $\alpha = \beta$. But 6.1.2 shows that the number of inequivalent modules $D^\beta$ equals the number of absolutely irreducible $FS_n$-modules. We have already proved that $D^\beta$ is absolutely irreducible, so the proof of part (i) is complete.

Now, $D^\beta = S^\beta/(S^\beta \cap S^{\beta\perp}) \simeq (S^\beta + S^{\beta\perp})/S^{\beta\perp}$, by the second isomorphism theorem, which in turn $\simeq$ dual of $S^\beta/(S^\beta \cap S^{\beta\perp})$, since always $A/B \simeq$ dual of $B^\perp/A^\perp$. Therefore, $D^\beta$ is self-dual, which finishes part (ii).

Next, 7.1.15 shows that the only possible composition factors of $M^\beta$ have the form $D^\alpha$ with $\alpha \trianglerighteq \beta$. Furthermore, if $A$ is an $FS_n$-submodule of $M^\beta$ and $D^\beta \simeq M^\beta/A$ then there is a homomorphism from $S^\beta$ to $M^\beta/A$:

$$S^\beta \overset{\text{canon.}}{\rightarrow} D^\beta \overset{\text{iso.}}{\rightarrow} M^\beta/A,$$

and 7.1.15 and 7.1.5(ii) prove that the image of this homomorphism is $(S^\beta + A)/A$. Therefore no composition factor of $M^\beta/S^\beta$ or of $S^\beta \cap S^{\beta\perp}$ is isomorphic to $D^\beta$, and this completes the proof of part (iii). ■

Compare the ease with which we have the next result, with the struggle to get 6.3.50:

**7.1.16** COROLLARY. *The matrix recording the composition factors of Specht modules over a field of characteristic p has the form*

$$\begin{array}{r c}
 & \overbrace{\hphantom{1 \quad 1 \quad \ddots \quad 1}}^{D^\beta\ (\beta\ p\text{-regular})} \\
S^\beta\ (\beta\ p\text{-regular}) \left\{ \right. & \left( \begin{array}{cccc} 1 & & & 0 \\ & 1 & & \\ & & \ddots & \\ & & & 1 \\ & * & & \end{array} \right) \\
S^\beta\ (\beta\ p\text{-singular}) \left\{ \right. &
\end{array}$$

*when the p-regular partitions are placed in lexicographic order before all the p-singular partitions.*

*Proof.* Recall that $D^\beta := S^\beta/(S^\beta \cap S^{\beta\perp})$, and use part (iii) of the theorem. ■

We have carefully avoided saying that the above matrix is the decomposition matrix of $S_n$, although we shall see in the next section that this is indeed the case. The problem is that we are not yet certain that $S^\beta$ defined over the field of $p$ elements is the reduction modulo $p$ of $S^\beta$ defined over the rational field.

## 7.2 The Standard Basis of the Specht Module

Our objective in this section is to prove that

$$\{e_t \mid t \text{ is a standard } \beta\text{-tableau}\}$$

is a basis for $S^\beta$, whatever field we work over.

We start by finding elements in the group algebra $FS_n$ which annihilate a given polytabloid $e_t$. There are some elements of $FS_n$ which obviously annihilate $e_t$, namely

**7.2.1** $\qquad 1-(\operatorname{sgn}\pi)\pi \quad$ *annihilates* $e_t \qquad$ if $\quad \pi \in V^t$.

*Proof.* $\mathcal{V}^t = (\operatorname{sgn}\pi)\pi\mathcal{V}^t$, so $(1-(\operatorname{sgn}\pi)\pi)\mathcal{V}^t\{t\} = (1-(\operatorname{sgn}\pi)\pi)e_t = 0$. ∎

Now let $X$ be a subset of the $i$th column, and $Y$ be a subset of the $j$th column of $t$, with $i<j$. Let $\sigma_1, \ldots, \sigma_k$ be coset representatives of $S_X \times S_Y$ in $S_{X\cup Y}$, and let

$$G_{X,Y} := \sum_{e=1}^{k} (\operatorname{sgn}\sigma_e)\sigma_e.$$

$G_{X,Y}$ is called a *Garnir element* for $X\cup Y$. The permutations $\sigma_1, \ldots, \sigma_k$ are, of course, not unique, but for practical purposes note that we may take $\sigma_1, \ldots, \sigma_k$ so that $\sigma_1 t, \ldots, \sigma_k t$ are all the tableaux which agree with $t$ except in the positions occupied by $X\cup Y$, and whose entries increase vertically downwards in the positions occupied by $X\cup Y$.

**7.2.2** Example. If

$$t = \begin{matrix} 1 & 2 \\ 4 & 3 \\ 5 & \end{matrix}, \qquad X=\{4,\ 5\} \quad \text{and} \quad Y=\{2,\ 3\},$$

then we may take $\sigma_1 t, \ldots, \sigma_k t$ to be

$$\begin{matrix} 1 & 2 \\ 4 & 3 \\ 5 & \end{matrix},\quad \begin{matrix} 1 & 2 \\ 3 & 4 \\ 5 & \end{matrix},\quad \begin{matrix} 1 & 2 \\ 3 & 5 \\ 4 & \end{matrix},\quad \begin{matrix} 1 & 3 \\ 2 & 4 \\ 5 & \end{matrix},\quad \begin{matrix} 1 & 3 \\ 2 & 5 \\ 4 & \end{matrix},\quad \begin{matrix} 1 & 4 \\ 2 & 5 \\ 3 & \end{matrix}$$

With this choice of coset representatives, we have

$$G_{X,Y}=1-(34)+(354)+(234)-(2354)+(24)(35).$$

**7.2.3 THEOREM.** *Let t be a $\beta$-tableau, X a subset of the i th column of t, and Y a subset of the j th column of t, with $i<j$. Let $G_{X,Y}$ be a Garnir element for $X\cup Y$. If $|X\cup Y|>\beta_i'$ (where $\beta'$ is the partition associated with $\beta$), then $G_{X,Y}e_t=0$.*

*Proof.* Write

$$S_X^- S_Y^- \quad \text{for} \quad \Sigma\{(\operatorname{sgn}\sigma)\sigma \mid \sigma\in S_X\times S_Y\},$$

and

$$S_{X\cup Y}^- \quad \text{for} \quad \Sigma\{(\operatorname{sgn}\sigma)\sigma \mid \sigma\in S_{X\cup Y}\}.$$

Since $|X\cup Y|>\beta_i'$, for every $\tau\in V^t$ some pair of numbers in $X\cup Y$ are in the same row of $\tau t$. Hence

$$S_{X\cup Y}^-\{\tau t\}=0.$$

Therefore,

$$S_{X\cup Y}^-\mathcal{V}^t\{t\}=0.$$

Now, $S_X^- S_Y^-$ is a factor of $\mathcal{V}^t$ and

$$S_{X\cup Y}^-=G_{X,Y}S_X^- S_Y^- .$$

Therefore,

$$0=S_{X\cup Y}^-\mathcal{V}^t\{t\}=|X|!|Y|!G_{X,Y}\mathcal{V}^t\{t\}.$$

Thus, $G_{X,Y}\mathcal{V}^t\{t\}=0$ when the field is the field of rational numbers. But all the tabloid coefficients in $G_{X,Y}\mathcal{V}^t\{t\}$ are integers in this special case, and so $G_{X,Y}\mathcal{V}^t\{t\}=0$, whatever the field. ∎

In fact, the left ideal of $FS_n$ annihilating $e_t$ is generated by the elements of the group algebra described in 7.2.1, together with all Garnir elements for $X\cup Y$, with $X, Y$ ranging over subsets of adjacent pairs of columns of $t$ (say $X$ a subset of the $i$th and $Y$ a subset of the $(i+1)$th) with the property that $|X\cup Y|=\beta_i'+1$; but we do not require this result.

**7.2.4** EXAMPLE. If

$$t = \begin{array}{cc} 1 & 2 \\ 4 & 3 \\ 5 & \end{array}$$

and $t_1, \ldots, t_6$ are the six tableaux listed in Example 7.2.2, then

$$0 = G_{X,Y} e_t = e_{t_1} - e_{t_2} + e_{t_3} + e_{t_4} - e_{t_5} + e_{t_6},$$

so

$$e_t = e_{t_2} - e_{t_3} - e_{t_4} + e_{t_5} - e_{t_6}$$

In this way, we have written $e_t$ (where $t$ is not standard) as a linear combination of polytabloids $e_{t_i}$ for which $t_i$ is standard. Our plan is to do this in general.

Remembering that a tableau $t$ is standard if the numbers increase along the rows and down the columns of $t$, we define $e_t$ to be a *standard polytabloid* if $t$ is standard.

Now totally order the set of $\beta$-tabloids by saying

**7.2.5.** $\{t_1\} < \{t_2\}$ if for some $i$

(i) when $j > i$, $j$ is in the same row of $\{t_1\}$ and $\{t_2\}$, and
(ii) $i$ is in a higher row of $\{t_1\}$ than $\{t_2\}$.

For example,

$$\begin{array}{ccc}\hline 3 & 4 & 5 \\ \hline 1 & 2 & \\ \hline\end{array} < \begin{array}{ccc}\hline 2 & 4 & 5 \\ \hline 1 & 3 & \\ \hline\end{array} < \begin{array}{ccc}\hline 1 & 4 & 5 \\ \hline 2 & 3 & \\ \hline\end{array} < \begin{array}{ccc}\hline 2 & 3 & 5 \\ \hline 1 & 4 & \\ \hline\end{array} < \begin{array}{ccc}\hline 1 & 3 & 5 \\ \hline 2 & 4 & \\ \hline\end{array}$$

$$< \begin{array}{ccc}\hline 1 & 2 & 5 \\ \hline 3 & 4 & \\ \hline\end{array} < \begin{array}{ccc}\hline 2 & 3 & 4 \\ \hline 1 & 5 & \\ \hline\end{array} < \begin{array}{ccc}\hline 1 & 3 & 4 \\ \hline 2 & 5 & \\ \hline\end{array} < \begin{array}{ccc}\hline 1 & 2 & 4 \\ \hline 3 & 5 & \\ \hline\end{array} < \begin{array}{ccc}\hline 1 & 2 & 3 \\ \hline 4 & 5 & \\ \hline\end{array}.$$

It is clear (cf. Example 7.1.2) that

**7.2.6.** *If the numbers increase down the columns of $t$, then $\{t\}$ is the last tabloid involved in $e_t$.*

The basis of $S^\beta$ described in the next theorem is called the *standard basis* of $S^\beta$.

**7.2.7** THEOREM. $\{e_t \mid t$ *is a standard $\beta$-tableau*$\}$ *is a basis for* $S^\beta$.

*Proof.* By 7.2.6, the standard polytabloids are linearly independent. It is therefore sufficient to prove that any polytabloid can be written as a linear combination of standard polytabloids.

Let $[t]$ denote the column equivalence class of the tableau $t$; that is, $[t]=\{t_1 | t_1=\pi t \text{ for some } \pi \in V^t\}$. Totally order the set of column equivalence classes in a way similar to the total order on the set of tabloids.

Suppose that $t$ is not standard. By induction, we may assume that $e_{t'}$ can be written as a linear combination of standard polytabloids when $[t']<[t]$. We want to prove the same result for $e_t$. (There is no need to "get the induction started", since we are assuming nothing when there is no $t'$ with $[t']<[t]$.) Since $\pi e_t=(\operatorname{sgn}\pi)e_t$ when $\pi\in V^t$, we may suppose that the entries in $t$ are in increasing order down columns. Since $t$ is not standard, some adjacent pair of columns, say the $i$th and $(i+1)$th columns, have entries $a_1<a_2<\cdots<a_r$ and $b_1<b_2<\cdots<b_s$ with $a_q>b_q$ for some $q$:

$$
\begin{array}{ccc}
a_1 & & b_1 \\
\cdot & & \wedge \\
\cdot & & \cdot \\
\cdot & & \cdot \\
\cdot & & \cdot \\
\cdot & & \wedge \\
a_q & > & b_q \\
\wedge & & \cdot \\
\cdot & & \cdot \\
\cdot & & \cdot \\
\cdot & & b_s \\
\wedge & & \\
a_r & &
\end{array}
$$

Let $X=\{a_q,\ldots,a_r\}$ and $Y=\{b_1,\ldots,b_q\}$, and consider a Garnir element for $X\cup Y$, say $G_{X,Y}=\Sigma(\operatorname{sgn}\sigma)\sigma$. We may choose our coset representatives to include $\sigma=1$. Then 7.2.3 gives

$$\sum(\operatorname{sgn}\sigma)e_{\sigma t}=\sum(\operatorname{sgn}\sigma)\sigma e_t=0.$$

Because $b_1<\cdots<b_q<a_q<\cdots<a_r$, $[\sigma t]<[t]$ for $\sigma\neq 1$.

Since $e_t=-\Sigma_{\sigma\neq 1}(\operatorname{sgn}\sigma)e_{\sigma t}$, our induction hypothesis shows that $e_t$ can be written as a linear combination of standard polytabloids, as we wished to show. ∎

**7.2.8** COROLLARY. *The dimension of the Specht module $S^\beta$ is independent of the field, and equals the number of standard $\beta$-tableaux.*

We say that a tabloid $\{\hat{t}\}$ is standard if there exists a standard tableau in the row equivalence class $\{\hat{t}\}$.

By 7.2.6, different standard polytabloids involve different last tabloids. Since the standard polytabloids span the Specht module, we have:

**7.2.9** COROLLARY. *The last tabloid involved in each nonzero element of the Specht module is standard.*

The "converse" of this result looks rather artificial, but its proof displays a useful technique:

**7.2.10** LEMMA. *Suppose that $A$ is a subspace of $M^\beta$ containing $S^\beta$. Assume that the last tabloid involved in each nonzero element of $A$ is standard. Then $A=S^\beta$.*

*Proof.* Let $0\neq a\in A$, and let $\{\hat{t}\}$ be the last tabloid involved in $a$, with coefficient $c$, say. By hypothesis, there is a standard tableau $t$ such that $\{t\}=\{\hat{t}\}$. Now 7.2.6 shows that the last tabloid involved in $a-ce_t$ is earlier than $\{t\}$, so, by induction, $a-ce_t$ is a linear combination of standard polytabloids. Therefore, $a$ is a linear combination of standard polytabloids, and $a\in S^\beta$. ■

A similar proof, combined with 7.2.9, gives:

**7.2.11.** *If $F=\mathbb{Q}$, and $v\in S^\beta$ has the property that the coefficients of all the tabloids involved in $v$ are integers, then $v$ is an integral linear combination of standard polytabloids.*

**7.2.12** THEOREM. *The entries in the matrix representing a permutation with respect to the standard basis of $S^\beta$, defined over $\mathbb{Q}$, are all integers. If the matrix entries are reduced modulo $p$, we obtain the matrix representing the same permutation with respect to the standard basis of $S^\beta$, defined over a field of characteristic $p$.*

*Proof.* $\pi e_t=e_{\pi t}$, and all the tabloids involved in $e_{\pi t}$ have coefficient $\pm 1$. When $F=\mathbb{Q}$, 7.2.11 gives $e_{\pi t}=\Sigma c_i e_{t_i}$ ($c_i\in\mathbb{Z}$, summing over the standard polytabloids $e_{t_i}$). This gives the first sentence of the theorem.

Let $\bar{c}_i$ denote the integer $c_i$, reduced modulo $p$. The equation $\pi e_t=e_{\pi t}=\Sigma\bar{c}_i e_{t_i}$ shows how to write $\pi e_t$ as a linear combination of standard polytabloids when $\operatorname{char} F=p$, and proves the second sentence of the theorem. ■

The representation arising from $S^\beta$ defined over $\mathbb{Q}$ is just Young's natural representation.

**7.2.13** COROLLARY. *$S^\beta$ defined over the field of $p$ elements is the $p$-modular representation of $S_n$ obtained from $S^\beta$ defined over $\mathbb{Q}$.*

It follows that the matrix defined at the end of Section 7.1 is indeed the decomposition matrix of $S_n$.

Let $t_1, t_2, \ldots$ be the standard $\beta$-tableaux, ordered by 3.1.16, and $e_1, e_2, \ldots$ be the corresponding standard polytabloids.

**7.2.14** THEOREM. *The dimension of the p-modular irreducible representation $D^\beta$ of $S_n$ equals the p-rank of the Gram matrix (whose $(i, j)$th entry is $\Phi(e_i, e_j)$), with respect to the standard basis of $S^\beta$.*

*Proof.* $D^\beta$, by definition, equals $S^\beta/(S^\beta \cap S^{\beta\perp})$, and the result follows from elementary algebra. ∎

Although this algorithm for finding $\dim D^\beta$ seems daunting, it has been used in practice to solve problems which do not, as yet, yield to other methods. The largest such example we know of is the calculation of $\dim D^{(7,4,2)}$ for $p=2$, performed by D. Stockhofe, using a computer; this involved calculating the 2-rank of a $6006 \times 6006$ matrix. Some theoretical results involving the $p$-rank of the Gram matrix appear in the next section.

## 7.3 On the Role of Hook Lengths

We now investigate the properties of an element $R_m$ of the group algebra of $S_n$, introduced by G. E. Murphy. This operator does much to illuminate both the Nakayama conjecture and the appearance of "axial distances" in Young's seminormal matrices.

Let $t_1, t_2, \ldots$ be the standard $\beta$-tableaux, in order, and let $e_1, e_2, \ldots$ be the corresponding standard polytabloids. When we wish to distinguish the partition $\beta$ with which we are working, we write $t_1^\beta, t_2^\beta, \ldots$ and $e_1^\beta, e_2^\beta, \ldots$.

For every standard tableau except the first one (the tableau with $1, 2, \ldots, n$ in order down successive columns), there are numbers $l < m$ where $l$ is in a later column of the tableau than $m$. Let us examine the action of the transposition $(lm)$ on such a standard tableau:

**7.3.1** LEMMA. *Let $t_i$ be a standard $\beta$-tableau, and assume that $l<m$ and the column of $t_i$ containing $l$ is later than the column containing $m$. Then $(lm)e_i$ is a linear combination of standard polytabloids earlier than $e_i$.*

*Proof.* Rearrange the numbers in each column of $(lm)t_i$ so that the numbers increase down the columns, and call the new tableau $\hat{t}$. Since $t_i$ is standard, $m$ is in a higher row in $\hat{t}$ than in $t_i$, and since $l<m$, the positions occupied by $m+1, m+2, \ldots,$ and $n$ are the same in $\hat{t}$ and $t_i$. Therefore $\{\hat{t}\} < \{t_i\}$.

Now, $(lm)e_i = \pm e_{\hat{t}}$, and by 7.2.6 and 7.2.9, there is a standard tableau $t^*$ such that $\{t^*\} = \{\hat{t}\} < \{t_i\}$. The technique used to prove 7.2.10 shows that

$e_{\hat{t}} = e_{t^*} +$(a linear combination of standard polytabloids earlier than $e_{t^*}$), and the result follows. ■

**7.3.2.** If $m$ lies in the $u$th row and $v$th column of the $i$th standard $\beta$-tableau $t_i^\beta$, define $a_i^\beta(m) := v - u \in F$.

Thus, $a_i^\beta(m)$ is simply the $p$-residue ($p$ prime or $=\infty$, depending on the characteristic of the field $F$) of the node which $m$ replaces in $t_i^\beta$.

**7.3.3** EXAMPLE. If $\beta = (3, 2)$, then

$$t_1^\beta = \begin{matrix} 1 & 3 & 5 \\ 2 & 4 & \end{matrix}, \quad t_2^\beta = \begin{matrix} 1 & 2 & 5 \\ 3 & 4 & \end{matrix}, \quad t_3^\beta = \begin{matrix} 1 & 3 & 4 \\ 2 & 5 & \end{matrix}$$

$$t_4^\beta = \begin{matrix} 1 & 2 & 4 \\ 3 & 5 & \end{matrix}, \quad t_5^\beta = \begin{matrix} 1 & 2 & 3 \\ 4 & 5 & \end{matrix}.$$

Therefore, the sequence $a_i^\beta(1), a_i^\beta(2), \ldots, a_i^\beta(5)$ is

$$\begin{array}{rrrrrl} 0, & -1, & 1, & 0, & 2 & \text{when } i=1, \\ 0, & 1, & -1, & 0, & 2 & \text{when } i=2, \\ 0, & -1, & 1, & 2, & 0 & \text{when } i=3, \\ 0, & 1, & -1, & 2, & 0 & \text{when } i=4, \\ 0, & 1, & 2, & -1, & 0 & \text{when } i=5. \end{array}$$

Consider the element $R_m$ of the group algebra $FS_n$ defined by

**7.3.4**
$$R_m := \sum_{i=1}^{m-1} (i, m).$$

It is clear that $(jk)R_m = R_m(jk)$ when $j$ and $k$ are either both less than $m$ or both greater than $m$. Therefore, $R_l$ and $R_m$ commute $(1 \leq l, m \leq n)$.

Murphy's key result is:

**7.3.5** THEOREM. *Let $e_1, e_2, \ldots$ be the standard basis of $S^\beta$. Then*

$$R_m e_i = a_i^\beta(m) e_i + \sum_{j=1}^{i-1} c_j e_j \qquad (c_j \in F, \textit{possibly zero}).$$

*Proof.* Let $m$ be in the $(uv)$th place of $t_i$. If $v' < v$, the Garnir relations (7.2.3) show that $e_i$ is fixed by the sum of all the transpositions of the form $(xm)$ with $x$ in the $v'$th column of $t_i$. On the other hand, $(xm)e_i = -e_i$ when $x \neq m$ and $x$ lies in the $v$th column of $t_i$.

Acting on $e_i$ with the sum of all the transpositions $(xm)$ where $x$ lies in the $(u', v')$th place of $t_i$ with $v' < v$ or $v' = v$ and $u' < u$, we therefore obtain the element

$$((v-1)-(u-1))e_i = a_i^\beta(m)e_i.$$

But our sum of transpositions is

$$R_m + \Sigma_1(ym) - \Sigma_2(zm)$$

where $\Sigma_1$ is over all $y>m$ lying in a column of $t_i$ before the $v$th and $\Sigma_2$ is over all $z<m$ lying in a column of $t_i$ after the $v$th. By Lemma 7.3.1,

$$(\Sigma_1(ym) - \Sigma_2(zm))e_i$$

is a linear combination of standard polytabloids earlier than $e_i$. Therefore,

$$\begin{aligned} R_m e_i &= a_i^\beta(m)e_i - (\Sigma_1(ym) - \Sigma_2(zm))e_i \\ &= a_i^\beta(m)e_i + \sum_{j=1}^{i-1} c_j e_j. \end{aligned}$$ ■

We illustrate the power of Theorem 7.3.5 by proving at once a result giving more information than the useful part of the Nakayama conjecture:

**7.3.6** THEOREM. *Suppose $\alpha, \beta \vdash n$. Assume that for every $i$, there exists an $m$ with $1 \leqslant m \leqslant n$ such that $a_1^\alpha(m) \neq a_i^\beta(m)$. Then every $FS_n$-homomorphism from $S^\alpha$ into a quotient of $S^\beta$ is zero.*

*Proof.* Assume, by way of contradiction, that $\theta$ is a nonzero $FS_n$-homomorphism from $S^\alpha$ into $S^\beta/A$, where $A$ is an $FS_n$-submodule of $S^\beta$. Then we may write

$$\theta(e_1^\alpha) = b_1 e_1^\beta + \cdots + b_r e_r^\beta + A \qquad (b_i \in F, \quad b_r \neq 0).$$

Choose the coset representative $b_1 e_1^\beta + \cdots + b_r e_r^\beta$ in such a way that $(b_1 e_1^\beta + \cdots + b_r e_r^\beta) - (b_1' e_1^\beta + \cdots + b_s' e_s^\beta) \in A$ and $b_s' \neq 0$ implies $s \geqslant r$.

Now, the hypothesis of the theorem means that for some $m$, $a_1^\alpha(m) \neq a_r^\beta(m)$. Therefore,

$$\left(R_m - a_r^\beta(m)\right)e_1^\alpha = \left(a_1^\alpha(m) - a_r^\beta(m)\right)e_1^\alpha \neq 0,$$

while

$$\left(R_m - a_r^\beta(m)\right)\left(b_1 e_1^\beta + \cdots + b_r e_r^\beta\right)$$

is a linear combination of $e_1^\beta, e_2^\beta, \ldots, e_{r-1}^\beta$, by 7.3.5. Thus we may apply $(a_1^\alpha(m)-a_r^\beta(m))^{-1}(R_m-a_r^\beta(m))$ to $\theta(e_1^\alpha)$ and contradict our choice of coset representative. ■

One half of the Nakayama conjecture now follows. For if $[\alpha]$ and $[\beta]$ have different $p$-cores, then their $p$-contents differ, by 2.7.41, so the hypothesis of the theorem holds, and we conclude that $D^\alpha$ is not a composition factor of $S^\beta$ when $\alpha$ is $p$-regular.

**7.3.7 Example.** If $p=2$ and $\alpha=(4,3)$, the sequence $a_1^\alpha(1), a_1^\alpha(2), \ldots, a_1^\alpha(7)$ is $0,1,1,0,0,1,1$. The only diagrams $[\beta]$ which can be built up by adding successively to the empty diagram nodes whose $p$-residues are $0,1,1,0,0,1,1$ are:

$$\begin{array}{llll} 0 & 1 & 0 & 1 \\ 1 & 0 & 1 & \end{array}, \quad \begin{array}{llll} 0 & 1 & 0 & 1 \\ 1 & & & \\ 0 & & & \\ 1 & & & \end{array}, \quad \begin{array}{ll} 0 & 1 \\ 1 & 0 \\ 0 & 1 \\ 1 & \end{array}$$

so $D^{(4,3)}$ can be a composition factor only of $S^{(4,3)}$, $S^{(4,1^3)}$, and $S^{(2^3,1)}$. This much is given immediately by the Nakayama conjecture. But consider now $p=2$ and $\alpha=(7,3)$. The sequence $a_1^\alpha(1), a_1^\alpha(2), \ldots, a_1^\alpha(10)$ is $0,1,1,0,0,1,1,0,1,0$. Any diagram which can be built by adding, in turn, nodes with these $p$-residues must contain $[4,3]$, $[4,1^3]$, or $[2^3,1]$. Therefore $D^{(7,3)}$ is not a composition factor of $S^{(6,2^2)}$, even though $[7,3]$ and $[6,2^2]$ both have empty 2-cores.

For the remainder of this section we shall be dealing with just one partition $\beta$, so we shall omit the superscript $\beta$ from $t_i^\beta$, $e_i^\beta$, and $a_i^\beta(m)$.

The next few results will be of interest mainly in the case where char $F=0$, but we shall make a slightly weaker assumption:

**7.3.8 Hypothesis.** Assume that for all $i$ and $j$ with $i\neq j$ there is an $m$ such that $a_i(m)\neq a_j(m)$.

This means that we can recover the standard $\beta$-tableau $t_i$ from the sequence

$$a_i(1), a_i(2), \ldots, a_i(n).$$

The hypothesis is certainly valid in the case where char $F=0$, since $a_i(m)\neq a_j(m)$ for the smallest $m$ which lies in different places in $t_i$ and $t_j$, and it also holds in some cases where char $F=p$ (e.g. $p=3$ and $\beta=(2,1)$).

The plan is to construct, under hypothesis 7.3.8, an orthogonal basis for $S^\beta$.

From 7.3.5,

$$\prod_{k=1}^{j}(R_m - a_k(m))e_j = 0.$$

Since $R_l$ and $R_m$ commute, there is no need to choose the same $m$ in each term of the product, and this, together with hypothesis 7.3.8, gives

**7.3.9.** *For* $j<i$, $G(i)e_j = 0$ *where*

$$G(i) := \prod_{\substack{1 \leqslant k \leqslant i-1 \\ 1 \leqslant m \leqslant n \\ a_i(m) \neq a_k(m)}} \frac{R_m - a_k(m)}{a_i(m) - a_k(m)}.$$

Now, $R_m - a_i(m)$ sends $e_i$ to a linear combination of earlier polytabloids, so 7.3.9 proves that

**7.3.10** $\qquad R_m f_i = a_i(m) f_i, \qquad$ *where* $\; f_i := G(i)e_i.$

Consider the inner product of $f_i$ and $f_j$:

$$a_i(m)\Phi(f_i, f_j) = \Phi(R_m f_i, f_j) = \Phi(f_i, R_m f_j) = a_j(m)\Phi(f_i, f_j),$$

and hypothesis 7.3.8 shows

**7.3.11** $\qquad \Phi(f_i, f_j) = 0 \qquad$ *when* $\; i \neq j.$

Next, $(R_m - a_k(m))e_i = (a_i(m) - a_k(m))e_i + \Sigma_{j=1}^{i-1} c_j e_j$, so

**7.3.12.** $f_i = e_i +$ (*a linear combination of earlier polytabloids*).

We can now give an easy alternative construction of Young's seminormal form:

**7.3.13 Theorem** (cf. 3.3.29). *Assume hypothesis* 7.3.8 (*e.g.* char $F=0$). *An orthogonal basis* $f_1, f_2, \ldots$ *for the Specht module can be obtained from the standard basis* $e_1, e_2, \ldots$ *by means of a unimodular transformation. For all* $m$ *with* $1 < m \leqslant n$,

$$(m-1, m) f_i = \rho_1 f_i + \rho_2 f_j$$

*where* $\rho_1^{-1} = a_i(m) - a_i(m-1)$ $(\neq 0)$, *and*

$$\rho_2 = \begin{cases} 0 & \text{if } (m-1,m)t_i \text{ is not standard,} \\ 1 & \text{if } i<j \text{ and } t_j := (m-1,m)t_i, \\ 1-\rho_1^2 & \text{if } i>j \text{ and } t_j := (m-1,m)t_i. \end{cases}$$

*Proof.* We have already proved that $f_1, f_2, \ldots$ is an orthogonal basis obtained from $e_1, e_2, \ldots$ by means of a unimodular transformation. Therefore, for some field coefficients $c_k$,

$$(m-1, m)f_i = \sum c_k f_k.$$

Operate on this equation by $R_l$ where $l \neq m-1, m$, and use 7.3.10:

$$a_i(l)(m-1, m)f_i = \sum_k a_k(l) c_k f_k = a_i(l) \sum_k c_k f_k.$$

Comparing coefficients, we deduce that if $c_k \neq 0$, then $a_k(l) = a_i(l)$ for all $l \neq m-1, m$. Hypothesis 7.3.8 now shows that $c_k \neq 0$ only if $t_k$ equals $t_i$ or $(m-1, m)t_i$. Therefore,

**7.3.14.** $(m-1, m)f_i = c_i f_i + c_j f_j$, *where* $t_j := (m-1, m)t_i$ *if* $(m-1, m)t_i$ *is standard, and* $c_j = 0$ *otherwise.*

Operate on this equation with $R_m$, noting that $R_m(m-1, m) = (m-1, m) R_{m-1} + 1$:

$$\begin{aligned} a_i(m)c_i f_i + a_j(m)c_j f_j &= (m-1, m)R_{m-1} f_i + f_i \\ &= a_i(m-1)(m-1, m)f_i + f_i \\ &= (a_i(m-1)c_i + 1)f_i + a_i(m-1)c_j f_j. \end{aligned}$$

The coefficient of $f_i$ shows that $c_i \neq 0$, and $c_i^{-1} = a_i(m) - a_i(m-1)$, which proves that $\rho_1$ in the statement of the theorem is correct.

We have already proved that $c_j = 0$ when $(m-1, m)t_i$ is not standard, so assume $(m-1, m)t_i = t_j$.

Consider first the case where $i < j$. By 7.3.12, there exist coefficients $d_k \in F$ such that $f_i = e_i + \sum_{k=1}^{i-1} d_k f_k$, and now 7.3.14 gives:

$$\begin{aligned} (m-1, m)f_i &= e_j + (\text{a linear combination of } f_k\text{'s with } k \neq i, j) \\ &= f_j + c_i f_i, \qquad \text{by 7.3.12 and 7.3.14.} \end{aligned}$$

This proves that $\rho_2 = 1$ when $i < j$.

Finally, apply $(m-1, m)$ to the last equation:

$$f_i = (m-1, m)f_j + c_i(f_j + c_i f_i),$$

so

$$(m-1, m)f_j = c_i f_j + (1 - c_i^2) f_i$$

Therefore, the stated value of $\rho_2$ is correct in this last case. ∎

We now want to work out the inner product $\Phi(f_i, f_i)$. This will be used to derive a formula for the determinant of the Gram matrix with respect to the standard basis of the Specht module.

A repeated application of 7.3.10 gives $G(i)f_i = f_i$. Therefore,

**7.3.15** $$\Phi(f_i, e_i) = \Phi(G(i)f_i, e_i) = \Phi(f_i, G(i)e_i) = \Phi(f_i, f_i).$$

On the other hand, by 7.3.9,

**7.3.16.** *If* $j<i$, $\Phi(f_i, e_j) = \Phi(G(i)f_i, e_j) = \Phi(e_i, G(i)e_j) = 0$.

If the nodes $x+1, x+2, \ldots, n$ are removed from the standard tableau $t_i$, let $h(x,k)$ denote the length of the hook in the resulting tableau with $x$ at the foot of the hook and the hand of the hook being in the $k$th row. Thus, $h(x,k)$ is defined only if $x$ is in a row lower than the $k$th.

For example if

$$t_i = \begin{matrix} 1 & 2 & 5 & 6 & 13 \\ 3 & 4 & 7 & 12 & 14 \\ 8 & 9 & 11 & & \\ 10 & 15 & & & \end{matrix}$$

then $h(9,1)=5$ and $h(9,2)=3$.

**7.3.17** THEOREM. *Assume* char $F=0$. *Then*

$$\Phi(f_i, f_i) = \prod_{x=1}^{n} \prod_{k} \frac{h(x,k)}{h(x,k)-1},$$

*where the second product is over all* $h(x,k)$ *defined above.*

*Proof.* We work by induction on $i$. $f_1 = e_1$, and $\Phi(e_1, e_1) = \Pi_a(\beta_a'!)$ where $\beta'$ is the partition associated with $\beta$. If $x$ is in the $j$th row of $t_1$, then $h(x,k) = j-k+1$, so that

$$\prod_k \frac{h(x,k)}{h(x,k)-1} = \prod_{k=1}^{j-1} \frac{j-k+1}{j-k} = j.$$

Therefore,

$$\prod_{x=1}^{n} \prod_{k} \frac{h(x,k)}{h(x,k)-1} = \prod_a (\beta_a'!),$$

as required.

If $i>1$, then some $m$ in $t_i$ is such that $m-1$ is in a later column than $m$. Now, $(m-1, m)t_i$ is standard, and equals $t_j$, say. Since $j<i$, we may assume the result is true for $f_j$. But, using 7.3.13, 7.3.15, and 7.3.16, we have

$$\begin{aligned}\Phi(f_i, f_i) &= \Phi(f_i, e_i) = \Phi(f_i,(m-1, m)e_j) = \Phi((m-1, m)f_i, e_j)\\ &= \Phi(\rho_1 f_i + \rho_2 f_j, e_j) = \rho_2 \Phi(f_j, f_j)\end{aligned}$$

and

$$\begin{aligned}\rho_2 &= \frac{a_i(m)-a_i(m-1)+1}{a_i(m)-a_i(m-1)} \cdot \frac{a_i(m)-a_i(m-1)-1}{a_i(m)-a_i(m-1)}\\ &= \frac{a_j(m)-a_j(m-1)-1}{a_j(m)-a_j(m-1)} \cdot \frac{a_i(m-1)-a_i(m)+1}{a_i(m-1)-a_i(m).}\end{aligned}$$

Now, $t_i$ and $t_j$ have the same "hooks" $h(x, k)$, except that the hook from $m$ to $m-1$ in $t_i$ is replaced by a hook shortened by one node in $t_j$, since $m$ is removed before $m-1$. The necessary adjustment to

$$\prod_{x=1}^{n} \prod_{k} \frac{h(x,k)}{h(x,k)-1}$$

is precisely multiplication by $\rho_2$, as above, and the theorem is proved. ■

Let $\det \beta$ denote the determinant of the Gram matrix with respect to the standard basis $e_1, e_2, \ldots$ of $S^\beta$, defined over $\mathbb{Q}$.

**7.3.18** THEOREM. $\det \beta$ *equals the product over all standard* $\beta$*-tableaux of the expressions*

$$\prod_{x=1}^{n} \prod_{k} \frac{h(x,k)}{h(x,k)-1}.$$

*If* $\beta$ *is p-regular, and* $p^a$ *divides* $\det \beta$ *but* $p^{a+1}$ *does not divide* $\det \beta$, *then*

$$\dim D^\beta \geqslant \dim S^\beta - a,$$

*where* $D^\beta$ *is the irreducible p-modular representation of* $S_n$ *corresponding to* $\beta$.

*Proof.* The orthogonal basis $f_1, f_2, \ldots$ for $S^\beta$, defined over $\mathbb{Q}$, has been constructed from the standard basis $e_1, e_2, \ldots$ by a unimodular transformation, so 7.3.17 gives the value of $\det \beta$. The second part of the theorem follows from the fact that $\dim D^\beta$ is the $p$-rank of the Gram matrix with respect to the standard basis. ■

**7.3.19** EXAMPLE. Under each standard (3,2)-tableau below we give the value of $\prod_{x=1}^{n}\prod_k h(x,k)/(h(x,k)-1)$:

$$\begin{array}{cccccc} & \begin{matrix}1&3&5\\2&4\end{matrix} & \begin{matrix}1&2&5\\3&4\end{matrix} & \begin{matrix}1&3&4\\2&5\end{matrix} & \begin{matrix}1&2&4\\3&5\end{matrix} & \begin{matrix}1&2&3\\4&5\end{matrix} \\ \displaystyle\prod_{x=1}^{n}\prod_k \frac{h(x,k)}{h(x,k)-1} = & \frac{2}{1}\cdot\frac{2}{1} & \frac{3}{2}\cdot\frac{2}{1} & \frac{2}{1}\cdot\frac{3}{2} & \frac{3}{2}\cdot\frac{3}{2} & \frac{4}{3}\cdot\frac{3}{2} \end{array}$$

Therefore $\det(3,2)=2\times 3^4$, and

$$\dim D^{(3,2)}\begin{cases} =5 & \text{if } \operatorname{char} F\neq 2,3,\\ \geqslant 4 & \text{if } \operatorname{char} F=2,\\ \geqslant 1 & \text{if } \operatorname{char} F=3.\end{cases}$$

In this case, the lower bounds for $\dim D^{(3,2)}$ are the correct dimensions.

A neater expression for $\det\beta$ can be obtained. To describe this, we need a certain function $d$ on finite sequences of integers (cf. the discussion preceding 2.7.7):

Let $\beta_1,\beta_2,\dots,\beta_s$ be integers. Let $d(\beta_1,\beta_2,\dots,\beta_s):=0$ if some $\beta_i$ is negative, or if some pair of $\beta_i$'s are equal. Otherwise, define $d(\beta_1,\beta_2,\dots,\beta_s):=$ (the dimension of the ordinary irreducible representation of $S_n$ corresponding to the diagram whose $\beta$-numbers are $\beta_1,\beta_2,\dots,\beta_s$) $\times$(the signature of the permutation $\pi$ of $1,2,\dots,s$ for which $\beta_{\pi 1}>\beta_{\pi 2}>\cdots>\beta_{\pi s}$).

**7.3.20** THEOREM. *Suppose that $\beta$ has $s$ nonzero parts. As usual, let $h_{ij}$ be the hook length of the $(i,j)$ node of $[\beta]$, and let $\beta'$ be the partition associated with $\beta$. Then*

$$\det\beta=\prod_{1\leqslant a<b\leqslant s}\ \prod_{c=1}^{\beta_b}\left(\frac{h_{ac}}{h_{bc}}\right)^{d(h_{11},h_{21},\dots,h_{a1}+h_{bc},\dots,h_{b1}-h_{bc},\dots,h_{s1})}$$

$$=\prod_{(b,c)\in[\beta]}\ \prod_{\substack{a\neq b\\ a\leqslant\beta'_c}} h_{bc}^{-d(h_{11},h_{21},\dots,\underset{\uparrow\ a\text{ th place}}{h_{a1}+h_{bc}},\dots,\underset{\uparrow\ b\text{ th place}}{h_{b1}-h_{bc}},\dots,h_{s1})}$$

*Proof.* The theorem is proved by applying an induction argument to the expression for $\det\beta$ given in the last theorem; the details can be found in the paper by James and Murphy [1979]. ■

To clarify the exponent described in the theorem, we explain how to obtain

$$d(h_{11},h_{21},\dots,h_{a1}+h_{bc},\dots,h_{b1}-h_{bc},\dots,h_{s1})$$

given $\beta$, $a$, $b$, and $c$. It is, to within a sign, the dimension of $S^\alpha$, where $[\alpha]$ is obtained from $[\beta]$ by removing the rim $(b, c)$-hook, and "wrapping on" a rim hook of the same length, whose foot node lies in the $a$th row. The sign is $+$ if and only if the sum of the leg lengths of the unwrapped and wrapped-on rim hooks is even. Of course, dim $S^\alpha$ is most easily evaluated using the Hook Formula 2.3.21.

**7.3.21 EXAMPLE.** $\beta = (11, 9, 6, 5, 4^2, 3)$, $a = 3$, $b = 5$, $c = 2$. The $(a, c)$ and $(b, c)$ nodes are marked below:

$$
[\beta] = 
\begin{array}{l}
\times\,\times\,\times\,\times\,\times\,\times\,\times\,\times\,\times\,\times\,\times \\
\times\,\times\,\times\,\times\,\times\,\times\,\times\,\times\,\times \\
\times\,\otimes\,\times\,\times\,\times\,\times \\
\times\,\times\,\times\,\times\,\times \\
\times\,\otimes\,\times\,\times \\
\times\,\times\,\times \\
\times\,\times\,\times
\end{array}
\qquad , \qquad
[\alpha] = 
\begin{array}{l}
\times\,\times\,\times\,\times\,\times\,\times\,\times\,\times\,\times\,\times\,\times \\
\times\,\times\,\times\,\times\,\times\,\times\,\times\,\times\,\times\,\times\,\times \\
\times\,\otimes\,\times\,\times\,\times\,\times\,\times\,\times\,\times\,\times \\
\times\,\times\,\times\,\times\,\times \\
\times\,\otimes\,\times \\
\times\,\times \\
\times
\end{array}
$$

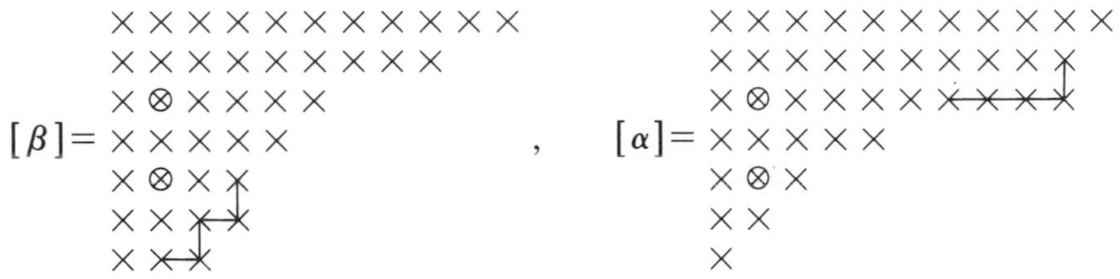

and $d(h_{11}, h_{21}, \ldots, h_{a1} + h_{bc}, \ldots, h_{b1} - h_{bc}, \ldots, h_{s1}) = -\dim(11, 10, 10, 5, 3, 2, 1)$, where we write dim $\alpha$ for dim $S^\alpha$.

**7.3.22 EXAMPLE.** If $\beta = (4, 3, 2)$, then the hook graph for $[\beta]$ is

$$
\begin{array}{cccc}
6 & 5 & 3 & 1 \\
4 & 3 & 1 & \\
2 & 1 & &
\end{array}
$$

and the two expressions for det $\beta$ given in the theorem are

$$
\det\beta = \left(\frac{6}{4}\right)^{d(10,0,2)} \cdot \left(\frac{6}{2}\right)^{d(8,4,0)} \cdot \left(\frac{4}{2}\right)^{d(6,6,0)} \cdot \left(\frac{5}{3}\right)^{d(9,1,2)}
$$
$$
\cdot \left(\frac{5}{1}\right)^{d(7,4,1)} \cdot \left(\frac{3}{1}\right)^{d(6,5,1)} \cdot \left(\frac{3}{1}\right)^{d(7,3,2)}
$$

and

$$
\det\beta = 6^{-d(0,10,2)-d(0,4,8)} \cdot 4^{-d(10,0,2)-d(6,0,6)}
$$
$$
\cdot 2^{-d(8,4,0)-d(6,6,0)} \cdot 5^{-d(1,9,2)-d(1,4,7)}
$$
$$
\cdot 3^{-d(9,1,2)-d(6,1,5)} \cdot 3^{-d(3,7,2)}.
$$

These numbers are clearly equal. (But it is far from transparent that they will turn out to be integers.) Evaluating the second expression, we get

$$
\det\beta = 6^{-\dim(8,1)+\dim(6,3)} \cdot 4^{\dim(8,1)} \cdot 2^{-\dim(6,3)}
$$
$$
\cdot 5^{-\dim(7,1,1)+\dim(5,3,1)} \cdot 3^{\dim(7,1,1)+\dim(4,4,1)+\dim(5,2,2)}
$$
$$
= 6^{-8+48} \cdot 4^8 \cdot 2^{-48} \cdot 5^{-28+162} \cdot 3^{28+84+120}
$$
$$
= 2^8 \cdot 3^{272} \cdot 5^{134}.
$$

Since $\dim S^\beta = 168$, we deduce that

$$\dim D^\beta \geqslant 160 \quad \text{when} \quad \operatorname{char} F = 2,$$
$$\dim D^\beta \geqslant 34 \quad \text{when} \quad \operatorname{char} F = 5.$$

In fact, $\dim D^\beta = 160$ when $\operatorname{char} F = 2$, and $\dim D^\beta = 34$ when $\operatorname{char} F = 5$.

In the above example, the theorem gives no information about $\dim D^\beta$ when $\operatorname{char} F = 3$. We do not know how frequently good bounds for $\dim D^\beta$ emerge, but for small $n$, and for special cases in blocks of arbitrarily large defect, the bound turns out to be the exact dimension.

We now classify those cases where $D^\beta = S^\beta$.

**7.3.23** THEOREM. *$\beta$ is $p$-regular and $S^\beta$ is irreducible over fields of characteristic $p$ if and only if, for all $a, b, c$ with $(a, c), (b, c) \in [\beta]$, the exponent of $p$ dividing the hook length $h_{ac}$ equals the exponent of $p$ dividing $h_{bc}$.*

*Proof.* If $z$ is an integer, let $\nu_p(z) :=$ the exponent of $p$ dividing $z$.

Suppose that $\nu_p(h_{ac}) = \nu_p(h_{bc})$ for all $a, b, c$. Then $\beta$ is obviously $p$-regular. Also $p$ does not divide $\det \beta$, by the first expression in Theorem 7.3.20. Therefore, the Gram matrix has full $p$-rank, and $S^\beta$ is irreducible.

Now suppose that $\beta$ is $p$-regular and $\nu_p(h_{ac}) > \nu_p(h_{ac})$ for some $a, b, c$. Consider the largest $c$, then the largest $a$ for which there exists a $b$ with $\nu_p(h_{ac}) > \nu_p(h_{bc})$. It is easy to see that the $(a, c)$ node does not lie at the foot of a column or at the end of a row. We claim that it is therefore sufficient to prove that $S^\beta$ is reducible in the case where

**7.3.24.** $\nu_p(h_{ac}) = \nu_p(h_{bc})$ for $c > 1$, and for $c = 1$ and $1 < a \leqslant b$, but $\nu_p(h_{11}) > \nu_p(h_{21})$.

The most satisfactory way of justifying this reduction would be to prove that $\det \mu$ divides $\det \lambda$ when $[\lambda]$ is obtained from $[\mu]$ by adding a new first column or a new first row, but we are not certain that this is true. However, when $[\lambda]$ is $p$-regular and is obtained from $[\mu]$ by adding a new first column, we can prove that $S^\lambda$ is reducible if $S^\mu$ is reducible, thus: $S^\mu$ reducible implies there is a submodule of $S^\mu$ isomorphic to $D^\nu$ for some $\nu \rhd \mu$. Therefore, $\operatorname{Hom}_{FS_n}(S^\nu, S^\mu) \neq 0$, and 8.4.4 below shows $\operatorname{Hom}_{\bar{U}_F}(W^\mu, W^\nu) \neq 0$. Add to $[\nu]$ a new first column, equal in length to the first column of $[\lambda]$, thereby obtaining $[\nu^*]$, say. By 8.1.22, $\operatorname{Hom}_{\bar{U}_F}(W^\lambda, W^{\nu^*}) \neq 0$, and applying 8.4.4 again, we get $\operatorname{Hom}_{FS_n}(S^{\nu^*}, S^\lambda) \neq 0$. Since $\nu^* \rhd \lambda$, $S^\lambda$ is reducible. When $[\lambda]$ is $p$-regular, and is obtained from $[\mu]$ by adding a new first row, it again turns out that $S^\lambda$ is reducible if $S^\mu$ is reducible, but the proof is more subtle, since there is no obvious way of exhibiting a composition factor; the reader is referred to James's paper [1978].

Finally, therefore, we must prove that $p$ divides $\det\beta$ when $[\beta]$ has the property described in 7.3.24. In the presence of condition 7.3.24, the first expression for $\det\beta$ in 7.3.20 shows that the exponent of $p$ dividing $\det\beta$ is a strictly positive multiple of

$$d(h_{11}+h_{21},0,h_{31},\ldots,h_{s1})+d(h_{11}+h_{31},h_{21},0,\ldots,h_{s1})+\cdots \\ +d(h_{11}+h_{s1},h_{21},h_{31},\ldots,0).$$

We shall be home if we prove this expression is strictly positive. In fact, we prove a slightly stronger result, namely:

**7.3.25.** *If* $h_1>h_2>\cdots>h_s\geqslant 0$, *then*

$$\sum_{i=2}^{s} d(h_1+h_i,h_2,\ldots,\underset{\substack{\uparrow \\ i\text{ th place}}}{0},\ldots,h_s)>0.$$

When $h_s=0$, $d(h_1,h_2,\ldots,h_{s-1},0)$ is the only nonzero term in the sum, so 7.3.25 is certainly true in this case.

Assume, therefore, that $h_s\geqslant 1$. By the branching theorem,

$$\begin{aligned}\sum_{i=2}^{s} d(h_1+h_i,h_2,\ldots,0,\ldots,h_s)=&\sum_{i=2}^{s} d(h_1+h_i-1,h_2,\ldots,0,\ldots,h_s)\\ &+\sum_{i=2}^{s} d(h_1+h_i,h_2-1,\ldots,0,\ldots,h_s)+\cdots\\ &+\sum_{i-2}^{s} d(h_1+h_i,h_2,\ldots,0,\ldots,h_s-1).\end{aligned}$$

Since $h_1>h_2>\cdots>h_s-1\geqslant 0$, the last sum is strictly positive, by induction. All the other sums are also positive, by induction, if we note that $\sum_{i=2}^{s} d(h_1+h_i,h_2,\ldots,0,\ldots,h_s)=0$ when $h_1>h_2>\cdots>h_j=h_{j+1}>\cdots>h_s\geqslant 0$. This completes the proof of 7.3.25, and we have finished the proof of the theorem. ■

**7.3.26** EXAMPLE. If $\beta=(8,5,2)$, then the hook lengths are

| | | | | | | | |
|---|---|---|---|---|---|---|---|
| 10 | 9 | 7 | 6 | 5 | 3 | 2 | 1 |
| 6 | 5 | 3 | 2 | 1 | | | |
| 2 | 1 | | | | | | |

There are two numbers $h_{ac}$ and $h_{bc}$ in the same column such that $\nu_p(h_{ac})\neq\nu_p(h_{bc})$ if and only if $p=3,5,7$. Therefore, $S^{(8,5,2)}$ is reducible if and only if char $F=3,5$, or 7.

# Exercises

7.1 Let $[t]$ denote the column equivalence class of the $\beta$-tableau $t$, and let $\overline{M}^\beta$ be the vector space over $F$ spanned by the $[t]$ as $t$ varies over $\beta$-tableaux. Turn $\overline{M}^\beta$ into an $FS_n$-module by defining

$$\pi[t]=(\operatorname{sgn}\pi)[\pi t] \qquad (\pi\in S_n).$$

Let $\overline{S}^\beta$ be the subspace of $\overline{M}^\beta$ spanned by the vectors of the form $\mathcal{K}'[t]$.

(a) Verify that for every result of Section 7.1 concerning $S^\beta$ there is a corresponding result for $\overline{S}^\beta$. In particular, show that when the partition $\beta'$ associated with $\beta$ is $p$-regular, $\overline{S}^\beta$ has a unique maximal submodule $\overline{S}^\beta\cap\overline{S}^{\beta\perp}$.

(b) Prove that

$$\overline{M}^\beta\simeq M^{\beta'}\otimes S^{(1^n)} \quad\text{and}\quad \overline{S}^\beta\simeq S^{\beta'}\otimes S^{(1^n)}.$$

(c) Let $t$ be a given $\beta$-tableau, and suppose that $\theta$ is the element of $\operatorname{Hom}_{FS_n}(M^\beta,\overline{S}^\beta)$ which sends $\{t\}$ to $\mathcal{K}'[t]$. Prove that $\operatorname{Ker}\theta=S^{\beta\perp}$ and deduce that $\overline{S}^\beta$ is isomorphic to the dual of $S^\beta$.

7.2 Prove that, as $\alpha$ varies over $p$-regular partitions of $n$, the left ideals $FS_n\mathcal{K}^\alpha\mathcal{V}^\alpha\mathcal{K}^\alpha$ of $FS_n$ vary over a complete set of inequivalent irreducible $FS_n$-modules. Also, as $\alpha'$ varies over $p$-regular partitions of $n$, $FS_n\mathcal{V}^\alpha\mathcal{K}^\alpha\mathcal{V}^\alpha$ varies over a complete set of inequivalent irreducible $FS_n$-modules.

7.3 Suppose that the entries in the tableau $t$ increase down each column. Prove that $\{t\}$ is a standard tabloid—that is, there exists a standard tableau which is row-equivalent to $t$.

7.4 Prove that $\operatorname{Hom}_{FS_n}(S^\beta,S^\beta)\simeq F$ unless $\operatorname{char}F=2$ and both $\beta$ and $\beta'$ are 2-singular.

7.5 Suppose that $\operatorname{Hom}_{FS_n}(S^\beta,S^\beta)\simeq F$. Prove that $S^\beta$ is irreducible if and only if $S^\beta$ is $FS_n$-isomorphic to its dual $\overline{S}^\beta$.

7.6 Let $F$ be the field of 2 elements. Let $V$ be the vector space over $F$ whose elements are graphs (without loops or multiple edges) on $n$ points, the sum of two graphs being defined to be their symmetric difference. Since $S_n$ permutes the points, $V$ can be regarded as an $FS_n$-module. Prove that $V$ is isomorphic to $M^{(n-2,2)}$ (assuming $n\geqslant 4$). Which subspace corresponds to $S^{(n-2,2)}$? Why is $S^{(n-2,2)}\subseteq S^{(n-2,2)\perp}$ if and only if $n=4$?

7.7 Verify that the Gram matrix with respect to the standard basis of $S^{(3,1^2)}$ is

$$\begin{bmatrix} 6 & 2 & -2 & 2 & -2 & 0\\ 2 & 6 & 2 & 2 & 0 & -2\\ -2 & 2 & 6 & 0 & 2 & -2\\ 2 & 2 & 0 & 6 & 2 & 2\\ -2 & 0 & 2 & 2 & 6 & 2\\ 0 & -2 & -2 & 2 & 2 & 6 \end{bmatrix}.$$

Prove that the determinant of this matrix is $2^6\cdot 5^3$, both by direct calculation and using Theorem 7.3.20.

CHAPTER 8

# *Representations of General Linear Groups*

If $F$ is an arbitrary field, there is a very close resemblance between the representation theory of $S_n$ over $F$, and that of $GL(m, F)$ over $F$. (Here $GL(m, F)$ is the group of nonsingular $m \times m$ matrices with entries from $F$.) So-called Weyl modules, $W^\alpha$, play a part analogous to that of Specht modules $S^\alpha$. We follow the approach of Carter and Lusztig, studying $W^\alpha$ and $S^\alpha$ as subspaces of tensor space, where information about Weyl modules sheds further light on the theory of Specht modules.

In this chapter we generalize some of the results of Section 5.2 on symmetrization. In particular, it will emerge that when $F = \mathbb{C}$, the Weyl module $W^\alpha$ affords the representation previously denoted by $\langle\alpha\rangle = \mathrm{id}_{GL(V)} \boxdot [\alpha]$.

The nature of $F$ affects the theory of Weyl modules in a way that did not occur for Specht modules. In some results, the case where $\operatorname{char} F = 2$ is tricky, and it is sometimes easier to assume that $F$ is infinite. In fact, we shall investigate $\bar{U}_F$-submodules of tensor space, where $\bar{U}_F$ is a certain associative $F$-algebra defined in Section 8.2; $\bar{U}_F$-submodules of tensor space coincide with $FGL(F)$-submodules when $F$ is infinite.

Every group $G$ which has a representation of degree $m$ over $F$ has, by definition, a homomorphic image inside $GL(m, F)$. Hence $GL(m, F)$-modules may be regarded as $G$-modules. In particular, many of the theorems in this chapter go through word for word when $GL(m, F)$ is replaced by the special linear group $SL(m, F)$. Throughout the chapter, the reader may care to refer to the remarkable list at the end which details results for Specht modules and the parallel results for Weyl modules.

ENCYCLOPEDIA OF MATHEMATICS and Its Applications, Gian-Carlo Rota (ed.). Vol. 16: G. D. James and A. Kerber, The Representation Theory of the Symmetric Group
ISBN 0-201-13515-9

## 8.1 Weyl Modules

Let $W^{(1)}$ be the $m$-dimensional vector space over $F$, with basis $w_1, w_2, \ldots, w_m$ on which $\mathrm{GL}(m, F)$ acts in the natural way; that is, the matrix $(g_{ij})$ belonging to $\mathrm{GL}(m, F)$ sends $w_i$ to $\Sigma_j g_{ji} w_j$. We denote by $L^{(n)}$ the $n$-fold tensor product $W^{(1)} \otimes \cdots \otimes W^{(1)}$. This has a basis consisting of all tensors

$$w_{i_1} \otimes w_{i_2} \otimes \cdots \otimes w_{i_n} \qquad (1 \leqslant i_j \leqslant m),$$

and $\mathrm{GL}(m, F)$ acts on $L^{(n)}$ by

$$(g_{ij}) : w_{i_1} \otimes \cdots \otimes w_{i_n} \mapsto \Sigma g_{j_1 i_1} g_{j_2 i_2} \cdots g_{j_n i_n} w_{j_1} \otimes \cdots \otimes w_{j_n},$$

where the sum is over all $j_1, \ldots, j_n$ with $1 \leqslant j_1 \leqslant m, \ldots, 1 \leqslant j_n \leqslant m$.

But $S_n$ also acts on $L^{(n)}$ by "place permutations of the subscripts":

$$\pi : w_{i_1} \otimes \cdots \otimes w_{i_n} \mapsto w_{i_{\pi^{-1}(1)}} \otimes \cdots \otimes w_{i_{\pi^{-1}(n)}}.$$

Comparing 4.3.7 and 4.3.11, we see that the representation of $S_n$ afforded this way is simply that which was called $P_n^m$. We are writing $L^{(n)}$, in place of $\otimes^n W^{(1)}$, to fit in with later notation.

If $\alpha$ is a partition of $n$ having at most $m$ nonzero parts, we embed the permutation module $M^\alpha$ (see 7.1.1) in $L^{(n)}$ as follows:

**8.1.1.** Suppose that $f(i)$ is the number of the row in which $i$ lies in the $\alpha$-tabloid $\{t\}$. Then map

$$\{t\} \mapsto w_{f(1)} \otimes w_{f(2)} \otimes \cdots \otimes w_{f(n)}.$$

Our copy of $M^\alpha$ (which we shall also call $M^\alpha$) clearly consists of the space spanned by those tensors $w_{i_1} \otimes \cdots \otimes w_{i_n}$ where, for each $i$, precisely $\alpha_i$ of the subscripts equal $i$. Note, too, that

**8.1.2.** *Standard tabloids correspond to tensors* $w_{i_1} \otimes \cdots \otimes w_{i_n}$ *for which* $i_1, i_2, \ldots, i_n$ *is a lattice permutation.*

Let $w^\alpha$ denote the tensor which corresponds to the first standard $\alpha$-tabloid. (We define $w^\alpha := 0$ if $\alpha$ has more than $m$ nonzero parts.) For example, if $\alpha = (3,2)$,

$$\begin{array}{ccc} \hline 1 & 3 & 5 \\ \hline 2 & 4 & \\ \hline \end{array} \mapsto w_1 \otimes w_2 \otimes w_1 \otimes w_2 \otimes w_1 = w^\alpha.$$

Throughout this chapter $\mathcal{V}^\alpha$ will be the signed column sum for the first standard $\alpha$-tableau. $FS_n\mathcal{V}^\alpha w^\alpha$ is an isomorphic copy of the Specht module $S^\alpha$ in $L^{(n)}$ (provided that $\alpha$ has at most $m$ nonzero parts). If $\alpha=(3,2)$, then

$$\mathcal{V}^\alpha w^\alpha = w_1\otimes w_2\otimes w_1\otimes w_2\otimes w_1 - w_2\otimes w_1\otimes w_1\otimes w_2\otimes w_1$$
$$- w_1\otimes w_2\otimes w_2\otimes w_1\otimes w_1 + w_2\otimes w_1\otimes w_2\otimes w_1\otimes w_1.$$

The action of $\mathrm{GL}(m, F)$ commutes with the action of $S_n$ on $L^{(n)}$ (cf. 5.3.1). Therefore, for every left ideal $I$ of the group algebra of $S_n$, the set of vectors in $L^{(n)}$ annihilated by all elements in $I$ is a $\mathrm{GL}(m, F)$-submodule of $L^{(n)}$. We shall examine certain modules of this form. In particular, we are interested in

8.1.3 $\quad \{w \mid w\in L^{(n)} \text{ and } sw=0 \text{ for all } s\in FS_n \text{ such that } s\mathcal{V}^\alpha w^\alpha=0\}.$

This is certainly a natural $\mathrm{GL}(m, F)$-module to study, especially since the Garnir elements for the first standard polytabloid annihilate $\mathcal{V}^\alpha w^\alpha$. It turns out that when $\operatorname{char} F=0$, 8.1.3 gives us an irreducible $\mathrm{GL}(m, F)$-module. We shall, however, work with an *arbitrary field F*, and we run into a snag when $\operatorname{char} F=2$. The problem is this. The ideal of $FS_2$ annihilating $\mathcal{V}^\alpha w^\alpha$ consists of multiples of $1+(12)$ when $\alpha=(1^2)$, and consists of multiples of $1-(12)$ when $\alpha=(2)$. 8.1.3 therefore gives the same $\mathrm{GL}(m, F)$-module in both cases if $\operatorname{char} F=2$—a possibility we want to avoid. This forces us to give a preliminary definition.

**8.1.4.** Let $L^{(1^r)}$ be the $F$-space spanned by the vectors of the form $\mathcal{V}^{(1^r)} w_{i_1}\otimes\cdots\otimes w_{i_r}$ with $1\leqslant i_1\leqslant m, \ldots, 1\leqslant i_r\leqslant m$.

Thus, $L^{(1^r)}$ is simply the antisymmetric part of $\otimes^r W^{(1)}$. More generally if $\alpha\vdash n$ and $\nu$ is the partition associated with $\alpha$, let

**8.1.5.** $L^\alpha := L^{(1^{\nu(1)})}\otimes L^{(1^{\nu(2)})}\otimes\cdots$; that is to say, $L^\alpha=\mathcal{V}^\alpha L^{(n)}$. (Here we have written $\nu_j$ as $\nu(j)$, for typographical convenience.) Notice that this is consistent with our earlier definition, $L^{(n)}=W^{(1)}\otimes W^{(1)}\otimes\cdots$.

In the representation theory of $\mathrm{GL}(m, F)$, $L^\alpha$ plays a rôle similar to that occupied by the permutation module $M^\alpha$ in the representation theory of $S_n$. We are now in a position to define the $\mathrm{GL}(m, F)$-modules $W^\alpha$, which we shall call *Weyl modules*; many properties of Weyl modules parallel those of Specht modules. Let $\alpha\vdash n$.

**8.1.6.**

(i) If $\operatorname{char} F=2$ and $\alpha$ is 2-singular, let

$$W^\alpha := \{w \mid w\in L^{(n)} \text{ and } sw=0 \text{ for all } s\in FS_n \text{ such that } s\mathcal{V}^\alpha w^\alpha=0\}\cap L^\alpha.$$

(ii) In all other cases, let

$$W^\alpha := \{w \mid w \in L^{(n)} \text{ and } sw = 0 \text{ for all } s \in FS_n \text{ such that } s\mathcal{V}^\alpha w^\alpha = 0\}.$$

Note that we have imposed the restriction that $W^\alpha$ lies in $L^\alpha$ only for the cases where char $F=2$ and $\alpha$ is 2-singular; in all other cases, it turns out that $W^\alpha$ automatically is contained in $L^\alpha$. It seems plausible at this stage that $W^\alpha = \mathcal{V}^\alpha \mathcal{H}^\alpha L^{(n)}$ when char $F=0$, but we postpone the proof of this to Section 8.3.

Our first objective is to construct the "semistandard basis" of $W^\alpha$, and the method is very similar to that for the standard basis of the Specht module. The construction uses generalized $\alpha$-tableaux with entries from $\{1,\ldots,m\}$, and throughout this chapter we shall denote such a tableau by a capital $T$. With each $\alpha$-tableau $T$ (in general containing repeated entries) we associate a tensor $w_T$ in $L^{(n)}$ as follows:

**8.1.7.** Let $T(1), T(2),\ldots, T(n)$ be the entries in $T$, reading down successive columns. Then define $w_T := w_{T(1)} \otimes w_{T(2)} \otimes \cdots \otimes w_{T(n)}$ (and $w_T := 0$ if some entry in $T$ is greater than $m$).

For example, if

$$T = \begin{matrix} 1 & 1 & 4 \\ 2 & 3 & 3 \\ 3 & & \end{matrix},$$

then $w_T = w_1 \otimes w_2 \otimes w_3 \otimes w_1 \otimes w_3 \otimes w_4 \otimes w_3$.

The construction ensures that

**8.1.8.** *$w^\alpha = w_T$, where $T$ is the $\alpha$-tableau in which, for each $i$, all the entries in the $i$th row equal $i$.*

Thus, for instance, $w^{(3,2)} = w_1 \otimes w_2 \otimes w_1 \otimes w_2 \otimes w_1 = w_T$ where

$$T = \begin{matrix} 1 & 1 & 1 \\ 2 & 2 & \end{matrix}.$$

It is clear that

**8.1.9.** *$\mathcal{V}^\alpha w_T = 0$ if and only if some column of $T$ contains two identical numbers.*

In particular,

**8.1.10.** *$\mathcal{V}^\alpha w_T = 0$ if the content of $T$ is $\beta$ and $\alpha \ntrianglerighteq \beta$.*

We say that two tableaux $T_1$ and $T_2$ are *row* (respectively *column*) *equivalent* if $T_2$ can be obtained from $T_1$ by permuting the order of appearance of the numbers in each row (respectively column) of $T_1$. For example,

$$\begin{matrix}1&1&2\\2&3&\end{matrix},\quad \begin{matrix}1&2&1\\2&3&\end{matrix},\quad \begin{matrix}2&1&1\\2&3&\end{matrix},\quad \begin{matrix}1&1&2\\3&2&\end{matrix},\quad \begin{matrix}1&2&1\\3&2&\end{matrix},\quad \begin{matrix}2&1&1\\3&2&\end{matrix}$$

forms a row-equivalence class.

**8.1.11.** If $T$ is a $\alpha$-tableau, let

$$E_T := \mathcal{V}^\alpha \sum \{w_{T_1} \mid T_1 \text{ is row-equivalent to } T\}.$$

The practical way of writing down $E_T$ is to permute the numbers in the columns of each $T_1$ which is row-equivalent to $T$, attaching the sign of the column permutation. This tells us which tableaux occur as subscripts in the expression for $E_T$. For example, if

$$T = \begin{matrix}1&1&2,\\2&3&\end{matrix}$$

we write down

$$\begin{matrix}
+\begin{matrix}1&1&2\\2&3&\end{matrix} & +\begin{matrix}1&2&1\\2&3&\end{matrix} & +\begin{matrix}2&1&1\\2&3&\end{matrix} & +\begin{matrix}1&1&2\\3&2&\end{matrix} & +\begin{matrix}1&2&1\\3&2&\end{matrix} & +\begin{matrix}2&1&1\\3&2&\end{matrix}\\
-\begin{matrix}2&1&2\\1&3&\end{matrix} & -\begin{matrix}2&2&1\\1&3&\end{matrix} & -\begin{matrix}2&1&1\\2&3&\end{matrix} & -\begin{matrix}3&1&2\\1&2&\end{matrix} & -\begin{matrix}3&2&1\\1&2&\end{matrix} & -\begin{matrix}3&1&1\\2&2&\end{matrix}\\
-\begin{matrix}1&3&2\\2&1&\end{matrix} & -\begin{matrix}1&3&1\\2&2&\end{matrix} & -\begin{matrix}2&3&1\\2&1&\end{matrix} & -\begin{matrix}1&2&2\\3&1&\end{matrix} & -\begin{matrix}1&2&1\\3&2&\end{matrix} & -\begin{matrix}2&2&1\\3&1&\end{matrix}\\
+\begin{matrix}2&3&2\\1&1&\end{matrix} & +\begin{matrix}2&3&1\\1&2&\end{matrix} & +\begin{matrix}2&3&1\\2&1&\end{matrix} & +\begin{matrix}3&2&2\\1&1&\end{matrix} & +\begin{matrix}3&2&1\\1&2&\end{matrix} & +\begin{matrix}3&2&1\\2&1&\end{matrix}
\end{matrix}$$

and $E_T = w_1 \otimes w_2 \otimes w_1 \otimes w_3 \otimes w_2 + w_1 \otimes w_2 \otimes w_2 \otimes w_3 \otimes w_1 + \cdots + w_3 \otimes w_2 \otimes w_2 \otimes w_1 \otimes w_1$.

We shall prove that $\{E_T \mid T \text{ is semistandard } \alpha\text{-tableau}\}$ is a basis for $W^\alpha$. Certainly $E_T$ belongs to $L^\alpha = \mathcal{V}^\alpha L^{(n)}$. To see that $E_T$ belongs to $W^\alpha$, note that, *given* $T$, the map which sends the first standard $\alpha$-tabloid to

$$\Sigma \{w_{T_1} \mid T_1 \text{ is row-equivalent to } T\}$$

extends to a well-defined $FS_n$-homomorphism from $M^\alpha$ into $L^{(n)}$ (this is *not* the isomorphism of 8.1.1). Therefore, $E_T$, being the image of the first standard $\alpha$-polytabloid under an $FS_n$-homomorphism, is annihilated by all elements of $FS_n$ which annihilate the first standard $\alpha$-polytabloid.

**8.1.12** LEMMA. *$\{E_T | T$ is a semistandard $\alpha$-tableau$\}$ is a linearly independent subset of $W^\alpha$.*

*Proof.* Put an equivalence relation on the set of $\alpha$-tableaux by saying that $T_1$ and $T_2$ are equivalent if, for every $j$, the sum of the numbers in the $j$th column of $T_1$ equals the sum of the numbers in the $j$th column of $T_2$. Let $[T]$ denote the equivalence class of $T$. Put a partial order $\trianglelefteq$ on the set of equivalence classes by defining $[T_1] \trianglelefteq [T_2]$ if, for all $j$, the sum of the entries in the first $j$ columns of $T_1$ is greater than or equal to the sum of the entries in the first $j$ columns of $T_2$. It is clear that

**8.1.13.** *If $T$ is semistandard and $T'$ is row-equivalent to $T$, then $[T'] \trianglelefteq [T]$.*

Now consider $\Sigma c_T E_T$ ($c_T \in F$, summing over semistandard tableaux $T$). If not all the coefficients $c_T$ are zero, choose $T_1$ such that $c_{T_1} \neq 0$ but $c_T = 0$ if $[T_1] \triangleleft [T]$. Let $T_1, T_2, \ldots, T_k$ be the semistandard tableaux such that $[T_i] = [T_1]$. Then, by 8.1.13 and the fact that $\mathcal{V}^\alpha$ preserves column equivalence classes,

$$\Sigma c_T E_T = \mathcal{V}^\alpha (c_{T_1} w_{T_1} + \cdots + c_{T_k} w_{T_k})$$
$$+ \text{ terms of the form } c_{T'} \mathcal{V}^\alpha w_{T'} \text{ with } [T_1] \not\trianglelefteq [T'].$$

Since $T_1$ is semistandard, $\mathcal{V}^\alpha w_{T_1} \neq 0$ by 8.1.9. Now, the fact that different semistandard $\alpha$-tableaux belong to different column equivalence classes, and $\mathcal{V}^\alpha$ preserves these classes, shows that $\Sigma c_T E_T$ is nonzero, as required. ■

**8.1.14** LEMMA. *If $\Sigma c_T w_T$ ($c_T \in F$) is a nonzero element of $W^\alpha$, then $c_{T_1} \neq 0$ for some semistandard tableau $T_1$.*

*Proof.* We prove first that

**8.1.15.** *$c_{T*} = 0$ for every tableau $T^*$ having a repeated entry in some column.*

Suppose that $i$ and $j$ are in the same column of the first standard $\alpha$-tableau $t_1^\alpha$, and $T^*(i) = T^*(j)$ (see Definition 8.1.7). We must prove that $c_{T*} = 0$.

Since $(1+(ij))\mathcal{V}^\alpha w^\alpha = 0$, $1+(ij)$ annihilates $\Sigma c_T w_T$. Because $(ij) w_{T*} = w_{T*}$, we conclude that $c_{T*} = 0$ when $\operatorname{char} F \neq 2$.

If $\operatorname{char} F = 2$ and $\alpha$ is 2-singular, then, by definition, $W^\alpha \subseteq L^\alpha$. Therefore, $\Sigma c_T w_T \in L^\alpha = \mathcal{V}^\alpha L^{(n)}$, and $c_{T*} = 0$, by 8.1.9.

If $\operatorname{char} F = 2$ and $\alpha$ is 2-regular, then, by Corollary 7.1.12, there is a $\pi \in S_n$ such that $1 - \mathcal{V}^\alpha \pi$ annihilates $\mathcal{V}^\alpha w^\alpha$. Therefore, $\Sigma c_T w_T = \mathcal{V}^\alpha \pi \Sigma c_T w_T \in \mathcal{V}^\alpha L^{(n)}$, and again we have $c_{T*} = 0$. This completes the proof of 8.1.15.

When $\pi$ belongs to the column stabilizer of $t_1^\alpha$, $1-(\operatorname{sgn}\pi)\pi$ annihilates $\mathcal{V}^\alpha w^\alpha$. Therefore, $\Sigma c_T w_T = \Sigma c_T(\operatorname{sgn}\pi)\pi w_T$, and so $c_{T_1} = \pm c_{T_2}$ when $T_1$ and $T_2$ are column-equivalent.

Since $\Sigma c_T w_T \neq 0$, we may choose a tableau $T_1$ such that $c_{T_1} \neq 0$, but $c_T = 0$ if $[T_1] \lhd [T]$. The previous paragraph and 8.1.15 show that we may assume that the numbers strictly increase down the columns of $T_1$.

We shall be home if we can derive a contradiction from assuming that, for some $j$, $a_1 < a_2 < \cdots < a_r$ are the entries in the $j$th column of $T_1$ $b_1 < b_2 < \cdots < b_s$ are the entries in the $(j+1)$th column of $T_1$ and $a_q > b_q$ for some $q$:

$$\begin{array}{ccc}
a_1 & & b_1 \\
\cdot & & \wedge \\
\cdot & & \cdot \\
\cdot & & \cdot \\
\cdot & & \cdot \\
a_q & > & b_q \\
\wedge & & \cdot \\
\cdot & & \cdot \\
\cdot & & \cdot \\
\cdot & & b_s \\
\wedge & & \\
a_r & &
\end{array}$$

Let $x_{ij}$ be the entry in the $(i, j)$th place of $t_1^\alpha$, and let $\Sigma(\operatorname{sgn}\sigma)\sigma$ be a Garnir element for the sets $\{x_{qj}, \ldots, x_{rj}\}$ and $\{x_{1,j+1}, \ldots, x_{q,j+1}\}$. Then $\Sigma(\operatorname{sgn}\sigma)\sigma$ annihilates $\mathcal{V}^\alpha w^\alpha$, so

$$\Sigma(\operatorname{sgn}\sigma)\sigma \Sigma c_T w_T = 0$$

For every tableau $T$, $\Sigma(\operatorname{sgn}\sigma)\sigma w_T$ is a linear combination of terms $w_{\hat{T}}$ where $\hat{T}$ agrees with $T$ on all except the $(1, j+1)$th, $(2, j+1)$th, $\ldots$, $(q, j+1)$th, $(q, j)$th, $\ldots$, $(r, j)$th places. All the terms involved in $\Sigma(\operatorname{sgn}\sigma)\sigma c_{T_1} w_{T_1}$ have coefficient $\pm c_{T_1} \neq 0$, and since $\Sigma(\operatorname{sgn}\sigma)\sigma\Sigma c_T w_T = 0$, there must be a tableau $T \neq T_1$ with $c_T \neq 0$ such that $T$ agrees with $T_1$ on all except the places described above. Because $b_1 < \cdots < b_q < a_q < \cdots a_r$, we must have $[T_1] \lhd [T]$, and this contradicts our choice of $T_1$. ■

**8.1.16** THEOREM. *$\{E_T \mid T$ is a semistandard $\alpha$-tableau$\}$ is a basis for $W^\alpha$.*

*Proof.* The given vectors are linearly independent, by 8.1.12.

If $\Sigma c_T w_T$ is a non-zero element of $W^\alpha$, then by the last lemma $c_{T_1} \neq 0$ for some semistandard tableau $T_1$. We may assume that $c_T = 0$ if $T$ is semistandard and $[T_1] \lhd [T]$. Let $T_1, T_2, \ldots, T_k$ be the semistandard tableaux such that $[T_i]=[T_1]$. Then, by 8.1.13, and using induction down the partial order $\trianglelefteq$, we find that

$$\Sigma c_T w_T - c_{T_1}E_{T_1} - c_{T_2}E_{T_2} - \cdots - c_{T_k}E_{T_k}$$

is a linear combination of terms $E_{T'}$ with $T'$ semistandard, and so the same is true of $\Sigma c_T w_T$. Therefore, the given vectors span $W^\alpha$. ■

**8.1.17** Corollary (cf. 5.2.14). *The dimension of $W^\alpha$ equals the number of semistandard $\alpha$-tableaux with entries from $\{1,2,\ldots,m\}$, which also equals $\prod_{i,j}(m-i+j)/h_{ij}^\alpha$.*

**8.1.18** Corollary. *The diagonal matrix in* GL$(m,F)$ *whose diagonal entries are $x_1, x_2,\ldots, x_m$ has trace*

$$\Sigma\{x^T \mid T \textit{ is a semistandard } \alpha\textit{-tableau}\}$$

*on $W^\alpha$, where $x^T := x_1^{\lambda_1}\cdot x_2^{\lambda_2}\cdots x_m^{\lambda_m}$ if $T$ has content $(\lambda_1, \lambda_2,\ldots, \lambda_m)$.*

*Proof.* Write diag$(x_1,\ldots, x_m)$ for the element of GL$(m, F)$ which is under discussion. If $T$ has content $\lambda$, then clearly

$$\operatorname{diag}(x_1,\ldots, x_m)w_T = (x_1^{\lambda_1}\cdot x_2^{\lambda_2}\cdots x_m^{\lambda_m})w_T.$$

Thus,

$$\operatorname{diag}(x_1,\ldots, x_m)E_T = x^T E_T,$$

and the corollary follows. ■

It turns out, although it is by no means obvious, that the expression

$$\Sigma\{x^T \mid T \text{ is a semistandard } \alpha\text{-tableau}\},$$

which gives the trace of a diagonal matrix, is just the Schur function, defined in 5.1.30.

Theorem 8.1.16 also gives:

**8.1.19.**

(i) $W^\alpha \subseteq \mathcal{V}^\alpha L^{(n)} = L^\alpha$.
(ii) $W^{(1^n)} = L^{(1^n)}$.
(iii) $W^\alpha = 0$ *if $\alpha$ has more than $m$ nonzero parts.*
(iv) $W^{(n)}$ *is the symmetric part of* $\otimes^n W^{(1)}$.

In particular, $W^{(1^m)} = L^{(1^m)}$, which affords the representation $g \mapsto \det g$ $(g\in \mathrm{GL}(m, F))$. It is convenient to call $W^{(1^m)}$ the *determinant module* $\Delta$.

Since $\Delta$ is 1-dimensional, $\Delta^d := $(the representation $g \mapsto (\det g)^d$) is a 1-dimensional representation of $\mathrm{GL}(m, F)$ for each integer $d$.

Let $T(\Delta)$ denote the tableau with one column of height $m$ in which, for each $i$, $i$ appears in the $i$th row. Then $\{E_{T(\Delta)}\}$ is a basis for $\Delta$. If $\alpha$ has at most $m$ nonzero parts, and $T$ is a semistandard $\alpha$-tableau, then adding $T(\Delta)$ as a new first column, we obtain a semistandard $(\alpha_1+1, \ldots, \alpha_m+1)$-tableau $T^*$. When $T'$ is row-equivalent to $T^*$ and $\mathcal{V}^{(\alpha_1+1,\ldots,\alpha_m+1)}T' \neq 0$, $T'$ has $i$ in the $(i, 1)$th place. Therefore, $E_{T*} = E_{T(\Delta)} \otimes E_T$. This proves

**8.1.20** $$W^{(\alpha_1+1,\ldots,\alpha_m+1)} = \Delta \otimes W^{(\alpha_1,\ldots,\alpha_m)}.$$

In particular,

**8.1.21.** $\Delta^d = W^{(d,d,\ldots,d)}$ *(where* $(d, d, \ldots, d) = (d^m)$*) when* $d \geq 0$.

Sometimes, the trick of tensoring with $\Delta$ or $\Delta^{-1}$ is useful:

**8.1.22** THEOREM. *Suppose that $\alpha$ and $\beta$ are partitions of $n$ into at most $n$ nonzero parts. Let $\alpha^+ = (\alpha_1+1, \ldots, \alpha_m+1)$ and $\beta^+ = (\beta_1+1, \ldots, \beta_m+1)$. Then* $\mathrm{Hom}_{\mathrm{GL}(m,F)}(W^\alpha, W^\beta) \neq 0$ *if and only if* $\mathrm{Hom}_{\mathrm{GL}(m,F)}(W^{\alpha^+}, W^{\beta^+}) \neq 0$.

*Proof.* Let $\iota$ denote the identity map $\Delta \to \Delta$. If $\theta$ is a nonzero $\mathrm{GL}(m, F)$-homomorphism from $W^\alpha$ into $W^\beta$, then $\iota \otimes \theta$ is a nonzero $\mathrm{GL}(m, F)$-homomorphism from $W^{\alpha^+} = \Delta \otimes W^\alpha$ into $W^{\beta^+} = \Delta \otimes W^\beta$. The other implication is obtained by tensoring with $\Delta^{-1}$. ■

## 8.2 The Hyperalgebra

When dealing with tensor space $L^{(n)}$, it is often easier to consider $\bar{U}_F$-modules, for a certain $F$-algebra $\bar{U}_F$, than to work with $\mathrm{GL}(m, F)$-modules. In many cases, for example if $F$ is infinite, $\mathrm{GL}(m, F)$-submodules of $L^{(n)}$ coincide with $\bar{U}_F$-submodules.

The ideas behind the construction of $\bar{U}_F$ are inspired by a well-known set of generators for $\mathrm{GL}(m, F)$. Let $E_{ij}$ be the $m \times m$ matrix with 1 in the $(i, j)$ place and zeros elsewhere. Then it can be verified that the set of all matrices of the forms

8.2.1 $$\begin{aligned} e_{ij}(f) &:= I + fE_{ij} \qquad (f \in F), \\ h_i(f) &:= I + fE_{ii} \qquad (-1 \neq f \in F). \end{aligned}$$

generates $\mathrm{GL}(m, F)$.

Recalling that $L^{(n)}$ is the $n$-fold tensor product of the $m$-dimensional $F$-vector space $W^{(1)}$ on which $\mathrm{GL}(m, F)$ acts naturally, we define $L^{(0)} = F \cdot 1$ and $L := \bigotimes_{n \geq 0} L^{(n)}$. Then $L$ is an $F$-algebra in the obvious way; indeed, we may regard $L$ as the algebra of polynomials in the noncommuting variables $w_1, w_2, \ldots, w_m$. A *derivation* of $L$ is an $F$-linear map $D: L \to L$ such that, for all $l_1, l_2 \in L$,

$$D(l_1 \otimes l_2) = D(l_1) \otimes l_2 + l_1 \otimes D(l_2).$$

It is easy to see that there is a unique derivation of $L$ such that $Dw_j = w_i$ and $Dw_k = 0$ $(k \neq j)$. Call this derivation $D_{ij}$.

**8.2.2** LEMMA. $D_{ij} D_{kl} - D_{kl} D_{ij} = \delta_{jk} D_{il} - \delta_{li} D_{kj}$.

(Here, as usual, $\delta_{jk} = 1$ if $j = k$, and $\delta_{jk} = 0$ if $j \neq k$.)

*Proof.* If $D_1$ and $D_2$ are derivations of $L$, then one readily checks that $D_1 D_2 - D_2 D_1$ is also a derivation. Hence, $D_{ij} D_{kl} - D_{kl} D_{ij}$ and $\delta_{jk} D_{il} - \delta_{li} D_{kj}$ are both derivations. It is straightforward to verify that they coincide on $W^{(1)}$, so they must be the same derivation. ■

Let $U$ be the associative $F$-algebra given by generators and relations as follows:

$$U = \langle 1, e_{ij} \mid 1 \leqslant i \leqslant m,\ 1 \leqslant j \leqslant m,\ e_{ij} e_{kl} - e_{kl} e_{ij} = \delta_{jk} e_{il} - \delta_{li} e_{kj} \rangle.$$

$U$ is, in fact, the universal enveloping algebra of the Lie algebra of $\mathrm{GL}(m, F)$ (the relations are those satisfied by the elementary matrices $E_{ij}$), but we shall not make use of this observation.

In view of Lemma 8.2.2, we can regard $L$ as a $U$-module by defining the action of $e_{ij}$ to be that of $D_{ij}$.

**8.2.3** EXAMPLES.

(i) $e_{12}(w_1 \otimes w_2 \otimes w_2 \otimes w_3 \otimes w_2)$
$= w_1 \otimes w_1 \otimes w_2 \otimes w_3 \otimes w_2 + w_1 \otimes w_2 \otimes w_1 \otimes w_3 \otimes w_2 + w_1 \otimes w_2 \otimes w_2 \otimes w_3 \otimes w_1.$

(ii) $e_{12}^3(w_1 \otimes w_2 \otimes w_2 \otimes w_3 \otimes w_2) = 3!\, w_1 \otimes w_1 \otimes w_1 \otimes w_3 \otimes w_1.$

(iii) $e_{22}(w_1 \otimes w_2 \otimes w_2 \otimes w_3 \otimes w_2) = 3 w_1 \otimes w_2 \otimes w_2 \otimes w_3 \otimes w_2.$

(iv) $e_{21}^2(w_1 \otimes w_1 \otimes w_2 \otimes w_1 \otimes w_4)$
$= 2(w_2 \otimes w_2 \otimes w_2 \otimes w_1 \otimes w_4 + w_2 \otimes w_1 \otimes w_2 \otimes w_2 \otimes w_4 + w_1 \otimes w_2 \otimes w_2 \otimes w_2 \otimes w_4).$

To motivate an "adjustment" we shall make to the definition of $U$, it is sensible to work for a short while with the case $F = \mathbb{Q}$. Now (cf. 8.2.3),

**8.2.4.** *If $i \neq j$, the effect of applying $\dfrac{e_{ij}^a}{a!}$ to $w_{i_1} \otimes \cdots \otimes w_{i_n}$ is to change $a$ of the subscripts $j$ to $i$ in all possible ways and sum. (The effect is to multiply by zero if fewer than $a$ subscripts equal $j$.)*

Let $h_i := e_{ii}$ and

$$\binom{h_i}{c} := \frac{1}{c!} h_i(h_i - 1) \cdots (h_i - c + 1).$$

Then

**8.2.5.** *If $N_i$ of the subscripts $i_1, i_2, \ldots, i_n$ equal $i$, then*

$$\binom{h_i}{c} w_{i_1} \otimes \cdots \otimes w_{i_n} = \binom{N_i}{c} w_{i_1} \otimes \cdots \otimes w_{i_n}.$$

To draw attention to the fact that our algebras $U$ and $L$ are temporarily defined over $\mathbb{Q}$, we shall write $U_{\mathbb{Q}}$ and $L_{\mathbb{Q}}$. Our aim is to devise a means of interpreting the following theorem for arbitrary "large" fields (even those of finite characteristic).

**8.2.6 Theorem.** *Assume that $F = \mathbb{Q}$. Then*

(i) *For $i \neq j$, $e_{ij}(f)$ and* exp$(fe_{ij})$ *act in the same way on $L_{\mathbb{Q}}^{(n)}$, where*

$$\exp(fe_{ij}) := \sum_{a=0}^{n} f^a \frac{e_{ij}^a}{a!}.$$

(ii) *$h_i(f)$ and $\Sigma_{c=0}^{n} f^c \binom{h_i}{c}$ act in the same way on $L_{\mathbb{Q}}^{(n)}$.*

(iii) *$e_{ij}^a / a!$ acts on $L_{\mathbb{Q}}^{(n)}$ in the same way as a certain linear combination of terms of the form $e_{ij}(f)$.*

(iv) *$\binom{h_i}{c}$ acts on $L_{\mathbb{Q}}^{(n)}$ in the same way as a certain linear combination of terms of the form $h_i(f)$.*

*Proof.* (i): Let $D = fe_{ij}$, and $l_1, l_2 \in \oplus_{i=0}^{n} L_{\mathbb{Q}}^{(i)}$. Then

$$(\exp D)(l_1 l_2) = \sum_{a=0}^{n} \frac{D^a}{a!}(l_1 l_2) = \sum_{a=0}^{n} \frac{1}{a!} \sum_{b=0}^{n} \binom{a}{b} (D^b l_1)(D^{a-b} l_2)$$

$$= \sum_{a=0}^{n} \sum_{\substack{b,c \\ b+c=a}} \left(\frac{D^b l_1}{b!}\right)\left(\frac{D^c l_2}{c!}\right) = \left(\sum_{b=0}^{n} \frac{D^b l_1}{b!}\right)\left(\sum_{c=0}^{n} \frac{D^c l_2}{c!}\right)$$

$$= (\exp Dl_1)(\exp Dl_2).$$

But $\exp D$ acts on $L_{\mathbb{Q}}^{(1)}$ in the same way as $1 + fe_{ij}$, and this is the way that $e_{ij}(f)$ acts. Therefore,

$$e_{ij}(f)(w_{i_1} \otimes \cdots \otimes w_{i_n}) = \exp Dw_{i_1} \otimes \cdots \otimes \exp Dw_{i_n}$$
$$= (\exp D)(w_{i_1} \otimes \cdots \otimes w_{i_n}),$$

as required.

(ii): Let $N$ be the number of subscripts in $w_{i_1}\otimes\cdots\otimes w_{i_n}$ which equal $i$. Then, by 8.2.5,

$$\begin{aligned}\sum_{c=0}^{n} f^c\binom{h_i}{c}w_{i_1}\otimes\cdots\otimes w_{i_n} &= \sum_{c=0}^{n} f^c\binom{N}{c}w_{i_1}\otimes\cdots\otimes w_{i_n}\\ &=(1+f)^N w_{i_1}\otimes\cdots\otimes w_{i_n}\\ &=h_i(f)w_{i_1}\otimes\cdots\otimes w_{i_n}.\end{aligned}$$

(iii): By part (i),

$$e_{ij}(f)=1+fe_{ij}+\frac{f^2}{2!}e_{ij}^2+\cdots+\frac{f^n}{n!}e_{ij}^n$$

in its action on $L_{\mathbf{Q}}^{(n)}$. Now,

$$\begin{vmatrix} 1 & f_1 & \cdots & f_1^n\\ 1 & f_2 & \cdots & f_2^n\\ \vdots & \vdots & & \vdots\\ 1 & f_{n+1} & \cdots & f_{n+1}^n\end{vmatrix} = \prod_{1\leqslant i<j\leqslant n+1}(f_j-f_i).$$

Since $\mathbf{Q}$ is infinite, we may choose $f_1,\ldots,f_{n+1}$ such that the determinant is nonzero. Then we may invert and write $e_{ij}^a/a!$ as a linear combination of terms of the form $e_{ij}(f)$.

(iv): The proof of part (iv) is similar to that of part (iii). ■

There is a natural bilinear form $\Phi_{\mathbf{Q}}$ on $L_{\mathbf{Q}}^{(n)}$, given by

8.2.7 $$\Phi_{\mathbf{Q}}\left(w_{i_1}\otimes\cdots\otimes w_{i_n}, w_{j_1}\otimes\cdots\otimes w_{j_n}\right):=\delta_{i_1j_1}\delta_{i_2j_2}\cdots\delta_{i_nj_n}.$$

Now,

$$\begin{aligned}&\Phi_{\mathbf{Q}}\left(e_{ij}\left(w_{i_1}\otimes\cdots\otimes w_{i_n}\right), w_{j_1}\otimes\cdots\otimes w_{j_n}\right)\\ &\qquad=\sum_{k=1}^{n}\left(\left(\delta_{ij_k}\delta_{i_kj}\right)\left(\delta_{i_1j_1}\delta_{i_2j_2}\cdots\hat{\delta}_{i_kj_k}\cdots\delta_{i_nj_n}\right)\right)\\ &\qquad=\Phi_{\mathbf{Q}}\left(w_{i_1}\otimes\cdots\otimes w_{i_n}, e_{ji}\left(w_{j_1}\otimes\cdots\otimes w_{j_n}\right)\right),\end{aligned}$$

and hence

**8.2.8.** *For all* $l_1, l_2\in L_{\mathbf{Q}}^{(n)}$, $\Phi_{\mathbf{Q}}(e_{ij}l_1, l_2)=\Phi_{\mathbf{Q}}(l_1, e_{ji}l_2)$.

At this stage, we transfer our attention to an *arbitrary field F*. Let $L_{\mathbf{Z}}^{(n)}$ be the set of $\mathbf{Z}$-linear combinations of tensors $w_{i_1}\otimes\cdots\otimes w_{i_n}$, and define $L_{\mathbf{Z}}^{\alpha}:=$

$L_{\mathbb{Q}}^{\alpha} \cap L_{\mathbb{Z}}^{(n)}$. Then $\Phi_{\mathbb{Q}}$ induces a bilinear form $\Phi_{\mathbb{Z}}$ on $L_{\mathbb{Z}}^{\alpha}$ whose image lies in $(\nu_1!\nu_2!\cdots)\mathbb{Z}$, where $\nu$ is the partition associated with $\alpha$. Define a bilinear form $\Phi^{\alpha}$ on $L_{\mathbb{Z}}^{\alpha} \otimes_{\mathbb{Z}} F$ by

**8.2.9**
$$\Phi^{\alpha}(l_1 \otimes 1, l_2 \otimes 1) := \frac{\Phi_{\mathbb{Z}}(l_1, l_2)}{\nu_1!\nu_2!\cdots} \otimes 1 \qquad (l_1, l_2 \in L_{\mathbb{Z}}^{\alpha})$$

We quote the following results without proof; they both follow from a careful application of the relations on the generators of $U$:

**8.2.10.**

(i) *The following is a basis for* $U_{\mathbb{Q}}$:

$$\left\{ \prod_{i>j} \frac{e_{ij}^{b_{ij}}}{b_{ij}!} \prod_i \binom{h_i}{c_i} \prod_{i<j} \frac{e_{ij}^{a_{ij}}}{a_{ij}!} \;\middle|\; \begin{array}{l} b_{ij}, c_i, a_{ij} \in \mathbb{Z}^+, \textit{and} \\ \textit{the factors in the first and third} \\ \textit{product are ordered lexicographically} \end{array} \right\}.$$

(ii) *Let* $U_{\mathbb{Z}} :=$ *the set of* $\mathbb{Z}$*-linear combinations of the basis elements described in part* (i). *Then* $U_{\mathbb{Z}}$ *is a subring of* $U_{\mathbb{Q}}$.

This is known as the *Kostant* $\mathbb{Z}$*-form* of $U$. By 8.2.4 and 8.2.5, $L_{\mathbb{Z}}^{(n)}$ is a $U_{\mathbb{Z}}$-module.

For our arbitrary field $F$, we write

$$\bar{U}_F = U_{\mathbb{Z}} \otimes_{\mathbb{Z}} F \quad \text{and} \quad \bar{L}^{(n)} = L_{\mathbb{Z}}^{(n)} \otimes_{\mathbb{Z}} F.$$

Remembering that $W^{(1)} = L^{(1)}$, we have, in particular, $\bar{W}^{(1)} = W_{\mathbb{Z}}^{(1)} \otimes_{\mathbb{Z}} F$. Writing $\bar{w}_i = w_i \otimes 1$, we shall identify $\bar{w}_{i_1} \otimes \cdots \otimes \bar{w}_{i_n}$ with $w_{i_1} \otimes \cdots \otimes w_{i_n} \otimes 1$. Then, as far as tensor space is concerned, we are (to within isomorphism) back where we started; that is, we are considering the tensor space of $W^{(1)}$ over an arbitrary field. We shall therefore omit the bars from the names of all subspaces of tensor space.

Let us examine what we have achieved by this maneuvre. We have Weyl modules $W^{\alpha}$ and $\mathrm{GL}(m, F)$-modules $L^{\alpha}$ as before, defined over the field $F$. All that has happened is that we now have an $F$-algebra $\bar{U}_F$, generated by elements called

$$1, \qquad \frac{e_{ij}^{a}}{a!} \quad (i \neq j), \quad \text{and} \quad \binom{h_i}{c},$$

and $\bar{U}_F$ acts on $L^{(n)}$ in the way described in 8.2.4 and 8.2.5 (where $\binom{N_i}{c}$ in 8.2.5 is now regarded as the field element $1+1+\cdots+1$, $\binom{N_i}{c}$ times). $\bar{U}_F$ is called the *hyperalgebra*.

For each partition $\alpha$ we have a symmetric bilinear form $\Phi^\alpha: L^\alpha \times L^\alpha \to F$, satisfying

$$\Phi^\alpha\left(\frac{e_{ij}^a}{a!}l_1, l_2\right)=\Phi^\alpha\left(l_1, \frac{e_{ji}^a}{a!}l_2\right)$$

and

$$\Phi^\alpha\left(\binom{h_i}{c}l_1, l_2\right)=\Phi^\alpha\left(l_1, \binom{h_i}{c}l_2\right).$$

Thus, if $u \mapsto \bar{u}$ is the anti-automorphism of $\overline{U}_F$ sending

$$\frac{e_{ij}^a}{a!} \text{ to } \frac{e_{ji}^a}{a!} \quad \text{and} \quad \binom{h_i}{c} \text{ to } \binom{h_i}{c},$$

then

**8.2.11** $\qquad \Phi^\alpha(ul_1, l_2)=\Phi^\alpha(l_1, \bar{u}l_2) \qquad (l_1, l_2 \in L^\alpha, \quad u \in \overline{U}_F).$

**8.2.12** THEOREM.

(i) *If* $g \in \mathrm{GL}(m, F)$, *then there exists* $u \in \overline{U}_F$ *such that* $g$ *and* $u$ *have the same action on* $L^{(n)}$.

(ii) *If* $|F| \geqslant n+2$, *and* $u \in \overline{U}_F$, *then there exists* $g \in F\mathrm{GL}(m, F)$, *the group algebra of* $\mathrm{GL}(m, F)$ *over* $F$, *such that* $u$ *and* $g$ *have the same action on* $L^{(n)}$.

*Proof.* (i): The proof of Theorem 8.2.6 shows that $e_{ij}(f) \in \mathrm{GL}(m, F)$ and $1+fe_{ij}+f^2e_{ij}^2/2!+\cdots+f^ne_{ij}^n/n! \in \overline{U}_F$ have the same action in $L^{(n)}$, and that $h_i(f) \in \mathrm{GL}(m, F)$ and

$$1+f\binom{h_i}{1}+f^2\binom{h_i}{2}+\cdots+f^n\binom{h_i}{n} \in \overline{U}_F$$

have the same action on $L^{(n)}$. But $\mathrm{GL}(m, F)$ is generated by matrices of the form $e_{ij}(f)$ and $h_i(f)$, so part (i) is true.

(ii): We can choose $f_1, \ldots, f_{n+1} \in F$ so that the determinant in the proof of part (iii) of Theorem 8.2.6 is nonzero, since $|F| \geqslant n+1$. Therefore, we can find an element of $F\mathrm{GL}(m, F)$ which acts on $L^{(n)}$ in the same way as $e_{ij}^a/a!$. Similarly, we can find an element of $F\mathrm{GL}(m, F)$ which acts on $L^{(n)}$ in the same way as $\binom{h_i}{c}$ (here we need $|F| \geqslant n+2$, since we are not allowed to take $f_j=-1$). Since $\overline{U}_F$ is generated by elements of the form

$$1, \qquad \frac{e_{ij}^a}{a!} \quad \text{and} \quad \binom{h_i}{c},$$

part (ii) of the theorem follows. ∎

As immediate corollaries, we have

**8.2.13** THEOREM.

(i) *Every* $\overline{U}_F$*-submodule of* $L^{(n)}$ *is a* $\mathrm{GL}(m, F)$*-submodule.*
(ii) *If* $|F|\geqslant n+2$, *every* $\mathrm{GL}(m, F)$*-submodule of* $L^{(n)}$ *is a* $\overline{U}_F$*-submodule.*

**8.2.14** THEOREM.

(i) *If* $W$ *is finitely generated* $\overline{U}_F$*-submodule of* $L^{(n)}$, *then every* $\overline{U}_F$*-homomorphism from* $W$ *into* $L$ *is a* $\mathrm{GL}(m, F)$*-homomorphism.*
(ii) *If* $|F|\geqslant \max(k, n)+2$ *and* $W$ *is a* $\mathrm{GL}(m, F)$*-submodule of* $L^{(n)}$, *then every* $\mathrm{GL}(m, F)$*-homomorphism from* $W$ *into* $L^{(k)}$ *is a* $\overline{U}_F$*-homomorphism.*

To illustrate how things can go wrong when $F$ is small, we give:

**8.2.15** EXAMPLE. Let $m=2$, and write $w_iw_jw_k$ for $w_i\otimes w_j\otimes w_k$. Then

$$\{w_1w_1w_1, w_2w_2w_2, w_1w_1w_2+w_1w_2w_1+w_2w_1w_1, w_1w_2w_2+w_2w_1w_2+w_2w_2w_1\}$$

is a basis for $W^{(3)}$.

Let $x=w_1w_1w_2+w_1w_2w_1+w_2w_1w_1$ and $y=w_1w_2w_2+w_2w_1w_2+w_2w_2w_1$. Then

$$\begin{pmatrix} 1 & f \\ 0 & 1 \end{pmatrix} \in \mathrm{GL}(2, F)$$

sends $x$ to

$$w_1w_1(fw_1+w_2)+w_1(fw_1+w_2)w_1+(fw_1+w_2)w_1w_1.$$

Therefore

$$\begin{aligned}
\begin{pmatrix} 1 & f \\ 0 & 1 \end{pmatrix} &: x\mapsto x+3f\,w_1w_1w_1, \quad \text{and similarly} \\
&\quad y\mapsto y+2fx+3f^2w_1w_1w_1, \\
\begin{pmatrix} 1 & 0 \\ f & 1 \end{pmatrix} &: x\mapsto x+2fy+3f^2w_2w_2w_2, \\
&\quad y\mapsto y+3f\,w_2w_2w_2, \\
e_{12} &: x\mapsto 3w_1w_1w_1, \\
&\quad y\mapsto 2x, \\
e_{21} &: x\mapsto 2y, \\
&\quad y\mapsto 3w_2w_2w_2.
\end{aligned}$$

Therefore, the space spanned by $x+y$ is a $\mathrm{GL}(2, F)$-module but not a

$\overline{U}_F$-module when $|F|=2$. In fact, when $|F|=2$, the space spanned by $x+y$ is the identity GL($m$, $F$)-module, so there is a GL($m$, $F$)-isomorphism from $W^{(0)}$ into $L^{(3)}$; this shows that we cannot let $|F|$ be too small in 8.2.14(ii).

## 8.3 Irreducible GL($m, F$)-modules over $F$

The title of this section is, perhaps, a misnomer, since we shall, in fact, construct irreducible $\overline{U}_F$-modules. However, we now know that $\overline{U}_F$-submodules and GL($m$, $F$)-submodules of $L^{(n)}$ usually coincide, and the irreducible $\overline{U}_F$-modules we construct will be irreducible GL($m$, $F$)-modules, if, for example, $F$ is infinite.

It is clear from 8.2.4 and 8.2.5 that $\overline{U}_F$ and $S_n$ commute in their action on $L^{(n)}$. The Weyl module, $W^\alpha$, is therefore a $\overline{U}_F$-module, lying in $L^\alpha$. Now, let us define

$$W^{\alpha\perp}=\{l\,|\,l\in L^\alpha \text{ and } \Phi^\alpha(l,w)=0 \text{ for all } w\in W^\alpha\}$$

In view of 8.2.11, $W^{\alpha\perp}$ is a $\overline{U}_F$-module. We shall prove that $W^\alpha/(W^\alpha\cap W^{\alpha\perp})$ is an irreducible $\overline{U}_F$-module; $W^\alpha/(W^\alpha\cap W^{\alpha\perp})$ is therefore a GL($m$, $F$)-module, which is irreducible when $|F|\geqslant n+2$.

**8.3.1** THEOREM. *$\mathcal{V}^\alpha w^\alpha$ generates $W^\alpha$ as a $\overline{U}_F$-module, and generates $S^\alpha$ as an $FS_n$-module.*

*Proof.* Recall from 8.1.8 that $w^\alpha=w_T$, where $T$ is the $\alpha$-tableau in which, for each $i$, all the entries in the $i$th row equal $i$. Now,

$$\prod_{i<j}\frac{e_{ij}^{a_{ij}}}{a_{ij}!}w^\alpha=\sum\{w_{T'}\,|\,T' \text{ is obtained from } T \text{ by changing } a_{ij}\ j\text{'s to } i\text{'s}\},$$

provided that we apply $e_{ij}^{a_{ij}}/a_{ij}!$ before $e_{xy}^{a_{xy}}/a_{xy}!$ when $i>x$ or $i=x$ and $j>y$. Therefore, if $T_1$ is a semistandard tableau whose $j$th row contains precisely $a_{ij}$ $i$'s, then

$$\prod_{i<j}\frac{e_{ij}^{a_{ij}}}{a_{ij}!}w^\alpha=\sum\{w_{T'}\,|\,T' \text{ is row-equivalent to } T_1\}.$$

For example, if

$$T=\begin{matrix}1&1&1&1\\2&2&2&2\\3&3&&\end{matrix},$$

then

$$\frac{e_{21}^2}{2!}\frac{e_{32}^2}{2!}e_{42}e_{43}w^\alpha=\sum\left\{w_{T'}\,|\,T' \text{ is row-equivalent to } \begin{matrix}1&1&2&2\\2&3&3&4\\3&4&&\end{matrix}\right\}.$$

Applying $\mathcal{V}^\alpha$, we have

$$\prod_{i<j} \frac{e_{ij}^{a_{ij}}}{a_{ij}!} \mathcal{V}^\alpha w^\alpha = \mathcal{V}^\alpha \sum \{w_{T'} | T' \text{ is row-equivalent to } T_1\}$$
$$= E_{T_1}.$$

Since $\{E_{T_1} | T_1 \text{ is semistandard}\}$ is a basis for $W^\alpha$, this proves that $\mathcal{V}^\alpha w^\alpha$ generates $W^\alpha$ as a $\overline{U}_F$-module.

We already know (7.1.3) that $S^\alpha$ is generated by $\mathcal{V}^\alpha w^\alpha$ as an $FS_n$-module. ■

Compare the next theorem with 7.1.7:

**8.3.2** THEOREM. *If $A$ is a $\overline{U}_F$-submodule of $L^\alpha$, then either $A \supseteq W^\alpha$ or $A \subseteq W^{\alpha\perp}$.*

*Proof.* Let $\alpha$ have $k$ nonzero parts. We may assume $k \leqslant m$; if $k > m$, $L^\alpha = W^\alpha = 0$.

Suppose first that $k < m$. Then

$$X := \frac{e_{k,k+1}^{\alpha_k}}{\alpha_k!} \cdots \frac{e_{23}^{\alpha_2}}{\alpha_2!} \frac{e_{12}^{\alpha_1}}{\alpha_1!}$$

sends $w_{i_1} \otimes \cdots \otimes w_{i_n}$ to zero or to $w_{j_1} \otimes \cdots \otimes w_{j_n}$, where for each $i$, $\alpha_i$ of the subscripts equal $i$. Therefore, $XL^\alpha$ is the 1-dimensional space spanned by $\mathcal{V}^\alpha w^\alpha$ (cf. 7.1.5(ii)). Thus, if $Xa \neq 0$ for some $a \in A$, then $\mathcal{V}^\alpha w^\alpha \in A$ and $A \supseteq W^\alpha$ by 8.3.1.

On the other hand, if $Xa = 0$ for every $a \in A$, then

$$0 = \Phi^\alpha(Xa, \mathcal{V}^\alpha w^\alpha) = \Phi^\alpha\left(a, \mathcal{V}^\alpha \frac{e_{21}^{\alpha_1}}{\alpha_1!} \cdots \frac{e_{k+1,k}^{\alpha_k}}{\alpha_k!} w^\alpha\right) \qquad \text{for all} \quad a \in A.$$

But

$$\mathcal{V}^\alpha \frac{e_{21}^{\alpha_1}}{\alpha_1!} \cdots \frac{e_{k+1,k}^{\alpha_k}}{\alpha_k!} w^\alpha = \mathcal{V}^\alpha w_{T_1},$$

where $T_1$ is the $\alpha$-tableau having $\alpha_1$ 2's in row 1, $\alpha_2$ 3's in row 2, and so on. Since $\mathcal{V}^\alpha w_{T_1}$ generates $W^\alpha$ as a $\overline{U}_F$-module (the proof of this being similar to the proof of 8.3.1), every $w \in W^\alpha$ can be written in the form $u\mathcal{V}^\alpha w_{T_1}$ for some $u \in \overline{U}_F$. Using 8.2.11,

$$\Phi^\alpha(a, w) = \Phi^\alpha(a, u\mathcal{V}^\alpha w_{T_1}) = \Phi^\alpha(\bar{u}a, \mathcal{V}^\alpha w_{T_1}) = 0,$$

so $A \subseteq W^{\alpha\perp}$.

Now suppose that $\alpha$ has $m$ nonzero parts. Then

$$L^\alpha = \Delta^{\alpha_m} \otimes L^{(\alpha_1 - \alpha_m, \ldots, \alpha_{m-1} - \alpha_m)},$$

where $\Delta$ is the determinant module. If there were a $\bar{U}_F$-submodule $A$ of $L^\alpha$ neither containing $W^\alpha$, nor contained in $W^{\alpha\perp}$, then $\Delta^{-\alpha_m}\otimes A$ would be a $\bar{U}_F$-submodule of $L^{(\alpha_1-\alpha_m,\ldots,\alpha_{m-1}-\alpha_m)}$ neither containing $W^{(\alpha_1-\alpha_m,\ldots,\alpha_{m-1}-\alpha_m)}$ nor contained in $W^{(\alpha_1-\alpha_m,\ldots,\alpha_{m-1}-\alpha_m)\perp}$, contradicting the part of the theorem we have already proved. ■

**8.3.3** COROLLARY. *$W^\alpha/(W^\alpha\cap W^{\alpha\perp})$ is an irreducible $\bar{U}_F$-module (or zero, if $\alpha$ has more than $m$ nonzero parts).*

*Proof.* $\Phi^\alpha(\mathcal{V}^\alpha w^\alpha,\mathcal{V}^\alpha w^\alpha)=1$, by construction, so $W^\alpha\cap W^{\alpha\perp}\neq W^\alpha$. The corollary is now immediate. ■

**8.3.4** COROLLARY. *If* char $F=0$, *then $W^\alpha$ is an irreducible $\bar{U}_F$-module, and an irreducible* GL$(m,F)$*-module (except that $W^\alpha=0$ if $\alpha$ has more than $m$ nonzero parts).*

*Proof.* Since $\bar{U}_F$-submodules and GL$(m,F)$-submodules of $L^{(n)}$ coincide when $|F|\geqslant n+2$, it is sufficient to prove that $W^\alpha\cap W^{\alpha\perp}=0$. When $F=\mathbb{Q}$, $\Phi^\alpha$ is an inner product, so the determinant of the Gram matrix with respect to the semistandard basis of $W^\alpha$ is nonzero. Since this determinant is an integer, $W^\alpha\cap W^{\alpha\perp}=0$ over all fields of characteristic 0. ■

Now let $F=\mathbb{C}$. Since $\mathcal{H}^\alpha w^\alpha=(\alpha_1!\alpha_2!\cdots)w^\alpha$ and $\mathcal{V}^\alpha w^\alpha$ generates $W^\alpha$, we have $W^\alpha\subseteq\mathcal{V}^\alpha\mathcal{H}^\alpha L^{(n)}$. But the latter is the irreducible GL$(m,\mathbb{C})$-module affording $\langle\alpha\rangle$ which we mentioned in the proof of 5.2.26, so $W^\alpha=\mathcal{V}^\alpha\mathcal{H}^\alpha L^{(n)}$. This implies that (cf. Exercise 8.3)

**8.3.5** $$W^\alpha=\mathcal{V}^\alpha\mathcal{H}^\alpha L^{(n)}\quad\textit{whenever}\quad \operatorname{char}F=0.$$

Thus, if $D$ is an ordinary representation of a group $G$, and we take $W^{(1)}$ to be a $G$-module affording $D$, then

**8.3.6.** *$W^\alpha$ is a $G$-module affording the symmetrized representation $D\boxdot[\alpha]$.*

Returning to the case where $F$ is arbitrary, note that

**8.3.7** THEOREM. *The dimension of $W^\alpha/(W^\alpha\cap W^{\alpha\perp})$ equals the rank of the Gram matrix with respect to the semistandard basis of $W^\alpha$.*

8.3.3, combined with the technique used in the proof of 7.1.8, now gives:

**8.3.8** THEOREM. *Suppose that $\alpha\vdash n$ has at most $m$ nonzero parts (so that $W^\alpha\neq 0$). $W^\alpha/(W^\alpha\cap W^{\alpha\perp})$ is an absolutely irreducible $\bar{U}_F$-module. If $|F|\geqslant n+2$, then $W^\alpha/(W^\alpha\cap W^{\alpha\perp})$ is an absolutely irreducible* GL$(m,F)$*-module.*

In Exercise 8.4, we outline how to construct all the irreducible $F\mathrm{GL}(m, F)$ modules when $F$ is finite.

**8.3.9** COROLLARY. *$W^{(1^n)}$ is an absolutely irreducible $\overline{U}_F$-module (if $n \leqslant m$).*

*Proof.* The Gram matrix with respect to the semistandard basis of $W^{(1^n)}$ is the identity matrix. ∎

$W^\alpha/(W^\alpha \cap W^{\alpha\perp})$ need not be an irreducible GL($m$, $F$)-module when $F$ is small.

**8.3.10** EXAMPLE. Let $m=2$. The semistandard basis of $W^{(3)}$ is $\{w_1w_1w_1, w_2w_2w_2, x, y\}$, where

$$x = w_1w_1w_2 + w_1w_2w_1 + w_2w_1w_1 \quad \text{and} \quad y = w_1w_2w_2 + w_2w_1w_2 + w_2w_2w_1$$

as in Example 8.2.15. The Gram matrix for this basis is

$$\begin{bmatrix} 1 & & & \\ & 1 & & \\ & & 3 & \\ & & & 3 \end{bmatrix}.$$

When char $F=2$, $W^{(3)}$ is therefore an irreducible $\overline{U}_F$-module, but we have seen that it is not an irreducible GL(2, $F$)-module when $|F|=2$.

The Gram matrix proves that the dimension of $W^{(3)}/(W^{(3)} \cap W^{(3)\perp})$ is 2 if char $F=3$ and $m=2$. We recommend that the reader verifies directly that $W^{(3)} \cap W^{(3)\perp}$, the space spanned by $x$ and $y$, is $\overline{U}_F$-invariant in this case.

We now prove that no two of the irreducible $\overline{U}_F$-modules we have found are equivalent. To do this, the elements $\binom{h_i}{c}$ of $\overline{U}_F$ are brought into play.

An elementary property of binomial coefficients is

**8.3.11**
$$\binom{x}{c} = \sum_{0 \leqslant b \leqslant c} \binom{a}{c-b}\binom{x-a}{b} \qquad (x, a, c \in \mathbb{N}_0).$$

This works even for $a > x$ if we let

$$\binom{x-a}{b} := \frac{(x-a)(x-a-1)\cdots(x-a+1-b)}{b!}$$

Now, for $a \geqslant 0$, define

$$\binom{h_i - a}{c} \in \overline{U}_F$$

as follows. Let

$$\binom{h_i-a}{0}:=1.$$

Assuming inductively that $\binom{h_i-a}{b}$ has been defined for $0\leqslant b\leqslant c-1$, let

$$\binom{h_i-a}{c}:=\binom{h_i}{c}-\sum_{0\leqslant b\leqslant c-1}\binom{a}{c-b}\binom{h_i-a}{b}.$$

Here, of course,

$$\binom{a}{c-b} \quad \text{means} \quad 1+1+\cdots+1\left(\binom{a}{c-b}\text{ times}\right)\in F.$$

8.2.5 and 8.3.11 give:

**8.3.12.** *If $N_i$ of the subscripts $i_1, i_2, \ldots, i_n$ equal $i$, then*

$$\binom{h_i-a}{c}w_{i_1}\otimes\cdots\otimes w_{i_n}=\binom{N_i-a}{c}w_{i_1}\otimes\cdots\otimes w_{i_n}.$$

**8.3.13 Lemma.**

$$\binom{h_i-a}{c}w_{i_1}\otimes\cdots\otimes w_{i_n}=0 \textit{ for all } c\geqslant 1$$

*if and only if precisely $a$ of the subscripts $i_1, i_2, \ldots, i_n$ are equal to $i$.*

*Proof.* Let $N_i$ of the subscripts equal $i$. Then

$$\binom{h_i-a}{c}w_{i_1}\otimes\cdots\otimes w_{i_n}=\binom{N_i-a}{c}w_{i_1}\otimes\cdots\otimes w_{i_n}.$$

If $N_i=a$, then

$$\binom{N_i-a}{c}=0 \qquad \text{for all} \quad c\geqslant 1.$$

If $N_i>a$, then

$$\binom{N_i-a}{c}=1 \qquad \text{for} \quad c=N_i-a.$$

If $N_i < a$, then

$$\binom{N_i - a}{c} = (-1)^c \binom{a - N_i + c - 1}{c}.$$

When $c=1$, this equals $N_i - a$, which is nonzero if char $F=0$. On the other hand if char $F=p$ and $p$ divides $\binom{a-N_i+c-1}{c}$ for every $c$ with $1 \leqslant c \leqslant a - N_i$, then $p$ divides $\binom{a-N_i}{a-N_i}$, a contradiction. (Consider Pascal's triangle.) ■

**8.3.14** Lemma. *Let $\alpha \vdash n'$, $\beta \vdash n$. Suppose that $A$ is a $\overline{U}_F$-submodule of $L^\beta$ and that there exists a nonzero $\overline{U}_F$-homomorphism $\theta$ from $W^\alpha$ into $L^\beta/A$. Then $n=n'$ and $\beta \trianglerighteq \alpha$. Furthermore, if $\alpha=\beta$, then $\operatorname{Im}\theta = (W^\alpha + A)/A$.*

*Proof.* Choose a basis $l_1, l_2, \ldots$ of $L^\beta$ in such a way that each $l_j$ has the form $\mathcal{V}^\beta w_{T_j}$ for some $\beta$-tableau $T_j$. Let

$$\theta(\mathcal{V}^\alpha w^\alpha) = b_1 l_1 + \cdots + b_r l_r + A \qquad (b_j \in F, b_r \neq 0).$$

We may assume that the coset representative $b_1 l_1 + \cdots + b_r l_r$ is such that no element of $A$ has the form $x_1 l_1 + \cdots + x_r l_r$ with $x_j \in F$, $x_r \neq 0$.

If the content of $T_r$ is not $\alpha$, then there exists an $i$ such that the tensors $w_{i_1} \otimes \cdots \otimes w_{i_n}$ involved in $l_r$ do not have precisely $\alpha_i$ subscripts equal to $i$. By Lemma 8.3.13, for some $c \geqslant 1$, the element

$$u := \binom{h_i - \alpha_i}{c} \in \overline{U}_F$$

has the property that $ul_r = y_r l_r$ with $0 \neq y_r \in F$. But 8.3.12 shows that $ul_j = y_j l_j$ ($y_j \in F$) for each $j$, and by Lemma 8.3.13 again, $u\theta(\mathcal{V}^\alpha w^\alpha) = 0$. Therefore, when we apply $u$ to our expression for $\theta(\mathcal{V}^\alpha w^\alpha)$, we deduce that $b_1 y_1 l_1 + \cdots + b_r y_r l_r \in A$, in contradiction to our choice of coset representative.

Thus, the content of $T_r$ is $\alpha$. But unless $n=n'$ and $\beta \trianglerighteq \alpha$, there is no $\beta$-tableau $T_r$ of content $\alpha$ such that $\mathcal{V}^\beta w_{T_r} \neq 0$ (see 8.1.10). This proves the first part of the lemma. When $\alpha=\beta$, the only nonzero vectors of the form $\mathcal{V}^\alpha w_T$ where $T$ is a $\beta$-tableau of content $\alpha$ are $\pm \mathcal{V}^\alpha w^\alpha$. Therefore, in this case, $r=1$ and $\theta(\mathcal{V}^\alpha w^\alpha) = b_1 \mathcal{V}^\alpha w^\alpha + A$, and from this the last statement of the lemma follows. ■

If $\alpha$ has at most $m$ nonzero parts, define

$$F^\alpha := W^\alpha / (W^\alpha \cap W^{\alpha\perp}).$$

This is both a $\overline{U}_F$-module and a $\mathrm{GL}(m, F)$-module; it is an irreducible $\overline{U}_F$-module.

**8.3.15** THEOREM. *If $F^\alpha$ is a composition factor of $L^\beta$ as a $\overline{U}_F$-module, then $\alpha$ and $\beta$ are partitions of the same integer and $\beta \trianglerighteq \alpha$. Furthermore, the multiplicity of $F^\alpha$ as a composition factor of $L^\alpha$ is one.*

*Proof.* Since $F^\alpha$ is a homomorphic image of $W^\alpha$, $F^\alpha$ is a composition factor of $L^\beta$ if and only if there exists a nonzero $\overline{U}_F$-homomorphism from $W^\alpha$ into a quotient of $L^\beta$. Now apply the last lemma. ∎

**8.3.16** COROLLARY.

(i) *If $F^\alpha$ and $F^\beta$ are isomorphic as $\overline{U}_F$-modules, then $\alpha = \beta$.*

(ii) *If $F$ is infinite, and $F^\alpha$ and $F^\beta$ are isomorphic as* $\mathrm{GL}(m, F)$*-modules, then $\alpha = \beta$.*

In particular,

**8.3.17** COROLLARY. *If* $\operatorname{char} F = 0$, *and* $W^\alpha \simeq W^\beta$ *(and not both are zero), then $\alpha = \beta$.*

When $F$ is algebraically closed of characteristic zero, it can be proved that the various nonzero Weyl modules give all the irreducible integral representations of $\mathrm{GL}(m, F)$; to obtain all the irreducible rational representations, simply tensor Weyl modules with $\Delta^{-d}$, a negative integral power of the determinant module.

The reader interested in constructing irreducible modules for $\mathrm{SL}(m, F)$ over $F$ (where $F$ is infinite) should note that an almost identical proof to that given shows that $F^\alpha$ is an irreducible $\mathrm{SL}(m, F)$-module. Our proof of inequivalence, though, used the elements $\binom{h_i - a}{c}$ of $\overline{U}_F$, which are not available in the special-linear-group case. However, the reader should have no difficulty (except when $\alpha \vdash n'$, $\beta \vdash n$, and $n' < n$) in proving Lemma 8.3.14 on the assumption that $\alpha$ has strictly less than $m$ nonzero parts, using the technique of 8.3.2. (The fact that $W^{(\alpha_1+1, \ldots, \alpha_m+1)} = \Delta \otimes W^{(\alpha_1, \ldots, \alpha_m)}$ shows that we can ignore partitions into $m$ nonzero parts when dealing with $\mathrm{SL}(m, F)$).

Now, it is possible to prove that every $\overline{U}_F$ composition factor of $L^\beta$ is isomorphic to some $F^\alpha$. (See Theorem (3.5a) in Green [1980].) Theorem 8.3.15 therefore gives (cf. 7.1.16):

**8.3.18** THEOREM. *The matrix recording the $\bar{U}_F$ composition factors of Weyl modules for partitions of n has the form*

$$W^\alpha \left\{ \overbrace{\begin{bmatrix} 1 & & & & \\ & 1 & & & \\ & & 1 & & 0 \\ & * & & \ddots & \\ & & & & 1 \end{bmatrix}}^{F^\alpha} \right.$$

*if the partitions are written down in reverse lexicographic order.*

**8.3.19** EXAMPLE. If char $F=3$, the $\bar{U}_F$ composition factors of Weyl modules for partitions of 3 are given by

$$\begin{array}{c} \\ W^{(1^3)} \\ W^{(2,1)} \\ W^{(3)} \end{array} \begin{array}{c} \begin{array}{ccc} F^{(1^3)} & F^{(2,1)} & F^{(3)} \end{array} \\ \begin{bmatrix} 1 & & \\ 1 & 1 & \\ & 1 & 1 \end{bmatrix} \end{array}.$$

It is known [James, 1980] that the matrix appearing in Theorem 8.3.18 contains the decomposition matrix of $S_n$ as a submatrix. To be explicit, let $\bar{S}^\beta$ be the dual of $S^\beta$, and when the partition $\beta'$ associated with $\beta$ is $p$-regular, let $\bar{D}^\beta$ be the quotient of $\bar{S}^\beta$ by its unique maximal $FS_n$-submodule (see Exercise 7.1). Then when $\beta'$ is $p$-regular, the multiplicity of $\bar{D}^\beta$ as a composition factor of $\bar{S}^\alpha$ equals the multiplicity of $F^\beta$ as a composition factor of $W^\alpha$. Compare the matrix in Example 8.3.19 with the decomposition matrix of $S_3$ for $p=3$, written in terms of $\bar{D}^\alpha$ and $\bar{S}^\alpha$:

$$\begin{array}{c} \\ \bar{S}^{(1^3)} \\ \bar{S}^{(2,1)} \\ \bar{S}^{(3)} \end{array} \begin{array}{c} \begin{array}{cc} \bar{D}^{(1^3)} & \bar{D}^{(2,1)} \end{array} \\ \begin{bmatrix} 1 & \\ 1 & 1 \\ & 1 \end{bmatrix} \end{array}.$$

## 8.4 Further Connections between Specht and Weyl Modules

We have already proved many results for Weyl modules which are analogous to those for Specht modules, and in this section we push the connection between these modules further, by examining the properties of the copy, $FS_n\mathcal{T}^\alpha w^\alpha$, of the Specht module $S^\alpha$ which lies in $L^{(n)}$.

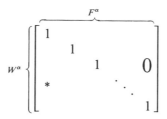

Recall that $M^\alpha$ is the space spanned by those $w_{i_1}\otimes\cdots\otimes w_{i_n}$ where, for each $i$, precisely $\alpha_i$ of the subscripts are $i$. By 8.3.13,

$$M^\alpha=\left\{w\,|\,w\in L^{(n)} \text{ and } \binom{h_i-\alpha_i}{c}w=0 \text{ for all } i \text{ and } c\geq 1\right\}.$$

We intend to prove that

$$S^\alpha=M^\alpha\cap\left\{w\,|\,w\in L^{(n)} \text{ and } \frac{e_{i,i+1}^a}{a!}w=0 \quad \text{for all } i \text{ and all } a \text{ with } 0<a\leq\alpha_{i+1}\right\}.$$

Now, we already know that $S^\alpha\subseteq M^\alpha$, and since $e_{i,i+1}^a/a!$ changes $a$ subscripts $i+1$ to $i$, 8.1.8 and 8.1.9 show that for $0<a\leq\alpha_{i+1}$, $e_{i,i+1}^a/a!$ annihilates $\mathcal{V}^\alpha w^\alpha$, and hence annihilates $S^\alpha$. To prove the other inclusion, we require

**8.4.1** Lemma. *Let $\alpha=(\alpha_1,\ldots,\alpha_k)$. Suppose $\alpha_k\neq 0$, $x<k$, $0\leq s\leq y<\alpha_{x+1}$, and $w_{i_1}\otimes\cdots\otimes w_{i_n}\in M^\alpha$. Choose $u_1,u_2,\ldots,u_y$ such that $i_{u_1}=i_{u_2}=\cdots=i_{u_s}=x+1$ and $i_{u_{s+1}}=\cdots=i_{u_y}=x$. Then the sum of all tensors $w_{j_1}\otimes\cdots\otimes w_{j_n}$ in $M^\alpha$ such that*

*(i) $j_k=i_k$ for $i_k\neq x,x+1$ and*
*(ii) $j_{u_r}=i_{u_r}$ for $1\leq r\leq y$*

*is a linear combination of vectors of the form $(e_{x+1,x}^a/a!)\,w$, with $0<a\leq\alpha_{x+1}$ and $w\in L^{(n)}$.*

*Proof.* Change to $x$ all the subscripts in $w_{i_1}\otimes\cdots\otimes w_{i_n}$ which are equal to $x+1$, except $i_{u_1},i_{u_2},\ldots,i_{u_s}$. Acting on the resulting tensor with

$$\frac{e_{x+1,x}^{(\alpha_{x+1}-s)}}{(\alpha_{x+1}-s)!},$$

we obtain the sum of all $w_{j_1}\otimes\cdots\otimes w_{j_n}$ in $M^\alpha$ such that

(i) $j_k=i_k$ for $i_k\neq x,x+1$,
(ii) $j_u=x+1$ for $u\in\{u_1,\ldots,u_s\}$.

This gives the required result when $s=y$.

If $s<y$, we may assume, by induction, that for every nonempty subset $B$ of $\{u_{s+1},\ldots,u_y\}$, the sum of all $w_{j_1}\otimes\cdots\otimes w_{j_n}$ in $M^\alpha$ such that

(i) $j_k=i_k$ for $i_k\neq x,x+1$
(ii) $j_u=x+1$ for $u\in\{u_1,\ldots,u_s\}\cup B$
(iii) $j_u=x$ for $u\in\{u_{s+1},\ldots,u_y\}\setminus B$

is a linear combination of vectors of the right form. Sum these vectors over all nonempty subsets $B$ of $\{u_{s+1},\ldots,u_y\}$ and subtract the sum from the vector obtained in the first paragraph to obtain the desired vector. ■

(The reader who has managed to thread his way through this horrible lemma is challenged to find and explain the connection with the Garnir relations.)

**8.4.2** THEOREM.

$$S^\alpha = M^\alpha \cap \left\{ w \mid w \in L^{(n)} \text{ and } \frac{e_{i,i+1}^a}{a!} w = 0 \quad \text{for all } i \text{ and all } a \text{ with } 0 < a \leqslant \alpha_{i+1} \right\}.$$

*Proof.* Let $A$ denote the right-hand side. We have already proved that $S^\alpha \subseteq A$.

Under the isomorphism 8.1.1, the total order 7.2.5 on the set of $\alpha$-tabloids induces a total order on $\{w_{i_1} \otimes \cdots \otimes w_{i_n} \in M^\alpha\}$. Although our field is arbitrary, we may define a bilinear form $\Phi$ on $L^{(n)}$ as in 8.2.7 (in fact, the restriction of $\Phi$ to $M^\alpha$ is just the bilinear form used in Chapter 7).

Now, suppose that $w$ is a nonzero element of $A$. We claim that the last $w_{i_1} \otimes \cdots \otimes w_{i_n}$ involved in $w$ is such that $i_1, i_2, \ldots, i_n$ is a lattice permutation. Suppose not. Then for some $x$ and $z$, $i_z = x+1$, and the number of subscripts before $i_z$ which equal $x+1$ is the same as the number of subscripts before $i_z$ which equal $x$. Let $i_{u_1}, i_{u_2}, \ldots, i_{u_s}$ be the subscripts after $i_z$ (excluding $i_z$) which equal $x+1$, and let $i_{u_{s+1}}, \ldots, i_{u_y}$ be the subscripts before $i_z$ which equal $x$. Then $y = \alpha_{x+1} - 1$.

Since $e_{x,x+1}^a/a!$ annihilates $w$, $w$ is orthogonal (under $\Phi$, defined above) to all vectors of the form $(e_{x+1,x}^a/a!)w'$ with $w' \in L^{(n)}$. In particular, by Lemma 8.4.1, $w$ is orthogonal to the sum $v$ of all tensors $w_{j_1} \otimes \cdots \otimes w_{j_n}$ in $M^\alpha$ such that

(i) $j_k = i_k$ for $i_k \neq x, x+1$ and
(ii) $j_{u_r} = i_{u_r}$ for $1 \leqslant r \leqslant y$.

But $w_{i_1} \otimes \cdots \otimes w_{i_n}$ is the first tensor satisfying these conditions, and by hypothesis $w_{i_1} \otimes \cdots \otimes w_{i_n}$ is the last tensor in $w$. This contradicts $\Phi(v, w) = 0$.

We have now proved that the last tensor involved in every nonzero element of $A$ has its subscripts in a lattice permutation. Therefore, $A = S^\alpha$, by 8.1.2 and 7.2.10. ■

Compare the next corollary with the definition of the Weyl module.

**8.4.3** COROLLARY.

$$S^\alpha = \{ w \mid w \in L^{(n)} \text{ and } uw = 0 \text{ for all } u \in \overline{U}_F \text{ such that } u \mathcal{V}^\alpha w^\alpha = 0 \}.$$

*Proof.* Since the action of $S_n$ on $L^{(n)}$ commutes with that of $\overline{U}_F$, and $\mathcal{V}^\alpha w^\alpha$ generates $S^\alpha$ as an $FS_n$-module, every element of $S^\alpha$ is certainly annihilated by those elements in $\overline{U}_F$ which annihilate $\mathcal{V}^\alpha w^\alpha$. This gives one inclusion, and the other follows from the descriptions of $M^\alpha$ and $S^\alpha$ we have given above. ■

Surprisingly, not only does tensor space contain Weyl modules and Specht modules, but it also contains copies of many vector spaces of module homomorphisms:

**8.4.4** THEOREM. *Suppose that $\alpha$ and $\beta$ have at most $m$ nonzero parts. Then there exist injective linear maps*

(i) *from* $W^\alpha \cap S^\beta$ *into* $\mathrm{Hom}_{FS_n}(S^\alpha, S^\beta)$,
(ii) *from* $W^\alpha \cap M^\beta$ *into* $\mathrm{Hom}_{FS_n}(S^\alpha, M^\beta)$,
(iii) *from* $W^\alpha \cap S^\beta$ *into* $\mathrm{Hom}_{\overline{U}_F}(W^\beta, W^\alpha)$, *and*
(iv) *from* $L^\alpha \cap S^\beta$ *into* $\mathrm{Hom}_{\overline{U}_F}(W^\beta, L^\alpha)$.

*The third and fourth maps are isomorphisms, and so are the first and second unless* $\mathrm{char}\, F=2$ *and $\alpha$ is 2-singular.*

*Proof.* (i): Since $\mathcal{V}^\alpha w^\alpha$ generates $W^\alpha$ as a $\overline{U}_F$-module, every element of $W^\alpha \cap S^\beta$ has the form $u\mathcal{V}^\alpha w^\alpha$ for some $u \in \overline{U}_F$. Let

$$\theta: s\mathcal{V}^\alpha w^\alpha \mapsto su\mathcal{V}^\alpha w^\alpha \qquad (s \in FS_n).$$

Then $\theta$ is well defined, since $s_1\mathcal{V}^\alpha w^\alpha = s_2\mathcal{V}^\alpha w^\alpha$ implies that $s_1 - s_2$ annihilates $\mathcal{V}^\alpha w^\alpha$, and so annihilates $u\mathcal{V}^\alpha w^\alpha$. Thus, we have associated a unique element of $\mathrm{Hom}_{FS_n}(S^\alpha, S^\beta)$ with each element of $W^\alpha \cap S^\beta$.

If $\varphi \in \mathrm{Hom}_{FS_n}(S^\alpha, S^\beta)$, then $\varphi\mathcal{V}^\alpha w^\alpha \in S^\beta$. Now, since $\varphi$ is well defined, $s\varphi\mathcal{V}^\alpha w^\alpha = 0$ for all $s \in FS_n$ such that $s\mathcal{V}^\alpha w^\alpha = 0$. Therefore, provided $\mathrm{char}\, F \neq 2$ or $\alpha$ is 2-regular, $\varphi\mathcal{V}^\alpha w^\alpha \in W^\alpha$ by Definition 8.1.6. This gives us an injection $\varphi \mapsto \varphi\mathcal{V}^\alpha w^\alpha$ from $\mathrm{Hom}_{FS_n}(S^\alpha, S^\beta)$ into $W^\alpha \cap S^\beta$. It is straightforward to check that this is the inverse of the map described in the first paragraph.

(ii), (iii), (iv): The proof of these parts are almost identical to the proof of part (i), except that in parts (iii) and (iv) we use 8.4.3 instead of 8.1.6. ■

To illustrate that the conditions on $F$ must appear in the statement of the theorem, we give

**8.4.5** EXAMPLE. Suppose $\mathrm{char}\, F=2$. Then

$$\theta: w_1 \otimes w_2 - w_2 \otimes w_1 \mapsto w_1 \otimes w_1$$

extends to a well-defined element of $\mathrm{Hom}_{FS_2}(S^{(1^2)}, S^{(2)})$. But $W^{(1^2)}$ is spanned by $w_1 \otimes w_2 - w_2 \otimes w_1$, so $W^{(1^2)} \cap S^{(2)} = 0$. Since $M^{(2)} = S^{(2)}$, this example simultaneously shows that the maps in parts (i) and (ii) of the theorem need not be isomorphisms when $\operatorname{char} F = 2$ and $\alpha$ is 2-singular.

Theorems 8.4.2 and 8.4.4 are very powerful when it comes to finding composition factors of Specht modules. To decide whether or not $\operatorname{Hom}_{FS_n}(S^\alpha, S^\beta)$ is zero, all we have to do (at least in the case $\operatorname{char} F \neq 2$ or $\alpha$ 2-regular) is see whether any element of $W^\alpha \cap M^\beta$ is annihilated by $e_{i,i+1}^{a}/a!$ for all $i$ and all $a$ with $0 < a \leqslant \alpha_{i+1}$.

Since 8.1.1 and 8.1.16 show that $\{E_T \mid T$ is a semistandard $\alpha$-tableau of content $\beta\}$ is a basis of $W^\alpha \cap M^\beta$, the problem reduces, for a given $\alpha$ and $\beta$, to that of seeing whether a certain set of linear equations has a nontrivial solution. Examples of the progress which can be made along these lines are to be found in James's book [1978e]. Unfortunately, sometimes there exist composition factors of $S^\beta$ (apart from $D^\beta$) which are not in the image of an earlier Specht module, and in these cases more work must be done.

In conclusion, we list some of the properties of $FS_n$-modules and $\overline{U}_F$-modules which we have proved; the similarities are impressive:

(1)(a) $\mathcal{V}^\alpha w^\alpha$ generates the Specht module $S^\alpha$ as an $FS_n$-module.
 (b) $\mathcal{V}^\alpha w^\alpha$ generates the Weyl module $W^\alpha$ as a $\overline{U}_F$-module.

(2)(a) $M^\alpha$ is the subset of $L^{(n)}$ annihilated by

$$\left\{ \begin{pmatrix} h_i - \alpha_i \\ c \end{pmatrix} \middle| i \geqslant 1,\ c \geqslant 1 \right\}.$$

 (b) $L^\alpha$ is the subset of $L^{(n)}$ annihilated by $\{1 - (\operatorname{sgn} \pi)\pi \mid \pi \in \mathcal{V}^\alpha\}$ (if $\operatorname{char} F \neq 2$).

(3)(a) $S^\alpha$ is the subset of $L^{(n)}$ annihilated by those $u \in \overline{U}_F$ which annihilate $\mathcal{V}^\alpha w^\alpha$.
 (b) $W^\alpha$ is the subset of $L^{(n)}$ annihilated by those $s \in FS_n$ which annihilate $\mathcal{V}^\alpha w^\alpha$ (unless $\operatorname{char} F = 2$ and $\alpha$ is 2-singular).

(4)(a) $S^\alpha = M^\alpha \cap \{w \mid w \in L^{(n)}$ and $w$ is annihilated by $e_{i,i+1}^{a}/a!$ for $i \geqslant 1$, $0 < a \leqslant \alpha_{i+1}\}$.
 (b) $W^\alpha = L^\alpha \cap \{w \mid w \in L^{(n)}$ and $w$ is annihilated by the Garnir elements for the first standard $\alpha$-tableau$\}$.

(5)(a) $\dim S^\alpha =$ the number of standard $\alpha$-tableaux.
 (b) $\dim W^\alpha =$ the number of semistandard $\alpha$-tableaux.

(6)(a) $S^\alpha$ has a unique top composition factor $D^\alpha$ ($\alpha$ $p$-regular).
 (b) $W^\alpha$ has a unique top composition factor $F^\alpha$.

(7)(a) dim $D^\alpha$ equals the rank of a certain Gram matrix.
(b) dim $F^\alpha$ equals the rank of a certain Gram matrix.

(8)(a) Every $FS_n$-submodule of $M^\alpha$ either contains $S^\alpha$ or is contained in $S^{\alpha\perp}$.
(b) Every $\overline{U}_F$-submodule of $L^\alpha$ either contains $W^\alpha$ or is contained in $W^{\alpha\perp}$.

(9)(a) Every composition factor of $S^\alpha$ has the form $D^\beta$ ($\beta$ $p$-regular), where $\beta \trianglerighteq \alpha$. If $\alpha$ is $p$-regular, $D^\alpha$ is a composition factor with multiplicity 1.
(b) Every composition factor of $W^\alpha$ has the form $F^\beta$, where $\alpha \trianglerighteq \beta$. $F^\alpha$ is a composition factor with multiplicity 1.

(10) $\operatorname{Hom}_{FS_n}(S^\alpha, S^\beta) \simeq W^\alpha \cap S^\beta \simeq \operatorname{Hom}_{\overline{U}_F}(W^\beta, W^\alpha)$ (the first isomorphism assumes char $F \neq 2$ or that $\alpha$ is 2-regular).

## Exercises

8.1 Work out $(123)w_1 \otimes w_1 \otimes w_2$. Then use the fact that $(123)=(12)(23)$ to see whether your have got the right answer.

8.2 Assume that $m \geqslant n$. Prove that

$$L^{(n)} = \overline{U}_F(w_1 \otimes w_2 \otimes \cdots \otimes w_n).$$

8.3 Suppose that the partition $\alpha'$ ($:=\beta$, for short) associated with $\alpha$ is $p$-regular, and that the underlying field has characteristic $p$. Consider

$$w := \left(w_{\beta_1} \otimes w_{\beta_1 - 1} \otimes \cdots \otimes w_1\right) \otimes \left(w_{\beta_2} \otimes w_{\beta_2 - 1} \otimes \cdots \otimes w_1\right)$$
$$\otimes \cdots \otimes \left(w_{\beta_r} \otimes w_{\beta_r - 1} \otimes \cdots \otimes w_1\right)$$

($r :=$ the number of nonzero parts of $\beta$). Prove that $\mathcal{V}^\alpha \mathcal{H}^\alpha w$ equals a non-zero multiple of $\mathcal{V}^\alpha w$ (cf. the proof of 7.1.10(ii)). Deduce that

$$\alpha' \text{ } p\text{-regular} \quad \Rightarrow \quad \mathcal{V}^\alpha \mathcal{H}^\alpha L^{(n)} = W^\alpha.$$

The converse of this statement is also true (see James [1980]).

8.4 Assume that $|F| = q$ is a power of $p$. The purpose of this exercise is to construct the inequivalent irreducible $p$-modular representations of $\mathrm{GL}(m, F)$.

(a) Prove that $\mathrm{GL}(m, F)$ has $q^{m-1}(q-1)$ $p$-regular conjugacy classes.

(b) Prove that

$$\sum_{x \in F} x^s = \begin{cases} -1 & \text{if } q-1 \text{ divides } s, \\ 0 & \text{otherwise.} \end{cases}$$

(c) Using (b) and 8.2.6(i), prove that if $0 \leqslant s \leqslant q-1$ there is an element of the group algebra of $\mathrm{GL}(m, F)$, over $F$, which acts on $L^{(n)}$ in the same way as the following element of $\overline{U}_F$:

$$u(i, j, s) := \frac{e_{ij}^{s}}{s!} + \frac{e_{ij}^{s+q-1}}{(s+q-1)!} + \frac{e_{ij}^{s+2q-2}}{(s+2q-2)!} + \cdots.$$

(d) Assume that $\alpha$ has $k$ nonzero parts, where $k<m$, and let

$$\gamma_i := u(i, k+1, \alpha_i - \alpha_{i+1})u(i, k, \alpha_{i+1} - \alpha_{i+2}) \cdots u(i, i+1, \alpha_k) \in \overline{U}_F.$$

Prove that $\gamma_k \gamma_{k-1} \cdots \gamma_2 \gamma_1 \mathcal{V}^{\alpha} L^{(n)}$ is the 1-dimensional space spanned by $\mathcal{V}^{\alpha} w^{\alpha}$. (To show that it is nonzero, consider the action on $w_T$, where $T$ is the tableau where every entry in the $i$th row equals $k-i+2$.)

(e) Deduce from (c) and (d), as in Theorem 8.3.2, that if $\alpha$ has fewer than $m$ nonzero parts, and the difference between successive parts is at most $q-1$, then $W^{\alpha}/(W^{\alpha} \cap W^{\alpha\perp})$ is an absolutely irreducible $F\mathrm{GL}(m, F)$-module.

(f) Using (a) and (e), construct all the inequivalent irreducible $p$-modular representations of $\mathrm{GL}(m, F)$.

8.5 Prove that the only identity $FS_n$-submodule of $M^{\alpha}$ is the sum $s$ of all the tensors $w_{i_1} \otimes \cdots \otimes w_{i_n}$ which have the property that for each $i$ precisely $\alpha_i$ of the subscripts equal $i$. What is the image of $s$ under $e_{ij}^{a}/a!$? Using the characterization of the Specht module $S^{\alpha}$ given by Theorem 8.4.2, prove that when $\alpha = (n-3, 3)$, $s$ belongs to $S^{\alpha}$ if and only if $\operatorname{char} F = p$ and $p$ divides

$$\binom{n}{3}, \quad \binom{n-1}{2}, \quad \text{and} \quad \binom{n-2}{1}.$$

8.6 Assume that $\dim W^{(1)} = m$. Prove that

$$\dim W^{(1^n)} = \binom{m}{n} \quad \text{and} \quad \dim W^{(n)} = \binom{m+n-1}{n}.$$

Suppose now that $\operatorname{char} F = p$ and $n = a_0 + a_1 p + \cdots + a_r p^r$ $(0 \leqslant a_i < p)$. Prove that

$$\dim F^{(n)} = \prod_{i=0}^{r} \binom{m+a_i-1}{a_i}.$$

APPENDIX I

# Tables

## I.A Character Tables

This section of Appendix I contains the character tables of the symmetric groups $S_n$ for $n \leqslant 10$. Above each character table the reader will find the numbering of the partitions and the numbering of the conjugacy classes, which is the reverse of the numbering of the partitions. Furthermore the order of the conjugacy classes is given, as well as the order of the corresponding centralizers. For both these orders their prime-number decomposition is indicated.

The program was implemented by G. Berard.

ENCYCLOPEDIA OF MATHEMATICS and Its Applications, Gian-Carlo Rota (ed.). Vol. 16: G. D. James and A. Kerber, The Representation Theory of the Symmetric Group

ISBN 0-201-13515-9

CHARACTER TABLE OF S2

| NO | OF PARTITION | NO | CLASS ORDER | ORDER OF CENTRALIZER |
|---|---|---|---|---|
| X1 | [2] | C2 | 1 = 1(0) | 2 = 2 |
| X2 | [1,1] | C1 | 1 = 1(0) | 2 = 2 |

CHARACTERS

| | C1 | C2 |
|---|---|---|
| X1 | 1 | 1 |
| X2 | 1 | -1 |

CHARACTER TABLE OF S3

| NO | OF PARTITION | NO | CLASS ORDER | ORDER OF CENTRALIZER |
|---|---|---|---|---|
| X1 | [3] | C3 | 2 = 2 | 3 = 3 |
| X2 | [2,1] | C2 | 3 = 3 | 2 = 2 |
| X3 | [1,1,1] | C1 | 1 = 1(0) | 6 = 2*3 |

CHARACTERS

| | C1 | C2 | C3 |
|---|---|---|---|
| X1 | 1 | 1 | 1 |
| X2 | 2 | 0 | -1 |
| X3 | 1 | -1 | 1 |

CHARACTER TABLE OF S4

| NO | OF PARTITION | NO | CLASS ORDER | ORDER OF CENTRALIZER |
|---|---|---|---|---|
| X1 | [4] | C5 | 6 = 2*3 | 4 = 2(2) |
| X2 | [3,1] | C4 | 8 = 2(3) | 3 = 3 |
| X3 | [2,2] | C3 | 3 = 3 | 8 = 2(3) |
| X4 | [2,1,1] | C2 | 6 = 2*3 | 4 = 2(2) |
| X5 | [1,1,1,1] | C1 | 1 = 1(0) | 24 = 2(3)*3 |

CHARACTERS

| | C1 | C2 | C3 | C4 | C5 |
|---|---|---|---|---|---|
| X1 | 1 | 1 | 1 | 1 | 1 |
| X2 | 3 | 1 | -1 | 0 | -1 |
| X3 | 2 | 0 | 2 | -1 | 0 |
| X4 | 3 | -1 | -1 | 0 | 1 |
| X5 | 1 | -1 | 1 | 1 | -1 |

CHARACTER TABLE OF S5

| NO | OF PARTITION | NO | CLASS ORDER | ORDER OF CENTRALIZER |
|---|---|---|---|---|
| X1 | [5] | C7 | 24 = 2(3)*3 | 5 = 5 |
| X2 | [4,1] | C6 | 30 = 2*3*5 | 4 = 2(2) |
| X3 | [3,2] | C5 | 20 = 2(2)*5 | 6 = 2*3 |
| X4 | [3,1,1] | C4 | 20 = 2(2)*5 | 6 = 2*3 |
| X5 | [2,2,1] | C3 | 15 = 3*5 | 8 = 2(3) |
| X6 | [2,1,1,1] | C2 | 10 = 2*5 | 12 = 2(2)*3 |
| X7 | [1,1,1,1,1] | C1 | 1 = 1(0) | 120 = 2(3)*3*5 |

CHARACTERS

| | C1 | C2 | C3 | C4 | C5 | C6 | C7 |
|---|---|---|---|---|---|---|---|
| X1 | 1 | 1 | 1 | 1 | 1 | 1 | 1 |
| X2 | 4 | 2 | 0 | 1 | -1 | 0 | -1 |
| X3 | 5 | 1 | 1 | -1 | 1 | -1 | 0 |
| X4 | 6 | 0 | -2 | 0 | 0 | 0 | 1 |
| X5 | 5 | -1 | 1 | -1 | -1 | 1 | 0 |
| X6 | 4 | -2 | 0 | 1 | 1 | 0 | -1 |
| X7 | 1 | -1 | 1 | 1 | -1 | -1 | 1 |

CHARACTER TABLE OF S6

| NO OF | PARTITION | NO | CLASS ORDER | ORDER OF CENTRALIZER |
|---|---|---|---|---|
| X1 | [6] | C11 | 120 = 2(3)*3*5 | 6 = 2*3 |
| X2 | [5,1] | C10 | 144 = 2(4)*3(2) | 5 = 5 |
| X3 | [4,2] | C9 | 90 = 2*3(2)*5 | 8 = 2(3) |
| X4 | [4,1,1] | C8 | 90 = 2*3(2)*5 | 8 = 2(3) |
| X5 | [3,3] | C7 | 40 = 2(3)*5 | 18 = 2*3(2) |
| X6 | [3,2,1] | C6 | 120 = 2(3)*3*5 | 6 = 2*3 |
| X7 | [3,1,1,1] | C5 | 40 = 2(3)*5 | 18 = 2*3(2) |
| X8 | [2,2,2] | C4 | 15 = 3*5 | 48 = 2(4)*3 |
| X9 | [2,2,1,1] | C3 | 45 = 3(2)*5 | 16 = 2(4) |
| X10 | [2,1,1,1,1] | C2 | 15 = 3*5 | 48 = 2(4)*3 |
| X11 | [1,1,1,1,1,1] | C1 | 1 = 1(0) | 720 = 2(4)*3(2)*5 |

CHARACTERS

| | C1 | C2 | C3 | C4 | C5 | C6 | C7 | C8 | C9 | C10 | C11 |
|---|---|---|---|---|---|---|---|---|---|---|---|
| X1 | 1 | 1 | 1 | 1 | 1 | 1 | 1 | 1 | 1 | 1 | 1 |
| X2 | 5 | 3 | 1 | -1 | 2 | 0 | -1 | 1 | -1 | 0 | -1 |
| X3 | 9 | 3 | 1 | 3 | 0 | 0 | 0 | -1 | 1 | -1 | 0 |
| X4 | 10 | 2 | -2 | -2 | 1 | -1 | 1 | 0 | 0 | 0 | 1 |
| X5 | 5 | 1 | 1 | -3 | -1 | 1 | 2 | -1 | -1 | 0 | 0 |
| X6 | 16 | 0 | 0 | 0 | -2 | 0 | -2 | 0 | 0 | 1 | 0 |
| X7 | 10 | -2 | -2 | 2 | 1 | 1 | 1 | 0 | 0 | 0 | -1 |
| X8 | 5 | -1 | 1 | 3 | -1 | -1 | 2 | 1 | -1 | 0 | 0 |
| X9 | 9 | -3 | 1 | -3 | 0 | 0 | 0 | 1 | 1 | -1 | 0 |
| X10 | 5 | -3 | 1 | 1 | 2 | 0 | -1 | -1 | -1 | 0 | 1 |
| X11 | 1 | -1 | 1 | -1 | 1 | -1 | 1 | -1 | 1 | 1 | -1 |

CHARACTER TABLE OF S7

| NO OF | PARTITION | NO | CLASS ORDER | ORDER OF CENTRALIZER |
|---|---|---|---|---|
| X1 | [7] | C15 | 720 = 2(4)*3(2)*5 | 7 = 7 |
| X2 | [6,1] | C14 | 840 = 2(3)*3*5*7 | 6 = 2*3 |
| X3 | [5,2] | C13 | 504 = 2(3)*3(2)*7 | 10 = 2*5 |
| X4 | [5,1,1] | C12 | 504 = 2(3)*3(2)*7 | 10 = 2*5 |
| X5 | [4,3] | C11 | 420 = 2(2)*3*5*7 | 12 = 2(2)*3 |
| X6 | [4,2,1] | C10 | 630 = 2*3(2)*5*7 | 8 = 2(3) |
| X7 | [4,1,1,1] | C9 | 210 = 2*3*5*7 | 24 = 2(3)*3 |
| X8 | [3,3,1] | C8 | 280 = 2(3)*5*7 | 18 = 2*3(2) |
| X9 | [3,2,2] | C7 | 210 = 2*3*5*7 | 24 = 2(3)*3 |
| X10 | [3,2,1,1] | C6 | 420 = 2(2)*3*5*7 | 12 = 2(2)*3 |
| X11 | [3,1,1,1,1] | C5 | 70 = 2*5*7 | 72 = 2(3)*3(2) |
| X12 | [2,2,2,1] | C4 | 105 = 3*5*7 | 48 = 2(4)*3 |
| X13 | [2,2,1,1,1] | C3 | 105 = 3*5*7 | 48 = 2(4)*3 |
| X14 | [2,1,1,1,1,1] | C2 | 21 = 3*7 | 240 = 2(4)*3*5 |
| X15 | [1,1,1,1,1,1,1] | C1 | 1 = 1(0) | 5040 = 2(4)*3(2)*5*7 |

CHARACTERS

| | C1 | C2 | C3 | C4 | C5 | C6 | C7 | C8 | C9 | C10 | C11 | C12 | C13 | C14 | C15 |
|---|---|---|---|---|---|---|---|---|---|---|---|---|---|---|---|
| X1 | 1 | 1 | 1 | 1 | 1 | 1 | 1 | 1 | 1 | 1 | 1 | 1 | 1 | 1 | 1 |
| X2 | 6 | 4 | 2 | 0 | 3 | 1 | -1 | 0 | 2 | 0 | -1 | 1 | -1 | 0 | -1 |
| X3 | 14 | 6 | 2 | 2 | 2 | 0 | 2 | -1 | 0 | 0 | 0 | -1 | 1 | -1 | 0 |
| X4 | 15 | 5 | -1 | -3 | 3 | -1 | -1 | 0 | 1 | -1 | 1 | 0 | 0 | 0 | 1 |
| X5 | 14 | 4 | 2 | 0 | -1 | 1 | -1 | 2 | -2 | 0 | 1 | -1 | -1 | 0 | 0 |
| X6 | 35 | 5 | -1 | 1 | -1 | -1 | -1 | -1 | -1 | 1 | -1 | 0 | 0 | 1 | 0 |
| X7 | 20 | 0 | -4 | 0 | 2 | 0 | 2 | 2 | 0 | 0 | 0 | 0 | 0 | 0 | -1 |
| X8 | 21 | 1 | 1 | -3 | -3 | 1 | 1 | 0 | -1 | -1 | -1 | 1 | 1 | 0 | 0 |
| X9 | 21 | -1 | 1 | 3 | -3 | -1 | 1 | 0 | 1 | -1 | 1 | 1 | -1 | 0 | 0 |
| X10 | 35 | -5 | -1 | -1 | -1 | 1 | -1 | -1 | 1 | 1 | 1 | 0 | 0 | -1 | 0 |
| X11 | 15 | -5 | -1 | 3 | 3 | 1 | -1 | 0 | -1 | -1 | -1 | 0 | 0 | 0 | 1 |
| X12 | 14 | -4 | 2 | 0 | -1 | -1 | -1 | 2 | 2 | 0 | -1 | -1 | 1 | 0 | 0 |
| X13 | 14 | -6 | 2 | -2 | 2 | 0 | 2 | -1 | 0 | 0 | 0 | -1 | -1 | 1 | 0 |
| X14 | 6 | -4 | 2 | 0 | 3 | -1 | -1 | 0 | -2 | 0 | 1 | 1 | 1 | 0 | -1 |
| X15 | 1 | -1 | 1 | -1 | 1 | -1 | 1 | 1 | -1 | 1 | -1 | 1 | -1 | -1 | 1 |

## CHARACTER TABLE OF S8

| NO OF | PARTITION | NO | CLASS ORDER | ORDER OF CENTRALIZER |
|---|---|---|---|---|
| X1 | [8] | C22 | 5040 = 2(4)*3(2)*5*7 | 8 = 2(3) |
| X2 | [7,1] | C21 | 5760 = 2(7)*3(2)*5 | 7 = 7 |
| X3 | [6,2] | C20 | 3360 = 2(5)*3*5*7 | 12 = 2(2)*3 |
| X4 | [6,1,1] | C19 | 3360 = 2(5)*3*5*7 | 12 = 2(2)*3 |
| X5 | [5,3] | C18 | 2688 = 2(7)*3*7 | 15 = 3*5 |
| X6 | [5,2,1] | C17 | 4032 = 2(6)*3(2)*7 | 10 = 2*5 |
| X7 | [5,1,1,1] | C16 | 1344 = 2(6)*3*7 | 30 = 2*3*5 |
| X8 | [4,4] | C15 | 1260 = 2(2)*3(2)*5*7 | 32 = 2(5) |
| X9 | [4,3,1] | C14 | 3360 = 2(5)*3*5*7 | 12 = 2(2)*3 |
| X10 | [4,2,2] | C13 | 1260 = 2(2)*3(2)*5*7 | 32 = 2(5) |
| X11 | [4,2,1,1] | C12 | 2520 = 2(3)*3(2)*5*7 | 16 = 2(4) |
| X12 | [4,1,1,1,1] | C11 | 420 = 2(2)*3*5*7 | 96 = 2(5)*3 |
| X13 | [3,3,2] | C10 | 1120 = 2(5)*5*7 | 36 = 2(2)*3(2) |
| X14 | [3,3,1,1] | C9 | 1120 = 2(5)*5*7 | 36 = 2(2)*3(2) |
| X15 | [3,2,2,1] | C8 | 1680 = 2(4)*3*5*7 | 24 = 2(3)*3 |
| X16 | [3,2,1,1,1] | C7 | 1120 = 2(5)*5*7 | 36 = 2(2)*3(2) |
| X17 | [3,1,1,1,1,1] | C6 | 112 = 2(4)*7 | 360 = 2(3)*3(2)*5 |
| X18 | [2,2,2,2] | C5 | 105 = 3*5*7 | 384 = 2(7)*3 |
| X19 | [2,2,2,1,1] | C4 | 420 = 2(2)*3*5*7 | 96 = 2(5)*3 |
| X20 | [2,2,1,1,1,1] | C3 | 210 = 2*3*5*7 | 192 = 2(6)*3 |
| X21 | [2,1,1,1,1,1,1] | C2 | 28 = 2(2)*7 | 1440 = 2(5)*3(2)*5 |
| X22 | [1,1,1,1,1,1,1,1] | C1 | 1 = 1(0) | 40320 = 2(7)*3(2)*5*7 |

## CHARACTERS

| | C1 | C2 | C3 | C4 | C5 | C6 | C7 | C8 | C9 | C10 | C11 | C12 | C13 | C14 | C15 | C16 | C17 | C18 | C19 | C20 | C21 | C22 |
|---|---|---|---|---|---|---|---|---|---|---|---|---|---|---|---|---|---|---|---|---|---|---|
| X1 | 1 | 1 | 1 | 1 | 1 | 1 | 1 | 1 | 1 | 1 | 1 | 1 | 1 | 1 | 1 | 1 | 1 | 1 | 1 | 1 | 1 | 1 |
| X2 | 7 | 5 | 3 | 1 | -1 | 4 | 2 | 0 | 1 | -1 | 3 | 1 | -1 | 0 | -1 | 2 | 0 | -1 | 1 | -1 | 0 | -1 |
| X3 | 20 | 10 | 4 | 2 | 4 | 5 | 1 | 1 | -1 | 1 | 2 | 0 | 2 | -1 | 0 | 0 | 0 | 0 | -1 | 1 | -1 | 0 |
| X4 | 21 | 9 | 1 | -3 | -3 | 6 | 0 | -2 | 0 | 0 | 3 | -1 | -1 | 0 | 1 | 1 | -1 | 1 | 0 | 0 | 0 | 1 |
| X5 | 28 | 10 | 4 | 2 | -4 | 1 | 1 | 1 | 1 | 1 | -2 | 0 | -2 | 1 | 0 | -2 | 0 | 1 | -1 | -1 | 0 | 0 |
| X6 | 64 | 16 | 0 | 0 | 0 | 4 | -2 | 0 | -2 | -2 | 0 | 0 | 0 | 0 | 0 | -1 | 1 | -1 | 0 | 0 | 1 | 0 |
| X7 | 35 | 5 | -5 | -3 | 3 | 5 | -1 | 1 | 2 | 2 | 1 | -1 | 1 | 1 | -1 | 0 | 0 | 0 | 0 | 0 | 0 | -1 |
| X8 | 14 | 4 | 2 | 0 | 6 | -1 | 1 | -1 | 2 | -2 | -2 | 0 | 2 | 1 | 2 | -1 | -1 | -1 | 0 | 0 | 0 | 0 |
| X9 | 70 | 10 | 2 | -2 | -2 | -5 | 1 | -1 | 1 | 1 | -4 | 0 | 0 | -1 | -2 | 0 | 0 | 0 | 1 | 1 | 0 | 0 |
| X10 | 56 | 4 | 0 | 4 | 8 | -4 | -2 | 0 | -1 | 1 | 0 | 0 | 0 | 0 | 0 | 1 | -1 | 1 | 1 | -1 | 0 | 0 |
| X11 | 90 | 0 | -6 | 0 | -6 | 0 | 0 | 0 | 0 | 0 | 0 | 2 | 0 | 0 | 2 | 0 | 0 | 0 | 0 | 0 | -1 | 0 |
| X12 | 35 | -5 | -5 | 3 | 3 | 5 | 1 | 1 | 2 | -2 | -1 | -1 | -1 | -1 | -1 | 0 | 0 | 0 | 0 | 0 | 0 | 1 |
| X13 | 42 | 0 | 2 | 0 | -6 | -6 | 0 | 2 | 0 | 0 | 0 | -2 | 0 | 0 | 2 | 2 | 0 | -1 | 0 | 0 | 0 | 0 |
| X14 | 56 | -4 | 0 | -4 | 8 | -4 | 2 | 0 | -1 | -1 | 0 | 0 | 0 | 0 | 0 | 1 | 1 | 1 | -1 | -1 | 0 | 0 |
| X15 | 70 | -10 | 2 | 2 | -2 | -5 | -1 | -1 | 1 | -1 | 4 | 0 | 0 | 1 | -2 | 0 | 0 | 0 | -1 | 1 | 0 | 0 |
| X16 | 64 | -16 | 0 | 0 | 0 | 4 | 2 | 0 | -2 | 2 | 0 | 0 | 0 | 0 | 0 | -1 | -1 | -1 | 0 | 0 | 1 | 0 |
| X17 | 21 | -9 | 1 | 3 | -3 | 6 | 0 | -2 | 0 | 0 | -3 | -1 | 1 | 0 | 1 | 1 | 1 | 1 | 0 | 0 | 0 | -1 |
| X18 | 14 | -4 | 2 | 0 | 6 | -1 | -1 | -1 | 2 | 2 | 2 | 0 | -2 | -1 | 2 | -1 | 1 | -1 | 0 | 0 | 0 | 0 |
| X19 | 28 | -10 | 4 | -2 | -4 | 1 | -1 | 1 | 1 | -1 | 2 | 0 | 2 | -1 | 0 | -2 | 0 | 1 | 1 | -1 | 0 | 0 |
| X20 | 20 | -10 | 4 | -2 | 4 | 5 | -1 | 1 | -1 | -1 | -2 | 0 | -2 | 1 | 0 | 0 | 0 | 0 | 1 | 1 | -1 | 0 |
| X21 | 7 | -5 | 3 | -1 | -1 | 4 | -2 | 0 | 1 | 1 | -3 | 1 | 1 | 0 | -1 | 2 | 0 | -1 | -1 | -1 | 0 | 1 |
| X22 | 1 | -1 | 1 | -1 | 1 | 1 | -1 | 1 | 1 | -1 | -1 | 1 | -1 | -1 | 1 | 1 | -1 | 1 | -1 | 1 | 1 | -1 |

## CHARACTER TABLE OF S9

| NO OF | PARTITION | NO | CLASS ORDER | ORDER OF CENTRALIZER |
|---|---|---|---|---|
| X1 | [9] | C30 | 40320 = 2(7)*3(2)*5*7 | 9 = 3(2) |
| X2 | [8,1] | C29 | 45360 = 2(4)*3(4)*5*7 | 8 = 2(3) |
| X3 | [7,2] | C28 | 25920 = 2(6)*3(4)*5 | 14 = 2*7 |
| X4 | [7,1,1] | C27 | 25920 = 2(6)*3(4)*5 | 14 = 2*7 |
| X5 | [6,3] | C26 | 20160 = 2(6)*3(2)*5*7 | 18 = 2*3(2) |
| X6 | [6,2,1] | C25 | 30240 = 2(5)*3(3)*5*7 | 12 = 2(2)*3 |
| X7 | [6,1,1,1] | C24 | 10080 = 2(5)*3(2)*5*7 | 36 = 2(2)*3(2) |
| X8 | [5,4] | C23 | 18144 = 2(5)*3(4)*7 | 20 = 2(2)*5 |
| X9 | [5,3,1] | C22 | 24192 = 2(7)*3(3)*7 | 15 = 3*5 |
| X10 | [5,2,2] | C21 | 9072 = 2(4)*3(4)*7 | 40 = 2(3)*5 |
| X11 | [5,2,1,1] | C20 | 18144 = 2(5)*3(4)*7 | 20 = 2(2)*5 |
| X12 | [5,1,1,1,1] | C19 | 3024 = 2(4)*3(3)*7 | 120 = 2(3)*3*5 |
| X13 | [4,4,1] | C18 | 11340 = 2(2)*3(4)*5*7 | 32 = 2(5) |
| X14 | [4,3,2] | C17 | 15120 = 2(4)*3(3)*5*7 | 24 = 2(3)*3 |
| X15 | [4,3,1,1] | C16 | 15120 = 2(4)*3(3)*5*7 | 24 = 2(3)*3 |
| X16 | [4,2,2,1] | C15 | 11340 = 2(2)*3(4)*5*7 | 32 = 2(5) |
| X17 | [4,2,1,1,1] | C14 | 7560 = 2(3)*3(3)*5*7 | 48 = 2(4)*3 |
| X18 | [4,1,1,1,1,1] | C13 | 756 = 2(2)*3(3)*7 | 480 = 2(5)*3*5 |
| X19 | [3,3,3] | C12 | 2240 = 2(6)*5*7 | 162 = 2*3(4) |
| X20 | [3,3,2,1] | C11 | 10080 = 2(5)*3(2)*5*7 | 36 = 2(2)*3(2) |
| X21 | [3,3,1,1,1] | C10 | 3360 = 2(5)*3*5*7 | 108 = 2(2)*3(3) |
| X22 | [3,2,2,2] | C9 | 2520 = 2(3)*3(2)*5*7 | 144 = 2(4)*3(2) |
| X23 | [3,2,2,1,1] | C8 | 7560 = 2(3)*3(3)*5*7 | 48 = 2(4)*3 |
| X24 | [3,2,1,1,1,1] | C7 | 2520 = 2(3)*3(2)*5*7 | 144 = 2(4)*3(2) |
| X25 | [3,1,1,1,1,1,1] | C6 | 168 = 2(3)*3*7 | 2160 = 2(4)*3(3)*5 |
| X26 | [2,2,2,2,1] | C5 | 945 = 3(3)*5*7 | 384 = 2(7)*3 |
| X27 | [2,2,2,1,1,1] | C4 | 1260 = 2(2)*3(2)*5*7 | 288 = 2(5)*3(2) |
| X28 | [2,2,1,1,1,1,1] | C3 | 378 = 2*3(3)*7 | 960 = 2(6)*3*5 |
| X29 | [2,1,1,1,1,1,1,1] | C2 | 36 = 2(2)*3(2) | 10080 = 2(5)*3(2)*5*7 |
| X30 | [1,1,1,1,1,1,1,1,1] | C1 | 1 = 1(0) | 362880 = 2(7)*3(4)*5*7 |

CHARACTERS

| | C1 | C2 | C3 | C4 | C5 | C6 | C7 | C8 | C9 | C10 | C11 | C12 | C13 | C14 | C15 | C16 | C17 | C18 | C19 | C20 | C21 | C22 | C23 | C24 | C25 | C26 | C27 | C28 | C29 | C30 |
|---|---|---|---|---|---|---|---|---|---|---|---|---|---|---|---|---|---|---|---|---|---|---|---|---|---|---|---|---|---|---|
| X1 | 1 | 1 | 1 | 1 | 1 | 1 | 1 | 1 | 1 | 1 | 1 | 1 | 1 | 1 | 1 | 1 | 1 | 1 | 1 | 1 | 1 | 1 | 1 | 1 | 1 | 1 | 1 | 1 | 1 | 1 |
| X2 | 8 | 6 | 4 | 2 | 0 | 5 | 3 | 1 | -1 | 2 | 0 | -1 | 4 | 2 | 0 | 1 | -1 | 0 | 3 | 1 | -1 | 0 | -1 | 2 | 0 | -1 | 1 | -1 | 0 | -1 |
| X3 | 27 | 15 | 7 | 3 | 3 | 9 | 3 | 1 | 3 | 0 | 0 | 0 | 5 | 1 | 1 | -1 | 1 | -1 | 2 | 0 | 2 | -1 | 0 | 0 | 0 | 0 | -1 | 1 | -1 | 0 |
| X4 | 28 | 14 | 4 | -2 | -4 | 10 | 2 | -2 | -2 | 1 | -1 | 1 | 6 | 0 | -2 | 0 | 0 | 0 | 3 | -1 | -1 | 0 | 1 | 1 | -1 | 1 | 0 | 0 | 0 | 1 |
| X5 | 48 | 20 | 8 | 4 | 0 | 6 | 2 | 2 | -2 | 0 | 2 | 3 | 0 | 0 | 0 | 0 | 0 | 0 | -2 | 0 | -2 | 1 | 0 | -2 | 0 | 1 | -1 | -1 | 0 | 0 |
| X6 | 105 | 35 | 5 | -1 | 1 | 15 | -1 | -1 | -1 | -3 | -1 | -3 | 5 | -1 | 1 | -1 | -1 | 1 | 0 | 0 | 0 | 0 | 0 | -1 | 1 | -1 | 0 | 0 | 1 | 0 |
| X7 | 56 | 14 | -4 | -6 | 0 | 11 | -1 | -1 | 3 | 2 | 2 | 2 | 4 | -2 | 0 | 1 | 1 | 0 | 1 | -1 | 1 | 1 | -1 | 0 | 0 | 0 | 0 | 0 | 0 | -1 |
| X8 | 42 | 14 | 6 | 2 | 2 | 0 | 2 | 0 | 2 | 3 | -1 | -3 | -4 | 0 | 0 | 2 | 0 | 2 | -3 | -1 | 1 | 0 | 1 | -1 | -1 | -1 | 0 | 0 | 0 | 0 |
| X9 | 162 | 36 | 6 | 0 | -6 | 0 | 0 | 0 | 0 | 0 | 0 | 0 | -6 | 0 | -2 | 0 | 0 | -2 | -3 | 1 | 1 | 0 | -1 | 0 | 0 | 0 | 1 | 1 | 0 | 0 |
| X10 | 120 | 20 | 0 | 4 | 8 | 0 | -4 | 0 | 4 | -3 | -1 | 3 | 0 | 0 | 0 | 0 | 0 | 0 | 0 | 0 | 0 | 0 | 0 | 1 | -1 | 1 | 1 | -1 | 0 | 0 |
| X11 | 189 | 21 | -11 | -3 | -3 | 9 | -3 | 1 | -3 | 0 | 0 | 0 | 1 | 1 | 1 | 1 | 1 | 1 | -1 | 1 | -1 | -1 | 1 | 0 | 0 | 0 | 0 | 0 | -1 | 0 |
| X12 | 70 | 0 | -10 | 0 | 6 | 10 | 0 | 2 | 0 | 4 | 0 | -2 | 0 | -2 | 0 | 0 | -2 | -2 | 0 | 0 | 0 | 0 | 0 | 0 | 0 | 0 | 0 | 0 | 0 | 1 |
| X13 | 84 | 14 | 4 | -2 | 4 | -6 | 2 | -2 | -2 | 3 | -1 | 3 | -6 | 0 | 2 | 0 | 0 | 0 | -1 | -1 | -1 | -1 | -1 | 1 | 1 | 1 | 0 | 0 | 0 | 0 |
| X14 | 168 | 14 | 4 | 2 | 0 | -15 | -1 | 1 | -1 | 0 | 2 | -3 | -4 | -2 | 0 | -1 | 1 | 0 | 3 | -1 | -1 | 0 | 1 | 2 | 0 | -1 | 0 | 0 | 0 | 0 |
| X15 | 216 | 6 | -4 | -6 | 0 | -9 | 3 | -1 | 3 | 0 | 0 | 0 | -4 | 2 | 0 | -1 | -1 | 0 | 1 | 1 | 1 | 1 | 1 | 0 | 0 | 0 | -1 | -1 | 0 | 0 |
| X16 | 216 | -6 | -4 | 6 | 0 | -9 | -3 | -1 | -3 | 0 | 0 | 0 | 4 | 2 | 0 | 1 | -1 | 0 | 1 | -1 | 1 | 1 | -1 | 0 | 0 | 0 | -1 | 1 | 0 | 0 |
| X17 | 189 | -21 | -11 | 3 | -3 | 9 | 3 | 1 | 3 | 0 | 0 | 0 | -1 | 1 | -1 | -1 | 1 | 1 | -1 | -1 | -1 | -1 | -1 | 0 | 0 | 0 | 0 | 0 | 1 | 0 |
| X18 | 56 | -14 | -4 | 6 | 0 | 11 | 1 | -1 | -3 | 2 | -2 | 2 | -4 | -2 | 0 | -1 | 1 | 0 | 1 | 1 | 1 | 1 | 1 | 0 | 0 | 0 | 0 | 0 | 0 | -1 |
| X19 | 42 | 0 | 2 | 0 | -6 | -6 | 0 | 2 | 0 | 0 | 0 | 6 | 0 | -2 | 0 | 0 | -2 | 2 | 2 | 0 | 2 | -1 | 0 | 0 | 0 | 0 | 0 | 0 | 0 | 0 |
| X20 | 168 | -14 | 4 | -2 | 0 | -15 | 1 | 1 | 1 | 0 | -2 | -3 | 4 | -2 | 0 | 1 | 1 | 0 | 3 | 1 | -1 | 0 | -1 | -2 | 0 | 1 | 0 | 0 | 0 | 0 |
| X21 | 120 | -20 | 0 | -4 | 8 | 0 | 4 | 0 | -4 | -3 | 1 | 3 | 0 | 0 | 0 | 0 | 0 | 0 | 0 | 0 | 0 | 0 | 0 | -1 | -1 | -1 | 1 | 1 | 0 | 0 |
| X22 | 84 | -14 | 4 | 2 | 4 | -6 | -2 | -2 | 2 | 3 | 1 | 3 | 6 | 0 | -2 | 0 | 0 | 0 | -1 | 1 | -1 | -1 | 1 | -1 | 1 | -1 | 0 | 0 | 0 | 0 |
| X23 | 162 | -36 | 6 | 0 | -6 | 0 | 0 | 0 | 0 | 0 | 0 | 0 | 6 | 0 | 2 | 0 | 0 | -2 | -3 | -1 | 1 | 0 | 1 | 0 | 0 | 0 | 1 | -1 | 0 | 0 |
| X24 | 105 | -35 | 5 | 1 | 1 | 15 | 1 | -1 | 1 | -3 | 1 | -3 | -5 | -1 | -1 | 1 | -1 | 1 | 0 | 0 | 0 | 0 | 0 | 1 | 1 | 1 | 0 | 0 | -1 | 0 |
| X25 | 28 | -14 | 4 | 2 | -4 | 10 | -2 | -2 | 2 | 1 | 1 | 1 | -6 | 0 | 2 | 0 | 0 | 0 | 3 | 1 | -1 | 0 | -1 | -1 | -1 | -1 | 0 | 0 | 0 | 1 |
| X26 | 42 | -14 | 6 | -2 | 2 | 0 | -2 | 0 | -2 | 3 | 1 | -3 | 4 | 0 | 0 | -2 | 0 | 2 | -3 | 1 | 1 | 0 | -1 | 1 | -1 | 1 | 0 | 0 | 0 | 0 |
| X27 | 48 | -20 | 8 | -4 | 0 | 6 | -2 | 2 | 2 | 0 | -2 | 3 | 0 | 0 | 0 | 0 | 0 | 0 | -2 | 0 | -2 | 1 | 0 | 2 | 0 | -1 | -1 | 1 | 0 | 0 |
| X28 | 27 | -15 | 7 | -3 | 3 | 9 | -3 | 1 | -3 | 0 | 0 | 0 | -5 | 1 | -1 | 1 | 1 | -1 | 2 | 0 | 2 | -1 | 0 | 0 | 0 | 0 | -1 | -1 | 1 | 0 |
| X29 | 8 | -6 | 4 | -2 | 0 | 5 | -3 | 1 | 1 | 2 | 0 | -1 | -4 | 2 | 0 | -1 | -1 | 0 | 3 | -1 | -1 | 0 | 1 | -2 | 0 | 1 | 1 | 1 | 0 | -1 |
| X30 | 1 | -1 | 1 | -1 | 1 | 1 | -1 | 1 | -1 | 1 | -1 | 1 | -1 | 1 | -1 | -1 | 1 | 1 | 1 | -1 | 1 | 1 | -1 | -1 | 1 | -1 | 1 | -1 | -1 | 1 |

CHARACTER TABLE OF S10

| NO OF | PARTITION | NO | CLASS ORDER | ORDER OF CENTRALIZER |
|---|---|---|---|---|
| X1 | [10] | C42 | 362880 = 2(7)*3(4)*5*7 | 10 = 2*5 |
| X2 | [9,1] | C41 | 403200 = 2(8)*3(2)*5(2)*7 | 9 = 3(2) |
| X3 | [8,2] | C40 | 226800 = 2(4)*3(4)*5(2)*7 | 16 = 2(4) |
| X4 | [8,1,1] | C39 | 226800 = 2(4)*3(4)*5(2)*7 | 16 = 2(4) |
| X5 | [7,3] | C38 | 172800 = 2(8)*3(3)*5(2) | 21 = 3*7 |
| X6 | [7,2,1] | C37 | 259200 = 2(7)*3(4)*5(2) | 14 = 2*7 |
| X7 | [7,1,1,1] | C36 | 86400 = 2(7)*3(3)*5(2) | 42 = 2*3*7 |
| X8 | [6,4] | C35 | 151200 = 2(5)*3(3)*5(2)*7 | 24 = 2(3)*3 |
| X9 | [6,3,1] | C34 | 201600 = 2(7)*3(2)*5(2)*7 | 18 = 2*3(2) |
| X10 | [6,2,2] | C33 | 75600 = 2(4)*3(3)*5(2)*7 | 48 = 2(4)*3 |
| X11 | [6,2,1,1] | C32 | 151200 = 2(5)*3(3)*5(2)*7 | 24 = 2(3)*3 |
| X12 | [6,1,1,1,1] | C31 | 25200 = 2(4)*3(2)*5(2)*7 | 144 = 2(4)*3(2) |
| X13 | [5,5] | C30 | 72576 = 2(7)*3(4)*7 | 50 = 2*5(2) |
| X14 | [5,4,1] | C29 | 181440 = 2(6)*3(4)*5*7 | 20 = 2(2)*5 |
| X15 | [5,3,2] | C28 | 120960 = 2(7)*3(3)*5*7 | 30 = 2*3*5 |
| X16 | [5,3,1,1] | C27 | 120960 = 2(7)*3(3)*5*7 | 30 = 2*3*5 |
| X17 | [5,2,2,1] | C26 | 90720 = 2(5)*3(4)*5*7 | 40 = 2(3)*5 |
| X18 | [5,2,1,1,1] | C25 | 60480 = 2(6)*3(3)*5*7 | 60 = 2(2)*3*5 |
| X19 | [5,1,1,1,1,1] | C24 | 6048 = 2(5)*3(3)*7 | 600 = 2(3)*3*5(2) |
| X20 | [4,4,2] | C23 | 56700 = 2(2)*3(4)*5(2)*7 | 64 = 2(6) |
| X21 | [4,4,1,1] | C22 | 56700 = 2(2)*3(4)*5(2)*7 | 64 = 2(6) |
| X22 | [4,3,3] | C21 | 50400 = 2(5)*3(2)*5(2)*7 | 72 = 2(3)*3(2) |
| X23 | [4,3,2,1] | C20 | 151200 = 2(5)*3(3)*5(2)*7 | 24 = 2(3)*3 |
| X24 | [4,3,1,1,1] | C19 | 50400 = 2(5)*3(2)*5(2)*7 | 72 = 2(3)*3(2) |
| X25 | [4,2,2,2] | C18 | 18900 = 2(2)*3(3)*5(2)*7 | 192 = 2(6)*3 |
| X26 | [4,2,2,1,1] | C17 | 56700 = 2(2)*3(4)*5(2)*7 | 64 = 2(6) |
| X27 | [4,2,1,1,1,1] | C16 | 18900 = 2(2)*3(3)*5(2)*7 | 192 = 2(6)*3 |
| X28 | [4,1,1,1,1,1,1] | C15 | 1260 = 2(2)*3(2)*5*7 | 2880 = 2(6)*3(2)*5 |
| X29 | [3,3,3,1] | C14 | 22400 = 2(7)*5(2)*7 | 162 = 2*3(4) |
| X30 | [3,3,2,2] | C13 | 25200 = 2(4)*3(2)*5(2)*7 | 144 = 2(4)*3(2) |
| X31 | [3,3,2,1,1] | C12 | 50400 = 2(5)*3(2)*5(2)*7 | 72 = 2(3)*3(2) |
| X32 | [3,3,1,1,1,1] | C11 | 8400 = 2(4)*3*5(2)*7 | 432 = 2(4)*3(3) |
| X33 | [3,2,2,2,1] | C10 | 25200 = 2(4)*3(2)*5(2)*7 | 144 = 2(4)*3(2) |
| X34 | [3,2,2,1,1,1] | C9 | 25200 = 2(4)*3(2)*5(2)*7 | 144 = 2(4)*3(2) |
| X35 | [3,2,1,1,1,1,1] | C8 | 5040 = 2(4)*3(2)*5*7 | 720 = 2(4)*3(2)*5 |
| X36 | [3,1,1,1,1,1,1,1] | C7 | 240 = 2(4)*3*5 | 15120 = 2(4)*3(3)*5*7 |
| X37 | [2,2,2,2,2] | C6 | 945 = 3(3)*5*7 | 3840 = 2(8)*3*5 |
| X38 | [2,2,2,2,1,1] | C5 | 4725 = 3(3)*5(2)*7 | 768 = 2(8)*3 |
| X39 | [2,2,2,1,1,1,1] | C4 | 3150 = 2*3(2)*5(2)*7 | 1152 = 2(7)*3(2) |
| X40 | [2,2,1,1,1,1,1,1] | C3 | 630 = 2*3(2)*5*7 | 5760 = 2(7)*3(2)*5 |
| X41 | [2,1,1,1,1,1,1,1,1] | C2 | 45 = 3(2)*5 | 80640 = 2(8)*3(2)*5*7 |
| X42 | [1,1,1,1,1,1,1,1,1,1] | C1 | 1 = 1(0) | 3628800 = 2(8)*3(4)*5(2)*7 |

CHARACTERS

| | C1 | C2 | C3 | C4 | C5 | C6 | C7 | C8 | C9 | C10 | C11 | C12 | C13 | C14 | C15 | C16 | C17 | C18 | C19 | C20 | C21 |
|---|---|---|---|---|---|---|---|---|---|---|---|---|---|---|---|---|---|---|---|---|---|
| X1 | 1 | 1 | 1 | 1 | 1 | 1 | 1 | 1 | 1 | 1 | 1 | 1 | 1 | 1 | 1 | 1 | 1 | 1 | 1 | 1 | 1 |
| X2 | 9 | 7 | 5 | 3 | 1 | -1 | 6 | 4 | 2 | 0 | 3 | 1 | -1 | 0 | 5 | 3 | 1 | -1 | 2 | 0 | -1 |
| X3 | 35 | 21 | 11 | 5 | 3 | 5 | 14 | 6 | 2 | 2 | 2 | 0 | 2 | -1 | 9 | 3 | 1 | 3 | 0 | 0 | 0 |
| X4 | 36 | 20 | 8 | 0 | -4 | -4 | 15 | 5 | -1 | -3 | 3 | -1 | -1 | 0 | 10 | 2 | -2 | -2 | 1 | -1 | 1 |
| X5 | 75 | 35 | 15 | 7 | 3 | -5 | 15 | 5 | 3 | 1 | 0 | 2 | 0 | 3 | 5 | 1 | 1 | -3 | -1 | 1 | 2 |
| X6 | 160 | 64 | 16 | 0 | 0 | 0 | 34 | 4 | -2 | 0 | -2 | -2 | -2 | -2 | 16 | 0 | 0 | 0 | -2 | 0 | -2 |
| X7 | 84 | 28 | 0 | -8 | -4 | 4 | 21 | 1 | -3 | 1 | 3 | 1 | 3 | 3 | 10 | -2 | -2 | 2 | 1 | 1 | 1 |
| X8 | 90 | 34 | 14 | 6 | 2 | 10 | 6 | 4 | 2 | 0 | 3 | 1 | -1 | 0 | -4 | 0 | 0 | 4 | 2 | 0 | -1 |
| X9 | 315 | 91 | 19 | 3 | -5 | -5 | 21 | 1 | 1 | -3 | -3 | 1 | 1 | 0 | -1 | -1 | -1 | -1 | -1 | -1 | -1 |
| X10 | 225 | 55 | 5 | 3 | 9 | 15 | 15 | -5 | -1 | 3 | -6 | -2 | 2 | 0 | 5 | -1 | 1 | 3 | -1 | -1 | 2 |
| X11 | 350 | 70 | -10 | -10 | -2 | -10 | 35 | -5 | -1 | -1 | -1 | 1 | -1 | -1 | 10 | -2 | 2 | -2 | 1 | 1 | 1 |
| X12 | 126 | 14 | -14 | -6 | 6 | 6 | 21 | -1 | 1 | 3 | 6 | 2 | -2 | 0 | 4 | -4 | 0 | 0 | 1 | -1 | -2 |
| X13 | 42 | 14 | 6 | 2 | 2 | -10 | 0 | 2 | 0 | 2 | 3 | -1 | 3 | -3 | -4 | 0 | 0 | -4 | 2 | 0 | -1 |
| X14 | 288 | 64 | 16 | 0 | 0 | 0 | -6 | 4 | -2 | 0 | 6 | -2 | -2 | 0 | -16 | 0 | 0 | 0 | 2 | 0 | 2 |
| X15 | 450 | 70 | 10 | 6 | 2 | -10 | -15 | -5 | 1 | 3 | -3 | 1 | 1 | 0 | -10 | -2 | -2 | -2 | -1 | 1 | -1 |
| X16 | 567 | 63 | -9 | -9 | -9 | 15 | 0 | 0 | 0 | 0 | 0 | 0 | 0 | 0 | -9 | 3 | -1 | 3 | 0 | 0 | 0 |
| X17 | 525 | 35 | -15 | 7 | 5 | -5 | 0 | -10 | 0 | -2 | -3 | -1 | -3 | 3 | 5 | 3 | 1 | -1 | 2 | 0 | -1 |
| X18 | 448 | 0 | -32 | 0 | 0 | 0 | 28 | 0 | 4 | 0 | 4 | 0 | 4 | -2 | 0 | 0 | 0 | 0 | 0 | 0 | 0 |
| X19 | 126 | -14 | -14 | 6 | 6 | -6 | 21 | 1 | 1 | -3 | 6 | -2 | -2 | 0 | -4 | -4 | 0 | 0 | -1 | -1 | 2 |
| X20 | 252 | 28 | 8 | 0 | 4 | 20 | -21 | 1 | -1 | -3 | 3 | 1 | -1 | 0 | -10 | -2 | 2 | 2 | -1 | 1 | -1 |
| X21 | 300 | 20 | 0 | -8 | 4 | -20 | -15 | 5 | -3 | 1 | 3 | -1 | 3 | 3 | -10 | 2 | 2 | -2 | -1 | -1 | -1 |
| X22 | 210 | 14 | 6 | 2 | -6 | -10 | -21 | -1 | 3 | -1 | 0 | 2 | 0 | 3 | -4 | -4 | 0 | 0 | -1 | -1 | 2 |
| X23 | 768 | 0 | 0 | 0 | 0 | 0 | -48 | 0 | 0 | 0 | 0 | 0 | 0 | -6 | 0 | 0 | 0 | 0 | 0 | 0 | 0 |
| X24 | 525 | -35 | -15 | -7 | 5 | 5 | 0 | 10 | 0 | 2 | -3 | 1 | -3 | 3 | -5 | 3 | -1 | -1 | -2 | 0 | 1 |
| X25 | 300 | -20 | 0 | 8 | 4 | 20 | -15 | -5 | -3 | -1 | 3 | 1 | 3 | 3 | 10 | 2 | -2 | -2 | 1 | -1 | 1 |
| X26 | 567 | -63 | -9 | 9 | -9 | -15 | 0 | 0 | 0 | 0 | 0 | 0 | 0 | 0 | 9 | 3 | 1 | 3 | 0 | 0 | 0 |
| X27 | 350 | -70 | -10 | 10 | -2 | 10 | 35 | 5 | -1 | 1 | -1 | -1 | -1 | -1 | -10 | -2 | -2 | -2 | -1 | 1 | -1 |
| X28 | 84 | -28 | 0 | 8 | -4 | -4 | 21 | -1 | -3 | -1 | 3 | -1 | 3 | 3 | -10 | -2 | 2 | 2 | -1 | 1 | -1 |
| X29 | 210 | -14 | 6 | -2 | -6 | 10 | -21 | 1 | 3 | 1 | 0 | -2 | 0 | 3 | 4 | -4 | 0 | 0 | 1 | -1 | -2 |
| X30 | 252 | -28 | 8 | 0 | 4 | -20 | -21 | -1 | -1 | 3 | 3 | -1 | -1 | 0 | 10 | -2 | -2 | 2 | 1 | 1 | 1 |
| X31 | 450 | -70 | 10 | -6 | 2 | 10 | -15 | 5 | 1 | -3 | -3 | -1 | 1 | 0 | 10 | -2 | 2 | -2 | 1 | 1 | 1 |
| X32 | 225 | -55 | 5 | -3 | 9 | -15 | 15 | 5 | -1 | -3 | -6 | 2 | 2 | 0 | -5 | -1 | -1 | 3 | 1 | -1 | -2 |
| X33 | 288 | -64 | 16 | 0 | 0 | 0 | -6 | -4 | -2 | 0 | 6 | 2 | -2 | 0 | 16 | 0 | 0 | 0 | -2 | 0 | -2 |
| X34 | 315 | -91 | 19 | -3 | -5 | 5 | 21 | -1 | 1 | 3 | -3 | -1 | 1 | 0 | 1 | -1 | 1 | -1 | 1 | -1 | 1 |
| X35 | 160 | -64 | 16 | 0 | 0 | 0 | 34 | -4 | -2 | 0 | -2 | 2 | -2 | -2 | -16 | 0 | 0 | 0 | 2 | 0 | 2 |
| X36 | 36 | -20 | 8 | 0 | -4 | 4 | 15 | -5 | -1 | 3 | 3 | 1 | -1 | 0 | -10 | 2 | 2 | -2 | -1 | -1 | -1 |
| X37 | 42 | -14 | 6 | -2 | 2 | 10 | 0 | -2 | 0 | -2 | 3 | 1 | 3 | -3 | 4 | 0 | 0 | -4 | -2 | 0 | 1 |
| X38 | 90 | -34 | 14 | -6 | 2 | -10 | 6 | -4 | 2 | 0 | 3 | -1 | -1 | 0 | 4 | 0 | 0 | 4 | -2 | 0 | 1 |
| X39 | 75 | -35 | 15 | -7 | 3 | 5 | 15 | -5 | 3 | -1 | 0 | -2 | 0 | 3 | -5 | 1 | -1 | -3 | 1 | 1 | -2 |
| X40 | 35 | -21 | 11 | -5 | 3 | -5 | 14 | -6 | 2 | -2 | 2 | 0 | 2 | -1 | -9 | 3 | -1 | 3 | 0 | 0 | 0 |
| X41 | 9 | -7 | 5 | -3 | 1 | 1 | 6 | -4 | 2 | 0 | 3 | -1 | -1 | 0 | -5 | 3 | -1 | -1 | -2 | 0 | 1 |
| X42 | 1 | -1 | 1 | -1 | 1 | -1 | 1 | -1 | 1 | -1 | 1 | -1 | 1 | 1 | -1 | 1 | -1 | 1 | -1 | 1 | -1 |

| | C22 | C23 | C24 | C25 | C26 | C27 | C28 | C29 | C30 | C31 | C32 | C33 | C34 | C35 | C36 | C37 | C38 | C39 | C40 | C41 | C42 |
|---|---|---|---|---|---|---|---|---|---|---|---|---|---|---|---|---|---|---|---|---|---|
| X1 | 1 | 1 | 1 | 1 | 1 | 1 | 1 | 1 | 1 | 1 | 1 | 1 | 1 | 1 | 1 | 1 | 1 | 1 | 1 | 1 | 1 |
| X2 | 1 | -1 | 4 | 2 | 0 | 1 | -1 | 0 | -1 | 3 | 1 | -1 | 0 | -1 | 2 | 0 | -1 | 1 | -1 | 0 | -1 |
| X3 | -1 | 1 | 5 | 1 | 1 | -1 | 1 | -1 | 0 | 2 | 0 | 2 | -1 | 0 | 0 | 0 | 0 | -1 | 1 | -1 | 0 |
| X4 | 0 | 0 | 6 | 0 | -2 | 0 | 0 | 0 | 1 | 3 | -1 | -1 | 0 | 1 | 1 | -1 | 1 | 0 | 0 | 0 | 1 |
| X5 | -1 | -1 | 0 | 0 | 0 | 0 | 0 | 0 | 0 | -2 | 0 | -2 | 1 | 0 | -2 | 0 | 1 | -1 | -1 | 0 | 0 |
| X6 | 0 | 0 | 5 | -1 | 1 | -1 | -1 | 1 | 0 | 0 | 0 | 0 | 0 | 0 | -1 | 1 | -1 | 0 | 0 | 1 | 0 |
| X7 | 0 | 0 | 4 | -2 | 0 | 1 | 1 | 0 | -1 | 1 | -1 | 1 | 1 | -1 | 0 | 0 | 0 | 0 | 0 | 0 | -1 |
| X8 | 2 | 2 | -5 | -1 | -1 | 1 | -1 | 1 | 0 | -3 | -1 | 1 | 0 | 1 | -1 | -1 | -1 | 0 | 0 | 0 | 0 |
| X9 | -1 | -1 | -5 | 1 | -1 | 1 | 1 | -1 | 0 | -3 | 1 | 1 | 0 | -1 | 0 | 0 | 0 | 1 | 1 | 0 | 0 |
| X10 | 1 | -1 | 0 | 0 | 0 | 0 | 0 | 0 | 0 | 0 | 0 | 0 | 0 | 0 | 1 | -1 | 1 | 1 | -1 | 0 | 0 |
| X11 | 2 | 2 | 0 | 0 | 0 | 0 | 0 | 0 | 0 | -1 | 1 | -1 | -1 | 1 | 0 | 0 | 0 | 0 | 0 | -1 | 0 |
| X12 | -2 | -2 | 1 | -1 | 1 | 1 | -1 | -1 | 1 | 0 | 0 | 0 | 0 | 0 | 0 | 0 | 0 | 0 | 0 | 0 | 1 |
| X13 | 2 | -2 | -3 | -1 | 1 | 0 | 2 | 1 | 2 | -1 | -1 | -1 | -1 | -1 | 0 | 0 | 0 | 0 | 0 | 0 | 0 |
| X14 | 0 | 0 | -7 | -1 | 1 | -1 | -1 | -1 | -2 | 0 | 0 | 0 | 0 | 0 | 1 | 1 | 1 | 0 | 0 | 0 | 0 |
| X15 | -2 | 2 | 0 | 0 | 0 | 0 | 0 | 0 | 0 | 3 | -1 | -1 | 0 | 1 | 2 | 0 | -1 | 0 | 0 | 0 | 0 |
| X16 | -1 | -1 | -3 | 3 | 1 | 0 | 0 | 1 | 2 | 0 | 0 | 0 | 0 | 0 | 0 | 0 | 0 | -1 | -1 | 0 | 0 |
| X17 | 1 | -1 | 0 | 0 | 0 | 0 | 0 | 0 | 0 | 1 | -1 | 1 | 1 | -1 | 0 | 0 | 0 | -1 | 1 | 0 | 0 |
| X18 | 0 | 0 | -2 | 0 | -2 | -2 | 0 | 0 | -2 | 0 | 0 | 0 | 0 | 0 | 0 | 0 | 0 | 0 | 0 | 1 | 0 |
| X19 | -2 | 2 | 1 | 1 | 1 | 1 | 1 | 1 | 1 | 0 | 0 | 0 | 0 | 0 | 0 | 0 | 0 | 0 | 0 | 0 | -1 |
| X20 | 0 | 0 | 2 | -2 | -2 | -1 | 1 | 0 | 2 | 3 | 1 | -1 | 0 | -1 | 0 | 0 | 0 | 0 | 0 | 0 | 0 |
| X21 | 0 | 0 | 0 | 0 | 0 | 0 | 0 | 0 | 0 | 1 | 1 | 1 | 1 | 1 | -1 | -1 | -1 | 0 | 0 | 0 | 0 |
| X22 | 2 | -2 | 5 | -1 | 1 | -1 | -1 | 1 | 0 | 2 | 0 | 2 | -1 | 0 | 0 | 0 | 0 | 0 | 0 | 0 | 0 |
| X23 | 0 | 0 | 8 | 0 | 0 | 2 | 0 | 0 | -2 | 0 | 0 | 0 | 0 | 0 | -2 | 0 | 1 | 0 | 0 | 0 | 0 |
| X24 | 1 | 1 | 0 | 0 | 0 | 0 | 0 | 0 | 0 | -1 | -1 | -1 | -1 | -1 | 0 | 0 | 0 | 1 | 1 | 0 | 0 |
| X25 | 0 | 0 | 0 | 0 | 0 | 0 | 0 | 0 | 0 | -1 | 1 | -1 | -1 | 1 | -1 | 1 | -1 | 0 | 0 | 0 | 0 |
| X26 | -1 | 1 | -3 | -3 | 1 | 0 | 0 | -1 | 2 | 0 | 0 | 0 | 0 | 0 | 0 | 0 | 0 | 1 | -1 | 0 | 0 |
| X27 | 2 | -2 | 0 | 0 | 0 | 0 | 0 | 0 | 0 | 1 | 1 | 1 | 1 | 1 | 0 | 0 | 0 | 0 | 0 | -1 | 0 |
| X28 | 0 | 0 | 4 | 2 | 0 | 1 | -1 | 0 | -1 | -1 | -1 | -1 | -1 | -1 | 0 | 0 | 0 | 0 | 0 | 0 | 1 |
| X29 | 2 | 2 | 5 | 1 | 1 | -1 | 1 | -1 | 0 | -2 | 0 | -2 | 1 | 0 | 0 | 0 | 0 | 0 | 0 | 0 | 0 |
| X30 | 0 | 0 | 2 | 2 | -2 | -1 | -1 | 0 | 2 | -3 | 1 | 1 | 0 | -1 | 0 | 0 | 0 | 0 | 0 | 0 | 0 |
| X31 | -2 | -2 | 0 | 0 | 0 | 0 | 0 | 0 | 0 | -3 | -1 | 1 | 0 | 1 | 2 | 0 | -1 | 0 | 0 | 0 | 0 |
| X32 | 1 | 1 | 0 | 0 | 0 | 0 | 0 | 0 | 0 | 0 | 0 | 0 | 0 | 0 | 1 | 1 | 1 | -1 | -1 | 0 | 0 |
| X33 | 0 | 0 | -7 | 1 | 1 | -1 | 1 | 1 | -2 | 0 | 0 | 0 | 0 | 0 | 1 | -1 | 1 | 0 | 0 | 0 | 0 |
| X34 | -1 | 1 | -5 | -1 | -1 | 1 | -1 | 1 | 0 | 3 | 1 | -1 | 0 | -1 | 0 | 0 | 0 | -1 | 1 | 0 | 0 |
| X35 | 0 | 0 | 5 | 1 | 1 | -1 | 1 | -1 | 0 | 0 | 0 | 0 | 0 | 0 | -1 | -1 | -1 | 0 | 0 | 1 | 0 |
| X36 | 0 | 0 | 6 | 0 | -2 | 0 | 0 | 0 | 1 | -3 | -1 | 1 | 0 | 1 | 1 | 1 | 1 | 0 | 0 | 0 | -1 |
| X37 | 2 | 2 | -3 | 1 | 1 | 0 | -2 | -1 | 2 | 1 | -1 | 1 | 1 | -1 | 0 | 0 | 0 | 0 | 0 | 0 | 0 |
| X38 | 2 | -2 | -5 | 1 | -1 | 1 | 1 | -1 | 0 | 3 | -1 | -1 | 0 | 1 | -1 | 1 | -1 | 0 | 0 | 0 | 0 |
| X39 | -1 | 1 | 0 | 0 | 0 | 0 | 0 | 0 | 0 | 2 | 0 | 2 | -1 | 0 | -2 | 0 | 1 | 1 | -1 | 0 | 0 |
| X40 | -1 | -1 | 5 | -1 | 1 | -1 | -1 | 1 | 0 | -2 | 0 | -2 | 1 | 0 | 0 | 0 | 0 | 1 | 1 | -1 | 0 |
| X41 | 1 | 1 | 4 | -2 | 0 | 1 | 1 | 0 | -1 | -3 | 1 | 1 | 0 | -1 | 2 | 0 | -1 | -1 | -1 | 0 | 1 |
| X42 | 1 | -1 | 1 | -1 | 1 | 1 | -1 | -1 | 1 | -1 | 1 | -1 | -1 | 1 | 1 | -1 | 1 | -1 | 1 | 1 | -1 |

## I.B Class Multiplication Coefficients

This section of Appendix I contains the class multiplication coefficients $a_{ijk}$ which satisfy

$$Ci * Cj = \sum_{k} a_{ijk} Ck.$$

Here $Ci$, $Cj$ and $Ck$ denote conjugacy classes of $S_n$, numbered in the same way as in Appendix I.A. $*$ indicates the complex product of the corresponding class sums. As $a_{ijk} = a_{jik}$, we need only show $a_{ijk}$ for $i \leqslant j$.

The program was implemented by G. Berard. The coefficients $a_{ijk}$, $i \leqslant j$, are tabulated for $n \leqslant 8$.

Class multiplication coefficients for $S_2$, $S_3$, $S_4$, and $S_5$

| | C1 | C2 |
|---|---|---|
| C1*C1 | 1 | 0 |
| C1*C2 | 0 | 1 |
| C2*C2 | 1 | 0 |

| | C1 | C2 | C3 |
|---|---|---|---|
| C1*C1 | 1 | 0 | 0 |
| C1*C2 | 0 | 1 | 0 |
| C1*C3 | 0 | 0 | 1 |
| C2*C2 | 3 | 0 | 3 |
| C2*C3 | 0 | 2 | 0 |
| C3*C3 | 2 | 0 | 1 |

| | C1 | C2 | C3 | C4 | C5 |
|---|---|---|---|---|---|
| C1*C1 | 1 | 0 | 0 | 0 | 0 |
| C1*C2 | 0 | 1 | 0 | 0 | 0 |
| C1*C3 | 0 | 0 | 1 | 0 | 0 |
| C1*C4 | 0 | 0 | 0 | 1 | 0 |
| C1*C5 | 0 | 0 | 0 | 0 | 1 |
| C2*C2 | 6 | 0 | 2 | 3 | 0 |
| C2*C3 | 0 | 1 | 0 | 0 | 2 |
| C2*C4 | 0 | 4 | 0 | 0 | 4 |
| C2*C5 | 0 | 0 | 4 | 3 | 0 |
| C3*C3 | 3 | 0 | 2 | 0 | 0 |
| C3*C4 | 0 | 0 | 0 | 3 | 0 |
| C3*C5 | 0 | 2 | 0 | 0 | 1 |
| C4*C4 | 8 | 0 | 8 | 4 | 0 |
| C4*C5 | 0 | 4 | 0 | 0 | 4 |
| C5*C5 | 6 | 0 | 2 | 3 | 0 |

| | C1 | C2 | C3 | C4 | C5 | C6 | C7 |
|---|---|---|---|---|---|---|---|
| C1*C1 | 1 | 0 | 0 | 0 | 0 | 0 | 0 |
| C1*C2 | 0 | 1 | 0 | 0 | 0 | 0 | 0 |
| C1*C3 | 0 | 0 | 1 | 0 | 0 | 0 | 0 |
| C1*C4 | 0 | 0 | 0 | 1 | 0 | 0 | 0 |
| C1*C5 | 0 | 0 | 0 | 0 | 1 | 0 | 0 |
| C1*C6 | 0 | 0 | 0 | 0 | 0 | 1 | 0 |
| C1*C7 | 0 | 0 | 0 | 0 | 0 | 0 | 1 |
| C2*C2 | 10 | 0 | 2 | 3 | 0 | 0 | 0 |
| C2*C3 | 0 | 3 | 0 | 0 | 3 | 2 | 0 |
| C2*C4 | 0 | 6 | 0 | 0 | 1 | 4 | 0 |
| C2*C5 | 0 | 0 | 4 | 1 | 0 | 0 | 5 |
| C2*C6 | 0 | 0 | 4 | 6 | 0 | 0 | 5 |
| C2*C7 | 0 | 0 | 0 | 0 | 6 | 4 | 0 |
| C3*C3 | 15 | 0 | 2 | 3 | 0 | 0 | 5 |
| C3*C4 | 0 | 0 | 4 | 6 | 0 | 0 | 5 |
| C3*C5 | 0 | 6 | 0 | 0 | 6 | 4 | 0 |
| C3*C6 | 0 | 6 | 0 | 0 | 6 | 9 | 0 |
| C3*C7 | 0 | 0 | 8 | 6 | 0 | 0 | 5 |
| C4*C4 | 20 | 0 | 8 | 7 | 0 | 0 | 5 |
| C4*C5 | 0 | 2 | 0 | 0 | 7 | 8 | 0 |
| C4*C6 | 0 | 12 | 0 | 0 | 12 | 8 | 0 |
| C4*C7 | 0 | 0 | 8 | 6 | 0 | 0 | 10 |
| C5*C5 | 20 | 0 | 8 | 7 | 0 | 0 | 5 |
| C5*C6 | 0 | 0 | 8 | 12 | 0 | 0 | 10 |
| C5*C7 | 0 | 12 | 0 | 0 | 6 | 8 | 0 |
| C6*C6 | 30 | 0 | 18 | 12 | 0 | 0 | 15 |
| C6*C7 | 0 | 12 | 0 | 0 | 12 | 12 | 0 |
| C7*C7 | 24 | 0 | 8 | 12 | 0 | 0 | 8 |

Class multiplication coefficients for $S_6$

| | C1 | C2 | C3 | C4 | C5 | C6 | C7 | C8 | C9 | C10 | C11 |
|---|---|---|---|---|---|---|---|---|---|---|---|
| C1*C1 | 1 | 0 | 0 | 0 | 0 | 0 | 0 | 0 | 0 | 0 | 0 |
| C1*C2 | 0 | 1 | 0 | 0 | 0 | 0 | 0 | 0 | 0 | 0 | 0 |
| C1*C3 | 0 | 0 | 1 | 0 | 0 | 0 | 0 | 0 | 0 | 0 | 0 |
| C1*C4 | 0 | 0 | 0 | 1 | 0 | 0 | 0 | 0 | 0 | 0 | 0 |
| C1*C5 | 0 | 0 | 0 | 0 | 1 | 0 | 0 | 0 | 0 | 0 | 0 |
| C1*C6 | 0 | 0 | 0 | 0 | 0 | 1 | 0 | 0 | 0 | 0 | 0 |
| C1*C7 | 0 | 0 | 0 | 0 | 0 | 0 | 1 | 0 | 0 | 0 | 0 |
| C1*C8 | 0 | 0 | 0 | 0 | 0 | 0 | 0 | 1 | 0 | 0 | 0 |
| C1*C9 | 0 | 0 | 0 | 0 | 0 | 0 | 0 | 0 | 1 | 0 | 0 |
| C1*C10 | 0 | 0 | 0 | 0 | 0 | 0 | 0 | 0 | 0 | 1 | 0 |
| C1*C11 | 0 | 0 | 0 | 0 | 0 | 0 | 0 | 0 | 0 | 0 | 1 |
| C2*C2 | 15 | 0 | 2 | 0 | 3 | 0 | 0 | 0 | 0 | 0 | 0 |
| C2*C3 | 0 | 6 | 0 | 3 | 0 | 3 | 0 | 2 | 0 | 0 | 0 |
| C2*C4 | 0 | 0 | 1 | 0 | 0 | 0 | 0 | 0 | 2 | 0 | 0 |
| C2*C5 | 0 | 8 | 0 | 0 | 0 | 1 | 0 | 4 | 0 | 0 | 0 |
| C2*C6 | 0 | 0 | 8 | 0 | 3 | 0 | 6 | 0 | 4 | 5 | 0 |
| C2*C7 | 0 | 0 | 0 | 0 | 0 | 2 | 0 | 0 | 0 | 0 | 3 |
| C2*C8 | 0 | 0 | 4 | 0 | 9 | 0 | 0 | 0 | 1 | 5 | 0 |
| C2*C9 | 0 | 0 | 0 | 12 | 0 | 3 | 0 | 1 | 0 | 0 | 6 |
| C2*C10 | 0 | 0 | 0 | 0 | 0 | 6 | 0 | 8 | 0 | 0 | 6 |
| C2*C11 | 0 | 0 | 0 | 0 | 0 | 0 | 9 | 0 | 8 | 5 | 0 |
| C3*C3 | 45 | 0 | 4 | 0 | 9 | 0 | 9 | 0 | 4 | 5 | 0 |
| C3*C4 | 0 | 3 | 0 | 6 | 0 | 0 | 0 | 2 | 0 | 0 | 3 |
| C3*C5 | 0 | 0 | 8 | 0 | 9 | 0 | 0 | 0 | 4 | 5 | 0 |
| C3*C6 | 0 | 24 | 0 | 0 | 0 | 15 | 0 | 12 | 0 | 0 | 18 |
| C3*C7 | 0 | 0 | 8 | 0 | 0 | 0 | 9 | 0 | 4 | 5 | 0 |
| C3*C8 | 0 | 12 | 0 | 12 | 0 | 9 | 0 | 17 | 0 | 0 | 9 |
| C3*C9 | 0 | 0 | 8 | 0 | 9 | 0 | 9 | 0 | 17 | 10 | 0 |
| C3*C10 | 0 | 0 | 16 | 0 | 18 | 0 | 18 | 0 | 16 | 20 | 0 |
| C3*C11 | 0 | 0 | 0 | 24 | 0 | 18 | 0 | 12 | 0 | 0 | 15 |
| C4*C4 | 15 | 0 | 2 | 0 | 0 | 0 | 3 | 0 | 0 | 0 | 0 |
| C4*C5 | 0 | 0 | 0 | 0 | 0 | 3 | 0 | 0 | 0 | 0 | 2 |
| C4*C6 | 0 | 0 | 0 | 0 | 9 | 0 | 0 | 0 | 8 | 5 | 0 |
| C4*C7 | 0 | 0 | 0 | 8 | 0 | 0 | 0 | 4 | 0 | 0 | 1 |
| C4*C8 | 0 | 0 | 4 | 0 | 0 | 0 | 9 | 0 | 1 | 5 | 0 |
| C4*C9 | 0 | 12 | 0 | 0 | 0 | 6 | 0 | 1 | 0 | 0 | 3 |
| C4*C10 | 0 | 0 | 0 | 0 | 0 | 6 | 0 | 8 | 0 | 0 | 6 |
| C4*C11 | 0 | 0 | 8 | 0 | 6 | 0 | 3 | 0 | 4 | 5 | 0 |

| | C1 | C2 | C3 | C4 | C5 | C6 | C7 | C8 | C9 | C10 | C11 |
|---|---|---|---|---|---|---|---|---|---|---|---|
| C5*C5 | 40 | 0 | 8 | 0 | 10 | 0 | 2 | 0 | 0 | 5 | 0 |
| C5*C6 | 0 | 8 | 0 | 24 | 0 | 12 | 0 | 16 | 0 | 0 | 12 |
| C5*C7 | 0 | 0 | 0 | 0 | 2 | 0 | 2 | 0 | 8 | 5 | 0 |
| C5*C8 | 0 | 24 | 0 | 0 | 0 | 12 | 0 | 12 | 0 | 0 | 6 |
| C5*C9 | 0 | 0 | 8 | 0 | 0 | 0 | 18 | 0 | 12 | 10 | 0 |
| C5*C10 | 0 | 0 | 16 | 0 | 18 | 0 | 18 | 0 | 16 | 15 | 0 |
| C5*C11 | 0 | 0 | 0 | 16 | 0 | 12 | 0 | 8 | 0 | 0 | 20 |
| C6*C6 | 120 | 0 | 40 | 0 | 36 | 0 | 60 | 0 | 32 | 40 | 0 |
| C6*C7 | 0 | 16 | 0 | 0 | 0 | 20 | 0 | 8 | 0 | 0 | 12 |
| C6*C8 | 0 | 0 | 24 | 0 | 36 | 0 | 18 | 0 | 36 | 30 | 0 |
| C6*C9 | 0 | 24 | 0 | 48 | 0 | 24 | 0 | 36 | 0 | 0 | 30 |
| C6*C10 | 0 | 48 | 0 | 48 | 0 | 48 | 0 | 48 | 0 | 0 | 48 |
| C6*C11 | 0 | 0 | 48 | 0 | 36 | 0 | 36 | 0 | 40 | 40 | 0 |
| C7*C7 | 40 | 0 | 8 | 0 | 2 | 0 | 10 | 0 | 0 | 5 | 0 |
| C7*C8 | 0 | 0 | 0 | 24 | 0 | 6 | 0 | 12 | 0 | 0 | 12 |
| C7*C9 | 0 | 0 | 8 | 0 | 18 | 0 | 0 | 0 | 12 | 10 | 0 |
| C7*C10 | 0 | 0 | 16 | 0 | 18 | 0 | 18 | 0 | 16 | 15 | 0 |
| C7*C11 | 0 | 24 | 0 | 8 | 0 | 12 | 0 | 16 | 0 | 0 | 12 |
| C8*C8 | 90 | 0 | 34 | 0 | 27 | 0 | 27 | 0 | 16 | 20 | 0 |
| C8*C9 | 0 | 6 | 0 | 6 | 0 | 27 | 0 | 16 | 0 | 0 | 27 |
| C8*C10 | 0 | 48 | 0 | 48 | 0 | 36 | 0 | 32 | 0 | 0 | 36 |
| C8*C11 | 0 | 0 | 24 | 0 | 18 | 0 | 36 | 0 | 36 | 30 | 0 |
| C9*C9 | 90 | 0 | 34 | 0 | 27 | 0 | 27 | 0 | 16 | 20 | 0 |
| C9*C10 | 0 | 0 | 32 | 0 | 36 | 0 | 36 | 0 | 32 | 40 | 0 |
| C9*C11 | 0 | 48 | 0 | 24 | 0 | 30 | 0 | 36 | 0 | 0 | 24 |
| C10*C10 | 144 | 0 | 64 | 0 | 54 | 0 | 54 | 0 | 64 | 53 | 0 |
| C10*C11 | 0 | 48 | 0 | 48 | 0 | 48 | 0 | 48 | 0 | 0 | 48 |
| C11*C11 | 120 | 0 | 40 | 0 | 60 | 0 | 36 | 0 | 32 | 40 | 0 |

## Class multiplication coefficients for $S_7$

| | C1 | C2 | C3 | C4 | C5 | C6 | C7 | C8 | C9 | C10 | C11 | C12 | C13 | C14 | C15 |
|---|---|---|---|---|---|---|---|---|---|---|---|---|---|---|---|
| C1*C1 | 1 | 0 | 0 | 0 | 0 | 0 | 0 | 0 | 0 | 0 | 0 | 0 | 0 | 0 | 0 |
| C1*C2 | 0 | 1 | 0 | 0 | 0 | 0 | 0 | 0 | 0 | 0 | 0 | 0 | 0 | 0 | 0 |
| C1*C3 | 0 | 0 | 1 | 0 | 0 | 0 | 0 | 0 | 0 | 0 | 0 | 0 | 0 | 0 | 0 |
| C1*C4 | 0 | 0 | 0 | 1 | 0 | 0 | 0 | 0 | 0 | 0 | 0 | 0 | 0 | 0 | 0 |
| C1*C5 | 0 | 0 | 0 | 0 | 1 | 0 | 0 | 0 | 0 | 0 | 0 | 0 | 0 | 0 | 0 |
| C1*C6 | 0 | 0 | 0 | 0 | 0 | 1 | 0 | 0 | 0 | 0 | 0 | 0 | 0 | 0 | 0 |
| C1*C7 | 0 | 0 | 0 | 0 | 0 | 0 | 1 | 0 | 0 | 0 | 0 | 0 | 0 | 0 | 0 |
| C1*C8 | 0 | 0 | 0 | 0 | 0 | 0 | 0 | 1 | 0 | 0 | 0 | 0 | 0 | 0 | 0 |
| C1*C9 | 0 | 0 | 0 | 0 | 0 | 0 | 0 | 0 | 1 | 0 | 0 | 0 | 0 | 0 | 0 |
| C1*C10 | 0 | 0 | 0 | 0 | 0 | 0 | 0 | 0 | 0 | 1 | 0 | 0 | 0 | 0 | 0 |
| C1*C11 | 0 | 0 | 0 | 0 | 0 | 0 | 0 | 0 | 0 | 0 | 1 | 0 | 0 | 0 | 0 |
| C1*C12 | 0 | 0 | 0 | 0 | 0 | 0 | 0 | 0 | 0 | 0 | 0 | 1 | 0 | 0 | 0 |
| C1*C13 | 0 | 0 | 0 | 0 | 0 | 0 | 0 | 0 | 0 | 0 | 0 | 0 | 1 | 0 | 0 |
| C1*C14 | 0 | 0 | 0 | 0 | 0 | 0 | 0 | 0 | 0 | 0 | 0 | 0 | 0 | 1 | 0 |
| C1*C15 | 0 | 0 | 0 | 0 | 0 | 0 | 0 | 0 | 0 | 0 | 0 | 0 | 0 | 0 | 1 |
| C2*C2 | 21 | 0 | 2 | 0 | 3 | 0 | 0 | 0 | 0 | 0 | 0 | 0 | 0 | 0 | 0 |
| C2*C3 | 0 | 10 | 0 | 3 | 0 | 3 | 0 | 0 | 2 | 0 | 0 | 0 | 0 | 0 | 0 |
| C2*C4 | 0 | 0 | 3 | 0 | 0 | 0 | 3 | 0 | 0 | 2 | 0 | 0 | 0 | 0 | 0 |
| C2*C5 | 0 | 10 | 0 | 0 | 0 | 1 | 0 | 0 | 4 | 0 | 0 | 0 | 0 | 0 | 0 |
| C2*C6 | 0 | 0 | 12 | 0 | 6 | 0 | 2 | 6 | 0 | 4 | 0 | 5 | 0 | 0 | 0 |
| C2*C7 | 0 | 0 | 0 | 6 | 0 | 1 | 0 | 0 | 0 | 0 | 2 | 0 | 5 | 0 | 0 |
| C2*C8 | 0 | 0 | 0 | 0 | 0 | 4 | 0 | 0 | 0 | 0 | 4 | 0 | 0 | 3 | 0 |
| C2*C9 | 0 | 0 | 4 | 0 | 12 | 0 | 0 | 0 | 0 | 1 | 0 | 5 | 0 | 0 | 0 |
| C2*C10 | 0 | 0 | 0 | 12 | 0 | 6 | 0 | 0 | 3 | 0 | 3 | 0 | 5 | 6 | 0 |
| C2*C11 | 0 | 0 | 0 | 0 | 0 | 0 | 4 | 6 | 0 | 2 | 0 | 0 | 0 | 0 | 7 |
| C2*C12 | 0 | 0 | 0 | 0 | 0 | 6 | 0 | 0 | 12 | 0 | 0 | 0 | 1 | 6 | 0 |
| C2*C13 | 0 | 0 | 0 | 0 | 0 | 0 | 12 | 0 | 0 | 4 | 0 | 1 | 0 | 0 | 7 |
| C2*C14 | 0 | 0 | 0 | 0 | 0 | 0 | 0 | 9 | 0 | 8 | 0 | 10 | 0 | 0 | 7 |
| C2*C15 | 0 | 0 | 0 | 0 | 0 | 0 | 0 | 0 | 0 | 0 | 12 | 0 | 10 | 6 | 0 |
| C3*C3 | 105 | 0 | 8 | 0 | 18 | 0 | 6 | 9 | 0 | 4 | 0 | 5 | 0 | 0 | 0 |
| C3*C4 | 0 | 15 | 0 | 6 | 0 | 3 | 0 | 0 | 6 | 0 | 6 | 0 | 5 | 3 | 0 |
| C3*C5 | 0 | 0 | 12 | 0 | 12 | 0 | 1 | 0 | 0 | 4 | 0 | 5 | 0 | 0 | 0 |
| C3*C6 | 0 | 60 | 0 | 12 | 0 | 25 | 0 | 0 | 24 | 0 | 14 | 0 | 10 | 18 | 0 |
| C3*C7 | 0 | 0 | 12 | 0 | 3 | 0 | 14 | 0 | 0 | 8 | 0 | 5 | 0 | 0 | 14 |
| C3*C8 | 0 | 0 | 24 | 0 | 0 | 0 | 0 | 15 | 0 | 12 | 0 | 10 | 0 | 0 | 14 |
| C3*C9 | 0 | 20 | 0 | 12 | 0 | 12 | 0 | 0 | 25 | 0 | 0 | 0 | 5 | 9 | 0 |
| C3*C10 | 0 | 0 | 24 | 0 | 36 | 0 | 24 | 27 | 0 | 25 | 0 | 25 | 0 | 0 | 28 |
| C3*C11 | 0 | 0 | 0 | 24 | 0 | 14 | 0 | 0 | 0 | 0 | 25 | 0 | 20 | 18 | 0 |
| C3*C12 | 0 | 0 | 24 | 0 | 36 | 0 | 12 | 18 | 0 | 20 | 0 | 35 | 0 | 0 | 14 |
| C3*C13 | 0 | 0 | 0 | 24 | 0 | 12 | 0 | 0 | 12 | 0 | 24 | 0 | 35 | 18 | 0 |
| C3*C14 | 0 | 0 | 0 | 24 | 0 | 36 | 0 | 0 | 36 | 0 | 36 | 0 | 30 | 39 | 0 |
| C3*C15 | 0 | 0 | 0 | 0 | 0 | 0 | 48 | 36 | 0 | 32 | 0 | 20 | 0 | 0 | 35 |

| | C1 | C2 | C3 | C4 | C5 | C6 | C7 | C8 | C9 | C10 | C11 | C12 | C13 | C14 | C15 |
|---|---|---|---|---|---|---|---|---|---|---|---|---|---|---|---|
| C4*C4 | 105 | 0 | 6 | 0 | 9 | 0 | 6 | 3 | 0 | 0 | 0 | 5 | 0 | 0 | 7 |
| C4*C5 | 0 | 0 | 0 | 6 | 0 | 6 | 0 | 0 | 0 | 0 | 0 | 0 | 5 | 2 | 0 |
| C4*C6 | 0 | 0 | 12 | 0 | 36 | 0 | 12 | 18 | 0 | 16 | 0 | 15 | 0 | 0 | 21 |
| C4*C7 | 0 | 30 | 0 | 12 | 0 | 6 | 0 | 0 | 12 | 0 | 12 | 0 | 10 | 6 | 0 |
| C4*C8 | 0 | 0 | 0 | 8 | 0 | 12 | 0 | 0 | 12 | 0 | 12 | 0 | 10 | 13 | 0 |
| C4*C9 | 0 | 0 | 12 | 0 | 0 | 0 | 12 | 9 | 0 | 9 | 0 | 10 | 0 | 0 | 7 |
| C4*C10 | 0 | 60 | 0 | 0 | 0 | 24 | 0 | 0 | 27 | 0 | 27 | 0 | 30 | 27 | 0 |
| C4*C11 | 0 | 0 | 24 | 0 | 0 | 0 | 24 | 18 | 0 | 18 | 0 | 20 | 0 | 0 | 14 |
| C4*C12 | 0 | 0 | 0 | 24 | 0 | 18 | 0 | 0 | 24 | 0 | 24 | 0 | 15 | 24 | 0 |
| C4*C13 | 0 | 0 | 24 | 0 | 36 | 0 | 24 | 18 | 0 | 24 | 0 | 15 | 0 | 0 | 21 |
| C4*C14 | 0 | 0 | 24 | 0 | 24 | 0 | 24 | 39 | 0 | 36 | 0 | 40 | 0 | 0 | 35 |
| C4*C15 | 0 | 0 | 0 | 48 | 0 | 36 | 0 | 0 | 24 | 0 | 24 | 0 | 30 | 30 | 0 |
| C5*C5 | 70 | 0 | 8 | 0 | 13 | 0 | 0 | 2 | 0 | 0 | 0 | 5 | 0 | 0 | 0 |
| C5*C6 | 0 | 20 | 0 | 24 | 0 | 17 | 0 | 0 | 24 | 0 | 4 | 0 | 5 | 12 | 0 |
| C5*C7 | 0 | 0 | 2 | 0 | 0 | 0 | 13 | 6 | 0 | 8 | 0 | 0 | 0 | 0 | 7 |
| C5*C8 | 0 | 0 | 0 | 0 | 8 | 0 | 8 | 8 | 0 | 8 | 0 | 10 | 0 | 0 | 7 |
| C5*C9 | 0 | 40 | 0 | 0 | 0 | 12 | 0 | 0 | 16 | 0 | 1 | 0 | 0 | 6 | 0 |
| C5*C10 | 0 | 0 | 24 | 0 | 0 | 0 | 24 | 18 | 0 | 18 | 0 | 20 | 0 | 0 | 14 |
| C5*C11 | 0 | 0 | 0 | 0 | 0 | 4 | 0 | 0 | 2 | 0 | 17 | 0 | 20 | 12 | 0 |
| C5*C12 | 0 | 0 | 24 | 0 | 36 | 0 | 0 | 18 | 0 | 16 | 0 | 20 | 0 | 0 | 7 |
| C5*C13 | 0 | 0 | 0 | 24 | 0 | 6 | 0 | 0 | 0 | 0 | 24 | 0 | 20 | 12 | 0 |
| C5*C14 | 0 | 0 | 0 | 16 | 0 | 24 | 0 | 0 | 24 | 0 | 24 | 0 | 20 | 26 | 0 |
| C5*C15 | 0 | 0 | 0 | 0 | 0 | 0 | 24 | 18 | 0 | 16 | 0 | 10 | 0 | 0 | 35 |
| C6*C6 | 420 | 0 | 100 | 0 | 102 | 0 | 50 | 78 | 0 | 64 | 0 | 80 | 0 | 0 | 63 |
| C6*C7 | 0 | 20 | 0 | 24 | 0 | 25 | 0 | 0 | 28 | 0 | 38 | 0 | 45 | 36 | 0 |
| C6*C8 | 0 | 80 | 0 | 48 | 0 | 52 | 0 | 0 | 32 | 0 | 52 | 0 | 40 | 48 | 0 |
| C6*C9 | 0 | 0 | 48 | 0 | 72 | 0 | 28 | 24 | 0 | 40 | 0 | 50 | 0 | 0 | 21 |
| C6*C10 | 0 | 120 | 0 | 96 | 0 | 96 | 0 | 0 | 120 | 0 | 120 | 0 | 100 | 102 | 0 |
| C6*C11 | 0 | 0 | 56 | 0 | 24 | 0 | 76 | 78 | 0 | 80 | 0 | 50 | 0 | 0 | 77 |
| C6*C12 | 0 | 120 | 0 | 72 | 0 | 96 | 0 | 0 | 120 | 0 | 60 | 0 | 80 | 84 | 0 |
| C6*C13 | 0 | 0 | 48 | 0 | 36 | 0 | 108 | 72 | 0 | 80 | 0 | 80 | 0 | 0 | 98 |
| C6*C14 | 0 | 0 | 144 | 0 | 144 | 0 | 144 | 144 | 0 | 136 | 0 | 140 | 0 | 0 | 140 |
| C6*C15 | 0 | 0 | 0 | 144 | 0 | 108 | 0 | 0 | 72 | 0 | 132 | 0 | 140 | 120 | 0 |
| C7*C7 | 210 | 0 | 28 | 0 | 39 | 0 | 26 | 18 | 0 | 16 | 0 | 15 | 0 | 0 | 14 |
| C7*C8 | 0 | 0 | 0 | 0 | 24 | 0 | 24 | 24 | 0 | 24 | 0 | 30 | 0 | 0 | 21 |
| C7*C9 | 0 | 0 | 0 | 24 | 0 | 14 | 0 | 0 | 0 | 0 | 25 | 0 | 20 | 18 | 0 |
| C7*C10 | 0 | 0 | 48 | 0 | 72 | 0 | 48 | 54 | 0 | 50 | 0 | 50 | 0 | 0 | 56 |
| C7*C11 | 0 | 40 | 0 | 48 | 0 | 38 | 0 | 0 | 50 | 0 | 25 | 0 | 30 | 36 | 0 |
| C7*C12 | 0 | 0 | 24 | 0 | 0 | 0 | 36 | 54 | 0 | 40 | 0 | 40 | 0 | 0 | 49 |
| C7*C13 | 0 | 120 | 0 | 48 | 0 | 54 | 0 | 0 | 48 | 0 | 36 | 0 | 40 | 36 | 0 |
| C7*C14 | 0 | 0 | 0 | 48 | 0 | 72 | 0 | 0 | 72 | 0 | 72 | 0 | 60 | 78 | 0 |
| C7*C15 | 0 | 0 | 96 | 0 | 72 | 0 | 48 | 54 | 0 | 64 | 0 | 70 | 0 | 0 | 49 |

## Class multiplication coefficients for $S_7$ (continued)

| | C1 | C2 | C3 | C4 | C5 | C6 | C7 | C8 | C9 | C10 | C11 | C12 | C13 | C14 | C15 |
|---|---|---|---|---|---|---|---|---|---|---|---|---|---|---|---|
| C8*C8 | 280 | 0 | 40 | 0 | 32 | 0 | 32 | 52 | 0 | 32 | 0 | 20 | 0 | 0 | 28 |
| C8*C9 | 0 | 0 | 0 | 24 | 0 | 16 | 0 | 0 | 36 | 0 | 16 | 0 | 30 | 24 | 0 |
| C8*C10 | 0 | 0 | 72 | 0 | 72 | 0 | 72 | 72 | 0 | 68 | 0 | 70 | 0 | 0 | 70 |
| C8*C11 | 0 | 80 | 0 | 48 | 0 | 52 | 0 | 0 | 32 | 0 | 52 | 0 | 40 | 48 | 0 |
| C8*C12 | 0 | 0 | 48 | 0 | 72 | 0 | 72 | 36 | 0 | 56 | 0 | 60 | 0 | 0 | 56 |
| C8*C13 | 0 | 0 | 0 | 48 | 0 | 48 | 0 | 0 | 72 | 0 | 48 | 0 | 60 | 60 | 0 |
| C8*C14 | 0 | 120 | 0 | 104 | 0 | 96 | 0 | 0 | 96 | 0 | 96 | 0 | 100 | 84 | 0 |
| C8*C15 | 0 | 0 | 96 | 0 | 72 | 0 | 72 | 72 | 0 | 80 | 0 | 80 | 0 | 0 | 84 |
| C9*C9 | 210 | 0 | 50 | 0 | 48 | 0 | 0 | 27 | 0 | 16 | 0 | 25 | 0 | 0 | 7 |
| C9*C10 | 0 | 30 | 0 | 54 | 0 | 60 | 0 | 0 | 48 | 0 | 48 | 0 | 55 | 51 | 0 |
| C9*C11 | 0 | 0 | 0 | 0 | 6 | 0 | 50 | 24 | 0 | 32 | 0 | 30 | 0 | 0 | 49 |
| C9*C12 | 0 | 120 | 0 | 48 | 0 | 60 | 0 | 0 | 60 | 0 | 36 | 0 | 20 | 42 | 0 |
| C9*C13 | 0 | 0 | 24 | 0 | 0 | 0 | 48 | 54 | 0 | 44 | 0 | 20 | 0 | 0 | 56 |
| C9*C14 | 0 | 0 | 72 | 0 | 72 | 0 | 72 | 72 | 0 | 68 | 0 | 70 | 0 | 0 | 70 |
| C9*C15 | 0 | 0 | 0 | 48 | 0 | 36 | 0 | 0 | 24 | 0 | 84 | 0 | 80 | 60 | 0 |
| C10*C10 | 630 | 0 | 150 | 0 | 162 | 0 | 150 | 153 | 0 | 144 | 0 | 165 | 0 | 0 | 168 |
| C10*C11 | 0 | 60 | 0 | 108 | 0 | 120 | 0 | 0 | 96 | 0 | 96 | 0 | 110 | 102 | 0 |
| C10*C12 | 0 | 0 | 120 | 0 | 144 | 0 | 120 | 126 | 0 | 132 | 0 | 120 | 0 | 0 | 126 |
| C10*C13 | 0 | 120 | 0 | 144 | 0 | 120 | 0 | 0 | 132 | 0 | 132 | 0 | 120 | 126 | 0 |
| C10*C14 | 0 | 240 | 0 | 216 | 0 | 204 | 0 | 0 | 204 | 0 | 204 | 0 | 210 | 216 | 0 |
| C10*C15 | 0 | 0 | 192 | 0 | 144 | 0 | 192 | 180 | 0 | 192 | 0 | 180 | 0 | 0 | 168 |
| C11*C11 | 420 | 0 | 100 | 0 | 102 | 0 | 50 | 78 | 0 | 64 | 0 | 80 | 0 | 0 | 63 |
| C11*C12 | 0 | 0 | 0 | 96 | 0 | 60 | 0 | 0 | 72 | 0 | 96 | 0 | 100 | 84 | 0 |
| C11*C13 | 0 | 0 | 96 | 0 | 144 | 0 | 72 | 72 | 0 | 88 | 0 | 100 | 0 | 0 | 70 |
| C11*C14 | 0 | 0 | 144 | 0 | 144 | 0 | 144 | 144 | 0 | 136 | 0 | 140 | 0 | 0 | 140 |
| C11*C15 | 0 | 240 | 0 | 96 | 0 | 132 | 0 | 0 | 168 | 0 | 108 | 0 | 100 | 120 | 0 |
| C12*C12 | 504 | 0 | 168 | 0 | 144 | 0 | 96 | 108 | 0 | 96 | 0 | 108 | 0 | 0 | 84 |
| C12*C13 | 0 | 24 | 0 | 72 | 0 | 96 | 0 | 0 | 48 | 0 | 120 | 0 | 108 | 108 | 0 |
| C12*C14 | 0 | 240 | 0 | 192 | 0 | 168 | 0 | 0 | 168 | 0 | 168 | 0 | 180 | 156 | 0 |
| C12*C15 | 0 | 0 | 96 | 0 | 72 | 0 | 168 | 144 | 0 | 144 | 0 | 120 | 0 | 0 | 168 |
| C13*C13 | 504 | 0 | 168 | 0 | 144 | 0 | 96 | 108 | 0 | 96 | 0 | 108 | 0 | 0 | 84 |
| C13*C14 | 0 | 0 | 144 | 0 | 144 | 0 | 144 | 180 | 0 | 168 | 0 | 180 | 0 | 0 | 168 |
| C13*C15 | 0 | 240 | 0 | 144 | 0 | 168 | 0 | 0 | 192 | 0 | 120 | 0 | 120 | 144 | 0 |
| C14*C14 | 840 | 0 | 312 | 0 | 312 | 0 | 312 | 252 | 0 | 288 | 0 | 260 | 0 | 0 | 280 |
| C14*C15 | 0 | 240 | 0 | 240 | 0 | 240 | 0 | 0 | 240 | 0 | 240 | 0 | 240 | 240 | 0 |
| C15*C15 | 720 | 0 | 240 | 0 | 360 | 0 | 168 | 216 | 0 | 192 | 0 | 240 | 0 | 0 | 180 |

Class multiplication coefficients for $S_8$

| | C1 | C2 | C3 | C4 | C5 | C6 | C7 | C8 | C9 | C10 | C11 | C12 | C13 | C14 | C15 | C16 | C17 | C18 | C19 | C20 | C21 | C22 |
|---|---|---|---|---|---|---|---|---|---|---|---|---|---|---|---|---|---|---|---|---|---|---|
| C1*C1 | 1 | 0 | 0 | 0 | 0 | 0 | 0 | 0 | 0 | 0 | 0 | 0 | 0 | 0 | 0 | 0 | 0 | 0 | 0 | 0 | 0 | 0 |
| C1*C2 | 0 | 1 | 0 | 0 | 0 | 0 | 0 | 0 | 0 | 0 | 0 | 0 | 0 | 0 | 0 | 0 | 0 | 0 | 0 | 0 | 0 | 0 |
| C1*C3 | 0 | 0 | 1 | 0 | 0 | 0 | 0 | 0 | 0 | 0 | 0 | 0 | 0 | 0 | 0 | 0 | 0 | 0 | 0 | 0 | 0 | 0 |
| C1*C4 | 0 | 0 | 0 | 1 | 0 | 0 | 0 | 0 | 0 | 0 | 0 | 0 | 0 | 0 | 0 | 0 | 0 | 0 | 0 | 0 | 0 | 0 |
| C1*C5 | 0 | 0 | 0 | 0 | 1 | 0 | 0 | 0 | 0 | 0 | 0 | 0 | 0 | 0 | 0 | 0 | 0 | 0 | 0 | 0 | 0 | 0 |
| C1*C6 | 0 | 0 | 0 | 0 | 0 | 1 | 0 | 0 | 0 | 0 | 0 | 0 | 0 | 0 | 0 | 0 | 0 | 0 | 0 | 0 | 0 | 0 |
| C1*C7 | 0 | 0 | 0 | 0 | 0 | 0 | 1 | 0 | 0 | 0 | 0 | 0 | 0 | 0 | 0 | 0 | 0 | 0 | 0 | 0 | 0 | 0 |
| C1*C8 | 0 | 0 | 0 | 0 | 0 | 0 | 0 | 1 | 0 | 0 | 0 | 0 | 0 | 0 | 0 | 0 | 0 | 0 | 0 | 0 | 0 | 0 |
| C1*C9 | 0 | 0 | 0 | 0 | 0 | 0 | 0 | 0 | 1 | 0 | 0 | 0 | 0 | 0 | 0 | 0 | 0 | 0 | 0 | 0 | 0 | 0 |
| C1*C10 | 0 | 0 | 0 | 0 | 0 | 0 | 0 | 0 | 0 | 1 | 0 | 0 | 0 | 0 | 0 | 0 | 0 | 0 | 0 | 0 | 0 | 0 |
| C1*C11 | 0 | 0 | 0 | 0 | 0 | 0 | 0 | 0 | 0 | 0 | 1 | 0 | 0 | 0 | 0 | 0 | 0 | 0 | 0 | 0 | 0 | 0 |
| C1*C12 | 0 | 0 | 0 | 0 | 0 | 0 | 0 | 0 | 0 | 0 | 0 | 1 | 0 | 0 | 0 | 0 | 0 | 0 | 0 | 0 | 0 | 0 |
| C1*C13 | 0 | 0 | 0 | 0 | 0 | 0 | 0 | 0 | 0 | 0 | 0 | 0 | 1 | 0 | 0 | 0 | 0 | 0 | 0 | 0 | 0 | 0 |
| C1*C14 | 0 | 0 | 0 | 0 | 0 | 0 | 0 | 0 | 0 | 0 | 0 | 0 | 0 | 1 | 0 | 0 | 0 | 0 | 0 | 0 | 0 | 0 |
| C1*C15 | 0 | 0 | 0 | 0 | 0 | 0 | 0 | 0 | 0 | 0 | 0 | 0 | 0 | 0 | 1 | 0 | 0 | 0 | 0 | 0 | 0 | 0 |
| C1*C16 | 0 | 0 | 0 | 0 | 0 | 0 | 0 | 0 | 0 | 0 | 0 | 0 | 0 | 0 | 0 | 1 | 0 | 0 | 0 | 0 | 0 | 0 |
| C1*C17 | 0 | 0 | 0 | 0 | 0 | 0 | 0 | 0 | 0 | 0 | 0 | 0 | 0 | 0 | 0 | 0 | 1 | 0 | 0 | 0 | 0 | 0 |
| C1*C18 | 0 | 0 | 0 | 0 | 0 | 0 | 0 | 0 | 0 | 0 | 0 | 0 | 0 | 0 | 0 | 0 | 0 | 1 | 0 | 0 | 0 | 0 |
| C1*C19 | 0 | 0 | 0 | 0 | 0 | 0 | 0 | 0 | 0 | 0 | 0 | 0 | 0 | 0 | 0 | 0 | 0 | 0 | 1 | 0 | 0 | 0 |
| C1*C20 | 0 | 0 | 0 | 0 | 0 | 0 | 0 | 0 | 0 | 0 | 0 | 0 | 0 | 0 | 0 | 0 | 0 | 0 | 0 | 1 | 0 | 0 |
| C1*C21 | 0 | 0 | 0 | 0 | 0 | 0 | 0 | 0 | 0 | 0 | 0 | 0 | 0 | 0 | 0 | 0 | 0 | 0 | 0 | 0 | 1 | 0 |
| C1*C22 | 0 | 0 | 0 | 0 | 0 | 0 | 0 | 0 | 0 | 0 | 0 | 0 | 0 | 0 | 0 | 0 | 0 | 0 | 0 | 0 | 0 | 1 |
| C2*C2 | 28 | 0 | 2 | 0 | 0 | 3 | 0 | 0 | 0 | 0 | 0 | 0 | 0 | 0 | 0 | 0 | 0 | 0 | 0 | 0 | 0 | 0 |
| C2*C3 | 0 | 15 | 0 | 3 | 0 | 0 | 3 | 0 | 0 | 0 | 2 | 0 | 0 | 0 | 0 | 0 | 0 | 0 | 0 | 0 | 0 | 0 |
| C2*C4 | 0 | 0 | 6 | 0 | 4 | 0 | 0 | 3 | 0 | 0 | 0 | 2 | 0 | 0 | 0 | 0 | 0 | 0 | 0 | 0 | 0 | 0 |
| C2*C5 | 0 | 0 | 0 | 1 | 0 | 0 | 0 | 0 | 0 | 0 | 0 | 0 | 2 | 0 | 0 | 0 | 0 | 0 | 0 | 0 | 0 | 0 |
| C2*C6 | 0 | 12 | 0 | 0 | 0 | 0 | 1 | 0 | 0 | 0 | 4 | 0 | 0 | 0 | 0 | 0 | 0 | 0 | 0 | 0 | 0 | 0 |
| C2*C7 | 0 | 0 | 16 | 0 | 0 | 10 | 0 | 2 | 6 | 0 | 0 | 4 | 0 | 0 | 0 | 5 | 0 | 0 | 0 | 0 | 0 | 0 |
| C2*C8 | 0 | 0 | 0 | 12 | 0 | 0 | 3 | 0 | 0 | 6 | 0 | 0 | 4 | 2 | 0 | 0 | 5 | 0 | 0 | 0 | 0 | 0 |
| C2*C9 | 0 | 0 | 0 | 0 | 0 | 0 | 6 | 0 | 0 | 1 | 0 | 0 | 0 | 4 | 0 | 0 | 0 | 0 | 3 | 0 | 0 | 0 |
| C2*C10 | 0 | 0 | 0 | 0 | 0 | 0 | 0 | 4 | 1 | 0 | 0 | 0 | 0 | 0 | 0 | 0 | 0 | 5 | 0 | 3 | 0 | 0 |
| C2*C11 | 0 | 0 | 4 | 0 | 0 | 15 | 0 | 0 | 0 | 0 | 0 | 1 | 0 | 0 | 0 | 5 | 0 | 0 | 0 | 0 | 0 | 0 |
| C2*C12 | 0 | 0 | 0 | 12 | 0 | 0 | 9 | 0 | 0 | 0 | 6 | 0 | 2 | 3 | 0 | 0 | 5 | 0 | 6 | 0 | 0 | 0 |
| C2*C13 | 0 | 0 | 0 | 0 | 24 | 0 | 0 | 3 | 0 | 0 | 0 | 1 | 0 | 0 | 4 | 0 | 0 | 0 | 0 | 6 | 0 | 0 |
| C2*C14 | 0 | 0 | 0 | 0 | 0 | 0 | 0 | 4 | 12 | 0 | 0 | 4 | 0 | 0 | 8 | 0 | 0 | 5 | 0 | 0 | 7 | 0 |
| C2*C15 | 0 | 0 | 0 | 0 | 0 | 0 | 0 | 0 | 0 | 0 | 0 | 0 | 4 | 3 | 0 | 0 | 0 | 0 | 0 | 0 | 0 | 4 |
| C2*C16 | 0 | 0 | 0 | 0 | 0 | 0 | 6 | 0 | 0 | 0 | 16 | 0 | 0 | 0 | 0 | 0 | 1 | 0 | 6 | 0 | 0 | 0 |
| C2*C17 | 0 | 0 | 0 | 0 | 0 | 0 | 0 | 12 | 0 | 0 | 0 | 8 | 0 | 0 | 0 | 3 | 0 | 3 | 0 | 6 | 7 | 0 |
| C2*C18 | 0 | 0 | 0 | 0 | 0 | 0 | 0 | 0 | 0 | 12 | 0 | 0 | 0 | 4 | 0 | 0 | 2 | 0 | 0 | 0 | 0 | 8 |
| C2*C19 | 0 | 0 | 0 | 0 | 0 | 0 | 0 | 0 | 9 | 0 | 0 | 8 | 0 | 0 | 0 | 15 | 0 | 0 | 0 | 1 | 7 | 0 |
| C2*C20 | 0 | 0 | 0 | 0 | 0 | 0 | 0 | 0 | 0 | 9 | 0 | 0 | 16 | 0 | 0 | 0 | 5 | 0 | 1 | 0 | 0 | 8 |
| C2*C21 | 0 | 0 | 0 | 0 | 0 | 0 | 0 | 0 | 0 | 0 | 0 | 0 | 0 | 12 | 0 | 0 | 10 | 0 | 12 | 0 | 0 | 8 |
| C2*C22 | 0 | 0 | 0 | 0 | 0 | 0 | 0 | 0 | 0 | 0 | 0 | 0 | 0 | 0 | 16 | 0 | 0 | 15 | 0 | 12 | 7 | 0 |

Class multiplication coefficients for $S_8$ (continued)

| | C1 | C2 | C3 | C4 | C5 | C6 | C7 | C8 | C9 | C10 | C11 | C12 | C13 | C14 | C15 | C16 | C17 | C18 | C19 | C20 | C21 | C22 |
|---|---|---|---|---|---|---|---|---|---|---|---|---|---|---|---|---|---|---|---|---|---|---|
| C3*C3 | 210 | 0 | 14 | 0 | 6 | 30 | 0 | 6 | 9 | 0 | 0 | 4 | 0 | 0 | 0 | 5 | 0 | 0 | 0 | 0 | 0 | 0 |
| C3*C4 | 0 | 45 | 0 | 9 | 0 | 0 | 9 | 0 | 0 | 9 | 12 | 0 | 6 | 6 | 0 | 0 | 5 | 0 | 3 | 0 | 0 | 0 |
| C3*C5 | 0 | 0 | 3 | 0 | 12 | 0 | 0 | 0 | 0 | 0 | 0 | 2 | 0 | 0 | 4 | 0 | 0 | 0 | 0 | 3 | 0 | 0 |
| C3*C6 | 0 | 0 | 16 | 0 | 0 | 15 | 0 | 1 | 0 | 0 | 0 | 4 | 0 | 0 | 0 | 5 | 0 | 0 | 0 | 0 | 0 | 0 |
| C3*C7 | 0 | 120 | 0 | 24 | 0 | 0 | 36 | 0 | 0 | 6 | 40 | 0 | 8 | 14 | 0 | 0 | 10 | 0 | 18 | 0 | 0 | 0 |
| C3*C8 | 0 | 0 | 48 | 0 | 0 | 15 | 0 | 29 | 6 | 0 | 0 | 20 | 0 | 0 | 16 | 15 | 0 | 20 | 0 | 18 | 14 | 0 |
| C3*C9 | 0 | 0 | 48 | 0 | 0 | 0 | 0 | 4 | 21 | 0 | 0 | 20 | 0 | 0 | 16 | 15 | 0 | 5 | 0 | 3 | 14 | 0 |
| C3*C10 | 0 | 0 | 0 | 24 | 0 | 0 | 6 | 0 | 0 | 21 | 0 | 0 | 8 | 4 | 0 | 0 | 15 | 0 | 3 | 0 | 0 | 20 |
| C3*C11 | 0 | 30 | 0 | 12 | 0 | 0 | 15 | 0 | 0 | 0 | 33 | 0 | 1 | 0 | 0 | 0 | 5 | 0 | 9 | 0 | 0 | 0 |
| C3*C12 | 0 | 0 | 48 | 0 | 48 | 90 | 0 | 30 | 45 | 0 | 0 | 34 | 0 | 0 | 4 | 45 | 0 | 15 | 0 | 15 | 28 | 0 |
| C3*C13 | 0 | 0 | 0 | 18 | 0 | 0 | 9 | 0 | 0 | 9 | 3 | 0 | 43 | 6 | 0 | 0 | 10 | 0 | 6 | 0 | 0 | 20 |
| C3*C14 | 0 | 0 | 0 | 48 | 0 | 0 | 42 | 0 | 0 | 12 | 0 | 0 | 16 | 48 | 0 | 0 | 30 | 0 | 36 | 0 | 0 | 40 |
| C3*C15 | 0 | 0 | 0 | 0 | 48 | 0 | 0 | 12 | 18 | 0 | 0 | 2 | 0 | 0 | 26 | 0 | 0 | 15 | 0 | 18 | 14 | 0 |
| C3*C16 | 0 | 0 | 32 | 0 | 0 | 60 | 0 | 12 | 18 | 0 | 0 | 24 | 0 | 0 | 0 | 50 | 0 | 0 | 0 | 6 | 14 | 0 |
| C3*C17 | 0 | 0 | 0 | 48 | 0 | 0 | 36 | 0 | 0 | 54 | 48 | 0 | 32 | 36 | 0 | 0 | 50 | 0 | 42 | 0 | 0 | 40 |
| C3*C18 | 0 | 0 | 0 | 0 | 0 | 0 | 0 | 32 | 12 | 0 | 0 | 16 | 0 | 0 | 32 | 0 | 0 | 50 | 0 | 36 | 28 | 0 |
| C3*C19 | 0 | 0 | 0 | 24 | 0 | 0 | 54 | 0 | 0 | 9 | 72 | 0 | 16 | 36 | 0 | 0 | 35 | 0 | 63 | 0 | 0 | 20 |
| C3*C20 | 0 | 0 | 0 | 0 | 96 | 0 | 0 | 36 | 9 | 0 | 0 | 20 | 0 | 0 | 48 | 15 | 0 | 45 | 0 | 63 | 28 | 0 |
| C3*C21 | 0 | 0 | 0 | 0 | 0 | 0 | 0 | 48 | 72 | 0 | 0 | 64 | 0 | 0 | 64 | 60 | 0 | 60 | 0 | 48 | 70 | 0 |
| C3*C22 | 0 | 0 | 0 | 0 | 0 | 0 | 0 | 0 | 0 | 90 | 0 | 0 | 80 | 60 | 0 | 0 | 50 | 0 | 30 | 0 | 0 | 70 |
| C4*C4 | 420 | 0 | 18 | 0 | 24 | 45 | 0 | 6 | 12 | 0 | 0 | 4 | 0 | 0 | 8 | 15 | 0 | 15 | 0 | 6 | 7 | 0 |
| C4*C5 | 0 | 15 | 0 | 6 | 0 | 0 | 0 | 0 | 0 | 3 | 6 | 0 | 4 | 0 | 0 | 0 | 0 | 0 | 3 | 0 | 0 | 4 |
| C4*C6 | 0 | 0 | 0 | 12 | 0 | 0 | 9 | 0 | 0 | 0 | 0 | 0 | 4 | 0 | 0 | 0 | 5 | 0 | 2 | 0 | 0 | 0 |
| C4*C7 | 0 | 0 | 48 | 0 | 0 | 90 | 0 | 30 | 36 | 0 | 0 | 28 | 0 | 0 | 16 | 30 | 0 | 15 | 0 | 20 | 21 | 0 |
| C4*C8 | 0 | 180 | 0 | 24 | 0 | 0 | 45 | 0 | 0 | 36 | 60 | 0 | 24 | 30 | 0 | 0 | 25 | 0 | 30 | 0 | 0 | 48 |
| C4*C9 | 0 | 0 | 0 | 32 | 0 | 0 | 36 | 0 | 0 | 9 | 24 | 0 | 8 | 24 | 0 | 0 | 25 | 0 | 25 | 0 | 0 | 24 |
| C4*C10 | 0 | 0 | 48 | 0 | 32 | 0 | 0 | 24 | 9 | 0 | 0 | 24 | 0 | 0 | 32 | 15 | 0 | 30 | 0 | 25 | 21 | 0 |
| C4*C11 | 0 | 0 | 24 | 0 | 24 | 0 | 0 | 15 | 9 | 0 | 0 | 17 | 0 | 0 | 0 | 15 | 0 | 0 | 0 | 9 | 7 | 0 |
| C4*C12 | 0 | 180 | 0 | 24 | 0 | 0 | 63 | 0 | 0 | 54 | 102 | 0 | 42 | 57 | 0 | 0 | 45 | 0 | 60 | 0 | 0 | 48 |
| C4*C13 | 0 | 0 | 36 | 0 | 48 | 45 | 0 | 18 | 9 | 0 | 0 | 21 | 0 | 0 | 36 | 30 | 0 | 30 | 0 | 39 | 21 | 0 |
| C4*C14 | 0 | 0 | 96 | 0 | 0 | 0 | 0 | 60 | 72 | 0 | 0 | 76 | 0 | 0 | 72 | 60 | 0 | 75 | 0 | 60 | 77 | 0 |
| C4*C15 | 0 | 0 | 0 | 24 | 0 | 0 | 18 | 0 | 0 | 36 | 0 | 0 | 36 | 27 | 0 | 0 | 30 | 0 | 24 | 0 | 0 | 24 |
| C4*C16 | 0 | 0 | 0 | 48 | 0 | 0 | 36 | 0 | 0 | 18 | 48 | 0 | 32 | 24 | 0 | 0 | 30 | 0 | 42 | 0 | 0 | 16 |
| C4*C17 | 0 | 0 | 96 | 0 | 0 | 180 | 0 | 60 | 90 | 0 | 0 | 72 | 0 | 0 | 96 | 90 | 0 | 90 | 0 | 90 | 84 | 0 |
| C4*C18 | 0 | 0 | 0 | 96 | 0 | 0 | 36 | 0 | 0 | 72 | 0 | 0 | 64 | 60 | 0 | 0 | 60 | 0 | 48 | 0 | 0 | 56 |
| C4*C19 | 0 | 0 | 48 | 0 | 96 | 60 | 0 | 60 | 75 | 0 | 0 | 80 | 0 | 0 | 64 | 105 | 0 | 60 | 0 | 51 | 77 | 0 |
| C4*C20 | 0 | 0 | 0 | 48 | 0 | 0 | 60 | 0 | 0 | 75 | 72 | 0 | 104 | 60 | 0 | 0 | 75 | 0 | 51 | 0 | 0 | 80 |
| C4*C21 | 0 | 0 | 0 | 96 | 0 | 0 | 108 | 0 | 0 | 108 | 96 | 0 | 96 | 132 | 0 | 0 | 120 | 0 | 132 | 0 | 0 | 120 |
| C4*C22 | 0 | 0 | 0 | 0 | 192 | 0 | 0 | 144 | 108 | 0 | 0 | 96 | 0 | 0 | 96 | 60 | 0 | 105 | 0 | 120 | 105 | 0 |

|  | C1 | C2 | C3 | C4 | C5 | C6 | C7 | C8 | C9 | C10 | C11 | C12 | C13 | C14 | C15 | C16 | C17 | C18 | C19 | C20 | C21 | C22 |
|---|---|---|---|---|---|---|---|---|---|---|---|---|---|---|---|---|---|---|---|---|---|---|
| C5*C5 | 105 | 0 | 6 | 0 | 12 | 0 | 0 | 0 | 3 | 0 | 0 | 0 | 0 | 0 | 4 | 0 | 0 | 0 | 0 | 0 | 0 | 0 |
| C5*C6 | 0 | 0 | 0 | 0 | 0 | 0 | 0 | 3 | 0 | 0 | 0 | 0 | 0 | 0 | 0 | 0 | 0 | 0 | 0 | 2 | 0 | 0 |
| C5*C7 | 0 | 0 | 0 | 0 | 0 | 0 | 9 | 0 | 0 | 0 | 0 | 0 | 16 | 6 | 0 | 0 | 5 | 0 | 2 | 0 | 0 | 8 |
| C5*C8 | 0 | 0 | 0 | 0 | 0 | 45 | 0 | 6 | 0 | 0 | 0 | 8 | 0 | 0 | 0 | 15 | 0 | 15 | 0 | 12 | 7 | 0 |
| C5*C9 | 0 | 0 | 0 | 0 | 32 | 0 | 0 | 0 | 18 | 0 | 0 | 4 | 0 | 0 | 16 | 0 | 0 | 0 | 0 | 7 | 7 | 0 |
| C5*C10 | 0 | 0 | 0 | 8 | 0 | 0 | 0 | 0 | 0 | 18 | 24 | 0 | 0 | 0 | 0 | 0 | 10 | 0 | 7 | 0 | 0 | 4 |
| C5*C11 | 0 | 0 | 0 | 6 | 0 | 0 | 0 | 0 | 0 | 9 | 0 | 0 | 1 | 0 | 0 | 0 | 5 | 0 | 0 | 0 | 0 | 2 |
| C5*C12 | 0 | 0 | 24 | 0 | 0 | 0 | 0 | 12 | 9 | 0 | 0 | 5 | 0 | 0 | 20 | 15 | 0 | 15 | 0 | 15 | 14 | 0 |
| C5*C13 | 0 | 90 | 0 | 12 | 0 | 0 | 18 | 0 | 0 | 0 | 3 | 0 | 18 | 12 | 0 | 0 | 0 | 0 | 3 | 0 | 0 | 6 |
| C5*C14 | 0 | 0 | 0 | 0 | 0 | 0 | 18 | 0 | 0 | 0 | 0 | 0 | 32 | 27 | 0 | 0 | 10 | 0 | 24 | 0 | 0 | 16 |
| C5*C15 | 0 | 0 | 24 | 0 | 48 | 0 | 0 | 0 | 18 | 0 | 0 | 10 | 0 | 0 | 13 | 0 | 0 | 0 | 0 | 6 | 7 | 0 |
| C5*C16 | 0 | 0 | 0 | 0 | 0 | 0 | 0 | 12 | 0 | 0 | 0 | 8 | 0 | 0 | 0 | 0 | 0 | 15 | 0 | 6 | 7 | 0 |
| C5*C17 | 0 | 0 | 0 | 0 | 0 | 0 | 18 | 0 | 0 | 36 | 48 | 0 | 0 | 12 | 0 | 0 | 30 | 0 | 18 | 0 | 0 | 24 |
| C5*C18 | 0 | 0 | 0 | 0 | 0 | 0 | 0 | 24 | 0 | 0 | 0 | 16 | 0 | 0 | 0 | 30 | 0 | 15 | 0 | 12 | 14 | 0 |
| C5*C19 | 0 | 0 | 0 | 24 | 0 | 0 | 6 | 0 | 0 | 21 | 0 | 0 | 8 | 24 | 0 | 0 | 15 | 0 | 18 | 0 | 0 | 20 |
| C5*C20 | 0 | 0 | 48 | 0 | 0 | 60 | 0 | 24 | 21 | 0 | 0 | 20 | 0 | 0 | 16 | 15 | 0 | 15 | 0 | 18 | 14 | 0 |
| C5*C21 | 0 | 0 | 0 | 0 | 0 | 0 | 0 | 24 | 36 | 0 | 0 | 32 | 0 | 0 | 32 | 30 | 0 | 30 | 0 | 24 | 35 | 0 |
| C5*C22 | 0 | 0 | 0 | 48 | 0 | 0 | 36 | 0 | 0 | 18 | 24 | 0 | 24 | 24 | 0 | 0 | 30 | 0 | 30 | 0 | 0 | 21 |
| C6*C6 | 112 | 0 | 8 | 0 | 0 | 16 | 0 | 0 | 2 | 0 | 0 | 0 | 0 | 0 | 0 | 5 | 0 | 0 | 0 | 0 | 0 | 0 |
| C6*C7 | 0 | 40 | 0 | 24 | 0 | 0 | 22 | 0 | 0 | 2 | 32 | 0 | 0 | 4 | 0 | 0 | 5 | 0 | 12 | 0 | 0 | 0 |
| C6*C8 | 0 | 0 | 8 | 0 | 48 | 0 | 0 | 20 | 12 | 0 | 0 | 16 | 0 | 0 | 0 | 0 | 0 | 5 | 0 | 12 | 7 | 0 |
| C6*C9 | 0 | 0 | 0 | 0 | 0 | 20 | 0 | 8 | 14 | 0 | 0 | 8 | 0 | 0 | 0 | 15 | 0 | 5 | 0 | 0 | 7 | 0 |
| C6*C10 | 0 | 0 | 0 | 0 | 0 | 0 | 2 | 0 | 0 | 14 | 0 | 0 | 16 | 8 | 0 | 0 | 5 | 0 | 0 | 0 | 0 | 8 |
| C6*C11 | 0 | 60 | 0 | 0 | 0 | 0 | 12 | 0 | 0 | 0 | 20 | 0 | 0 | 1 | 0 | 0 | 0 | 0 | 6 | 0 | 0 | 0 |
| C6*C12 | 0 | 0 | 48 | 0 | 0 | 0 | 0 | 24 | 18 | 0 | 0 | 24 | 0 | 0 | 8 | 30 | 0 | 0 | 0 | 6 | 14 | 0 |
| C6*C13 | 0 | 0 | 0 | 12 | 0 | 0 | 0 | 0 | 0 | 18 | 0 | 0 | 20 | 3 | 0 | 0 | 10 | 0 | 0 | 0 | 0 | 8 |
| C6*C14 | 0 | 0 | 0 | 0 | 0 | 0 | 12 | 0 | 0 | 24 | 8 | 0 | 8 | 24 | 0 | 0 | 20 | 0 | 24 | 0 | 0 | 16 |
| C6*C15 | 0 | 0 | 0 | 0 | 0 | 0 | 0 | 0 | 0 | 0 | 0 | 4 | 0 | 0 | 8 | 0 | 0 | 15 | 0 | 12 | 7 | 0 |
| C6*C16 | 0 | 0 | 32 | 0 | 0 | 60 | 0 | 0 | 18 | 0 | 0 | 16 | 0 | 0 | 0 | 25 | 0 | 1 | 0 | 0 | 7 | 0 |
| C6*C17 | 0 | 0 | 0 | 48 | 0 | 0 | 18 | 0 | 0 | 18 | 0 | 0 | 32 | 24 | 0 | 0 | 27 | 0 | 24 | 0 | 0 | 16 |
| C6*C18 | 0 | 0 | 0 | 0 | 0 | 0 | 0 | 8 | 12 | 0 | 0 | 0 | 0 | 0 | 32 | 2 | 0 | 26 | 0 | 24 | 14 | 0 |
| C6*C19 | 0 | 0 | 0 | 16 | 0 | 0 | 36 | 0 | 0 | 0 | 48 | 0 | 0 | 24 | 0 | 0 | 20 | 0 | 32 | 0 | 0 | 8 |
| C6*C20 | 0 | 0 | 0 | 0 | 64 | 0 | 0 | 24 | 0 | 0 | 0 | 8 | 0 | 0 | 32 | 0 | 0 | 30 | 0 | 32 | 14 | 0 |
| C6*C21 | 0 | 0 | 0 | 0 | 0 | 0 | 0 | 24 | 36 | 0 | 0 | 32 | 0 | 0 | 32 | 30 | 0 | 30 | 0 | 24 | 42 | 0 |
| C6*C22 | 0 | 0 | 0 | 0 | 0 | 0 | 0 | 0 | 0 | 36 | 0 | 0 | 32 | 24 | 0 | 0 | 20 | 0 | 12 | 0 | 0 | 56 |

Class multiplication coefficients for $S_8$ (continued)

| | C1 | C2 | C3 | C4 | C5 | C6 | C7 | C8 | C9 | C10 | C11 | C12 | C13 | C14 | C15 | C16 | C17 | C18 | C19 | C20 | C21 | C22 |
|---|---|---|---|---|---|---|---|---|---|---|---|---|---|---|---|---|---|---|---|---|---|---|
| C7*C7 | 1120 | 0 | 192 | 0 | 96 | 220 | 0 | 64 | 98 | 0 | 0 | 96 | 0 | 0 | 32 | 125 | 0 | 25 | 0 | 24 | 63 | 0 |
| C7*C8 | 0 | 120 | 0 | 120 | 0 | 0 | 96 | 0 | 0 | 108 | 112 | 0 | 80 | 84 | 0 | 0 | 100 | 0 | 84 | 0 | 0 | 96 |
| C7*C9 | 0 | 240 | 0 | 96 | 0 | 0 | 98 | 0 | 0 | 14 | 80 | 0 | 48 | 72 | 0 | 0 | 55 | 0 | 84 | 0 | 0 | 48 |
| C7*C10 | 0 | 0 | 32 | 0 | 0 | 20 | 0 | 72 | 14 | 0 | 0 | 48 | 0 | 0 | 64 | 25 | 0 | 85 | 0 | 84 | 63 | 0 |
| C7*C11 | 0 | 0 | 80 | 0 | 0 | 120 | 0 | 28 | 30 | 0 | 0 | 44 | 0 | 0 | 0 | 70 | 0 | 5 | 0 | 6 | 21 | 0 |
| C7*C12 | 0 | 360 | 0 | 168 | 0 | 0 | 216 | 0 | 0 | 108 | 264 | 0 | 104 | 144 | 0 | 0 | 140 | 0 | 180 | 0 | 0 | 96 |
| C7*C13 | 0 | 0 | 48 | 0 | 192 | 0 | 0 | 60 | 54 | 0 | 0 | 52 | 0 | 0 | 112 | 30 | 0 | 75 | 0 | 102 | 63 | 0 |
| C7*C14 | 0 | 0 | 224 | 0 | 192 | 120 | 0 | 168 | 216 | 0 | 0 | 192 | 0 | 0 | 224 | 160 | 0 | 180 | 0 | 168 | 196 | 0 |
| C7*C15 | 0 | 0 | 0 | 48 | 0 | 0 | 36 | 0 | 0 | 72 | 0 | 0 | 112 | 84 | 0 | 0 | 60 | 0 | 48 | 0 | 0 | 88 |
| C7*C16 | 0 | 240 | 0 | 96 | 0 | 0 | 150 | 0 | 0 | 30 | 224 | 0 | 32 | 64 | 0 | 0 | 85 | 0 | 120 | 0 | 0 | 32 |
| C7*C17 | 0 | 0 | 192 | 0 | 192 | 180 | 0 | 240 | 198 | 0 | 0 | 224 | 0 | 0 | 192 | 255 | 0 | 255 | 0 | 216 | 217 | 0 |
| C7*C18 | 0 | 0 | 0 | 96 | 0 | 0 | 60 | 0 | 0 | 204 | 32 | 0 | 160 | 144 | 0 | 0 | 170 | 0 | 96 | 0 | 0 | 192 |
| C7*C19 | 0 | 0 | 288 | 0 | 64 | 360 | 0 | 168 | 252 | 0 | 0 | 240 | 0 | 0 | 128 | 300 | 0 | 120 | 0 | 140 | 196 | 0 |
| C7*C20 | 0 | 0 | 0 | 160 | 0 | 0 | 72 | 0 | 0 | 252 | 48 | 0 | 272 | 168 | 0 | 0 | 180 | 0 | 140 | 0 | 0 | 240 |
| C7*C21 | 0 | 0 | 0 | 288 | 0 | 0 | 324 | 0 | 0 | 324 | 288 | 0 | 288 | 336 | 0 | 0 | 310 | 0 | 336 | 0 | 0 | 320 |
| C7*C22 | 0 | 0 | 0 | 0 | 384 | 0 | 0 | 288 | 216 | 0 | 0 | 192 | 0 | 0 | 352 | 120 | 0 | 360 | 0 | 360 | 280 | 0 |
| C8*C8 | 1680 | 0 | 232 | 0 | 96 | 300 | 0 | 104 | 168 | 0 | 0 | 96 | 0 | 0 | 192 | 160 | 0 | 155 | 0 | 144 | 133 | 0 |
| C8*C9 | 0 | 0 | 32 | 0 | 0 | 120 | 0 | 112 | 84 | 0 | 0 | 88 | 0 | 0 | 64 | 100 | 0 | 110 | 0 | 84 | 98 | 0 |
| C8*C10 | 0 | 240 | 0 | 96 | 0 | 0 | 108 | 0 | 0 | 84 | 80 | 0 | 128 | 112 | 0 | 0 | 80 | 0 | 84 | 0 | 0 | 88 |
| C8*C11 | 0 | 0 | 0 | 60 | 0 | 0 | 42 | 0 | 0 | 30 | 0 | 0 | 68 | 33 | 0 | 0 | 40 | 0 | 36 | 0 | 0 | 24 |
| C8*C12 | 0 | 0 | 240 | 0 | 192 | 360 | 0 | 144 | 198 | 0 | 0 | 216 | 0 | 0 | 216 | 210 | 0 | 240 | 0 | 210 | 210 | 0 |
| C8*C13 | 0 | 180 | 0 | 72 | 0 | 0 | 90 | 0 | 0 | 144 | 204 | 0 | 104 | 75 | 0 | 0 | 110 | 0 | 102 | 0 | 0 | 112 |
| C8*C14 | 0 | 240 | 0 | 240 | 0 | 0 | 252 | 0 | 0 | 336 | 264 | 0 | 200 | 288 | 0 | 0 | 300 | 0 | 288 | 0 | 0 | 272 |
| C8*C15 | 0 | 0 | 96 | 0 | 0 | 0 | 0 | 144 | 72 | 0 | 0 | 108 | 0 | 0 | 72 | 120 | 0 | 105 | 0 | 108 | 105 | 0 |
| C8*C16 | 0 | 0 | 96 | 0 | 192 | 0 | 0 | 128 | 120 | 0 | 0 | 112 | 0 | 0 | 128 | 120 | 0 | 80 | 0 | 120 | 112 | 0 |
| C8*C17 | 0 | 720 | 0 | 240 | 0 | 0 | 360 | 0 | 0 | 288 | 384 | 0 | 352 | 360 | 0 | 0 | 280 | 0 | 336 | 0 | 0 | 368 |
| C8*C18 | 0 | 0 | 256 | 0 | 384 | 120 | 0 | 248 | 264 | 0 | 0 | 256 | 0 | 0 | 224 | 160 | 0 | 200 | 0 | 216 | 224 | 0 |
| C8*C19 | 0 | 0 | 0 | 240 | 0 | 0 | 252 | 0 | 0 | 252 | 288 | 0 | 272 | 288 | 0 | 0 | 280 | 0 | 300 | 0 | 0 | 280 |
| C8*C20 | 0 | 0 | 288 | 0 | 384 | 360 | 0 | 288 | 252 | 0 | 0 | 280 | 0 | 0 | 288 | 300 | 0 | 270 | 0 | 300 | 266 | 0 |
| C8*C21 | 0 | 0 | 384 | 0 | 384 | 360 | 0 | 456 | 504 | 0 | 0 | 480 | 0 | 0 | 480 | 480 | 0 | 480 | 0 | 456 | 504 | 0 |
| C8*C22 | 0 | 0 | 0 | 576 | 0 | 0 | 432 | 0 | 0 | 396 | 288 | 0 | 448 | 408 | 0 | 0 | 460 | 0 | 420 | 0 | 0 | 392 |
| C9*C9 | 1120 | 0 | 112 | 0 | 192 | 140 | 0 | 56 | 112 | 0 | 0 | 64 | 0 | 0 | 80 | 55 | 0 | 35 | 0 | 54 | 63 | 0 |
| C9*C10 | 0 | 40 | 0 | 24 | 0 | 0 | 14 | 0 | 0 | 112 | 64 | 0 | 32 | 56 | 0 | 0 | 85 | 0 | 54 | 0 | 0 | 64 |
| C9*C11 | 0 | 0 | 0 | 24 | 0 | 0 | 30 | 0 | 0 | 24 | 72 | 0 | 0 | 20 | 0 | 0 | 30 | 0 | 36 | 0 | 0 | 12 |
| C9*C12 | 0 | 0 | 240 | 0 | 96 | 180 | 0 | 132 | 144 | 0 | 0 | 168 | 0 | 0 | 128 | 180 | 0 | 120 | 0 | 120 | 140 | 0 |
| C9*C13 | 0 | 0 | 0 | 24 | 0 | 0 | 54 | 0 | 0 | 36 | 0 | 0 | 120 | 96 | 0 | 0 | 50 | 0 | 72 | 0 | 0 | 76 |
| C9*C14 | 0 | 480 | 0 | 192 | 0 | 0 | 216 | 0 | 0 | 168 | 160 | 0 | 256 | 224 | 0 | 0 | 160 | 0 | 168 | 0 | 0 | 176 |
| C9*C15 | 0 | 0 | 96 | 0 | 192 | 0 | 0 | 48 | 90 | 0 | 0 | 64 | 0 | 0 | 128 | 30 | 0 | 60 | 0 | 78 | 70 | 0 |
| C9*C16 | 0 | 0 | 96 | 0 | 0 | 180 | 0 | 80 | 66 | 0 | 0 | 96 | 0 | 0 | 32 | 135 | 0 | 55 | 0 | 60 | 77 | 0 |
| C9*C17 | 0 | 0 | 0 | 240 | 0 | 0 | 198 | 0 | 0 | 306 | 288 | 0 | 160 | 192 | 0 | 0 | 245 | 0 | 228 | 0 | 0 | 224 |
| C9*C18 | 0 | 0 | 64 | 0 | 0 | 120 | 0 | 176 | 84 | 0 | 0 | 128 | 0 | 0 | 128 | 110 | 0 | 190 | 0 | 168 | 154 | 0 |
| C9*C19 | 0 | 360 | 0 | 200 | 0 | 0 | 252 | 0 | 0 | 162 | 288 | 0 | 192 | 168 | 0 | 0 | 190 | 0 | 210 | 0 | 0 | 160 |
| C9*C20 | 0 | 0 | 48 | 0 | 224 | 0 | 0 | 168 | 162 | 0 | 0 | 160 | 0 | 0 | 208 | 150 | 0 | 210 | 0 | 210 | 196 | 0 |
| C9*C21 | 0 | 0 | 384 | 0 | 384 | 360 | 0 | 336 | 324 | 0 | 0 | 320 | 0 | 0 | 320 | 330 | 0 | 330 | 0 | 336 | 294 | 0 |
| C9*C22 | 0 | 0 | 0 | 288 | 0 | 0 | 216 | 0 | 0 | 288 | 144 | 0 | 304 | 264 | 0 | 0 | 280 | 0 | 240 | 0 | 0 | 336 |

| | C1 | C2 | C3 | C4 | C5 | C6 | C7 | C8 | C9 | C10 | C11 | C12 | C13 | C14 | C15 | C16 | C17 | C18 | C19 | C20 | C21 | C22 |
|---|---|---|---|---|---|---|---|---|---|---|---|---|---|---|---|---|---|---|---|---|---|---|
| C10*C10 | 1120 | 0 | 112 | 0 | 192 | 140 | 0 | 56 | 112 | 0 | 0 | 64 | 0 | 0 | 80 | 55 | 0 | 35 | 0 | 54 | 63 | 0 |
| C10*C11 | 0 | 0 | 0 | 0 | 96 | 0 | 0 | 20 | 24 | 0 | 0 | 12 | 0 | 0 | 48 | 0 | 0 | 25 | 0 | 36 | 21 | 0 |
| C10*C12 | 0 | 0 | 0 | 144 | 0 | 0 | 108 | 0 | 0 | 144 | 72 | 0 | 152 | 132 | 0 | 0 | 140 | 0 | 120 | 0 | 0 | 168 |
| C10*C13 | 0 | 0 | 48 | 0 | 0 | 180 | 0 | 96 | 36 | 0 | 0 | 76 | 0 | 0 | 16 | 120 | 0 | 75 | 0 | 72 | 63 | 0 |
| C10*C14 | 0 | 0 | 64 | 0 | 0 | 240 | 0 | 224 | 168 | 0 | 0 | 176 | 0 | 0 | 128 | 200 | 0 | 220 | 0 | 168 | 196 | 0 |
| C10*C15 | 0 | 0 | 0 | 96 | 0 | 0 | 72 | 0 | 0 | 90 | 144 | 0 | 16 | 48 | 0 | 0 | 90 | 0 | 78 | 0 | 0 | 64 |
| C10*C16 | 0 | 0 | 0 | 48 | 0 | 0 | 30 | 0 | 0 | 66 | 0 | 0 | 128 | 80 | 0 | 0 | 65 | 0 | 60 | 0 | 0 | 96 |
| C10*C17 | 0 | 0 | 288 | 0 | 384 | 180 | 0 | 192 | 306 | 0 | 0 | 224 | 0 | 0 | 288 | 195 | 0 | 195 | 0 | 228 | 217 | 0 |
| C10*C18 | 0 | 480 | 0 | 192 | 0 | 0 | 204 | 0 | 0 | 84 | 160 | 0 | 160 | 176 | 0 | 0 | 130 | 0 | 168 | 0 | 0 | 128 |
| C10*C19 | 0 | 0 | 48 | 0 | 224 | 0 | 0 | 168 | 162 | 0 | 0 | 160 | 0 | 0 | 208 | 150 | 0 | 210 | 0 | 210 | 196 | 0 |
| C10*C20 | 0 | 360 | 0 | 200 | 0 | 0 | 252 | 0 | 0 | 162 | 288 | 0 | 192 | 168 | 0 | 0 | 190 | 0 | 210 | 0 | 0 | 160 |
| C10*C21 | 0 | 0 | 0 | 288 | 0 | 0 | 324 | 0 | 0 | 324 | 288 | 0 | 288 | 336 | 0 | 0 | 310 | 0 | 336 | 0 | 0 | 320 |
| C10*C22 | 0 | 0 | 480 | 0 | 192 | 360 | 0 | 264 | 288 | 0 | 0 | 336 | 0 | 0 | 256 | 360 | 0 | 240 | 0 | 240 | 280 | 0 |
| C11*C11 | 420 | 0 | 66 | 0 | 0 | 75 | 0 | 0 | 27 | 0 | 0 | 16 | 0 | 0 | 2 | 30 | 0 | 0 | 0 | 0 | 7 | 0 |
| C11*C12 | 0 | 90 | 0 | 102 | 0 | 0 | 99 | 0 | 0 | 27 | 96 | 0 | 36 | 51 | 0 | 0 | 60 | 0 | 75 | 0 | 0 | 24 |
| C11*C13 | 0 | 0 | 6 | 0 | 12 | 0 | 0 | 51 | 0 | 0 | 0 | 18 | 0 | 0 | 18 | 0 | 0 | 45 | 0 | 39 | 21 | 0 |
| C11*C14 | 0 | 0 | 0 | 0 | 0 | 30 | 0 | 66 | 60 | 0 | 0 | 68 | 0 | 0 | 40 | 90 | 0 | 75 | 0 | 72 | 77 | 0 |
| C11*C15 | 0 | 0 | 0 | 0 | 0 | 0 | 0 | 0 | 0 | 54 | 6 | 0 | 18 | 15 | 0 | 0 | 30 | 0 | 18 | 0 | 0 | 42 |
| C11*C16 | 0 | 240 | 0 | 48 | 0 | 0 | 84 | 0 | 0 | 0 | 96 | 0 | 0 | 36 | 0 | 0 | 20 | 0 | 48 | 0 | 0 | 8 |
| C11*C17 | 0 | 0 | 96 | 0 | 192 | 0 | 0 | 96 | 108 | 0 | 0 | 96 | 0 | 0 | 96 | 60 | 0 | 60 | 0 | 84 | 84 | 0 |
| C11*C18 | 0 | 0 | 0 | 0 | 0 | 0 | 12 | 0 | 0 | 60 | 0 | 0 | 96 | 60 | 0 | 0 | 40 | 0 | 36 | 0 | 0 | 88 |
| C11*C19 | 0 | 0 | 144 | 0 | 0 | 180 | 0 | 72 | 108 | 0 | 0 | 100 | 0 | 0 | 48 | 120 | 0 | 45 | 0 | 24 | 77 | 0 |
| C11*C20 | 0 | 0 | 0 | 72 | 0 | 0 | 18 | 0 | 0 | 108 | 0 | 0 | 104 | 72 | 0 | 0 | 70 | 0 | 24 | 0 | 0 | 100 |
| C11*C21 | 0 | 0 | 0 | 96 | 0 | 0 | 108 | 0 | 0 | 108 | 96 | 0 | 96 | 132 | 0 | 0 | 120 | 0 | 132 | 0 | 0 | 120 |
| C11*C22 | 0 | 0 | 0 | 0 | 96 | 0 | 0 | 72 | 54 | 0 | 0 | 48 | 0 | 0 | 168 | 30 | 0 | 165 | 0 | 150 | 105 | 0 |
| C12*C12 | 2520 | 0 | 408 | 0 | 120 | 540 | 0 | 324 | 378 | 0 | 0 | 320 | 0 | 0 | 320 | 450 | 0 | 270 | 0 | 270 | 308 | 0 |
| C12*C13 | 0 | 90 | 0 | 126 | 0 | 0 | 117 | 0 | 0 | 171 | 108 | 0 | 168 | 153 | 0 | 0 | 150 | 0 | 135 | 0 | 0 | 192 |
| C12*C14 | 0 | 360 | 0 | 456 | 0 | 0 | 432 | 0 | 0 | 396 | 408 | 0 | 408 | 408 | 0 | 0 | 420 | 0 | 420 | 0 | 0 | 432 |
| C12*C15 | 0 | 0 | 24 | 0 | 240 | 90 | 0 | 162 | 144 | 0 | 0 | 160 | 0 | 0 | 160 | 90 | 0 | 180 | 0 | 180 | 154 | 0 |
| C12*C16 | 0 | 0 | 288 | 0 | 192 | 360 | 0 | 168 | 216 | 0 | 0 | 240 | 0 | 0 | 96 | 240 | 0 | 120 | 0 | 120 | 168 | 0 |
| C12*C17 | 0 | 720 | 0 | 432 | 0 | 0 | 504 | 0 | 0 | 504 | 576 | 0 | 480 | 504 | 0 | 0 | 480 | 0 | 504 | 0 | 0 | 528 |
| C12*C18 | 0 | 0 | 192 | 0 | 384 | 0 | 0 | 384 | 288 | 0 | 0 | 288 | 0 | 0 | 384 | 240 | 0 | 360 | 0 | 384 | 336 | 0 |
| C12*C19 | 0 | 720 | 0 | 480 | 0 | 0 | 540 | 0 | 0 | 360 | 600 | 0 | 360 | 420 | 0 | 0 | 420 | 0 | 480 | 0 | 0 | 360 |
| C12*C20 | 0 | 0 | 240 | 0 | 480 | 180 | 0 | 420 | 360 | 0 | 0 | 360 | 0 | 0 | 480 | 300 | 0 | 480 | 0 | 480 | 420 | 0 |
| C12*C21 | 0 | 0 | 768 | 0 | 768 | 720 | 0 | 720 | 720 | 0 | 0 | 704 | 0 | 0 | 704 | 720 | 0 | 720 | 0 | 720 | 728 | 0 |
| C12*C22 | 0 | 0 | 0 | 576 | 0 | 0 | 432 | 0 | 0 | 756 | 288 | 0 | 768 | 648 | 0 | 0 | 660 | 0 | 540 | 0 | 0 | 672 |

## Class multiplication coefficients for $S_8$ (continued)

| | C1 | C2 | C3 | C4 | C5 | C6 | C7 | C8 | C9 | C10 | C11 | C12 | C13 | C14 | C15 | C16 | C17 | C18 | C19 | C20 | C21 | C22 |
|---|---|---|---|---|---|---|---|---|---|---|---|---|---|---|---|---|---|---|---|---|---|---|
| C13*C13 | 1260 | 0 | 258 | 0 | 216 | 225 | 0 | 78 | 135 | 0 | 0 | 84 | 0 | 0 | 114 | 60 | 0 | 60 | 0 | 72 | 63 | 0 |
| C13*C14 | 0 | 0 | 96 | 0 | 384 | 90 | 0 | 150 | 288 | 0 | 0 | 204 | 0 | 0 | 264 | 210 | 0 | 165 | 0 | 204 | 231 | 0 |
| C13*C15 | 0 | 180 | 0 | 108 | 0 | 0 | 126 | 0 | 0 | 18 | 54 | 0 | 114 | 99 | 0 | 0 | 60 | 0 | 90 | 0 | 0 | 66 |
| C13*C16 | 0 | 0 | 0 | 96 | 0 | 0 | 36 | 0 | 0 | 144 | 0 | 0 | 64 | 84 | 0 | 0 | 100 | 0 | 48 | 0 | 0 | 104 |
| C13*C17 | 0 | 0 | 192 | 0 | 0 | 360 | 0 | 264 | 180 | 0 | 0 | 240 | 0 | 0 | 192 | 300 | 0 | 300 | 0 | 252 | 252 | 0 |
| C13*C18 | 0 | 0 | 0 | 192 | 0 | 0 | 180 | 0 | 0 | 180 | 288 | 0 | 128 | 132 | 0 | 0 | 200 | 0 | 204 | 0 | 0 | 136 |
| C13*C19 | 0 | 0 | 96 | 0 | 96 | 0 | 0 | 204 | 216 | 0 | 0 | 180 | 0 | 0 | 240 | 120 | 0 | 255 | 0 | 204 | 231 | 0 |
| C13*C20 | 0 | 720 | 0 | 312 | 0 | 0 | 306 | 0 | 0 | 216 | 312 | 0 | 192 | 204 | 0 | 0 | 210 | 0 | 204 | 0 | 0 | 180 |
| C13*C21 | 0 | 0 | 0 | 288 | 0 | 0 | 324 | 0 | 0 | 324 | 288 | 0 | 288 | 396 | 0 | 0 | 360 | 0 | 396 | 0 | 0 | 360 |
| C13*C22 | 0 | 0 | 480 | 0 | 288 | 360 | 0 | 336 | 342 | 0 | 0 | 384 | 0 | 0 | 264 | 390 | 0 | 255 | 0 | 270 | 315 | 0 |
| C14*C14 | 3360 | 0 | 768 | 0 | 864 | 720 | 0 | 576 | 672 | 0 | 0 | 544 | 0 | 0 | 608 | 480 | 0 | 540 | 0 | 576 | 532 | 0 |
| C14*C15 | 0 | 360 | 0 | 216 | 0 | 0 | 252 | 0 | 0 | 144 | 120 | 0 | 264 | 228 | 0 | 0 | 180 | 0 | 216 | 0 | 0 | 216 |
| C14*C16 | 0 | 0 | 0 | 192 | 0 | 0 | 192 | 0 | 0 | 240 | 288 | 0 | 224 | 192 | 0 | 0 | 240 | 0 | 240 | 0 | 0 | 224 |
| C14*C17 | 0 | 0 | 576 | 0 | 384 | 720 | 0 | 720 | 576 | 0 | 0 | 672 | 0 | 0 | 576 | 720 | 0 | 720 | 0 | 672 | 672 | 0 |
| C14*C18 | 0 | 480 | 0 | 480 | 0 | 0 | 432 | 0 | 0 | 528 | 480 | 0 | 352 | 432 | 0 | 0 | 480 | 0 | 432 | 0 | 0 | 448 |
| C14*C19 | 0 | 0 | 576 | 0 | 768 | 720 | 0 | 576 | 504 | 0 | 0 | 560 | 0 | 0 | 576 | 600 | 0 | 540 | 0 | 600 | 532 | 0 |
| C14*C20 | 0 | 0 | 0 | 480 | 0 | 0 | 504 | 0 | 0 | 504 | 576 | 0 | 544 | 576 | 0 | 0 | 560 | 0 | 600 | 0 | 0 | 560 |
| C14*C21 | 0 | 1440 | 0 | 1056 | 0 | 0 | 1008 | 0 | 0 | 1008 | 1056 | 0 | 1056 | 912 | 0 | 0 | 960 | 0 | 912 | 0 | 0 | 960 |
| C14*C22 | 0 | 0 | 960 | 0 | 768 | 720 | 0 | 816 | 792 | 0 | 0 | 864 | 0 | 0 | 864 | 840 | 0 | 840 | 0 | 840 | 840 | 0 |
| C15*C15 | 1260 | 0 | 156 | 0 | 156 | 90 | 0 | 54 | 144 | 0 | 0 | 80 | 0 | 0 | 116 | 90 | 0 | 45 | 0 | 72 | 77 | 0 |
| C15*C16 | 0 | 0 | 0 | 0 | 0 | 0 | 0 | 96 | 36 | 0 | 0 | 48 | 0 | 0 | 96 | 60 | 0 | 120 | 0 | 108 | 84 | 0 |
| C15*C17 | 0 | 0 | 0 | 288 | 0 | 0 | 216 | 0 | 0 | 324 | 288 | 0 | 192 | 216 | 0 | 0 | 300 | 0 | 252 | 0 | 0 | 240 |
| C15*C18 | 0 | 0 | 192 | 0 | 0 | 360 | 0 | 168 | 144 | 0 | 0 | 192 | 0 | 0 | 96 | 240 | 0 | 180 | 0 | 144 | 168 | 0 |
| C15*C19 | 0 | 0 | 0 | 192 | 0 | 0 | 144 | 0 | 0 | 234 | 144 | 0 | 240 | 216 | 0 | 0 | 210 | 0 | 174 | 0 | 0 | 240 |
| C15*C20 | 0 | 0 | 288 | 0 | 192 | 360 | 0 | 216 | 234 | 0 | 0 | 240 | 0 | 0 | 192 | 270 | 0 | 180 | 0 | 174 | 210 | 0 |
| C15*C21 | 0 | 0 | 384 | 0 | 384 | 360 | 0 | 360 | 360 | 0 | 0 | 352 | 0 | 0 | 352 | 360 | 0 | 360 | 0 | 360 | 364 | 0 |
| C15*C22 | 0 | 720 | 0 | 288 | 0 | 0 | 396 | 0 | 0 | 288 | 504 | 0 | 264 | 324 | 0 | 0 | 300 | 0 | 360 | 0 | 0 | 276 |
| C16*C16 | 1344 | 0 | 320 | 0 | 0 | 300 | 0 | 96 | 162 | 0 | 0 | 128 | 0 | 0 | 64 | 173 | 0 | 45 | 0 | 24 | 91 | 0 |
| C16*C17 | 0 | 144 | 0 | 288 | 0 | 0 | 306 | 0 | 0 | 234 | 192 | 0 | 320 | 288 | 0 | 0 | 263 | 0 | 264 | 0 | 0 | 256 |
| C16*C18 | 0 | 0 | 0 | 0 | 384 | 24 | 0 | 128 | 132 | 0 | 0 | 128 | 0 | 0 | 256 | 90 | 0 | 218 | 0 | 240 | 182 | 0 |
| C16*C19 | 0 | 720 | 0 | 336 | 0 | 0 | 360 | 0 | 0 | 180 | 384 | 0 | 128 | 240 | 0 | 0 | 220 | 0 | 276 | 0 | 0 | 160 |
| C16*C20 | 0 | 0 | 96 | 0 | 192 | 0 | 0 | 240 | 180 | 0 | 0 | 160 | 0 | 0 | 288 | 60 | 0 | 300 | 0 | 276 | 224 | 0 |
| C16*C21 | 0 | 0 | 384 | 0 | 384 | 360 | 0 | 384 | 396 | 0 | 0 | 384 | 0 | 0 | 384 | 390 | 0 | 390 | 0 | 384 | 378 | 0 |
| C16*C22 | 0 | 0 | 0 | 192 | 0 | 0 | 144 | 0 | 0 | 432 | 96 | 0 | 416 | 336 | 0 | 0 | 320 | 0 | 240 | 0 | 0 | 448 |

| | C1 | C2 | C3 | C4 | C5 | C6 | C7 | C8 | C9 | C10 | C11 | C12 | C13 | C14 | C15 | C16 | C17 | C18 | C19 | C20 | C21 | C22 |
|---|---|---|---|---|---|---|---|---|---|---|---|---|---|---|---|---|---|---|---|---|---|---|
| C17*C17 | 4032 | 0 | 960 | 0 | 1152 | 972 | 0 | 672 | 882 | 0 | 0 | 768 | 0 | 0 | 960 | 789 | 0 | 789 | 0 | 792 | 819 | 0 |
| C17*C18 | 0 | 288 | 0 | 576 | 0 | 0 | 612 | 0 | 0 | 468 | 384 | 0 | 640 | 576 | 0 | 0 | 526 | 0 | 528 | 0 | 0 | 512 |
| C17*C19 | 0 | 0 | 672 | 0 | 576 | 720 | 0 | 672 | 684 | 0 | 0 | 672 | 0 | 0 | 672 | 660 | 0 | 660 | 0 | 684 | 672 | 0 |
| C17*C20 | 0 | 720 | 0 | 720 | 0 | 0 | 648 | 0 | 0 | 684 | 672 | 0 | 672 | 672 | 0 | 0 | 660 | 0 | 684 | 0 | 0 | 672 |
| C17*C21 | 0 | 1440 | 0 | 1152 | 0 | 0 | 1116 | 0 | 0 | 1116 | 1152 | 0 | 1152 | 1152 | 0 | 0 | 1170 | 0 | 1152 | 0 | 0 | 1152 |
| C17*C22 | 0 | 0 | 960 | 0 | 1152 | 720 | 0 | 1104 | 1008 | 0 | 0 | 1056 | 0 | 0 | 960 | 960 | 0 | 960 | 0 | 1008 | 1008 | 0 |
| C18*C18 | 2688 | 0 | 640 | 0 | 384 | 624 | 0 | 320 | 456 | 0 | 0 | 384 | 0 | 0 | 384 | 436 | 0 | 308 | 0 | 288 | 364 | 0 |
| C18*C19 | 0 | 0 | 0 | 384 | 0 | 0 | 288 | 0 | 0 | 504 | 288 | 0 | 544 | 432 | 0 | 0 | 440 | 0 | 408 | 0 | 0 | 512 |
| C18*C20 | 0 | 0 | 576 | 0 | 384 | 720 | 0 | 432 | 504 | 0 | 0 | 512 | 0 | 0 | 384 | 600 | 0 | 360 | 0 | 408 | 448 | 0 |
| C18*C21 | 0 | 0 | 768 | 0 | 768 | 720 | 0 | 768 | 792 | 0 | 0 | 768 | 0 | 0 | 768 | 780 | 0 | 780 | 0 | 768 | 756 | 0 |
| C18*C22 | 0 | 1440 | 0 | 672 | 0 | 0 | 864 | 0 | 0 | 576 | 1056 | 0 | 544 | 672 | 0 | 0 | 640 | 0 | 768 | 0 | 0 | 608 |
| C19*C19 | 3360 | 0 | 1008 | 0 | 576 | 960 | 0 | 600 | 630 | 0 | 0 | 640 | 0 | 0 | 464 | 690 | 0 | 510 | 0 | 486 | 532 | 0 |
| C19*C20 | 0 | 120 | 0 | 408 | 0 | 0 | 420 | 0 | 0 | 630 | 192 | 0 | 544 | 600 | 0 | 0 | 570 | 0 | 486 | 0 | 0 | 640 |
| C19*C21 | 0 | 1440 | 0 | 1056 | 0 | 0 | 1008 | 0 | 0 | 1008 | 1056 | 0 | 1056 | 912 | 0 | 0 | 960 | 0 | 912 | 0 | 0 | 960 |
| C19*C22 | 0 | 0 | 480 | 0 | 960 | 360 | 0 | 840 | 720 | 0 | 0 | 720 | 0 | 0 | 960 | 600 | 0 | 960 | 0 | 960 | 840 | 0 |
| C20*C20 | 3360 | 0 | 1008 | 0 | 576 | 960 | 0 | 600 | 630 | 0 | 0 | 640 | 0 | 0 | 464 | 690 | 0 | 510 | 0 | 486 | 532 | 0 |
| C20*C21 | 0 | 0 | 768 | 0 | 768 | 720 | 0 | 912 | 1008 | 0 | 0 | 960 | 0 | 0 | 960 | 960 | 0 | 960 | 0 | 912 | 1008 | 0 |
| C20*C22 | 0 | 1440 | 0 | 960 | 0 | 0 | 1080 | 0 | 0 | 720 | 1200 | 0 | 720 | 840 | 0 | 0 | 840 | 0 | 960 | 0 | 0 | 720 |
| C21*C21 | 5760 | 0 | 1920 | 0 | 1920 | 2160 | 0 | 1728 | 1512 | 0 | 0 | 1664 | 0 | 0 | 1664 | 1620 | 0 | 1620 | 0 | 1728 | 1580 | 0 |
| C21*C22 | 0 | 1440 | 0 | 1440 | 0 | 0 | 1440 | 0 | 0 | 1440 | 1440 | 0 | 1440 | 1440 | 0 | 0 | 1440 | 0 | 1440 | 0 | 0 | 1440 |
| C22*C22 | 5040 | 0 | 1680 | 0 | 1008 | 2520 | 0 | 1176 | 1512 | 0 | 0 | 1344 | 0 | 0 | 1104 | 1680 | 0 | 1140 | 0 | 1080 | 1260 | 0 |

## I.C Representing Matrices

This section of Appendix I contains the matrices which represent the generators (12), $(1 \ldots n)$ of $S_n$ and the matrix representing $(1 \ldots n)^{-1}$, so that the reader can get the matrix representing an arbitrary element of $S_n$ by carrying out suitable multiplications and conjugations. The representing matrices are the ones which correspond to the ordinary irreducible representations in Young's so-called natural form (see Chapter 3); the algorithm used is the one which is described in the first edition of the book by H. Boerner [1955], and these calculations were implemented by W. Ulbrich. The tables are for $n \leq 7$, and give the representations corresponding to all partitions, except $(n)$ and $(1^n)$.

Matrices representing (1 2), $(1 \dots n)$, and $(1 \dots n)^{-1}$ for $n=4$ and 5

```
----------------
*    2 *  3 1
----------------

 1 0-1
 0 1-1
 0 0-1

-1 1 0
-1 0 1
-1 0 0

 0 0-1
 1 0-1
 0 1-1
----------------
*    3 *  2 2
----------------

 1-1
 0-1

-1 0
-1 1

-1 0
-1 1
----------------
*    4 *  2 1 1
----------------

 1-1 1
 0-1 0
 0 0-1

 1-1 1
 1 0 0
 0 1 0

 0 1 0
 0 0 1
 1-1 1
----------------
```

```
----------------
*    2 *  4 1
----------------

 1 0 0-1
 0 1 0-1
 0 0 1-1
 0 0 0-1

-1 1 0 0
-1 0 1 0
-1 0 0 1
-1 0 0 0

 0 0 0-1
 1 0 0-1
 0 1 0-1
 0 0 1-1
----------------
*    3 *  3 2
----------------

 1 0 0-1-1
 0 1 0-1 0
 0 0 1 0-1
 0 0 0-1 0
 0 0 0 0-1

-1-1 1 1 0
-1 0 0 0 1
 0-1 0 0 0
-1 0 0 1 0
 0-1 0 1 0

 0 0-1-1 1
 0 0-1 0 0
 1 0-1-1 0
 0 0-1 0 1
 0 1-1-1 1
```

```
----------------
*    4 *  3 1 1
----------------

 1 0 0-1 1 0
 0 1 0-1 0 1
 0 0 1 0-1 1
 0 0 0-1 0 0
 0 0 0 0-1 0
 0 0 0 0 0-1

 1-1 1 0 0 0
 1 0 0-1 1 0
 0 1 0-1 0 1
 1 0 0 0 0 0
 0 1 0 0 0 0
 0 0 0 1 0 0

 0 0 0 1 0 0
 0 0 0 0 1 0
 1 0 0-1 1 0
 0 0 0 0 0 1
 0 1 0-1 0 1
 0 0 1 0-1 1
----------------
*    5 *  2 2 1
----------------

 1 0-1 0 1
 0 1 0-1 1
 0 0-1 0 0
 0 0 0-1 0
 0 0 0 0-1

 1-1-1 0 0
 1 0-1 0 0
 1-1-1 1 0
 1 0 0 0 1
 0 0 1 0 0

 0 1 0 0 1
-1 1 0 0 0
 0 0 0 0 1
-1 0 1 0 0
 0-1 0 1-1
```

```
------------------
*    6 *  2 1 1 1
------------------

 1-1 1-1
 0-1 0 0
 0 0-1 0
 0 0 0-1

-1 1-1 1
-1 0 0 0
 0-1 0 0
 0 0-1 0

 0-1 0 0
 0 0-1 0
 0 0 0-1
 1-1 1-1
------------------
```

## Matrices representing (1 2), (1…6), and $(1\ldots6)^{-1}$ for $S_6$

```
--------------------
*     2 *  5 1
--------------------

 1 0 0 0-1
 0 1 0 0-1
 0 0 1 0-1
 0 0 0 1-1
 0 0 0 0-1

-1 1 0 0 0
-1 0 1 0 0
-1 0 0 1 0
-1 0 0 0 1
-1 0 0 0 0

 0 0 0 0-1
 1 0 0 0-1
 0 1 0 0-1
 0 0 1 0-1
 0 0 0 1-1
--------------------
*     3 *  - 2
--------------------

 1 0 0 0 0 0-1-1 0
 0 1 0 0 0 0-1 0-1
 0 0 1 0 0 0 0-1-1
 0 0 0 1 0 0-1 0 0
 0 0 0 0 1 0 0-1 0
 0 0 0 0 0 1 0 0-1
 0 0 0 0 0 0-1 0 0
 0 0 0 0 0 0 0-1 0
 0 0 0 0 0 0 0 0-1

-1-1 1 0 0 0 1 0 0
-1 0 0-1 1 0 1 0 0
 0-1 0-1 0 1 1 0 0
-1 0 0 0 0 0 0 1 0
 0-1 0 0 0 0 0 0 1
 0 0 0-1 0 0 0 0 0
-1 0 0 0 0 0 1 0 0
 0-1 0 0 0 0 1 0 0
 0 0 0-1 0 0 1 0 0

 0 0 0 0 0-1-1 0 1
 0 0 0 0 0-1 0-1 1
 1 0 0 0 0-1-1-1 1
 0 0 0 0 0-1 0 0 0
 0 1 0 0 0-1-1 0 0
 0 0 1 0 0-1 0-1 0
 0 0 0 0 0-1 0 0 1
 0 0 0 1 0-1-1 0 1
 0 0 0 0 1-1 0-1 1
```

```
--------------------
*     4 *  4 1 1
--------------------

 1 0 0 0 0 0-1 1 0 0
 0 1 0 0 0 0-1 0 1 0
 0 0 1 0 0 0 0-1 1 0
 0 0 0 1 0 0-1 0 0 1
 0 0 0 0 1 0 0-1 0 1
 0 0 0 0 0 1 0 0-1 1
 0 0 0 0 0 0-1 0 0 0
 0 0 0 0 0 0 0-1 0 0
 0 0 0 0 0 0 0 0-1 0
 0 0 0 0 0 0 0 0 0-1

 1-1 1 0 0 0 0 0 0 0
 1 0 0-1 1 0 0 0 0 0
 0 1 0-1 0 1 0 0 0 0
 1 0 0 0 0 0-1 1 0 0
 0 1 0 0 0 0-1 0 1 0
 0 0 0 1 0 0-1 0 0 1
 1 0 0 0 0 0 0 0 0 0
 0 1 0 0 0 0 0 0 0 0
 0 0 0 1 0 0 0 0 0 0
 0 0 0 0 0 0 1 0 0 0

 0 0 0 0 0 0 1 0 0 0
 0 0 0 0 0 0 0 1 0 0
 1 0 0 0 0 0-1 1 0 0
 0 0 0 0 0 0 0 0 1 0
 0 1 0 0 0 0-1 0 1 0
 0 0 1 0 0 0 0-1 1 0
 0 0 0 0 0 0 0 0 0 1
 0 0 0 1 0 0-1 0 0 1
 0 0 0 0 1 0 0-1 0 1
 0 0 0 0 0 1 0 0-1 1
--------------------
*     5 *  3 3
--------------------

 1 0 0-1-1
 0 1 0-1 0
 0 0 1 0-1
 0 0 0-1 0
 0 0 0 0-1

-1-1 1 1 0
-1 0 1 0 0
 0-1 1 0 0
-1 0 1 1 0
 0-1 1 1-1

-1-1 1 1 0
-1 0 0 1 0
-1 0 1 1 0
 0-1 0 1 0
 0-1 1 1-1
```

```
--------------------------------
*     6 *  3 2 1
--------------------------------

 1 0 0 0 0 0 0 0-1 0-1 1 0 1 0 0
 0 1 0 0 0 0 0 0 0-1 0 1-1 1 0 0
 0 0 1 0 0 0 0 0-1 0 0 0 0 0 1 1
 0 0 0 1 0 0 0 0 0-1 0 0 0 0 1 1
 0 0 0 0 1 0 0 0 0 0-1 0 0 0 1 0
 0 0 0 0 0 1 0 0 0 0 0-1 0 0 1 0
 0 0 0 0 0 0 1 0 0 0 0 0-1 0 0 1
 0 0 0 0 0 0 0 1 0 0 0 0 0-1 0 1
 0 0 0 0 0 0 0 0-1 0 0 0 0 0 0 0
 0 0 0 0 0 0 0 0 0-1 0 0 0 0 0 0
 0 0 0 0 0 0 0 0 0 0-1 0 0 0 0 0
 0 0 0 0 0 0 0 0 0 0 0-1 0 0 0 0
 0 0 0 0 0 0 0 0 0 0 0 0-1 0 0 0
 0 0 0 0 0 0 0 0 0 0 0 0 0-1 0 0
 0 0 0 0 0 0 0 0 0 0 0 0 0 0-1 0
 0 0 0 0 0 0 0 0 0 0 0 0 0 0 0-1

 1-1 0-1-1 0 1 0-1 1 0 0 0 0-1 0
 1 0-1 0 0-1 0 1 0 0-1 1 0 0-1 0
 1-1 0 0 0 0 0 0-1 0-1 0 1 0 0 0
 1 0 0 0 0 0 0 0-1 0-1 0 0 1 0 0
 0 0 1-1 0 0 0 0-1 0 0 0 0 0 0 0
 0 0 1 0 0 0 0 0-1 0 0 0 0 0 0 1
 0 0 0 0 1-1 0 0 0 0-1 0 0 0 0 0
 0 0 0 0 1 0 0 0 0 0-1 0 0 0 0 0
 1-1 0 0 0 0 0 0-1 1 0 0 0 0 0 0
 1 0 0 0 0 0 0 0 0 0-1 1 0 0 0 0
 0 0 1-1 0 0 0 0-1 1 0 0 0 0 0 0
 0 0 1 0 0 0 0 0 0 0 0 0 0 0 1 0
 0 0 0 0 1-1 0 0 0 0-1 1 0 0 0 0
 0 0 0 0 1 0 0 0 0 0 0 0 0 0 1 0
 0 0 0 0 0 0 0 0 1 0 0 0 0 0 0 0
 0 0 0 0 0 0 0 0 0 0 1 0 0 0 0 0

 0 0 0 0 0 0 1 0 0 1 0 0-1 0 0 1
 0 0 0 0-1 0 1 0-1 1 1 0-1 0-1 1
 0 0 0 0 0 0 0 1 0 0 0 1 0-1 0 1
 0 0 0 0-1 0 0 1 0 0 0 1 0-1-1 1
 0 0 0 0 0 0 0 1 0 0 0 0 0 0 0 1
 0 0 0 0 0 0-1 1 0 0 0 0 0 0 0 0
 1 0 0 0-1 0 0 1-1 0 0 1 0 0-1 1
 0 1 0 0 0 0-1 1 0-1 0 1 0 0 0 0
 0 0 0 0 0 0 0 0 0 0 0 0 0 0 1 0
 0 0 0 0-1 0 0 0 0 0 1 0 0 0 0 0
 0 0 0 0 0 0 0 0 0 0 0 0 0 0 0 1
 0 0 0 0 0 0-1 0 0 0 0 0 1 0 0 0
 0 0 1 0-1 0 0 0-1 0 1 0 0 0 0 1
 0 0 0 1 0 0-1 0 0-1 0 0 1 0 1 0
 0 0 0 0 0 0 0-1 0 0 0 0 0 1 0-1
 0 0 0 0 0 1 0-1 0 0 0-1 0 1 1-1
```

```
--------------------
*     7 *  3 1 1 1
--------------------

 1 0 0 0-1 1-1 0 0 0
 0 1 0 0-1 0 0 1-1 0
 0 0 1 0 0-1 0 1 0-1
 0 0 0 1 0 0-1 0 1-1
 0 0 0 0-1 0 0 0 0 0
 0 0 0 0 0-1 0 0 0 0
 0 0 0 0 0 0-1 0 0 0
 0 0 0 0 0 0 0-1 0 0
 0 0 0 0 0 0 0 0-1 0
 0 0 0 0 0 0 0 0 0-1

-1 1-1 1 0 0 0 0 0 0
-1 0 0 0 1-1 1 0 0 0
 0-1 0 0 1 0 0-1 1 0
 0 0-1 0 0 1 0-1 0 1
-1 0 0 0 0 0 0 0 0 0
 0-1 0 0 0 0 0 0 0 0
 0 0-1 0 0 0 0 0 0 0
 0 0 0 0-1 0 0 0 0 0
 0 0 0 0 0-1 0 0 0 0
 0 0 0 0 0 0 0-1 0 0

 0 0 0 0-1 0 0 0 0 0
 0 0 0 0 0-1 0 0 0 0
 0 0 0 0 0 0-1 0 0 0
 1 0 0 0-1 1-1 0 0 0
 0 0 0 0 0 0 0-1 0 0
 0 0 0 0 0 0 0 0-1 0
 0 1 0 0-1 0 0 1-1 0
 0 0 0 0 0 0 0 0 0-1
 0 0 1 0 0-1 0 1 0-1
 0 0 0 1 0 0-1 0 1-1
--------------------
*     8 *  2 2 2
--------------------

 1 0-1 0 1
 0 1 0-1 1
 0 0-1 0 0
 0 0 0-1 0
 0 0 0 0-1

 1 0-1 1 0
 1-1-1 1 0
 1 0-1 0 0
 1-1 0 0 0
 0 0 1-1 1

 1-1 0 1 0
 1-1 0 0 0
 1-1-1 1 0
 1 0-1 0 0
 0 1 0-1 1
```

```
--------------------
*     9 *  2 2 1 1
--------------------

 1 0 0-1 0 0 1-1 0
 0 1 0 0-1 0 1 0-1
 0 0 1 0 0-1 0 1-1
 0 0 0-1 0 0 0 0 0
 0 0 0 0-1 0 0 0 0
 0 0 0 0 0-1 0 0 0
 0 0 0 0 0 0-1 0 0
 0 0 0 0 0 0 0-1 0
 0 0 0 0 0 0 0 0-1

-1 1-1 1-1 0 0 0 0
-1 0 0 1 0 0-1 0 0
 0-1 0 0 1 0-1 0 0
-1 1-1 1-1 1 0 0 0
-1 0 0 0 0 0-1 1 0
 0-1 0 0 0 0 0 0 1
 0 0 0-1 0 0 0 0 0
 0 0 0 0-1 0 0 0 0
 0 0 0 0 0 0-1 0 0

 0-1 0 0 0 0-1 0 1
 0 0-1 0 0 0 0-1 1
-1 1-1 0 0 0 0 0 0
 0 0 0 0 0 0-1 0 0
 0 0 0 0 0 0 0-1 0
-1 0 0 1 0 0 0 0 0
 0 0 0 0 0 0 0 0-1
 0-1 0 0 1 0-1 0 0
 0 0-1 0 0 1 0-1 1
--------------------
*    10 *  2 1 1 1 1
--------------------

 1-1 1-1 1
 0-1 0 0 0
 0 0-1 0 0
 0 0 0-1 0
 0 0 0 0-1

 1-1 1-1 1
 1 0 0 0 0
 0 1 0 0 0
 0 0 1 0 0
 0 0 0 1 0

 0 1 0 0 0
 0 0 1 0 0
 0 0 0 1 0
 0 0 0 0 1
 1-1 1-1 1
--------------------
```

Matrices representing (1 2), (1 ... 7), and $(1 \ldots 7)^{-1}$ for the partitions (6,1), (5,2), (5,1,1), and (4,3) of 7

```
---------------
 *    2 *  6 1
---------------

 1 0 0 0 0-1
 0 1 0 0 0-1
 0 0 1 0 0-1
 0 0 0 1 0-1
 0 0 0 0 1-1
 0 0 0 0 0-1

-1 1 0 0 0 0
-1 0 1 0 0 0
-1 0 0 1 0 0
-1 0 0 0 1 0
-1 0 0 0 0 1
-1 0 0 0 0 0

 0 0 0 0 0-1
 1 0 0 0 0-1
 0 1 0 0 0-1
 0 0 1 0 0-1
 0 0 0 1 0-1
 0 0 0 0 1-1
```

```
----------------------------
 *    3 *  5 2
----------------------------

 1 0 0 0 0 0 0 0 0 0-1-1 0 0
 0 1 0 0 0 0 0 0 0 0-1 0-1 0
 0 0 1 0 0 0 0 0 0 0 0-1-1 0
 0 0 0 1 0 0 0 0 0 0-1 0 0-1
 0 0 0 0 1 0 0 0 0 0 0-1 0-1
 0 0 0 0 0 1 0 0 0 0 0 0-1-1
 0 0 0 0 0 0 1 0 0 0-1 0 0 0
 0 0 0 0 0 0 0 1 0 0 0-1 0 0
 0 0 0 0 0 0 0 0 1 0 0 0-1 0
 0 0 0 0 0 0 0 0 0 1 0 0 0-1
 0 0 0 0 0 0 0 0 0 0-1 0 0 0
 0 0 0 0 0 0 0 0 0 0 0-1 0 0
 0 0 0 0 0 0 0 0 0 0 0 0-1 0
 0 0 0 0 0 0 0 0 0 0 0 0 0-1

-1-1 1 0 0 0 0 0 0 0 1 0 0 0
-1 0 0-1 1 0 0 0 0 0 1 0 0 0
 0-1 0-1 0 1 0 0 0 0 1 0 0 0
-1 0 0 0 0 0-1 1 0 0 1 0 0 0
 0-1 0 0 0 0-1 0 1 0 1 0 0 0
 0 0 0-1 0 0-1 0 0 1 1 0 0 0
-1 0 0 0 0 0 0 0 0 0 0 1 0 0
 0-1 0 0 0 0 0 0 0 0 0 0 1 0
 0 0 0-1 0 0 0 0 0 0 0 0 0 1
 0 0 0 0 0 0-1 0 0 0 0 0 0 0
-1 0 0 0 0 0 0 0 0 0 1 0 0 0
 0-1 0 0 0 0 0 0 0 0 1 0 0 0
 0 0 0-1 0 0 0 0 0 0 1 0 0 0
 0 0 0 0 0 0-1 0 0 0 1 0 0 0

 0 0 0 0 0 0 0 0 0-1-1 0 0 1
 0 0 0 0 0 0 0 0 0-1 0-1 0 1
 1 0 0 0 0 0 0 0 0-1-1-1 0 1
 0 0 0 0 0 0 0 0 0-1 0 0-1 1
 0 1 0 0 0 0 0 0 0-1-1 0-1 1
 0 0 1 0 0 0 0 0 0-1 0-1-1 1
 0 0 0 0 0 0 0 0 0-1 0 0 0 0
 0 0 0 1 0 0 0 0 0-1-1 0 0 0
 0 0 0 0 1 0 0 0 0-1 0-1 0 0
 0 0 0 0 0 1 0 0 0-1 0 0-1 0
 0 0 0 0 0 0 0 0 0-1 0 0 0 1
 0 0 0 0 0 0 1 0 0-1-1 0 0 1
 0 0 0 0 0 0 0 1 0-1 0-1 0 1
 0 0 0 0 0 0 0 0 1-1 0 0-1 1
```

```
-------------------------------
 *    4 *  5 1 1
-------------------------------

 1 0 0 0 0 0 0 0 0 0-1 1 0 0 0
 0 1 0 0 0 0 0 0 0 0-1 0 1 0 0
 0 0 1 0 0 0 0 0 0 0 0-1 1 0 0
 0 0 0 1 0 0 0 0 0 0-1 0 0 1 0
 0 0 0 0 1 0 0 0 0 0 0-1 0 1 0
 0 0 0 0 0 1 0 0 0 0 0 0-1 1 0
 0 0 0 0 0 0 1 0 0 0-1 0 0 0 1
 0 0 0 0 0 0 0 1 0 0 0-1 0 0 1
 0 0 0 0 0 0 0 0 1 0 0 0-1 0 1
 0 0 0 0 0 0 0 0 0 1 0 0 0-1 1
 0 0 0 0 0 0 0 0 0 0-1 0 0 0 0
 0 0 0 0 0 0 0 0 0 0 0-1 0 0 0
 0 0 0 0 0 0 0 0 0 0 0 0-1 0 0
 0 0 0 0 0 0 0 0 0 0 0 0 0-1 0
 0 0 0 0 0 0 0 0 0 0 0 0 0 0-1

 1-1 1 0 0 0 0 0 0 0 0 0 0 0 0
 1 0 0-1 1 0 0 0 0 0 0 0 0 0 0
 0 1 0-1 0 1 0 0 0 0 0 0 0 0 0
 1 0 0 0 0 0-1 1 0 0 0 0 0 0 0
 0 1 0 0 0 0-1 0 1 0 0 0 0 0 0
 0 0 0 1 0 0-1 0 0 1 0 0 0 0 0
 1 0 0 0 0 0 0 0 0 0-1 1 0 0 0
 0 1 0 0 0 0 0 0 0 0-1 0 1 0 0
 0 0 0 1 0 0 0 0 0 0-1 0 0 1 0
 0 0 0 0 0 0 1 0 0 0-1 0 0 0 1
 1 0 0 0 0 0 0 0 0 0 0 0 0 0 0
 0 1 0 0 0 0 0 0 0 0 0 0 0 0 0
 0 0 0 1 0 0 0 0 0 0 0 0 0 0 0
 0 0 0 0 0 0 1 0 0 0 0 0 0 0 0
 0 0 0 0 0 0 0 0 0 0 1 0 0 0 0

 0 0 0 0 0 0 0 0 0 0 1 0 0 0 0
 0 0 0 0 0 0 0 0 0 0 0 1 0 0 0
 1 0 0 0 0 0 0 0 0 0-1 1 0 0 0
 0 0 0 0 0 0 0 0 0 0 0 0 1 0 0
 0 1 0 0 0 0 0 0 0 0-1 0 1 0 0
 0 0 1 0 0 0 0 0 0 0 0-1 1 0 0
 0 0 0 0 0 0 0 0 0 0 0 0 0 1 0
 0 0 0 1 0 0 0 0 0 0-1 0 0 1 0
 0 0 0 0 1 0 0 0 0 0 0-1 0 1 0
 0 0 0 0 0 1 0 0 0 0 0 0-1 1 0
 0 0 0 0 0 0 0 0 0 0 0 0 0 0 1
 0 0 0 0 0 0 1 0 0 0-1 0 0 0 1
 0 0 0 0 0 0 0 1 0 0 0-1 0 0 1
 0 0 0 0 0 0 0 0 1 0 0 0-1 0 1
 0 0 0 0 0 0 0 0 0 1 0 0 0-1 1
```

```
----------------------------
 *    5 *  4 3
----------------------------

 1 0 0 0 0 0 0 0 0-1-1-1 0 0
 0 1 0 0 0 0 0 0 0-1 0 0-1-1
 0 0 1 0 0 0 0 0 0 0-1 0-1 0
 0 0 0 1 0 0 0 0 0 0 0-1 0-1
 0 0 0 0 1 0 0 0 0-1 0 0 0 0
 0 0 0 0 0 1 0 0 0 0-1 0 0 0
 0 0 0 0 0 0 1 0 0 0 0-1 0 0
 0 0 0 0 0 0 0 1 0 0 0 0-1 0
 0 0 0 0 0 0 0 0 1 0 0 0 0-1
 0 0 0 0 0 0 0 0 0-1 0 0 0 0
 0 0 0 0 0 0 0 0 0 0-1 0 0 0
 0 0 0 0 0 0 0 0 0 0 0-1 0 0
 0 0 0 0 0 0 0 0 0 0 0 0-1 0
 0 0 0 0 0 0 0 0 0 0 0 0 0-1

-1-1-1 1 0 0 0 2 0 1 1 0 0 0
-1 0 0 0-1-1 1 2 0 1 1 0-1 0
 0-1 0 0-1 0 0 1 1 1 0 0 0 0
 0 0-1 0 0-1 0 1 0 0 1 0 0 0
-1 0 0 0 0 0 0 1 0 0 0 1 0 0
 0-1 0 0 0 0 0 1 0 0 0 0 0 1
 0 0-1 0 0 0 0 1 0 0 0 0 0 0
 0 0 0 0-1 0 0 1 0 0 0 0 0 0
 0 0 0 0 0-1 0 1 0 0 0 0 0 0
-1 0 0 0 0 0 0 1 0 1 1 0-1 0
 0-1 0 0 0 0 0 1 0 1 0 0 0 0
 0 0-1 0 0 0 0 1 0 0 1 0 0 0
 0 0 0 0-1 0 0 1 0 1 0 0-1 0
 0 0 0 0 0-1 0 1 0 0 1 0-1 0

 0 0 0-1 0 0-1-1 1-1 0 2 1 0
 0 0 0-1 0 0-1-1 2 0-1 2 1-1
 0 0 0-1 0 0-1 0 1 0 0 1 0 0
 1 0 0-1 0 0-1-1 1-1-1 1 1 0
 0 0 0-1 0 0 0-1 1 0 0 1 0 0
 0 0 0-1 0 0 0 0 0 0 0 1 0 0
 0 1 0-1 0 0 0-1 0-1 0 1 0 0
 0 0 0-1 0 0 0 0 1 0 0 1 0 0
 0 0 1-1 0 0 0-1 1 0-1 1 0 0
 0 0 0 0 0 0-1-1 1 0 0 1 1-1
 0 0 0 0 0 0-1 0 0 0 0 1 0 0
 0 0 0 0 1 0-1-1 0-1 0 1 1 0
 0 0 0 0 0 0-1 0 1 0 0 1 0-1
 0 0 0 0 0 1-1-1 1 0-1 1 1-1
```

Matrices representing (12) and (1 . . . 7) for the partition (4, 2, 1) of 7

* 6 * 4 2 1

---

| | | | | | | | | | | | | | | | | | | | | | | | | | | | | | | | | | | |
|---|---|---|---|---|---|---|---|---|---|---|---|---|---|---|---|---|---|---|---|---|---|---|---|---|---|---|---|---|---|---|---|---|---|---|
| 1 | 0 | 0 | 0 | 0 | 0 | 0 | 0 | 0 | 0 | 0 | 0 | 0 | 0 | 0 | 0 | 0 | 0 | 0 | 0 | -1 | 0 | -1 | 1 | 0 | 1 | 0 | 0 | 0 | 0 | 0 | 0 | 0 | 0 | 0 |
| 0 | 1 | 0 | 0 | 0 | 0 | 0 | 0 | 0 | 0 | 0 | 0 | 0 | 0 | 0 | 0 | 0 | 0 | 0 | 0 | 0 | -1 | 0 | 1 | -1 | 1 | 0 | 0 | 0 | 0 | 0 | 0 | 0 | 0 | 0 |
| 0 | 0 | 1 | 0 | 0 | 0 | 0 | 0 | 0 | 0 | 0 | 0 | 0 | 0 | 0 | 0 | 0 | 0 | 0 | 0 | -1 | 0 | 0 | 0 | 0 | 0 | -1 | 1 | 0 | 1 | 0 | 0 | 0 | 0 | 0 |
| 0 | 0 | 0 | 1 | 0 | 0 | 0 | 0 | 0 | 0 | 0 | 0 | 0 | 0 | 0 | 0 | 0 | 0 | 0 | 0 | 0 | -1 | 0 | 0 | 0 | 0 | 0 | 1 | -1 | 1 | 0 | 0 | 0 | 0 | 0 |
| 0 | 0 | 0 | 0 | 1 | 0 | 0 | 0 | 0 | 0 | 0 | 0 | 0 | 0 | 0 | 0 | 0 | 0 | 0 | 0 | 0 | 0 | -1 | 0 | 0 | 0 | -1 | 1 | 0 | 0 | 0 | 1 | 0 | 0 | 0 |
| 0 | 0 | 0 | 0 | 0 | 1 | 0 | 0 | 0 | 0 | 0 | 0 | 0 | 0 | 0 | 0 | 0 | 0 | 0 | 0 | 0 | 0 | 0 | -1 | 0 | 0 | 0 | 1 | 0 | 0 | -1 | 1 | 0 | 0 | 0 |
| 0 | 0 | 0 | 0 | 0 | 0 | 1 | 0 | 0 | 0 | 0 | 0 | 0 | 0 | 0 | 0 | 0 | 0 | 0 | 0 | 0 | 0 | 0 | 0 | -1 | 0 | 0 | 0 | -1 | 1 | 0 | 1 | 0 | 0 | 0 |
| 0 | 0 | 0 | 0 | 0 | 0 | 0 | 1 | 0 | 0 | 0 | 0 | 0 | 0 | 0 | 0 | 0 | 0 | 0 | 0 | 0 | 0 | 0 | 0 | 0 | -1 | 0 | 0 | 0 | 1 | -1 | 1 | 0 | 0 | 0 |
| 0 | 0 | 0 | 0 | 0 | 0 | 0 | 0 | 1 | 0 | 0 | 0 | 0 | 0 | 0 | 0 | 0 | 0 | 0 | 0 | -1 | 0 | 0 | 0 | 0 | 0 | 0 | 0 | 0 | 0 | 0 | 0 | 1 | 1 | 0 |
| 0 | 0 | 0 | 0 | 0 | 0 | 0 | 0 | 0 | 1 | 0 | 0 | 0 | 0 | 0 | 0 | 0 | 0 | 0 | 0 | 0 | -1 | 0 | 0 | 0 | 0 | 0 | 0 | 0 | 0 | 0 | 0 | 1 | 1 | 0 |
| 0 | 0 | 0 | 0 | 0 | 0 | 0 | 0 | 0 | 0 | 1 | 0 | 0 | 0 | 0 | 0 | 0 | 0 | 0 | 0 | 0 | 0 | -1 | 0 | 0 | 0 | 0 | 0 | 0 | 0 | 0 | 0 | 1 | 0 | 1 |
| 0 | 0 | 0 | 0 | 0 | 0 | 0 | 0 | 0 | 0 | 0 | 1 | 0 | 0 | 0 | 0 | 0 | 0 | 0 | 0 | 0 | 0 | 0 | -1 | 0 | 0 | 0 | 0 | 0 | 0 | 0 | 0 | 1 | 0 | 1 |
| 0 | 0 | 0 | 0 | 0 | 0 | 0 | 0 | 0 | 0 | 0 | 0 | 1 | 0 | 0 | 0 | 0 | 0 | 0 | 0 | 0 | 0 | 0 | 0 | -1 | 0 | 0 | 0 | 0 | 0 | 0 | 0 | 0 | 1 | 1 |
| 0 | 0 | 0 | 0 | 0 | 0 | 0 | 0 | 0 | 0 | 0 | 0 | 0 | 1 | 0 | 0 | 0 | 0 | 0 | 0 | 0 | 0 | 0 | 0 | 0 | -1 | 0 | 0 | 0 | 0 | 0 | 0 | 0 | 1 | 1 |
| 0 | 0 | 0 | 0 | 0 | 0 | 0 | 0 | 0 | 0 | 0 | 0 | 0 | 0 | 1 | 0 | 0 | 0 | 0 | 0 | 0 | 0 | 0 | 0 | 0 | 0 | -1 | 0 | 0 | 0 | 0 | 0 | 1 | 0 | 0 |
| 0 | 0 | 0 | 0 | 0 | 0 | 0 | 0 | 0 | 0 | 0 | 0 | 0 | 0 | 0 | 1 | 0 | 0 | 0 | 0 | 0 | 0 | 0 | 0 | 0 | 0 | 0 | -1 | 0 | 0 | 0 | 0 | 1 | 0 | 0 |
| 0 | 0 | 0 | 0 | 0 | 0 | 0 | 0 | 0 | 0 | 0 | 0 | 0 | 0 | 0 | 0 | 1 | 0 | 0 | 0 | 0 | 0 | 0 | 0 | 0 | 0 | 0 | 0 | -1 | 0 | 0 | 0 | 0 | 1 | 0 |
| 0 | 0 | 0 | 0 | 0 | 0 | 0 | 0 | 0 | 0 | 0 | 0 | 0 | 0 | 0 | 0 | 0 | 1 | 0 | 0 | 0 | 0 | 0 | 0 | 0 | 0 | 0 | 0 | 0 | -1 | 0 | 0 | 0 | 1 | 0 |
| 0 | 0 | 0 | 0 | 0 | 0 | 0 | 0 | 0 | 0 | 0 | 0 | 0 | 0 | 0 | 0 | 0 | 0 | 1 | 0 | 0 | 0 | 0 | 0 | 0 | 0 | 0 | 0 | 0 | 0 | -1 | 0 | 0 | 0 | 1 |
| 0 | 0 | 0 | 0 | 0 | 0 | 0 | 0 | 0 | 0 | 0 | 0 | 0 | 0 | 0 | 0 | 0 | 0 | 0 | 1 | 0 | 0 | 0 | 0 | 0 | 0 | 0 | 0 | 0 | 0 | 0 | -1 | 0 | 0 | 1 |
| 0 | 0 | 0 | 0 | 0 | 0 | 0 | 0 | 0 | 0 | 0 | 0 | 0 | 0 | 0 | 0 | 0 | 0 | 0 | 0 | -1 | 0 | 0 | 0 | 0 | 0 | 0 | 0 | 0 | 0 | 0 | 0 | 0 | 0 | 0 |
| 0 | 0 | 0 | 0 | 0 | 0 | 0 | 0 | 0 | 0 | 0 | 0 | 0 | 0 | 0 | 0 | 0 | 0 | 0 | 0 | 0 | -1 | 0 | 0 | 0 | 0 | 0 | 0 | 0 | 0 | 0 | 0 | 0 | 0 | 0 |
| 0 | 0 | 0 | 0 | 0 | 0 | 0 | 0 | 0 | 0 | 0 | 0 | 0 | 0 | 0 | 0 | 0 | 0 | 0 | 0 | 0 | 0 | -1 | 0 | 0 | 0 | 0 | 0 | 0 | 0 | 0 | 0 | 0 | 0 | 0 |
| 0 | 0 | 0 | 0 | 0 | 0 | 0 | 0 | 0 | 0 | 0 | 0 | 0 | 0 | 0 | 0 | 0 | 0 | 0 | 0 | 0 | 0 | 0 | -1 | 0 | 0 | 0 | 0 | 0 | 0 | 0 | 0 | 0 | 0 | 0 |
| 0 | 0 | 0 | 0 | 0 | 0 | 0 | 0 | 0 | 0 | 0 | 0 | 0 | 0 | 0 | 0 | 0 | 0 | 0 | 0 | 0 | 0 | 0 | 0 | -1 | 0 | 0 | 0 | 0 | 0 | 0 | 0 | 0 | 0 | 0 |
| 0 | 0 | 0 | 0 | 0 | 0 | 0 | 0 | 0 | 0 | 0 | 0 | 0 | 0 | 0 | 0 | 0 | 0 | 0 | 0 | 0 | 0 | 0 | 0 | 0 | -1 | 0 | 0 | 0 | 0 | 0 | 0 | 0 | 0 | 0 |
| 0 | 0 | 0 | 0 | 0 | 0 | 0 | 0 | 0 | 0 | 0 | 0 | 0 | 0 | 0 | 0 | 0 | 0 | 0 | 0 | 0 | 0 | 0 | 0 | 0 | 0 | -1 | 0 | 0 | 0 | 0 | 0 | 0 | 0 | 0 |
| 0 | 0 | 0 | 0 | 0 | 0 | 0 | 0 | 0 | 0 | 0 | 0 | 0 | 0 | 0 | 0 | 0 | 0 | 0 | 0 | 0 | 0 | 0 | 0 | 0 | 0 | 0 | -1 | 0 | 0 | 0 | 0 | 0 | 0 | 0 |
| 0 | 0 | 0 | 0 | 0 | 0 | 0 | 0 | 0 | 0 | 0 | 0 | 0 | 0 | 0 | 0 | 0 | 0 | 0 | 0 | 0 | 0 | 0 | 0 | 0 | 0 | 0 | 0 | -1 | 0 | 0 | 0 | 0 | 0 | 0 |
| 0 | 0 | 0 | 0 | 0 | 0 | 0 | 0 | 0 | 0 | 0 | 0 | 0 | 0 | 0 | 0 | 0 | 0 | 0 | 0 | 0 | 0 | 0 | 0 | 0 | 0 | 0 | 0 | 0 | -1 | 0 | 0 | 0 | 0 | 0 |
| 0 | 0 | 0 | 0 | 0 | 0 | 0 | 0 | 0 | 0 | 0 | 0 | 0 | 0 | 0 | 0 | 0 | 0 | 0 | 0 | 0 | 0 | 0 | 0 | 0 | 0 | 0 | 0 | 0 | 0 | -1 | 0 | 0 | 0 | 0 |
| 0 | 0 | 0 | 0 | 0 | 0 | 0 | 0 | 0 | 0 | 0 | 0 | 0 | 0 | 0 | 0 | 0 | 0 | 0 | 0 | 0 | 0 | 0 | 0 | 0 | 0 | 0 | 0 | 0 | 0 | 0 | -1 | 0 | 0 | 0 |
| 0 | 0 | 0 | 0 | 0 | 0 | 0 | 0 | 0 | 0 | 0 | 0 | 0 | 0 | 0 | 0 | 0 | 0 | 0 | 0 | 0 | 0 | 0 | 0 | 0 | 0 | 0 | 0 | 0 | 0 | 0 | 0 | -1 | 0 | 0 |
| 0 | 0 | 0 | 0 | 0 | 0 | 0 | 0 | 0 | 0 | 0 | 0 | 0 | 0 | 0 | 0 | 0 | 0 | 0 | 0 | 0 | 0 | 0 | 0 | 0 | 0 | 0 | 0 | 0 | 0 | 0 | 0 | 0 | -1 | 0 |
| 0 | 0 | 0 | 0 | 0 | 0 | 0 | 0 | 0 | 0 | 0 | 0 | 0 | 0 | 0 | 0 | 0 | 0 | 0 | 0 | 0 | 0 | 0 | 0 | 0 | 0 | 0 | 0 | 0 | 0 | 0 | 0 | 0 | 0 | -1 |

| | | | | | | | | | | | | | | | | | | | | | | | | | | | | | | | | | | |
|---|---|---|---|---|---|---|---|---|---|---|---|---|---|---|---|---|---|---|---|---|---|---|---|---|---|---|---|---|---|---|---|---|---|---|
| 1 | -1 | 0 | -1 | -1 | 0 | 1 | 0 | 0 | 0 | 0 | 0 | 0 | 0 | 0 | 0 | 0 | 0 | 0 | 0 | -1 | 1 | 0 | 0 | 0 | 0 | 1 | -1 | 0 | 0 | 0 | 0 | 0 | 0 | 0 |
| 1 | 0 | -1 | 0 | 0 | -1 | 0 | 1 | 0 | 0 | 0 | 0 | 0 | 0 | 0 | 0 | 0 | 0 | 0 | 0 | 0 | 0 | -1 | 1 | 0 | 0 | 1 | -1 | 0 | 0 | 0 | 0 | 0 | 0 | 0 |
| 1 | -1 | 0 | 0 | 0 | 0 | 0 | 0 | 0 | -1 | -1 | 0 | 1 | 0 | 0 | 0 | 0 | 0 | 0 | 0 | -1 | 1 | 0 | 0 | 0 | 0 | 0 | 0 | 0 | 0 | 0 | 0 | -1 | 0 | 0 |
| 1 | 0 | 0 | 0 | 0 | 0 | 0 | 0 | -1 | 0 | 0 | -1 | 0 | 1 | 0 | 0 | 0 | 0 | 0 | 0 | 0 | 0 | -1 | 1 | 0 | 0 | 0 | 0 | 0 | 0 | 0 | 0 | -1 | 0 | 0 |
| 0 | 0 | 1 | -1 | 0 | 0 | 0 | 0 | 0 | -1 | 0 | 0 | 0 | 0 | -1 | 0 | 1 | 0 | 0 | 0 | -1 | 1 | 0 | 0 | 0 | 0 | 0 | 0 | 0 | 0 | 0 | 0 | -1 | 0 | 0 |
| 0 | 0 | 1 | 0 | 0 | 0 | 0 | 0 | -1 | 0 | 0 | 0 | 0 | 0 | 0 | -1 | 0 | 1 | 0 | 0 | 0 | 0 | 0 | 0 | 0 | 0 | -1 | 1 | 0 | 0 | 0 | 0 | -1 | 0 | 0 |
| 0 | 0 | 0 | 0 | 1 | -1 | 0 | 0 | 0 | 0 | 0 | -1 | 0 | 0 | -1 | 0 | 0 | 0 | 1 | 0 | 0 | 0 | -1 | 1 | 0 | 0 | 0 | 0 | 0 | 0 | 0 | 0 | -1 | 0 | 0 |
| 0 | 0 | 0 | 0 | 1 | 0 | 0 | 0 | 0 | 0 | -1 | 0 | 0 | 0 | 0 | -1 | 0 | 0 | 0 | 1 | 0 | 0 | 0 | 0 | 0 | 0 | -1 | 1 | 0 | 0 | 0 | 0 | -1 | 0 | 0 |
| 1 | -1 | 0 | 0 | 0 | 0 | 0 | 0 | 0 | 0 | 0 | 0 | 0 | 0 | 0 | 0 | 0 | 0 | 0 | 0 | -1 | 0 | -1 | 0 | 1 | 0 | 0 | 0 | 0 | 0 | 0 | 0 | 0 | 0 | 0 |
| 1 | 0 | 0 | 0 | 0 | 0 | 0 | 0 | 0 | 0 | 0 | 0 | 0 | 0 | 0 | 0 | 0 | 0 | 0 | 0 | -1 | 0 | -1 | 0 | 0 | 1 | 0 | 0 | 0 | 0 | 0 | 0 | 0 | 0 | 0 |
| 0 | 0 | 1 | -1 | 0 | 0 | 0 | 0 | 0 | 0 | 0 | 0 | 0 | 0 | 0 | 0 | 0 | 0 | 0 | 0 | -1 | 0 | 0 | 0 | 0 | 0 | -1 | 0 | 1 | 0 | 0 | 0 | 0 | 0 | 0 |
| 0 | 0 | 1 | 0 | 0 | 0 | 0 | 0 | 0 | 0 | 0 | 0 | 0 | 0 | 0 | 0 | 0 | 0 | 0 | 0 | -1 | 0 | 0 | 0 | 0 | 0 | -1 | 0 | 0 | 1 | 0 | 0 | 0 | 0 | 0 |
| 0 | 0 | 0 | 0 | 1 | -1 | 0 | 0 | 0 | 0 | 0 | 0 | 0 | 0 | 0 | 0 | 0 | 0 | 0 | 0 | 0 | 0 | -1 | 0 | 0 | 0 | -1 | 0 | 0 | 0 | 1 | 0 | 0 | 0 | 0 |
| 0 | 0 | 0 | 0 | 1 | 0 | 0 | 0 | 0 | 0 | 0 | 0 | 0 | 0 | 0 | 0 | 0 | 0 | 0 | 0 | 0 | 0 | -1 | 0 | 0 | 0 | -1 | 0 | 0 | 0 | 0 | 1 | 0 | 0 | 0 |
| 0 | 0 | 0 | 0 | 0 | 0 | 0 | 0 | 1 | -1 | 0 | 0 | 0 | 0 | 0 | 0 | 0 | 0 | 0 | 0 | -1 | 0 | 0 | 0 | 0 | 0 | 0 | 0 | 0 | 0 | 0 | 0 | 0 | 0 | 0 |
| 0 | 0 | 0 | 0 | 0 | 0 | 0 | 0 | 1 | 0 | 0 | 0 | 0 | 0 | 0 | 0 | 0 | 0 | 0 | 0 | -1 | 0 | 0 | 0 | 0 | 0 | 0 | 0 | 0 | 0 | 0 | 0 | 0 | 1 | 0 |
| 0 | 0 | 0 | 0 | 0 | 0 | 0 | 0 | 0 | 0 | 1 | -1 | 0 | 0 | 0 | 0 | 0 | 0 | 0 | 0 | 0 | 0 | -1 | 0 | 0 | 0 | 0 | 0 | 0 | 0 | 0 | 0 | 0 | 0 | 0 |
| 0 | 0 | 0 | 0 | 0 | 0 | 0 | 0 | 0 | 0 | 1 | 0 | 0 | 0 | 0 | 0 | 0 | 0 | 0 | 0 | 0 | 0 | -1 | 0 | 0 | 0 | 0 | 0 | 0 | 0 | 0 | 0 | 0 | 0 | 1 |
| 0 | 0 | 0 | 0 | 0 | 0 | 0 | 0 | 0 | 0 | 0 | 0 | 0 | 0 | 1 | -1 | 0 | 0 | 0 | 0 | 0 | 0 | 0 | 0 | 0 | 0 | -1 | 0 | 0 | 0 | 0 | 0 | 0 | 0 | 0 |
| 0 | 0 | 0 | 0 | 0 | 0 | 0 | 0 | 0 | 0 | 0 | 0 | 0 | 0 | 1 | 0 | 0 | 0 | 0 | 0 | 0 | 0 | 0 | 0 | 0 | 0 | -1 | 0 | 0 | 0 | 0 | 0 | 0 | 0 | 0 |
| 1 | -1 | 0 | 0 | 0 | 0 | 0 | 0 | 0 | 0 | 0 | 0 | 0 | 0 | 0 | 0 | 0 | 0 | 0 | 0 | -1 | 1 | 0 | 0 | 0 | 0 | 0 | 0 | 0 | 0 | 0 | 0 | 0 | 0 | 0 |
| 1 | 0 | 0 | 0 | 0 | 0 | 0 | 0 | 0 | 0 | 0 | 0 | 0 | 0 | 0 | 0 | 0 | 0 | 0 | 0 | 0 | 0 | -1 | 1 | 0 | 0 | 0 | 0 | 0 | 0 | 0 | 0 | 0 | 0 | 0 |
| 0 | 0 | 1 | -1 | 0 | 0 | 0 | 0 | 0 | 0 | 0 | 0 | 0 | 0 | 0 | 0 | 0 | 0 | 0 | 0 | -1 | 1 | 0 | 0 | 0 | 0 | 0 | 0 | 0 | 0 | 0 | 0 | 0 | 0 | 0 |
| 0 | 0 | 1 | 0 | 0 | 0 | 0 | 0 | 0 | 0 | 0 | 0 | 0 | 0 | 0 | 0 | 0 | 0 | 0 | 0 | 0 | 0 | 0 | 0 | 0 | 0 | -1 | 1 | 0 | 0 | 0 | 0 | 0 | 0 | 0 |
| 0 | 0 | 0 | 0 | 1 | -1 | 0 | 0 | 0 | 0 | 0 | 0 | 0 | 0 | 0 | 0 | 0 | 0 | 0 | 0 | 0 | 0 | -1 | 1 | 0 | 0 | 0 | 0 | 0 | 0 | 0 | 0 | 0 | 0 | 0 |
| 0 | 0 | 0 | 0 | 1 | 0 | 0 | 0 | 0 | 0 | 0 | 0 | 0 | 0 | 0 | 0 | 0 | 0 | 0 | 0 | 0 | 0 | 0 | 0 | 0 | 0 | -1 | 1 | 0 | 0 | 0 | 0 | 0 | 0 | 0 |
| 0 | 0 | 0 | 0 | 0 | 0 | 0 | 0 | 1 | -1 | 0 | 0 | 0 | 0 | 0 | 0 | 0 | 0 | 0 | 0 | -1 | 1 | 0 | 0 | 0 | 0 | 0 | 0 | 0 | 0 | 0 | 0 | 0 | 0 | 0 |
| 0 | 0 | 0 | 0 | 0 | 0 | 0 | 0 | 1 | 0 | 0 | 0 | 0 | 0 | 0 | 0 | 0 | 0 | 0 | 0 | 0 | 0 | 0 | 0 | 0 | 0 | 0 | 0 | 0 | 0 | 0 | 0 | 1 | 0 | 0 |
| 0 | 0 | 0 | 0 | 0 | 0 | 0 | 0 | 0 | 0 | 1 | -1 | 0 | 0 | 0 | 0 | 0 | 0 | 0 | 0 | 0 | 0 | -1 | 1 | 0 | 0 | 0 | 0 | 0 | 0 | 0 | 0 | 0 | 0 | 0 |
| 0 | 0 | 0 | 0 | 0 | 0 | 0 | 0 | 0 | 0 | 1 | 0 | 0 | 0 | 0 | 0 | 0 | 0 | 0 | 0 | 0 | 0 | 0 | 0 | 0 | 0 | 0 | 0 | 0 | 0 | 0 | 0 | 1 | 0 | 0 |
| 0 | 0 | 0 | 0 | 0 | 0 | 0 | 0 | 0 | 0 | 0 | 0 | 0 | 0 | 1 | -1 | 0 | 0 | 0 | 0 | 0 | 0 | 0 | 0 | 0 | 0 | -1 | 1 | 0 | 0 | 0 | 0 | 0 | 0 | 0 |
| 0 | 0 | 0 | 0 | 0 | 0 | 0 | 0 | 0 | 0 | 0 | 0 | 0 | 0 | 1 | 0 | 0 | 0 | 0 | 0 | 0 | 0 | 0 | 0 | 0 | 0 | 0 | 0 | 0 | 0 | 0 | 0 | 1 | 0 | 0 |
| 0 | 0 | 0 | 0 | 0 | 0 | 0 | 0 | 0 | 0 | 0 | 0 | 0 | 0 | 0 | 0 | 0 | 0 | 0 | 0 | 1 | 0 | 0 | 0 | 0 | 0 | 0 | 0 | 0 | 0 | 0 | 0 | 0 | 0 | 0 |
| 0 | 0 | 0 | 0 | 0 | 0 | 0 | 0 | 0 | 0 | 0 | 0 | 0 | 0 | 0 | 0 | 0 | 0 | 0 | 0 | 0 | 0 | 1 | 0 | 0 | 0 | 0 | 0 | 0 | 0 | 0 | 0 | 0 | 0 | 0 |
| 0 | 0 | 0 | 0 | 0 | 0 | 0 | 0 | 0 | 0 | 0 | 0 | 0 | 0 | 0 | 0 | 0 | 0 | 0 | 0 | 0 | 0 | 0 | 0 | 0 | 0 | 1 | 0 | 0 | 0 | 0 | 0 | 0 | 0 | 0 |

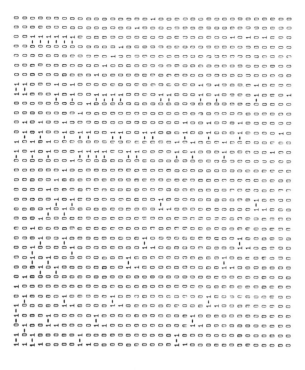

Matrix representing $(1 \ldots 7)^{-1}$ for the partition $(4,2,1)$ of 7

| | | | | | | | | | | | | | | | | | | | | | | | | | | | | | | | | | | |
|---|---|---|---|---|---|---|---|---|---|---|---|---|---|---|---|---|---|---|---|---|---|---|---|---|---|---|---|---|---|---|---|---|---|---|
| 0 | 0 | 0 | 0 | 0 | 0 | 0 | 0 | 0 | 0 | 0 | 0 | 0 | 0 | 0 | 0 | 1 | 0 | 0 | 0 | 0 | 1 | 0 | 0 | 0 | 0 | 0 | 0 | -1 | 0 | 0 | 0 | 0 | 1 | 0 |
| 0 | 0 | 0 | 0 | 0 | 0 | 0 | 0 | 0 | 0 | 0 | 0 | 0 | 0 | -1 | 0 | 1 | 0 | 0 | 0 | -1 | 1 | 0 | 0 | 0 | 0 | 1 | 0 | -1 | 0 | 0 | 0 | -1 | 1 | 0 |
| 0 | 0 | 0 | 0 | 0 | 0 | 0 | 0 | 0 | 0 | 0 | 0 | 0 | 0 | 0 | 0 | 0 | 0 | 1 | 0 | 0 | 0 | 0 | 1 | 0 | 0 | 0 | 0 | 0 | 0 | -1 | 0 | 0 | 0 | 1 |
| 0 | 0 | 0 | 0 | 0 | 0 | 0 | 0 | 0 | 0 | 0 | 0 | 0 | 0 | -1 | 0 | 0 | 0 | 1 | 0 | 0 | 0 | -1 | 1 | 0 | 0 | 1 | 0 | 0 | 0 | -1 | 0 | -1 | 0 | 1 |
| 0 | 0 | 0 | 0 | 0 | 0 | 0 | 0 | 0 | 0 | 0 | 0 | 0 | 0 | 0 | 0 | 0 | 0 | 1 | 0 | 0 | 0 | 0 | 0 | 0 | 1 | 0 | 0 | 0 | 0 | -1 | 0 | 0 | 0 | 1 |
| 0 | 0 | 0 | 0 | 0 | 0 | 0 | 0 | 0 | 0 | 0 | 0 | 0 | 0 | 0 | 0 | -1 | 0 | 1 | 0 | 0 | 0 | 0 | 0 | -1 | 1 | 0 | 0 | 1 | 0 | -1 | 0 | 0 | -1 | 1 |
| 1 | 0 | 0 | 0 | 0 | 0 | 0 | 0 | 0 | 0 | 0 | 0 | 0 | 0 | -1 | 0 | 0 | 0 | 1 | 0 | -1 | 0 | -1 | 1 | 0 | 1 | 1 | 0 | 0 | 0 | -1 | 0 | -1 | 0 | 1 |
| 0 | 1 | 0 | 0 | 0 | 0 | 0 | 0 | 0 | 0 | 0 | 0 | 0 | 0 | 0 | 0 | -1 | 0 | 1 | 0 | 0 | -1 | 0 | 1 | -1 | 1 | 0 | 0 | 1 | 0 | -1 | 0 | 0 | -1 | 1 |
| 0 | 0 | 0 | 0 | 0 | 0 | 0 | 0 | 0 | 0 | 0 | 0 | 0 | 0 | 0 | 0 | 0 | 0 | 0 | 1 | 0 | 0 | 0 | 0 | 0 | 0 | 0 | 1 | 0 | 0 | 0 | -1 | 0 | 0 | 1 |
| 0 | 0 | 0 | 0 | 0 | 0 | 0 | 0 | 0 | 0 | 0 | 0 | 0 | 0 | -1 | 0 | 0 | 0 | 0 | 1 | 0 | 0 | 0 | 0 | 0 | 0 | 0 | 1 | 0 | 0 | 0 | -1 | -1 | 0 | 1 |
| 0 | 0 | 0 | 0 | 0 | 0 | 0 | 0 | 0 | 0 | 0 | 0 | 0 | 0 | 0 | 0 | 0 | 0 | 0 | 1 | 0 | 0 | 0 | 0 | 0 | 0 | 0 | 0 | 0 | 1 | 0 | -1 | 0 | 0 | 1 |
| 0 | 0 | 0 | 0 | 0 | 0 | 0 | 0 | 0 | 0 | 0 | 0 | 0 | 0 | 0 | 0 | -1 | 0 | 0 | 1 | 0 | 0 | 0 | 0 | 0 | 0 | 0 | 0 | 0 | 1 | 0 | -1 | 0 | -1 | 1 |
| 0 | 0 | 1 | 0 | 0 | 0 | 0 | 0 | 0 | 0 | 0 | 0 | 0 | 0 | -1 | 0 | 0 | 0 | 0 | 1 | -1 | 0 | 0 | 0 | 0 | 0 | 0 | 1 | 0 | 1 | 0 | -1 | -1 | 0 | 1 |
| 0 | 0 | 0 | 1 | 0 | 0 | 0 | 0 | 0 | 0 | 0 | 0 | 0 | 0 | 0 | 0 | -1 | 0 | 0 | 1 | 0 | -1 | 0 | 0 | 0 | 0 | 0 | 1 | 0 | 1 | 0 | -1 | 0 | -1 | 1 |
| 0 | 0 | 0 | 0 | 0 | 0 | 0 | 0 | 0 | 0 | 0 | 0 | 0 | 0 | 0 | 0 | 0 | 0 | 0 | 1 | 0 | 0 | 0 | 0 | 0 | 0 | 0 | 0 | 0 | 0 | 0 | 0 | 0 | 0 | 1 |
| 0 | 0 | 0 | 0 | 0 | 0 | 0 | 0 | 0 | 0 | 0 | 0 | 0 | 0 | 0 | 0 | 0 | 0 | -1 | 1 | 0 | 0 | 0 | 0 | 0 | 0 | 0 | 0 | 0 | 0 | 0 | 0 | 0 | 0 | 0 |
| 0 | 0 | 0 | 0 | 1 | 0 | 0 | 0 | 0 | 0 | 0 | 0 | 0 | 0 | -1 | 0 | 0 | 0 | 0 | 1 | 0 | 0 | -1 | 0 | 0 | 0 | 0 | 1 | 0 | 0 | 0 | 0 | -1 | 0 | 1 |
| 0 | 0 | 0 | 0 | 0 | 1 | 0 | 0 | 0 | 0 | 0 | 0 | 0 | 0 | 0 | 0 | 0 | 0 | -1 | 1 | 0 | 0 | 0 | -1 | 0 | 0 | 0 | 1 | 0 | 0 | 0 | 0 | 0 | 0 | 0 |
| 0 | 0 | 0 | 0 | 0 | 0 | 1 | 0 | 0 | 0 | 0 | 0 | 0 | 0 | 0 | 0 | -1 | 0 | 0 | 1 | 0 | 0 | 0 | 0 | -1 | 0 | 0 | 0 | 0 | 1 | 0 | 0 | 0 | -1 | 1 |
| 0 | 0 | 0 | 0 | 0 | 0 | 0 | 1 | 0 | 0 | 0 | 0 | 0 | 0 | 0 | 0 | 0 | 0 | -1 | 1 | 0 | 0 | 0 | 0 | 0 | -1 | 0 | 0 | 0 | 1 | 0 | 0 | 0 | 0 | 0 |
| 0 | 0 | 0 | 0 | 0 | 0 | 0 | 0 | 0 | 0 | 0 | 0 | 0 | 0 | 0 | 0 | 0 | 0 | 0 | 0 | 0 | 0 | 0 | 0 | 0 | 0 | 0 | 0 | 0 | 0 | 0 | 0 | 1 | 0 | 0 |
| 0 | 0 | 0 | 0 | 0 | 0 | 0 | 0 | 0 | 0 | 0 | 0 | 0 | 0 | -1 | 0 | 0 | 0 | 0 | 0 | 0 | 0 | 0 | 0 | 0 | 0 | 1 | 0 | 0 | 0 | 0 | 0 | 0 | 0 | 0 |
| 0 | 0 | 0 | 0 | 0 | 0 | 0 | 0 | 0 | 0 | 0 | 0 | 0 | 0 | 0 | 0 | 0 | 0 | 0 | 0 | 0 | 0 | 0 | 0 | 0 | 0 | 0 | 0 | 0 | 0 | 0 | 0 | 0 | 1 | 0 |
| 0 | 0 | 0 | 0 | 0 | 0 | 0 | 0 | 0 | 0 | 0 | 0 | 0 | 0 | 0 | 0 | -1 | 0 | 0 | 0 | 0 | 0 | 0 | 0 | 0 | 0 | 0 | 0 | 1 | 0 | 0 | 0 | 0 | 0 | 0 |
| 0 | 0 | 0 | 0 | 0 | 0 | 0 | 0 | 1 | 0 | 0 | 0 | 0 | 0 | -1 | 0 | 0 | 0 | 0 | 0 | -1 | 0 | 0 | 0 | 0 | 0 | 1 | 0 | 0 | 0 | 0 | 0 | 0 | 1 | 0 |
| 0 | 0 | 0 | 0 | 0 | 0 | 0 | 0 | 0 | 1 | 0 | 0 | 0 | 0 | 0 | 0 | -1 | 0 | 0 | 0 | 0 | -1 | 0 | 0 | 0 | 0 | 0 | 0 | 1 | 0 | 0 | 0 | 1 | 0 | 0 |
| 0 | 0 | 0 | 0 | 0 | 0 | 0 | 0 | 0 | 0 | 0 | 0 | 0 | 0 | 0 | 0 | 0 | 0 | 0 | 0 | 0 | 0 | 0 | 0 | 0 | 0 | 0 | 0 | 0 | 0 | 0 | 0 | 0 | 0 | 1 |
| 0 | 0 | 0 | 0 | 0 | 0 | 0 | 0 | 0 | 0 | 0 | 0 | 0 | 0 | 0 | 0 | 0 | 0 | -1 | 0 | 0 | 0 | 0 | 0 | 0 | 0 | 0 | 0 | 0 | 0 | 1 | 0 | 0 | 0 | 0 |
| 0 | 0 | 0 | 0 | 0 | 0 | 0 | 0 | 0 | 0 | 1 | 0 | 0 | 0 | -1 | 0 | 0 | 0 | 0 | 0 | 0 | 0 | -1 | 0 | 0 | 0 | 1 | 0 | 0 | 0 | 0 | 0 | 0 | 0 | 1 |
| 0 | 0 | 0 | 0 | 0 | 0 | 0 | 0 | 0 | 0 | 0 | 1 | 0 | 0 | 0 | 0 | 0 | 0 | -1 | 0 | 0 | 0 | 0 | -1 | 0 | 0 | 0 | 0 | 0 | 0 | 1 | 0 | 1 | 0 | 0 |
| 0 | 0 | 0 | 0 | 0 | 0 | 0 | 0 | 0 | 0 | 0 | 0 | 1 | 0 | 0 | 0 | -1 | 0 | 0 | 0 | 0 | 0 | 0 | 0 | -1 | 0 | 0 | 0 | 1 | 0 | 0 | 0 | 0 | 0 | 1 |
| 0 | 0 | 0 | 0 | 0 | 0 | 0 | 0 | 0 | 0 | 0 | 0 | 0 | 1 | 0 | 0 | 0 | 0 | -1 | 0 | 0 | 0 | 0 | 0 | 0 | -1 | 0 | 0 | 0 | 0 | 1 | 0 | 0 | 1 | 0 |
| 0 | 0 | 0 | 0 | 0 | 0 | 0 | 0 | 0 | 0 | 0 | 0 | 0 | 0 | 0 | 0 | 0 | 0 | 0 | -1 | 0 | 0 | 0 | 0 | 0 | 0 | 0 | 0 | 0 | 0 | 0 | 1 | 0 | 0 | -1 |
| 0 | 0 | 0 | 0 | 0 | 0 | 0 | 0 | 0 | 0 | 0 | 0 | 0 | 0 | 0 | 1 | 0 | 0 | 0 | -1 | 0 | 0 | 0 | 0 | 0 | 0 | 0 | -1 | 0 | 0 | 0 | 1 | 1 | 0 | -1 |
| 0 | 0 | 0 | 0 | 0 | 0 | 0 | 0 | 0 | 0 | 0 | 0 | 0 | 0 | 0 | 0 | 0 | 1 | 0 | -1 | 0 | 0 | 0 | 0 | 0 | 0 | 0 | 0 | 0 | -1 | 0 | 1 | 0 | 1 | -1 |

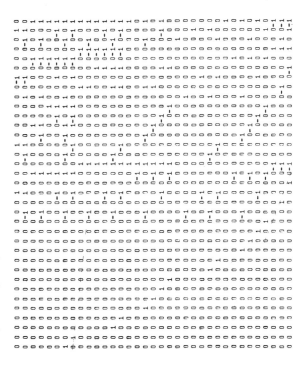

Matrices representing (1 2), $(1\ldots7)$, and $(1\ldots7)^{-1}$ for the partition $(4,1^3)$ of 7

```
-----------------------------------------
*     7 *   4 1 1 1
-----------------------------------------

 1 0 0 0 0 0 0 0 0 0-1 1-1 0 0 0 0 0 0 0
 0 1 0 0 0 0 0 0 0 0-1 0 0 1-1 0 0 0 0 0
 0 0 1 0 0 0 0 0 0 0 0-1 0 1 0-1 0 0 0 0
 0 0 0 1 0 0 0 0 0 0 0 0-1 0 1-1 0 0 0 0
 0 0 0 0 1 0 0 0 0 0-1 0 0 0 0 0 1-1 0 0
 0 0 0 0 0 1 0 0 0 0 0-1 0 0 0 0 1 0-1 0
 0 0 0 0 0 0 1 0 0 0 0 0-1 0 0 0 0 1-1 0
 0 0 0 0 0 0 0 1 0 0 0 0 0-1 0 0 1 0 0-1
 0 0 0 0 0 0 0 0 1 0 0 0 0 0-1 0 0 1 0-1
 0 0 0 0 0 0 0 0 0 1 0 0 0 0 0-1 0 0 1-1
 0 0 0 0 0 0 0 0 0 0-1 0 0 0 0 0 0 0 0 0
 0 0 0 0 0 0 0 0 0 0 0-1 0 0 0 0 0 0 0 0
 0 0 0 0 0 0 0 0 0 0 0 0-1 0 0 0 0 0 0 0
 0 0 0 0 0 0 0 0 0 0 0 0 0-1 0 0 0 0 0 0
 0 0 0 0 0 0 0 0 0 0 0 0 0 0-1 0 0 0 0 0
 0 0 0 0 0 0 0 0 0 0 0 0 0 0 0-1 0 0 0 0
 0 0 0 0 0 0 0 0 0 0 0 0 0 0 0 0-1 0 0 0
 0 0 0 0 0 0 0 0 0 0 0 0 0 0 0 0 0-1 0 0
 0 0 0 0 0 0 0 0 0 0 0 0 0 0 0 0 0 0-1 0
 0 0 0 0 0 0 0 0 0 0 0 0 0 0 0 0 0 0 0-1
```

```
-1 1-1 1 0 0 0 0 0 0 0 0 0 0 0 0 0 0 0 0
-1 0 0 0 1-1 1 0 0 0 0 0 0 0 0 0 0 0 0 0
 0-1 0 0 1 0 0-1 1 0 0 0 0 0 0 0 0 0 0 0
 0 0-1 0 0 1 0-1 0 1 0 0 0 0 0 0 0 0 0 0
-1 0 0 0 0 0 0 0 0 0 1-1 1 0 0 0 0 0 0 0
 0-1 0 0 0 0 0 0 0 0 1 0 0-1 1 0 0 0 0 0
 0 0-1 0 0 0 0 0 0 0 0 1 0-1 0 1 0 0 0 0
 0 0 0 0-1 0 0 0 0 0 1 0 0 0 0 0-1 1 0 0
 0 0 0 0 0-1 0 0 0 0 0 1 0 0 0 0-1 0 1 0
 0 0 0 0 0 0 0-1 0 0 0 0 0 1 0 0-1 0 0 1
-1 0 0 0 0 0 0 0 0 0 0 0 0 0 0 0 0 0 0 0
 0-1 0 0 0 0 0 0 0 0 0 0 0 0 0 0 0 0 0 0
 0 0-1 0 0 0 0 0 0 0 0 0 0 0 0 0 0 0 0 0
 0 0 0 0-1 0 0 0 0 0 0 0 0 0 0 0 0 0 0 0
 0 0 0 0 0-1 0 0 0 0 0 0 0 0 0 0 0 0 0 0
 0 0 0 0 0 0 0-1 0 0 0 0 0 0 0 0 0 0 0 0
 0 0 0 0 0 0 0 0 0 0-1 0 0 0 0 0 0 0 0 0
 0 0 0 0 0 0 0 0 0 0 0-1 0 0 0 0 0 0 0 0
 0 0 0 0 0 0 0 0 0 0 0 0 0-1 0 0 0 0 0 0
 0 0 0 0 0 0 0 0 0 0 0 0 0 0 0 0-1 0 0 0
```

```
 0 0 0 0 0 0 0 0 0 0-1 0 0 0 0 0 0 0 0 0
 0 0 0 0 0 0 0 0 0 0 0-1 0 0 0 0 0 0 0 0
 0 0 0 0 0 0 0 0 0 0 0 0-1 0 0 0 0 0 0 0
 1 0 0 0 0 0 0 0 0 0-1 1-1 0 0 0 0 0 0 0
 0 0 0 0 0 0 0 0 0 0 0 0 0-1 0 0 0 0 0 0
 0 0 0 0 0 0 0 0 0 0 0 0 0 0-1 0 0 0 0 0
 0 1 0 0 0 0 0 0 0 0-1 0 0 1-1 0 0 0 0 0
 0 0 0 0 0 0 0 0 0 0 0 0 0 0 0-1 0 0 0 0
 0 0 1 0 0 0 0 0 0 0 0-1 0 1 0-1 0 0 0 0
 0 0 0 1 0 0 0 0 0 0 0 0-1 0 1-1 0 0 0 0
 0 0 0 0 0 0 0 0 0 0 0 0 0 0 0 0-1 0 0 0
 0 0 0 0 0 0 0 0 0 0 0 0 0 0 0 0 0-1 0 0
 0 0 0 0 1 0 0 0 0 0-1 0 0 0 0 0 1-1 0 0
 0 0 0 0 0 0 0 0 0 0 0 0 0 0 0 0 0 0-1 0
 0 0 0 0 0 1 0 0 0 0 0-1 0 0 0 0 1 0-1 0
 0 0 0 0 0 0 1 0 0 0 0 0-1 0 0 0 0 1-1 0
 0 0 0 0 0 0 0 0 0 0 0 0 0 0 0 0 0 0 0-1
 0 0 0 0 0 0 0 1 0 0 0 0 0-1 0 0 1 0 0-1
 0 0 0 0 0 0 0 0 1 0 0 0 0 0-1 0 0 1 0-1
 0 0 0 0 0 0 0 0 0 1 0 0 0 0 0-1 0 0 1-1
```

Matrices representing (1 2), (1 . . . 7), and $(1 \ldots 7)^{-1}$ for the partition (3, 3, 1) of 7

```
-----------------------------------------
*     8 *   3 3 1
-----------------------------------------

1 0 0 0 0 0 0 0 0 0 0-1 0 0-1 0 1 0 1 0-1
0 1 0 0 0 0 0 0 0 0 0 0-1 0 0-1 1 0 1 0 0
0 0 1 0 0 0 0 0 0 0 0 0 0-1 0 0 1-1 1 0 0
0 0 0 1 0 0 0 0 0 0 0-1 0 0 0 0 0 0 0 1 0
0 0 0 0 1 0 0 0 0 0 0 0-1 0 0 0 0 0 0 1 1
0 0 0 0 0 1 0 0 0 0 0 0 0-1 0 0 0 0 0 1 1
0 0 0 0 0 0 1 0 0 0 0 0 0 0-1 0 0 0 0 1-1
0 0 0 0 0 0 0 1 0 0 0 0 0 0 0-1 0 0 0 1 0
0 0 0 0 0 0 0 0 1 0 0 0 0 0 0 0-1 0 0 1 0
0 0 0 0 0 0 0 0 0 1 0 0 0 0 0 0 0-1 0 0 1
0 0 0 0 0 0 0 0 0 0 1 0 0 0 0 0 0 0-1 0 1
0 0 0 0 0 0 0 0 0 0 0-1 0 0 0 0 0 0 0 0 0
0 0 0 0 0 0 0 0 0 0 0 0-1 0 0 0 0 0 0 0 0
0 0 0 0 0 0 0 0 0 0 0 0 0-1 0 0 0 0 0 0 0
0 0 0 0 0 0 0 0 0 0 0 0 0 0-1 0 0 0 0 0 0
0 0 0 0 0 0 0 0 0 0 0 0 0 0 0-1 0 0 0 0 0
0 0 0 0 0 0 0 0 0 0 0 0 0 0 0 0-1 0 0 0 0
0 0 0 0 0 0 0 0 0 0 0 0 0 0 0 0 0-1 0 0 0
0 0 0 0 0 0 0 0 0 0 0 0 0 0 0 0 0 0-1 0 0
0 0 0 0 0 0 0 0 0 0 0 0 0 0 0 0 0 0 0-1 0
0 0 0 0 0 0 0 0 0 0 0 0 0 0 0 0 0 0 0 0-1

1-1 0 0-1 0-2 1 0 0 0-1 1 0 1 0 0 0 0-1 0
1 0-1 0 0-1-1 0 0 1 0-1 0 1 0 0 0 0 0-1 0
1 0 0-1 0 0 0 0-1 0 1 0 0 0-1 0 1 0 0-1 0
1-1 0 0 0 0-1 1 0 0 0-1 0 0 0 0 0 0 0 0 0
1 0-1 0 0 0 0 0 0 1 0-1 0 0-1 0 0 0 0 0 0
1 0 0 0 0 0 0 0 0 0 1-1 0 0-1 0 0 0 0 0 0
0 0 0 1-1 0-1 1 0 0 0-1 0 0 1 0 0 0 0 0 0
0 0 0 1 0-1 0 0 0 1 0-1 0 0 0 0 0 0 0 0 0
0 0 0 1 0 0 0 0 0 0 0-1 0 0 0 0 0 0 0 0 0
0 0 0 0 0 0 1 0-1 0 1 0 0 0-1 0 0 0 0 0 0
0 0 0 0 0 0 1 0 0 0 0 0 0 0-1 0 0 0 0 0 0
1-1 0 0 0 0-1 1 0 0 0-1 1 0 0 0 0 0 0 0 0
1 0-1 0 0 0 0 0 0 1 0-1 0 1 0 0 0 0 0 0 0
1 0 0 0 0 0 0 0 0 0 1 0 0 0-1 0 1 0 0 0 0
0 0 0 1-1 0-1 1 0 0 0-1 1 0 1-1 0 0 0 0 0
0 0 0 1 0-1 0 0 0 1 0-1 0 1 0 0 0-1 0 0 0
0 0 0 1 0 0 0 0 0 0 0 0 0 0 0 0 0 0 0 1 0
0 0 0 0 0 0 1 0-1 0 1 0 0 0-1 0 1 0-1 0 0
0 0 0 0 0 0 1 0 0 0 0 0 0 0 0 0 0 0 0 1-1
0 0 0 0 0 0 0 0 0 0 0 1 0 0 0 0 0 0 0 0 0
0 0 0 0 0 0 0 0 0 0 0 0 0 1 0 0 0 0 0 0 0

 0 0 1 0 0 1 0 0 0-1 1 0 0-1 0 0 1 0 0 1 1
-1 0 1-1 0 1 1 0 0-1 0 1 0-1 0 0 0 0 0 1 0
 0-1 1 0-1 1 0 1 0-1 0 0 1-1 0 0 0 0 0 1-1
 0 0 0 0 0 0 0 0 1 0 0 0 0 0 0 0 0 0 0 1 0
-1 0 0 0 0 0 0 0 1 0-1 1 0 0 0 0-1 0 0 1 0
 0-1 0 0 0 0 0 0 1 0-1 0 1 0 0 0-1 0 0 1-1
 0 0 0 0 0 0 0 0 0 0 1 0 0 0 0 0 0 0 0 0 1
-1 0 0 0 0 0 1 0 0 0 0 1 0 0 0 0-1 0 0 1 0
 0 0-1 0 0 0 0 0 0 0 0 0 0 1 0 0-1 0 0 0 0
 0-1 0 0 0 0 0 1 0 0-1 0 1 0 0 0-1 0 0 1-1
 0 0-1 0 0 0 0 0 0 1-1 0 0 1 0 0-1 0 0 0 0
 0 0 0 0 0 0 0 0 0 0 0 0 0 0 0 0 0 0 0 1 0
 0 0 0-1 0 0 0 0 0 0 0 1 0 0 0 0 0 0 0 0 0
 0 0 0 0-1 0 0 0 0 0 0 0 1 0 0 0 0 0 0 0-1
 0 0 0 0 0 0 0 0 0 0 0 0 0 0 0 0 0 0 0 0 1
 0 0 0-1 0 0 1 0 0 0 0 1 0 0-1 0 0 0 0 0 0
 0 0 0 0 0-1 0 0 0 0 0 0 0 1 0 0 0 0 0-1 0
 0 0 0 0-1 0 0 1 0 0 0 0 1 0 0-1 0 0 0 0-1
 0 0 0 0 0-1 0 0 0 1 0 0 0 1 0 0 0-1 0-1 0
 0 0 0 0 0 0 0 0-1 0 0 0 0 0 0 0 1 0 0-1 0
 0 0 0 0 0 0 0 0-1 0 1 0 0 0 0 0 1 0-1-1 1
```

## Matrices representing (1 2), (1 . . . 7), and $(1\ldots7)^{-1}$ for the partition (3, 2, 2) of 7

```
------------------------------------------
*     9 *   3 2 2
------------------------------------------

1 0 0 0 0 0 0 0 0 0-1 0-1 1 0 1 1 0 0 0 0
0 1 0 0 0 0 0 0 0 0 0-1 0 1-1 1 1-1 0 0 0
0 0 1 0 0 0 0 0 0 0-1 0 0 0 0 0 0 0 1 1 0
0 0 0 1 0 0 0 0 0 0 0-1 0 0 0 0 0 0 1 1-1
0 0 0 0 1 0 0 0 0 0 0 0-1 0 0 0 0 0 1 0 0
0 0 0 0 0 1 0 0 0 0 0 0 0-1 0 0 0 0 1 0 0
0 0 0 0 0 0 1 0 0 0 0 0 0 0-1 0 0 0 0 1 0
0 0 0 0 0 0 0 1 0 0 0 0 0 0 0-1 0 0 0 1 0
0 0 0 0 0 0 0 0 1 0 0 0 0 0 0 0-1 0 0 0 1
0 0 0 0 0 0 0 0 0 1 0 0 0 0 0 0 0-1 0 0 1
0 0 0 0 0 0 0 0 0 0-1 0 0 0 0 0 0 0 0 0 0
0 0 0 0 0 0 0 0 0 0 0-1 0 0 0 0 0 0 0 0 0
0 0 0 0 0 0 0 0 0 0 0 0-1 0 0 0 0 0 0 0 0
0 0 0 0 0 0 0 0 0 0 0 0 0-1 0 0 0 0 0 0 0
0 0 0 0 0 0 0 0 0 0 0 0 0 0-1 0 0 0 0 0 0
0 0 0 0 0 0 0 0 0 0 0 0 0 0 0-1 0 0 0 0 0
0 0 0 0 0 0 0 0 0 0 0 0 0 0 0 0-1 0 0 0 0
0 0 0 0 0 0 0 0 0 0 0 0 0 0 0 0 0-1 0 0 0
0 0 0 0 0 0 0 0 0 0 0 0 0 0 0 0 0 0-1 0 0
0 0 0 0 0 0 0 0 0 0 0 0 0 0 0 0 0 0 0-1 0
0 0 0 0 0 0 0 0 0 0 0 0 0 0 0 0 0 0 0 0-1
```

```
1 0 0 1-1 1-1 0 1 0-1 0 0 0 1 0 0 0 0 1 0
1-1-1 1 0 1 0-1 0 1 0 0-1 0 1 0 0 0 0 0 0
1 0 0 0 0 0 0 0 0 0-1 1-1 1 0 0 1 0-1 1 0
1-1 0 0 0 0 0 0 0 0-1 1-1 1 1-1 0 1-1 1 0
0 0 1 0 0 0 0 0 0 0-1 1 0 0 0 0 0 0 0 1 0
0 0 1-1 0 0 0 0 0 0-1 1 0 0 0 0 0 0 0 0 1
0 0 0 0 1 0 0 0 0 0 0 0-1 1 0 0 0 0 0 0 0
0 0 0 0 1-1 0 0 0 0 0 0-1 1 0 0 0 0 0 0 0
0 0 0 0 0 0 1 0 0 0 0 0 0 0-1 1 0 0 0 0 0
0 0 0 0 0 0 1-1 0 0 0 0 0 0-1 1 0 0 0 0 0
1 0 0 0 0 0 0 0 0 0-1 0 0 0 1 0 0 0 0 1 0
1-1 0 0 0 0 0 0 0 0 0 0-1 0 1 0 0 0 0 0 0
0 0 1 0 0 0 0 0 0 0-1 0 0 0 0 0 0 0 0 1 0
0 0 1-1 0 0 0 0 0 0 0 0 0 0 0 0 0 0 0 0 0
0 0 0 0 1 0 0 0 0 0 0 0-1 0 0 0 0 0 0 0 0
0 0 0 0 1-1 0 0 0 0 0 0 0 0 0 0 0 0 0 0 0
0 0 0 0 0 0 1 0 0 0 0 0 0 0-1 0 0 0 0 0 0
0 0 0 0 0 0 1-1 0 0 0 0 0 0 0 0 0 0 0 0 0
0 0 0 0 0 0 0 0 0 0 1-1 0 0 0 0 0 0 1-1 0
0 0 0 0 0 0 0 0 0 0 0 0 1-1 0 0 0 0 1 0 0
0 0 0 0 0 0 0 0 0 0 0 0 0 0 1-1 0 0 0 1 0
```

```
0 0 0 0 1 0 0-1-1 1 1 0-1 0 0 1 1-1 1-1 0
0 0 0 0 1 0-1 0 0 0 1-1-1 0 1 0 0 0 1-1 0
0 0 0 0 1 0 0-1 0 0 0 0 0 0 0 1 0 0 1-1 0
0 0 0 0 1 0 0-1 0 0 0 0 0-1 0 1 0 0 1-1 0
0 0 0 0 0 0 1-1 0 0 0 0 0 0 0 1 0 0 0 0 0
0 0 0 0 0 0 1-1 0 0 0 0 0 0 0 0 0 0 0 0 0
0 0 0 0 0 0 0 0 1-1 0 0 0 0 0 0 0 1 0 0 0
0 0 0 0 0 0 0 0 1-1 0 0 0 0 0 0 0 0 0 0 0
1 0 0 0-1 0 0 1 1-1-1 0 0 1 0 0 0 1-1 1 0
0 1 0 0 0 0-1 1 1-1 0-1 0 1 0 0 0 0 0 0 0
0 0 0 0 1 0 0-1 0 1 0 0-1 0 0 1 0-1 1-1 1
0 0 0 0 1 0 0 0 0 0 0 0-1 0 0 0 0 0 0 0 0
0 0 0 0 0 0 1-1 0 0 0 0 0 0-1 1 0 0 0 0 0
0 0 0 0 0 0 1 0 0 0 0 0 0 0-1 0 0 0 0 0 0
0 0 0 0 0 0 0 0 1-1 0 0 0 0 0 0-1 1 0 0 0
0 0 0 0 0 0 0 0 1 0 0 0 0 0 0 0-1 0 0 0 0
0 0 1 0-1 0 0 0 1-1-1 0 1 0 0 0-1 1 0 1 0
0 0 0 1 0 0-1 0 1 0 0-1 0 0 1 0-1 0 1 0 0
0 0 0 0 0 0 0 1 0 0 0 0 0 0 0-1 0 0 0 1 0
0 0 0 0 0 0 0 0 0 1 0 0 0 0 0 0 0-1 0 0 1
0 0 0 0 0 1 0-1 0 1 0 0 0-1 0 1 0-1 1-1 1
```

```
------------------------------------------------------------------------
*   10 *   3 2 1 1
------------------------------------------------------------------------

1 0 0 0 0 0 0 0 0 0 0 0 0 0 0-1 0 0-1 1-1 0 1 0 0-1 0 0 0 0 0 0 0 0 0
0 1 0 0 0 0 0 0 0 0 0 0 0 0 0 0-1 0 0 1 0-1 1-1 0 0-1 0 0 0 0 0 0 0 0
0 0 1 0 0 0 0 0 0 0 0 0 0 0 0 0 0-1 0 0 1 0 0-1-1 1-1 0 0 0 0 0 0 0 0
0 0 0 1 0 0 0 0 0 0 0 0 0 0 0-1 0 0 0 0 0 0 0 0 0 0 0 1-1 1 0-1 0 0 0
0 0 0 0 1 0 0 0 0 0 0 0 0 0 0 0-1 0 0 0 0 0 0 0 0 0 0 1 0 1-1 0-1 0 0
0 0 0 0 0 1 0 0 0 0 0 0 0 0 0 0 0-1 0 0 0 0 0 0 0 0 0 0 1 0-1 1-1 0 0
0 0 0 0 0 0 1 0 0 0 0 0 0 0 0 0 0 0-1 0 0 0 0 0 0 0 0 1-1 0 0 0 0 0 0
0 0 0 0 0 0 0 1 0 0 0 0 0 0 0 0 0 0 0-1 0 0 0 0 0 0 0 1 0 0 0 0 0-1-1
0 0 0 0 0 0 0 0 1 0 0 0 0 0 0 0 0 0 0 0-1 0 0 0 0 0 0 0 1 0 0 0 0-1-1
0 0 0 0 0 0 0 0 0 1 0 0 0 0 0 0 0 0 0 0 0-1 0 0 0 0 0 0 0 1-1 0 0 0 0
0 0 0 0 0 0 0 0 0 0 1 0 0 0 0 0 0 0 0 0 0 0-1 0 0 0 0 0 0 1 0 0 0-1 0
0 0 0 0 0 0 0 0 0 0 0 1 0 0 0 0 0 0 0 0 0 0 0-1 0 0 0 0 0 0 1 0 0-1 0
0 0 0 0 0 0 0 0 0 0 0 0 1 0 0 0 0 0 0 0 0 0 0 0-1 0 0 0 0 0 0 1-1 0 0
0 0 0 0 0 0 0 0 0 0 0 0 0 1 0 0 0 0 0 0 0 0 0 0 0-1 0 0 0 0 0 1 0 0-1
0 0 0 0 0 0 0 0 0 0 0 0 0 0 1 0 0 0 0 0 0 0 0 0 0 0-1 0 0 0 0 0 1 0-1
0 0 0 0 0 0 0 0 0 0 0 0 0 0 0-1 0 0 0 0 0 0 0 0 0 0 0 0 0 0 0 0 0 0 0
0 0 0 0 0 0 0 0 0 0 0 0 0 0 0 0-1 0 0 0 0 0 0 0 0 0 0 0 0 0 0 0 0 0 0
0 0 0 0 0 0 0 0 0 0 0 0 0 0 0 0 0-1 0 0 0 0 0 0 0 0 0 0 0 0 0 0 0 0 0
0 0 0 0 0 0 0 0 0 0 0 0 0 0 0 0 0 0-1 0 0 0 0 0 0 0 0 0 0 0 0 0 0 0 0
0 0 0 0 0 0 0 0 0 0 0 0 0 0 0 0 0 0 0-1 0 0 0 0 0 0 0 0 0 0 0 0 0 0 0
0 0 0 0 0 0 0 0 0 0 0 0 0 0 0 0 0 0 0 0-1 0 0 0 0 0 0 0 0 0 0 0 0 0 0
0 0 0 0 0 0 0 0 0 0 0 0 0 0 0 0 0 0 0 0 0-1 0 0 0 0 0 0 0 0 0 0 0 0 0
0 0 0 0 0 0 0 0 0 0 0 0 0 0 0 0 0 0 0 0 0 0-1 0 0 0 0 0 0 0 0 0 0 0 0
0 0 0 0 0 0 0 0 0 0 0 0 0 0 0 0 0 0 0 0 0 0 0-1 0 0 0 0 0 0 0 0 0 0 0
0 0 0 0 0 0 0 0 0 0 0 0 0 0 0 0 0 0 0 0 0 0 0 0-1 0 0 0 0 0 0 0 0 0 0
0 0 0 0 0 0 0 0 0 0 0 0 0 0 0 0 0 0 0 0 0 0 0 0 0-1 0 0 0 0 0 0 0 0 0
0 0 0 0 0 0 0 0 0 0 0 0 0 0 0 0 0 0 0 0 0 0 0 0 0 0-1 0 0 0 0 0 0 0 0
0 0 0 0 0 0 0 0 0 0 0 0 0 0 0 0 0 0 0 0 0 0 0 0 0 0 0-1 0 0 0 0 0 0 0
0 0 0 0 0 0 0 0 0 0 0 0 0 0 0 0 0 0 0 0 0 0 0 0 0 0 0 0-1 0 0 0 0 0 0
0 0 0 0 0 0 0 0 0 0 0 0 0 0 0 0 0 0 0 0 0 0 0 0 0 0 0 0 0-1 0 0 0 0 0
0 0 0 0 0 0 0 0 0 0 0 0 0 0 0 0 0 0 0 0 0 0 0 0 0 0 0 0 0 0-1 0 0 0 0
0 0 0 0 0 0 0 0 0 0 0 0 0 0 0 0 0 0 0 0 0 0 0 0 0 0 0 0 0 0 0-1 0 0 0
0 0 0 0 0 0 0 0 0 0 0 0 0 0 0 0 0 0 0 0 0 0 0 0 0 0 0 0 0 0 0 0-1 0 0
0 0 0 0 0 0 0 0 0 0 0 0 0 0 0 0 0 0 0 0 0 0 0 0 0 0 0 0 0 0 0 0 0-1 0
0 0 0 0 0 0 0 0 0 0 0 0 0 0 0 0 0 0 0 0 0 0 0 0 0 0 0 0 0 0 0 0 0 0-1
```

```
-1 1-1 0 0-1 1 0 0-1 0 0 1 0 0 1-1 1 0 0 0 0 0 0 0 0 0 1-1-1 1 0 0 0 0
-1 0 0 1 0 0 0 0-1 0-1 0 0 1 0 0 0 0 1-1 1 0 0 0 0 0 0 1-1 0 0 0 0 1 0
 0-1 0 0 1 0 0-1 0 0 0-1 0 0 1 0 0 0 0 0 0 1-1 1 0 0 0 0 0 1-1 0 0 1 0
-1 1-1 0 0 0 0 0 0 0 0 0 0 0 0 1-1 0 1 0 0-1 0 0 1 0 0 0 0 0 0 0 0 0 0
-1 0 0 0 0 0 0 0 0 0 0 0 0 0 0 1 0 0 1-1 0 0-1 0 0 1 0 0 0 0 0 0 0 0 0
 0-1 0 0 0 0 0 0 0 0 0 0 0 0 0 0 1 0 0-1 0 1-1 0 0 0 1 0 0 0 0 0 0 0 0
 0 0 0-1 1-1 0 0 0 0 0 0 0 0 0 1-1 0 0 0 0 0 0 0 0 0 0 0 0 0 0 0 0 0 0
 0 0 0-1 0 0 0 0 0 0 0 0 0 0 0 1 0 0 0 0 0 0 0 0 0 0 0-1 0-1 0 1 0 0 0
 0 0 0 0-1 0 0 0 0 0 0 0 0 0 0 0 1 0 0 0 0 0 0 0 0 0 0-1 0-1 0 0 1 0 0
 0 0 0 0 0 0-1 1-1 0 0 0 0 0 0 0 0 0 1-1 0 0 0 0 0 0 0 0 0 0 0 0 0 0 0
 0 0 0 0 0 0-1 0 0 0 0 0 0 0 0 0 0 0 1 0 0 0 0 0 0 0 0-1 0 0 0 0 0 0 0
 0 0 0 0 0 0 0-1 0 0 0 0 0 0 0 0 0 0 0 1 0 0 0 0 0 0 0-1 0 0 0 0 0 0 1
 0 0 0 0 0 0 0 0 0-1 1-1 0 0 0 0 0 0 0 0 0 1-1 0 0 0 0 0 0 0 0 0 0 0 0
 0 0 0 0 0 0 0 0 0-1 0 0 0 0 0 0 0 0 0 0 0 1 0 0 0 0 0 0 0-1 0 0 0 0 0
 0 0 0 0 0 0 0 0 0 0-1 0 0 0 0 0 0 0 0 0 0 0 1 0 0 0 0 0 0-1 0 0 0 0 0
-1 1-1 0 0 0 0 0 0 0 0 0 0 0 0 1-1 1 0 0 0 0 0 0 0 0 0 0 0 0 0 0 0 0 0
-1 0 0 0 0 0 0 0 0 0 0 0 0 0 0 0 0 0 1-1 1 0 0 0 0 0 0 0 0 0 0 0 0 0 0
 0-1 0 0 0 0 0 0 0 0 0 0 0 0 0 0 0 0 0 0 0 1-1 1 0 0 0 0 0 0 0 0 0 0 0
 0 0 0-1 1-1 0 0 0 0 0 0 0 0 0 1-1 1 0 0 0 0 0 0 0 0 0 0 0 0 0 0 0 0 0
 0 0 0-1 0 0 0 0 0 0 0 0 0 0 0 0 0 0 0 0 0 0 0 0 0 0 0-1 1 0 0 0 0 0 0
 0 0 0 0-1 0 0 0 0 0 0 0 0 0 0 0 0 0 0 0 0 0 0 0 0 0 0 0 0-1 1 0 0 0 0
 0 0 0 0 0 0-1 1-1 0 0 0 0 0 0 0 0 0 1-1 1 0 0 0 0 0 0 0 0 0 0 0 0 0 0
 0 0 0 0 0 0-1 0 0 0 0 0 0 0 0 0 0 0 0 0 0 0 0 0 0 0 0-1 1 0 0 0 0 0 0
 0 0 0 0 0 0 0-1 0 0 0 0 0 0 0 0 0 0 0 0 0 0 0 0 0 0 0 0 0 0 0 0 0 1 0
 0 0 0 0 0 0 0 0 0-1 1-1 0 0 0 0 0 0 0 0 0 1-1 1 0 0 0 0 0 0 0 0 0 0 0
 0 0 0 0 0 0 0 0 0-1 0 0 0 0 0 0 0 0 0 0 0 0 0 0 0 0 0 0 0-1 1 0 0 0 0
 0 0 0 0 0 0 0 0 0 0-1 0 0 0 0 0 0 0 0 0 0 0 0 0 0 0 0 0 0 0 0 0 0 1 0
 0 0 0 0 0 0 0 0 0 0 0 0 0 0 0-1 0 0 0 0 0 0 0 0 0 0 0 0 0 0 0 0 0 0 0
 0 0 0 0 0 0 0 0 0 0 0 0 0 0 0 0-1 0 0 0 0 0 0 0 0 0 0 0 0 0 0 0 0 0 0
 0 0 0 0 0 0 0 0 0 0 0 0 0 0 0 0 0 0-1 0 0 0 0 0 0 0 0 0 0 0 0 0 0 0 0
 0 0 0 0 0 0 0 0 0 0 0 0 0 0 0 0 0 0 0-1 0 0 0 0 0 0 0 0 0 0 0 0 0 0 0
 0 0 0 0 0 0 0 0 0 0 0 0 0 0 0 0 0 0 0 0 0-1 0 0 0 0 0 0 0 0 0 0 0 0 0
 0 0 0 0 0 0 0 0 0 0 0 0 0 0 0 0 0 0 0 0 0 0-1 0 0 0 0 0 0 0 0 0 0 0 0
 0 0 0 0 0 0 0 0 0 0 0 0 0 0 0 0 0 0 0 0 0 0 0 0 0 0 0-1 0 0 0 0 0 0 0
 0 0 0 0 0 0 0 0 0 0 0 0 0 0 0 0 0 0 0 0 0 0 0 0 0 0 0 0 0-1 0 0 0 0 0
```

Matrix representing $(1 \ldots 7)^{-1}$ for the partition (3, 2, 1, 1) of 7

| | | | | | | | | | | | | | | | | | | | | | | | | | | | | | | | | | | |
|---|---|---|---|---|---|---|---|---|---|---|---|---|---|---|---|---|---|---|---|---|---|---|---|---|---|---|---|---|---|---|---|---|---|---|
| 0 | 0 | 0 | 0 | 0 | 0 | 0 | 0 | 0 | -1 | 0 | 0 | 0 | 0 | 0 | 0 | -1 | 0 | 0 | 0 | 0 | 1 | 0 | 0 | 0 | 0 | 0 | 0 | 0 | -1 | 1 | 0 | 0 | 0 | 0 |
| 0 | 0 | 0 | 0 | 0 | 0 | 0 | 0 | 0 | 0 | 0 | 0 | -1 | 0 | 0 | 0 | 0 | -1 | 0 | 0 | 0 | 0 | 0 | 0 | 1 | 0 | 0 | 0 | 0 | 0 | 0 | -1 | 1 | 0 | 0 |
| 0 | 0 | 0 | 0 | 0 | 0 | -1 | 0 | 0 | 1 | 0 | 0 | -1 | 0 | 0 | -1 | 1 | -1 | 1 | 0 | 0 | -1 | 0 | 0 | 1 | 0 | 0 | -1 | 1 | 1 | -1 | -1 | 1 | 0 | 0 |
| 0 | 0 | 0 | 0 | 0 | 0 | 0 | 0 | 0 | 0 | -1 | 0 | 0 | 0 | 0 | 0 | 0 | 0 | 0 | -1 | 0 | 0 | 1 | 0 | 0 | 0 | 0 | 0 | 0 | -1 | 0 | 0 | 0 | 1 | 0 |
| 0 | 0 | 0 | 0 | 0 | 0 | 0 | 0 | 0 | 0 | 0 | 0 | 0 | -1 | 0 | 0 | 0 | 0 | 0 | 0 | -1 | 0 | 0 | 0 | 0 | 1 | 0 | 0 | 0 | 0 | 0 | -1 | 0 | 0 | 1 |
| 0 | 0 | 0 | 0 | 0 | 0 | -1 | 0 | 0 | 0 | 1 | 0 | 0 | -1 | 0 | 0 | 0 | 0 | 0 | 1 | -1 | 0 | -1 | 0 | 0 | 1 | 0 | -1 | 1 | 1 | 0 | -1 | 0 | -1 | 1 |
| 0 | 0 | 0 | 0 | 0 | 0 | 0 | 0 | 0 | 0 | -1 | 0 | 0 | 0 | 0 | 0 | 0 | 0 | 0 | 0 | 0 | 0 | 0 | 0 | 0 | 0 | 0 | 0 | 0 | -1 | 0 | 0 | 0 | 1 | 0 |
| 0 | 0 | 0 | 0 | 0 | 0 | 0 | 0 | 0 | 0 | 0 | 0 | 0 | 0 | -1 | 0 | 0 | 0 | 0 | 0 | 0 | 0 | 0 | -1 | 0 | 0 | 1 | 0 | 0 | 0 | 0 | 0 | -1 | 0 | 1 |
| 0 | 0 | 0 | 0 | 0 | 0 | 0 | 0 | 0 | -1 | 1 | 0 | 0 | 0 | -1 | 0 | 0 | 0 | 0 | 0 | 0 | 0 | 0 | -1 | 0 | 0 | 1 | 0 | 0 | 0 | 1 | 0 | -1 | -1 | 1 |
| 0 | 0 | 0 | 0 | 0 | 0 | 0 | 0 | 0 | 0 | 0 | 0 | 0 | -1 | 0 | 0 | 0 | 0 | 0 | 0 | 0 | 0 | 0 | 0 | 0 | 0 | 0 | 0 | 0 | 0 | 0 | -1 | 0 | 0 | 1 |
| 0 | 0 | 0 | 0 | 0 | 0 | 0 | 0 | 0 | 0 | 0 | 0 | 0 | 0 | -1 | 0 | 0 | 0 | 0 | 0 | 0 | 0 | 0 | 0 | 0 | 0 | 0 | 0 | 0 | 0 | 0 | 0 | -1 | 0 | 1 |
| 0 | 0 | 0 | 0 | 0 | 0 | 0 | 0 | 0 | 0 | 0 | 0 | -1 | 1 | -1 | 0 | 0 | 0 | 0 | 0 | 0 | 0 | 0 | 0 | 0 | 0 | 0 | 0 | 0 | 0 | 0 | 0 | 0 | 0 | 0 |
| 1 | 0 | 0 | 0 | 0 | 0 | -1 | 0 | 0 | 0 | 1 | 0 | 0 | -1 | 0 | -1 | 0 | 0 | 0 | 1 | -1 | 0 | 0 | 0 | 0 | 0 | 0 | -1 | 1 | 1 | 0 | -1 | 0 | -1 | 1 |
| 0 | 1 | 0 | 0 | 0 | 0 | 0 | 0 | 0 | -1 | 1 | 0 | 0 | 0 | -1 | 0 | -1 | 0 | 0 | 1 | 0 | 0 | 0 | -1 | 0 | 0 | 0 | 0 | 0 | 0 | 1 | 0 | -1 | -1 | 1 |
| 0 | 0 | 1 | 0 | 0 | 0 | 0 | 0 | 0 | 0 | 0 | 0 | -1 | 1 | -1 | 0 | 0 | -1 | 0 | 0 | 1 | 0 | 0 | -1 | 0 | 0 | 0 | 0 | 0 | 0 | 0 | 0 | 0 | 0 | 0 |
| 0 | 0 | 0 | 0 | 0 | 0 | 0 | 0 | 0 | 0 | 0 | 0 | 0 | 0 | 0 | 0 | 0 | 0 | 0 | 0 | 0 | 0 | 0 | 0 | 0 | 0 | 0 | -1 | 0 | 0 | 0 | 0 | 0 | 0 | 0 |
| 0 | 0 | 0 | 0 | 0 | 0 | 0 | 0 | 0 | 0 | 0 | 0 | 0 | 0 | 0 | 0 | 0 | 0 | 0 | 0 | 0 | 0 | 0 | 0 | 0 | 0 | 0 | 0 | -1 | 0 | 0 | 0 | 0 | 0 | 0 |
| 0 | 0 | 0 | 0 | 0 | 0 | -1 | 0 | 0 | 0 | 0 | 0 | 0 | 0 | 0 | 0 | 0 | 0 | 1 | 0 | 0 | 0 | 0 | 0 | 0 | 0 | 0 | 0 | 0 | 0 | 0 | 0 | 0 | 0 | 0 |
| 0 | 0 | 0 | 0 | 0 | 0 | 0 | 0 | 0 | 0 | 0 | 0 | 0 | 0 | 0 | 0 | 0 | 0 | 0 | 0 | 0 | 0 | 0 | 0 | 0 | 0 | 0 | 0 | 0 | -1 | 0 | 0 | 0 | 0 | 0 |
| 0 | 0 | 0 | 0 | 0 | 0 | 0 | 0 | 0 | 0 | 0 | 0 | 0 | 0 | 0 | 0 | 0 | 0 | 0 | 0 | 0 | 0 | 0 | 0 | 0 | 0 | 0 | 0 | 0 | 0 | -1 | 0 | 0 | 0 | 0 |
| 0 | 0 | 0 | 0 | 0 | 0 | 0 | 0 | 0 | -1 | 0 | 0 | 0 | 0 | 0 | 0 | 0 | 0 | 0 | 0 | 0 | 1 | 0 | 0 | 0 | 0 | 0 | 0 | 0 | 0 | 0 | 0 | 0 | 0 | 0 |
| 0 | 0 | 0 | 0 | 0 | 0 | 0 | 0 | 0 | 0 | 0 | 0 | 0 | 0 | 0 | 0 | 0 | 0 | 0 | 0 | 0 | 0 | 0 | 0 | 0 | 0 | 0 | 0 | 0 | 0 | 0 | -1 | 0 | 0 | 0 |
| 0 | 0 | 0 | 0 | 0 | 0 | 0 | 0 | 0 | 0 | 0 | 0 | 0 | 0 | 0 | 0 | 0 | 0 | 0 | 0 | 0 | 0 | 0 | 0 | 0 | 0 | 0 | 0 | 0 | 0 | 0 | 0 | -1 | 0 | 0 |
| 0 | 0 | 0 | 0 | 0 | 0 | 0 | 0 | 0 | 0 | 0 | 0 | -1 | 0 | 0 | 0 | 0 | 0 | 0 | 0 | 0 | 0 | 0 | 0 | 1 | 0 | 0 | 0 | 0 | 0 | 0 | 0 | 0 | 0 | 0 |
| 0 | 0 | 0 | 1 | 0 | 0 | -1 | 0 | 0 | 0 | 0 | 0 | 0 | 0 | 0 | -1 | 0 | 0 | 1 | 0 | 0 | 0 | 0 | 0 | 0 | 0 | 0 | 0 | 0 | 1 | 0 | -1 | 0 | 0 | 0 |
| 0 | 0 | 0 | 0 | 1 | 0 | 0 | 0 | 0 | -1 | 0 | 0 | 0 | 0 | 0 | 0 | -1 | 0 | 0 | 0 | 0 | 1 | 0 | 0 | 0 | 0 | 0 | 1 | 0 | 0 | 0 | 0 | -1 | 0 | 0 |
| 0 | 0 | 0 | 0 | 0 | 1 | 0 | 0 | 0 | 0 | 0 | 0 | -1 | 0 | 0 | 0 | 0 | -1 | 0 | 0 | 0 | 0 | 0 | 0 | 1 | 0 | 0 | 0 | 1 | 0 | -1 | 0 | 0 | 0 | 0 |
| 0 | 0 | 0 | 0 | 0 | 0 | 0 | 0 | 0 | 0 | 0 | 0 | 0 | 0 | 0 | 0 | 0 | 0 | 0 | 0 | 0 | 0 | 0 | 0 | 0 | 0 | 0 | 0 | 0 | 0 | 0 | 0 | 0 | -1 | 0 |
| 0 | 0 | 0 | 0 | 0 | 0 | 0 | 0 | 0 | 0 | -1 | 0 | 0 | 0 | 0 | 0 | 0 | 0 | 0 | 0 | 0 | 0 | 1 | 0 | 0 | 0 | 0 | 0 | 0 | -1 | 0 | 0 | 0 | 0 | 0 |
| 0 | 0 | 0 | 0 | 0 | 0 | 0 | 0 | 0 | 0 | 0 | 0 | 0 | 0 | 0 | 0 | 0 | 0 | 0 | 0 | 0 | 0 | 0 | 0 | 0 | 0 | 0 | 0 | 0 | 0 | 0 | 0 | 0 | 0 | -1 |
| 0 | 0 | 0 | 0 | 0 | 0 | 0 | 0 | 0 | 0 | 0 | 0 | 0 | -1 | 0 | 0 | 0 | 0 | 0 | 0 | 0 | 0 | 0 | 0 | 0 | 1 | 0 | 0 | 0 | 0 | 0 | -1 | 0 | 0 | 0 |
| 0 | 0 | 0 | 0 | 0 | 0 | 0 | 1 | 0 | 0 | -1 | 0 | 0 | 0 | 0 | 0 | 0 | 0 | 0 | -1 | 0 | 0 | 1 | 0 | 0 | 0 | 0 | 1 | 0 | -1 | 0 | 0 | 0 | 0 | -1 |
| 0 | 0 | 0 | 0 | 0 | 0 | 0 | 0 | 1 | 0 | 0 | 0 | 0 | -1 | 0 | 0 | 0 | 0 | 0 | 0 | -1 | 0 | 0 | 0 | 0 | 1 | 0 | 0 | 1 | 0 | 0 | -1 | 0 | -1 | 0 |
| 0 | 0 | 0 | 0 | 0 | 0 | 0 | 0 | 0 | 0 | 0 | 0 | 0 | 0 | -1 | 0 | 0 | 0 | 0 | 0 | 0 | 0 | 0 | 0 | 0 | 0 | 1 | 0 | 0 | 0 | 0 | 0 | -1 | 0 | 1 |
| 0 | 0 | 0 | 0 | 0 | 0 | 0 | 0 | 0 | 0 | 0 | 1 | 0 | 0 | -1 | 0 | 0 | 0 | 0 | 0 | 0 | 0 | 0 | -1 | 0 | 0 | 1 | 0 | 0 | 0 | 1 | 0 | -1 | -1 | 1 |

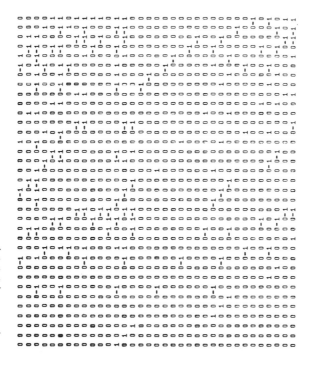

## Matrices representing (1 2), (1 … 7), and $(1 \ldots 7)^{-1}$ for the partitions $(3,1^4)$, $(2^3,1)$, $(2^2,1^3)$, and $(2,1^5)$ of 7

```
------------------------------------
*   11 *   3 1 1 1 1
------------------------------------

 1 0 0 0 0-1 1-1 1 0 0 0 0 0 0
 0 1 0 0 0-1 0 0 0 1-1 1 0 0 0
 0 0 1 0 0 0-1 0 0 1 0 0-1 1 0
 0 0 0 1 0 0 0-1 0 0 1 0-1 0 1
 0 0 0 0 1 0 0 0-1 0 0 1 0-1 1
 0 0 0 0 0-1 0 0 0 0 0 0 0 0 0
 0 0 0 0 0 0-1 0 0 0 0 0 0 0 0
 0 0 0 0 0 0 0-1 0 0 0 0 0 0 0
 0 0 0 0 0 0 0 0-1 0 0 0 0 0 0
 0 0 0 0 0 0 0 0 0-1 0 0 0 0 0
 0 0 0 0 0 0 0 0 0 0-1 0 0 0 0
 0 0 0 0 0 0 0 0 0 0 0-1 0 0 0
 0 0 0 0 0 0 0 0 0 0 0 0-1 0 0
 0 0 0 0 0 0 0 0 0 0 0 0 0-1 0
 0 0 0 0 0 0 0 0 0 0 0 0 0 0-1

 1-1 1-1 1 0 0 0 0 0 0 0 0 0 0
 1 0 0 0 0-1 1-1 1 0 0 0 0 0 0
 0 1 0 0 0-1 0 0 0 1-1 1 0 0 0
 0 0 1 0 0 0-1 0 0 1 0 0-1 1 0
 0 0 0 1 0 0 0-1 0 0 1 0-1 0 1
 1 0 0 0 0 0 0 0 0 0 0 0 0 0 0
 0 1 0 0 0 0 0 0 0 0 0 0 0 0 0
 0 0 1 0 0 0 0 0 0 0 0 0 0 0 0
 0 0 0 1 0 0 0 0 0 0 0 0 0 0 0
 0 0 0 0 0 1 0 0 0 0 0 0 0 0 0
 0 0 0 0 0 0 1 0 0 0 0 0 0 0 0
 0 0 0 0 0 0 0 1 0 0 0 0 0 0 0
 0 0 0 0 0 0 0 0 0 1 0 0 0 0 0
 0 0 0 0 0 0 0 0 0 0 1 0 0 0 0
 0 0 0 0 0 0 0 0 0 0 0 0 1 0 0

 0 0 0 0 0 1 0 0 0 0 0 0 0 0 0
 0 0 0 0 0 0 1 0 0 0 0 0 0 0 0
 0 0 0 0 0 0 0 1 0 0 0 0 0 0 0
 0 0 0 0 0 0 0 0 1 0 0 0 0 0 0
 1 0 0 0 0-1 1-1 1 0 0 0 0 0 0
 0 0 0 0 0 0 0 0 0 1 0 0 0 0 0
 0 0 0 0 0 0 0 0 0 0 1 0 0 0 0
 0 0 0 0 0 0 0 0 0 0 0 1 0 0 0
 0 1 0 0 0-1 0 0 0 1-1 1 0 0 0
 0 0 0 0 0 0 0 0 0 0 0 0 1 0 0
 0 0 0 0 0 0 0 0 0 0 0 0 0 1 0
 0 0 1 0 0 0-1 0 0 1 0 0-1 1 0
 0 0 0 0 0 0 0 0 0 0 0 0 0 0 1
 0 0 0 1 0 0 0-1 0 0 1 0-1 0 1
 0 0 0 0 1 0 0 0-1 0 0 1 0-1 1
```

```
------------------------------------
*   12 *   2 2 2 1
------------------------------------

 1 0 0 0 0-1 0 0 0 0 1 0-1 1
 0 1 0 0 0 0-1 0 0 0 0 1-1 0
 0 0 1 0 0 0 0-1 0 0 1 0 0 0
 0 0 0 1 0 0 0 0-1 0 0 1 0-1
 0 0 0 0 1 0 0 0 0-1 0 0 1-1
 0 0 0 0 0-1 0 0 0 0 0 0 0 0
 0 0 0 0 0 0-1 0 0 0 0 0 0 0
 0 0 0 0 0 0 0-1 0 0 0 0 0 0
 0 0 0 0 0 0 0 0-1 0 0 0 0 0
 0 0 0 0 0 0 0 0 0-1 0 0 0 0
 0 0 0 0 0 0 0 0 0 0-1 0 0 0
 0 0 0 0 0 0 0 0 0 0 0-1 0 0
 0 0 0 0 0 0 0 0 0 0 0 0-1 0
 0 0 0 0 0 0 0 0 0 0 0 0 0-1

-1 1 1 0 0 1-1-2 1 0 1 0 0 0
-1 0 1 0 0 1 0-1 0 1 0 0 0 0
-1 1 1-1 0 1-1-1 1 0 0 0 0 0
-1 0 0 0-1 1 0 0 0 1-1 0 0 0
 0 0-1 0 0 0 0 1 0 0-1 0 0 0
-1 1 1 0 0 1-1-1 0 0 0 0 0 0
-1 0 1 0 0 1 0-1 0 0 0 0 0 0
-1 1 1-1 0 0 0 0 0 0-1 0 0 0
-1 0 0 0-1 0 0 0 0 0-1 0 0 0
 0 0-1 0 0 0 0 0 0 0 0 0 0 0
 0 0 0 0 0-1 1 1-1 0-1 1 0 0
 0 0 0 0 0-1 0 0 0-1 0 0 1 0
 0 0 0 0 0 0 0-1 0 0 0 0 0 1
 0 0 0 0 0 0 0 0 0 0-1 0 0 0

 0-1 0 1-1 0 0 0-1 0 0 0 0 1
 1-1-1 1 0 0 0 1-1 0 0 0 0 0
 0 0 0 0 0 0 0 0 0-1 0 0 0 0
 1 0-1 0 1 0 0 0 0-1 0 0 0 0
 0 1 0-1 1 0 0 0 0 0 0 0 0 0
 0-1 0 1 0 0 1 0-1 0 0 0 0 0
 1-1-1 1 0-1 1 1-1 0 0 0 0 0
 0 0 0 0 1 0 0 0 0-1 0 0 0-1
 1 0 0 0 1-1 0 0 0-1 0 0 0 0
 0 1 0 0 0 0-1 0 0 0 0 0 0 0
 0 0 0 0 0 0 0 0 0 0 0 0 0-1
 0 0 1 0 0 0 0-1 0 0 1 0 0 0
 0 0 0 1 0 0 0 0-1 0 0 1 0 0
 0 0 0 0 1 0 0 0 0-1 0 0 1-1
```

```
------------------------------------
*   13 *   2 2 1 1 1
------------------------------------

 1 0 0 0-1 0 0 0 1-1 1 0 0 0
 0 1 0 0 0-1 0 0 1 0 0-1 1 0
 0 0 1 0 0 0-1 0 0 1 0-1 0 1
 0 0 0 1 0 0 0-1 0 0 1 0-1 1
 0 0 0 0-1 0 0 0 0 0 0 0 0 0
 0 0 0 0 0-1 0 0 0 0 0 0 0 0
 0 0 0 0 0 0-1 0 0 0 0 0 0 0
 0 0 0 0 0 0 0-1 0 0 0 0 0 0
 0 0 0 0 0 0 0 0-1 0 0 0 0 0
 0 0 0 0 0 0 0 0 0-1 0 0 0 0
 0 0 0 0 0 0 0 0 0 0-1 0 0 0
 0 0 0 0 0 0 0 0 0 0 0-1 0 0
 0 0 0 0 0 0 0 0 0 0 0 0-1 0
 0 0 0 0 0 0 0 0 0 0 0 0 0-1

 1-1 1-1-1 1-1 0 0 0 0 0 0 0
 1 0 0 0-1 0 0 0 1-1 0 0 0 0
 0 1 0 0 0-1 0 0 1 0 0-1 0 0
 0 0 1 0 0 0-1 0 0 1 0-1 0 0
 1-1 1-1-1 1-1 1 0 0 0 0 0 0
 1 0 0 0 0 0 0 0 1-1 1 0 0 0
 0 1 0 0 0 0 0 0 0 0 0-1 1 0
 0 0 1 0 0 0 0 0 0 0 0 0 0 1
 0 0 0 0 1 0 0 0 0 0 0 0 0 0
 0 0 0 0 0 1 0 0 0 0 0 0 0 0
 0 0 0 0 0 0 1 0 0 0 0 0 0 0
 0 0 0 0 0 0 0 0 1 0 0 0 0 0
 0 0 0 0 0 0 0 0 0 1 0 0 0 0
 0 0 0 0 0 0 0 0 0 0 0 1 0 0

 0 1 0 0 0 0 0 0 1 0 0-1 1 0
 0 0 1 0 0 0 0 0 0 1 0-1 0 1
 0 0 0 1 0 0 0 0 0 0 1 0-1 1
-1 1-1 1 0 0 0 0 0 0 0 0 0 0
 0 0 0 0 0 0 0 0 1 0 0 0 0 0
 0 0 0 0 0 0 0 0 0 1 0 0 0 0
 0 0 0 0 0 0 0 0 0 0 1 0 0 0
-1 0 0 0 1 0 0 0 0 0 0 0 0 0
 0 0 0 0 0 0 0 0 0 0 0 1 0 0
 0 0 0 0 0 0 0 0 0 0 0 0 1 0
 0-1 0 0 0 1 0 0-1 0 0 0 0 0
 0 0 0 0 0 0 0 0 0 0 0 0 0 1
 0 0-1 0 0 0 1 0 0-1 0 1 0 0
 0 0 0-1 0 0 0 1 0 0-1 0 1-1
```

```
------------------------------------
*   14 *   2 1 1 1 1 1
------------------------------------

 1-1 1-1 1-1
 0-1 0 0 0 0
 0 0-1 0 0 0
 0 0 0-1 0 0
 0 0 0 0-1 0
 0 0 0 0 0-1

-1 1-1 1-1 1
-1 0 0 0 0 0
 0-1 0 0 0 0
 0 0-1 0 0 0
 0 0 0-1 0 0
 0 0 0 0-1 0

 0-1 0 0 0 0
 0 0-1 0 0 0
 0 0 0-1 0 0
 0 0 0 0-1 0
 0 0 0 0 0-1
 1-1 1-1 1-1
------------------------------------
```

## I.D Decompositions of Symmetrizations and Permutrizations

The present section contains matrices

$$S([\alpha], n) := (s_{\beta\gamma} := ([\alpha] \triangle_n [\gamma], [\beta]))$$

(cf. 5.3.7). The representations are denoted by X$i$ and numbered in the same way as in Appendix I.A. The upper left hand-entry (X$i$, $m$), $n$ of each one of the following tables means that the table in question shows the decompositions involving $[\alpha] := \mathrm{X}i$ (see Appendix I.A), which is a partition of $m$. The $j$th row gives the decomposition of $[\alpha] \boxdot \mathrm{X}j$, X$j$ the $j$th partition of $n$, while the $k$th column gives the decomposition of $[\alpha] \triangle_n \mathrm{X}k$, X$k$ the $k$th partition of $m$.

The tables are due to N. Esper, and show his results for $m \leqslant 8$, $n \leqslant 5$.

| (X1,2),2 | X1 | X2 |
|---|---|---|
| X1 | 1 | 0 |
| X2 | 0 | 0 |

| (X2,2),2 | X1 | X2 |
|---|---|---|
| X1 | 1 | 0 |
| X2 | 0 | 0 |

| (X1,2),3 | X1 | X2 |
|---|---|---|
| X1 | 1 | 0 |
| X2 | 0 | 0 |
| X3 | 0 | 0 |

| (X2,2),3 | X1 | X2 |
|---|---|---|
| X1 | 0 | 1 |
| X2 | 0 | 0 |
| X3 | 0 | 0 |

| (X1,2),4 | X1 | X2 |
|---|---|---|
| X1 | 1 | 0 |
| X2 | 0 | 0 |
| X3 | 0 | 0 |
| X4 | 0 | 0 |
| X5 | 0 | 0 |

| (X2,2),4 | X1 | X2 |
|---|---|---|
| X1 | 1 | 0 |
| X2 | 0 | 0 |
| X3 | 0 | 0 |
| X4 | 0 | 0 |
| X5 | 0 | 0 |

| (X1,2),5 | X1 | X2 |
|---|---|---|
| X1 | 1 | 0 |
| X2 | 0 | 0 |
| X3 | 0 | 0 |
| X4 | 0 | 0 |
| X5 | 0 | 0 |
| X6 | 0 | 0 |
| X7 | 0 | 0 |

| (X2,2),5 | X1 | X2 |
|---|---|---|
| X1 | 0 | 1 |
| X2 | 0 | 0 |
| X3 | 0 | 0 |
| X4 | 0 | 0 |
| X5 | 0 | 0 |
| X6 | 0 | 0 |
| X7 | 0 | 0 |

| (X1,3),2 | X1 | X2 | X3 |
|---|---|---|---|
| X1 | 1 | 0 | 0 |
| X2 | 0 | 0 | 0 |

| (X2,3),2 | X1 | X2 | X3 |
|---|---|---|---|
| X1 | 1 | 1 | 0 |
| X2 | 0 | 0 | 1 |

| (X3,3),2 | X1 | X2 | X3 |
|---|---|---|---|
| X1 | 1 | 0 | 0 |
| X2 | 0 | 0 | 0 |

| (X1,3),3 | X1 | X2 | X3 |
|---|---|---|---|
| X1 | 1 | 0 | 0 |
| X2 | 0 | 0 | 0 |
| X3 | 0 | 0 | 0 |

| (X2,3),3 | X1 | X2 | X3 |
|---|---|---|---|
| X1 | 1 | 1 | 1 |
| X2 | 0 | 1 | 0 |
| X3 | 0 | 0 | 0 |

| (X3,3),3 | X1 | X2 | X3 |
|---|---|---|---|
| X1 | 0 | 0 | 1 |
| X2 | 0 | 0 | 0 |
| X3 | 0 | 0 | 0 |

| (X1,3),4 | X1 | X2 | X3 |
|---|---|---|---|
| X1 | 1 | 0 | 0 |
| X2 | 0 | 0 | 0 |
| X3 | 0 | 0 | 0 |
| X4 | 0 | 0 | 0 |
| X5 | 0 | 0 | 0 |

| (X2,3),4 | X1 | X2 | X3 |
|---|---|---|---|
| X1 | 1 | 2 | 0 |
| X2 | 0 | 1 | 1 |
| X3 | 1 | 0 | 0 |
| X4 | 0 | 0 | 0 |
| X5 | 0 | 0 | 0 |

| (X3,3),4 | X1 | X2 | X3 |
|---|---|---|---|
| X1 | 1 | 0 | 0 |
| X2 | 0 | 0 | 0 |
| X3 | 0 | 0 | 0 |
| X4 | 0 | 0 | 0 |
| X5 | 0 | 0 | 0 |

| (X1,3),5 | X1 | X2 | X3 |
|---|---|---|---|
| X1 | 1 | 0 | 0 |
| X2 | 0 | 0 | 0 |
| X3 | 0 | 0 | 0 |
| X4 | 0 | 0 | 0 |
| X5 | 0 | 0 | 0 |
| X6 | 0 | 0 | 0 |
| X7 | 0 | 0 | 0 |

| (X2,3),5 | X1 | X2 | X3 |
|---|---|---|---|
| X1 | 1 | 2 | 1 |
| X2 | 1 | 1 | 1 |
| X3 | 0 | 1 | 0 |
| X4 | 0 | 0 | 0 |
| X5 | 0 | 0 | 0 |
| X6 | 0 | 0 | 0 |
| X7 | 0 | 0 | 0 |

| (X3,3),5 | X1 | X2 | X3 |
|---|---|---|---|
| X1 | 0 | 0 | 1 |
| X2 | 0 | 0 | 0 |
| X3 | 0 | 0 | 0 |
| X4 | 0 | 0 | 0 |
| X5 | 0 | 0 | 0 |
| X6 | 0 | 0 | 0 |
| X7 | 0 | 0 | 0 |

| (X1,4),2 | X1 | X2 | X3 | X4 | X5 |
|---|---|---|---|---|---|
| X1 | 1 | 0 | 0 | 0 | 0 |
| X2 | 0 | 0 | 0 | 0 | 0 |

| (X2,4),2 | X1 | X2 | X3 | X4 | X5 |
|---|---|---|---|---|---|
| X1 | 1 | 1 | 1 | 0 | 0 |
| X2 | 0 | 0 | 0 | 1 | 0 |

| (X3,4),2 | X1 | X2 | X3 | X4 | X5 |
|---|---|---|---|---|---|
| X1 | 1 | 0 | 1 | 0 | 0 |
| X2 | 0 | 0 | 0 | 0 | 1 |

| (X4,4),2 | X1 | X2 | X3 | X4 | X5 |
|---|---|---|---|---|---|
| X1 | 1 | 1 | 1 | 0 | 0 |
| X2 | 0 | 0 | 0 | 1 | 0 |

| (X5,4),2 | X1 | X2 | X3 | X4 | X5 |
|---|---|---|---|---|---|
| X1 | 1 | 0 | 0 | 0 | 0 |
| X2 | 0 | 0 | 0 | 0 | 0 |

| (X1,4),3 | X1 | X2 | X3 | X4 | X5 |
|---|---|---|---|---|---|
| X1 | 1 | 0 | 0 | 0 | 0 |
| X2 | 0 | 0 | 0 | 0 | 0 |
| X3 | 0 | 0 | 0 | 0 | 0 |

| (X2,4),3 | X1 | X2 | X3 | X4 | X5 |
|---|---|---|---|---|---|
| X1 | 1 | 2 | 0 | 1 | 0 |
| X2 | 0 | 1 | 1 | 1 | 0 |
| X3 | 0 | 0 | 0 | 0 | 1 |

| (X3,4),3 | X1 | X2 | X3 | X4 | X5 |
|---|---|---|---|---|---|
| X1 | 1 | 0 | 1 | 0 | 1 |
| X2 | 0 | 0 | 1 | 0 | 0 |
| X3 | 0 | 0 | 0 | 0 | 0 |

| (X4,4),3 | X1 | X2 | X3 | X4 | X5 |
|---|---|---|---|---|---|
| X1 | 0 | 1 | 0 | 2 | 1 |
| X2 | 0 | 1 | 1 | 1 | 0 |
| X3 | 1 | 0 | 0 | 0 | 0 |

| (X5,4),3 | X1 | X2 | X3 | X4 | X5 |
|---|---|---|---|---|---|
| X1 | 0 | 0 | 0 | 0 | 1 |
| X2 | 0 | 0 | 0 | 0 | 0 |
| X3 | 0 | 0 | 0 | 0 | 0 |

| (X1,4),4 | X1 | X2 | X3 | X4 | X5 |
|---|---|---|---|---|---|
| X1 | 1 | 0 | 0 | 0 | 0 |
| X2 | 0 | 0 | 0 | 0 | 0 |
| X3 | 0 | 0 | 0 | 0 | 0 |
| X4 | 0 | 0 | 0 | 0 | 0 |
| X5 | 0 | 0 | 0 | 0 | 0 |

| (X2,4),4 | X1 | X2 | X3 | X4 | X5 |
|---|---|---|---|---|---|
| X1 | 2 | 2 | 2 | 1 | 0 |
| X2 | 0 | 2 | 1 | 2 | 1 |
| X3 | 1 | 1 | 1 | 0 | 0 |
| X4 | 0 | 0 | 0 | 1 | 0 |
| X5 | 0 | 0 | 0 | 0 | 0 |

| (X3,4),4 | X1 | X2 | X3 | X4 | X5 |
|---|---|---|---|---|---|
| X1 | 1 | 0 | 2 | 0 | 0 |
| X2 | 0 | 0 | 1 | 0 | 1 |
| X3 | 1 | 0 | 0 | 0 | 0 |
| X4 | 0 | 0 | 0 | 0 | 0 |
| X5 | 0 | 0 | 0 | 0 | 0 |

| (X4,4),4 | X1 | X2 | X3 | X4 | X5 |
|---|---|---|---|---|---|
| X1 | 2 | 2 | 2 | 1 | 0 |
| X2 | 0 | 2 | 1 | 2 | 1 |
| X3 | 1 | 1 | 1 | 0 | 0 |
| X4 | 0 | 0 | 0 | 1 | 0 |
| X5 | 0 | 0 | 0 | 0 | 0 |

| (X5,4),4 | X1 | X2 | X3 | X4 | X5 |
|---|---|---|---|---|---|
| X1 | 1 | 0 | 0 | 0 | 0 |
| X2 | 0 | 0 | 0 | 0 | 0 |
| X3 | 0 | 0 | 0 | 0 | 0 |
| X4 | 0 | 0 | 0 | 0 | 0 |
| X5 | 0 | 0 | 0 | 0 | 0 |

| (X1,4),5 | X1 | X2 | X3 | X4 | X5 |
|---|---|---|---|---|---|
| X1 | 1 | 0 | 0 | 0 | 0 |
| X2 | 0 | 0 | 0 | 0 | 0 |
| X3 | 0 | 0 | 0 | 0 | 0 |
| X4 | 0 | 0 | 0 | 0 | 0 |
| X5 | 0 | 0 | 0 | 0 | 0 |
| X6 | 0 | 0 | 0 | 0 | 0 |
| X7 | 0 | 0 | 0 | 0 | 0 |

| (X2,4),5 | X1 | X2 | X3 | X4 | X5 |
|---|---|---|---|---|---|
| X1 | 1 | 4 | 1 | 2 | 0 |
| X2 | 1 | 3 | 2 | 3 | 1 |
| X3 | 1 | 2 | 1 | 2 | 0 |
| X4 | 0 | 0 | 1 | 1 | 1 |
| X5 | 0 | 1 | 0 | 0 | 0 |
| X6 | 0 | 0 | 0 | 0 | 0 |
| X7 | 0 | 0 | 0 | 0 | 0 |

| (X3,4),5 | X1 | X2 | X3 | X4 | X5 |
|---|---|---|---|---|---|
| X1 | 1 | 0 | 2 | 0 | 1 |
| X2 | 1 | 0 | 1 | 0 | 1 |
| X3 | 0 | 0 | 1 | 0 | 0 |
| X4 | 0 | 0 | 0 | 0 | 0 |
| X5 | 0 | 0 | 0 | 0 | 0 |
| X6 | 0 | 0 | 0 | 0 | 0 |
| X7 | 0 | 0 | 0 | 0 | 0 |

| (X4,4),5 | X1 | X2 | X3 | X4 | X5 |
|---|---|---|---|---|---|
| X1 | 0 | 2 | 1 | 4 | 1 |
| X2 | 1 | 3 | 2 | 3 | 1 |
| X3 | 0 | 2 | 1 | 2 | 1 |
| X4 | 1 | 1 | 1 | 0 | 0 |
| X5 | 0 | 0 | 0 | 1 | 0 |
| X6 | 0 | 0 | 0 | 0 | 0 |
| X7 | 0 | 0 | 0 | 0 | 0 |

| (X5,4),5 | X1 | X2 | X3 | X4 | X5 |
|---|---|---|---|---|---|
| X1 | 0 | 0 | 0 | 0 | 1 |
| X2 | 0 | 0 | 0 | 0 | 0 |
| X3 | 0 | 0 | 0 | 0 | 0 |
| X4 | 0 | 0 | 0 | 0 | 0 |
| X5 | 0 | 0 | 0 | 0 | 0 |
| X6 | 0 | 0 | 0 | 0 | 0 |
| X7 | 0 | 0 | 0 | 0 | 0 |

| (X1,5),2 | I | X1 | X2 | X3 | X4 | X5 | X6 | X7 |
|---|---|---|---|---|---|---|---|---|
| | I | | | | | | | |
| X1 | I | 1 | 0 | 0 | 0 | 0 | 0 | 0 |
| X2 | I | 0 | 0 | 0 | 0 | 0 | 0 | 0 |

| (X2,5),2 | I | X1 | X2 | X3 | X4 | X5 | X6 | X7 |
|---|---|---|---|---|---|---|---|---|
| | I | | | | | | | |
| X1 | I | 1 | 1 | 1 | 0 | 0 | 0 | 0 |
| X2 | I | 0 | 0 | 0 | 1 | 0 | 0 | 0 |

| (X3,5),2 | I | X1 | X2 | X3 | X4 | X5 | X6 | X7 |
|---|---|---|---|---|---|---|---|---|
| | I | | | | | | | |
| X1 | I | 1 | 1 | 1 | 0 | 1 | 0 | 0 |
| X2 | I | 0 | 0 | 0 | 1 | 0 | 1 | 0 |

| (X4,5),2 | I | X1 | X2 | X3 | X4 | X5 | X6 | X7 |
|---|---|---|---|---|---|---|---|---|
| | I | | | | | | | |
| X1 | I | 1 | 1 | 2 | 0 | 1 | 0 | 1 |
| X2 | I | 0 | 0 | 0 | 1 | 1 | 1 | 0 |

| (X5,5),2 | I | X1 | X2 | X3 | X4 | X5 | X6 | X7 |
|---|---|---|---|---|---|---|---|---|
| | I | | | | | | | |
| X1 | I | 1 | 1 | 1 | 0 | 1 | 0 | 0 |
| X2 | I | 0 | 0 | 0 | 1 | 0 | 1 | 0 |

| (X6,5),2 | I | X1 | X2 | X3 | X4 | X5 | X6 | X7 |
|---|---|---|---|---|---|---|---|---|
| | I | | | | | | | |
| X1 | I | 1 | 1 | 1 | 0 | 0 | 0 | 0 |
| X2 | I | 0 | 0 | 0 | 1 | 0 | 0 | 0 |

| (X7,5),2 | I | X1 | X2 | X3 | X4 | X5 | X6 | X7 |
|---|---|---|---|---|---|---|---|---|
| | I | | | | | | | |
| X1 | I | 1 | 0 | 0 | 0 | 0 | 0 | 0 |
| X2 | I | 0 | 0 | 0 | 0 | 0 | 0 | 0 |

| (X1,5),3 | I | X1 | X2 | X3 | X4 | X5 | X6 | X7 |
|---|---|---|---|---|---|---|---|---|
| | I | | | | | | | |
| X1 | I | 1 | 0 | 0 | 0 | 0 | 0 | 0 |
| X2 | I | 0 | 0 | 0 | 0 | 0 | 0 | 0 |
| X3 | I | 0 | 0 | 0 | 0 | 0 | 0 | 0 |

| (X2,5),3 | I | X1 | X2 | X3 | X4 | X5 | X6 | X7 |
|---|---|---|---|---|---|---|---|---|
| | I | | | | | | | |
| X1 | I | 1 | 2 | 1 | 1 | 0 | 0 | 0 |
| X2 | I | 0 | 1 | 1 | 1 | 1 | 0 | 0 |
| X3 | I | 0 | 0 | 0 | 0 | 0 | 1 | 0 |

| (X3,5),3 | I | X1 | X2 | X3 | X4 | X5 | X6 | X7 |
|---|---|---|---|---|---|---|---|---|
| | I | | | | | | | |
| X1 | I | 1 | 2 | 2 | 1 | 1 | 1 | 1 |
| X2 | I | 0 | 1 | 2 | 2 | 2 | 1 | 0 |
| X3 | I | 0 | 0 | 0 | 1 | 0 | 1 | 0 |

| (X4,5),3 | I | X1 | X2 | X3 | X4 | X5 | X6 | X7 |
|---|---|---|---|---|---|---|---|---|
| | I | | | | | | | |
| X1 | I | 0 | 2 | 1 | 5 | 1 | 2 | 0 |
| X2 | I | 0 | 2 | 3 | 4 | 3 | 2 | 0 |
| X3 | I | 1 | 1 | 1 | 0 | 1 | 1 | 1 |

| (X5,5),3 | I | X1 | X2 | X3 | X4 | X5 | X6 | X7 |
|---|---|---|---|---|---|---|---|---|
| | I | | | | | | | |
| X1 | I | 1 | 1 | 1 | 1 | 2 | 2 | 1 |
| X2 | I | 0 | 1 | 2 | 2 | 2 | 1 | 0 |
| X3 | I | 0 | 1 | 0 | 1 | 0 | 0 | 0 |

| (X6,5),3 | I | X1 | X2 | X3 | X4 | X5 | X6 | X7 |
|---|---|---|---|---|---|---|---|---|
| | I | | | | | | | |
| X1 | I | 0 | 0 | 0 | 1 | 1 | 2 | 1 |
| X2 | I | 0 | 0 | 1 | 1 | 1 | 1 | 0 |
| X3 | I | 0 | 1 | 0 | 0 | 0 | 0 | 0 |

| (X7,5),3 | I | X1 | X2 | X3 | X4 | X5 | X6 | X7 |
|---|---|---|---|---|---|---|---|---|
| | I | | | | | | | |
| X1 | I | 0 | 0 | 0 | 0 | 0 | 0 | 1 |
| X2 | I | 0 | 0 | 0 | 0 | 0 | 0 | 0 |
| X3 | I | 0 | 0 | 0 | 0 | 0 | 0 | 0 |

| (X1,5),4 | X1 | X2 | X3 | X4 | X5 | X6 | X7 |
|---|---|---|---|---|---|---|---|
| X1 | 1 | 0 | 0 | 0 | 0 | 0 | 0 |
| X2 | 0 | 0 | 0 | 0 | 0 | 0 | 0 |
| X3 | 0 | 0 | 0 | 0 | 0 | 0 | 0 |
| X4 | 0 | 0 | 0 | 0 | 0 | 0 | 0 |
| X5 | 0 | 0 | 0 | 0 | 0 | 0 | 0 |

| (X2,5),4 | X1 | X2 | X3 | X4 | X5 | X6 | X7 |
|---|---|---|---|---|---|---|---|
| X1 | 2 | 3 | 2 | 1 | 1 | 0 | 0 |
| X2 | 0 | 2 | 2 | 3 | 1 | 1 | 0 |
| X3 | 1 | 1 | 2 | 0 | 1 | 0 | 0 |
| X4 | 0 | 0 | 0 | 1 | 1 | 1 | 0 |
| X5 | 0 | 0 | 0 | 0 | 0 | 0 | 1 |

| (X3,5),4 | X1 | X2 | X3 | X4 | X5 | X6 | X7 |
|---|---|---|---|---|---|---|---|
| X1 | 2 | 3 | 4 | 2 | 4 | 1 | 0 |
| X2 | 0 | 3 | 4 | 6 | 4 | 4 | 1 |
| X3 | 2 | 3 | 3 | 1 | 2 | 1 | 1 |
| X4 | 0 | 1 | 1 | 3 | 2 | 2 | 0 |
| X5 | 0 | 0 | 1 | 0 | 0 | 0 | 0 |

| (X4,5),4 | X1 | X2 | X3 | X4 | X5 | X6 | X7 |
|---|---|---|---|---|---|---|---|
| X1 | 4 | 5 | 7 | 3 | 7 | 3 | 2 |
| X2 | 1 | 6 | 9 | 10 | 9 | 8 | 3 |
| X3 | 3 | 5 | 6 | 3 | 5 | 2 | 1 |
| X4 | 0 | 3 | 3 | 7 | 4 | 4 | 0 |
| X5 | 0 | 1 | 1 | 1 | 0 | 0 | 0 |

| (X5,5),4 | X1 | X2 | X3 | X4 | X5 | X6 | X7 |
|---|---|---|---|---|---|---|---|
| X1 | 2 | 3 | 4 | 2 | 4 | 1 | 0 |
| X2 | 0 | 3 | 4 | 6 | 4 | 4 | 1 |
| X3 | 2 | 3 | 3 | 1 | 2 | 1 | 1 |
| X4 | 0 | 1 | 1 | 3 | 2 | 2 | 0 |
| X5 | 0 | 0 | 1 | 0 | 0 | 0 | 0 |

| (X6,5),4 | X1 | X2 | X3 | X4 | X5 | X6 | X7 |
|---|---|---|---|---|---|---|---|
| X1 | 2 | 3 | 2 | 1 | 1 | 0 | 0 |
| X2 | 0 | 2 | 2 | 3 | 1 | 1 | 0 |
| X3 | 1 | 1 | 2 | 0 | 1 | 0 | 0 |
| X4 | 0 | 0 | 0 | 1 | 1 | 1 | 0 |
| X5 | 0 | 0 | 0 | 0 | 0 | 0 | 1 |

| (X7,5),4 | X1 | X2 | X3 | X4 | X5 | X6 | X7 |
|---|---|---|---|---|---|---|---|
| X1 | 1 | 0 | 0 | 0 | 0 | 0 | 0 |
| X2 | 0 | 0 | 0 | 0 | 0 | 0 | 0 |
| X3 | 0 | 0 | 0 | 0 | 0 | 0 | 0 |
| X4 | 0 | 0 | 0 | 0 | 0 | 0 | 0 |
| X5 | 0 | 0 | 0 | 0 | 0 | 0 | 0 |

| (X1,5),5 | X1 | X2 | X3 | X4 | X5 | X6 | X7 |
|---|---|---|---|---|---|---|---|
| X1 | 1 | 0 | 0 | 0 | 0 | 0 | 0 |
| X2 | 0 | 0 | 0 | 0 | 0 | 0 | 0 |
| X3 | 0 | 0 | 0 | 0 | 0 | 0 | 0 |
| X4 | 0 | 0 | 0 | 0 | 0 | 0 | 0 |
| X5 | 0 | 0 | 0 | 0 | 0 | 0 | 0 |
| X6 | 0 | 0 | 0 | 0 | 0 | 0 | 0 |
| X7 | 0 | 0 | 0 | 0 | 0 | 0 | 0 |

| (X2,5),5 | X1 | X2 | X3 | X4 | X5 | X6 | X7 |
|---|---|---|---|---|---|---|---|
| X1 | 2 | 4 | 3 | 3 | 1 | 0 | 0 |
| X2 | 1 | 4 | 4 | 4 | 3 | 2 | 0 |
| X3 | 1 | 3 | 3 | 3 | 2 | 1 | 0 |
| X4 | 0 | 0 | 1 | 2 | 2 | 2 | 1 |
| X5 | 0 | 1 | 1 | 1 | 1 | 0 | 0 |
| X6 | 0 | 0 | 0 | 0 | 0 | 1 | 0 |
| X7 | 0 | 0 | 0 | 0 | 0 | 0 | 0 |

| (X3,5),5 | X1 | X2 | X3 | X4 | X5 | X6 | X7 |
|---|---|---|---|---|---|---|---|
| X1 | 2 | 5 | 7 | 5 | 5 | 3 | 2 |
| X2 | 2 | 8 | 9 | 11 | 9 | 8 | 2 |
| X3 | 2 | 6 | 8 | 8 | 8 | 5 | 1 |
| X4 | 0 | 3 | 4 | 8 | 5 | 5 | 1 |
| X5 | 1 | 3 | 4 | 3 | 3 | 2 | 1 |
| X6 | 0 | 1 | 1 | 1 | 1 | 1 | 0 |
| X7 | 1 | 0 | 0 | 0 | 0 | 0 | 0 |

| (X4,5),5 | X1 | X2 | X3 | X4 | X5 | X6 | X7 |
|---|---|---|---|---|---|---|---|
| X1 | 0 | 8 | 8 | 18 | 8 | 8 | 0 |
| X2 | 3 | 17 | 20 | 27 | 20 | 17 | 3 |
| X3 | 2 | 14 | 16 | 24 | 16 | 14 | 2 |
| X4 | 5 | 11 | 16 | 13 | 16 | 11 | 5 |
| X5 | 1 | 7 | 8 | 12 | 8 | 7 | 1 |
| X6 | 1 | 3 | 4 | 3 | 4 | 3 | 1 |
| X7 | 0 | 0 | 0 | 1 | 0 | 0 | 0 |

| (X5,5),5 | X1 | X2 | X3 | X4 | X5 | X6 | X7 |
|---|---|---|---|---|---|---|---|
| X1 | 2 | 3 | 5 | 5 | 7 | 5 | 2 |
| X2 | 2 | 8 | 9 | 11 | 9 | 8 | 2 |
| X3 | 1 | 5 | 8 | 8 | 8 | 6 | 2 |
| X4 | 1 | 5 | 5 | 8 | 4 | 3 | 0 |
| X5 | 1 | 2 | 3 | 3 | 4 | 3 | 1 |
| X6 | 0 | 1 | 1 | 1 | 1 | 1 | 0 |
| X7 | 0 | 0 | 0 | 0 | 0 | 0 | 1 |

| (X6,5),5 | X1 | X2 | X3 | X4 | X5 | X6 | X7 |
|---|---|---|---|---|---|---|---|
| X1 | 0 | 0 | 1 | 3 | 3 | 4 | 2 |
| X2 | 0 | 2 | 3 | 4 | 4 | 4 | 1 |
| X3 | 0 | 1 | 2 | 3 | 3 | 3 | 1 |
| X4 | 1 | 2 | 2 | 2 | 1 | 0 | 0 |
| X5 | 0 | 0 | 1 | 1 | 1 | 1 | 0 |
| X6 | 0 | 1 | 0 | 0 | 0 | 0 | 0 |
| X7 | 0 | 0 | 0 | 0 | 0 | 0 | 0 |

| (X7,5),5 | X1 | X2 | X3 | X4 | X5 | X6 | X7 |
|---|---|---|---|---|---|---|---|
| X1 | 0 | 0 | 0 | 0 | 0 | 0 | 1 |
| X2 | 0 | 0 | 0 | 0 | 0 | 0 | 0 |
| X3 | 0 | 0 | 0 | 0 | 0 | 0 | 0 |
| X4 | 0 | 0 | 0 | 0 | 0 | 0 | 0 |
| X5 | 0 | 0 | 0 | 0 | 0 | 0 | 0 |
| X6 | 0 | 0 | 0 | 0 | 0 | 0 | 0 |
| X7 | 0 | 0 | 0 | 0 | 0 | 0 | 0 |

| (X1,6),2 | X1 | X2 | X3 | X4 | X5 | X6 | X7 | X8 | X9 | X10 | X11 |
|---|---|---|---|---|---|---|---|---|---|---|---|
| X1 | 1 | 0 | 0 | 0 | 0 | 0 | 0 | 0 | 0 | 0 | 0 |
| X2 | 0 | 0 | 0 | 0 | 0 | 0 | 0 | 0 | 0 | 0 | 0 |

| (X2,6),2 | X1 | X2 | X3 | X4 | X5 | X6 | X7 | X8 | X9 | X10 | X11 |
|---|---|---|---|---|---|---|---|---|---|---|---|
| X1 | 1 | 1 | 1 | 0 | 0 | 0 | 0 | 0 | 0 | 0 | 0 |
| X2 | 0 | 0 | 0 | 1 | 0 | 0 | 0 | 0 | 0 | 0 | 0 |

| (X3,6),2 | X1 | X2 | X3 | X4 | X5 | X6 | X7 | X8 | X9 | X10 | X11 |
|---|---|---|---|---|---|---|---|---|---|---|---|
| X1 | 1 | 1 | 2 | 0 | 0 | 1 | 0 | 1 | 0 | 0 | 0 |
| X2 | 0 | 0 | 0 | 1 | 0 | 1 | 1 | 0 | 0 | 0 | 0 |

| (X4,6),2 | X1 | X2 | X3 | X4 | X5 | X6 | X7 | X8 | X9 | X10 | X11 |
|---|---|---|---|---|---|---|---|---|---|---|---|
| X1 | 1 | 1 | 2 | 0 | 1 | 1 | 0 | 1 | 0 | 1 | 0 |
| X2 | 0 | 0 | 0 | 1 | 0 | 1 | 1 | 0 | 1 | 0 | 0 |

| (X5,6),2 | X1 | X2 | X3 | X4 | X5 | X6 | X7 | X8 | X9 | X10 | X11 |
|---|---|---|---|---|---|---|---|---|---|---|---|
| X1 | 1 | 0 | 1 | 0 | 0 | 0 | 0 | 1 | 0 | 0 | 0 |
| X2 | 0 | 0 | 0 | 0 | 0 | 0 | 1 | 0 | 0 | 0 | 0 |

| (X6,6),2 | X1 | X2 | X3 | X4 | X5 | X6 | X7 | X8 | X9 | X10 | X11 |
|---|---|---|---|---|---|---|---|---|---|---|---|
| X1 | 1 | 2 | 3 | 1 | 1 | 3 | 1 | 2 | 1 | 1 | 1 |
| X2 | 0 | 0 | 0 | 3 | 1 | 2 | 3 | 0 | 2 | 1 | 0 |

| (X7,6),2 | X1 | X2 | X3 | X4 | X5 | X6 | X7 | X8 | X9 | X10 | X11 |
|---|---|---|---|---|---|---|---|---|---|---|---|
| X1 | 1 | 1 | 2 | 0 | 1 | 1 | 0 | 1 | 0 | 1 | 0 |
| X2 | 0 | 0 | 0 | 1 | 0 | 1 | 1 | 0 | 1 | 0 | 0 |

| (X8,6),2 | X1 | X2 | X3 | X4 | X5 | X6 | X7 | X8 | X9 | X10 | X11 |
|---|---|---|---|---|---|---|---|---|---|---|---|
| X1 | 1 | 0 | 1 | 0 | 0 | 0 | 0 | 1 | 0 | 0 | 0 |
| X2 | 0 | 0 | 0 | 0 | 0 | 0 | 1 | 0 | 0 | 0 | 0 |

| (X9,6),2 | X1 | X2 | X3 | X4 | X5 | X6 | X7 | X8 | X9 | X10 | X11 |
|---|---|---|---|---|---|---|---|---|---|---|---|
| X1 | 1 | 1 | 2 | 0 | 0 | 1 | 0 | 1 | 0 | 0 | 0 |
| X2 | 0 | 0 | 0 | 1 | 0 | 1 | 1 | 0 | 0 | 0 | 0 |

| (X10,6),2 | X1 | X2 | X3 | X4 | X5 | X6 | X7 | X8 | X9 | X10 | X11 |
|---|---|---|---|---|---|---|---|---|---|---|---|
| X1 | 1 | 1 | 1 | 0 | 0 | 0 | 0 | 0 | 0 | 0 | 0 |
| X2 | 0 | 0 | 0 | 1 | 0 | 0 | 0 | 0 | 0 | 0 | 0 |

| (X11,6),2 | X1 | X2 | X3 | X4 | X5 | X6 | X7 | X8 | X9 | X10 | X11 |
|---|---|---|---|---|---|---|---|---|---|---|---|
| X1 | 1 | 0 | 0 | 0 | 0 | 0 | 0 | 0 | 0 | 0 | 0 |
| X2 | 0 | 0 | 0 | 0 | 0 | 0 | 0 | 0 | 0 | 0 | 0 |

(X1,6),3

| | X1 | X2 | X3 | X4 | X5 | X6 | X7 | X8 | X9 | X10 | X11 |
|---|---|---|---|---|---|---|---|---|---|---|---|
| X1 | 1 | 0 | 0 | 0 | 0 | 0 | 0 | 0 | 0 | 0 | 0 |
| X2 | 0 | 0 | 0 | 0 | 0 | 0 | 0 | 0 | 0 | 0 | 0 |
| X3 | 0 | 0 | 0 | 0 | 0 | 0 | 0 | 0 | 0 | 0 | 0 |

(X2,6),3

| | X1 | X2 | X3 | X4 | X5 | X6 | X7 | X8 | X9 | X10 | X11 |
|---|---|---|---|---|---|---|---|---|---|---|---|
| X1 | 1 | 2 | 1 | 1 | 1 | 0 | 0 | 0 | 0 | 0 | 0 |
| X2 | 0 | 1 | 1 | 1 | 0 | 1 | 0 | 0 | 0 | 0 | 0 |
| X3 | 0 | 0 | 0 | 0 | 0 | 0 | 1 | 0 | 0 | 0 | 0 |

(X3,6),3

| | X1 | X2 | X3 | X4 | X5 | X6 | X7 | X8 | X9 | X10 | X11 |
|---|---|---|---|---|---|---|---|---|---|---|---|
| X1 | 2 | 2 | 5 | 2 | 1 | 3 | 2 | 2 | 0 | 1 | 0 |
| X2 | 0 | 2 | 4 | 3 | 1 | 6 | 3 | 2 | 2 | 1 | 0 |
| X3 | 0 | 0 | 0 | 2 | 1 | 1 | 2 | 0 | 2 | 1 | 0 |

(X4,6),3

| | X1 | X2 | X3 | X4 | X5 | X6 | X7 | X8 | X9 | X10 | X11 |
|---|---|---|---|---|---|---|---|---|---|---|---|
| X1 | 0 | 2 | 2 | 6 | 2 | 4 | 4 | 0 | 2 | 0 | 0 |
| X2 | 0 | 2 | 4 | 5 | 2 | 8 | 4 | 2 | 4 | 2 | 0 |
| X3 | 1 | 1 | 2 | 0 | 1 | 2 | 2 | 2 | 2 | 2 | 1 |

(X5,6),3

| | X1 | X2 | X3 | X4 | X5 | X6 | X7 | X8 | X9 | X10 | X11 |
|---|---|---|---|---|---|---|---|---|---|---|---|
| X1 | 0 | 1 | 0 | 1 | 2 | 0 | 0 | 0 | 1 | 0 | 1 |
| X2 | 0 | 0 | 0 | 1 | 1 | 1 | 0 | 0 | 1 | 0 | 0 |
| X3 | 0 | 0 | 0 | 0 | 0 | 0 | 1 | 0 | 0 | 0 | 0 |

(X6,6),3

| | X1 | X2 | X3 | X4 | X5 | X6 | X7 | X8 | X9 | X10 | X11 |
|---|---|---|---|---|---|---|---|---|---|---|---|
| X1 | 2 | 6 | 10 | 12 | 6 | 17 | 12 | 6 | 10 | 6 | 2 |
| X2 | 1 | 9 | 17 | 18 | 9 | 32 | 18 | 9 | 17 | 9 | 1 |
| X3 | 1 | 4 | 7 | 8 | 4 | 12 | 8 | 4 | 7 | 4 | 1 |

(X7,6),3

| | X1 | X2 | X3 | X4 | X5 | X6 | X7 | X8 | X9 | X10 | X11 |
|---|---|---|---|---|---|---|---|---|---|---|---|
| X1 | 0 | 0 | 2 | 4 | 0 | 4 | 6 | 2 | 2 | 2 | 0 |
| X2 | 0 | 2 | 4 | 4 | 2 | 8 | 5 | 2 | 4 | 2 | 0 |
| X3 | 1 | 2 | 2 | 2 | 2 | 2 | 0 | 1 | 2 | 1 | 1 |

(X8,6),3

| | X1 | X2 | X3 | X4 | X5 | X6 | X7 | X8 | X9 | X10 | X11 |
|---|---|---|---|---|---|---|---|---|---|---|---|
| X1 | 1 | 0 | 1 | 0 | 0 | 0 | 1 | 2 | 0 | 1 | 0 |
| X2 | 0 | 0 | 1 | 0 | 0 | 1 | 1 | 1 | 0 | 0 | 0 |
| X3 | 0 | 0 | 0 | 1 | 0 | 0 | 0 | 0 | 0 | 0 | 0 |

(X9,6),3

| | X1 | X2 | X3 | X4 | X5 | X6 | X7 | X8 | X9 | X10 | X11 |
|---|---|---|---|---|---|---|---|---|---|---|---|
| X1 | 0 | 1 | 0 | 2 | 2 | 3 | 2 | 1 | 5 | 2 | 2 |
| X2 | 0 | 1 | 2 | 3 | 2 | 6 | 3 | 1 | 4 | 2 | 0 |
| X3 | 0 | 1 | 2 | 2 | 0 | 1 | 2 | 1 | 0 | 0 | 0 |

(X10,6),3

| | X1 | X2 | X3 | X4 | X5 | X6 | X7 | X8 | X9 | X10 | X11 |
|---|---|---|---|---|---|---|---|---|---|---|---|
| X1 | 0 | 0 | 0 | 0 | 0 | 0 | 1 | 1 | 1 | 2 | 1 |
| X2 | 0 | 0 | 0 | 0 | 0 | 1 | 1 | 0 | 1 | 1 | 0 |
| X3 | 0 | 0 | 0 | 1 | 0 | 0 | 0 | 0 | 0 | 0 | 0 |

(X11,6),3

| | X1 | X2 | X3 | X4 | X5 | X6 | X7 | X8 | X9 | X10 | X11 |
|---|---|---|---|---|---|---|---|---|---|---|---|
| X1 | 0 | 0 | 0 | 0 | 0 | 0 | 0 | 0 | 0 | 0 | 1 |
| X2 | 0 | 0 | 0 | 0 | 0 | 0 | 0 | 0 | 0 | 0 | 0 |
| X3 | 0 | 0 | 0 | 0 | 0 | 0 | 0 | 0 | 0 | 0 | 0 |

| (X1,6),4 | X1 | X2 | X3 | X4 | X5 | X6 | X7 | X8 | X9 | X10 | X11 |
|---|---|---|---|---|---|---|---|---|---|---|---|
| X1 | 1 | 0 | 0 | 0 | 0 | 0 | 0 | 0 | 0 | 0 | 0 |
| X2 | 0 | 0 | 0 | 0 | 0 | 0 | 0 | 0 | 0 | 0 | 0 |
| X3 | 0 | 0 | 0 | 0 | 0 | 0 | 0 | 0 | 0 | 0 | 0 |
| X4 | 0 | 0 | 0 | 0 | 0 | 0 | 0 | 0 | 0 | 0 | 0 |
| X5 | 0 | 0 | 0 | 0 | 0 | 0 | 0 | 0 | 0 | 0 | 0 |

| (X2,6),4 | X1 | X2 | X3 | X4 | X5 | X6 | X7 | X8 | X9 | X10 | X11 |
|---|---|---|---|---|---|---|---|---|---|---|---|
| X1 | 2 | 3 | 3 | 1 | 0 | 1 | 0 | 0 | 0 | 0 | 0 |
| X2 | 0 | 2 | 2 | 3 | 1 | 2 | 1 | 0 | 0 | 0 | 0 |
| X3 | 1 | 1 | 2 | 0 | 1 | 1 | 0 | 1 | 0 | 0 | 0 |
| X4 | 0 | 0 | 0 | 1 | 0 | 1 | 1 | 0 | 1 | 0 | 0 |
| X5 | 0 | 0 | 0 | 0 | 0 | 0 | 0 | 0 | 0 | 1 | 0 |

| (X3,6),4 | X1 | X2 | X3 | X4 | X5 | X6 | X7 | X8 | X9 | X10 | X11 |
|---|---|---|---|---|---|---|---|---|---|---|---|
| X1 | 4 | 6 | 12 | 5 | 2 | 11 | 5 | 6 | 3 | 2 | 0 |
| X2 | 1 | 7 | 14 | 15 | 6 | 22 | 15 | 7 | 9 | 6 | 0 |
| X3 | 3 | 6 | 11 | 5 | 4 | 12 | 5 | 6 | 5 | 4 | 1 |
| X4 | 0 | 3 | 5 | 10 | 4 | 14 | 10 | 3 | 10 | 4 | 1 |
| X5 | 0 | 1 | 1 | 1 | 2 | 3 | 1 | 1 | 2 | 2 | 1 |

| (X4,6),4 | X1 | X2 | X3 | X4 | X5 | X6 | X7 | X8 | X9 | X10 | X11 |
|---|---|---|---|---|---|---|---|---|---|---|---|
| X1 | 5 | 8 | 14 | 7 | 5 | 15 | 7 | 8 | 8 | 5 | 2 |
| X2 | 1 | 10 | 18 | 21 | 11 | 33 | 21 | 10 | 18 | 11 | 2 |
| X3 | 4 | 8 | 15 | 8 | 6 | 19 | 8 | 8 | 9 | 6 | 1 |
| X4 | 0 | 5 | 9 | 16 | 6 | 22 | 16 | 5 | 14 | 6 | 1 |
| X5 | 0 | 2 | 3 | 3 | 2 | 4 | 3 | 2 | 2 | 2 | 1 |

| (X5,6),4 | X1 | X2 | X3 | X4 | X5 | X6 | X7 | X8 | X9 | X10 | X11 |
|---|---|---|---|---|---|---|---|---|---|---|---|
| X1 | 2 | 0 | 3 | 0 | 0 | 1 | 1 | 3 | 0 | 0 | 0 |
| X2 | 0 | 0 | 2 | 1 | 0 | 2 | 3 | 2 | 0 | 1 | 0 |
| X3 | 1 | 1 | 2 | 0 | 0 | 1 | 0 | 1 | 0 | 1 | 0 |
| X4 | 0 | 0 | 0 | 1 | 0 | 1 | 1 | 0 | 1 | 0 | 0 |
| X5 | 0 | 0 | 0 | 0 | 1 | 0 | 0 | 0 | 0 | 0 | 0 |

| (X6,6),4 | X1 | X2 | X3 | X4 | X5 | X6 | X7 | X8 | X9 | X10 | X11 |
|---|---|---|---|---|---|---|---|---|---|---|---|
| X1 | 9 | 30 | 55 | 48 | 26 | 89 | 48 | 30 | 47 | 26 | 5 |
| X2 | 8 | 60 | 108 | 132 | 64 | 204 | 132 | 60 | 116 | 64 | 12 |
| X3 | 16 | 45 | 80 | 69 | 40 | 118 | 69 | 45 | 64 | 40 | 10 |
| X4 | 8 | 47 | 84 | 103 | 48 | 158 | 103 | 47 | 92 | 48 | 10 |
| X5 | 4 | 15 | 28 | 21 | 14 | 42 | 21 | 15 | 20 | 14 | 2 |

| (X7,6),4 | X1 | X2 | X3 | X4 | X5 | X6 | X7 | X8 | X9 | X10 | X11 |
|---|---|---|---|---|---|---|---|---|---|---|---|
| X1 | 5 | 8 | 14 | 7 | 5 | 15 | 7 | 8 | 8 | 5 | 2 |
| X2 | 1 | 10 | 18 | 21 | 11 | 33 | 21 | 10 | 18 | 11 | 2 |
| X3 | 4 | 8 | 15 | 8 | 6 | 19 | 8 | 8 | 9 | 6 | 1 |
| X4 | 0 | 5 | 9 | 16 | 6 | 22 | 16 | 5 | 14 | 6 | 1 |
| X5 | 0 | 2 | 3 | 3 | 2 | 4 | 3 | 2 | 2 | 2 | 1 |

| X8,6),4 | X1 | X2 | X3 | X4 | X5 | X6 | X7 | X8 | X9 | X10 | X11 |
|---|---|---|---|---|---|---|---|---|---|---|---|
| X1 | 2 | 0 | 3 | 0 | 0 | 1 | 1 | 3 | 0 | 0 | 0 |
| X2 | 0 | 0 | 2 | 1 | 0 | 2 | 3 | 2 | 0 | 1 | 0 |
| X3 | 1 | 1 | 2 | 0 | 0 | 1 | 0 | 1 | 0 | 1 | 0 |
| X4 | 0 | 0 | 0 | 1 | 0 | 1 | 1 | 0 | 1 | 0 | 0 |
| X5 | 0 | 0 | 0 | 0 | 1 | 0 | 0 | 0 | 0 | 0 | 0 |

| X9,6),4 | X1 | X2 | X3 | X4 | X5 | X6 | X7 | X8 | X9 | X10 | X11 |
|---|---|---|---|---|---|---|---|---|---|---|---|
| X1 | 4 | 6 | 12 | 5 | 2 | 11 | 5 | 6 | 3 | 2 | 0 |
| X2 | 1 | 7 | 14 | 15 | 6 | 22 | 15 | 7 | 9 | 6 | 0 |
| X3 | 3 | 6 | 11 | 5 | 4 | 12 | 5 | 6 | 5 | 4 | 1 |
| X4 | 0 | 3 | 5 | 10 | 4 | 14 | 10 | 3 | 10 | 4 | 1 |
| X5 | 0 | 1 | 1 | 1 | 2 | 3 | 1 | 1 | 2 | 2 | 1 |

| X10,6),4 | X1 | X2 | X3 | X4 | X5 | X6 | X7 | X8 | X9 | X10 | X11 |
|---|---|---|---|---|---|---|---|---|---|---|---|
| X1 | 2 | 3 | 3 | 1 | 0 | 1 | 0 | 0 | 0 | 0 | 0 |
| X2 | 0 | 2 | 2 | 3 | 1 | 2 | 1 | 0 | 0 | 0 | 0 |
| X3 | 1 | 1 | 2 | 0 | 1 | 1 | 0 | 1 | 0 | 0 | 0 |
| X4 | 0 | 0 | 0 | 1 | 0 | 1 | 1 | 0 | 1 | 0 | 0 |
| X5 | 0 | 0 | 0 | 0 | 0 | 0 | 0 | 0 | 0 | 1 | 0 |

| X11,6),4 | X1 | X2 | X3 | X4 | X5 | X6 | X7 | X8 | X9 | X10 | X11 |
|---|---|---|---|---|---|---|---|---|---|---|---|
| X1 | 1 | 0 | 0 | 0 | 0 | 0 | 0 | 0 | 0 | 0 | 0 |
| X2 | 0 | 0 | 0 | 0 | 0 | 0 | 0 | 0 | 0 | 0 | 0 |
| X3 | 0 | 0 | 0 | 0 | 0 | 0 | 0 | 0 | 0 | 0 | 0 |
| X4 | 0 | 0 | 0 | 0 | 0 | 0 | 0 | 0 | 0 | 0 | 0 |
| X5 | 0 | 0 | 0 | 0 | 0 | 0 | 0 | 0 | 0 | 0 | 0 |

| (X1,6),5 | X1 | X2 | X3 | X4 | X5 | X6 | X7 | X8 | X9 | X10 | X11 |
|---|---|---|---|---|---|---|---|---|---|---|---|
| X1 | 1 | 0 | 0 | 0 | 0 | 0 | 0 | 0 | 0 | 0 | 0 |
| X2 | 0 | 0 | 0 | 0 | 0 | 0 | 0 | 0 | 0 | 0 | 0 |
| X3 | 0 | 0 | 0 | 0 | 0 | 0 | 0 | 0 | 0 | 0 | 0 |
| X4 | 0 | 0 | 0 | 0 | 0 | 0 | 0 | 0 | 0 | 0 | 0 |
| X5 | 0 | 0 | 0 | 0 | 0 | 0 | 0 | 0 | 0 | 0 | 0 |
| X6 | 0 | 0 | 0 | 0 | 0 | 0 | 0 | 0 | 0 | 0 | 0 |
| X7 | 0 | 0 | 0 | 0 | 0 | 0 | 0 | 0 | 0 | 0 | 0 |

| (X2,6),5 | X1 | X2 | X3 | X4 | X5 | X6 | X7 | X8 | X9 | X10 | X11 |
|---|---|---|---|---|---|---|---|---|---|---|---|
| X1 | 2 | 5 | 3 | 3 | 2 | 2 | 0 | 0 | 0 | 0 | 0 |
| X2 | 1 | 4 | 5 | 5 | 2 | 4 | 2 | 1 | 1 | 0 | 0 |
| X3 | 1 | 3 | 4 | 3 | 2 | 4 | 1 | 0 | 1 | 0 | 0 |
| X4 | 0 | 0 | 1 | 2 | 0 | 3 | 3 | 1 | 1 | 1 | 0 |
| X5 | 0 | 1 | 1 | 1 | 1 | 2 | 0 | 1 | 1 | 0 | 0 |
| X6 | 0 | 0 | 0 | 0 | 0 | 0 | 1 | 0 | 1 | 1 | 0 |
| X7 | 0 | 0 | 0 | 0 | 0 | 0 | 0 | 0 | 0 | 0 | 1 |

| (X3,6),5 | X1 | X2 | X3 | X4 | X5 | X6 | X7 | X8 | X9 | X10 | X11 |
|---|---|---|---|---|---|---|---|---|---|---|---|
| X1 | 6 | 12 | 26 | 16 | 7 | 29 | 16 | 12 | 8 | 7 | 1 |
| X2 | 6 | 24 | 46 | 44 | 20 | 70 | 44 | 24 | 34 | 20 | 2 |
| X3 | 7 | 24 | 44 | 40 | 19 | 66 | 40 | 24 | 31 | 19 | 2 |
| X4 | 1 | 16 | 27 | 41 | 20 | 62 | 41 | 16 | 39 | 20 | 5 |
| X5 | 4 | 14 | 27 | 25 | 13 | 42 | 25 | 14 | 22 | 13 | 3 |
| X6 | 0 | 6 | 10 | 14 | 8 | 22 | 14 | 6 | 16 | 8 | 2 |
| X7 | 1 | 2 | 2 | 1 | 1 | 3 | 1 | 2 | 1 | 1 | 0 |

| (X4,6),5 | X1 | X2 | X3 | X4 | X5 | X6 | X7 | X8 | X9 | X10 | X11 |
|---|---|---|---|---|---|---|---|---|---|---|---|
| X1 | 1 | 15 | 22 | 37 | 15 | 44 | 30 | 8 | 22 | 8 | 1 |
| X2 | 6 | 36 | 64 | 74 | 36 | 114 | 72 | 34 | 64 | 34 | 6 |
| X3 | 5 | 35 | 60 | 75 | 35 | 110 | 70 | 30 | 60 | 30 | 5 |
| X4 | 10 | 32 | 62 | 56 | 32 | 106 | 64 | 40 | 62 | 40 | 10 |
| X5 | 3 | 23 | 40 | 50 | 23 | 74 | 46 | 20 | 40 | 20 | 3 |
| X6 | 3 | 13 | 24 | 24 | 13 | 40 | 26 | 14 | 24 | 14 | 3 |
| X7 | 0 | 2 | 2 | 6 | 2 | 6 | 4 | 0 | 2 | 0 | 0 |

| (X5,6),5 | X1 | X2 | X3 | X4 | X5 | X6 | X7 | X8 | X9 | X10 | X11 |
|---|---|---|---|---|---|---|---|---|---|---|---|
| X1 | 0 | 2 | 0 | 3 | 5 | 2 | 0 | 0 | 3 | 0 | 2 |
| X2 | 0 | 2 | 1 | 5 | 4 | 4 | 2 | 0 | 5 | 1 | 1 |
| X3 | 0 | 2 | 1 | 3 | 3 | 4 | 1 | 0 | 4 | 0 | 1 |
| X4 | 0 | 0 | 1 | 2 | 0 | 3 | 3 | 1 | 1 | 1 | 0 |
| X5 | 0 | 1 | 1 | 1 | 1 | 2 | 0 | 0 | 1 | 1 | 0 |
| X6 | 0 | 0 | 1 | 0 | 0 | 0 | 1 | 1 | 0 | 0 | 0 |
| X7 | 1 | 0 | 0 | 0 | 0 | 0 | 0 | 0 | 0 | 0 | 0 |

| (X6,6),5 | X1 | X2 | X3 | X4 | X5 | X6 | X7 | X8 | X9 | X10 | X11 |
|---|---|---|---|---|---|---|---|---|---|---|---|
| X1 | 23 | 108 | 193 | 216 | 108 | 344 | 216 | 108 | 193 | 108 | 23 |
| X2 | 66 | 324 | 582 | 648 | 324 | 1029 | 648 | 324 | 582 | 324 | 66 |
| X3 | 66 | 339 | 612 | 678 | 339 | 1092 | 678 | 339 | 612 | 339 | 66 |
| X4 | 72 | 357 | 642 | 714 | 357 | 1143 | 714 | 357 | 642 | 357 | 72 |
| X5 | 52 | 264 | 476 | 528 | 264 | 848 | 528 | 264 | 476 | 264 | 52 |
| X6 | 34 | 172 | 310 | 344 | 172 | 549 | 344 | 172 | 310 | 172 | 34 |
| X7 | 8 | 31 | 54 | 62 | 31 | 95 | 62 | 31 | 54 | 31 | 8 |

| (X7,6),5 | X1 | X2 | X3 | X4 | X5 | X6 | X7 | X8 | X9 | X10 | X11 |
|---|---|---|---|---|---|---|---|---|---|---|---|
| X1 | 1 | 8 | 22 | 30 | 8 | 44 | 37 | 15 | 22 | 15 | 1 |
| X2 | 6 | 34 | 64 | 72 | 34 | 114 | 74 | 36 | 64 | 36 | 6 |
| X3 | 5 | 30 | 60 | 70 | 30 | 110 | 75 | 35 | 60 | 35 | 5 |
| X4 | 10 | 40 | 62 | 64 | 40 | 106 | 56 | 32 | 62 | 32 | 10 |
| X5 | 3 | 20 | 40 | 46 | 20 | 74 | 50 | 23 | 40 | 23 | 3 |
| X6 | 3 | 14 | 24 | 26 | 14 | 40 | 24 | 13 | 24 | 13 | 3 |
| X7 | 0 | 0 | 2 | 4 | 0 | 6 | 6 | 2 | 2 | 2 | 0 |

| (X8,6),5 | X1 | X2 | X3 | X4 | X5 | X6 | X7 | X8 | X9 | X10 | X11 |
|---|---|---|---|---|---|---|---|---|---|---|---|
| X1 | 2 | 0 | 3 | 0 | 0 | 2 | 3 | 5 | 0 | 2 | 0 |
| X2 | 1 | 1 | 5 | 2 | 0 | 4 | 5 | 4 | 1 | 2 | 0 |
| X3 | 1 | 0 | 4 | 1 | 0 | 4 | 3 | 3 | 1 | 2 | 0 |
| X4 | 0 | 1 | 1 | 3 | 1 | 3 | 2 | 0 | 1 | 0 | 0 |
| X5 | 0 | 1 | 1 | 0 | 0 | 2 | 1 | 1 | 1 | 1 | 0 |
| X6 | 0 | 0 | 0 | 1 | 1 | 0 | 0 | 0 | 1 | 0 | 0 |
| X7 | 0 | 0 | 0 | 0 | 0 | 0 | 0 | 0 | 0 | 0 | 1 |

| (X9,6),5 | X1 | X2 | X3 | X4 | X5 | X6 | X7 | X8 | X9 | X10 | X11 |
|---|---|---|---|---|---|---|---|---|---|---|---|
| X1 | 1 | 7 | 8 | 16 | 12 | 29 | 16 | 7 | 26 | 12 | 6 |
| X2 | 2 | 20 | 34 | 44 | 24 | 70 | 44 | 20 | 46 | 24 | 6 |
| X3 | 2 | 19 | 31 | 40 | 24 | 66 | 40 | 19 | 44 | 24 | 7 |
| X4 | 5 | 20 | 39 | 41 | 16 | 62 | 41 | 20 | 27 | 16 | 1 |
| X5 | 3 | 13 | 22 | 25 | 14 | 42 | 25 | 13 | 27 | 14 | 4 |
| X6 | 2 | 8 | 16 | 14 | 6 | 22 | 14 | 8 | 10 | 6 | 0 |
| X7 | 0 | 1 | 1 | 1 | 2 | 3 | 1 | 1 | 2 | 2 | 1 |

| (X10,6),5 | X1 | X2 | X3 | X4 | X5 | X6 | X7 | X8 | X9 | X10 | X11 |
|---|---|---|---|---|---|---|---|---|---|---|---|
| X1 | 0 | 0 | 0 | 0 | 0 | 2 | 3 | 2 | 3 | 5 | 2 |
| X2 | 0 | 0 | 1 | 2 | 1 | 4 | 5 | 2 | 5 | 4 | 1 |
| X3 | 0 | 0 | 1 | 1 | 0 | 4 | 3 | 2 | 4 | 3 | 1 |
| X4 | 0 | 1 | 1 | 3 | 1 | 3 | 2 | 0 | 1 | 0 | 0 |
| X5 | 0 | 0 | 1 | 0 | 1 | 2 | 1 | 1 | 1 | 1 | 0 |
| X6 | 0 | 1 | 1 | 1 | 0 | 0 | 0 | 0 | 0 | 0 | 0 |
| X7 | 1 | 0 | 0 | 0 | 0 | 0 | 0 | 0 | 0 | 0 | 0 |

| (X11,6),5 | X1 | X2 | X3 | X4 | X5 | X6 | X7 | X8 | X9 | X10 | X11 |
|---|---|---|---|---|---|---|---|---|---|---|---|
| X1 | 0 | 0 | 0 | 0 | 0 | 0 | 0 | 0 | 0 | 0 | 1 |
| X2 | 0 | 0 | 0 | 0 | 0 | 0 | 0 | 0 | 0 | 0 | 0 |
| X3 | 0 | 0 | 0 | 0 | 0 | 0 | 0 | 0 | 0 | 0 | 0 |
| X4 | 0 | 0 | 0 | 0 | 0 | 0 | 0 | 0 | 0 | 0 | 0 |
| X5 | 0 | 0 | 0 | 0 | 0 | 0 | 0 | 0 | 0 | 0 | 0 |
| X6 | 0 | 0 | 0 | 0 | 0 | 0 | 0 | 0 | 0 | 0 | 0 |
| X7 | 0 | 0 | 0 | 0 | 0 | 0 | 0 | 0 | 0 | 0 | 0 |

| (X1,7),2 | X1 | X2 | X3 | X4 | X5 | X6 | X7 | X8 | X9 | X10 | X11 | X12 | X13 | X14 | X15 |
|---|---|---|---|---|---|---|---|---|---|---|---|---|---|---|---|
| X1 | 1 | 0 | 0 | 0 | 0 | 0 | 0 | 0 | 0 | 0 | 0 | 0 | 0 | 0 | 0 |
| X2 | 0 | 0 | 0 | 0 | 0 | 0 | 0 | 0 | 0 | 0 | 0 | 0 | 0 | 0 | 0 |

| (X2,7),2 | X1 | X2 | X3 | X4 | X5 | X6 | X7 | X8 | X9 | X10 | X11 | X12 | X13 | X14 | X15 |
|---|---|---|---|---|---|---|---|---|---|---|---|---|---|---|---|
| X1 | 1 | 1 | 1 | 0 | 0 | 0 | 0 | 0 | 0 | 0 | 0 | 0 | 0 | 0 | 0 |
| X2 | 0 | 0 | 0 | 1 | 0 | 0 | 0 | 0 | 0 | 0 | 0 | 0 | 0 | 0 | 0 |

| (X3,7),2 | X1 | X2 | X3 | X4 | X5 | X6 | X7 | X8 | X9 | X10 | X11 | X12 | X13 | X14 | X15 |
|---|---|---|---|---|---|---|---|---|---|---|---|---|---|---|---|
| X1 | 1 | 1 | 2 | 0 | 1 | 1 | 0 | 0 | 1 | 0 | 0 | 0 | 0 | 0 | 0 |
| X2 | 0 | 0 | 0 | 1 | 0 | 1 | 1 | 1 | 0 | 0 | 0 | 0 | 0 | 0 | 0 |

| (X4,7),2 | X1 | X2 | X3 | X4 | X5 | X6 | X7 | X8 | X9 | X10 | X11 | X12 | X13 | X14 | X15 |
|---|---|---|---|---|---|---|---|---|---|---|---|---|---|---|---|
| X1 | 1 | 1 | 2 | 0 | 1 | 1 | 0 | 0 | 1 | 0 | 1 | 0 | 0 | 0 | 0 |
| X2 | 0 | 0 | 0 | 1 | 0 | 1 | 1 | 0 | 0 | 1 | 0 | 0 | 0 | 0 | 0 |

| (X5,7),2 | X1 | X2 | X3 | X4 | X5 | X6 | X7 | X8 | X9 | X10 | X11 | X12 | X13 | X14 | X15 |
|---|---|---|---|---|---|---|---|---|---|---|---|---|---|---|---|
| X1 | 1 | 1 | 1 | 0 | 1 | 1 | 0 | 0 | 1 | 0 | 0 | 1 | 0 | 0 | 0 |
| X2 | 0 | 0 | 0 | 1 | 0 | 0 | 1 | 1 | 0 | 1 | 0 | 0 | 0 | 0 | 0 |

| (X6,7),2 | X1 | X2 | X3 | X4 | X5 | X6 | X7 | X8 | X9 | X10 | X11 | X12 | X13 | X14 | X15 |
|---|---|---|---|---|---|---|---|---|---|---|---|---|---|---|---|
| X1 | 1 | 2 | 4 | 1 | 3 | 5 | 1 | 2 | 4 | 3 | 2 | 2 | 1 | 1 | 0 |
| X2 | 0 | 0 | 0 | 3 | 1 | 4 | 4 | 3 | 1 | 5 | 1 | 1 | 2 | 0 | 0 |

| (X7,7),2 | X1 | X2 | X3 | X4 | X5 | X6 | X7 | X8 | X9 | X10 | X11 | X12 | X13 | X14 | X15 |
|---|---|---|---|---|---|---|---|---|---|---|---|---|---|---|---|
| X1 | 1 | 1 | 2 | 0 | 2 | 1 | 0 | 1 | 2 | 0 | 1 | 1 | 1 | 1 | 0 |
| X2 | 0 | 0 | 0 | 1 | 0 | 1 | 1 | 1 | 0 | 2 | 0 | 1 | 1 | 0 | 1 |

| (X8,7),2 | X1 | X2 | X3 | X4 | X5 | X6 | X7 | X8 | X9 | X10 | X11 | X12 | X13 | X14 | X15 |
|---|---|---|---|---|---|---|---|---|---|---|---|---|---|---|---|
| X1 | 1 | 1 | 2 | 0 | 1 | 2 | 0 | 0 | 2 | 1 | 1 | 1 | 0 | 1 | 0 |
| X2 | 0 | 0 | 0 | 1 | 0 | 1 | 2 | 1 | 0 | 2 | 1 | 0 | 1 | 0 | 0 |

| (X9,7),2 | X1 | X2 | X3 | X4 | X5 | X6 | X7 | X8 | X9 | X10 | X11 | X12 | X13 | X14 | X15 |
|---|---|---|---|---|---|---|---|---|---|---|---|---|---|---|---|
| X1 | 1 | 1 | 2 | 0 | 1 | 2 | 0 | 0 | 2 | 1 | 1 | 1 | 0 | 1 | 0 |
| X2 | 0 | 0 | 0 | 1 | 0 | 1 | 2 | 1 | 0 | 2 | 1 | 0 | 1 | 0 | 0 |

| (X10,7),2 | X1 | X2 | X3 | X4 | X5 | X6 | X7 | X8 | X9 | X10 | X11 | X12 | X13 | X14 | X15 |
|---|---|---|---|---|---|---|---|---|---|---|---|---|---|---|---|
| X1 | 1 | 2 | 4 | 1 | 3 | 5 | 1 | 2 | 4 | 3 | 2 | 2 | 1 | 1 | 0 |
| X2 | 0 | 0 | 0 | 3 | 1 | 4 | 4 | 3 | 1 | 5 | 1 | 1 | 2 | 0 | 0 |

| (X11,7),2 | X1 | X2 | X3 | X4 | X5 | X6 | X7 | X8 | X9 | X10 | X11 | X12 | X13 | X14 | X15 |
|---|---|---|---|---|---|---|---|---|---|---|---|---|---|---|---|
| X1 | 1 | 1 | 2 | 0 | 1 | 1 | 0 | 0 | 1 | 0 | 1 | 0 | 0 | 0 | 0 |
| X2 | 0 | 0 | 0 | 1 | 0 | 1 | 1 | 0 | 0 | 1 | 0 | 0 | 0 | 0 | 0 |

| (X12,7),2 | X1 | X2 | X3 | X4 | X5 | X6 | X7 | X8 | X9 | X10 | X11 | X12 | X13 | X14 | X15 |
|---|---|---|---|---|---|---|---|---|---|---|---|---|---|---|---|
| X1 | 1 | 1 | 1 | 0 | 1 | 1 | 0 | 0 | 1 | 0 | 0 | 1 | 0 | 0 | 0 |
| X2 | 0 | 0 | 0 | 1 | 0 | 0 | 1 | 1 | 0 | 1 | 0 | 0 | 0 | 0 | 0 |

| (X13,7),2 | I | X1 | X2 | X3 | X4 | X5 | X6 | X7 | X8 | X9 | X10 | X11 | X12 | X13 | X14 | X15 |
|---|---|---|---|---|---|---|---|---|---|---|---|---|---|---|---|---|
| X1 | I | 1 | 1 | 2 | 0 | 1 | 1 | 0 | 0 | 1 | 0 | 0 | 0 | 0 | 0 | 0 |
| X2 | I | 0 | 0 | 0 | 1 | 0 | 1 | 1 | 1 | 0 | 0 | 0 | 0 | 0 | 0 | 0 |

| (X14,7),2 | I | X1 | X2 | X3 | X4 | X5 | X6 | X7 | X8 | X9 | X10 | X11 | X12 | X13 | X14 | X15 |
|---|---|---|---|---|---|---|---|---|---|---|---|---|---|---|---|---|
| X1 | I | 1 | 1 | 1 | 0 | 0 | 0 | 0 | 0 | 0 | 0 | 0 | 0 | 0 | 0 | 0 |
| X2 | I | 0 | 0 | 0 | 1 | 0 | 0 | 0 | 0 | 0 | 0 | 0 | 0 | 0 | 0 | 0 |

| (X15,7),2 | I | X1 | X2 | X3 | X4 | X5 | X6 | X7 | X8 | X9 | X10 | X11 | X12 | X13 | X14 | X15 |
|---|---|---|---|---|---|---|---|---|---|---|---|---|---|---|---|---|
| X1 | I | 1 | 0 | 0 | 0 | 0 | 0 | 0 | 0 | 0 | 0 | 0 | 0 | 0 | 0 | 0 |
| X2 | I | 0 | 0 | 0 | 0 | 0 | 0 | 0 | 0 | 0 | 0 | 0 | 0 | 0 | 0 | 0 |

| (X1,7),3 | I | X1 | X2 | X3 | X4 | X5 | X6 | X7 | X8 | X9 | X10 | X11 | X12 | X13 | X14 | X15 |
|---|---|---|---|---|---|---|---|---|---|---|---|---|---|---|---|---|
| X1 | I | 1 | 0 | 0 | 0 | 0 | 0 | 0 | 0 | 0 | 0 | 0 | 0 | 0 | 0 | 0 |
| X2 | I | 0 | 0 | 0 | 0 | 0 | 0 | 0 | 0 | 0 | 0 | 0 | 0 | 0 | 0 | 0 |
| X3 | I | 0 | 0 | 0 | 0 | 0 | 0 | 0 | 0 | 0 | 0 | 0 | 0 | 0 | 0 | 0 |

| (X2,7),3 | I | X1 | X2 | X3 | X4 | X5 | X6 | X7 | X8 | X9 | X10 | X11 | X12 | X13 | X14 | X15 |
|---|---|---|---|---|---|---|---|---|---|---|---|---|---|---|---|---|
| X1 | I | 1 | 2 | 1 | 1 | 1 | 0 | 0 | 0 | 0 | 0 | 0 | 0 | 0 | 0 | 0 |
| X2 | I | 0 | 1 | 1 | 1 | 0 | 1 | 0 | 0 | 0 | 0 | 0 | 0 | 0 | 0 | 0 |
| X3 | I | 0 | 0 | 0 | 0 | 0 | 0 | 1 | 0 | 0 | 0 | 0 | 0 | 0 | 0 | 0 |

| (X3,7),3 | I | X1 | X2 | X3 | X4 | X5 | X6 | X7 | X8 | X9 | X10 | X11 | X12 | X13 | X14 | X15 |
|---|---|---|---|---|---|---|---|---|---|---|---|---|---|---|---|---|
| X1 | I | 2 | 3 | 5 | 2 | 4 | 5 | 2 | 2 | 3 | 1 | 1 | 1 | 0 | 0 | 0 |
| X2 | I | 0 | 2 | 5 | 4 | 3 | 8 | 3 | 4 | 4 | 5 | 1 | 1 | 1 | 0 | 0 |
| X3 | I | 0 | 0 | 0 | 2 | 1 | 2 | 3 | 2 | 0 | 3 | 1 | 1 | 1 | 0 | 0 |

| (X4,7),3 | I | X1 | X2 | X3 | X4 | X5 | X6 | X7 | X8 | X9 | X10 | X11 | X12 | X13 | X14 | X15 |
|---|---|---|---|---|---|---|---|---|---|---|---|---|---|---|---|---|
| X1 | I | 0 | 2 | 2 | 6 | 3 | 5 | 4 | 4 | 0 | 4 | 0 | 1 | 1 | 0 | 1 |
| X2 | I | 0 | 2 | 4 | 5 | 3 | 9 | 4 | 5 | 4 | 7 | 2 | 2 | 3 | 1 | 0 |
| X3 | I | 1 | 1 | 2 | 0 | 2 | 2 | 2 | 1 | 3 | 3 | 3 | 2 | 1 | 1 | 0 |

| (X5,7),3 | I | X1 | X2 | X3 | X4 | X5 | X6 | X7 | X8 | X9 | X10 | X11 | X12 | X13 | X14 | X15 |
|---|---|---|---|---|---|---|---|---|---|---|---|---|---|---|---|---|
| X1 | I | 1 | 2 | 3 | 2 | 4 | 4 | 2 | 3 | 2 | 2 | 1 | 2 | 1 | 1 | 1 |
| X2 | I | 0 | 1 | 3 | 3 | 3 | 7 | 3 | 4 | 4 | 6 | 2 | 2 | 2 | 1 | 0 |
| X3 | I | 0 | 0 | 0 | 1 | 1 | 2 | 3 | 1 | 1 | 3 | 2 | 1 | 1 | 0 | 0 |

| (X6,7),3 | I | X1 | X2 | X3 | X4 | X5 | X6 | X7 | X8 | X9 | X10 | X11 | X12 | X13 | X14 | X15 |
|---|---|---|---|---|---|---|---|---|---|---|---|---|---|---|---|---|
| | I | | | | | | | | | | | | | | | |
| X1 | I | 3 | 11 | 24 | 25 | 24 | 56 | 34 | 31 | 32 | 51 | 23 | 20 | 17 | 7 | 1 |
| X2 | I | 2 | 17 | 41 | 43 | 39 | 101 | 55 | 60 | 60 | 99 | 41 | 38 | 39 | 16 | 2 |
| X3 | I | 2 | 8 | 16 | 19 | 19 | 43 | 26 | 28 | 26 | 46 | 20 | 21 | 21 | 10 | 3 |

| (X7,7),3 | I | X1 | X2 | X3 | X4 | X5 | X6 | X7 | X8 | X9 | X10 | X11 | X12 | X13 | X14 | X15 |
|---|---|---|---|---|---|---|---|---|---|---|---|---|---|---|---|---|
| | I | | | | | | | | | | | | | | | |
| X1 | I | 0 | 0 | 2 | 6 | 3 | 11 | 12 | 5 | 5 | 11 | 6 | 3 | 2 | 0 | 0 |
| X2 | I | 0 | 2 | 7 | 8 | 6 | 19 | 12 | 11 | 11 | 19 | 8 | 6 | 7 | 2 | 0 |
| X3 | I | 1 | 3 | 4 | 3 | 5 | 7 | 3 | 5 | 5 | 7 | 3 | 5 | 4 | 3 | 1 |

| (X8,7),3 | I | X1 | X2 | X3 | X4 | X5 | X6 | X7 | X8 | X9 | X10 | X11 | X12 | X13 | X14 | X15 |
|---|---|---|---|---|---|---|---|---|---|---|---|---|---|---|---|---|
| | I | | | | | | | | | | | | | | | |
| X1 | I | 0 | 3 | 4 | 8 | 6 | 11 | 7 | 10 | 5 | 12 | 3 | 6 | 6 | 3 | 2 |
| X2 | I | 0 | 3 | 8 | 9 | 8 | 22 | 11 | 14 | 13 | 22 | 8 | 8 | 9 | 3 | 0 |
| X3 | I | 1 | 1 | 4 | 2 | 4 | 9 | 7 | 4 | 7 | 9 | 6 | 4 | 2 | 1 | 0 |

| (X9,7),3 | I | X1 | X2 | X3 | X4 | X5 | X6 | X7 | X8 | X9 | X10 | X11 | X12 | X13 | X14 | X15 |
|---|---|---|---|---|---|---|---|---|---|---|---|---|---|---|---|---|
| | I | | | | | | | | | | | | | | | |
| X1 | I | 2 | 3 | 6 | 3 | 6 | 12 | 7 | 5 | 10 | 11 | 8 | 6 | 4 | 3 | 0 |
| X2 | I | 0 | 3 | 9 | 8 | 8 | 22 | 11 | 13 | 14 | 22 | 9 | 3 | 8 | 3 | 0 |
| X3 | I | 0 | 1 | 2 | 6 | 4 | 9 | 7 | 7 | 4 | 9 | 2 | 4 | 4 | 1 | 1 |

| (X10,7),3 | I | X1 | X2 | X3 | X4 | X5 | X6 | X7 | X8 | X9 | X10 | X11 | X12 | X13 | X14 | X15 |
|---|---|---|---|---|---|---|---|---|---|---|---|---|---|---|---|---|
| | I | | | | | | | | | | | | | | | |
| X1 | I | 1 | 7 | 17 | 23 | 20 | 51 | 34 | 32 | 31 | 56 | 25 | 24 | 24 | 11 | 3 |
| X2 | I | 2 | 16 | 39 | 41 | 38 | 99 | 55 | 60 | 60 | 101 | 43 | 39 | 41 | 17 | 2 |
| X3 | I | 3 | 10 | 21 | 20 | 21 | 46 | 26 | 26 | 28 | 43 | 19 | 19 | 16 | 8 | 2 |

| (X11,7),3 | I | X1 | X2 | X3 | X4 | X5 | X6 | X7 | X8 | X9 | X10 | X11 | X12 | X13 | X14 | X15 |
|---|---|---|---|---|---|---|---|---|---|---|---|---|---|---|---|---|
| | I | | | | | | | | | | | | | | | |
| X1 | I | 1 | 0 | 1 | 0 | 1 | 4 | 4 | 0 | 4 | 5 | 6 | 3 | 2 | 2 | 0 |
| X2 | I | 0 | 1 | 3 | 2 | 2 | 7 | 4 | 4 | 5 | 9 | 5 | 3 | 4 | 2 | 0 |
| X3 | I | 0 | 1 | 1 | 3 | 2 | 3 | 2 | 3 | 1 | 2 | 0 | 2 | 2 | 1 | 1 |

| (X12,7),3 | I | X1 | X2 | X3 | X4 | X5 | X6 | X7 | X8 | X9 | X10 | X11 | X12 | X13 | X14 | X15 |
|---|---|---|---|---|---|---|---|---|---|---|---|---|---|---|---|---|
| | I | | | | | | | | | | | | | | | |
| X1 | I | 1 | 1 | 1 | 1 | 2 | 2 | 2 | 2 | 3 | 4 | 2 | 4 | 3 | 2 | 1 |
| X2 | I | 0 | 1 | 2 | 2 | 2 | 6 | 3 | 4 | 4 | 7 | 3 | 3 | 3 | 1 | 0 |
| X3 | I | 0 | 0 | 1 | 2 | 1 | 3 | 3 | 1 | 1 | 2 | 1 | 1 | 0 | 0 | 0 |

| (X13,7),3 | I | X1 | X2 | X3 | X4 | X5 | X6 | X7 | X8 | X9 | X10 | X11 | X12 | X13 | X14 | X15 |
|---|---|---|---|---|---|---|---|---|---|---|---|---|---|---|---|---|
| | I | | | | | | | | | | | | | | | |
| X1 | I | 0 | 0 | 0 | 1 | 1 | 1 | 2 | 3 | 2 | 5 | 2 | 4 | 5 | 3 | 2 |
| X2 | I | 0 | 0 | 1 | 1 | 1 | 5 | 3 | 4 | 4 | 8 | 4 | 3 | 5 | 2 | 0 |
| X3 | I | 0 | 0 | 1 | 1 | 1 | 3 | 3 | 0 | 2 | 2 | 2 | 1 | 0 | 0 | 0 |

| (X14,7),3 | I | X1 | X2 | X3 | X4 | X5 | X6 | X7 | X8 | X9 | X10 | X11 | X12 | X13 | X14 | X15 |
|---|---|---|---|---|---|---|---|---|---|---|---|---|---|---|---|---|
| | I | | | | | | | | | | | | | | | |
| X1 | I | 0 | 0 | 0 | 0 | 0 | 0 | 0 | 0 | 0 | 0 | 1 | 1 | 1 | 2 | 1 |
| X2 | I | 0 | 0 | 0 | 0 | 0 | 0 | 0 | 0 | 0 | 1 | 1 | 0 | 1 | 1 | 0 |
| X3 | I | 0 | 0 | 0 | 0 | 0 | 0 | 1 | 0 | 0 | 0 | 0 | 0 | 0 | 0 | 0 |

| (X15,7),3 | I | X1 | X2 | X3 | X4 | X5 | X6 | X7 | X8 | X9 | X10 | X11 | X12 | X13 | X14 | X15 |
|---|---|---|---|---|---|---|---|---|---|---|---|---|---|---|---|---|
| | I | | | | | | | | | | | | | | | |
| X1 | I | 0 | 0 | 0 | 0 | 0 | 0 | 0 | 0 | 0 | 0 | 0 | 0 | 0 | 0 | 1 |
| X2 | I | 0 | 0 | 0 | 0 | 0 | 0 | 0 | 0 | 0 | 0 | 0 | 0 | 0 | 0 | 0 |
| X3 | I | 0 | 0 | 0 | 0 | 0 | 0 | 0 | 0 | 0 | 0 | 0 | 0 | 0 | 0 | 0 |

| (X1,7),4 | X1 | X2 | X3 | X4 | X5 | X6 | X7 | X8 | X9 | X10 | X11 | X12 | X13 | X14 | X15 |
|---|---|---|---|---|---|---|---|---|---|---|---|---|---|---|---|
| X1 | 1 | 0 | 0 | 0 | 0 | 0 | 0 | 0 | 0 | 0 | 0 | 0 | 0 | 0 | 0 |
| X2 | 0 | 0 | 0 | 0 | 0 | 0 | 0 | 0 | 0 | 0 | 0 | 0 | 0 | 0 | 0 |
| X3 | 0 | 0 | 0 | 0 | 0 | 0 | 0 | 0 | 0 | 0 | 0 | 0 | 0 | 0 | 0 |
| X4 | 0 | 0 | 0 | 0 | 0 | 0 | 0 | 0 | 0 | 0 | 0 | 0 | 0 | 0 | 0 |
| X5 | 0 | 0 | 0 | 0 | 0 | 0 | 0 | 0 | 0 | 0 | 0 | 0 | 0 | 0 | 0 |

| (X2,7),4 | X1 | X2 | X3 | X4 | X5 | X6 | X7 | X8 | X9 | X10 | X11 | X12 | X13 | X14 | X15 |
|---|---|---|---|---|---|---|---|---|---|---|---|---|---|---|---|
| X1 | 2 | 3 | 3 | 1 | 1 | 1 | 0 | 0 | 0 | 0 | 0 | 0 | 0 | 0 | 0 |
| X2 | 0 | 2 | 2 | 3 | 1 | 2 | 1 | 1 | 0 | 0 | 0 | 0 | 0 | 0 | 0 |
| X3 | 1 | 1 | 2 | 0 | 1 | 1 | 0 | 0 | 1 | 0 | 0 | 0 | 0 | 0 | 0 |
| X4 | 0 | 0 | 0 | 1 | 0 | 1 | 1 | 0 | 0 | 1 | 0 | 0 | 0 | 0 | 0 |
| X5 | 0 | 0 | 0 | 0 | 0 | 0 | 0 | 0 | 0 | 0 | 1 | 0 | 0 | 0 | 0 |

| (X3,7),4 | X1 | X2 | X3 | X4 | X5 | X6 | X7 | X8 | X9 | X10 | X11 | X12 | X13 | X14 | X15 |
|---|---|---|---|---|---|---|---|---|---|---|---|---|---|---|---|
| X1 | 4 | 9 | 17 | 8 | 11 | 21 | 6 | 8 | 12 | 11 | 4 | 4 | 2 | 1 | 0 |
| X2 | 1 | 10 | 21 | 23 | 19 | 43 | 23 | 25 | 21 | 32 | 11 | 10 | 9 | 2 | 0 |
| X3 | 4 | 8 | 16 | 8 | 14 | 25 | 9 | 12 | 17 | 17 | 8 | 9 | 5 | 3 | 0 |
| X4 | 0 | 4 | 9 | 15 | 10 | 29 | 19 | 18 | 14 | 30 | 10 | 10 | 11 | 3 | 1 |
| X5 | 0 | 1 | 3 | 2 | 2 | 6 | 3 | 4 | 5 | 7 | 5 | 3 | 4 | 3 | 0 |

| (X4,7),4 | X1 | X2 | X3 | X4 | X5 | X6 | X7 | X8 | X9 | X10 | X11 | X12 | X13 | X14 | X15 |
|---|---|---|---|---|---|---|---|---|---|---|---|---|---|---|---|
| X1 | 5 | 9 | 18 | 8 | 12 | 25 | 10 | 8 | 16 | 16 | 10 | 7 | 4 | 3 | 0 |
| X2 | 1 | 10 | 23 | 25 | 21 | 53 | 31 | 29 | 29 | 46 | 21 | 16 | 15 | 6 | 0 |
| X3 | 4 | 9 | 19 | 9 | 16 | 31 | 12 | 16 | 22 | 25 | 13 | 12 | 9 | 5 | 0 |
| X4 | 0 | 5 | 11 | 19 | 13 | 37 | 24 | 24 | 19 | 41 | 14 | 15 | 17 | 6 | 2 |
| X5 | 0 | 2 | 4 | 4 | 4 | 8 | 4 | 6 | 6 | 9 | 5 | 5 | 6 | 4 | 1 |

| (X5,7),4 | X1 | X2 | X3 | X4 | X5 | X6 | X7 | X8 | X9 | X10 | X11 | X12 | X13 | X14 | X15 |
|---|---|---|---|---|---|---|---|---|---|---|---|---|---|---|---|
| X1 | 4 | 7 | 11 | 6 | 11 | 18 | 7 | 9 | 12 | 13 | 5 | 9 | 4 | 2 | 1 |
| X2 | 0 | 7 | 15 | 20 | 16 | 38 | 23 | 25 | 21 | 37 | 14 | 14 | 14 | 5 | 1 |
| X3 | 3 | 7 | 15 | 7 | 11 | 24 | 8 | 11 | 17 | 19 | 10 | 9 | 8 | 5 | 0 |
| X4 | 0 | 3 | 8 | 13 | 9 | 28 | 19 | 17 | 15 | 31 | 12 | 11 | 12 | 4 | 1 |
| X5 | 1 | 1 | 3 | 1 | 4 | 6 | 4 | 4 | 6 | 6 | 5 | 4 | 2 | 2 | 0 |

| (X6,7),4 | X1 | X2 | X3 | X4 | X5 | X6 | X7 | X8 | X9 | X10 | X11 | X12 | X13 | X14 | X15 |
|---|---|---|---|---|---|---|---|---|---|---|---|---|---|---|---|
| X1 | 24 | 102 | 230 | 213 | 217 | 523 | 277 | 301 | 320 | 499 | 216 | 205 | 195 | 87 | 13 |
| X2 | 32 | 231 | 541 | 606 | 547 | 1378 | 799 | 834 | 813 | 1378 | 582 | 541 | 547 | 228 | 38 |
| X3 | 40 | 168 | 381 | 354 | 367 | 874 | 472 | 509 | 544 | 847 | 378 | 355 | 338 | 156 | 24 |
| X4 | 29 | 200 | 471 | 532 | 479 | 1222 | 714 | 740 | 724 | 1241 | 524 | 490 | 498 | 208 | 37 |
| X5 | 14 | 66 | 156 | 144 | 150 | 363 | 195 | 214 | 231 | 358 | 163 | 150 | 147 | 69 | 9 |

| (X7,7),4 | X1 | X2 | X3 | X4 | X5 | X6 | X7 | X8 | X9 | X10 | X11 | X12 | X13 | X14 | X15 |
|---|---|---|---|---|---|---|---|---|---|---|---|---|---|---|---|
| X1 | 9 | 19 | 34 | 21 | 33 | 60 | 25 | 35 | 42 | 56 | 25 | 32 | 27 | 16 | 5 |
| X2 | 3 | 27 | 60 | 67 | 63 | 148 | 83 | 97 | 90 | 152 | 63 | 64 | 67 | 30 | 7 |
| X3 | 8 | 22 | 49 | 32 | 42 | 95 | 42 | 52 | 64 | 87 | 40 | 39 | 36 | 18 | 2 |
| X4 | 1 | 17 | 43 | 57 | 45 | 125 | 79 | 75 | 70 | 129 | 53 | 47 | 49 | 18 | 3 |
| X5 | 2 | 6 | 15 | 14 | 15 | 35 | 22 | 17 | 22 | 31 | 18 | 13 | 9 | 5 | 0 |

| (X8,7),4 | X1 | X2 | X3 | X4 | X5 | X6 | X7 | X8 | X9 | X10 | X11 | X12 | X13 | X14 | X15 |
|---|---|---|---|---|---|---|---|---|---|---|---|---|---|---|---|
| X1 | 7 | 16 | 38 | 23 | 33 | 77 | 36 | 39 | 53 | 70 | 34 | 32 | 25 | 13 | 1 |
| X2 | 3 | 28 | 70 | 80 | 71 | 184 | 110 | 110 | 110 | 186 | 82 | 72 | 73 | 31 | 4 |
| X3 | 9 | 27 | 58 | 42 | 51 | 114 | 56 | 62 | 75 | 105 | 52 | 47 | 43 | 23 | 3 |
| X4 | 2 | 24 | 54 | 71 | 57 | 151 | 92 | 94 | 84 | 157 | 62 | 60 | 64 | 25 | 6 |
| X5 | 2 | 9 | 20 | 17 | 20 | 41 | 20 | 27 | 27 | 39 | 17 | 17 | 17 | 8 | 1 |

(X9,7),4

| | I | X1 | X2 | X3 | X4 | X5 | X6 | X7 | X8 | X9 | X10 | X11 | X12 | X13 | X14 | X15 |
|---|---|---|---|---|---|---|---|---|---|---|---|---|---|---|---|---|
| X1 | I | 7 | 16 | 38 | 23 | 33 | 77 | 36 | 39 | 53 | 70 | 34 | 32 | 25 | 13 | 1 |
| X2 | I | 3 | 28 | 70 | 80 | 71 | 184 | 110 | 110 | 110 | 186 | 82 | 72 | 73 | 31 | 4 |
| X3 | I | 9 | 27 | 58 | 42 | 51 | 114 | 56 | 62 | 75 | 105 | 52 | 47 | 43 | 23 | 3 |
| X4 | I | 2 | 24 | 54 | 71 | 57 | 151 | 92 | 94 | 84 | 157 | 62 | 60 | 64 | 25 | 6 |
| X5 | I | 2 | 9 | 20 | 17 | 20 | 41 | 20 | 27 | 27 | 39 | 17 | 17 | 17 | 8 | 1 |

(X10,7),4

| | I | X1 | X2 | X3 | X4 | X5 | X6 | X7 | X8 | X9 | X10 | X11 | X12 | X13 | X14 | X15 |
|---|---|---|---|---|---|---|---|---|---|---|---|---|---|---|---|---|
| X1 | I | 24 | 102 | 230 | 213 | 217 | 523 | 277 | 301 | 320 | 499 | 216 | 205 | 195 | 37 | 13 |
| X2 | I | 32 | 231 | 541 | 606 | 547 | 1378 | 799 | 834 | 813 | 1378 | 582 | 541 | 547 | 228 | 38 |
| X3 | I | 40 | 168 | 381 | 354 | 367 | 374 | 472 | 509 | 544 | 847 | 378 | 355 | 338 | 156 | 24 |
| X4 | I | 29 | 200 | 471 | 532 | 479 | 1222 | 714 | 740 | 724 | 1241 | 524 | 490 | 498 | 208 | 37 |
| X5 | I | 14 | 66 | 156 | 144 | 150 | 363 | 135 | 214 | 231 | 358 | 163 | 150 | 147 | 69 | 9 |

(X11,7),4

| | I | X1 | X2 | X3 | X4 | X5 | X6 | X7 | X8 | X9 | X10 | X11 | X12 | X13 | X14 | X15 |
|---|---|---|---|---|---|---|---|---|---|---|---|---|---|---|---|---|
| X1 | I | 5 | 9 | 18 | 8 | 12 | 25 | 10 | 8 | 16 | 16 | 10 | 7 | 4 | 3 | 0 |
| X2 | I | 1 | 10 | 23 | 25 | 21 | 53 | 31 | 29 | 29 | 46 | 21 | 16 | 15 | 6 | 0 |
| X3 | I | 4 | 9 | 19 | 9 | 16 | 31 | 12 | 16 | 22 | 25 | 13 | 12 | 9 | 5 | 0 |
| X4 | I | 0 | 5 | 11 | 19 | 13 | 37 | 24 | 24 | 19 | 41 | 14 | 15 | 17 | 6 | 2 |
| X5 | I | 0 | 2 | 4 | 4 | 4 | 8 | 4 | 6 | 6 | 9 | 5 | 5 | 6 | 4 | 1 |

(X12,7),4

| | I | X1 | X2 | X3 | X4 | X5 | X6 | X7 | X8 | X9 | X10 | X11 | X12 | X13 | X14 | X15 |
|---|---|---|---|---|---|---|---|---|---|---|---|---|---|---|---|---|
| X1 | I | 4 | 7 | 11 | 6 | 11 | 18 | 7 | 9 | 12 | 13 | 5 | 9 | 4 | 2 | 1 |
| X2 | I | 0 | 7 | 15 | 20 | 16 | 38 | 23 | 25 | 21 | 37 | 14 | 14 | 14 | 5 | 1 |
| X3 | I | 3 | 7 | 15 | 7 | 11 | 24 | 8 | 11 | 17 | 19 | 10 | 9 | 8 | 5 | 0 |
| X4 | I | 0 | 3 | 8 | 13 | 9 | 28 | 19 | 17 | 15 | 31 | 12 | 11 | 12 | 4 | 1 |
| X5 | I | 1 | 1 | 3 | 1 | 4 | 6 | 4 | 4 | 6 | 6 | 5 | 4 | 2 | 2 | 0 |

(X13,7),4

| | I | X1 | X2 | X3 | X4 | X5 | X6 | X7 | X8 | X9 | X10 | X11 | X12 | X13 | X14 | X15 |
|---|---|---|---|---|---|---|---|---|---|---|---|---|---|---|---|---|
| X1 | I | 4 | 9 | 17 | 8 | 11 | 21 | 6 | 8 | 12 | 11 | 4 | 4 | 2 | 1 | 0 |
| X2 | I | 1 | 10 | 21 | 23 | 19 | 43 | 23 | 25 | 21 | 32 | 11 | 10 | 9 | 2 | 0 |
| X3 | I | 4 | 8 | 16 | 8 | 14 | 25 | 9 | 12 | 17 | 17 | 8 | 9 | 5 | 3 | 0 |
| X4 | I | 0 | 4 | 9 | 15 | 10 | 29 | 19 | 18 | 14 | 30 | 10 | 10 | 11 | 3 | 1 |
| X5 | I | 0 | 1 | 3 | 2 | 2 | 6 | 3 | 4 | 5 | 7 | 5 | 3 | 4 | 3 | 0 |

(X14,7),4

| | I | X1 | X2 | X3 | X4 | X5 | X6 | X7 | X8 | X9 | X10 | X11 | X12 | X13 | X14 | X15 |
|---|---|---|---|---|---|---|---|---|---|---|---|---|---|---|---|---|
| X1 | I | 2 | 3 | 3 | 1 | 1 | 1 | 0 | 0 | 0 | 0 | 0 | 0 | 0 | 0 | 0 |
| X2 | I | 0 | 2 | 2 | 3 | 1 | 2 | 1 | 1 | 0 | 0 | 0 | 0 | 0 | 0 | 0 |
| X3 | I | 1 | 1 | 2 | 0 | 1 | 1 | 0 | 0 | 1 | 0 | 0 | 0 | 0 | 0 | 0 |
| X4 | I | 0 | 0 | 0 | 1 | 0 | 1 | 1 | 0 | 0 | 1 | 0 | 0 | 0 | 0 | 0 |
| X5 | I | 0 | 0 | 0 | 0 | 0 | 0 | 0 | 0 | 0 | 0 | 1 | 0 | 0 | 0 | 0 |

(X15,7),4

| | I | X1 | X2 | X3 | X4 | X5 | X6 | X7 | X8 | X9 | X10 | X11 | X12 | X13 | X14 | X15 |
|---|---|---|---|---|---|---|---|---|---|---|---|---|---|---|---|---|
| X1 | I | 1 | 0 | 0 | 0 | 0 | 0 | 0 | 0 | 0 | 0 | 0 | 0 | 0 | 0 | 0 |
| X2 | I | 0 | 0 | 0 | 0 | 0 | 0 | 0 | 0 | 0 | 0 | 0 | 0 | 0 | 0 | 0 |
| X3 | I | 0 | 0 | 0 | 0 | 0 | 0 | 0 | 0 | 0 | 0 | 0 | 0 | 0 | 0 | 0 |
| X4 | I | 0 | 0 | 0 | 0 | 0 | 0 | 0 | 0 | 0 | 0 | 0 | 0 | 0 | 0 | 0 |
| X5 | I | 0 | 0 | 0 | 0 | 0 | 0 | 0 | 0 | 0 | 0 | 0 | 0 | 0 | 0 | 0 |

| (X1,7),5 | X1 | X2 | X3 | X4 | X5 | X6 | X7 | X8 | X9 | X10 | X11 | X12 | X13 | X14 | X15 |
|---|---|---|---|---|---|---|---|---|---|---|---|---|---|---|---|
| X1 | 1 | 0 | 0 | 0 | 0 | 0 | 0 | 0 | 0 | 0 | 0 | 0 | 0 | 0 | 0 |
| X2 | 0 | 0 | 0 | 0 | 0 | 0 | 0 | 0 | 0 | 0 | 0 | 0 | 0 | 0 | 0 |
| X3 | 0 | 0 | 0 | 0 | 0 | 0 | 0 | 0 | 0 | 0 | 0 | 0 | 0 | 0 | 0 |
| X4 | 0 | 0 | 0 | 0 | 0 | 0 | 0 | 0 | 0 | 0 | 0 | 0 | 0 | 0 | 0 |
| X5 | 0 | 0 | 0 | 0 | 0 | 0 | 0 | 0 | 0 | 0 | 0 | 0 | 0 | 0 | 0 |
| X6 | 0 | 0 | 0 | 0 | 0 | 0 | 0 | 0 | 0 | 0 | 0 | 0 | 0 | 0 | 0 |
| X7 | 0 | 0 | 0 | 0 | 0 | 0 | 0 | 0 | 0 | 0 | 0 | 0 | 0 | 0 | 0 |

| (X2,7),5 | X1 | X2 | X3 | X4 | X5 | X6 | X7 | X8 | X9 | X10 | X11 | X12 | X13 | X14 | X15 |
|---|---|---|---|---|---|---|---|---|---|---|---|---|---|---|---|
| X1 | 2 | 5 | 4 | 3 | 2 | 2 | 0 | 1 | 0 | 0 | 0 | 0 | 0 | 0 | 0 |
| X2 | 1 | 4 | 5 | 5 | 3 | 5 | 2 | 1 | 1 | 1 | 0 | 0 | 0 | 0 | 0 |
| X3 | 1 | 3 | 4 | 3 | 3 | 4 | 1 | 2 | 1 | 1 | 0 | 0 | 0 | 0 | 0 |
| X4 | 0 | 0 | 1 | 2 | 0 | 3 | 3 | 1 | 1 | 2 | 1 | 0 | 0 | 0 | 0 |
| X5 | 0 | 1 | 1 | 1 | 1 | 2 | 0 | 1 | 1 | 1 | 0 | 1 | 0 | 0 | 0 |
| X6 | 0 | 0 | 0 | 0 | 0 | 0 | 1 | 0 | 0 | 1 | 1 | 0 | 1 | 0 | 0 |
| X7 | 0 | 0 | 0 | 0 | 0 | 0 | 0 | 0 | 0 | 0 | 0 | 0 | 0 | 1 | 0 |

| (X3,7),5 | X1 | X2 | X3 | X4 | X5 | X6 | X7 | X8 | X9 | X10 | X11 | X12 | X13 | X14 | X15 |
|---|---|---|---|---|---|---|---|---|---|---|---|---|---|---|---|
| X1 | 8 | 23 | 44 | 31 | 35 | 70 | 29 | 35 | 39 | 45 | 18 | 17 | 11 | 5 | 0 |
| X2 | 9 | 43 | 90 | 89 | 83 | 188 | 98 | 105 | 101 | 154 | 58 | 57 | 50 | 17 | 2 |
| X3 | 10 | 44 | 93 | 85 | 84 | 191 | 96 | 107 | 108 | 161 | 64 | 61 | 54 | 21 | 2 |
| X4 | 2 | 29 | 67 | 88 | 70 | 184 | 112 | 114 | 101 | 184 | 72 | 67 | 70 | 26 | 5 |
| X5 | 6 | 27 | 61 | 57 | 58 | 137 | 72 | 79 | 83 | 127 | 54 | 52 | 48 | 21 | 3 |
| X6 | 1 | 11 | 28 | 33 | 30 | 79 | 49 | 49 | 49 | 88 | 40 | 36 | 39 | 17 | 3 |
| X7 | 2 | 3 | 7 | 3 | 6 | 13 | 6 | 7 | 11 | 13 | 8 | 8 | 6 | 5 | 1 |

| (X4,7),5 | X1 | X2 | X3 | X4 | X5 | X6 | X7 | X8 | X9 | X10 | X11 | X12 | X13 | X14 | X15 |
|---|---|---|---|---|---|---|---|---|---|---|---|---|---|---|---|
| X1 | 1 | 21 | 35 | 55 | 36 | 85 | 50 | 57 | 36 | 76 | 21 | 24 | 28 | 8 | 3 |
| X2 | 7 | 49 | 104 | 117 | 102 | 245 | 137 | 147 | 135 | 230 | 90 | 98 | 89 | 35 | 7 |
| X3 | 6 | 48 | 103 | 122 | 105 | 255 | 144 | 157 | 138 | 243 | 91 | 91 | 93 | 35 | 8 |
| X4 | 12 | 45 | 109 | 97 | 133 | 257 | 143 | 146 | 166 | 256 | 122 | 107 | 102 | 49 | 6 |
| X5 | 3 | 32 | 74 | 98 | 76 | 191 | 110 | 118 | 111 | 189 | 75 | 74 | 74 | 29 | 5 |
| X6 | 4 | 19 | 47 | 46 | 46 | 118 | 70 | 68 | 75 | 124 | 60 | 51 | 52 | 24 | 3 |
| X7 | 0 | 3 | 6 | 12 | 7 | 21 | 14 | 14 | 9 | 22 | 7 | 7 | 9 | 4 | 2 |

| (X5,7),5 | X1 | X2 | X3 | X4 | X5 | X6 | X7 | X8 | X9 | X10 | X11 | X12 | X13 | X14 | X15 |
|---|---|---|---|---|---|---|---|---|---|---|---|---|---|---|---|
| X1 | 5 | 16 | 31 | 28 | 33 | 62 | 30 | 39 | 36 | 52 | 20 | 24 | 20 | 11 | 4 |
| X2 | 6 | 33 | 74 | 77 | 74 | 175 | 98 | 104 | 102 | 167 | 70 | 67 | 65 | 27 | 5 |
| X3 | 7 | 36 | 78 | 79 | 76 | 180 | 96 | 108 | 107 | 172 | 70 | 70 | 68 | 29 | 5 |
| X4 | 3 | 25 | 65 | 77 | 66 | 181 | 112 | 107 | 108 | 187 | 83 | 71 | 72 | 30 | 4 |
| X5 | 4 | 25 | 57 | 57 | 55 | 134 | 71 | 81 | 80 | 131 | 55 | 52 | 54 | 24 | 4 |
| X6 | 2 | 12 | 30 | 34 | 33 | 81 | 50 | 49 | 50 | 85 | 38 | 36 | 35 | 15 | 3 |
| X7 | 2 | 5 | 7 | 6 | 7 | 13 | 6 | 9 | 9 | 13 | 5 | 7 | 6 | 3 | 1 |

| (X6,7),5 | I | X1 | X2 | X3 | X4 | X5 | X6 | X7 | X8 | X9 | X10 | X11 | X12 | X13 | X14 | X15 |
|---|---|---|---|---|---|---|---|---|---|---|---|---|---|---|---|---|
| | I | | | | | | | | | | | | | | | |
| X1 | I | 121 | 699 | 1630 | 1731 | 1613 | 4033 | 2299 | 2390 | 2400 | 3973 | 1701 | 1570 | 1553 | 659 | 184 |
| X2 | I | 402 | 2404 | 5598 | 5993 | 5593 | 13959 | 7969 | 8368 | 8361 | 13925 | 5959 | 5565 | 5557 | 2376 | 395 |
| X3 | I | 462 | 2766 | 6452 | 6900 | 6433 | 16084 | 9184 | 9625 | 9635 | 16027 | 6867 | 6394 | 6378 | 2724 | 445 |
| X4 | I | 514 | 3104 | 7230 | 7768 | 7254 | 18121 | 10353 | 10918 | 10891 | 18191 | 7787 | 7300 | 7324 | 3150 | 538 |
| X5 | I | 412 | 2456 | 5740 | 6139 | 5730 | 14336 | 8194 | 8579 | 8593 | 14307 | 6137 | 5711 | 5701 | 2440 | 482 |
| X6 | I | 296 | 1785 | 4175 | 4476 | 4180 | 10460 | 5984 | 6280 | 6285 | 10484 | 4500 | 4199 | 4204 | 1804 | 381 |
| X7 | I | 67 | 389 | 910 | 970 | 902 | 2266 | 1301 | 1344 | 1358 | 2245 | 971 | 891 | 880 | 375 | 57 |

| (X7,7),5 | I | X1 | X2 | X3 | X4 | X5 | X6 | X7 | X8 | X9 | X10 | X11 | X12 | X13 | X14 | X15 |
|---|---|---|---|---|---|---|---|---|---|---|---|---|---|---|---|---|
| | I | | | | | | | | | | | | | | | |
| X1 | I | 3 | 37 | 102 | 135 | 105 | 301 | 202 | 169 | 169 | 301 | 135 | 105 | 102 | 37 | 3 |
| X2 | I | 22 | 150 | 364 | 406 | 365 | 939 | 556 | 555 | 555 | 939 | 406 | 365 | 364 | 150 | 22 |
| X3 | I | 22 | 160 | 393 | 442 | 392 | 1023 | 608 | 603 | 603 | 1023 | 442 | 392 | 393 | 160 | 22 |
| X4 | I | 39 | 203 | 453 | 463 | 453 | 1090 | 598 | 666 | 666 | 1090 | 463 | 453 | 453 | 203 | 39 |
| X5 | I | 20 | 136 | 328 | 358 | 325 | 835 | 484 | 497 | 497 | 835 | 358 | 325 | 328 | 136 | 20 |
| X6 | I | 20 | 104 | 232 | 240 | 235 | 561 | 312 | 341 | 341 | 561 | 240 | 235 | 232 | 104 | 20 |
| X7 | I | 2 | 16 | 39 | 48 | 40 | 109 | 69 | 63 | 63 | 109 | 48 | 40 | 39 | 16 | 2 |

| (X8,7),5 | I | X1 | X2 | X3 | X4 | X5 | X6 | X7 | X8 | X9 | X10 | X11 | X12 | X13 | X14 | X15 |
|---|---|---|---|---|---|---|---|---|---|---|---|---|---|---|---|---|
| | I | | | | | | | | | | | | | | | |
| X1 | I | 7 | 68 | 142 | 175 | 151 | 363 | 206 | 239 | 207 | 370 | 142 | 148 | 159 | 67 | 17 |
| X2 | I | 32 | 204 | 470 | 514 | 472 | 1178 | 676 | 712 | 700 | 1182 | 502 | 472 | 478 | 204 | 36 |
| X3 | I | 34 | 223 | 510 | 564 | 520 | 1288 | 732 | 791 | 764 | 1295 | 535 | 519 | 525 | 220 | 42 |
| X4 | I | 44 | 234 | 566 | 581 | 554 | 1409 | 810 | 819 | 860 | 1400 | 625 | 558 | 545 | 238 | 32 |
| X5 | I | 28 | 184 | 424 | 464 | 431 | 1064 | 604 | 652 | 635 | 1067 | 446 | 428 | 432 | 183 | 33 |
| X6 | I | 22 | 125 | 296 | 309 | 293 | 733 | 418 | 434 | 443 | 730 | 318 | 293 | 290 | 125 | 19 |
| X7 | I | 5 | 29 | 56 | 67 | 58 | 139 | 78 | 91 | 78 | 142 | 53 | 57 | 62 | 26 | 8 |

| (X9,7),5 | I | X1 | X2 | X3 | X4 | X5 | X6 | X7 | X8 | X9 | X10 | X11 | X12 | X13 | X14 | X15 |
|---|---|---|---|---|---|---|---|---|---|---|---|---|---|---|---|---|
| | I | | | | | | | | | | | | | | | |
| X1 | I | 17 | 67 | 159 | 142 | 148 | 370 | 206 | 207 | 239 | 363 | 175 | 151 | 142 | 68 | 7 |
| X2 | I | 36 | 204 | 478 | 502 | 472 | 1182 | 676 | 700 | 712 | 1178 | 514 | 472 | 470 | 204 | 32 |
| X3 | I | 42 | 220 | 525 | 535 | 519 | 1295 | 732 | 764 | 791 | 1288 | 564 | 520 | 510 | 223 | 34 |
| X4 | I | 32 | 238 | 545 | 625 | 558 | 1400 | 810 | 860 | 819 | 1409 | 581 | 554 | 566 | 234 | 44 |
| X5 | I | 33 | 183 | 432 | 446 | 428 | 1067 | 604 | 635 | 652 | 1064 | 464 | 431 | 424 | 184 | 28 |
| X6 | I | 19 | 125 | 290 | 318 | 293 | 730 | 418 | 443 | 434 | 733 | 309 | 293 | 296 | 125 | 22 |
| X7 | I | 8 | 26 | 62 | 53 | 57 | 142 | 78 | 78 | 91 | 139 | 67 | 58 | 56 | 29 | 5 |

| (X10,7),5 | I | X1 | X2 | X3 | X4 | X5 | X6 | X7 | X8 | X9 | X10 | X11 | X12 | X13 | X14 | X15 |
|---|---|---|---|---|---|---|---|---|---|---|---|---|---|---|---|---|
| X1 | I | 104 | 659 | 1553 | 1701 | 1570 | 3973 | 2299 | 2400 | 2390 | 4033 | 1731 | 1613 | 1630 | 699 | 121 |
| X2 | I | 395 | 2376 | 5557 | 5959 | 5565 | 13925 | 7969 | 8361 | 8368 | 13959 | 5993 | 5593 | 5598 | 2404 | 402 |
| X3 | I | 445 | 2724 | 6378 | 6867 | 6394 | 16027 | 9184 | 9635 | 9625 | 16084 | 6900 | 6433 | 6452 | 2766 | 462 |
| X4 | I | 538 | 3150 | 7324 | 7787 | 7300 | 18191 | 10353 | 10891 | 10918 | 18121 | 7768 | 7254 | 7230 | 3104 | 514 |
| X5 | I | 402 | 2440 | 5701 | 6137 | 5711 | 14307 | 8194 | 8593 | 8579 | 14336 | 6139 | 5730 | 5740 | 2456 | 412 |
| X6 | I | 301 | 1804 | 4204 | 4500 | 4199 | 10484 | 5984 | 6285 | 6280 | 10460 | 4476 | 4180 | 4175 | 1785 | 296 |
| X7 | I | 57 | 375 | 880 | 971 | 891 | 2245 | 1301 | 1358 | 1344 | 2266 | 970 | 902 | 910 | 389 | 67 |

| (X11,7),5 | I | X1 | X2 | X3 | X4 | X5 | X6 | X7 | X8 | X9 | X10 | X11 | X12 | X13 | X14 | X15 |
|---|---|---|---|---|---|---|---|---|---|---|---|---|---|---|---|---|
| X1 | I | 3 | 8 | 28 | 21 | 24 | 76 | 50 | 36 | 57 | 85 | 55 | 36 | 35 | 21 | 1 |
| X2 | I | 7 | 35 | 89 | 90 | 88 | 230 | 137 | 135 | 147 | 245 | 117 | 102 | 104 | 49 | 7 |
| X3 | I | 8 | 35 | 93 | 91 | 91 | 243 | 144 | 138 | 157 | 255 | 122 | 105 | 103 | 48 | 6 |
| X4 | I | 6 | 49 | 102 | 122 | 107 | 256 | 143 | 166 | 146 | 257 | 97 | 103 | 109 | 45 | 12 |
| X5 | I | 5 | 29 | 74 | 75 | 74 | 189 | 110 | 111 | 118 | 191 | 88 | 76 | 74 | 32 | 3 |
| X6 | I | 3 | 24 | 52 | 60 | 51 | 124 | 70 | 75 | 68 | 118 | 46 | 46 | 47 | 19 | 4 |
| X7 | I | 2 | 4 | 9 | 7 | 7 | 22 | 14 | 9 | 14 | 21 | 12 | 7 | 6 | 3 | 0 |

| (X12,7),5 | I | X1 | X2 | X3 | X4 | X5 | X6 | X7 | X8 | X9 | X10 | X11 | X12 | X13 | X14 | X15 |
|---|---|---|---|---|---|---|---|---|---|---|---|---|---|---|---|---|
| X1 | I | 4 | 11 | 20 | 20 | 24 | 52 | 30 | 36 | 39 | 62 | 28 | 33 | 31 | 16 | 5 |
| X2 | I | 5 | 27 | 65 | 70 | 67 | 167 | 98 | 102 | 104 | 175 | 77 | 74 | 74 | 33 | 6 |
| X3 | I | 5 | 29 | 68 | 70 | 70 | 172 | 96 | 107 | 108 | 180 | 79 | 76 | 78 | 36 | 7 |
| X4 | I | 4 | 30 | 72 | 83 | 71 | 187 | 112 | 108 | 107 | 181 | 77 | 66 | 65 | 25 | 3 |
| X5 | I | 4 | 24 | 54 | 55 | 52 | 131 | 71 | 80 | 81 | 134 | 57 | 55 | 57 | 25 | 4 |
| X6 | I | 3 | 15 | 35 | 38 | 36 | 85 | 50 | 50 | 49 | 81 | 34 | 33 | 30 | 12 | 2 |
| X7 | I | 1 | 3 | 6 | 5 | 7 | 13 | 6 | 9 | 9 | 13 | 6 | 7 | 7 | 5 | 2 |

| (X13,7),5 | I | X1 | X2 | X3 | X4 | X5 | X6 | X7 | X8 | X9 | X10 | X11 | X12 | X13 | X14 | X15 |
|---|---|---|---|---|---|---|---|---|---|---|---|---|---|---|---|---|
| X1 | I | 0 | 5 | 11 | 18 | 17 | 45 | 29 | 39 | 35 | 70 | 31 | 35 | 44 | 23 | 8 |
| X2 | I | 2 | 17 | 50 | 58 | 57 | 154 | 98 | 101 | 105 | 188 | 89 | 83 | 90 | 43 | 9 |
| X3 | I | 2 | 21 | 54 | 64 | 61 | 161 | 96 | 108 | 107 | 191 | 85 | 84 | 93 | 44 | 10 |
| X4 | I | 5 | 26 | 70 | 72 | 67 | 184 | 112 | 101 | 114 | 184 | 88 | 70 | 67 | 29 | 2 |
| X5 | I | 3 | 21 | 48 | 54 | 52 | 127 | 72 | 83 | 79 | 137 | 57 | 58 | 61 | 27 | 6 |
| X6 | I | 3 | 17 | 39 | 40 | 36 | 98 | 49 | 49 | 49 | 79 | 33 | 30 | 28 | 11 | 1 |
| X7 | I | 1 | 5 | 6 | 8 | 8 | 13 | 6 | 11 | 7 | 13 | 3 | 6 | 7 | 3 | 2 |

| (X14,7),5 | I | X1 | X2 | X3 | X4 | X5 | X6 | X7 | X8 | X9 | X10 | X11 | X12 | X13 | X14 | X15 |
|---|---|---|---|---|---|---|---|---|---|---|---|---|---|---|---|---|
| X1 | I | 0 | 0 | 0 | 0 | 0 | 0 | 0 | 0 | 1 | 2 | 3 | 2 | 4 | 5 | 2 |
| X2 | I | 0 | 0 | 0 | 0 | 0 | 1 | 2 | 1 | 1 | 5 | 5 | 3 | 5 | 4 | 1 |
| X3 | I | 0 | 0 | 0 | 0 | 0 | 1 | 1 | 1 | 2 | 4 | 3 | 3 | 4 | 3 | 1 |
| X4 | I | 0 | 0 | 0 | 1 | 0 | 2 | 3 | 1 | 1 | 3 | 2 | 0 | 1 | 0 | 0 |
| X5 | I | 0 | 0 | 0 | 0 | 1 | 1 | 0 | 1 | 1 | 2 | 1 | 1 | 1 | 1 | 0 |
| X6 | I | 0 | 0 | 1 | 1 | 0 | 1 | 1 | 0 | 0 | 0 | 0 | 0 | 0 | 0 | 0 |
| X7 | I | 0 | 1 | 0 | 0 | 0 | 0 | 0 | 0 | 0 | 0 | 0 | 0 | 0 | 0 | 0 |

| (X15,7),5 | I | X1 | X2 | X3 | X4 | X5 | X6 | X7 | X8 | X9 | X10 | X11 | X12 | X13 | X14 | X15 |
|---|---|---|---|---|---|---|---|---|---|---|---|---|---|---|---|---|
| X1 | I | 0 | 0 | 0 | 0 | 0 | 0 | 0 | 0 | 0 | 0 | 0 | 0 | 0 | 0 | 1 |
| X2 | I | 0 | 0 | 0 | 0 | 0 | 0 | 0 | 0 | 0 | 0 | 0 | 0 | 0 | 0 | 0 |
| X3 | I | 0 | 0 | 0 | 0 | 0 | 0 | 0 | 0 | 0 | 0 | 0 | 0 | 0 | 0 | 0 |
| X4 | I | 0 | 0 | 0 | 0 | 0 | 0 | 0 | 0 | 0 | 0 | 0 | 0 | 0 | 0 | 0 |
| X5 | I | 0 | 0 | 0 | 0 | 0 | 0 | 0 | 0 | 0 | 0 | 0 | 0 | 0 | 0 | 0 |
| X6 | I | 0 | 0 | 0 | 0 | 0 | 0 | 0 | 0 | 0 | 0 | 0 | 0 | 0 | 0 | 0 |
| X7 | I | 0 | 0 | 0 | 0 | 0 | 0 | 0 | 0 | 0 | 0 | 0 | 0 | 0 | 0 | 0 |

| (X1,8),2 | X1 | X2 | X3 | X4 | X5 | X6 | X7 | X8 | X9 | X10 | X11 | X12 | X13 | X14 | X15 | X16 | X17 | X18 | X19 | X20 | X21 | X22 |
|---|---|---|---|---|---|---|---|---|---|---|---|---|---|---|---|---|---|---|---|---|---|---|
| X1 | 1 | 0 | 0 | 0 | 0 | 0 | 0 | 0 | 0 | 0 | 0 | 0 | 0 | 0 | 0 | 0 | 0 | 0 | 0 | 0 | 0 | 0 |
| X2 | 0 | 0 | 0 | 0 | 0 | 0 | 0 | 0 | 0 | 0 | 0 | 0 | 0 | 0 | 0 | 0 | 0 | 0 | 0 | 0 | 0 | 0 |

| (X2,8),2 | X1 | X2 | X3 | X4 | X5 | X6 | X7 | X8 | X9 | X10 | X11 | X12 | X13 | X14 | X15 | X16 | X17 | X18 | X19 | X20 | X21 | X22 |
|---|---|---|---|---|---|---|---|---|---|---|---|---|---|---|---|---|---|---|---|---|---|---|
| X1 | 1 | 1 | 1 | 0 | 0 | 0 | 0 | 0 | 0 | 0 | 0 | 0 | 0 | 0 | 0 | 0 | 0 | 0 | 0 | 0 | 0 | 0 |
| X2 | 0 | 0 | 0 | 1 | 0 | 0 | 0 | 0 | 0 | 0 | 0 | 0 | 0 | 0 | 0 | 0 | 0 | 0 | 0 | 0 | 0 | 0 |

| (X3,8),2 | X1 | X2 | X3 | X4 | X5 | X6 | X7 | X8 | X9 | X10 | X11 | X12 | X13 | X14 | X15 | X16 | X17 | X18 | X19 | X20 | X21 | X22 |
|---|---|---|---|---|---|---|---|---|---|---|---|---|---|---|---|---|---|---|---|---|---|---|
| X1 | 1 | 1 | 2 | 0 | 1 | 1 | 0 | 1 | 0 | 1 | 0 | 0 | 0 | 0 | 0 | 0 | 0 | 0 | 0 | 0 | 0 | 0 |
| X2 | 0 | 0 | 0 | 1 | 0 | 1 | 1 | 0 | 1 | 0 | 0 | 0 | 0 | 0 | 0 | 0 | 0 | 0 | 0 | 0 | 0 | 0 |

| (X4,8),2 | X1 | X2 | X3 | X4 | X5 | X6 | X7 | X8 | X9 | X10 | X11 | X12 | X13 | X14 | X15 | X16 | X17 | X18 | X19 | X20 | X21 | X22 |
|---|---|---|---|---|---|---|---|---|---|---|---|---|---|---|---|---|---|---|---|---|---|---|
| X1 | 1 | 1 | 2 | 0 | 1 | 1 | 0 | 0 | 0 | 1 | 0 | 1 | 0 | 0 | 0 | 0 | 0 | 0 | 0 | 0 | 0 | 0 |
| X2 | 0 | 0 | 0 | 1 | 0 | 1 | 1 | 0 | 0 | 0 | 1 | 0 | 0 | 0 | 0 | 0 | 0 | 0 | 0 | 0 | 0 | 0 |

| (X5,8),2 | X1 | X2 | X3 | X4 | X5 | X6 | X7 | X8 | X9 | X10 | X11 | X12 | X13 | X14 | X15 | X16 | X17 | X18 | X19 | X20 | X21 | X22 |
|---|---|---|---|---|---|---|---|---|---|---|---|---|---|---|---|---|---|---|---|---|---|---|
| X1 | 1 | 1 | 2 | 0 | 1 | 1 | 0 | 1 | 1 | 2 | 0 | 0 | 0 | 0 | 1 | 0 | 0 | 0 | 0 | 0 | 0 | 0 |
| X2 | 0 | 0 | 0 | 1 | 0 | 1 | 1 | 0 | 1 | 0 | 1 | 0 | 1 | 1 | 0 | 0 | 0 | 0 | 0 | 0 | 0 | 0 |

| (X6,8),2 | X1 | X2 | X3 | X4 | X5 | X6 | X7 | X8 | X9 | X10 | X11 | X12 | X13 | X14 | X15 | X16 | X17 | X18 | X19 | X20 | X21 | X22 |
|---|---|---|---|---|---|---|---|---|---|---|---|---|---|---|---|---|---|---|---|---|---|---|
| X1 | 1 | 2 | 4 | 1 | 4 | 5 | 1 | 2 | 4 | 5 | 3 | 2 | 2 | 2 | 3 | 2 | 1 | 1 | 0 | 0 | 0 | 0 |
| X2 | 0 | 0 | 0 | 3 | 1 | 4 | 4 | 0 | 5 | 1 | 6 | 1 | 2 | 3 | 2 | 2 | 0 | 0 | 1 | 0 | 0 | 0 |

| (X7,8),2 | X1 | X2 | X3 | X4 | X5 | X6 | X7 | X8 | X9 | X10 | X11 | X12 | X13 | X14 | X15 | X16 | X17 | X18 | X19 | X20 | X21 | X22 |
|---|---|---|---|---|---|---|---|---|---|---|---|---|---|---|---|---|---|---|---|---|---|---|
| X1 | 1 | 1 | 2 | 0 | 2 | 1 | 0 | 1 | 1 | 2 | 0 | 1 | 1 | 0 | 1 | 1 | 1 | 1 | 0 | 1 | 0 | 0 |
| X2 | 0 | 0 | 0 | 1 | 0 | 1 | 1 | 0 | 1 | 0 | 2 | 0 | 0 | 1 | 1 | 1 | 0 | 0 | 1 | 0 | 1 | 0 |

| (X8,8),2 | X1 | X2 | X3 | X4 | X5 | X6 | X7 | X8 | X9 | X10 | X11 | X12 | X13 | X14 | X15 | X16 | X17 | X18 | X19 | X20 | X21 | X22 |
|---|---|---|---|---|---|---|---|---|---|---|---|---|---|---|---|---|---|---|---|---|---|---|
| X1 | 1 | 0 | 1 | 0 | 0 | 0 | 0 | 1 | 0 | 1 | 0 | 0 | 0 | 0 | 0 | 0 | 0 | 1 | 0 | 0 | 0 | 0 |
| X2 | 0 | 0 | 0 | 0 | 0 | 0 | 1 | 0 | 0 | 0 | 0 | 0 | 0 | 1 | 0 | 0 | 0 | 0 | 0 | 0 | 0 | 0 |

| (X9,8),2 | X1 | X2 | X3 | X4 | X5 | X6 | X7 | X8 | X9 | X10 | X11 | X12 | X13 | X14 | X15 | X16 | X17 | X18 | X19 | X20 | X21 | X22 |
|---|---|---|---|---|---|---|---|---|---|---|---|---|---|---|---|---|---|---|---|---|---|---|
| X1 | 1 | 2 | 4 | 1 | 3 | 5 | 1 | 2 | 4 | 6 | 4 | 2 | 2 | 2 | 5 | 3 | 2 | 2 | 1 | 1 | 0 | 0 |
| X2 | 0 | 0 | 0 | 3 | 1 | 4 | 4 | 0 | 4 | 1 | 7 | 2 | 3 | 4 | 3 | 4 | 0 | 0 | 2 | 1 | 0 | 0 |

| (X10,8),2 | X1 | X2 | X3 | X4 | X5 | X6 | X7 | X8 | X9 | X10 | X11 | X12 | X13 | X14 | X15 | X16 | X17 | X18 | X19 | X20 | X21 | X22 |
|---|---|---|---|---|---|---|---|---|---|---|---|---|---|---|---|---|---|---|---|---|---|---|
| X1 | 1 | 1 | 3 | 0 | 2 | 3 | 0 | 2 | 2 | 5 | 2 | 2 | 1 | 2 | 3 | 2 | 2 | 2 | 0 | 1 | 0 | 0 |
| X2 | 0 | 0 | 0 | 1 | 0 | 2 | 3 | 0 | 3 | 1 | 4 | 2 | 1 | 3 | 2 | 3 | 0 | 0 | 1 | 1 | 0 | 0 |

| (X11,8),2 | X1 | X2 | X3 | X4 | X5 | X6 | X7 | X8 | X9 | X10 | X11 | X12 | X13 | X14 | X15 | X16 | X17 | X18 | X19 | X20 | X21 | X22 |
|---|---|---|---|---|---|---|---|---|---|---|---|---|---|---|---|---|---|---|---|---|---|---|
| X1 | 1 | 2 | 5 | 1 | 5 | 7 | 1 | 3 | 7 | 9 | 6 | 4 | 4 | 5 | 8 | 6 | 3 | 3 | 2 | 3 | 1 | 0 |
| X2 | 0 | 0 | 0 | 3 | 1 | 6 | 5 | 1 | 7 | 3 | 11 | 2 | 4 | 7 | 6 | 7 | 1 | 1 | 4 | 2 | 1 | 1 |

| (X12,8),2 | X1 | X2 | X3 | X4 | X5 | X6 | X7 | X8 | X9 | X10 | X11 | X12 | X13 | X14 | X15 | X16 | X17 | X18 | X19 | X20 | X21 | X22 |
|---|---|---|---|---|---|---|---|---|---|---|---|---|---|---|---|---|---|---|---|---|---|---|
| X1 | 1 | 1 | 2 | 0 | 2 | 1 | 0 | 1 | 1 | 2 | 0 | 1 | 1 | 0 | 1 | 1 | 1 | 1 | 0 | 1 | 0 | 0 |
| X2 | 0 | 0 | 0 | 1 | 0 | 1 | 1 | 0 | 1 | 0 | 2 | 0 | 0 | 1 | 1 | 1 | 0 | 0 | 1 | 0 | 1 | 0 |

| (X13,8),2 | X1 | X2 | X3 | X4 | X5 | X6 | X7 | X8 | X9 | X10 | X11 | X12 | X13 | X14 | X15 | X16 | X17 | X18 | X19 | X20 | X21 | X22 |
|---|---|---|---|---|---|---|---|---|---|---|---|---|---|---|---|---|---|---|---|---|---|---|
| X1 | 1 | 1 | 2 | 0 | 1 | 2 | 0 | 1 | 1 | 3 | 1 | 1 | 0 | 1 | 2 | 1 | 1 | 1 | 0 | 1 | 1 | 0 |
| X2 | 0 | 0 | 0 | 1 | 0 | 1 | 2 | 0 | 1 | 0 | 3 | 1 | 1 | 2 | 0 | 2 | 0 | 0 | 1 | 1 | 0 | 1 |

| (X14,8),2 | X1 | X2 | X3 | X4 | X5 | X6 | X7 | X8 | X9 | X10 | X11 | X12 | X13 | X14 | X15 | X16 | X17 | X18 | X19 | X20 | X21 | X22 |
|---|---|---|---|---|---|---|---|---|---|---|---|---|---|---|---|---|---|---|---|---|---|---|
| X1 | 1 | 1 | 3 | 0 | 2 | 3 | 0 | 2 | 2 | 5 | 2 | 2 | 1 | 2 | 3 | 2 | 2 | 2 | 0 | 1 | 0 | 0 |
| X2 | 0 | 0 | 0 | 1 | 0 | 2 | 3 | 0 | 3 | 1 | 4 | 2 | 1 | 3 | 2 | 3 | 0 | 0 | 1 | 1 | 0 | 0 |

| (X15,8),2 | X1 | X2 | X3 | X4 | X5 | X6 | X7 | X8 | X9 | X10 | X11 | X12 | X13 | X14 | X15 | X16 | X17 | X18 | X19 | X20 | X21 | X22 |
|---|---|---|---|---|---|---|---|---|---|---|---|---|---|---|---|---|---|---|---|---|---|---|
| X1 | 1 | 2 | 4 | 1 | 3 | 5 | 1 | 2 | 4 | 6 | 4 | 2 | 2 | 2 | 5 | 3 | 2 | 2 | 1 | 1 | 0 | 0 |
| X2 | 0 | 0 | 0 | 3 | 1 | 4 | 4 | 0 | 4 | 1 | 7 | 2 | 3 | 4 | 3 | 4 | 0 | 0 | 2 | 1 | 0 | 0 |

| (X16,8),2 | X1 | X2 | X3 | X4 | X5 | X6 | X7 | X8 | X9 | X10 | X11 | X12 | X13 | X14 | X15 | X16 | X17 | X18 | X19 | X20 | X21 | X22 |
|---|---|---|---|---|---|---|---|---|---|---|---|---|---|---|---|---|---|---|---|---|---|---|
| X1 | 1 | 2 | 4 | 1 | 4 | 5 | 1 | 2 | 4 | 5 | 3 | 2 | 2 | 2 | 3 | 2 | 1 | 1 | 0 | 0 | 0 | 0 |
| X2 | 0 | 0 | 0 | 3 | 1 | 4 | 4 | 0 | 5 | 1 | 6 | 1 | 2 | 3 | 2 | 2 | 0 | 0 | 1 | 0 | 0 | 0 |

| (X17,8),2 | X1 | X2 | X3 | X4 | X5 | X6 | X7 | X8 | X9 | X10 | X11 | X12 | X13 | X14 | X15 | X16 | X17 | X18 | X19 | X20 | X21 | X22 |
|---|---|---|---|---|---|---|---|---|---|---|---|---|---|---|---|---|---|---|---|---|---|---|
| X1 | 1 | 1 | 2 | 0 | 1 | 1 | 0 | 0 | 0 | 1 | 0 | 1 | 0 | 0 | 0 | 0 | 0 | 0 | 0 | 0 | 0 | 0 |
| X2 | 0 | 0 | 0 | 1 | 0 | 1 | 1 | 0 | 0 | 0 | 1 | 0 | 0 | 0 | 0 | 0 | 0 | 0 | 0 | 0 | 0 | 0 |

| (X18,8),2 | X1 | X2 | X3 | X4 | X5 | X6 | X7 | X8 | X9 | X10 | X11 | X12 | X13 | X14 | X15 | X16 | X17 | X18 | X19 | X20 | X21 | X22 |
|---|---|---|---|---|---|---|---|---|---|---|---|---|---|---|---|---|---|---|---|---|---|---|
| X1 | 1 | 0 | 1 | 0 | 0 | 0 | 0 | 1 | 0 | 1 | 0 | 0 | 0 | 0 | 0 | 0 | 0 | 1 | 0 | 0 | 0 | 0 |
| X2 | 0 | 0 | 0 | 0 | 0 | 0 | 1 | 0 | 0 | 0 | 0 | 0 | 0 | 1 | 0 | 0 | 0 | 0 | 0 | 0 | 0 | 0 |

| (X19,8),2 | X1 | X2 | X3 | X4 | X5 | X6 | X7 | X8 | X9 | X10 | X11 | X12 | X13 | X14 | X15 | X16 | X17 | X18 | X19 | X20 | X21 | X22 |
|---|---|---|---|---|---|---|---|---|---|---|---|---|---|---|---|---|---|---|---|---|---|---|
| X1 | 1 | 1 | 2 | 0 | 1 | 1 | 0 | 1 | 1 | 2 | 0 | 0 | 0 | 0 | 1 | 0 | 0 | 0 | 0 | 0 | 0 | 0 |
| X2 | 0 | 0 | 0 | 1 | 0 | 1 | 1 | 0 | 1 | 0 | 1 | 0 | 1 | 1 | 0 | 0 | 0 | 0 | 0 | 0 | 0 | 0 |

| (X20,8),2 | X1 | X2 | X3 | X4 | X5 | X6 | X7 | X8 | X9 | X10 | X11 | X12 | X13 | X14 | X15 | X16 | X17 | X18 | X19 | X20 | X21 | X22 |
|---|---|---|---|---|---|---|---|---|---|---|---|---|---|---|---|---|---|---|---|---|---|---|
| X1 | 1 | 1 | 2 | 0 | 1 | 1 | 0 | 1 | 0 | 1 | 0 | 0 | 0 | 0 | 0 | 0 | 0 | 0 | 0 | 0 | 0 | 0 |
| X2 | 0 | 0 | 0 | 1 | 0 | 1 | 1 | 0 | 1 | 0 | 0 | 0 | 0 | 0 | 0 | 0 | 0 | 0 | 0 | 0 | 0 | 0 |

| (X21,8),2 | X1 | X2 | X3 | X4 | X5 | X6 | X7 | X8 | X9 | X10 | X11 | X12 | X13 | X14 | X15 | X16 | X17 | X18 | X19 | X20 | X21 | X22 |
|---|---|---|---|---|---|---|---|---|---|---|---|---|---|---|---|---|---|---|---|---|---|---|
| X1 | 1 | 1 | 1 | 0 | 0 | 0 | 0 | 0 | 0 | 0 | 0 | 0 | 0 | 0 | 0 | 0 | 0 | 0 | 0 | 0 | 0 | 0 |
| X2 | 0 | 0 | 0 | 1 | 0 | 0 | 0 | 0 | 0 | 0 | 0 | 0 | 0 | 0 | 0 | 0 | 0 | 0 | 0 | 0 | 0 | 0 |

| (X22,8),2 | X1 | X2 | X3 | X4 | X5 | X6 | X7 | X8 | X9 | X10 | X11 | X12 | X13 | X14 | X15 | X16 | X17 | X18 | X19 | X20 | X21 | X22 |
|---|---|---|---|---|---|---|---|---|---|---|---|---|---|---|---|---|---|---|---|---|---|---|
| X1 | 1 | 0 | 0 | 0 | 0 | 0 | 0 | 0 | 0 | 0 | 0 | 0 | 0 | 0 | 0 | 0 | 0 | 0 | 0 | 0 | 0 | 0 |
| X2 | 0 | 0 | 0 | 0 | 0 | 0 | 0 | 0 | 0 | 0 | 0 | 0 | 0 | 0 | 0 | 0 | 0 | 0 | 0 | 0 | 0 | 0 |

| (X1,8),3 | X1 | X2 | X3 | X4 | X5 | X6 | X7 | X8 | X9 | X10 | X11 | X12 | X13 | X14 | X15 | X16 | X17 | X18 | X19 | X20 | X21 | X22 |
|---|---|---|---|---|---|---|---|---|---|---|---|---|---|---|---|---|---|---|---|---|---|---|
| X1 | 1 | 0 | 0 | 0 | 0 | 0 | 0 | 0 | 0 | 0 | 0 | 0 | 0 | 0 | 0 | 0 | 0 | 0 | 0 | 0 | 0 | 0 |
| X2 | 0 | 0 | 0 | 0 | 0 | 0 | 0 | 0 | 0 | 0 | 0 | 0 | 0 | 0 | 0 | 0 | 0 | 0 | 0 | 0 | 0 | 0 |
| X3 | 0 | 0 | 0 | 0 | 0 | 0 | 0 | 0 | 0 | 0 | 0 | 0 | 0 | 0 | 0 | 0 | 0 | 0 | 0 | 0 | 0 | 0 |

| (X2,8),3 | X1 | X2 | X3 | X4 | X5 | X6 | X7 | X8 | X9 | X10 | X11 | X12 | X13 | X14 | X15 | X16 | X17 | X18 | X19 | X20 | X21 | X22 |
|---|---|---|---|---|---|---|---|---|---|---|---|---|---|---|---|---|---|---|---|---|---|---|
| X1 | 1 | 2 | 1 | 1 | 1 | 0 | 0 | 0 | 0 | 0 | 0 | 0 | 0 | 0 | 0 | 0 | 0 | 0 | 0 | 0 | 0 | 0 |
| X2 | 0 | 1 | 1 | 1 | 0 | 1 | 0 | 0 | 0 | 0 | 0 | 0 | 0 | 0 | 0 | 0 | 0 | 0 | 0 | 0 | 0 | 0 |
| X3 | 0 | 0 | 0 | 0 | 0 | 0 | 1 | 0 | 0 | 0 | 0 | 0 | 0 | 0 | 0 | 0 | 0 | 0 | 0 | 0 | 0 | 0 |

| (X3,8),3 | X1 | X2 | X3 | X4 | X5 | X6 | X7 | X8 | X9 | X10 | X11 | X12 | X13 | X14 | X15 | X16 | X17 | X18 | X19 | X20 | X21 | X22 |
|---|---|---|---|---|---|---|---|---|---|---|---|---|---|---|---|---|---|---|---|---|---|---|
| X1 | 2 | 3 | 6 | 2 | 4 | 5 | 2 | 3 | 4 | 4 | 1 | 1 | 1 | 1 | 1 | 0 | 0 | 1 | 0 | 0 | 0 | 0 |
| X2 | 0 | 2 | 5 | 4 | 4 | 9 | 3 | 2 | 6 | 5 | 5 | 1 | 2 | 3 | 2 | 1 | 0 | 0 | 0 | 0 | 0 | 0 |
| X3 | 0 | 0 | 0 | 2 | 1 | 2 | 3 | 0 | 3 | 0 | 4 | 1 | 1 | 1 | 1 | 1 | 0 | 0 | 0 | 0 | 0 | 0 |

| (X4,8),3 | X1 | X2 | X3 | X4 | X5 | X6 | X7 | X8 | X9 | X10 | X11 | X12 | X13 | X14 | X15 | X16 | X17 | X18 | X19 | X20 | X21 | X22 |
|---|---|---|---|---|---|---|---|---|---|---|---|---|---|---|---|---|---|---|---|---|---|---|
| X1 | 0 | 2 | 2 | 6 | 3 | 5 | 4 | 1 | 5 | 0 | 4 | 0 | 1 | 2 | 1 | 1 | 0 | 0 | 1 | 0 | 1 | 0 |
| X2 | 0 | 2 | 4 | 5 | 3 | 9 | 4 | 1 | 6 | 4 | 7 | 2 | 2 | 3 | 3 | 3 | 1 | 0 | 1 | 1 | 0 | 0 |
| X3 | 1 | 1 | 2 | 0 | 2 | 2 | 2 | 1 | 1 | 3 | 3 | 3 | 1 | 1 | 2 | 2 | 1 | 1 | 0 | 0 | 0 | 0 |

| (X5,8),3 | I | X1 | X2 | X3 | X4 | X5 | X6 | X7 | X8 | X9 | X10 | X11 | X12 | X13 | X14 | X15 | X16 | X17 | X18 | X19 | X20 | X21 | X22 |
|---|---|---|---|---|---|---|---|---|---|---|---|---|---|---|---|---|---|---|---|---|---|---|---|
| X1 | I | 1 | 4 | 5 | 4 | 9 | 9 | 2 | 2 | 10 | 6 | 8 | 2 | 6 | 2 | 6 | 3 | 2 | 1 | 2 | 0 | 1 | 0 |
| X2 | I | 0 | 2 | 6 | 6 | 8 | 16 | 6 | 3 | 15 | 11 | 16 | 4 | 8 | 9 | 11 | 8 | 2 | 1 | 3 | 2 | 0 | 0 |
| X3 | I | 0 | 0 | 1 | 2 | 1 | 5 | 6 | 2 | 6 | 4 | 8 | 4 | 2 | 6 | 4 | 5 | 0 | 2 | 2 | 1 | 0 | 0 |

| (X6,8),3 | I | X1 | X2 | X3 | X4 | X5 | X6 | X7 | X8 | X9 | X10 | X11 | X12 | X13 | X14 | X15 | X16 | X17 | X18 | X19 | X20 | X21 | X22 |
|---|---|---|---|---|---|---|---|---|---|---|---|---|---|---|---|---|---|---|---|---|---|---|---|
| X1 | I | 3 | 13 | 31 | 32 | 41 | 85 | 46 | 20 | 88 | 66 | 102 | 37 | 47 | 59 | 71 | 59 | 17 | 14 | 24 | 14 | 5 | 1 |
| X2 | I | 2 | 19 | 53 | 54 | 69 | 156 | 79 | 33 | 160 | 126 | 195 | 70 | 91 | 118 | 142 | 124 | 37 | 25 | 51 | 35 | 10 | 1 |
| X3 | I | 2 | 9 | 22 | 23 | 31 | 67 | 38 | 16 | 74 | 57 | 93 | 37 | 43 | 57 | 71 | 64 | 21 | 15 | 28 | 19 | 7 | 1 |

| (X7,8),3 | I | X1 | X2 | X3 | X4 | X5 | X6 | X7 | X8 | X9 | X10 | X11 | X12 | X13 | X14 | X15 | X16 | X17 | X18 | X19 | X20 | X21 | X22 |
|---|---|---|---|---|---|---|---|---|---|---|---|---|---|---|---|---|---|---|---|---|---|---|---|
| X1 | I | 0 | 0 | 2 | 6 | 3 | 13 | 14 | 3 | 14 | 9 | 20 | 9 | 5 | 13 | 10 | 11 | 1 | 3 | 3 | 3 | 1 | 1 |
| X2 | I | 0 | 2 | 7 | 8 | 9 | 24 | 14 | 4 | 24 | 20 | 33 | 13 | 14 | 20 | 23 | 23 | 7 | 4 | 9 | 7 | 2 | 0 |
| X3 | I | 1 | 3 | 5 | 3 | 7 | 9 | 3 | 3 | 11 | 9 | 13 | 6 | 8 | 6 | 14 | 11 | 7 | 3 | 6 | 3 | 2 | 0 |

| (X8,8),3 | I | X1 | X2 | X3 | X4 | X5 | X6 | X7 | X8 | X9 | X10 | X11 | X12 | X13 | X14 | X15 | X16 | X17 | X18 | X19 | X20 | X21 | X22 |
|---|---|---|---|---|---|---|---|---|---|---|---|---|---|---|---|---|---|---|---|---|---|---|---|
| X1 | I | 1 | 0 | 2 | 0 | 0 | 1 | 1 | 3 | 1 | 2 | 0 | 1 | 0 | 2 | 0 | 0 | 0 | 2 | 0 | 1 | 0 | 1 |
| X2 | I | 0 | 0 | 1 | 0 | 0 | 2 | 1 | 2 | 1 | 3 | 1 | 1 | 0 | 3 | 1 | 1 | 0 | 1 | 0 | 1 | 0 | 0 |
| X3 | I | 0 | 0 | 0 | 0 | 0 | 0 | 1 | 0 | 1 | 0 | 1 | 1 | 0 | 0 | 1 | 1 | 0 | 0 | 0 | 0 | 0 | 0 |

| (X9,8),3 | I | X1 | X2 | X3 | X4 | X5 | X6 | X7 | X8 | X9 | X10 | X11 | X12 | X13 | X14 | X15 | X16 | X17 | X18 | X19 | X20 | X21 | X22 |
|---|---|---|---|---|---|---|---|---|---|---|---|---|---|---|---|---|---|---|---|---|---|---|---|
| X1 | I | 2 | 13 | 32 | 37 | 47 | 99 | 55 | 23 | 111 | 79 | 132 | 47 | 64 | 82 | 99 | 88 | 27 | 20 | 42 | 28 | 12 | 3 |
| X2 | I | 2 | 19 | 58 | 60 | 81 | 186 | 98 | 40 | 202 | 161 | 255 | 96 | 120 | 159 | 196 | 178 | 57 | 37 | 76 | 55 | 18 | 2 |
| X3 | I | 2 | 9 | 26 | 24 | 37 | 83 | 48 | 20 | 93 | 79 | 123 | 54 | 55 | 75 | 97 | 88 | 31 | 21 | 36 | 25 | 9 | 1 |

| (X10,8),3 | I | X1 | X2 | X3 | X4 | X5 | X6 | X7 | X8 | X9 | X10 | X11 | X12 | X13 | X14 | X15 | X16 | X17 | X18 | X19 | X20 | X21 | X22 |
|---|---|---|---|---|---|---|---|---|---|---|---|---|---|---|---|---|---|---|---|---|---|---|---|
| X1 | I | 4 | 7 | 22 | 11 | 22 | 49 | 27 | 17 | 51 | 54 | 64 | 34 | 27 | 43 | 54 | 47 | 1[illegible] | 16 | 15 | 15 | 3 | 1 |
| X2 | I | 1 | 9 | 31 | 28 | 39 | 94 | 50 | 22 | 100 | 87 | 128 | 51 | 59 | 85 | 101 | 93 | 2[illegible] | 22 | 38 | 30 | 9 | 1 |
| X3 | I | 0 | 4 | 9 | 18 | 18 | 42 | 27 | 8 | 51 | 31 | 64 | 21 | 31 | 39 | 48 | 44 | 1[illegible] | 9 | 24 | 15 | 7 | 2 |

| (X11,8),3 | X1 | X2 | X3 | X4 | X5 | X6 | X7 | X8 | X9 | X10 | X11 | X12 | X13 | X14 | X15 | X16 | X17 | X18 | X19 | X20 | X21 | X22 |
|---|---|---|---|---|---|---|---|---|---|---|---|---|---|---|---|---|---|---|---|---|---|---|
| X1 | 1 | 19 | 52 | 67 | 85 | 198 | 117 | 37 | 217 | 167 | 294 | 117 | 132 | 167 | 217 | 198 | 67 | 37 | 85 | 52 | 19 | 1 |
| X2 | 5 | 40 | 119 | 126 | 167 | 387 | 210 | 81 | 421 | 337 | 546 | 210 | 254 | 337 | 421 | 387 | 126 | 81 | 167 | 119 | 40 | 5 |
| X3 | 6 | 25 | 65 | 61 | 85 | 185 | 99 | 49 | 205 | 167 | 253 | 99 | 120 | 167 | 205 | 185 | 61 | 49 | 85 | 65 | 25 | 6 |

| (X12,8),3 | X1 | X2 | X3 | X4 | X5 | X6 | X7 | X8 | X9 | X10 | X11 | X12 | X13 | X14 | X15 | X16 | X17 | X18 | X19 | X20 | X21 | X22 |
|---|---|---|---|---|---|---|---|---|---|---|---|---|---|---|---|---|---|---|---|---|---|---|
| X1 | 1 | 1 | 3 | 1 | 3 | 11 | 9 | 3 | 10 | 13 | 20 | 14 | 5 | 9 | 14 | 13 | 6 | 3 | 3 | 2 | 0 | 0 |
| X2 | 0 | 2 | 7 | 7 | 9 | 23 | 13 | 4 | 23 | 20 | 33 | 14 | 14 | 20 | 24 | 24 | 8 | 4 | 9 | 7 | 2 | 0 |
| X3 | 0 | 2 | 3 | 7 | 6 | 11 | 6 | 3 | 14 | 6 | 13 | 3 | 8 | 9 | 11 | 9 | 3 | 3 | 7 | 5 | 3 | 1 |

| (X13,8),3 | X1 | X2 | X3 | X4 | X5 | X6 | X7 | X8 | X9 | X10 | X11 | X12 | X13 | X14 | X15 | X16 | X17 | X18 | X19 | X20 | X21 | X22 |
|---|---|---|---|---|---|---|---|---|---|---|---|---|---|---|---|---|---|---|---|---|---|---|
| X1 | 0 | 4 | 5 | 9 | 12 | 20 | 10 | 3 | 25 | 15 | 30 | 10 | 17 | 15 | 25 | 20 | 9 | 3 | 12 | 5 | 4 | 0 |
| X2 | 0 | 3 | 11 | 12 | 17 | 39 | 19 | 7 | 44 | 34 | 56 | 19 | 29 | 34 | 44 | 39 | 12 | 7 | 17 | 11 | 3 | 0 |
| X3 | 1 | 1 | 5 | 4 | 7 | 17 | 12 | 6 | 19 | 18 | 26 | 12 | 12 | 18 | 19 | 17 | 4 | 6 | 7 | 5 | 1 | 1 |

| (X14,8),3 | X1 | X2 | X3 | X4 | X5 | X6 | X7 | X8 | X9 | X10 | X11 | X12 | X13 | X14 | X15 | X16 | X17 | X18 | X19 | X20 | X21 | X22 |
|---|---|---|---|---|---|---|---|---|---|---|---|---|---|---|---|---|---|---|---|---|---|---|
| X1 | 1 | 3 | 15 | 17 | 15 | 47 | 34 | 16 | 54 | 43 | 64 | 27 | 27 | 54 | 51 | 49 | 11 | 17 | 22 | 22 | 7 | 4 |
| X2 | 1 | 9 | 30 | 29 | 38 | 93 | 51 | 22 | 101 | 85 | 128 | 50 | 59 | 87 | 100 | 94 | 28 | 22 | 39 | 31 | 9 | 1 |
| X3 | 2 | 7 | 15 | 13 | 24 | 44 | 21 | 9 | 48 | 39 | 64 | 27 | 31 | 31 | 51 | 42 | 18 | 8 | 18 | 9 | 4 | 0 |

| (X15,8),3 | X1 | X2 | X3 | X4 | X5 | X6 | X7 | X8 | X9 | X10 | X11 | X12 | X13 | X14 | X15 | X16 | X17 | X18 | X19 | X20 | X21 | X22 |
|---|---|---|---|---|---|---|---|---|---|---|---|---|---|---|---|---|---|---|---|---|---|---|
| X1 | 3 | 12 | 28 | 27 | 42 | 88 | 47 | 20 | 99 | 82 | 132 | 55 | 64 | 79 | 111 | 99 | 37 | 23 | 47 | 32 | 13 | 2 |
| X2 | 2 | 18 | 55 | 57 | 76 | 178 | 96 | 37 | 196 | 159 | 255 | 98 | 120 | 161 | 202 | 186 | 60 | 40 | 81 | 58 | 19 | 2 |
| X3 | 1 | 9 | 25 | 31 | 36 | 88 | 54 | 21 | 97 | 75 | 123 | 48 | 55 | 79 | 93 | 83 | 24 | 20 | 37 | 26 | 9 | 2 |

| (X16,8),3 | X1 | X2 | X3 | X4 | X5 | X6 | X7 | X8 | X9 | X10 | X11 | X12 | X13 | X14 | X15 | X16 | X17 | X18 | X19 | X20 | X21 | X22 |
|---|---|---|---|---|---|---|---|---|---|---|---|---|---|---|---|---|---|---|---|---|---|---|
| X1 | 1 | 5 | 14 | 17 | 24 | 59 | 37 | 14 | 71 | 59 | 102 | 46 | 47 | 66 | 88 | 85 | 32 | 20 | 41 | 31 | 13 | 3 |
| X2 | 1 | 10 | 35 | 37 | 51 | 124 | 70 | 25 | 142 | 118 | 195 | 79 | 91 | 126 | 160 | 156 | 54 | 33 | 69 | 53 | 19 | 2 |
| X3 | 1 | 7 | 19 | 21 | 26 | 64 | 37 | 15 | 71 | 57 | 93 | 38 | 43 | 57 | 74 | 67 | 23 | 16 | 31 | 22 | 9 | 2 |

| (X17,8),3 | X1 | X2 | X3 | X4 | X5 | X6 | X7 | X8 | X9 | X10 | X11 | X12 | X13 | X14 | X15 | X16 | X17 | X18 | X19 | X20 | X21 | X22 |
|---|---|---|---|---|---|---|---|---|---|---|---|---|---|---|---|---|---|---|---|---|---|---|
| X1 | 0 | 1 | 0 | 0 | 1 | 1 | 0 | 0 | 1 | 2 | 4 | 4 | 1 | 0 | 5 | 5 | 6 | 1 | 3 | 2 | 2 | 0 |
| X2 | 0 | 0 | 1 | 1 | 1 | 3 | 2 | 0 | 3 | 3 | 7 | 4 | 2 | 4 | 6 | 9 | 5 | 1 | 3 | 4 | 2 | 0 |
| X3 | 0 | 0 | 0 | 1 | 0 | 2 | 3 | 1 | 2 | 1 | 3 | 2 | 1 | 3 | 1 | 2 | 0 | 1 | 2 | 2 | 1 | 1 |

| (X18,8),3 | X1 | X2 | X3 | X4 | X5 | X6 | X7 | X8 | X9 | X10 | X11 | X12 | X13 | X14 | X15 | X16 | X17 | X18 | X19 | X20 | X21 | X22 |
|---|---|---|---|---|---|---|---|---|---|---|---|---|---|---|---|---|---|---|---|---|---|---|
| X1 | 1 | 0 | 1 | 0 | 0 | 0 | 1 | 2 | 0 | 2 | 0 | 1 | 0 | 2 | 1 | 1 | 0 | 3 | 0 | 2 | 0 | 1 |
| X2 | 0 | 0 | 1 | 0 | 0 | 1 | 1 | 1 | 1 | 3 | 1 | 1 | 0 | 3 | 1 | 2 | 0 | 2 | 0 | 1 | 0 | 0 |
| X3 | 0 | 0 | 0 | 0 | 0 | 1 | 1 | 0 | 1 | 0 | 1 | 1 | 0 | 0 | 1 | 0 | 0 | 0 | 0 | 0 | 0 | 0 |

| (X19,8),3 | X1 | X2 | X3 | X4 | X5 | X6 | X7 | X8 | X9 | X10 | X11 | X12 | X13 | X14 | X15 | X16 | X17 | X18 | X19 | X20 | X21 | X22 |
|---|---|---|---|---|---|---|---|---|---|---|---|---|---|---|---|---|---|---|---|---|---|---|
| X1 | 0 | 1 | 0 | 2 | 2 | 3 | 2 | 1 | 6 | 2 | 8 | 2 | 6 | 6 | 10 | 9 | 4 | 2 | 9 | 5 | 4 | 1 |
| X2 | 0 | 0 | 2 | 2 | 3 | 8 | 4 | 1 | 11 | 9 | 16 | 6 | 8 | 11 | 15 | 16 | 6 | 3 | 8 | 6 | 2 | 0 |
| X3 | 0 | 0 | 1 | 0 | 2 | 5 | 4 | 2 | 4 | 6 | 8 | 6 | 2 | 4 | 6 | 5 | 2 | 2 | 1 | 1 | 0 | 0 |

| (X20,8),3 | X1 | X2 | X3 | X4 | X5 | X6 | X7 | X8 | X9 | X10 | X11 | X12 | X13 | X14 | X15 | X16 | X17 | X18 | X19 | X20 | X21 | X22 |
|---|---|---|---|---|---|---|---|---|---|---|---|---|---|---|---|---|---|---|---|---|---|---|
| X1 | 0 | 0 | 0 | 0 | 0 | 0 | 1 | 1 | 1 | 1 | 1 | 2 | 1 | 4 | 4 | 5 | 2 | 3 | 4 | 6 | 3 | 2 |
| X2 | 0 | 0 | 0 | 0 | 0 | 1 | 1 | 0 | 2 | 3 | 5 | 3 | 2 | 5 | 6 | 9 | 4 | 2 | 4 | 5 | 2 | 0 |
| X3 | 0 | 0 | 0 | 0 | 0 | 1 | 1 | 0 | 1 | 1 | 4 | 3 | 1 | 0 | 3 | 2 | 2 | 0 | 1 | 0 | 0 | 0 |

| (X21,8),3 | X1 | X2 | X3 | X4 | X5 | X6 | X7 | X8 | X9 | X10 | X11 | X12 | X13 | X14 | X15 | X16 | X17 | X18 | X19 | X20 | X21 | X22 |
|---|---|---|---|---|---|---|---|---|---|---|---|---|---|---|---|---|---|---|---|---|---|---|
| X1 | 0 | 0 | 0 | 0 | 0 | 0 | 0 | 0 | 0 | 0 | 0 | 0 | 0 | 0 | 0 | 0 | 1 | 0 | 1 | 1 | 2 | 1 |
| X2 | 0 | 0 | 0 | 0 | 0 | 0 | 0 | 0 | 0 | 0 | 0 | 0 | 0 | 0 | 0 | 1 | 1 | 0 | 0 | 1 | 1 | 0 |
| X3 | 0 | 0 | 0 | 0 | 0 | 0 | 0 | 0 | 0 | 0 | 0 | 1 | 0 | 0 | 0 | 0 | 0 | 0 | 0 | 0 | 0 | 0 |

| (X22,8),3 | X1 | X2 | X3 | X4 | X5 | X6 | X7 | X8 | X9 | X10 | X11 | X12 | X13 | X14 | X15 | X16 | X17 | X18 | X19 | X20 | X21 | X22 |
|---|---|---|---|---|---|---|---|---|---|---|---|---|---|---|---|---|---|---|---|---|---|---|
| X1 | 0 | 0 | 0 | 0 | 0 | 0 | 0 | 0 | 0 | 0 | 0 | 0 | 0 | 0 | 0 | 0 | 0 | 0 | 0 | 0 | 0 | 1 |
| X2 | 0 | 0 | 0 | 0 | 0 | 0 | 0 | 0 | 0 | 0 | 0 | 0 | 0 | 0 | 0 | 0 | 0 | 0 | 0 | 0 | 0 | 0 |
| X3 | 0 | 0 | 0 | 0 | 0 | 0 | 0 | 0 | 0 | 0 | 0 | 0 | 0 | 0 | 0 | 0 | 0 | 0 | 0 | 0 | 0 | 0 |

| (X1,8),4 | X1 | X2 | X3 | X4 | X5 | X6 | X7 | X8 | X9 | X10 | X11 | X12 | X13 | X14 | X15 | X16 | X17 | X18 | X19 | X20 | X21 | X22 |
|---|---|---|---|---|---|---|---|---|---|---|---|---|---|---|---|---|---|---|---|---|---|---|
| X1 | 1 | 0 | 0 | 0 | 0 | 0 | 0 | 0 | 0 | 0 | 0 | 0 | 0 | 0 | 0 | 0 | 0 | 0 | 0 | 0 | 0 | 0 |
| X2 | 0 | 0 | 0 | 0 | 0 | 0 | 0 | 0 | 0 | 0 | 0 | 0 | 0 | 0 | 0 | 0 | 0 | 0 | 0 | 0 | 0 | 0 |
| X3 | 0 | 0 | 0 | 0 | 0 | 0 | 0 | 0 | 0 | 0 | 0 | 0 | 0 | 0 | 0 | 0 | 0 | 0 | 0 | 0 | 0 | 0 |
| X4 | 0 | 0 | 0 | 0 | 0 | 0 | 0 | 0 | 0 | 0 | 0 | 0 | 0 | 0 | 0 | 0 | 0 | 0 | 0 | 0 | 0 | 0 |
| X5 | 0 | 0 | 0 | 0 | 0 | 0 | 0 | 0 | 0 | 0 | 0 | 0 | 0 | 0 | 0 | 0 | 0 | 0 | 0 | 0 | 0 | 0 |

| (X2,8),4 | X1 | X2 | X3 | X4 | X5 | X6 | X7 | X8 | X9 | X10 | X11 | X12 | X13 | X14 | X15 | X16 | X17 | X18 | X19 | X20 | X21 | X22 |
|---|---|---|---|---|---|---|---|---|---|---|---|---|---|---|---|---|---|---|---|---|---|---|
| X1 | 2 | 3 | 3 | 1 | 1 | 1 | 0 | 1 | 0 | 0 | 0 | 0 | 0 | 0 | 0 | 0 | 0 | 0 | 0 | 0 | 0 | 0 |
| X2 | 0 | 2 | 2 | 3 | 1 | 2 | 1 | 0 | 1 | 0 | 0 | 0 | 0 | 0 | 0 | 0 | 0 | 0 | 0 | 0 | 0 | 0 |
| X3 | 1 | 1 | 2 | 0 | 1 | 1 | 0 | 0 | 0 | 1 | 0 | 0 | 0 | 0 | 0 | 0 | 0 | 0 | 0 | 0 | 0 | 0 |
| X4 | 0 | 0 | 0 | 1 | 0 | 1 | 1 | 0 | 0 | 0 | 1 | 0 | 0 | 0 | 0 | 0 | 0 | 0 | 0 | 0 | 0 | 0 |
| X5 | 0 | 0 | 0 | 0 | 0 | 0 | 0 | 0 | 0 | 0 | 0 | 1 | 0 | 0 | 0 | 0 | 0 | 0 | 0 | 0 | 0 | 0 |

| (X3,8),4 | X1 | X2 | X3 | X4 | X5 | X6 | X7 | X8 | X9 | X10 | X11 | X12 | X13 | X14 | X15 | X16 | X17 | X18 | X19 | X20 | X21 | X22 |
|---|---|---|---|---|---|---|---|---|---|---|---|---|---|---|---|---|---|---|---|---|---|---|
| X1 | 5 | 9 | 21 | 9 | 16 | 26 | 7 | 10 | 19 | 20 | 14 | 4 | 7 | 9 | 9 | 5 | 1 | 3 | 0 | 1 | 0 | 0 |
| X2 | 1 | 11 | 25 | 25 | 27 | 55 | 27 | 13 | 50 | 34 | 45 | 13 | 21 | 27 | 25 | 17 | 2 | 4 | 5 | 3 | 0 | 0 |
| X3 | 4 | 9 | 19 | 9 | 19 | 31 | 10 | 11 | 27 | 27 | 25 | 10 | 13 | 14 | 19 | 11 | 4 | 5 | 3 | 2 | 0 | 0 |
| X4 | 0 | 4 | 10 | 17 | 15 | 37 | 22 | 6 | 36 | 22 | 45 | 13 | 18 | 23 | 24 | 21 | 4 | 2 | 8 | 3 | 1 | 0 |
| X5 | 0 | 1 | 3 | 2 | 4 | 8 | 4 | 0 | 8 | 7 | 11 | 6 | 5 | 5 | 8 | 9 | 4 | 1 | 2 | 2 | 0 | 0 |

| (X4,8),4 | X1 | X2 | X3 | X4 | X5 | X6 | X7 | X8 | X9 | X10 | X11 | X12 | X13 | X14 | X15 | X16 | X17 | X18 | X19 | X20 | X21 | X22 |
|---|---|---|---|---|---|---|---|---|---|---|---|---|---|---|---|---|---|---|---|---|---|---|
| X1 | 5 | 9 | 19 | 8 | 16 | 27 | 10 | 8 | 18 | 22 | 21 | 11 | 8 | 9 | 13 | 10 | 5 | 3 | 1 | 2 | 0 | 0 |
| X2 | 1 | 10 | 23 | 25 | 26 | 58 | 32 | 10 | 51 | 38 | 60 | 24 | 24 | 31 | 35 | 31 | 9 | 5 | 10 | 7 | 2 | 0 |
| X3 | 4 | 9 | 20 | 9 | 20 | 33 | 12 | 10 | 29 | 30 | 31 | 15 | 16 | 18 | 25 | 20 | 8 | 6 | 6 | 6 | 1 | 0 |
| X4 | 0 | 5 | 11 | 19 | 15 | 40 | 25 | 7 | 40 | 25 | 53 | 16 | 21 | 30 | 33 | 32 | 9 | 5 | 15 | 9 | 4 | 1 |
| X5 | 0 | 2 | 4 | 4 | 5 | 9 | 4 | 2 | 10 | 8 | 11 | 6 | 6 | 8 | 11 | 11 | 5 | 3 | 5 | 5 | 2 | 0 |

| (X5,8),4 | X1 | X2 | X3 | X4 | X5 | X6 | X7 | X8 | X9 | X10 | X11 | X12 | X13 | X14 | X15 | X16 | X17 | X18 | X19 | X20 | X21 | X22 |
|---|---|---|---|---|---|---|---|---|---|---|---|---|---|---|---|---|---|---|---|---|---|---|
| X1 | 7 | 15 | 36 | 19 | 35 | 63 | 24 | 22 | 61 | 57 | 61 | 22 | 31 | 39 | 50 | 36 | 12 | 14 | 14 | 10 | 3 | 1 |
| X2 | 2 | 20 | 53 | 57 | 69 | 151 | 80 | 34 | 158 | 117 | 180 | 63 | 86 | 113 | 129 | 110 | 29 | 24 | 46 | 31 | 9 | 1 |
| X3 | 6 | 17 | 43 | 26 | 48 | 93 | 37 | 26 | 92 | 87 | 105 | 42 | 52 | 65 | 87 | 70 | 26 | 20 | 27 | 21 | 6 | 0 |
| X4 | 0 | 11 | 31 | 44 | 48 | 118 | 70 | 22 | 129 | 92 | 168 | 59 | 75 | 99 | 117 | 108 | 31 | 19 | 47 | 29 | 10 | 2 |
| X5 | 1 | 4 | 10 | 8 | 15 | 30 | 16 | 6 | 34 | 30 | 45 | 22 | 22 | 26 | 38 | 34 | 14 | 8 | 13 | 10 | 4 | 0 |

| (X6,8),4 | I | X1 | X2 | X3 | X4 | X5 | X6 | X7 | X8 | X9 | X10 | X11 | X12 | X13 | X14 | X15 | X16 | X17 | X18 | X19 | X20 | X21 | X22 |
|---|---|---|---|---|---|---|---|---|---|---|---|---|---|---|---|---|---|---|---|---|---|---|---|
| | I | | | | | | | | | | | | | | | | | | | | | | |
| X1 | I | 35 | 172 | 463 | 429 | 598 | 1304 | 663 | 300 | 1369 | 1118 | 1689 | 644 | 793 | 1033 | 1290 | 1137 | 369 | 256 | 476 | 336 | 109 | 13 |
| X2 | I | 50 | 404 | 1126 | 1219 | 1568 | 3563 | 1945 | 768 | 3845 | 3002 | 4848 | 1832 | 2255 | 2981 | 3658 | 3302 | 1039 | 704 | 1424 | 987 | 333 | 46 |
| X3 | I | 60 | 290 | 790 | 732 | 1049 | 2285 | 1187 | 532 | 2458 | 2022 | 3072 | 1206 | 1452 | 1902 | 2410 | 2146 | 716 | 497 | 916 | 657 | 223 | 28 |
| X4 | I | 46 | 354 | 1008 | 1097 | 1419 | 3264 | 1800 | 704 | 3557 | 2800 | 4560 | 1742 | 2117 | 2821 | 3488 | 3184 | 1021 | 683 | 1392 | 976 | 340 | 50 |
| X5 | I | 21 | 118 | 336 | 311 | 451 | 1004 | 524 | 227 | 1089 | 912 | 1392 | 562 | 659 | 874 | 1120 | 1019 | 352 | 233 | 440 | 330 | 114 | 14 |

| (X7,8),4 | I | X1 | X2 | X3 | X4 | X5 | X6 | X7 | X8 | X9 | X10 | X11 | X12 | X13 | X14 | X15 | X16 | X17 | X18 | X19 | X20 | X21 | X22 |
|---|---|---|---|---|---|---|---|---|---|---|---|---|---|---|---|---|---|---|---|---|---|---|---|
| | I | | | | | | | | | | | | | | | | | | | | | | |
| X1 | I | 11 | 25 | 58 | 35 | 66 | 121 | 52 | 37 | 127 | 117 | 150 | 63 | 75 | 94 | 132 | 113 | 45 | 34 | 51 | 40 | 17 | 3 |
| X2 | I | 3 | 37 | 100 | 110 | 142 | 317 | 173 | 68 | 348 | 270 | 437 | 168 | 208 | 275 | 342 | 314 | 104 | 69 | 141 | 103 | 38 | 6 |
| X3 | I | 10 | 31 | 83 | 56 | 99 | 203 | 92 | 52 | 214 | 194 | 264 | 109 | 129 | 167 | 222 | 196 | 71 | 49 | 82 | 64 | 21 | 2 |
| X4 | I | 1 | 24 | 74 | 98 | 112 | 278 | 165 | 55 | 306 | 231 | 408 | 153 | 183 | 250 | 302 | 284 | 38 | 56 | 126 | 86 | 31 | 5 |
| X5 | I | 2 | 10 | 27 | 25 | 37 | 82 | 47 | 19 | 89 | 76 | 116 | 52 | 52 | 70 | 92 | 83 | 30 | 20 | 34 | 25 | 9 | 1 |

| (X8,8),4 | I | X1 | X2 | X3 | X4 | X5 | X6 | X7 | X8 | X9 | X10 | X11 | X12 | X13 | X14 | X15 | X16 | X17 | X18 | X19 | X20 | X21 | X22 |
|---|---|---|---|---|---|---|---|---|---|---|---|---|---|---|---|---|---|---|---|---|---|---|---|
| | I | | | | | | | | | | | | | | | | | | | | | | |
| X1 | I | 3 | 1 | 6 | 0 | 1 | 4 | 2 | 7 | 3 | 9 | 2 | 3 | 0 | 6 | 3 | 2 | 0 | 6 | 0 | 2 | 0 | 1 |
| X2 | I | 0 | 0 | 5 | 2 | 1 | 9 | 9 | 6 | 9 | 12 | 8 | 6 | 2 | 14 | 7 | 8 | 0 | 6 | 1 | 5 | 0 | 1 |
| X3 | I | 2 | 1 | 6 | 0 | 3 | 6 | 1 | 4 | 4 | 10 | 4 | 3 | 2 | 5 | 5 | 5 | 2 | 4 | 0 | 3 | 0 | 0 |
| X4 | I | 0 | 0 | 1 | 2 | 1 | 6 | 5 | 1 | 7 | 5 | 10 | 4 | 3 | 7 | 7 | 7 | 1 | 1 | 3 | 2 | 1 | 0 |
| X5 | I | 0 | 1 | 0 | 0 | 2 | 1 | 0 | 0 | 2 | 1 | 2 | 2 | 2 | 0 | 3 | 1 | 2 | 0 | 1 | 0 | 0 | 0 |

| (X9,8),4 | I | X1 | X2 | X3 | X4 | X5 | X6 | X7 | X8 | X9 | X10 | X11 | X12 | X13 | X14 | X15 | X16 | X17 | X18 | X19 | X20 | X21 | X22 |
|---|---|---|---|---|---|---|---|---|---|---|---|---|---|---|---|---|---|---|---|---|---|---|---|
| | I | | | | | | | | | | | | | | | | | | | | | | |
| X1 | I | 45 | 216 | 596 | 559 | 796 | 1756 | 919 | 407 | 1898 | 1564 | 2400 | 941 | 1131 | 1490 | 1891 | 1694 | 566 | 391 | 731 | 523 | 180 | 24 |
| X2 | I | 66 | 532 | 1518 | 1644 | 2143 | 4926 | 2717 | 1061 | 5383 | 4258 | 6912 | 2666 | 3218 | 4291 | 5329 | 4873 | 1577 | 1051 | 2132 | 1510 | 525 | 75 |
| X3 | I | 74 | 385 | 1074 | 1016 | 1441 | 3201 | 1684 | 730 | 3464 | 2859 | 4408 | 1747 | 2075 | 2746 | 3488 | 3156 | 1066 | 719 | 1363 | 994 | 344 | 45 |
| X4 | I | 62 | 489 | 1407 | 1536 | 1994 | 4616 | 2560 | 993 | 5059 | 4003 | 6541 | 2524 | 3037 | 4064 | 5050 | 4640 | 1507 | 999 | 2041 | 1446 | 509 | 76 |
| X5 | I | 31 | 169 | 475 | 455 | 651 | 1445 | 769 | 330 | 1581 | 1304 | 2019 | 808 | 955 | 1265 | 1612 | 1462 | 496 | 335 | 635 | 465 | 162 | 21 |

| (X10,8),4 | I | X1 | X2 | X3 | X4 | X5 | X6 | X7 | X8 | X9 | X10 | X11 | X12 | X13 | X14 | X15 | X16 | X17 | X18 | X19 | X20 | X21 | X22 |
|---|---|---|---|---|---|---|---|---|---|---|---|---|---|---|---|---|---|---|---|---|---|---|---|
| | I | | | | | | | | | | | | | | | | | | | | | | |
| X1 | I | 26 | 89 | 268 | 210 | 325 | 730 | 375 | 188 | 776 | 694 | 983 | 410 | 456 | 636 | 795 | 715 | 242 | 184 | 292 | 230 | 72 | 10 |
| X2 | I | 28 | 210 | 628 | 657 | 862 | 2019 | 1129 | 451 | 2204 | 1782 | 2835 | 1121 | 1306 | 1795 | 2199 | 2022 | 650 | 453 | 868 | 639 | 214 | 33 |
| X3 | I | 38 | 164 | 458 | 404 | 597 | 1309 | 679 | 314 | 1412 | 1194 | 1788 | 724 | 844 | 1124 | 1434 | 1292 | 444 | 309 | 552 | 413 | 141 | 20 |
| X4 | I | 21 | 195 | 558 | 636 | 806 | 1875 | 1048 | 395 | 2064 | 1604 | 2671 | 1016 | 1240 | 1655 | 2049 | 1888 | 607 | 397 | 840 | 587 | 211 | 32 |
| X5 | I | 10 | 72 | 190 | 191 | 272 | 585 | 301 | 125 | 642 | 512 | 811 | 311 | 394 | 497 | 645 | 582 | 200 | 125 | 260 | 183 | 66 | 7 |

| (X11,8),4 | I | X1 | X2 | X3 | X4 | X5 | X6 | X7 | X8 | X9 | X10 | X11 | X12 | X13 | X14 | X15 | X16 | X17 | X18 | X19 | X20 | X21 | X22 |
|---|---|---|---|---|---|---|---|---|---|---|---|---|---|---|---|---|---|---|---|---|---|---|---|
| | I | | | | | | | | | | | | | | | | | | | | | | |
| X1 | I | 102 | 547 | 1539 | 1489 | 2074 | 4646 | 2470 | 1069 | 5062 | 4141 | 6432 | 2525 | 3030 | 4049 | 5093 | 4623 | 1541 | 1061 | 2021 | 1477 | 518 | 77 |
| X2 | I | 192 | 1439 | 4125 | 4393 | 5794 | 13294 | 7308 | 2903 | 14570 | 11595 | 18708 | 7253 | 8734 | 11687 | 14539 | 13317 | 4341 | 2911 | 5847 | 4187 | 1468 | 217 |
| X3 | I | 176 | 1005 | 2835 | 2791 | 3874 | 8700 | 4650 | 1960 | 9473 | 7727 | 12105 | 4762 | 5683 | 7547 | 9530 | 8655 | 2892 | 1948 | 3763 | 2721 | 949 | 130 |
| X4 | I | 178 | 1358 | 3900 | 4204 | 5522 | 12716 | 7016 | 2741 | 13923 | 11058 | 17979 | 6959 | 8359 | 11146 | 13897 | 12738 | 4155 | 2745 | 5580 | 3952 | 1385 | 199 |
| X5 | I | 75 | 460 | 1301 | 1306 | 1806 | 4067 | 2186 | 895 | 4425 | 3598 | 5690 | 2243 | 2661 | 3510 | 4451 | 4045 | 1355 | 891 | 1748 | 1249 | 433 | 54 |

| (X12,8),4 | I | X1 | X2 | X3 | X4 | X5 | X6 | X7 | X8 | X9 | X10 | X11 | X12 | X13 | X14 | X15 | X16 | X17 | X18 | X19 | X20 | X21 | X22 |
|---|---|---|---|---|---|---|---|---|---|---|---|---|---|---|---|---|---|---|---|---|---|---|---|
| | I | | | | | | | | | | | | | | | | | | | | | | |
| X1 | I | 11 | 25 | 58 | 35 | 66 | 121 | 52 | 37 | 127 | 117 | 150 | 63 | 75 | 94 | 132 | 113 | 45 | 34 | 51 | 40 | 17 | 3 |
| X2 | I | 3 | 37 | 100 | 110 | 142 | 317 | 173 | 68 | 348 | 270 | 437 | 168 | 208 | 275 | 342 | 314 | 104 | 69 | 141 | 103 | 38 | 6 |
| X3 | I | 10 | 31 | 83 | 56 | 99 | 203 | 92 | 52 | 214 | 194 | 264 | 109 | 129 | 167 | 222 | 196 | 71 | 49 | 82 | 64 | 21 | 2 |
| X4 | I | 1 | 24 | 74 | 98 | 112 | 278 | 165 | 55 | 306 | 231 | 408 | 153 | 183 | 250 | 302 | 284 | 88 | 56 | 126 | 86 | 31 | 5 |
| X5 | I | 2 | 10 | 27 | 25 | 37 | 82 | 47 | 19 | 89 | 76 | 116 | 52 | 52 | 70 | 92 | 83 | 30 | 20 | 34 | 25 | 9 | 1 |

| (X13,8),4 | I | X1 | X2 | X3 | X4 | X5 | X6 | X7 | X8 | X9 | X10 | X11 | X12 | X13 | X14 | X15 | X16 | X17 | X18 | X19 | X20 | X21 | X22 |
|---|---|---|---|---|---|---|---|---|---|---|---|---|---|---|---|---|---|---|---|---|---|---|---|
| | I | | | | | | | | | | | | | | | | | | | | | | |
| X1 | I | 11 | 32 | 93 | 66 | 107 | 238 | 119 | 67 | 255 | 232 | 316 | 131 | 148 | 212 | 262 | 233 | 78 | 65 | 96 | 78 | 25 | 5 |
| X2 | I | 6 | 65 | 195 | 217 | 270 | 645 | 369 | 142 | 707 | 561 | 908 | 357 | 415 | 581 | 700 | 650 | 205 | 144 | 281 | 210 | 72 | 12 |
| X3 | I | 17 | 59 | 159 | 125 | 197 | 418 | 208 | 104 | 442 | 388 | 557 | 233 | 264 | 350 | 454 | 409 | 147 | 101 | 172 | 134 | 46 | 6 |
| X4 | I | 6 | 61 | 172 | 203 | 250 | 588 | 333 | 119 | 641 | 494 | 843 | 320 | 386 | 512 | 636 | 592 | 193 | 120 | 264 | 182 | 67 | 11 |
| X5 | I | 6 | 25 | 64 | 55 | 90 | 180 | 87 | 39 | 193 | 162 | 245 | 100 | 122 | 144 | 198 | 176 | 65 | 38 | 76 | 54 | 19 | 1 |

| (X14,8),4 | I | X1 | X2 | X3 | X4 | X5 | X6 | X7 | X8 | X9 | X10 | X11 | X12 | X13 | X14 | X15 | X16 | X17 | X18 | X19 | X20 | X21 | X22 |
|---|---|---|---|---|---|---|---|---|---|---|---|---|---|---|---|---|---|---|---|---|---|---|---|
| | I | | | | | | | | | | | | | | | | | | | | | | |
| X1 | I | 26 | 89 | 268 | 210 | 325 | 730 | 375 | 188 | 776 | 694 | 983 | 410 | 456 | 636 | 795 | 715 | 242 | 184 | 292 | 230 | 72 | 10 |
| X2 | I | 28 | 210 | 628 | 657 | 862 | 2019 | 1129 | 451 | 2204 | 1782 | 2835 | 1121 | 1306 | 1795 | 2199 | 2022 | 650 | 453 | 868 | 639 | 214 | 33 |
| X3 | I | 38 | 164 | 458 | 404 | 597 | 1309 | 679 | 314 | 1412 | 1194 | 1788 | 724 | 844 | 1124 | 1434 | 1292 | 444 | 309 | 552 | 413 | 141 | 20 |
| X4 | I | 21 | 195 | 558 | 636 | 806 | 1875 | 1048 | 395 | 2064 | 1604 | 2671 | 1016 | 1240 | 1655 | 2049 | 1888 | 607 | 397 | 840 | 587 | 211 | 32 |
| X5 | I | 10 | 72 | 190 | 191 | 272 | 585 | 301 | 125 | 642 | 512 | 811 | 311 | 394 | 497 | 645 | 582 | 200 | 125 | 260 | 183 | 66 | 7 |

| (X15,8),4 | I | X1 | X2 | X3 | X4 | X5 | X6 | X7 | X8 | X9 | X10 | X11 | X12 | X13 | X14 | X15 | X16 | X17 | X18 | X19 | X20 | X21 | X22 |
|---|---|---|---|---|---|---|---|---|---|---|---|---|---|---|---|---|---|---|---|---|---|---|---|
| | I | | | | | | | | | | | | | | | | | | | | | | |
| X1 | I | 45 | 216 | 596 | 559 | 796 | 1756 | 919 | 407 | 1898 | 1564 | 2400 | 941 | 1131 | 1490 | 1891 | 1694 | 566 | 391 | 731 | 523 | 180 | 24 |
| X2 | I | 66 | 532 | 1518 | 1644 | 2143 | 4926 | 2717 | 1061 | 5383 | 4258 | 6912 | 2666 | 3218 | 4291 | 5329 | 4873 | 1577 | 1051 | 2132 | 1510 | 525 | 75 |
| X3 | I | 74 | 385 | 1074 | 1016 | 1441 | 3201 | 1684 | 730 | 3464 | 2859 | 4408 | 1747 | 2075 | 2746 | 3488 | 3156 | 1066 | 719 | 1363 | 994 | 344 | 45 |
| X4 | I | 62 | 489 | 1407 | 1536 | 1994 | 4616 | 2560 | 993 | 5059 | 4003 | 6541 | 2524 | 3037 | 4064 | 5050 | 4640 | 1507 | 999 | 2041 | 1446 | 509 | 76 |
| X5 | I | 31 | 169 | 475 | 455 | 651 | 1445 | 769 | 330 | 1581 | 1304 | 2019 | 808 | 955 | 1265 | 1612 | 1462 | 496 | 335 | 635 | 465 | 162 | 21 |

| (X16,8),4 | X1 | X2 | X3 | X4 | X5 | X6 | X7 | X8 | X9 | X10 | X11 | X12 | X13 | X14 | X15 | X16 | X17 | X18 | X19 | X20 | X21 | X22 |
|---|---|---|---|---|---|---|---|---|---|---|---|---|---|---|---|---|---|---|---|---|---|---|
| X1 | 35 | 172 | 463 | 429 | 598 | 1304 | 663 | 300 | 1369 | 1118 | 1689 | 644 | 793 | 1033 | 1290 | 1137 | 369 | 256 | 476 | 336 | 109 | 13 |
| X2 | 50 | 404 | 1126 | 1219 | 1568 | 3563 | 1945 | 768 | 3845 | 3002 | 4848 | 1832 | 2255 | 2981 | 3658 | 3302 | 1039 | 704 | 1424 | 967 | 333 | 46 |
| X3 | 60 | 290 | 790 | 732 | 1049 | 2285 | 1187 | 532 | 2458 | 2022 | 3072 | 1206 | 1452 | 1902 | 2410 | 2146 | 716 | 497 | 916 | 657 | 223 | 28 |
| X4 | 46 | 354 | 1008 | 1097 | 1419 | 3264 | 1800 | 704 | 3557 | 2800 | 4560 | 1742 | 2117 | 2821 | 3488 | 3184 | 1021 | 683 | 1392 | 976 | 340 | 50 |
| X5 | 21 | 118 | 336 | 311 | 451 | 1004 | 524 | 227 | 1089 | 912 | 1392 | 562 | 659 | 874 | 1120 | 1019 | 352 | 233 | 440 | 330 | 114 | 14 |

| (X17,8),4 | X1 | X2 | X3 | X4 | X5 | X6 | X7 | X8 | X9 | X10 | X11 | X12 | X13 | X14 | X15 | X16 | X17 | X18 | X19 | X20 | X21 | X22 |
|---|---|---|---|---|---|---|---|---|---|---|---|---|---|---|---|---|---|---|---|---|---|---|
| X1 | 5 | 9 | 19 | 8 | 16 | 27 | 10 | 8 | 18 | 22 | 21 | 11 | 8 | 9 | 13 | 10 | 5 | 3 | 1 | 2 | 0 | 0 |
| X2 | 1 | 10 | 23 | 25 | 26 | 58 | 32 | 10 | 51 | 38 | 60 | 24 | 24 | 31 | 35 | 31 | 9 | 5 | 10 | 7 | 2 | 0 |
| X3 | 4 | 9 | 20 | 9 | 20 | 33 | 12 | 10 | 29 | 30 | 31 | 15 | 16 | 18 | 25 | 20 | 8 | 6 | 6 | 6 | 1 | 0 |
| X4 | 0 | 5 | 11 | 19 | 15 | 40 | 25 | 7 | 40 | 25 | 53 | 16 | 21 | 30 | 33 | 32 | 9 | 5 | 15 | 9 | 4 | 1 |
| X5 | 0 | 2 | 4 | 4 | 5 | 9 | 4 | 2 | 10 | 8 | 11 | 6 | 6 | 8 | 11 | 11 | 5 | 3 | 5 | 5 | 2 | 0 |

| (X18,8),4 | X1 | X2 | X3 | X4 | X5 | X6 | X7 | X8 | X9 | X10 | X11 | X12 | X13 | X14 | X15 | X16 | X17 | X18 | X19 | X20 | X21 | X22 |
|---|---|---|---|---|---|---|---|---|---|---|---|---|---|---|---|---|---|---|---|---|---|---|
| X1 | 3 | 1 | 6 | 0 | 1 | 4 | 2 | 7 | 3 | 9 | 2 | 3 | 0 | 6 | 3 | 2 | 0 | 6 | 0 | 2 | 0 | 1 |
| X2 | 0 | 0 | 5 | 2 | 1 | 9 | 9 | 6 | 9 | 12 | 8 | 6 | 2 | 14 | 7 | 8 | 0 | 6 | 1 | 5 | 0 | 1 |
| X3 | 2 | 1 | 6 | 0 | 3 | 6 | 1 | 4 | 4 | 10 | 4 | 3 | 2 | 5 | 5 | 5 | 2 | 4 | 0 | 3 | 0 | 0 |
| X4 | 0 | 0 | 1 | 2 | 1 | 6 | 5 | 1 | 7 | 5 | 10 | 4 | 3 | 7 | 7 | 7 | 1 | 1 | 3 | 2 | 1 | 0 |
| X5 | 0 | 1 | 0 | 0 | 2 | 1 | 0 | 0 | 2 | 1 | 2 | 2 | 2 | 0 | 3 | 1 | 2 | 0 | 1 | 0 | 0 | 0 |

| (X19,8),4 | X1 | X2 | X3 | X4 | X5 | X6 | X7 | X8 | X9 | X10 | X11 | X12 | X13 | X14 | X15 | X16 | X17 | X18 | X19 | X20 | X21 | X22 |
|---|---|---|---|---|---|---|---|---|---|---|---|---|---|---|---|---|---|---|---|---|---|---|
| X1 | 7 | 15 | 36 | 19 | 35 | 63 | 24 | 22 | 61 | 57 | 61 | 22 | 31 | 39 | 50 | 36 | 12 | 14 | 14 | 10 | 3 | 1 |
| X2 | 2 | 20 | 53 | 57 | 69 | 151 | 80 | 34 | 158 | 117 | 180 | 63 | 86 | 113 | 129 | 110 | 29 | 24 | 46 | 31 | 9 | 1 |
| X3 | 6 | 17 | 43 | 26 | 48 | 93 | 37 | 26 | 92 | 87 | 105 | 42 | 52 | 65 | 87 | 70 | 26 | 20 | 27 | 21 | 6 | 0 |
| X4 | 0 | 11 | 31 | 44 | 48 | 118 | 70 | 22 | 128 | 92 | 168 | 59 | 75 | 99 | 117 | 108 | 31 | 19 | 47 | 29 | 10 | 2 |
| X5 | 1 | 4 | 10 | 8 | 15 | 30 | 16 | 6 | 34 | 30 | 45 | 22 | 22 | 26 | 38 | 34 | 14 | 8 | 13 | 10 | 4 | 0 |

| (X20,8),4 | X1 | X2 | X3 | X4 | X5 | X6 | X7 | X8 | X9 | X10 | X11 | X12 | X13 | X14 | X15 | X16 | X17 | X18 | X19 | X20 | X21 | X22 |
|---|---|---|---|---|---|---|---|---|---|---|---|---|---|---|---|---|---|---|---|---|---|---|
| X1 | 5 | 9 | 21 | 9 | 16 | 26 | 7 | 10 | 19 | 20 | 14 | 4 | 7 | 9 | 9 | 5 | 1 | 3 | 0 | 1 | 0 | 0 |
| X2 | 1 | 11 | 25 | 25 | 27 | 55 | 27 | 13 | 50 | 34 | 45 | 13 | 21 | 27 | 25 | 17 | 2 | 4 | 5 | 3 | 0 | 0 |
| X3 | 4 | 9 | 19 | 9 | 19 | 31 | 10 | 11 | 27 | 27 | 25 | 10 | 13 | 14 | 19 | 11 | 4 | 5 | 3 | 2 | 0 | 0 |
| X4 | 0 | 4 | 10 | 17 | 15 | 37 | 22 | 6 | 36 | 22 | 45 | 13 | 18 | 23 | 24 | 21 | 4 | 2 | 8 | 3 | 1 | 0 |
| X5 | 0 | 1 | 3 | 2 | 4 | 8 | 4 | 0 | 8 | 7 | 11 | 6 | 5 | 5 | 8 | 9 | 4 | 1 | 2 | 2 | 0 | 0 |

| (X21,8),4 | X1 | X2 | X3 | X4 | X5 | X6 | X7 | X8 | X9 | X10 | X11 | X12 | X13 | X14 | X15 | X16 | X17 | X18 | X19 | X20 | X21 | X22 |
|---|---|---|---|---|---|---|---|---|---|---|---|---|---|---|---|---|---|---|---|---|---|---|
| X1 | 2 | 3 | 3 | 1 | 1 | 1 | 0 | 1 | 0 | 0 | 0 | 0 | 0 | 0 | 0 | 0 | 0 | 0 | 0 | 0 | 0 | 0 |
| X2 | 0 | 2 | 2 | 3 | 1 | 2 | 1 | 0 | 1 | 0 | 0 | 0 | 0 | 0 | 0 | 0 | 0 | 0 | 0 | 0 | 0 | 0 |
| X3 | 1 | 1 | 2 | 0 | 1 | 1 | 0 | 0 | 0 | 1 | 0 | 0 | 0 | 0 | 0 | 0 | 0 | 0 | 0 | 0 | 0 | 0 |
| X4 | 0 | 0 | 0 | 1 | 0 | 1 | 1 | 0 | 0 | 0 | 1 | 0 | 0 | 0 | 0 | 0 | 0 | 0 | 0 | 0 | 0 | 0 |
| X5 | 0 | 0 | 0 | 0 | 0 | 0 | 0 | 0 | 0 | 0 | 0 | 1 | 0 | 0 | 0 | 0 | 0 | 0 | 0 | 0 | 0 | 0 |

| (X22,8),4 | X1 | X2 | X3 | X4 | X5 | X6 | X7 | X8 | X9 | X10 | X11 | X12 | X13 | X14 | X15 | X16 | X17 | X18 | X19 | X20 | X21 | X22 |
|---|---|---|---|---|---|---|---|---|---|---|---|---|---|---|---|---|---|---|---|---|---|---|
| X1 | 1 | 0 | 0 | 0 | 0 | 0 | 0 | 0 | 0 | 0 | 0 | 0 | 0 | 0 | 0 | 0 | 0 | 0 | 0 | 0 | 0 | 0 |
| X2 | 0 | 0 | 0 | 0 | 0 | 0 | 0 | 0 | 0 | 0 | 0 | 0 | 0 | 0 | 0 | 0 | 0 | 0 | 0 | 0 | 0 | 0 |
| X3 | 0 | 0 | 0 | 0 | 0 | 0 | 0 | 0 | 0 | 0 | 0 | 0 | 0 | 0 | 0 | 0 | 0 | 0 | 0 | 0 | 0 | 0 |
| X4 | 0 | 0 | 0 | 0 | 0 | 0 | 0 | 0 | 0 | 0 | 0 | 0 | 0 | 0 | 0 | 0 | 0 | 0 | 0 | 0 | 0 | 0 |
| X5 | 0 | 0 | 0 | 0 | 0 | 0 | 0 | 0 | 0 | 0 | 0 | 0 | 0 | 0 | 0 | 0 | 0 | 0 | 0 | 0 | 0 | 0 |

| (X1,8),5 | I | X1 | X2 | X3 | X4 | X5 | X6 | X7 | X8 | X9 | X10 | X11 | X12 | X13 | X14 | X15 | X16 | X17 | X18 | X19 | X20 | X21 | X22 |
|---|---|---|---|---|---|---|---|---|---|---|---|---|---|---|---|---|---|---|---|---|---|---|---|
| | I | | | | | | | | | | | | | | | | | | | | | | |
| X1 | I | 1 | 0 | 0 | 0 | 0 | 0 | 0 | 0 | 0 | 0 | 0 | 0 | 0 | 0 | 0 | 0 | 0 | 0 | 0 | 0 | 0 | 0 |
| X2 | I | 0 | 0 | 0 | 0 | 0 | 0 | 0 | 0 | 0 | 0 | 0 | 0 | 0 | 0 | 0 | 0 | 0 | 0 | 0 | 0 | 0 | 0 |
| X3 | I | 0 | 0 | 0 | 0 | 0 | 0 | 0 | 0 | 0 | 0 | 0 | 0 | 0 | 0 | 0 | 0 | 0 | 0 | 0 | 0 | 0 | 0 |
| X4 | I | 0 | 0 | 0 | 0 | 0 | 0 | 0 | 0 | 0 | 0 | 0 | 0 | 0 | 0 | 0 | 0 | 0 | 0 | 0 | 0 | 0 | 0 |
| X5 | I | 0 | 0 | 0 | 0 | 0 | 0 | 0 | 0 | 0 | 0 | 0 | 0 | 0 | 0 | 0 | 0 | 0 | 0 | 0 | 0 | 0 | 0 |
| X6 | I | 0 | 0 | 0 | 0 | 0 | 0 | 0 | 0 | 0 | 0 | 0 | 0 | 0 | 0 | 0 | 0 | 0 | 0 | 0 | 0 | 0 | 0 |
| X7 | I | 0 | 0 | 0 | 0 | 0 | 0 | 0 | 0 | 0 | 0 | 0 | 0 | 0 | 0 | 0 | 0 | 0 | 0 | 0 | 0 | 0 | 0 |

| (X2,8),5 | I | X1 | X2 | X3 | X4 | X5 | X6 | X7 | X8 | X9 | X10 | X11 | X12 | X13 | X14 | X15 | X16 | X17 | X18 | X19 | X20 | X21 | X22 |
|---|---|---|---|---|---|---|---|---|---|---|---|---|---|---|---|---|---|---|---|---|---|---|---|
| | I | | | | | | | | | | | | | | | | | | | | | | |
| X1 | I | 2 | 5 | 4 | 3 | 3 | 2 | 0 | 0 | 1 | 0 | 0 | 0 | 0 | 0 | 0 | 0 | 0 | 0 | 0 | 0 | 0 | 0 |
| X2 | I | 1 | 4 | 5 | 5 | 3 | 5 | 2 | 1 | 2 | 1 | 1 | 0 | 0 | 0 | 0 | 0 | 0 | 0 | 0 | 0 | 0 | 0 |
| X3 | I | 1 | 3 | 4 | 3 | 3 | 4 | 1 | 1 | 2 | 1 | 1 | 0 | 1 | 0 | 0 | 0 | 0 | 0 | 0 | 0 | 0 | 0 |
| X4 | I | 0 | 0 | 1 | 2 | 0 | 3 | 3 | 0 | 1 | 1 | 2 | 1 | 0 | 1 | 0 | 0 | 0 | 0 | 0 | 0 | 0 | 0 |
| X5 | I | 0 | 1 | 1 | 1 | 1 | 2 | 0 | 0 | 1 | 1 | 1 | 0 | 0 | 0 | 1 | 0 | 0 | 0 | 0 | 0 | 0 | 0 |
| X6 | I | 0 | 0 | 0 | 0 | 0 | 0 | 1 | 0 | 0 | 0 | 1 | 1 | 0 | 0 | 0 | 1 | 0 | 0 | 0 | 0 | 0 | 0 |
| X7 | I | 0 | 0 | 0 | 0 | 0 | 0 | 0 | 0 | 0 | 0 | 0 | 0 | 0 | 0 | 0 | 0 | 1 | 0 | 0 | 0 | 0 | 0 |

| (X3,8),5 | I | X1 | X2 | X3 | X4 | X5 | X6 | X7 | X8 | X9 | X10 | X11 | X12 | X13 | X14 | X15 | X16 | X17 | X18 | X19 | X20 | X21 | X22 |
|---|---|---|---|---|---|---|---|---|---|---|---|---|---|---|---|---|---|---|---|---|---|---|---|
| | I | | | | | | | | | | | | | | | | | | | | | | |
| X1 | I | 10 | 28 | 63 | 40 | 58 | 105 | 41 | 33 | 92 | 78 | 78 | 26 | 40 | 51 | 53 | 34 | 8 | 13 | 10 | 8 | 1 | 0 |
| X2 | I | 11 | 53 | 126 | 116 | 144 | 292 | 140 | 72 | 278 | 209 | 289 | 93 | 134 | 168 | 185 | 141 | 32 | 34 | 50 | 30 | 6 | 1 |
| X3 | I | 12 | 54 | 130 | 112 | 151 | 302 | 140 | 76 | 294 | 230 | 311 | 105 | 149 | 188 | 213 | 166 | 42 | 41 | 60 | 40 | 8 | 1 |
| X4 | I | 2 | 36 | 92 | 117 | 130 | 297 | 165 | 56 | 303 | 214 | 369 | 125 | 164 | 208 | 239 | 207 | 55 | 35 | 82 | 46 | 13 | 1 |
| X5 | I | 7 | 33 | 86 | 77 | 105 | 222 | 109 | 53 | 226 | 181 | 261 | 94 | 124 | 157 | 191 | 158 | 47 | 37 | 64 | 41 | 12 | 1 |
| X6 | I | 1 | 14 | 38 | 46 | 57 | 132 | 76 | 25 | 141 | 108 | 189 | 74 | 84 | 109 | 137 | 129 | 42 | 23 | 54 | 36 | 12 | 1 |
| X7 | I | 2 | 4 | 10 | 5 | 12 | 22 | 10 | 6 | 24 | 25 | 31 | 16 | 15 | 20 | 30 | 27 | 13 | 9 | 11 | 12 | 5 | 1 |

| (X4,8),5 | I | X1 | X2 | X3 | X4 | X5 | X6 | X7 | X8 | X9 | X10 | X11 | X12 | X13 | X14 | X15 | X16 | X17 | X18 | X19 | X20 | X21 | X22 |
|---|---|---|---|---|---|---|---|---|---|---|---|---|---|---|---|---|---|---|---|---|---|---|---|
| | I | | | | | | | | | | | | | | | | | | | | | | |
| X1 | I | 1 | 22 | 42 | 62 | 53 | 115 | 63 | 21 | 111 | 61 | 120 | 31 | 54 | 69 | 68 | 61 | 14 | 9 | 29 | 17 | 7 | 2 |
| X2 | I | 7 | 51 | 121 | 132 | 150 | 328 | 174 | 69 | 321 | 237 | 381 | 135 | 168 | 218 | 255 | 225 | 67 | 44 | 91 | 61 | 21 | 3 |
| X3 | I | 6 | 50 | 120 | 137 | 157 | 344 | 183 | 73 | 354 | 252 | 417 | 142 | 190 | 248 | 287 | 254 | 73 | 49 | 109 | 71 | 26 | 4 |
| X4 | I | 12 | 47 | 127 | 109 | 156 | 346 | 183 | 77 | 349 | 302 | 449 | 189 | 200 | 265 | 336 | 302 | 104 | 67 | 116 | 86 | 26 | 2 |
| X5 | I | 3 | 33 | 85 | 99 | 117 | 263 | 142 | 56 | 279 | 209 | 343 | 122 | 161 | 210 | 254 | 227 | 70 | 49 | 101 | 69 | 24 | 4 |
| X6 | I | 4 | 20 | 55 | 52 | 72 | 164 | 91 | 36 | 175 | 147 | 234 | 99 | 106 | 144 | 184 | 175 | 61 | 38 | 75 | 57 | 20 | 3 |
| X7 | I | 0 | 3 | 7 | 14 | 12 | 31 | 19 | 5 | 36 | 21 | 46 | 14 | 21 | 30 | 35 | 35 | 12 | 6 | 20 | 15 | 9 | 2 |

| (X5,8),5 | X1 | X2 | X3 | X4 | X5 | X6 | X7 | X8 | X9 | X10 | X11 | X12 | X13 | X14 | X15 | X16 | X17 | X18 | X19 | X20 | X21 | X22 |
|---|---|---|---|---|---|---|---|---|---|---|---|---|---|---|---|---|---|---|---|---|---|---|
| X1 | 12 | 65 | 140 | 133 | 197 | 368 | 162 | 78 | 386 | 289 | 440 | 148 | 226 | 245 | 332 | 270 | 93 | 56 | 121 | 72 | 29 | 2 |
| X2 | 23 | 152 | 397 | 408 | 540 | 1169 | 606 | 257 | 1241 | 961 | 1512 | 552 | 716 | 913 | 1129 | 989 | 308 | 212 | 425 | 284 | 95 | 12 |
| X3 | 28 | 171 | 442 | 446 | 605 | 1299 | 666 | 284 | 1387 | 1083 | 1704 | 628 | 814 | 1025 | 1295 | 1137 | 368 | 243 | 496 | 331 | 116 | 13 |
| X4 | 18 | 142 | 430 | 467 | 583 | 1391 | 790 | 309 | 1502 | 1195 | 1920 | 744 | 872 | 1211 | 1444 | 1323 | 405 | 292 | 558 | 403 | 129 | 21 |
| X5 | 20 | 131 | 354 | 361 | 494 | 1086 | 570 | 236 | 1174 | 933 | 1480 | 564 | 700 | 903 | 1144 | 1026 | 339 | 222 | 447 | 312 | 109 | 14 |
| X6 | 10 | 74 | 225 | 238 | 315 | 741 | 418 | 164 | 814 | 660 | 1064 | 424 | 488 | 668 | 828 | 767 | 250 | 170 | 332 | 241 | 82 | 12 |
| X7 | 5 | 21 | 52 | 46 | 74 | 151 | 72 | 32 | 167 | 137 | 216 | 84 | 108 | 127 | 181 | 161 | 62 | 34 | 76 | 50 | 21 | 1 |

| (X6,8),5 | X1 | X2 | X3 | X4 | X5 | X6 | X7 | X8 | X9 | X10 | X11 | X12 | X13 | X14 | X15 | X16 | X17 | X18 | X19 | X20 | X21 | X22 |
|---|---|---|---|---|---|---|---|---|---|---|---|---|---|---|---|---|---|---|---|---|---|---|
| X1 | 289 | 1953 | 5451 | 5683 | 7517 | 16990 | 9192 | 3730 | 18372 | 14585 | 23268 | 8915 | 10855 | 14365 | 17817 | 16105 | 5185 | 3509 | 6962 | 4896 | 1675 | 232 |
| X2 | 1015 | 6984 | 19713 | 20618 | 27380 | 62196 | 33816 | 13636 | 67616 | 53870 | 86229 | 33264 | 40236 | 53428 | 66516 | 60439 | 19628 | 13198 | 26280 | 18613 | 6436 | 905 |
| X3 | 1192 | 8259 | 23387 | 24479 | 32536 | 74031 | 40288 | 16216 | 80585 | 64269 | 102960 | 39783 | 48049 | 63864 | 79573 | 72407 | 23567 | 15809 | 31524 | 22375 | 7752 | 1092 |
| X4 | 1356 | 9436 | 26849 | 28154 | 37437 | 85514 | 46672 | 18718 | 93343 | 74574 | 119691 | 46422 | 55856 | 74375 | 92842 | 84713 | 27704 | 18518 | 36986 | 26348 | 9185 | 1305 |
| X5 | 1093 | 7621 | 21695 | 22749 | 30293 | 69107 | 37712 | 15126 | 75432 | 60267 | 96720 | 37507 | 45139 | 60101 | 75025 | 68449 | 22381 | 14963 | 29886 | 21288 | 7419 | 1054 |
| X6 | 815 | 5712 | 16337 | 17162 | 22892 | 52356 | 28656 | 11452 | 57304 | 45862 | 73749 | 28712 | 34412 | 45908 | 57412 | 52535 | 17260 | 11494 | 23000 | 16445 | 5764 | 825 |
| X7 | 189 | 1311 | 3750 | 3944 | 5264 | 12048 | 6608 | 2636 | 13205 | 10574 | 17019 | 6639 | 7941 | 10600 | 13269 | 12152 | 4003 | 2664 | 5328 | 3814 | 1344 | 194 |

| (X7,8),5 | X1 | X2 | X3 | X4 | X5 | X6 | X7 | X8 | X9 | X10 | X11 | X12 | X13 | X14 | X15 | X16 | X17 | X18 | X19 | X20 | X21 | X22 |
|---|---|---|---|---|---|---|---|---|---|---|---|---|---|---|---|---|---|---|---|---|---|---|
| X1 | 6 | 81 | 266 | 320 | 365 | 928 | 568 | 200 | 1002 | 787 | 1313 | 520 | 574 | 829 | 961 | 901 | 258 | 193 | 373 | 271 | 85 | 15 |
| X2 | 46 | 340 | 987 | 1055 | 1380 | 3199 | 1779 | 693 | 3481 | 2784 | 4505 | 1763 | 2075 | 2791 | 3463 | 3181 | 1035 | 689 | 1375 | 982 | 338 | 48 |
| X3 | 48 | 380 | 1125 | 1219 | 1572 | 3684 | 2063 | 801 | 4016 | 3202 | 5191 | 2022 | 2385 | 3240 | 3976 | 3657 | 1170 | 795 | 1576 | 1131 | 383 | 58 |
| X4 | 79 | 485 | 1327 | 1342 | 1864 | 4137 | 2195 | 904 | 4530 | 3646 | 5810 | 2262 | 2750 | 3579 | 4586 | 4168 | 1416 | 912 | 1844 | 1309 | 475 | 62 |
| X5 | 44 | 344 | 1012 | 1082 | 1410 | 3280 | 1820 | 716 | 3583 | 2859 | 4610 | 1790 | 2139 | 2888 | 3561 | 3267 | 1051 | 712 | 1422 | 1020 | 349 | 51 |
| X6 | 42 | 273 | 759 | 785 | 1061 | 2387 | 1291 | 528 | 2617 | 2092 | 3350 | 1303 | 1575 | 2085 | 2630 | 2401 | 800 | 531 | 1063 | 761 | 274 | 40 |
| X7 | 5 | 49 | 153 | 179 | 210 | 520 | 308 | 115 | 566 | 444 | 732 | 284 | 328 | 470 | 546 | 511 | 153 | 114 | 219 | 163 | 54 | 12 |

| (X8,8),5 | X1 | X2 | X3 | X4 | X5 | X6 | X7 | X8 | X9 | X10 | X11 | X12 | X13 | X14 | X15 | X16 | X17 | X18 | X19 | X20 | X21 | X22 |
|---|---|---|---|---|---|---|---|---|---|---|---|---|---|---|---|---|---|---|---|---|---|---|
| X1 | 4 | 1 | 13 | 2 | 3 | 15 | 11 | 16 | 14 | 23 | 10 | 9 | 3 | 22 | 10 | 10 | 1 | 13 | 1 | 9 | 1 | 3 |
| X2 | 3 | 3 | 22 | 8 | 11 | 41 | 28 | 23 | 40 | 51 | 43 | 27 | 14 | 50 | 37 | 37 | 6 | 21 | 9 | 19 | 2 | 3 |
| X3 | 3 | 4 | 22 | 10 | 13 | 43 | 26 | 21 | 42 | 50 | 45 | 25 | 17 | 49 | 40 | 38 | 7 | 20 | 10 | 20 | 2 | 3 |
| X4 | 0 | 3 | 11 | 11 | 14 | 40 | 26 | 8 | 44 | 38 | 59 | 27 | 24 | 39 | 46 | 43 | 13 | 9 | 16 | 13 | 4 | 0 |
| X5 | 1 | 3 | 13 | 9 | 12 | 32 | 16 | 11 | 32 | 31 | 39 | 16 | 18 | 31 | 31 | 30 | 9 | 9 | 12 | 12 | 3 | 1 |
| X6 | 0 | 2 | 3 | 7 | 10 | 17 | 10 | 1 | 22 | 12 | 30 | 10 | 15 | 12 | 23 | 20 | 8 | 2 | 11 | 4 | 3 | 0 |
| X7 | 0 | 2 | 1 | 2 | 3 | 3 | 1 | 0 | 4 | 2 | 4 | 1 | 3 | 2 | 4 | 3 | 1 | 1 | 2 | 1 | 1 | 0 |

| (X9,8),5 | I | X1 | X2 | X3 | X4 | X5 | X6 | X7 | X8 | X9 | X10 | X11 | X12 | X13 | X14 | X15 | X16 | X17 | X18 | X19 | X20 | X21 | X22 |
|---|---|---|---|---|---|---|---|---|---|---|---|---|---|---|---|---|---|---|---|---|---|---|---|
| | I | | | | | | | | | | | | | | | | | | | | | | |
| X1 | I | 398 | 2855 | 8069 | 8532 | 11295 | 25718 | 14031 | 5623 | 28111 | 22324 | 35934 | 13857 | 16809 | 22361 | 27863 | 25423 | 8290 | 5546 | 11162 | 7945 | 2798 | 409 |
| X2 | I | 1500 | 10485 | 29910 | 31393 | 41832 | 95538 | 52210 | 20904 | 104413 | 83479 | 134109 | 52103 | 62580 | 83402 | 104195 | 95205 | 31204 | 20817 | 41623 | 29706 | 10385 | 1482 |
| X3 | I | 1787 | 12562 | 35813 | 37658 | 50132 | 114482 | 62558 | 25046 | 125186 | 99992 | 160731 | 62368 | 75058 | 100008 | 124915 | 114128 | 37379 | 24959 | 49965 | 35644 | 12477 | 1791 |
| X4 | I | 2096 | 14528 | 41614 | 43566 | 58222 | 133171 | 72852 | 29142 | 145609 | 116731 | 187425 | 73056 | 87387 | 116567 | 145808 | 133331 | 43813 | 29181 | 58219 | 41615 | 14531 | 2056 |
| X5 | I | 1677 | 11826 | 33748 | 35493 | 47275 | 108004 | 59045 | 23622 | 118166 | 94407 | 151797 | 58943 | 70884 | 94483 | 118048 | 107908 | 35370 | 23593 | 47260 | 33741 | 11825 | 1697 |
| X6 | I | 1286 | 9012 | 25777 | 27071 | 36112 | 82576 | 45182 | 18063 | 90359 | 72314 | 116250 | 45242 | 54246 | 72352 | 90472 | 82755 | 27177 | 18107 | 36220 | 25887 | 9069 | 1297 |
| X7 | I | 295 | 2119 | 6004 | 6367 | 8422 | 19224 | 10509 | 4199 | 21053 | 16739 | 27000 | 10432 | 12635 | 16829 | 20996 | 19203 | 6281 | 4196 | 8457 | 6033 | 2130 | 315 |

| (X10,8),5 | I | X1 | X2 | X3 | X4 | X5 | X6 | X7 | X8 | X9 | X10 | X11 | X12 | X13 | X14 | X15 | X16 | X17 | X18 | X19 | X20 | X21 | X22 |
|---|---|---|---|---|---|---|---|---|---|---|---|---|---|---|---|---|---|---|---|---|---|---|---|
| | I | | | | | | | | | | | | | | | | | | | | | | |
| X1 | I | 174 | 962 | 2829 | 2750 | 3784 | 8692 | 4735 | 1996 | 9421 | 7799 | 12096 | 4845 | 5593 | 7623 | 9476 | 8648 | 2849 | 1985 | 3674 | 2719 | 907 | 130 |
| X2 | I | 520 | 3472 | 10014 | 10372 | 13864 | 31794 | 17416 | 7048 | 34726 | 27959 | 44608 | 17450 | 20764 | 27906 | 34742 | 31782 | 10402 | 7044 | 13830 | 9980 | 3454 | 506 |
| X3 | I | 626 | 4130 | 11905 | 12271 | 16480 | 37733 | 20621 | 8360 | 41189 | 33227 | 52928 | 20725 | 24647 | 33059 | 41242 | 37690 | 12365 | 8350 | 16376 | 11801 | 4079 | 585 |
| X4 | I | 636 | 4743 | 13503 | 14432 | 19054 | 43604 | 23914 | 9465 | 47798 | 37951 | 61440 | 23743 | 28712 | 38226 | 47712 | 43673 | 14278 | 9482 | 19225 | 13674 | 4828 | 704 |
| X5 | I | 568 | 3851 | 11028 | 11469 | 15376 | 35125 | 19179 | 7717 | 38377 | 30827 | 49344 | 19252 | 23013 | 30709 | 38414 | 35095 | 11535 | 7710 | 15303 | 10955 | 3815 | 539 |
| X6 | I | 400 | 2909 | 8257 | 8764 | 11654 | 26580 | 14526 | 5759 | 29102 | 23165 | 37440 | 14497 | 17500 | 23210 | 29088 | 26591 | 8738 | 5762 | 11683 | 8286 | 2924 | 412 |
| X7 | I | 112 | 680 | 1939 | 1955 | 2676 | 6077 | 3293 | 1344 | 6609 | 5376 | 8512 | 3359 | 3963 | 5269 | 6642 | 6050 | 2015 | 1338 | 2610 | 1873 | 647 | 85 |

| (X11,8),5 | I | X1 | X2 | X3 | X4 | X5 | X6 | X7 | X8 | X9 | X10 | X11 | X12 | X13 | X14 | X15 | X16 | X17 | X18 | X19 | X20 | X21 | X22 |
|---|---|---|---|---|---|---|---|---|---|---|---|---|---|---|---|---|---|---|---|---|---|---|---|
| | I | | | | | | | | | | | | | | | | | | | | | | |
| X1 | I | 1304 | 9428 | 26977 | 28609 | 38034 | 87128 | 47790 | 18853 | 95252 | 76071 | 122910 | 47790 | 57218 | 76071 | 95252 | 87128 | 28609 | 18853 | 38034 | 26977 | 9428 | 1304 |
| X2 | I | 5128 | 36045 | 103010 | 108282 | 144330 | 329977 | 180519 | 72093 | 360894 | 288657 | 464190 | 180519 | 216564 | 288657 | 360894 | 329977 | 108282 | 72093 | 144330 | 103010 | 36045 | 5128 |
| X3 | I | 6171 | 43554 | 124491 | 130998 | 174552 | 399165 | 218442 | 87108 | 436548 | 349104 | 561690 | 218442 | 261996 | 349104 | 436548 | 399165 | 130998 | 87108 | 174552 | 124491 | 43554 | 6171 |
| X4 | I | 7409 | 51380 | 146728 | 153649 | 205026 | 468353 | 255918 | 102757 | 512324 | 410055 | 658068 | 255918 | 307298 | 410055 | 512324 | 468353 | 153649 | 102757 | 205026 | 146728 | 51380 | 7409 |
| X5 | I | 5934 | 41700 | 119166 | 125292 | 166992 | 381810 | 208884 | 83400 | 417576 | 333984 | 537156 | 208884 | 250584 | 333984 | 417576 | 381810 | 125292 | 83400 | 166992 | 119166 | 41700 | 5934 |
| X6 | I | 4636 | 32331 | 92360 | 96876 | 129210 | 295267 | 161421 | 64665 | 322962 | 258417 | 415086 | 161421 | 193752 | 258417 | 322962 | 295267 | 96876 | 64665 | 129210 | 92360 | 32331 | 4636 |
| X7 | I | 1054 | 7574 | 21665 | 22903 | 30474 | 69760 | 38232 | 15145 | 76280 | 60951 | 98299 | 38232 | 45806 | 60951 | 76280 | 69760 | 22903 | 15145 | 30474 | 21665 | 7574 | 1054 |

| (X12,8),5 | I | X1 | X2 | X3 | X4 | X5 | X6 | X7 | X8 | X9 | X10 | X11 | X12 | X13 | X14 | X15 | X16 | X17 | X18 | X19 | X20 | X21 | X22 |
|---|---|---|---|---|---|---|---|---|---|---|---|---|---|---|---|---|---|---|---|---|---|---|---|
| | I | | | | | | | | | | | | | | | | | | | | | | |
| X1 | I | 15 | 85 | 271 | 268 | 373 | 901 | 520 | 193 | 961 | 829 | 1313 | 568 | 574 | 787 | 1002 | 928 | 320 | 200 | 365 | 266 | 81 | 6 |
| X2 | I | 48 | 338 | 982 | 1035 | 1375 | 3181 | 1753 | 689 | 3463 | 2791 | 4505 | 1779 | 2075 | 2784 | 3481 | 3199 | 1055 | 693 | 1380 | 987 | 340 | 46 |
| X3 | I | 58 | 383 | 1131 | 1170 | 1576 | 3657 | 2022 | 795 | 3976 | 3240 | 5191 | 2063 | 2385 | 3202 | 4016 | 3684 | 1219 | 801 | 1572 | 1125 | 360 | 48 |
| X4 | I | 62 | 475 | 1309 | 1416 | 1844 | 4168 | 2262 | 912 | 4586 | 3579 | 5810 | 2195 | 2750 | 3646 | 4530 | 4137 | 1342 | 904 | 1864 | 1327 | 485 | 79 |
| X5 | I | 51 | 349 | 1020 | 1051 | 1422 | 3267 | 1790 | 712 | 3561 | 2888 | 4610 | 1820 | 2139 | 2859 | 3583 | 3280 | 1082 | 716 | 1410 | 1012 | 344 | 44 |
| X6 | I | 40 | 274 | 761 | 800 | 1063 | 2401 | 1303 | 531 | 2630 | 2085 | 3350 | 1291 | 1575 | 2092 | 2617 | 2387 | 785 | 528 | 1061 | 759 | 273 | 42 |
| X7 | I | 12 | 54 | 163 | 153 | 219 | 511 | 284 | 114 | 546 | 470 | 732 | 308 | 328 | 444 | 566 | 520 | 179 | 115 | 210 | 153 | 49 | 5 |

| (X13,8),5 | I | X1 | X2 | X3 | X4 | X5 | X6 | X7 | X8 | X9 | X10 | X11 | X12 | X13 | X14 | X15 | X16 | X17 | X18 | X19 | X20 | X21 | X22 |
|---|---|---|---|---|---|---|---|---|---|---|---|---|---|---|---|---|---|---|---|---|---|---|---|
| | I | | | | | | | | | | | | | | | | | | | | | | |
| X1 | I | 31 | 253 | 670 | 741 | 986 | 2178 | 1163 | 446 | 2392 | 1854 | 3078 | 1163 | 1471 | 1854 | 2392 | 2178 | 741 | 446 | 986 | 670 | 253 | 31 |
| X2 | I | 117 | 856 | 2410 | 2564 | 3414 | 7760 | 4231 | 1669 | 8491 | 6748 | 10934 | 4231 | 5114 | 6748 | 8491 | 7760 | 2564 | 1669 | 3414 | 2410 | 856 | 117 |
| X3 | I | 137 | 999 | 2806 | 2979 | 3992 | 9044 | 4911 | 1948 | 9924 | 7876 | 12750 | 4911 | 5987 | 7876 | 9924 | 9044 | 2979 | 1948 | 3992 | 2806 | 999 | 137 |
| X4 | I | 161 | 1109 | 3232 | 3367 | 4481 | 10339 | 5692 | 2289 | 11298 | 9101 | 14536 | 5692 | 6747 | 9101 | 11298 | 10339 | 3367 | 2289 | 4481 | 3232 | 1109 | 161 |
| X5 | I | 127 | 899 | 2562 | 2699 | 3616 | 8220 | 4473 | 1794 | 9014 | 7188 | 11570 | 4473 | 5434 | 7188 | 9014 | 8220 | 2699 | 1794 | 3616 | 2562 | 899 | 127 |
| X6 | I | 98 | 661 | 1918 | 1991 | 2661 | 6109 | 3349 | 1356 | 6681 | 5375 | 8578 | 3349 | 3997 | 5375 | 6681 | 6109 | 1991 | 1356 | 2661 | 1918 | 661 | 98 |
| X7 | I | 24 | 158 | 426 | 448 | 605 | 1351 | 725 | 289 | 1482 | 1169 | 1898 | 725 | 897 | 1169 | 1482 | 1351 | 448 | 289 | 605 | 426 | 158 | 24 |

| (X14,8),5 | I | X1 | X2 | X3 | X4 | X5 | X6 | X7 | X8 | X9 | X10 | X11 | X12 | X13 | X14 | X15 | X16 | X17 | X18 | X19 | X20 | X21 | X22 |
|---|---|---|---|---|---|---|---|---|---|---|---|---|---|---|---|---|---|---|---|---|---|---|---|
| | I | | | | | | | | | | | | | | | | | | | | | | |
| X1 | I | 130 | 907 | 2719 | 2849 | 3674 | 8648 | 4845 | 1985 | 9476 | 7623 | 12096 | 4735 | 5593 | 7793 | 9421 | 8692 | 2750 | 1996 | 3784 | 2829 | 962 | 174 |
| X2 | I | 506 | 3454 | 9980 | 10402 | 13830 | 31782 | 17450 | 7344 | 34742 | 27906 | 44608 | 17416 | 20764 | 27959 | 34726 | 31794 | 10372 | 7048 | 13864 | 10014 | 3472 | 520 |
| X3 | I | 585 | 4079 | 11801 | 12365 | 16376 | 37690 | 20725 | 8350 | 41242 | 33059 | 52928 | 20621 | 24647 | 33227 | 41189 | 37733 | 12271 | 8360 | 16480 | 11905 | 4130 | 626 |
| X4 | I | 704 | 4828 | 13674 | 14278 | 19225 | 43673 | 23743 | 9482 | 47712 | 38226 | 61440 | 23914 | 28712 | 37951 | 47798 | 43604 | 14432 | 9465 | 19054 | 13503 | 4743 | 636 |
| X5 | I | 539 | 3815 | 10955 | 11535 | 15303 | 35095 | 19252 | 7710 | 38414 | 30709 | 49344 | 19179 | 23013 | 30827 | 38377 | 35125 | 11469 | 7717 | 15376 | 11028 | 3851 | 568 |
| X6 | I | 412 | 2924 | 8286 | 8738 | 11683 | 26591 | 14497 | 5762 | 29088 | 23210 | 37440 | 14526 | 17500 | 23165 | 29102 | 26580 | 8764 | 5759 | 11654 | 8257 | 2909 | 400 |
| X7 | I | 85 | 647 | 1873 | 2015 | 2610 | 6050 | 3359 | 1338 | 6642 | 5269 | 8512 | 3293 | 3963 | 5376 | 6609 | 6077 | 1955 | 1344 | 2676 | 1939 | 680 | 112 |

| (X15,8),5 | I | X1 | X2 | X3 | X4 | X5 | X6 | X7 | X8 | X9 | X10 | X11 | X12 | X13 | X14 | X15 | X16 | X17 | X18 | X19 | X20 | X21 | X22 |
|---|---|---|---|---|---|---|---|---|---|---|---|---|---|---|---|---|---|---|---|---|---|---|---|
| | I | | | | | | | | | | | | | | | | | | | | | | |
| X1 | I | 409 | 2798 | 7945 | 8290 | 11162 | 25423 | 13857 | 5546 | 27863 | 22361 | 35934 | 14031 | 16809 | 22324 | 28111 | 25718 | 8532 | 5623 | 11295 | 8069 | 2855 | 398 |
| X2 | I | 1482 | 10385 | 29706 | 31204 | 41623 | 95205 | 52103 | 20817 | 104195 | 83402 | 134109 | 52210 | 62580 | 83479 | 104413 | 95538 | 31393 | 20904 | 41832 | 29910 | 10485 | 1500 |
| X3 | I | 1791 | 12477 | 35644 | 37379 | 49965 | 114128 | 62368 | 24959 | 124915 | 100008 | 160731 | 62558 | 75058 | 99992 | 125186 | 114482 | 37658 | 25046 | 50132 | 35813 | 12562 | 1787 |
| X4 | I | 2056 | 14531 | 41615 | 43813 | 58219 | 133331 | 73056 | 29181 | 145808 | 116567 | 187425 | 72852 | 87387 | 116731 | 145609 | 133171 | 43566 | 29142 | 58222 | 41614 | 14528 | 2096 |
| X5 | I | 1697 | 11825 | 33741 | 35370 | 47260 | 107908 | 58943 | 23593 | 118048 | 94483 | 151797 | 59045 | 70884 | 94407 | 118166 | 108004 | 35493 | 23622 | 47275 | 33748 | 11826 | 1677 |
| X6 | I | 1297 | 9069 | 25887 | 27177 | 36220 | 82755 | 45242 | 18107 | 90472 | 72352 | 116250 | 45182 | 54246 | 72314 | 90359 | 82576 | 27071 | 18063 | 36112 | 25777 | 9012 | 1286 |
| X7 | I | 315 | 2130 | 6033 | 6281 | 8457 | 19203 | 10432 | 4196 | 20996 | 16829 | 27000 | 10509 | 12635 | 16739 | 21053 | 19224 | 6367 | 4199 | 8422 | 6004 | 2119 | 295 |

| (X16,8),5 | I | X1 | X2 | X3 | X4 | X5 | X6 | X7 | X8 | X9 | X10 | X11 | X12 | X13 | X14 | X15 | X16 | X17 | X18 | X19 | X20 | X21 | X22 |
|---|---|---|---|---|---|---|---|---|---|---|---|---|---|---|---|---|---|---|---|---|---|---|---|
| | I | | | | | | | | | | | | | | | | | | | | | | |
| X1 | I | 232 | 1675 | 4896 | 5185 | 6962 | 16105 | 8915 | 3509 | 17817 | 14365 | 23268 | 9192 | 10855 | 14585 | 18372 | 16990 | 5683 | 3730 | 7517 | 5451 | 1953 | 289 |
| X2 | I | 905 | 6436 | 18613 | 19628 | 26280 | 60439 | 33264 | 13198 | 66516 | 53428 | 86229 | 33816 | 40236 | 53870 | 67616 | 62196 | 20618 | 13636 | 27380 | 19713 | 6984 | 1015 |
| X3 | I | 1092 | 7752 | 22375 | 23567 | 31524 | 72407 | 39783 | 15809 | 79573 | 63864 | 102960 | 40288 | 48049 | 64269 | 80585 | 74031 | 24479 | 16216 | 32536 | 23387 | 8259 | 1192 |
| X4 | I | 1305 | 9185 | 26348 | 27704 | 36986 | 84713 | 46422 | 18518 | 92842 | 74375 | 119691 | 46672 | 55856 | 74574 | 93343 | 85514 | 28154 | 18718 | 37487 | 26849 | 9436 | 1356 |
| X5 | I | 1054 | 7419 | 21288 | 22381 | 29886 | 68449 | 37507 | 14963 | 75025 | 60101 | 96720 | 37712 | 45139 | 60267 | 75432 | 69107 | 22749 | 15126 | 30293 | 21695 | 7621 | 1093 |
| X6 | I | 825 | 5764 | 16445 | 17260 | 23000 | 52535 | 28712 | 11494 | 57412 | 45908 | 73749 | 28656 | 34412 | 45862 | 57304 | 52356 | 17162 | 11452 | 22892 | 16337 | 5712 | 815 |
| X7 | I | 194 | 1344 | 3814 | 4003 | 5328 | 12152 | 6639 | 2664 | 13269 | 10600 | 17019 | 6608 | 7941 | 10574 | 13205 | 12048 | 3944 | 2636 | 5264 | 3750 | 1311 | 189 |

| (X17,8),5 | I | X1 | X2 | X3 | X4 | X5 | X6 | X7 | X8 | X9 | X10 | X11 | X12 | X13 | X14 | X15 | X16 | X17 | X18 | X19 | X20 | X21 | X22 |
|---|---|---|---|---|---|---|---|---|---|---|---|---|---|---|---|---|---|---|---|---|---|---|---|
| | I | | | | | | | | | | | | | | | | | | | | | | |
| X1 | I | 2 | 7 | 17 | 14 | 29 | 61 | 31 | 9 | 68 | 69 | 120 | 63 | 54 | 61 | 111 | 115 | 62 | 21 | 53 | 42 | 22 | 1 |
| X2 | I | 3 | 21 | 61 | 67 | 91 | 225 | 135 | 44 | 255 | 218 | 381 | 174 | 168 | 237 | 321 | 328 | 132 | 69 | 150 | 121 | 51 | 7 |
| X3 | I | 4 | 26 | 71 | 73 | 109 | 254 | 142 | 49 | 287 | 248 | 417 | 183 | 190 | 252 | 354 | 344 | 137 | 73 | 157 | 120 | 50 | 6 |
| X4 | I | 2 | 26 | 86 | 104 | 116 | 302 | 189 | 67 | 336 | 265 | 449 | 183 | 200 | 302 | 349 | 346 | 109 | 77 | 156 | 127 | 47 | 12 |
| X5 | I | 4 | 24 | 69 | 70 | 101 | 227 | 122 | 49 | 254 | 210 | 343 | 142 | 161 | 209 | 279 | 263 | 99 | 56 | 117 | 85 | 33 | 3 |
| X6 | I | 3 | 20 | 57 | 61 | 75 | 175 | 99 | 38 | 184 | 144 | 234 | 91 | 106 | 147 | 175 | 164 | 52 | 36 | 72 | 55 | 20 | 4 |
| X7 | I | 2 | 9 | 15 | 12 | 20 | 35 | 14 | 6 | 35 | 30 | 46 | 19 | 21 | 21 | 36 | 31 | 14 | 5 | 12 | 7 | 3 | 0 |

| (X18,8),5 | I | X1 | X2 | X3 | X4 | X5 | X6 | X7 | X8 | X9 | X10 | X11 | X12 | X13 | X14 | X15 | X16 | X17 | X18 | X19 | X20 | X21 | X22 |
|---|---|---|---|---|---|---|---|---|---|---|---|---|---|---|---|---|---|---|---|---|---|---|---|
| | I | | | | | | | | | | | | | | | | | | | | | | |
| X1 | I | 3 | 1 | 9 | 1 | 1 | 10 | 9 | 13 | 10 | 22 | 10 | 11 | 3 | 23 | 14 | 15 | 2 | 16 | 3 | 13 | 1 | 4 |
| X2 | I | 3 | 2 | 19 | 6 | 9 | 37 | 27 | 21 | 37 | 50 | 43 | 28 | 14 | 51 | 40 | 41 | 8 | 23 | 11 | 22 | 3 | 3 |
| X3 | I | 3 | 2 | 20 | 7 | 10 | 38 | 25 | 20 | 40 | 49 | 45 | 26 | 17 | 50 | 42 | 43 | 10 | 21 | 13 | 22 | 4 | 3 |
| X4 | I | 0 | 4 | 13 | 13 | 16 | 43 | 27 | 9 | 46 | 39 | 59 | 26 | 24 | 38 | 44 | 40 | 11 | 8 | 14 | 11 | 3 | 0 |
| X5 | I | 1 | 3 | 12 | 9 | 12 | 30 | 16 | 9 | 31 | 31 | 39 | 16 | 18 | 31 | 32 | 32 | 9 | 11 | 12 | 13 | 3 | 1 |
| X6 | I | 0 | 3 | 4 | 8 | 11 | 20 | 10 | 2 | 23 | 12 | 30 | 10 | 15 | 12 | 22 | 17 | 7 | 1 | 10 | 3 | 2 | 0 |
| X7 | I | 0 | 1 | 1 | 1 | 2 | 3 | 1 | 1 | 4 | 2 | 4 | 1 | 3 | 2 | 4 | 3 | 2 | 0 | 3 | 1 | 2 | 0 |

| (X19,8),5 | I | X1 | X2 | X3 | X4 | X5 | X6 | X7 | X8 | X9 | X10 | X11 | X12 | X13 | X14 | X15 | X16 | X17 | X18 | X19 | X20 | X21 | X22 |
|---|---|---|---|---|---|---|---|---|---|---|---|---|---|---|---|---|---|---|---|---|---|---|---|
| | I | | | | | | | | | | | | | | | | | | | | | | |
| X1 | I | 2 | 29 | 72 | 93 | 121 | 270 | 148 | 56 | 332 | 245 | 440 | 162 | 226 | 289 | 386 | 368 | 133 | 78 | 197 | 140 | 65 | 12 |
| X2 | I | 12 | 95 | 284 | 308 | 425 | 989 | 552 | 212 | 1129 | 913 | 1512 | 606 | 716 | 961 | 1241 | 1169 | 408 | 257 | 540 | 397 | 152 | 23 |
| X3 | I | 13 | 116 | 331 | 368 | 496 | 1137 | 628 | 243 | 1295 | 1025 | 1704 | 666 | 814 | 1083 | 1387 | 1299 | 446 | 284 | 605 | 442 | 171 | 28 |
| X4 | I | 21 | 129 | 403 | 405 | 558 | 1323 | 744 | 292 | 1444 | 1211 | 1920 | 790 | 872 | 1195 | 1502 | 1391 | 467 | 309 | 583 | 430 | 142 | 18 |
| X5 | I | 14 | 109 | 312 | 339 | 447 | 1026 | 564 | 222 | 1144 | 903 | 1480 | 570 | 700 | 933 | 1174 | 1086 | 361 | 236 | 494 | 354 | 131 | 20 |
| X6 | I | 12 | 82 | 241 | 250 | 332 | 767 | 424 | 170 | 828 | 668 | 1064 | 418 | 488 | 660 | 814 | 741 | 238 | 164 | 315 | 225 | 74 | 10 |
| X7 | I | 1 | 21 | 50 | 62 | 76 | 161 | 84 | 34 | 181 | 127 | 216 | 72 | 108 | 137 | 167 | 151 | 46 | 32 | 74 | 52 | 21 | 5 |

| (X20,8),5 | I | X1 | X2 | X3 | X4 | X5 | X6 | X7 | X8 | X9 | X10 | X11 | X12 | X13 | X14 | X15 | X16 | X17 | X18 | X19 | X20 | X21 | X22 |
|---|---|---|---|---|---|---|---|---|---|---|---|---|---|---|---|---|---|---|---|---|---|---|---|
| | I | | | | | | | | | | | | | | | | | | | | | | |
| X1 | I | 0 | 1 | 8 | 8 | 10 | 34 | 26 | 13 | 53 | 51 | 78 | 41 | 40 | 78 | 92 | 105 | 40 | 33 | 58 | 63 | 28 | 10 |
| X2 | I | 1 | 6 | 30 | 32 | 50 | 141 | 93 | 34 | 185 | 168 | 289 | 140 | 134 | 209 | 278 | 292 | 116 | 72 | 144 | 126 | 53 | 11 |
| X3 | I | 1 | 8 | 40 | 42 | 60 | 166 | 105 | 41 | 213 | 188 | 311 | 140 | 149 | 230 | 294 | 302 | 112 | 76 | 151 | 130 | 54 | 12 |
| X4 | I | 1 | 13 | 46 | 55 | 82 | 207 | 125 | 35 | 239 | 208 | 369 | 165 | 164 | 214 | 303 | 297 | 117 | 56 | 130 | 92 | 36 | 2 |
| X5 | I | 1 | 12 | 41 | 47 | 64 | 158 | 94 | 37 | 191 | 157 | 261 | 109 | 124 | 181 | 226 | 222 | 77 | 53 | 105 | 86 | 33 | 7 |
| X6 | I | 1 | 12 | 36 | 42 | 54 | 129 | 74 | 23 | 137 | 109 | 189 | 76 | 84 | 108 | 141 | 132 | 46 | 25 | 57 | 38 | 14 | 1 |
| X7 | I | 1 | 5 | 12 | 13 | 11 | 27 | 16 | 9 | 30 | 20 | 31 | 10 | 15 | 25 | 24 | 22 | 5 | 6 | 12 | 10 | 4 | 2 |

(X21,8),5

| | I | X1 | X2 | X3 | X4 | X5 | X6 | X7 | X8 | X9 | X10 | X11 | X12 | X13 | X14 | X15 | X16 | X17 | X18 | X19 | X20 | X21 | X22 |
|---|---|---|---|---|---|---|---|---|---|---|---|---|---|---|---|---|---|---|---|---|---|---|---|
| | I | | | | | | | | | | | | | | | | | | | | | | |
| X1 | I | 0 | 0 | 0 | 0 | 0 | 0 | 0 | 0 | 0 | 0 | 0 | 0 | 0 | 0 | 1 | 2 | 3 | 0 | 3 | 4 | 5 | 2 |
| X2 | I | 0 | 0 | 0 | 0 | 0 | 0 | 0 | 0 | 0 | 0 | 1 | 2 | 0 | 1 | 2 | 5 | 5 | 1 | 3 | 5 | 4 | 1 |
| X3 | I | 0 | 0 | 0 | 0 | 0 | 0 | 0 | 0 | 0 | 0 | 1 | 1 | 1 | 1 | 2 | 4 | 3 | 1 | 3 | 4 | 3 | 1 |
| X4 | I | 0 | 0 | 0 | 0 | 0 | 0 | 1 | 0 | 0 | 1 | 2 | 3 | 0 | 1 | 1 | 3 | 2 | 0 | 0 | 1 | 0 | 0 |
| X5 | I | 0 | 0 | 0 | 0 | 0 | 0 | 0 | 0 | 1 | 0 | 1 | 0 | 0 | 1 | 1 | 2 | 1 | 0 | 1 | 1 | 1 | 0 |
| X6 | I | 0 | 0 | 0 | 0 | 0 | 1 | 1 | 0 | 0 | 0 | 1 | 1 | 0 | 0 | 0 | 0 | 0 | 0 | 0 | 0 | 0 | 0 |
| X7 | I | 0 | 0 | 0 | 1 | 0 | 0 | 0 | 0 | 0 | 0 | 0 | 0 | 0 | 0 | 0 | 0 | 0 | 0 | 0 | 0 | 0 | 0 |

(X22,8),5

| | I | X1 | X2 | X3 | X4 | X5 | X6 | X7 | X8 | X9 | X10 | X11 | X12 | X13 | X14 | X15 | X16 | X17 | X18 | X19 | X20 | X21 | X22 |
|---|---|---|---|---|---|---|---|---|---|---|---|---|---|---|---|---|---|---|---|---|---|---|---|
| | I | | | | | | | | | | | | | | | | | | | | | | |
| X1 | I | 0 | 0 | 0 | 0 | 0 | 0 | 0 | 0 | 0 | 0 | 0 | 0 | 0 | 0 | 0 | 0 | 0 | 0 | 0 | 0 | 0 | 1 |
| X2 | I | 0 | 0 | 0 | 0 | 0 | 0 | 0 | 0 | 0 | 0 | 0 | 0 | 0 | 0 | 0 | 0 | 0 | 0 | 0 | 0 | 0 | 0 |
| X3 | I | 0 | 0 | 0 | 0 | 0 | 0 | 0 | 0 | 0 | 0 | 0 | 0 | 0 | 0 | 0 | 0 | 0 | 0 | 0 | 0 | 0 | 0 |
| X4 | I | 0 | 0 | 0 | 0 | 0 | 0 | 0 | 0 | 0 | 0 | 0 | 0 | 0 | 0 | 0 | 0 | 0 | 0 | 0 | 0 | 0 | 0 |
| X5 | I | 0 | 0 | 0 | 0 | 0 | 0 | 0 | 0 | 0 | 0 | 0 | 0 | 0 | 0 | 0 | 0 | 0 | 0 | 0 | 0 | 0 | 0 |
| X6 | I | 0 | 0 | 0 | 0 | 0 | 0 | 0 | 0 | 0 | 0 | 0 | 0 | 0 | 0 | 0 | 0 | 0 | 0 | 0 | 0 | 0 | 0 |
| X7 | I | 0 | 0 | 0 | 0 | 0 | 0 | 0 | 0 | 0 | 0 | 0 | 0 | 0 | 0 | 0 | 0 | 0 | 0 | 0 | 0 | 0 | 0 |

## I.E Decomposition Numbers

We have deliberately presented these $p$-modular decomposition matrices of $S_n$, $n \leqslant 13$, $p = 2, 3$, without sorting the characters into blocks. This makes it easier to spot patterns which might hold in general. For $n \geqslant 8$, we have inserted numbers at the bottom of each table which indicate the block to which the characters belong. Self-associated characters are marked with an asterisk.

## The decomposition matrices of $S_n$ for the prime 2

**n = 0**

| | | (0) 1 |
|---|---|---|
| *1 | (0) | 1 |

**n = 1**

| | | (1) 1 |
|---|---|---|
| *1 | (1) | 1 |

**n = 2**

| | | (2) 1 |
|---|---|---|
| 1 | (2) | 1 |
| 1 | $(1^2)$ | 1 |

**n = 3**

| | | (3) 1 | (2,1) 2 |
|---|---|---|---|
| 1 | (3) | 1 | |
| *2 | (2,1) | | 1 |
| 1 | $(1^3)$ | 1 | |

**n = 4**

| | | (4) 1 | (3,1) 2 |
|---|---|---|---|
| 1 | (4) | 1 | |
| 3 | (3,1) | 1 | 1 |
| *2 | $(2^2)$ | | 1 |
| 3 | $(21^2)$ | 1 | 1 |
| 1 | $(1^4)$ | 1 | |

**n = 5**

| | | (5) 1 | (4,1) 4 | (3,2) 4 |
|---|---|---|---|---|
| 1 | (5) | 1 | | |
| 4 | (4,1) | | 1 | |
| 5 | (3,2) | 1 | | 1 |
| *6 | $(31^2)$ | 2 | | 1 |
| 5 | $(2^21)$ | 1 | | 1 |
| 4 | $(21^3)$ | | 1 | |
| 1 | $(1^5)$ | 1 | | |

**n = 6**

| | | (6) 1 | (5,1) 4 | (4,2) 4 | (321) 16 |
|---|---|---|---|---|---|
| 1 | (6) | 1 | | | |
| 5 | (5,1) | 1 | 1 | | |
| 9 | (4,2) | 1 | 1 | 1 | |
| *16 | (321) | | | | 1 |
| 10 | $(41^2)$ | 2 | 1 | 1 | |
| 5 | $(3^2)$ | 1 | | 1 | |
| 10 | $(31^3)$ | 2 | 1 | 1 | |
| 5 | $(2^3)$ | 1 | | 1 | |
| 9 | $(2^21^2)$ | 1 | 1 | 1 | |
| 5 | $(21^4)$ | 1 | 1 | | |
| 1 | $(1^6)$ | 1 | | | |

**n = 7**

| | | (7) 1 | (6,1) 6 | (5,2) 14 | (4,3) 8 | (421) 20 |
|---|---|---|---|---|---|---|
| 1 | (7) | 1 | | | | |
| 6 | (6,1) | | 1 | | | |
| 14 | (5,2) | | | 1 | | |
| 14 | (4,3) | | 1 | | 1 | |
| 35 | (421) | 1 | | 1 | | 1 |
| 15 | $(51^2)$ | 1 | | 1 | | |
| 21 | $(3^21)$ | 1 | | | | 1 |
| 21 | $(32^2)$ | 1 | | | | 1 |
| *20 | $(41^3)$ | | 2 | | 1 | |
| 35 | $(321^2)$ | 1 | | 1 | | 1 |
| 14 | $(2^31)$ | | 1 | | 1 | |
| 15 | $(31^4)$ | 1 | | 1 | | |
| 14 | $(2^21^3)$ | | | 1 | | |
| 6 | $(21^5)$ | | 1 | | | |
| 1 | $(1^7)$ | 1 | | | | |

The decomposition matrix of $S_8$ for the prime 2

| | | | (8) 1 | (7,1) 6 | (6,2) 14 | (5,3) 8 | (521) 64 | (431) 40 |
|---|---|---|---|---|---|---|---|---|
| 1 | (8) | $(1^8)$ | 1 | | | | | |
| 7 | (7,1) | $(21^6)$ | 1 | 1 | | | | |
| 20 | (6,2) | $(2^21^4)$ | | 1 | 1 | | | |
| 28 | (5,3) | $(2^31^2)$ | | 1 | 1 | 1 | | |
| 64 | (521) | $(321^3)$ | | | | | 1 | |
| 70 | (431) | $(32^21)$ | 2 | 1 | 1 | 1 | | 1 |
| 14 | $(4^2)$ | $(2^4)$ | | 1 | | 1 | | |
| 21 | $(61^2)$ | $(31^5)$ | 1 | 1 | 1 | | | |
| 56 | $(42^2)$ | $(3^21^2)$ | 2 | | 1 | | | 1 |
| 42 | *$(3^22)$ | | 2 | | | | | 1 |
| 35 | $(51^3)$ | $(41^4)$ | 1 | 2 | 1 | 1 | | |
| 90 | *$(421^2)$ | | 2 | 2 | 2 | 1 | | 1 |
| | Block number: | | 1 | 1 | 1 | 1 | 2 | 1 |

The decomposition matrix of $S_9$ for the prime 2

| | | | (9) 1 | (8,1) 8 | (7,2) 26 | (6,3) 48 | (5,4) 16 | (621) 78 | (531) 40 | (432) 160 |
|---|---|---|---|---|---|---|---|---|---|---|
| 1 | (9) | $(1^9)$ | 1 | | | | | | | |
| 8 | (8,1) | $(21^7)$ | | 1 | | | | | | |
| 27 | (7,2) | $(2^21^5)$ | 1 | | 1 | | | | | |
| 48 | (6,3) | $(2^31^3)$ | | | | 1 | | | | |
| 42 | (5,4) | $(2^41)$ | | | 1 | | 1 | | | |
| 105 | (621) | $(321^4)$ | 1 | | 1 | | | 1 | | |
| 162 | (531) | $(32^21^2)$ | 2 | | 1 | | 1 | 1 | 1 | |
| 168 | (432) | $(3^221)$ | | 1 | | | | | | 1 |
| 28 | $(71^2)$ | $(31^6)$ | 2 | | 1 | | | | | |
| 84 | $(4^21)$ | $(32^3)$ | 2 | | 1 | | 1 | | 1 | |
| 120 | $(52^2)$ | $(3^21^3)$ | 2 | | | | | 1 | 1 | |
| 42 | *$(3^3)$ | | 2 | | | | | | 1 | |
| 56 | $(61^3)$ | $(41^5)$ | | 1 | | 1 | | | | |
| 189 | $(521^2)$ | $(421^3)$ | 3 | | 2 | | 1 | 1 | 1 | |
| 216 | $(431^2)$ | $(42^21)$ | | 1 | | 1 | | | | 1 |
| 70 | *$(51^4)$ | | 2 | | 2 | | 1 | | | |
| | Block number: | | 1 | 2 | 1 | 2 | 1 | 1 | 1 | 2 |

The decomposition matrix of $S_{10}$ for the prime 2

| | | | 1 | 8 | 26 | 48 | 16 | 160 | 198 | 128 | 200 | 768 |
|---|---|---|---|---|---|---|---|---|---|---|---|---|
| | | | (10) | (9,1) | (8,2) | (7,3) | (6,4) | (721) | (631) | (541) | (532) | (4321) |
| 1 | (10) | $(1^{10})$ | 1 | | | | | | | | | |
| 9 | (9,1) | $(21^8)$ | 1 | 1 | | | | | | | | |
| 35 | (8,2) | $(2^2 1^6)$ | 1 | 1 | 1 | | | | | | | |
| 75 | (7,3) | $(2^3 1^4)$ | 1 | | 1 | 1 | | | | | | |
| 90 | (6,4) | $(2^4 1^2)$ | | | 1 | 1 | 1 | | | | | |
| 160 | (721) | $(321^5)$ | | | | | | 1 | | | | |
| 315 | (631) | $(32^2 1^3)$ | 1 | | 2 | 1 | 1 | | 1 | | | |
| 288 | (541) | $(32^3 1)$ | | | | | | 1 | | 1 | | |
| 450 | (532) | $(3^2 21^2)$ | 2 | 1 | 1 | | 1 | | 1 | | 1 | |
| 768 | *(4321) | | | | | | | | | | | 1 |
| 42 | $(5^2)$ | $(2^5)$ | | | 1 | | 1 | | | | | |
| 36 | $(81^2)$ | $(31^7)$ | 2 | 1 | 1 | | | | | | | |
| 225 | $(62^2)$ | $(3^2 1^4)$ | 1 | | 1 | | | | 1 | | | |
| 252 | $(4^2 2)$ | $(3^2 2^2)$ | 2 | 1 | 1 | | 1 | | | | 1 | |
| 210 | $(43^2)$ | $(3^3 1)$ | 2 | 1 | | | | | | | 1 | |
| 84 | $(71^3)$ | $(41^6)$ | 2 | 1 | 1 | 1 | | | | | | |
| 350 | $(621^2)$ | $(421^4)$ | 2 | 1 | 3 | 1 | 1 | | 1 | | | |
| 567 | $(531^2)$ | $(42^2 1^2)$ | 3 | 1 | 3 | 1 | 2 | | 1 | | 1 | |
| 300 | $(4^2 1^2)$ | $(42^3)$ | 2 | 1 | 1 | 1 | 1 | | | | 1 | |
| 525 | $(52^2 1)$ | $(431^3)$ | 3 | 1 | 2 | 1 | 1 | | 1 | | 1 | |
| 126 | $(61^4)$ | $(51^5)$ | 2 | 1 | 2 | 1 | 1 | | | | | |
| 448 | $*(521^3)$ | | | | | | | 2 | | 1 | | |
| | Block number: | | 1 | 1 | 1 | 1 | 1 | 2 | 1 | 2 | 1 | 3 |

The decomposition matrix $S_{11}$ for the prime 2

| | | | (11) 1 | (10,1) 10 | (9,2) 44 | (8,3) 100 | (7,4) 164 | (6,5) 32 | (821) 186 | (731) 198 | (641) 144 | (632) 848 | (542) 416 | (5321) 1168 |
|---|---|---|---|---|---|---|---|---|---|---|---|---|---|---|
| 1 | (11) | $(1^{11})$ | 1 | | | | | | | | | | | |
| 10 | (10,1) | $(21^8)$ | | 1 | | | | | | | | | | |
| 44 | (9,2) | $(2^21^7)$ | | | 1 | | | | | | | | | |
| 110 | (8,3) | $(2^31^5)$ | | 1 | | 1 | | | | | | | | |
| 165 | (7,4) | $(2^41^3)$ | 1 | | | | 1 | | | | | | | |
| 132 | (6,5) | $(2^51)$ | | | | 1 | | 1 | | | | | | |
| 231 | (821) | $(321^6)$ | 1 | | 1 | | | | 1 | | | | | |
| 550 | (731) | $(32^21^4)$ | 2 | | | | 1 | | 1 | 1 | | | | |
| 693 | (641) | $(32^31^2)$ | 1 | | | | 1 | | 1 | 1 | 1 | | | |
| 990 | (632) | $(3^221^3)$ | | 1 | | 1 | | 1 | | | | 1 | | |
| 990 | (542) | $(3^22^21)$ | 2 | | 1 | | | | 1 | 1 | 1 | | 1 | |
| 2310 | (5321) | $(4321^2)$ | | 3 | | 2 | | 2 | | | | 1 | | 1 |
| 45 | $(91^2)$ | $(31^8)$ | 1 | | 1 | | | | | | | | | |
| 330 | $(5^21)$ | $(32^4)$ | | | | | | | 1 | | 1 | | | |
| 385 | $(72^2)$ | $(3^21^5)$ | 1 | | | | | | 1 | 1 | | | | |
| 660 | $(53^2)$ | $(3^31^2)$ | 2 | | 1 | | | | | 1 | | | 1 | |
| 462 | $(4^23)$ | $(3^32)$ | 2 | | 1 | | | | | | | | 1 | |
| 120 | $(81^3)$ | $(41^7)$ | | 2 | | 1 | | | | | | | | |
| 594 | $(721^2)$ | $(421^5)$ | 2 | | 1 | | 1 | | 1 | 1 | | | | |
| 1232 | $(631^2)$ | $(42^21^3)$ | | 2 | | 3 | | 2 | | | | 1 | | |
| 1155 | $(541^2)$ | $(42^31)$ | 3 | | 1 | | 1 | | 1 | 1 | 1 | | 1 | |
| 1100 | $(62^21)$ | $(431^4)$ | | 2 | | 2 | | 1 | | | | 1 | | |
| 1320 | $(4^221)$ | $(432^2)$ | | 2 | | 1 | | 1 | | | | | | 1 |
| 1188 | *$(43^21)$ | | | 2 | | | | | | | | | | 1 |
| 825 | $(52^3)$ | $(4^21^3)$ | 3 | | 1 | | 1 | | | 1 | | | 1 | |
| 210 | $(71^4)$ | $(51^6)$ | 2 | | 1 | | 1 | | | | | | | |
| 924 | $(621^3)$ | $(521^4)$ | 2 | | 1 | | 1 | | 2 | 1 | 1 | | | |
| 1540 | $(531^3)$ | $(52^21^2)$ | 4 | | 1 | | 1 | | 2 | 2 | 1 | | 1 | |
| 252 | *$(61^5)$ | | | 2 | | 2 | | 1 | | | | | | |
| | Block number: | | 1 | 2 | 1 | 2 | 1 | 2 | 1 | 1 | 1 | 2 | 1 | 2 |

## The decomposition matrix of $S_{12}$ for the prime 2

| | | | (12) 1 | (11,1) 10 | (10,2) 44 | (9,3) 100 | (8,4) 164 | (7,5) 32 | (921) 320 | (831) 570 | (741) 1408 | (651) 288 | (732) 1046 | (642) 416 | (543) 1792 | (6321) 5632 | (5421) 2368 |
|---|---|---|---|---|---|---|---|---|---|---|---|---|---|---|---|---|---|
| 1 | (12) | $(1^{12})$ | 1 | | | | | | | | | | | | | | |
| 11 | (11,1) | $(21^{10})$ | 1 | 1 | | | | | | | | | | | | | |
| 54 | (10,2) | $(2^2 1^8)$ | | 1 | 1 | | | | | | | | | | | | |
| 154 | (9,3) | $(2^3 1^6)$ | | 1 | 1 | 1 | | | | | | | | | | | |
| 275 | (8,4) | $(2^4 1^4)$ | 1 | 1 | | 1 | 1 | | | | | | | | | | |
| 297 | (7,5) | $(2^5 1^2)$ | 1 | | | 1 | 1 | 1 | | | | | | | | | |
| 320 | (921) | $(321^7)$ | | | | | | | 1 | | | | | | | | |
| 891 | (831) | $(32^2 1^5)$ | 3 | 1 | 1 | 1 | 1 | | | 1 | | | | | | | |
| 1408 | (741) | $(32^2 1^3)$ | | | | | | | | | 1 | | | | | | |
| 1155 | (651) | $(32^4 1)$ | 1 | | | 1 | 1 | 1 | | 1 | | 1 | | | | | |
| 1925 | (732) | $(2^5 1^2)$ | 3 | 1 | | 1 | 1 | 1 | | 1 | | | 1 | | | | |
| 2673 | (642) | $(3^2 2^2 1^2)$ | 3 | 1 | 1 | 1 | 1 | 1 | | 1 | | 1 | 1 | 1 | | | |
| 2112 | (543) | $(3^3 21)$ | | | | | | | 1 | | | | | | 1 | | |
| 5632 | (6321) | $(4321^3)$ | | | | | | | | | | | | | | 1 | |
| 5775 | (5421) | $(432^2 1)$ | 5 | 5 | 2 | 3 | 1 | 2 | | 1 | | 1 | 1 | 2 | | | 1 |
| 132 | $(6^2)$ | $(2^6)$ | | | | 1 | | 1 | | | | | | | | | |
| 55 | $(101^2)$ | $(31^9)$ | 1 | 1 | 1 | | | | | | | | | | | | |
| 616 | $(82^2)$ | $(3^2 1^6)$ | 2 | | 1 | | | | | 1 | | | | | | | |
| 1320 | $(5^2 2)$ | $(3^2 2^3)$ | 2 | | 1 | | | | | 1 | | 1 | | 1 | | | |
| 1650 | $(63^2)$ | $(3^3 1^3)$ | 2 | 1 | 1 | 1 | | 1 | | | | | 1 | 1 | | | |
| 462 | $(4^3)$ | $(3^4)$ | 2 | | 1 | | | | | | | | | 1 | | | |
| 165 | $(91^3)$ | $(41^8)$ | 1 | 2 | 1 | 1 | | | | | | | | | | | |
| 945 | $(821^2)$ | $(421^6)$ | 3 | 2 | 2 | 1 | 1 | | | 1 | | | | | | | |
| 2376 | $(731^2)$ | $(42^2 1^4)$ | 4 | 2 | 1 | 3 | 2 | 2 | | 1 | | | 1 | | | | |
| 3080 | $(641^2)$ | $(42^3 1^2)$ | 4 | 2 | 1 | 3 | 2 | 2 | | 1 | | 1 | 1 | 1 | | | |
| 1485 | $(5^2 1^2)$ | $(42^4)$ | 3 | | 1 | | 1 | | | 1 | | 1 | | 1 | | | |
| 2079 | $(72^2 1)$ | $(431^5)$ | 3 | 2 | 1 | 2 | 1 | 1 | | 1 | | | 1 | | | | |
| 4158 | $(53^2 1)$ | $(43^2 1^2)$ | 2 | 5 | 1 | 2 | | 1 | | | | | 1 | 1 | | | 1 |
| 2970 | $(4^2 31)$ | $(43^2 2)$ | 2 | 4 | 1 | 1 | | | | | | | | 1 | | | 1 |
| 1925 | $(62^3)$ | $(4^2 1^4)$ | 3 | 2 | 1 | 2 | 1 | 1 | | | | | 1 | 1 | | | |
| 4455 | $(532^2)$ | $(4^2 21^2)$ | 3 | 5 | 1 | 3 | 1 | 2 | | | | | 1 | 1 | | | 1 |
| 2640 | $*(4^2 2^2)$ | | | 4 | | 2 | | 1 | | | | | | | | | 1 |
| 330 | $(81^4)$ | $(51^7)$ | 2 | 2 | 1 | 1 | 1 | | | | | | | | | | |
| 3696 | $(631^3)$ | $(52^2 1^3)$ | 6 | 2 | 2 | 3 | 2 | 2 | | 2 | | 1 | 1 | 1 | | | |
| 3520 | $(541^3)$ | $(52^3 1)$ | | | | | | | 1 | | 1 | | | | 1 | | |
| 3564 | $(62^2 1^2)$ | $(531^4)$ | 6 | 2 | 2 | 2 | 2 | 1 | | 2 | | 1 | 1 | 1 | | | |
| 7700 | $*(5321^2)$ | | 8 | 6 | 2 | 4 | 2 | 3 | | 2 | | 1 | 2 | 2 | | | 1 |
| 462 | $(71^5)$ | $(61^6)$ | 2 | 2 | 1 | 2 | 1 | 1 | | | | | | | | | |
| 2100 | $*(621^4)$ | | 4 | 2 | 2 | 2 | 2 | 1 | | 2 | | 1 | | | | | |
| 1728 | $(721^3)$ | $(521^5)$ | | | | | | | 1 | | 1 | | | | | | |
| | Block number: | | 1 | 1 | 1 | 1 | 1 | 1 | 2 | 1 | 2 | 1 | 1 | 1 | 2 | 3 | 1 |

The decomposition matrix of $S_{13}$ for the prime 2

| | | | 1 (13) | 12 (12,1) | 64 (11,2) | 208 (10,3) | 364 (9,4) | 560 (8,5) | 64 (7,6) | 364 (1021) | 570 (931) | 1572 (841) | 288 (751) | 2848 (832) | 2510 (742) | 1728 (652) | 2208 (643) | 8008 (7321) | 3200 (6421) | 8448 (5431) |
|---|---|---|---|---|---|---|---|---|---|---|---|---|---|---|---|---|---|---|---|---|
| 1 | (13) | $(1^{13})$ | 1 | | | | | | | | | | | | | | | | | |
| 12 | (12,1) | $(21^{11})$ | | 1 | | | | | | | | | | | | | | | | |
| 65 | (11,2) | $(2^21^9)$ | 1 | | 1 | | | | | | | | | | | | | | | |
| 208 | (10,3) | $(2^31^7)$ | | | | 1 | | | | | | | | | | | | | | |
| 429 | (9,4) | $(2^41^5)$ | 1 | | 1 | | 1 | | | | | | | | | | | | | |
| 572 | (8,5) | $(2^51^3)$ | | 1 | | | | 1 | | | | | | | | | | | | |
| 429 | (7,6) | $(2^61)$ | 1 | | | | 1 | | 1 | | | | | | | | | | | |
| 429 | (1021) | $(321^8)$ | 1 | | 1 | | | | | 1 | | | | | | | | | | |
| 1365 | (931) | $(32^21^6)$ | 3 | | 1 | | 1 | | | 1 | 1 | | | | | | | | | |
| 2574 | (841) | $(32^31^4)$ | 4 | | 1 | | 1 | | | | 1 | 1 | | | | | | | | |
| 2860 | (751) | $(32^41^2)$ | 2 | | | | 1 | | 1 | | 1 | 1 | 1 | | | | | | | |
| 3432 | (832) | $(3^221^5)$ | | 2 | | | | 1 | | | | | | 1 | | | | | | |
| 6006 | (742) | $(3^22^21^3)$ | 4 | | 1 | | 1 | | 1 | | 2 | 1 | 1 | | 1 | | | | | |
| 5148 | (652) | $(3^22^31)$ | | 1 | | | | 1 | | | | | | 1 | | 1 | | | | |
| 6435 | (643) | $(3^321^2)$ | 3 | | 1 | | 1 | | 1 | 1 | 1 | | 1 | | 1 | | 1 | | | |
| 12012 | (7321) | $(4321^4)$ | | 3 | | | | 2 | | | | | | 1 | | | | 1 | | |
| 17160 | (6421) | $(432^21^2)$ | | 4 | | 1 | | 2 | | | | | | 1 | | 1 | | 1 | 1 | |
| 15015 | (5431) | $(43^221)$ | 7 | | 3 | | 1 | | 1 | 1 | 1 | | 1 | | 1 | | 1 | | | 1 |
| 66 | $(111^2)$ | $(31^{10})$ | 2 | | 1 | | | | | | | | | | | | | | | |
| 1287 | $(6^21)$ | $(32^5)$ | 1 | | | | 1 | | 1 | | 1 | | 1 | | | | | | | |
| 936 | $(92^2)$ | $(3^21^7)$ | 2 | | | | | | | 1 | 1 | | | | | | | | | |
| 3575 | $(73^2)$ | $(3^31^4)$ | 3 | | 1 | | 1 | | 1 | | 1 | | | | 1 | | | | | |
| 3432 | $(5^23)$ | $(3^32^2)$ | 2 | | | | | | | 1 | 1 | | 1 | | | | 1 | | | |
| 2574 | $(54^2)$ | $(3^41)$ | 2 | | | | | | | 1 | | | | | | | 1 | | | |
| 220 | $(101^3)$ | $(41^9)$ | | 1 | | 1 | | | | | | | | | | | | | | |
| 1430 | $(921^2)$ | $(421^7)$ | 4 | | 2 | | 1 | | | 1 | 1 | | | | | | | | | |
| 4212 | $(831^2)$ | $(42^21^5)$ | | 3 | | 1 | | 2 | | | | | | 1 | | | | | | |
| 6864 | $(741^2)$ | $(42^31^3)$ | 6 | | 2 | | 3 | | 2 | | 2 | 1 | 1 | | 1 | | | | | |
| 5720 | $(651^2)$ | $(42^41)$ | | 2 | | | | 2 | | | | | | 1 | | 1 | | | | |
| 3640 | $(82^21)$ | $(431^6)$ | | 2 | | 1 | | 1 | | | | | | 1 | | | | | | |
| 8580 | $(5^221)$ | $(432^3)$ | | 3 | | 1 | | 1 | | | | | | 1 | | 1 | | | 1 | |
| 11440 | $(63^21)$ | $(43^21^3)$ | | 2 | | 1 | | | | | | | | | | | | 1 | 1 | |
| 3432 | $(4^31)$ | $(43^3)$ | | 2 | | 1 | | | | | | | | | | | | | 1 | |
| 4004 | $(72^3)$ | $(4^21^5)$ | 4 | | 2 | | 2 | | 1 | | 1 | | | | 1 | | | | | |
| 12012 | $(632^2)$ | $(4^221^3)$ | | 3 | | 1 | | 1 | | | | | | | | | | 1 | 1 | |
| 12870 | $(542^2)$ | $(4^22^21)$ | 6 | | 3 | | 2 | | 2 | | 1 | | 1 | | 1 | | | | | 1 |
| 11583 | $(53^22)$ | $(4^231^2)$ | 5 | | 3 | | 1 | | 1 | | | | | | 1 | | | | | 1 |
| 8580 | | $*(4^232)$ | 4 | | 2 | | | | | | | | | | | | | | | 1 |
| 495 | $(91^4)$ | $(51^8)$ | 3 | | 2 | | 1 | | | | | | | | | | | | | |
| 3003 | $(821^3)$ | $(521^6)$ | 5 | | 2 | | 1 | | | 1 | 1 | 1 | | | | | | | | |
| 7800 | $(731^3)$ | $(52^21^4)$ | 8 | | 2 | | 3 | | 2 | 1 | 3 | 1 | 1 | | 1 | | | | | |
| 10296 | $(641^3)$ | $(52^31^2)$ | 8 | | 2 | | 3 | | 2 | 1 | 3 | 1 | 2 | | 1 | | 1 | | | |
| 5005 | $(5^21^3)$ | $(52^4)$ | 3 | | | | | | | 1 | 1 | 1 | 1 | | | | 1 | | | |
| 7371 | $(72^21^2)$ | $(531^5)$ | 7 | | 2 | | 2 | | 1 | 1 | 3 | 1 | 1 | | 1 | | | | | |
| 20592 | $(6321^2)$ | $(5321^3)$ | | 6 | | 1 | | 3 | | | | | | 2 | | 1 | | 1 | 1 | |
| 21450 | $(5421^2)$ | $(532^21)$ | 12 | | 4 | | 3 | | 3 | 1 | 3 | 1 | 2 | | 2 | | 1 | | | 1 |
| 16016 | | $*(53^21^2)$ | 8 | | 4 | | 2 | | 2 | | 2 | | 1 | | 2 | | | | | 1 |
| 9009 | $(62^31)$ | $(541^4)$ | 7 | | 2 | | 2 | | 1 | 1 | 2 | 1 | 1 | | 1 | | 1 | | | |
| 729 | $(81^5)$ | $(61^7)$ | | 2 | | 1 | | 1 | | | | | | | | | | | | |
| 4290 | $(721^4)$ | $(621^5)$ | 6 | | 2 | | 2 | | 1 | 1 | 2 | 1 | 1 | | | | | | | |
| 9360 | $(631^4)$ | $(62^21^3)$ | | 4 | | 1 | | 3 | | | | | | 2 | | 1 | | | | |
| 924 | | $*(71^6)$ | 4 | | 2 | | 2 | | 1 | | | | | | | | | | | |
| | | Block number: | 1 | 2 | 1 | 2 | 1 | 2 | 1 | 1 | 1 | 1 | 1 | 2 | 1 | 2 | 1 | 2 | 2 | 1 |

## The decomposition matrices of $S_n$ for the prime 3

**n = 0**

| | | (0) 1 |
|---|---|---|
| *1 | (0) | 1 |

**n = 1**

| | | (1) 1 |
|---|---|---|
| *1 | (1) | 1 |

**n = 2**

| | | (2) 1 | $(1^2)$ 1 |
|---|---|---|---|
| 1 | (2) | 1 | |
| 1 | $(1^2)$ | | 1 |

**n = 3**

| | | (3) 1 | (2,1) 1 |
|---|---|---|---|
| 1 | (3) | 1 | |
| *2 | (2,1) | 1 | 1 |
| 1 | $(1^3)$ | | 1 |

**n = 4**

| | | (4) 1 | (3,1) 3 | $(2^2)$ 1 | $(21^2)$ 3 |
|---|---|---|---|---|---|
| 1 | (4) | 1 | | | |
| 3 | (3,1) | | 1 | | |
| *2 | $(2^2)$ | 1 | | 1 | |
| 3 | $(21^2)$ | | | | 1 |
| 1 | $(1^4)$ | | | 1 | |

**n = 5**

| | | (5) 1 | (4,1) 4 | (3,2) 1 | $(31^2)$ 6 | $(2^21)$ 4 |
|---|---|---|---|---|---|---|
| 1 | (5) | 1 | | | | |
| 4 | (4,1) | | 1 | | | |
| 5 | (3,2) | | 1 | 1 | | |
| *6 | $(31^2)$ | | | | 1 | |
| 5 | $(2^21)$ | 1 | | | | 1 |
| 4 | $(21^3)$ | | | | | 1 |
| 1 | $(1^5)$ | | | 1 | | |

**n = 6**

| | | (6) 1 | (5,1) 4 | (4,2) 9 | $(3^2)$ 1 | $(41^2)$ 6 | (321) 4 | $(2^21^2)$ 9 |
|---|---|---|---|---|---|---|---|---|
| 1 | (6) | 1 | | | | | | |
| 5 | (5,1) | 1 | 1 | | | | | |
| 9 | (4,2) | | | 1 | | | | |
| 5 | $(3^2)$ | | 1 | | 1 | | | |
| 10 | $(41^2)$ | | 1 | | | 1 | | |
| *16 | (321) | 1 | 1 | | 1 | 1 | 1 | |
| 9 | $(2^21^2)$ | | | | | | | 1 |
| 5 | $(2^3)$ | 1 | | | | | 1 | |
| 10 | $(31^3)$ | | | | | 1 | 1 | |
| 5 | $(21^4)$ | | | | 1 | | 1 | |
| 1 | $(1^6)$ | | | | 1 | | | |

**n = 7**

| | | (7) 1 | (6,1) 6 | (5,2) 13 | (4,3) 1 | $(51^2)$ 15 | (421) 20 | $(3^21)$ 6 | $(32^2)$ 15 | $(321^2)$ 13 |
|---|---|---|---|---|---|---|---|---|---|---|
| 1 | (7) | 1 | | | | | | | | |
| 6 | (6,1) | | 1 | | | | | | | |
| 14 | (5,2) | 1 | | 1 | | | | | | |
| 14 | (4,3) | | | 1 | 1 | | | | | |
| 15 | $(51^2)$ | | | | | 1 | | | | |
| 35 | (421) | 1 | | 1 | 1 | | 1 | | | |
| 21 | $(3^21)$ | | | | | 1 | | 1 | | |
| 21 | $(32^2)$ | | 1 | | | | | | 1 | |
| 35 | $(321^2)$ | 1 | | | 1 | | 1 | | | 1 |
| *20 | $(41^3)$ | | | | | | 1 | | | |
| 14 | $(2^31)$ | 1 | | | | | | | | 1 |
| 15 | $(31^4)$ | | | | | | | | 1 | |
| 14 | $(2^21^3)$ | | | | 1 | | | | | 1 |
| 6 | $(21^5)$ | | | | | | | 1 | | |
| 1 | $(1^7)$ | | | | 1 | | | | | |

The decomposition matrix of $S_8$ for the prime 3

| | | 1 | 7 | 13 | 28 | 1 | 21 | 35 | 7 | 35 | 21 | 90 | 13 | 28 |
|---|---|---|---|---|---|---|---|---|---|---|---|---|---|---|
| | | (8) | (7,1) | (6,2) | (5,3) | $(4^2)$ | $(61^2)$ | (521) | (431) | $(42^2)$ | $(3^2 2)$ | $(421^2)$ | $(3^2 1^2)$ | $(32^2 1)$ |
| 1 | (8) | 1 | | | | | | | | | | | | |
| 7 | (7,1) | | 1 | | | | | | | | | | | |
| 20 | (6,2) | | 1 | 1 | | | | | | | | | | |
| 28 | (5,3) | | | | 1 | | | | | | | | | |
| 14 | $(4^2)$ | | | 1 | | 1 | | | | | | | | |
| 21 | $(61^2)$ | | | | | | 1 | | | | | | | |
| 64 | (521) | 1 | | | 1 | | | 1 | | | | | | |
| 70 | (431) | | | | 1 | | | 1 | 1 | | | | | |
| 56 | $(42^2)$ | | 1 | 1 | | 1 | | | | 1 | | | | |
| *42 | $(3^2 2)$ | | | | | | 1 | | | | 1 | | | |
| *90 | $(421^2)$ | | | | | | | | | | | 1 | | |
| 56 | $(3^2 1^2)$ | 1 | | | | | | 1 | 1 | | | | 1 | |
| 70 | $(32^2 1)$ | | 1 | | | | | | | 1 | | | | 1 |
| 35 | $(51^3)$ | | | | | | | 1 | | | | | | |
| 14 | $(2^4)$ | 1 | | | | | | | | | | | 1 | |
| 35 | $(41^4)$ | | | | | | | | | 1 | | | | |
| 64 | $(321^3)$ | | | | | 1 | | | | 1 | | | | 1 |
| 28 | $(2^3 1^2)$ | | | | | | | | | | | | | 1 |
| 21 | $(31^5)$ | | | | | | | | | | 1 | | | |
| 20 | $(2^2 1^4)$ | | | | | | | | 1 | | | | 1 | |
| 7 | $(21^6)$ | | | | | | | | 1 | | | | | |
| 1 | $(1^8)$ | | | | | 1 | | | | | | | | |
| Block number: | | 1 | 2 | 2 | 1 | 2 | 3 | 1 | 1 | 2 | 3 | 4 | 1 | 2 |

## The decomposition matrix of $S_9$ for the prime 3

| dim | $\lambda$ | $(9)$ | $(8,1)$ | $(7,2)$ | $(6,3)$ | $(5,4)$ | $(71^2)$ | $(621)$ | $(531)$ | $(4^21)$ | $(52^2)$ | $(432)$ | $(521^2)$ | $(431^2)$ | $(42^21)$ | $(3^221)$ | $(32^21^2)$ |
|---|---|---|---|---|---|---|---|---|---|---|---|---|---|---|---|---|---|
| | dim → | 1 | 7 | 27 | 41 | 1 | 21 | 35 | 162 | 7 | 35 | 21 | 189 | 27 | 189 | 41 | 162 |
| 1 | $(9)$ | 1 | | | | | | | | | | | | | | | |
| 8 | $(8,1)$ | 1 | 1 | | | | | | | | | | | | | | |
| 27 | $(7,2)$ | | | 1 | | | | | | | | | | | | | |
| 48 | $(6,3)$ | | 1 | | 1 | | | | | | | | | | | | |
| 42 | $(5,4)$ | | | | 1 | 1 | | | | | | | | | | | |
| 28 | $(71^2)$ | | 1 | | | | 1 | | | | | | | | | | |
| 105 | $(621)$ | 1 | 1 | | 1 | | 1 | 1 | | | | | | | | | |
| 162 | $(531)$ | | | | | | | | 1 | | | | | | | | |
| 84 | $(4^21)$ | | | | 1 | 1 | | 1 | | 1 | | | | | | | |
| 120 | $(52^2)$ | 1 | 1 | | 1 | 1 | | 1 | | | 1 | | | | | | |
| 168 | $(432)$ | | 1 | | 1 | 1 | 1 | 1 | | 1 | 1 | 1 | | | | | |
| 189 | $(521^2)$ | | | | | | | | | | | | 1 | | | | |
| 216 | $(431^2)$ | | | | | | | | | | | | 1 | 1 | | | |
| 216 | $(42^21)$ | | | 1 | | | | | | | | | | | 1 | | |
| 168 | $(3^221)$ | 1 | 1 | | | | 1 | 1 | | 1 | 1 | 1 | | | | 1 | |
| 162 | $(32^21^2)$ | | | | | | | | | | | | | | | | 1 |
| *42 | $(3^3)$ | | | | | | 1 | | | | | 1 | | | | | |
| 56 | $(61^3)$ | | | | | | 1 | 1 | | | | | | | | | |
| 84 | $(32^3)$ | 1 | 1 | | | | | | | | 1 | | | | | 1 | |
| *70 | $(51^4)$ | | | | | | | 1 | | | 1 | | | | | | |
| 189 | $(421^3)$ | | | | | | | | | | | | | | 1 | | |
| 120 | $(3^21^3)$ | 1 | | | | 1 | | 1 | | 1 | 1 | | | | | 1 | |
| 42 | $(2^41)$ | 1 | | | | | | | | | | | | | | 1 | |
| 56 | $(41^5)$ | | | | | | | | | | 1 | 1 | | | | | |
| 105 | $(321^4)$ | | | | | 1 | | | | 1 | 1 | 1 | | | | 1 | |
| 48 | $(2^31^3)$ | | | | | | | | | 1 | | | | | | 1 | |
| 28 | $(31^6)$ | | | | | | | | | 1 | | 1 | | | | | |
| 27 | $(2^21^5)$ | | | | | | | | | | | | | 1 | | | |
| 8 | $(21^7)$ | | | | | 1 | | | | 1 | | | | | | | |
| 1 | $(1^9)$ | | | | | 1 | | | | | | | | | | | |
| | Block number: | 1 | 1 | 2 | 1 | 1 | 1 | 1 | 3 | 1 | 1 | 1 | 4 | 4 | 2 | 1 | 5 |

The decomposition matrix of $S_{10}$ for the prime 3

| | | $(10)$ | $(9,1)$ | $(8,2)$ | $(7,3)$ | $(6,4)$ | $(5^2)$ | $(81^2)$ | $(721)$ | $(631)$ | $(541)$ | $(62^2)$ | $(532)$ | $(4^22)$ | $(43^2)$ | $(621^2)$ | $(531^2)$ | $(4^21^2)$ | $(52^21)$ | $(4321)$ | $(3^22^2)$ | $(42^21^2)$ | $(3^221^2)$ |
|---|---|---|---|---|---|---|---|---|---|---|---|---|---|---|---|---|---|---|---|---|---|---|---|
| | | 1 | 9 | 34 | 41 | 90 | 1 | 36 | 84 | 279 | 9 | 126 | 126 | 36 | 84 | 224 | 567 | 34 | 224 | 41 | 90 | 567 | 279 |
| 1 | $(10)$ | 1 | | | | | | | | | | | | | | | | | | | | | |
| 9 | $(9,1)$ | | 1 | | | | | | | | | | | | | | | | | | | | |
| 35 | $(8,2)$ | 1 | | 1 | | | | | | | | | | | | | | | | | | | |
| 75 | $(7,3)$ | | | 1 | 1 | | | | | | | | | | | | | | | | | | |
| 90 | $(6,4)$ | | | | | 1 | | | | | | | | | | | | | | | | | |
| 42 | $(5^2)$ | | | | 1 | | 1 | | | | | | | | | | | | | | | | |
| 36 | $(81^2)$ | | | | | | | 1 | | | | | | | | | | | | | | | |
| 160 | $(721)$ | 1 | | 1 | 1 | | | | 1 | | | | | | | | | | | | | | |
| 315 | $(631)$ | | | | | | | 1 | | 1 | | | | | | | | | | | | | |
| 288 | $(541)$ | | | | | | | | | 1 | 1 | | | | | | | | | | | | |
| 225 | $(62^2)$ | | 1 | | | 1 | | | | | | 1 | | | | | | | | | | | |
| 450 | $(532)$ | | | | | | | 1 | | 1 | 1 | | 1 | | | | | | | | | | |
| 252 | $(4^22)$ | | | | | 1 | | | | | | 1 | | 1 | | | | | | | | | |
| 210 | $(43^2)$ | | | | 1 | | 1 | | 1 | | | | | | 1 | | | | | | | | |
| 350 | $(621^2)$ | 1 | | | 1 | | | | 1 | | | | | | | 1 | | | | | | | |
| 567 | $(531^2)$ | | | | | | | | | | | | | | | | 1 | | | | | | |
| 300 | $(4^21^2)$ | | | | 1 | | 1 | | | | | | | | | 1 | | 1 | | | | | |
| 525 | $(52^21)$ | 1 | | 1 | 1 | | 1 | | | | | | | | | 1 | | | 1 | | | | |
| *768 | $(4321)$ | 1 | | 1 | 1 | | 1 | | 1 | | | | | | 1 | 1 | | 1 | 1 | 1 | | | |
| 252 | $(3^22^2)$ | | | | | | | 1 | | | | | 1 | | | | | | | | 1 | | |
| 567 | $(42^21^2)$ | | | | | | | | | | | | | | | | | | | | | 1 | |
| 450 | $(3^221^2)$ | | 1 | | | | | | | | | 1 | | 1 | | | | | | | | | 1 |
| 84 | $(71^3)$ | | | | | | | | 1 | | | | | | | | | | | | | | |
| 210 | $(3^31)$ | 1 | | | | | | | 1 | | | | | | 1 | | | | | 1 | | | |
| 300 | $(42^3)$ | 1 | | 1 | | | | | | | | | | | | | | | 1 | 1 | | | |
| 126 | $(61^4)$ | | | | | | | | | | | 1 | | | | | | | | | | | |
| *448 | $(521^3)$ | | | | | | | | | | | | | | | 1 | | | 1 | | | | |
| 525 | $(431^3)$ | 1 | | | | | 1 | | | | | | | | | 1 | | 1 | 1 | 1 | | | |
| 288 | $(32^31)$ | | 1 | | | | | | | | | | | | | | | | | | | | 1 |
| 42 | $(2^5)$ | 1 | | | | | | | | | | | | | | | | | | 1 | | | |
| 126 | $(51^5)$ | | | | | | | | | | | | 1 | | | | | | | | | | |
| 350 | $(421^4)$ | | | | | | 1 | | | | | | | | 1 | | | | 1 | 1 | | | |
| 225 | $(3^21^4)$ | | | | | | | | | | 1 | | 1 | | | | | | | | 1 | | |
| 315 | $(32^21^3)$ | | | | | | | | | | | | | 1 | | | | | | | | | 1 |
| 90 | $(2^41^2)$ | | | | | | | | | | | | | | | | | | | | 1 | | |
| 84 | $(41^6)$ | | | | | | | | | | | | | | 1 | | | | | | | | |
| 160 | $(321^5)$ | | | | | | 1 | | | | | | | | 1 | | | 1 | | 1 | | | |
| 75 | $(2^31^4)$ | | | | | | | | | | | | | | | | | 1 | | 1 | | | |
| 36 | $(31^7)$ | | | | | | | | | | | | | 1 | | | | | | | | | |
| 35 | $(2^21^6)$ | | | | | | 1 | | | | | | | | | | | 1 | | | | | |
| 9 | $(21^8)$ | | | | | | | | | | 1 | | | | | | | | | | | | |
| 1 | $(1^{10})$ | | | | | | 1 | | | | | | | | | | | | | | | | |
| Block numbers: | | 1 | 2 | 1 | 1 | 2 | 1 | 3 | 1 | 3 | 3 | 2 | 3 | 2 | 1 | 1 | 4 | 1 | 1 | 1 | 3 | 5 | 2 |

The decomposition matrix of $S_{11}$ for the prime 3

| | | (11) | (10,1) | (9,2) | (8,3) | (7,4) | (6,5) | (91²) | (821) | (731) | (641) | (5²1) | (72²) | (632) | (542) | (53²) | (4²3) | (721²) | (631²) | (541²) | (62²1) | (5321) | (4²21) | (43²1) | (432²) | (52²1²) | (4321²) | (3²2²1) |
|---|---|---|---|---|---|---|---|---|---|---|---|---|---|---|---|---|---|---|---|---|---|---|---|---|---|---|---|---|
| | | 1 | 10 | 34 | 109 | 131 | 1 | 45 | 120 | 320 | 693 | 10 | 210 | 252 | 45 | 210 | 120 | 594 | 791 | 34 | 714 | 714 | 109 | 594 | 131 | 791 | 320 | 693 |
| 1 | (11) | 1 | | | | | | | | | | | | | | | | | | | | | | | | | | |
| 10 | (10,1) | | 1 | | | | | | | | | | | | | | | | | | | | | | | | | |
| 44 | (9,2) | | 1 | 1 | | | | | | | | | | | | | | | | | | | | | | | | |
| 110 | (8,3) | 1 | | | 1 | | | | | | | | | | | | | | | | | | | | | | | |
| 165 | (7,4) | | | 1 | | 1 | | | | | | | | | | | | | | | | | | | | | | |
| 132 | (6,5) | | | | | 1 | 1 | | | | | | | | | | | | | | | | | | | | | |
| 45 | (91²) | | | | | | | 1 | | | | | | | | | | | | | | | | | | | | |
| 231 | (821) | 2 | | | 1 | | | | 1 | | | | | | | | | | | | | | | | | | | |
| 550 | (731) | 1 | | | 1 | | | | 1 | 1 | | | | | | | | | | | | | | | | | | |
| 693 | (641) | | | | | | | | | | 1 | | | | | | | | | | | | | | | | | |
| 330 | (5²1) | | | | | | | | | 1 | | 1 | | | | | | | | | | | | | | | | |
| 385 | (72²) | | 1 | 1 | | 1 | | | | | | | 1 | | | | | | | | | | | | | | | |
| 990 | (632) | | | | | | | 1 | | | 1 | | | 1 | | | | | | | | | | | | | | |
| 990 | (542) | | | | | | | | | | 1 | | | 1 | 1 | | | | | | | | | | | | | |
| 660 | (53²) | | | | | | | | 1 | 1 | | 1 | | | | 1 | | | | | | | | | | | | |
| 462 | (4²3) | | | | | 1 | 1 | | | | | | 1 | | | | 1 | | | | | | | | | | | |
| 594 | (721²) | | | | | | | | | | | | | | | | | 1 | | | | | | | | | | |
| 1232 | (631²) | 1 | | | | | | | 1 | 1 | | | | | | | | | 1 | | | | | | | | | |
| 1155 | (541²) | | | | | | | | | 1 | | 1 | | | | | | | 1 | 1 | | | | | | | | |
| 1100 | (62²1) | | 1 | 1 | | 1 | 1 | | | | | | 1 | | | | | | | | 1 | | | | | | | |
| 2310 | (5321) | 2 | | | 1 | | | | 1 | 1 | | 1 | | | | 1 | | | 1 | 1 | | 1 | | | | | | |
| 1320 | (4²21) | | | 1 | | 1 | 2 | | | | | | 1 | | | | 1 | | | | 1 | | 1 | | | | | |
| *1188 | (43²1) | | | | | | | | | | | | | | | | | 1 | | | | | | 1 | | | | |
| 1320 | (432²) | 2 | | | 1 | | | | 1 | | | | | | | 1 | | | | 1 | | 1 | | | 1 | | | |
| 1540 | (52²1²) | | | 1 | | | 1 | | | | | | | | | | | | | | 1 | | | | | 1 | | |
| 2310 | (4321²) | | 1 | 1 | | | 2 | | | | | | 1 | | | | 1 | | | | 1 | | 1 | | | 1 | 1 | |
| 990 | (3²2²1) | | | | | | | 1 | | | | | | 1 | | | | | | | | | | | | | | 1 |
| 120 | (81³) | | | | | | | | 1 | | | | | | | | | | | | | | | | | | | |
| 825 | (52³) | 2 | | | 1 | | | | | | | | | | | | | | | | | 1 | | | | | | |
| 462 | (3³2) | 1 | | | | | | | 1 | | | | | | | 1 | | | | | | | | | 1 | | | |
| 210 | (71⁴) | | | | | | | | | | | | 1 | | | | | | | | | | | | | | | |
| 924 | (621³) | | | | | | | | | | | | 1 | | | | | | | | 1 | | | | | | | |
| 1540 | (531³) | 1 | | | | | | | | | | | | | | | | | 1 | 1 | | 1 | | | | | | |
| 825 | (4²1³) | | | | | | 2 | | | | | | | | | | | | | | 1 | | 1 | | | | | |
| 660 | (3³1²) | | 1 | | | | | | | | | | 1 | | | | 1 | | | | | | | | | | 1 | |
| 1155 | (42³1) | | 1 | 1 | | | | | | | | | | | | | | | | | | | | | | 1 | 1 | |
| 330 | (32⁴) | | 1 | | | | | | | | | | | | | | | | | | | | | | | | 1 | |
| *252 | (61⁵) | | | | | | | | | | | | | 1 | | | | | | | | | | | | | | |
| 924 | (521⁴) | | | | | | | | | | | | | | | 1 | | | | | | 1 | | | | | | |
| 1100 | (431⁴) | 1 | | | | | | | | | | 1 | | | | 1 | | | | 1 | | 1 | | | 1 | | | |
| 1232 | (42²1³) | | | | | | 1 | | | | | | | | | | 1 | | | | | | | | | 1 | 1 | |
| 990 | (3²21³) | | | | | | | | | | | | | 1 | 1 | | | | | | | | | | | | | 1 |
| 693 | (32³1²) | | | | | | | | | | | | | | | | | | | | | | | | | | | 1 |
| 132 | (2⁵1) | 1 | | | | | | | | | | | | | | | | | | | | | | | 1 | | | |
| 210 | (51⁶) | | | | | | | | | | | | | | | 1 | | | | | | | | | | | | |
| 594 | (421⁵) | | | | | | | | | | | | | | | | | | | | | | | 1 | | | | |
| 385 | (3²1⁵) | | | | | | | | | | | 1 | | | | 1 | | | | 1 | | | | | 1 | | | |
| 550 | (32²1⁴) | | | | | | 1 | | | | | | | | | | 1 | | | | | | 1 | | | | 1 | |
| 165 | (2⁴1³) | | | | | | | | | | | | | | | | | | | 1 | | | | | 1 | | | |
| 120 | (41⁷) | | | | | | | | | | | | | | | | 1 | | | | | | | | | | | |
| 231 | (321⁶) | | | | | | 2 | | | | | | | | | | 1 | | | | | | 1 | | | | | |
| 110 | (2³1⁵) | | | | | | 1 | | | | | | | | | | | | | | | | 1 | | | | | |
| 45 | (31⁸) | | | | | | | | | | | | | | 1 | | | | | | | | | | | | | |
| 44 | (2²1⁷) | | | | | | | | | | | 1 | | | | | | | | 1 | | | | | | | | |
| 10 | (21⁹) | | | | | | | | | | | 1 | | | | | | | | | | | | | | | | |
| 1 | (1¹¹) | | | | | | 1 | | | | | | | | | | | | | | | | | | | | | |
| Block numbers: | | 1 | 2 | 2 | 1 | 2 | 2 | 3 | 1 | 1 | 3 | 1 | 2 | 3 | 3 | 1 | 2 | 4 | 1 | 1 | 2 | 1 | 2 | 4 | 1 | 2 | 2 | 3 |

The 3-regular part of the decomposition matrix of $S_{12}$ for the prime 3

| | | 1 | 10 | 54 | 143 | 131 | 297 | 1 | 45 | 120 | 891 | 1013 | 10 | 210 | 252 | 2673 | 45 | 210 | 120 | 945 | 1431 | 1936 | 54 | 1728 | 1428 | 143 | 1728 | 945 | 3564 | 131 | 297 | 3564 | 1936 | 891 | 1431 | 1013 | 2673 |
|---|---|---|---|---|---|---|---|---|---|---|---|---|---|---|---|---|---|---|---|---|---|---|---|---|---|---|---|---|---|---|---|---|---|---|---|---|---|
| 1 | (12) | 1 | | | | | | | | | | | | | | | | | | | | | | | | | | | | | | | | | | | |
| 11 | (11,1) | 1 | 1 | | | | | | | | | | | | | | | | | | | | | | | | | | | | | | | | | | |
| 54 | (10,2) | | | 1 | | | | | | | | | | | | | | | | | | | | | | | | | | | | | | | | | |
| 154 | (9,3) | 1 | 1 | | 1 | | | | | | | | | | | | | | | | | | | | | | | | | | | | | | | | |
| 275 | (8,4) | 1 | | | 1 | 1 | | | | | | | | | | | | | | | | | | | | | | | | | | | | | | | |
| 297 | (7,5) | | | | | | 1 | | | | | | | | | | | | | | | | | | | | | | | | | | | | | | |
| 132 | ($6^2$) | | | | | 1 | | 1 | | | | | | | | | | | | | | | | | | | | | | | | | | | | | |
| 55 | ($10\,1^2$) | | 1 | | | | | | 1 | | | | | | | | | | | | | | | | | | | | | | | | | | | | |
| 320 | (921) | 2 | 1 | | 1 | | | | 1 | 1 | | | | | | | | | | | | | | | | | | | | | | | | | | | |
| 891 | (831) | | | | | | | | | | 1 | | | | | | | | | | | | | | | | | | | | | | | | | | |
| 1408 | (741) | 1 | | | 1 | 1 | | | | 1 | | 1 | | | | | | | | | | | | | | | | | | | | | | | | | |
| 1155 | (651) | | | | | 1 | | 1 | | | | 1 | 1 | | | | | | | | | | | | | | | | | | | | | | | | |
| 616 | ($82^2$) | 2 | 1 | | 1 | 1 | | | | 1 | | | | 1 | | | | | | | | | | | | | | | | | | | | | | | |
| 1925 | (732) | 1 | 1 | | 1 | 1 | | | 1 | 1 | | 1 | | 1 | 1 | | | | | | | | | | | | | | | | | | | | | | |
| 2673 | (642) | | | | | | | | | | | | | | | 1 | | | | | | | | | | | | | | | | | | | | | |
| 1320 | ($5^22$) | | | | | | | | | | | 1 | 1 | | 1 | | 1 | | | | | | | | | | | | | | | | | | | | |
| 1650 | ($63^2$) | | | | | | | | 1 | 1 | | 1 | 1 | | 1 | | | 1 | | | | | | | | | | | | | | | | | | | |
| 2112 | (543) | | | | | 1 | | 1 | | 1 | | 1 | 1 | 1 | 1 | | 1 | 1 | 1 | | | | | | | | | | | | | | | | | | |
| 945 | ($821^2$) | | | | | | | | | | | | | | | | | | | 1 | | | | | | | | | | | | | | | | | |
| 2376 | ($731^2$) | | | | | | | | | | | | | | | | | | | 1 | 1 | | | | | | | | | | | | | | | | |
| 3080 | ($641^2$) | 1 | | | | | | | | 1 | | 1 | 1 | | | | | | | | | 1 | | | | | | | | | | | | | | | |
| 1485 | ($5^21^2$) | | | | | | | | | | | | | | | | | | | | 1 | | 1 | | | | | | | | | | | | | | |
| 2079 | ($72^21$) | | | 1 | | | 1 | | | | | | | | | | | | | | | | | 1 | | | | | | | | | | | | | |
| 5632 | (6321) | 3 | 1 | | 1 | 1 | | 1 | 1 | 2 | | 1 | 1 | 1 | 1 | | | 1 | | | | 1 | | | 1 | | | | | | | | | | | | |
| 5775 | (5421) | 2 | | | 1 | 1 | | 2 | | 1 | | 1 | 2 | 1 | 1 | | 1 | 1 | 1 | | | 1 | | | 1 | 1 | | | | | | | | | | | |
| 4158 | ($53^21$) | | | | | | | | | | | | | | | | | | | 1 | 1 | | 1 | | | | 1 | | | | | | | | | | |
| 2970 | ($4^231$) | | | | | | 1 | | | | | | | | | | | | | | | | | 1 | | | | 1 | | | | | | | | | |
| 4455 | ($532^2$) | | | | | | | | | | 1 | | | | | | | | | | | | | | | | | | 1 | | | | | | | | |
| *2640 | ($4^22^2$) | 2 | | | 1 | 1 | | 2 | | 1 | | | | 1 | | | | 1 | 1 | | | | | | 1 | 1 | | | | 1 | | | | | | | |
| 2970 | ($43^22$) | | | | | | | | | | | | | | | | | | | 1 | | | | | | | 1 | | | | 1 | | | | | | |
| 3564 | ($62^21^2$) | | | | | | | | | | | | | | | | | | | | | | | | | | | | | | | 1 | | | | | |
| *7700 | ($5321^2$) | 3 | 1 | | 1 | | | 3 | | 1 | | | 1 | 1 | | | | 1 | 1 | | | 1 | | | 2 | 1 | | | | | | | 1 | | | | |
| 4455 | ($4^221^2$) | | | | | | | | | | | | | | | | | | | | | | | | | | | | | | | 1 | | 1 | | | |
| 4158 | ($43^21^2$) | | | 1 | | | | | | | | | | | | | | | | | | | | 1 | | | | 1 | | | | | | | 1 | | |
| 5775 | ($432^21$) | 2 | 2 | | 1 | | | 2 | 1 | 1 | | | | 1 | 1 | | | 1 | 1 | | | | | | 1 | 1 | | | | 1 | | | 1 | | | 1 | |
| 2673 | ($3^22^21^2$) | | | | | | | | | | | | | | | | | | | | | | | | | | | | | | | | | | | | 1 |

The 3-singular part of the decomposition matrix of $S_{12}$ for the prime 3

| | | 1 | 10 | 54 | 143 | 131 | 297 | 1 | 45 | 120 | 891 | 1013 | 10 | 210 | 252 | 2673 | 45 | 210 | 120 | 945 | 1431 | 1936 | 54 | 1728 | 1428 | 143 | 1728 | 945 | 3564 | 131 | 297 | 3564 | 1936 | 891 | 1431 | 1013 | 2673 |
|---|---|---|---|---|---|---|---|---|---|---|---|---|---|---|---|---|---|---|---|---|---|---|---|---|---|---|---|---|---|---|---|---|---|---|---|---|---|
| 462 | $(4^3)$ | | | | | 1 | | 1 | | | | | | 1 | | | | | 1 | | | | | | | | | | | | | | | | | | |
| 165 | $(91^3)$ | | | | | | | | 1 | 1 | | | | | | | | | | | | | | | | | | | | | | | | | | | |
| 1925 | $(62^3)$ | 2 | 1 | | 1 | | | 1 | | | | | | 1 | | | | | | | | | | | 1 | | | | | | | | | | | | |
| 462 | $(3^4)$ | 1 | | | | | | | | 1 | | | | | | | | 1 | | | | | | | | | | | | 1 | | | | | | | |
| 330 | $(81^4)$ | | | | | | | | | 1 | | | | 1 | | | | | | | | | | | | | | | | | | | | | | | |
| 1728 | $(721^3)$ | | | | | | | | | | | | | | | | | | | | | | | 1 | | | | | | | | | | | | | |
| 3696 | $(631^3)$ | 2 | | | | | | | | 1 | | | | 1 | | | | | | | | 1 | | | 1 | | | | | | | | | | | | |
| 3520 | $(541^3)$ | 1 | | | | | | 2 | | | | | 1 | | | | | | | | | 1 | | | 1 | 1 | | | | | | | | | | | |
| 3520 | $(52^31)$ | 2 | 1 | | 1 | | | 1 | | | | | | | | | | | | | | | | | 1 | | | | | | | | 1 | | | | |
| 2112 | $(3^321)$ | 1 | 1 | | | | | | 1 | 1 | | | | 1 | 1 | | | 1 | 1 | | | | | | | | | | | 1 | | | | | | 1 | |
| 1485 | $(42^4)$ | | | 1 | | | | | | | | | | | | | | | | | | | | | | | | | | | | | | | 1 | | |
| 1320 | $(3^22^3)$ | | 1 | | | | | | 1 | | | | | | 1 | | | | | | | | | | | | | | | | | | | | | 1 | |
| 462 | $(71^5)$ | | | | | | | | | | | | | 1 | 1 | | | | | | | | | | | | | | | | | | | | | | |
| *2100 | $(621^4)$ | | | | | | | | | | | | | 1 | 1 | | | 1 | | | | | | | 1 | | | | | | | | | | | | |
| 3564 | $(531^4)$ | | | | | | | | | | | | | | | | | | | | | | | | | | | | 1 | | | | | | | | |
| 1925 | $(4^21^4)$ | 1 | | | | | | 2 | | | | | 1 | | | | | 1 | | | | | | | 1 | 1 | | | | 1 | | | | | | | |
| 3696 | $(52^21^3)$ | | | | | | | 2 | | | | | | | | | | 1 | 1 | | | | | | 1 | | | | | | | | 1 | | | | |
| 5632 | $(4321^3)$ | 1 | 1 | | | | | 3 | | | | | 1 | 1 | 1 | | 1 | 1 | 2 | | | | | | 1 | 1 | | | | 1 | | | 1 | | | 1 | |
| 1650 | $(3^31^3)$ | | 1 | | | | | | | | | | | 1 | 1 | | 1 | | 1 | | | | | | | | | | | | | | | | | 1 | |
| 3080 | $(42^31^2)$ | | 1 | | | | | 1 | | | | | | | | | | | 1 | | | | | | | | | | | | | | 1 | | | 1 | |
| 1155 | $(32^41)$ | 1 | 1 | | | | | | | | | | | | | | | | | | | | | | | | | | | 1 | | | | | | 1 | |
| 132 | $(2^6)$ | 1 | | | | | | | | | | | | | | | | | | | | | | | | | | | | 1 | | | | | | | |
| 462 | $(61^6)$ | | | | | | | | | | | | | | 1 | | | 1 | | | | | | | | | | | | | | | | | | | |
| 1728 | $(521^5)$ | | | | | | | | | | | | | | | | | | | | | | | | | | 1 | | | | | | | | | | |
| 2079 | $(431^5)$ | | | | | | | | | | | | | | | | | | | | | | 1 | | | | 1 | | | | 1 | | | | | | |
| 2376 | $(42^21^4)$ | | | | | | | | | | | | | | | | | | | | | | | | | | | 1 | | | | | | | 1 | | |
| 1925 | $(3^221^4)$ | | | | | | | 1 | | | | | 1 | | 1 | | 1 | 1 | 1 | | | | | | | 1 | | | | 1 | | | | | | 1 | |
| 1408 | $(32^31^3)$ | | | | | | | 1 | | | | | | | | | | | 1 | | | | | | | 1 | | | | 1 | | | | | | 1 | |
| 297 | $(2^51^2)$ | | | | | | | | | | | | | | | | | | | | | | | | | | | | | | 1 | | | | | | |
| 330 | $(51^7)$ | | | | | | | | | | | | | | | | | 1 | 1 | | | | | | | | | | | | | | | | | | |
| 945 | $(421^6)$ | | | | | | | | | | | | | | | | | | | | | | | | | | | 1 | | | | | | | | | |
| 616 | $(3^21^6)$ | | | | | | | 2 | | | | | 1 | | | | | 1 | 1 | | | | | | | 1 | | | | 1 | | | | | | | |
| 891 | $(32^21^5)$ | | | | | | | | | | | | | | | | | | | | | | | | | | | | | | | | | 1 | | | |
| 275 | $(2^41^4)$ | | | | | | | 1 | | | | | | | | | | | | | | | | | | 1 | | | | 1 | | | | | | | |
| 165 | $(41^8)$ | | | | | | | | | | | | | | | | 1 | | 1 | | | | | | | | | | | | | | | | | | |
| 320 | $(321^7)$ | | | | | | | 2 | | | | | 1 | | | | 1 | | 1 | | | | | | | 1 | | | | | | | | | | | |
| 154 | $(2^31^6)$ | | | | | | | 1 | | | | | 1 | | | | | | | | | | | | | 1 | | | | | | | | | | | |
| 55 | $(31^9)$ | | | | | | | | | | | | 1 | | | | 1 | | | | | | | | | | | | | | | | | | | | |
| 54 | $(2^21^8)$ | | | | | | | | | | | | | | | | | | | | | | 1 | | | | | | | | | | | | | | |
| 11 | $(21^{10})$ | | | | | | | 1 | | | | | 1 | | | | | | | | | | | | | | | | | | | | | | | | |
| 1 | $(1^{12})$ | | | | | | | 1 | | | | | | | | | | | | | | | | | | | | | | | | | | | | | |
| Block number: | | 1 | 1 | 2 | 1 | 1 | 2 | 1 | 1 | 1 | 3 | 1 | 1 | 1 | 1 | 4 | 1 | 1 | 1 | 5 | 5 | 1 | 5 | 2 | 1 | 1 | 5 | 2 | 3 | 1 | 5 | 6 | 1 | 6 | 2 | 1 | 7 |

The 3-regular part of the decomposition matrix of $S_{13}$ for the prime 3

| | | 1 | 12 | 64 | 143 | 417 | 428 | 1 | 66 | 220 | 1299 | 1275 | 2287 | 12 | 495 | 792 | 5082 | 66 | 924 | 792 | 220 | 495 | 1065 | 4212 | 3367 | 64 | 1938 | 1428 | 10296 | 143 | 1938 | 1065 | 8568 | 417 | 7371 | 428 | 7371 | 8568 | 1299 | 3367 | 4212 | 10296 | 1275 | 2287 | 5082 |
|---|---|---|---|---|---|---|---|---|---|---|---|---|---|---|---|---|---|---|---|---|---|---|---|---|---|---|---|---|---|---|---|---|---|---|---|---|---|---|---|---|---|---|---|---|---|
| 1 | $(13)$ |1||||||||||||||||||||||||||||||||||||||||||||
| 12 | $(12,1)$ ||1|||||||||||||||||||||||||||||||||||||||||||
| 65 | $(11,2)$ |1||1|||||||||||||||||||||||||||||||||||||||||
| 208 | $(10,3)$ |1||1|1||||||||||||||||||||||||||||||||||||||||
| 429 | $(9,4)$ ||1|||1|||||||||||||||||||||||||||||||||||||||
| 572 | $(8,5)$ |1|||1||1||||||||||||||||||||||||||||||||||||||
| 429 | $(7,6)$ ||||||1|1|||||||||||||||||||||||||||||||||||||
| 66 | $(11,1^2)$ ||||||||1||||||||||||||||||||||||||||||||||||||
| 429 | $(10,2,1)$ |2||1|1|||||1|||||||||||||||||||||||||||||||||||||
| 1365 | $(931)$ ||||||||1||1|||||||||||||||||||||||||||||||||||||
| 2574 | $(841)$ ||||||||||1|1|||||||||||||||||||||||||||||||||||
| 2860 | $(751)$ |1|||1||1|1|||||1|||||||||||||||||||||||||||||||||||
| 1287 | $(6^21)$ |||||||||||1||1||||||||||||||||||||||||||||||||||
| 936 | $(92^2)$ ||2|||1|||||||||1||||||||||||||||||||||||||||||||
| 3424 | $(832)$ ||||||||1||1|1||||1|||||||||||||||||||||||||||||||
| 6006 | $(742)$ ||1|||1|||||||||1||1|||||||||||||||||||||||||||||||
| 5148 | $(652)$ |||||||||||||||||1|1|||||||||||||||||||||||||||||
| 3575 | $(73^2)$ |1|||1|||||1|||1||||||1|||||||||||||||||||||||||||
| 6435 | $(643)$ |||||||||||||||1||1|1||1|||||||||||||||||||||||||||
| 3432 | $(5^23)$ |||||||1|||||1||||||1||1|||||||||||||||||||||||||||
| 2574 | $(54^2)$ |||||||||||1||1||1||||||1||||||||||||||||||||||||||
| 1430 | $(921^2)$ |2|||1|||||1||||||||||||||1||||||||||||||||||||||||
| 4212 | $(831^2)$ ||||||||||||||||||||||||1|||||||||||||||||||||||
| 6864 | $(741^2)$ |2|||1||||||||1||||||||||1||1||||||||||||||||||||||
| 5720 | $(651^2)$ |1||||||1|||||1||||||||||||1|1|||||||||||||||||||||
| 3640 | $(82^21)$ |2||1|1||1||||||||||||||||1||||1|||||||||||||||||||||
| 12012 | $(7321)$ |4||1|2||1|1||1|||1||||||1||||1||1||1|1|||||||||||||||||||
| 17160 | $(6421)$ ||1|||1|||||||||1||1|1||1|||||||||1|||||||||||||||||
| 8580 | $(5^221)$ |2|||1|||2|||||1||||||1||1||||1|1||1||1|||||||||||||||||
| 11440 | $(63^21)$ |3|||1|||1||1|||1||||||1||||1||1|1||1|||1|||||||||||||||
| 15015 | $(5431)$ |2|||1||1|3|||||1||||||1||1||1||1|1|1|1||1|1|1|||||||||||||
| 12012 | $(632^2)$ ||||||||1||1|1||1||1|||||||||||||||||1||||||||||||||
| 12870 | $(542^2)$ ||||||||||1|1||2||1||||||1|||||||||||||1|1||||||||||||
| 11583 | $(53^22)$ ||||||||||||||||||||||||1|||||||||||1|||||||||||
| *8580 | $(4^232)$ |2|||1||1|2|||||||||||||||1||||1|1||1|1|1||||1||||||||||
| 7371 | $(72^21^2)$ |||||||||||||||||||||||||||||||||||||1|||||||||
| 20592 | $(6321^2)$ ||2|||1|||||||||1|||||1|||||||||1|||||||||1||||||||
| 21450 | $(5421^2)$ ||1|||1||||||||||||1||1|||||||||1|||||||||1|1|||||||
| *16016 | $(53^21^2)$ |3||1|1|||3||||||||||||||1|||1|1|1|2||1|1|1|||||||||1||||||
| 11583 | $(4^231^2)$ |||||||||||||||||||||||||||||||||||||1||||1|||||
| 21450 | $(532^21)$ ||||||||1||1|||1||1|||||||||||||||||1|1||||||||1||||
| 12870 | $(4^22^21)$ ||2|||1|||||||||1|||||1|||||||||||||||||||1|1||||1|||
| 15015 | $(43^221)$ |3||1|1|||2||1|||||||||1||||1||||||1|1||1|1|1||||1||||1||||1||
| 17160 | $(432^21^2)$ ||||||||1|||||1||1||||||1|||||||||||||||1|||||||1|||1|

The decomposition matrix of $S_{13}$ for the prime 3 (continued)

| | | 1 | 12 | 64 | 143 | 417 | 428 | 1 | 66 | 220 | 1299 | 1275 | 2287 | 12 | 495 | 792 | 5082 | 66 | 924 | 792 | 220 | 495 | 1065 | 4212 | 3367 | 64 | 1938 | 1428 | 10296 | 143 | 1938 | 1065 | 8568 | 417 | 7371 | 428 | 7371 | 8568 | 1299 | 3367 | 4212 | 10296 | 1275 | 2287 | 5082 |
|---|---|---|---|---|---|---|---|---|---|---|---|---|---|---|---|---|---|---|---|---|---|---|---|---|---|---|---|---|---|---|---|---|---|---|---|---|---|---|---|---|---|---|---|---|---|
| 220 | $(101^3)$ | | | | | | | | | 1 | | | | | | | | | | | | | | | | | | | | | | | | | | | | | | | | | | | |
| 3432 | $(4^31)$ | | | | | | 1 | 1 | | | | | | | | | | | | | | | | | | | 1 | | | | | 1 | | | | | | | | | | | | | |
| 4004 | $(72^3)$ | 2 | | 1 | 1 | | 1 | 1 | | | | | | | | | | | | | | | | | | | 1 | 1 | | | | | | | | | | | | | | | | | |
| 2574 | $(43^3)$ | | | | | | | | | | | | | | | | | | | | | | 1 | | | | | | | | 1 | | | | | 1 | | | | | | | | | |
| 495 | $(91^4)$ | | | | | | | | | | | | | | 1 | | | | | | | | | | | | | | | | | | | | | | | | | | | | | | |
| 3003 | $(821^3)$ | | | | | | | | | | | | | | | | | | | | | | 1 | | | | 1 | | | | | | | | | | | | | | | | | | |
| 7800 | $(731^3)$ | 2 | | | | | | | | | | | | | | | | | | | | | 1 | | 1 | | 1 | 1 | | | | | | | | | | | | | | | | | |
| 10296 | $(641^3)$ | | | | | | | | | | | | | | | | | | | | | | | | | | | | 1 | | | | | | | | | | | | | | | | |
| 5005 | $(5^21^3)$ | 1 | | | | | | 2 | | | | | | | | | | | | | | | | | 1 | 1 | | 1 | | 1 | | | | | | | | | | | | | | | |
| 9009 | $(62^31)$ | | 2 | | | 1 | | | | | | | | | | | | | | | | | | | | | | | | | | | | | | | | 1 | | | | | | | |
| 2574 | $(3^41)$ | | 1 | | | | | | | | | | | | 1 | | | | | 1 | | | | | | | | | | | | | | | | | | | | | | | 1 | | |
| 5005 | $(52^4)$ | 2 | | 1 | 1 | | | 1 | | | | | | | | | | | | | | | | | | | | 1 | | | | | | | | | | | | 1 | | | | | |
| 8580 | $(432^3)$ | 2 | | 1 | 1 | | | 2 | | 1 | | | | | | | | | 1 | | | | | | | | | 1 | | 1 | | | | | | | | | | 1 | | | | 1 | |
| 3432 | $(3^32^2)$ | 1 | | | | | | | | 1 | | | | | | | | | 1 | | | | | | | | | | | | | | | | | | | | | | | | | 1 | |
| 792 | $(81^5)$ | | | | | | | | | | | | | | | 1 | | | | | | | | | | | | | | | | | | | | | | | | | | | | | |
| 4290 | $(721^4)$ | | | | | | | | | | | | | | | | | | 1 | | | | | | | | 1 | 1 | | | | | | | | | | | | | | | | | |
| 9360 | $(631^4)$ | | | | | | | | | | | | | | | 1 | | | | | | | | | | | | | | | | | 1 | | | | | | | | | | | | |
| 9009 | $(541^4)$ | | | | | | | | | | | | | 2 | | | | | | | | | | | | | | | | | | | 1 | 1 | | | | | | | | | | | |
| 9360 | $(62^21^3)$ | | | | | | | | | | | | | | | | | | | 1 | | | | | | | | | | | | | | | | | | 1 | | | | | | | |
| 20592 | $(5321^3)$ | | | | | | | | | | | | | 2 | | 1 | | | | | | 1 | | | | | | | | | | | 1 | 1 | | | | | | | | 1 | | | |
| 12012 | $(4^221^3)$ | | 1 | | | | | | | | | | | | | | | 1 | | 1 | | | | | | | | | | | | | | | | | | 1 | 1 | | | | 1 | | |
| 11440 | $(43^21^3)$ | 1 | | 1 | | | | 3 | | | | | | | | | | | 1 | | 1 | | | | | | 1 | 1 | | 1 | | 1 | | | | | | | | 1 | | | | 1 | |
| 10296 | $(52^31^2)$ | | | | | | | | | | | | | | | | | | | | | | | | | | | | | | | | | | | | | | | | | 1 | | | |
| 6435 | $(3^321^2)$ | | | | | | | | 1 | | | | | | | 1 | | | | | | 1 | | | | | | | | | | | | | | | | | | | | | | | 1 |
| 5720 | $(42^41)$ | 1 | | 1 | | | | 1 | | | | | | | | | | | | | | | | | | | | | | | | | | | | | | | | 1 | | | | 1 | |
| 5148 | $(3^22^31)$ | | | | | | | | 1 | | | | | | | | | | | | | | | | | | | | | | | | | | | | | | | | | | | | 1 |
| 1287 | $(32^5)$ | | 1 | | | | | | | | | | | | | | | | | | | | | | | | | | | | | | | | | | | | | | | | 1 | | |
| *924 | $(71^6)$ | | | | | | | | | | | | | | | | | | 1 | | | | | | | | | | | | | | | | | | | | | | | | | | |

The decomposition matrix of $S_{13}$ for the prime 3 (continued)

| | | 1 | 12 | 64 | 143 | 417 | 428 | 1 | 66 | 220 | 1299 | 1275 | 2287 | 12 | 495 | 792 | 5082 | 66 | 924 | 792 | 220 | 495 | 1065 | 4212 | 3367 | 64 | 1938 | 1428 | 10296 | 143 | 1938 | 1065 | 8568 | 417 | 7371 | 428 | 7371 | 8568 | 1299 | 3367 | 4212 | 10296 | 1275 | 2287 | 5082 |
|---|---|---|---|---|---|---|---|---|---|---|---|---|---|---|---|---|---|---|---|---|---|---|---|---|---|---|---|---|---|---|---|---|---|---|---|---|---|---|---|---|---|---|---|---|---|
| 4290 | $(621^5)$ | | | | | | | | | | | | | | | | | | 1 | | | | | | | | | 1 | | | 1 | | | | | | | | | | | | | | |
| 7371 | $(531^5)$ | | | | | | | | | | | | | | | | | | | | | | | | | | | | | | | | | | 1 | | | | | | | | | | |
| 4004 | $(4^21^5)$ | 1 | | | | | | 2 | | | | | | | | | | | | | | | | | | 1 | | 1 | | 1 | 1 | | | | | 1 | | | | | | | | | |
| 7800 | $(52^21^4)$ | | | | | | | 2 | | | | | | | | | | | | | | | | | | | | 1 | | | 1 | 1 | | | | | | | | 1 | | | | | |
| 12012 | $(4321^4)$ | 1 | | | | | | 4 | | | | | | | | | | | 1 | | 1 | | | | | 1 | | 1 | | 2 | 1 | 1 | | | | 1 | | | | 1 | | | | 1 | |
| 3575 | $(3^31^4)$ | | | | | | | 1 | | | | | | | | | | | 1 | | 1 | | | | | | | | | 1 | | | | | | | | | | | | | | 1 | |
| 6864 | $(42^31^3)$ | | | | | | | 2 | | | | | | | | | | | | | | | | | | | | | | 1 | | 1 | | | | | | | | 1 | | | | 1 | |
| 6006 | $(3^22^21^3)$ | | | | | | | | | | | | | 1 | | | | | | | | 1 | | | | | | | | | | | | 1 | | | | | | | | | | | 1 |
| 2860 | $(32^41^2)$ | 1 | | | | | | 1 | | | | | | | | | | | | | | | | | | | | | | 1 | | | | | | 1 | | | | | | | | 1 | |
| 429 | $(2^61)$ | 1 | | | | | | | | | | | | | | | | | | | | | | | | | | | | | | | | | | 1 | | | | | | | | | |
| 792 | $(61^7)$ | | | | | | | | | | | | | | | | | | | 1 | | | | | | | | | | | | | | | | | | | | | | | | | |
| 3003 | $(521^6)$ | | | | | | | | | | | | | | | | | | | | | | | | | | | | | | 1 | 1 | | | | | | | | | | | | | |
| 3640 | $(431^6)$ | | | | | | | 2 | | | | | | | | | | | | | | | | | | 1 | | | | 1 | 1 | 1 | | | | 1 | | | | | | | | | |
| 4212 | $(42^21^5)$ | | | | | | | | | | | | | | | | | | | | | | | | | | | | | | | | | | | | | | | | 1 | | | | |
| 3432 | $(3^221^5)$ | | | | | | | | | | | | | | | | | 1 | | 1 | | | | | | | | | | | | | | | | | | | 1 | | | | 1 | | |
| 2574 | $(32^31^4)$ | | | | | | | | | | | | | | | | | | | | | | | | | | | | | | | | | | | | | | 1 | | | | 1 | | |
| 572 | $(2^51^3)$ | | | | | | | 1 | | | | | | | | | | | | | | | | | | | | | | 1 | | | | | | 1 | | | | | | | | | |
| 495 | $(51^8)$ | | | | | | | | | | | | | | | | | | | | | 1 | | | | | | | | | | | | | | | | | | | | | | | |
| 1430 | $(421^7)$ | | | | | | | 2 | | | | | | | | | | | | | 1 | | | | | | | | | 1 | | 1 | | | | | | | | | | | | | |
| 936 | $(3^21^7)$ | | | | | | | | | | | | | 2 | | | | | | | | 1 | | | | | | | | | | | | 1 | | | | | | | | | | | |
| 1365 | $(32^21^6)$ | | | | | | | | | | | | | | | | | 1 | | | | | | | | | | | | | | | | | | | | | 1 | | | | | | |
| 429 | $(2^41^5)$ | | | | | | | | | | | | | 1 | | | | | | | | | | | | | | | | | | | | 1 | | | | | | | | | | | |
| 220 | $(41^9)$ | | | | | | | | | | | | | | | | | | | | 1 | | | | | | | | | | | | | | | | | | | | | | | | |
| 429 | $(321^8)$ | | | | | | | 2 | | | | | | | | | | | | | 1 | | | | | 1 | | | | 1 | | | | | | | | | | | | | | | |
| 208 | $(2^31^7)$ | | | | | | | 1 | | | | | | | | | | | | | | | | | | 1 | | | | 1 | | | | | | | | | | | | | | | |
| 66 | $(31^{10})$ | | | | | | | | | | | | | | | | | 1 | | | | | | | | | | | | | | | | | | | | | | | | | | | |
| 65 | $(2^21^9)$ | | | | | | | 1 | | | | | | | | | | | | | | | | | | 1 | | | | | | | | | | | | | | | | | | | |
| 12 | $(21^{11})$ | | | | | | | | | | | | | 1 | | | | | | | | | | | | | | | | | | | | | | | | | | | | | | | |
| 1 | $(1^{13})$ | | | | | | | 1 | | | | | | | | | | | | | | | | | | | | | | | | | | | | | | | | | | | | | |
| Block numbers: | | 1 | 2 | 1 | 1 | 2 | 1 | 1 | 3 | 1 | 3 | 3 | 1 | 3 | 2 | 3 | 2 | 2 | 1 | 2 | 1 | 3 | 1 | 4 | 1 | 1 | 1 | 1 | 2 | 1 | 1 | 1 | 3 | 3 | 4 | 1 | 5 | 2 | 2 | 1 | 5 | 3 | 2 | 1 | 3 |

## I.F Irreducible Brauer Characters

The ordinary irreducible characters of $S_n$ together with the $p$-modular decomposition numbers of $S_n$ yield the irreducible $p$-modular Brauer characters of $S_n$. We show them for $p=2,3$ and $n \leq 10$. The tables are due to D. Stockhofe.

The irreducible Brauer characters of $S_n$ for the prime 2

| $n=1$ | $(1)$ |
|---|---|
| $(1)$ | 1 |

| $n=2$ | $(1^2)$ |
|---|---|
| $(2)$ | 1 |

| $n=3$ | $(1^3)$ | $(3)$ |
|---|---|---|
| $(3)$ | 1 | 1 |
| $(2,1)$ | 2 | $-1$ |

| $n=4$ | $(1^4)$ | $(3,1)$ |
|---|---|---|
| $(4)$ | 1 | 1 |
| $(3,1)$ | 2 | $-1$ |

| $n=5$ | $(1^5)$ | $(3,1^2)$ | $(5)$ |
|---|---|---|---|
| $(5)$ | 1 | 1 | 1 |
| $(4,1)$ | 4 | 1 | $-1$ |
| $(3,2)$ | 4 | $-2$ | $-1$ |

| $n=6$ | $(1^6)$ | $(3,1^3)$ | $(3^2)$ | $(5,1)$ |
|---|---|---|---|---|
| $(6)$ | 1 | 1 | 1 | 1 |
| $(5,1)$ | 4 | 1 | $-2$ | $-1$ |
| $(4,2)$ | 4 | $-2$ | 1 | $-1$ |
| $(3,2,1)$ | 16 | $-2$ | $-2$ | 1 |

| $n=7$ | $(1^7)$ | $(3,1^4)$ | $(3^2,1)$ | $(5,1^2)$ | $(7)$ |
|---|---|---|---|---|---|
| $(7)$ | 1 | 1 | 1 | 1 | 1 |
| $(6,1)$ | 6 | 3 | 0 | 1 | $-1$ |
| $(5,2)$ | 14 | 2 | $-1$ | $-1$ | 0 |
| $(4,3)$ | 8 | $-4$ | 2 | $-2$ | 1 |
| $(4,2,1)$ | 20 | $-4$ | $-1$ | 0 | $-1$ |

| $n=8$ | $(1^8)$ | $(3,1^5)$ | $(3^2,1^2)$ | $(5,1^3)$ | $(5,3)$ | $(7,1)$ |
|---|---|---|---|---|---|---|
| (8) | 1 | 1 | 1 | 1 | 1 | 1 |
| (7,1) | 6 | 3 | 0 | 1 | −2 | −1 |
| (6,2) | 14 | 2 | −1 | −1 | 2 | 0 |
| (5,3) | 8 | −4 | 2 | −2 | 1 | 1 |
| (5,2,1) | 64 | 4 | −2 | −1 | −1 | 1 |
| (4,3,1) | 40 | −8 | −2 | 0 | −3 | −2 |

| $n=9$ | $(1^9)$ | $(3,1^6)$ | $(3^2,1^3)$ | $(3^3)$ | $(5,1^4)$ | $(5,3,1)$ | $(7,1^2)$ | $(9)$ |
|---|---|---|---|---|---|---|---|---|
| (9) | 1 | 1 | 1 | 1 | 1 | 1 | 1 | 1 |
| (8,1) | 8 | 5 | 2 | −1 | 3 | 0 | 1 | −1 |
| (7,2) | 26 | 8 | −1 | −1 | 1 | −2 | −2 | −1 |
| (6,3) | 48 | 6 | 0 | 3 | −2 | 1 | −1 | 0 |
| (6,2,1) | 78 | 6 | −3 | −3 | −2 | 1 | 1 | 0 |
| (5,4) | 16 | −8 | 4 | −2 | −4 | 2 | 2 | 1 |
| (5,3,1) | 40 | −8 | −2 | 4 | 0 | −3 | −2 | −2 |
| (4,3,2) | 160 | −20 | −2 | −2 | 0 | 0 | −1 | 1 |

| $n=10$ | $(1^{10})$ | $(3,1^7)$ | $(3^2,1^4)$ | $(3^3,1)$ | $(5,1^5)$ | $(5,3,1^2)$ | $(5^2)$ | $(7,1^3)$ | $(7,3)$ | $(9,1)$ |
|---|---|---|---|---|---|---|---|---|---|---|
| (10) | 1 | 1 | 1 | 1 | 1 | 1 | 1 | 1 | 1 | 1 |
| (9,1) | 8 | 5 | 2 | −1 | 3 | 0 | −2 | 1 | −2 | −1 |
| (8,2) | 26 | 8 | −1 | −1 | 1 | −2 | 1 | −2 | 1 | −1 |
| (7,3) | 48 | 6 | 0 | 3 | −2 | 1 | −2 | −1 | −1 | 1 |
| (7,2,1) | 160 | 34 | −2 | −2 | 5 | −1 | 0 | −1 | −1 | 0 |
| (6,4) | 16 | −8 | 4 | −2 | −4 | 2 | 1 | 2 | 0 | −1 |
| (6,3,1) | 198 | 6 | −6 | 0 | −2 | 1 | −2 | 2 | −1 | 0 |
| (5,4,1) | 128 | −40 | 8 | 2 | −12 | 0 | −2 | 2 | −1 | 1 |
| (5,3,2) | 200 | −28 | −4 | 2 | 0 | −3 | 0 | −3 | 1 | 0 |
| (4,3,2,1) | 768 | −48 | 0 | −6 | −8 | 2 | −2 | −2 | 2 | −1 |

The irreducible Brauer characters of $S_n$ for the prime 3

| $n=1$ | $(1)$ |
|---|---|
| $(1)$ | 1 |

| $n=2$ | $(1^2)$ | $(2)$ |
|---|---|---|
| $(2)$ | 1 | 1 |
| $(1^2)$ | 1 | $-1$ |

| $n=3$ | $(1^3)$ | $(2,1)$ |
|---|---|---|
| $(3)$ | 1 | 1 |
| $(2,1)$ | 1 | $-1$ |

| $n=4$ | $(1^4)$ | $(2,1^2)$ | $(2^2)$ | $(4)$ |
|---|---|---|---|---|
| $(4)$ | 1 | 1 | 1 | 1 |
| $(3,1)$ | 3 | 1 | $-1$ | $-1$ |
| $(2^2)$ | 1 | $-1$ | 1 | $-1$ |
| $(2,1^2)$ | 3 | $-1$ | $-1$ | 1 |

| $n=5$ | $(1^5)$ | $(2,1^3)$ | $(2^2,1)$ | $(4,1)$ | $(5)$ |
|---|---|---|---|---|---|
| $(5)$ | 1 | 1 | 1 | 1 | 1 |
| $(4,1)$ | 4 | 2 | 0 | 0 | $-1$ |
| $(3,2)$ | 1 | $-1$ | 1 | $-1$ | 1 |
| $(3,1^2)$ | 6 | 0 | $-2$ | 0 | 1 |
| $(2^2,1)$ | 4 | $-2$ | 0 | 0 | $-1$ |

| $n=6$ | $(1^6)$ | $(2,1^4)$ | $(2^2,1^2)$ | $(2^3)$ | $(4,1^2)$ | $(4,2)$ | $(5,1)$ |
|---|---|---|---|---|---|---|---|
| $(6)$ | 1 | 1 | 1 | 1 | 1 | 1 | 1 |
| $(5,1)$ | 4 | 2 | 0 | $-2$ | 0 | $-2$ | $-1$ |
| $(4,2)$ | 9 | 3 | 1 | 3 | $-1$ | 1 | $-1$ |
| $(4,1^2)$ | 6 | 0 | $-2$ | 0 | 0 | 2 | 1 |
| $(3^2)$ | 1 | $-1$ | 1 | $-1$ | $-1$ | 1 | 1 |
| $(3,2,1)$ | 4 | $-2$ | 0 | 2 | 0 | $-2$ | $-1$ |
| $(2^2,1^2)$ | 9 | $-3$ | 1 | $-3$ | 1 | 1 | $-1$ |

| $n=7$ | $(1^7)$ | $(2,1^5)$ | $(2^2,1^3)$ | $(2^3,1)$ | $(4,1^3)$ | $(4,2,1)$ | $(5,1^2)$ | $(5,2)$ | $(7)$ |
|---|---|---|---|---|---|---|---|---|---|
| $(7)$ | 1 | 1 | 1 | 1 | 1 | 1 | 1 | 1 | 1 |
| $(6,1)$ | 6 | 4 | 2 | 0 | 2 | 0 | 1 | 1 | $-1$ |
| $(5,2)$ | 13 | 5 | 1 | 1 | 1 | $-1$ | $-2$ | 0 | $-1$ |
| $(5,1^2)$ | 15 | 5 | $-1$ | $-3$ | 1 | $-1$ | 0 | 0 | 1 |
| $(4,3)$ | 1 | $-1$ | 1 | $-1$ | $-1$ | 1 | 1 | $-1$ | 1 |
| $(4,2,1)$ | 20 | 0 | $-4$ | 0 | 0 | 0 | 0 | 0 | $-1$ |
| $(3^2,1)$ | 6 | $-4$ | 2 | 0 | $-2$ | 0 | 1 | 1 | $-1$ |
| $(3,2^2)$ | 15 | $-5$ | $-1$ | 3 | $-1$ | $-1$ | 0 | 0 | 1 |
| $(3,2,1^2)$ | 13 | $-5$ | 1 | $-1$ | 1 | $-1$ | $-2$ | 0 | $-1$ |

| $n=8$ | $(1^8)$ | $(2,1^6)$ | $(2^2,1^4)$ | $(2^3,1^2)$ | $(2^4)$ | $(4,1^4)$ | $(4,2,1^2)$ | $(4,2^2)$ | $(4^2)$ | $(5,1^3)$ | $(5,2,1)$ | $(7,1)$ | $(8)$ |
|---|---|---|---|---|---|---|---|---|---|---|---|---|---|
| $(8)$ | 1 | 1 | 1 | 1 | 1 | 1 | 1 | 1 | 1 | 1 | 1 | 1 | 1 |
| $(7,1)$ | 7 | 5 | 3 | 1 | −1 | 3 | 1 | −1 | −1 | 2 | 0 | 0 | −1 |
| $(6,2)$ | 13 | 5 | 1 | 1 | 5 | −1 | −1 | 3 | 1 | −2 | 0 | −1 | 1 |
| $(6,1^2)$ | 21 | 9 | 1 | −3 | −3 | 3 | −1 | −1 | 1 | 1 | −1 | 0 | 1 |
| $(5,3)$ | 28 | 10 | 4 | 2 | −4 | −2 | 0 | −2 | 0 | −2 | 0 | 0 | 0 |
| $(5,2,1)$ | 35 | 5 | −5 | −3 | 3 | 1 | −1 | 1 | −1 | 0 | 0 | 0 | −1 |
| $(4^2)$ | 1 | −1 | 1 | −1 | 1 | −1 | 1 | −1 | 1 | 1 | −1 | 1 | −1 |
| $(4,3,1)$ | 7 | −5 | 3 | −1 | −1 | −3 | 1 | 1 | −1 | 2 | 0 | 0 | 1 |
| $(4,2^2)$ | 35 | −5 | −5 | 3 | 3 | −1 | −1 | −1 | −1 | 0 | 0 | 0 | 1 |
| $(4,2,1^2)$ | 90 | 0 | −6 | 0 | −6 | 0 | 2 | 0 | 2 | 0 | 0 | −1 | 0 |
| $(3^2,2)$ | 21 | −9 | 1 | 3 | −3 | −3 | −1 | 1 | 1 | 1 | 1 | 0 | −1 |
| $(3^2,1^2)$ | 13 | −5 | 1 | −1 | 5 | 1 | −1 | −3 | 1 | −2 | 0 | −1 | −1 |
| $(3,2^2,1)$ | 28 | −10 | 4 | −2 | −4 | 2 | 0 | 2 | 0 | −2 | 0 | 0 | 0 |

| $n=9$ | $(1^9)$ | $(2,1^7)$ | $(2^2,1^5)$ | $(2^3,1^3)$ | $(2^4,1)$ | $(4,1^5)$ | $(4,2,1^3)$ | $(4,2^2,1)$ | $(4^2,1)$ | $(5,1^5)$ | $(5,2,1^2)$ | $(5,2^2)$ | $(5,4)$ | $(7,1^2)$ | $(7,2)$ | $(9)$ |
|---|---|---|---|---|---|---|---|---|---|---|---|---|---|---|---|---|
| $(9)$ | 1 | 1 | 1 | 1 | 1 | 1 | 1 | 1 | 1 | 1 | 1 | 1 | 1 | 1 | 1 | 1 |
| $(8,1)$ | 7 | 5 | 3 | 1 | −1 | 3 | 1 | −1 | −1 | 2 | 0 | −2 | −2 | 0 | −2 | −1 |
| $(7,2)$ | 27 | 15 | 7 | 3 | 3 | 5 | 1 | 1 | −1 | 2 | 0 | 2 | 0 | −1 | 1 | −1 |
| $(7,1^2)$ | 21 | 9 | 1 | −3 | −3 | 3 | −1 | −1 | 1 | 1 | −1 | 1 | 3 | 0 | 2 | 1 |
| $(6,3)$ | 41 | 15 | 5 | 3 | 1 | −3 | −1 | 1 | 1 | −4 | 0 | 0 | 2 | −1 | 1 | 1 |
| $(6,2,1)$ | 35 | 5 | −5 | −3 | 3 | 1 | −1 | 1 | −1 | 0 | 0 | 0 | −4 | 0 | −2 | −1 |
| $(5,4)$ | 1 | −1 | 1 | −1 | 1 | −1 | 1 | −1 | 1 | 1 | −1 | 1 | −1 | 1 | −1 | −1 |
| $(5,3,1)$ | 162 | 36 | 6 | 0 | −6 | −6 | 0 | −2 | −2 | −3 | 1 | 1 | −1 | 1 | 1 | 0 |
| $(5,2^2)$ | 35 | −5 | −5 | 3 | 3 | −1 | −1 | −1 | −1 | 0 | 0 | 0 | 4 | 0 | 2 | 1 |
| $(5,2,1^2)$ | 189 | 21 | −11 | −3 | −3 | 1 | 1 | 1 | 1 | −1 | 1 | −1 | 1 | 0 | 0 | −1 |
| $(4^2,1)$ | 7 | −5 | 3 | −1 | −1 | −3 | 1 | 1 | −1 | 2 | 0 | −2 | 2 | 0 | 2 | 1 |
| $(4,3,2)$ | 21 | −9 | 1 | 3 | −3 | −3 | −1 | 1 | 1 | 1 | 1 | 1 | −3 | 0 | −2 | −1 |
| $(4,3,1^2)$ | 27 | −15 | 7 | −3 | 3 | −5 | 1 | −1 | −1 | 2 | 0 | 2 | 0 | −1 | −1 | 1 |
| $(4,2^2,1)$ | 189 | −21 | −11 | 3 | −3 | −1 | 1 | −1 | 1 | −1 | −1 | −1 | −1 | 0 | 0 | 1 |
| $(3^2,2,1)$ | 41 | −15 | 5 | −3 | 1 | 3 | −1 | −1 | 1 | −4 | 0 | 0 | −2 | −1 | −1 | −1 |
| $(3,2^2,1^2)$ | 162 | −36 | 6 | 0 | −6 | 6 | 0 | 2 | −2 | −3 | −1 | 1 | 1 | 1 | −1 | 0 |

| $n=10$ | $(1^{10})$ | $(2,1^8)$ | $(2^2,1^6)$ | $(2^3,1^4)$ | $(2^4,1^2)$ | $(2^5)$ | $(4,1^6)$ | $(4,2,1^4)$ | $(4,2^2,1^2)$ | $(4,2^3)$ | $(4^2,1^2)$ | $(4^2,2)$ | $(5,1^5)$ | $(5,2,1^3)$ | $(5,2^2,1)$ | $(5,4,1)$ | $(5^2)$ | $(7,1^3)$ | $(7,2,1)$ | $(8,1^2)$ | $(8,2)$ | $(10)$ |
|---|---|---|---|---|---|---|---|---|---|---|---|---|---|---|---|---|---|---|---|---|---|---|
| $(10)$ | 1 | 1 | 1 | 1 | 1 | 1 | 1 | 1 | 1 | 1 | 1 | 1 | 1 | 1 | 1 | 1 | 1 | 1 | 1 | 1 | 1 | 1 |
| $(9,1)$ | 9 | 7 | 5 | 3 | 1 | −1 | 5 | 3 | 1 | −1 | 1 | −1 | 4 | 2 | 0 | 0 | −1 | 2 | 0 | 1 | −1 | −1 |
| $(8,2)$ | 34 | 20 | 10 | 4 | 2 | 4 | 8 | 2 | 0 | 2 | −2 | 0 | 4 | 0 | 0 | −2 | −1 | −1 | −1 | −2 | 0 | −1 |
| $(8,1^2)$ | 36 | 20 | 8 | 0 | −4 | −4 | 10 | 2 | −2 | −2 | 0 | 0 | 6 | 0 | −2 | 0 | 1 | 1 | −1 | 0 | 0 | 1 |
| $(7,3)$ | 41 | 15 | 5 | 3 | 1 | −9 | −3 | −1 | 1 | −5 | 1 | −1 | −4 | 0 | 0 | 2 | 1 | −1 | 1 | 1 | −1 | 1 |
| $(7,2,1)$ | 84 | 28 | 0 | −8 | −4 | 4 | 10 | −2 | −2 | 2 | 0 | 0 | 4 | −2 | 0 | 0 | −1 | 0 | 0 | 0 | 0 | −1 |
| $(6,4)$ | 90 | 34 | 14 | 6 | 2 | 10 | −4 | 0 | 0 | 4 | 2 | 2 | −5 | −1 | −1 | 1 | 0 | −1 | −1 | 0 | 0 | 0 |
| $(6,3,1)$ | 279 | 71 | 11 | 3 | −1 | −1 | −11 | −3 | 1 | 1 | −1 | −1 | −11 | 1 | 1 | −1 | −1 | −1 | 1 | 1 | 1 | −1 |
| $(6,2^2)$ | 126 | 14 | −14 | −6 | 6 | 6 | 4 | −4 | 0 | 0 | −2 | −2 | 1 | −1 | 1 | −1 | 1 | 0 | 0 | 0 | 0 | 1 |
| $(6,2,1^2)$ | 224 | 26 | −16 | −6 | 0 | −6 | 2 | 0 | 2 | 0 | 0 | 2 | −1 | 1 | −1 | −3 | −1 | 0 | −2 | −2 | 0 | −1 |
| $(5^2)$ | 1 | −1 | 1 | −1 | 1 | −1 | −1 | 1 | −1 | 1 | 1 | −1 | 1 | −1 | 1 | −1 | 1 | 1 | −1 | −1 | 1 | −1 |
| $(5,4,1)$ | 9 | −7 | 5 | −3 | 1 | 1 | −5 | 3 | −1 | −1 | 1 | 1 | 4 | −2 | 0 | 0 | −1 | 2 | 0 | −1 | −1 | 1 |
| $(5,3,2)$ | 126 | −14 | −14 | 6 | 6 | −6 | −4 | −4 | 0 | 0 | −2 | 2 | 1 | 1 | 1 | 1 | 1 | 0 | 0 | 0 | 0 | −1 |
| $(5,3,1^2)$ | 567 | 63 | −9 | −9 | −9 | 15 | −9 | 3 | −1 | 3 | −1 | −1 | −3 | 3 | 1 | 1 | 2 | 0 | 0 | −1 | −1 | 0 |
| $(5,2^2,1)$ | 224 | −26 | −16 | 6 | 0 | 6 | −2 | 0 | −2 | 0 | 0 | −2 | −1 | −1 | −1 | 3 | −1 | 0 | 2 | 2 | 0 | 1 |
| $(4^2,2)$ | 36 | −20 | 8 | 0 | −4 | 4 | −10 | 2 | 2 | −2 | 0 | 0 | 6 | 0 | −2 | 0 | 1 | 1 | 1 | 0 | 0 | −1 |
| $(4^2,1^2)$ | 34 | −20 | 10 | −4 | 2 | −4 | −8 | 2 | 0 | 2 | −2 | 0 | 4 | 0 | 0 | 2 | −1 | −1 | 1 | 2 | 0 | 1 |
| $(4,3^2)$ | 84 | −28 | 0 | 8 | −4 | −4 | −10 | −2 | 2 | 2 | 0 | 0 | 4 | 2 | 0 | 0 | −1 | 0 | 0 | 0 | 0 | 1 |
| $(4,3,2,1)$ | 41 | −15 | 5 | −3 | 1 | 9 | 3 | −1 | −1 | −5 | 1 | 1 | −4 | 0 | 0 | −2 | 1 | −1 | −1 | −1 | −1 | −1 |
| $(4,2^2,1^2)$ | 567 | −63 | −9 | 9 | −9 | −15 | 9 | 3 | 1 | 3 | −1 | 1 | −3 | −3 | 1 | −1 | 2 | 0 | 0 | 1 | −1 | 0 |
| $(3^2,2^2)$ | 90 | −34 | 14 | −6 | 2 | −10 | 4 | 0 | 0 | 4 | 2 | −2 | −5 | 1 | −1 | −1 | 0 | −1 | 1 | 0 | 0 | 0 |
| $(3^2,2,1^2)$ | 279 | −71 | 11 | −3 | −1 | 1 | 11 | −3 | −1 | 1 | −1 | 1 | −11 | −1 | 1 | 1 | −1 | −1 | −1 | −1 | 1 | 1 |

# I.G Littlewood-Richardson Coefficients

This section shows tables of multiplicities of the following kind:

$$([\alpha]\cdot[\beta],[\gamma]):=([\alpha]\#[\beta]\uparrow S_{m+n},[\gamma]),$$

where $\alpha \vdash m$, $\beta \vdash n$, so that $\gamma \vdash m+n$. These multiplicities can be obtained from the Littlewood-Richardson rule.

The tables are taken from F. D. Murnaghan's paper in Amer. J. Math. **59** (1937), 437–488. They were checked with the aid of a computer program written by F. Sänger.

The arrows show how use is made of the fact that

$$([\alpha][\beta],[\gamma])=([\alpha'][\beta'],[\gamma']).$$

*1. m+n=2:*

| | [2] | $[1^2]$ |
|---|---|---|
| [1][1] | 1 | 1 |

*2. m+n=3:*

| | [3] | [21] | $[1^3]$ | |
|---|---|---|---|---|
| [2][1] | 1 | 1 | | $[1^2][1]$ |
| | $[1^3]$ | [21] | [3] | |

*3. m+n=4:*

| | [4] | [31] | $[2^2]$ | $[21^2]$ | $[1^4]$ | |
|---|---|---|---|---|---|---|
| [3][1] | 1 | 1 | | | | $[1^3][1]$ |
| [21][1] | | 1 | 1 | 1 | | [21][1] |
| [2][2] | 1 | 1 | 1 | | | $[1^2][1^2]$ |
| $[2][1^2]$ | | 1 | | 1 | | $[1^2][2]$ |
| | $[1^4]$ | $[21^2]$ | $[2^2]$ | $[3^1]$ | [4] | |

*4. m+n=5:*

| | [5] | [41] | [32] | $[31^2]$ | $[2^21]$ | $[21^3]$ | $[1^5]$ | |
|---|---|---|---|---|---|---|---|---|
| [4][1] | 1 | 1 | | | | | | $[1^4][1]$ |
| [31][1] | | 1 | 1 | 1 | | | | $[21^2][1]$ |
| $[2^2][1]$ | | | 1 | | 1 | | | $[2^2][1]$ |
| [3][2] | 1 | 1 | 1 | | | | | $[1^3][1^2]$ |
| [21][2] | | 1 | 1 | 1 | 1 | | | $[21][1^2]$ |
| $[1^3][2]$ | | | | 1 | | 1 | | $[3][1^2]$ |
| | $[1^5]$ | $[21^3]$ | $[2^21]$ | $[31^2]$ | [32] | [41] | [5] | |

*5. m+n=6:*

| | [6] | [51] | [42] | $[41^2]$ | $[3^2]$ | [321] | $[31^3]$ | $[2^3]$ | $[2^21^2]$ | $[21^4]$ | $[1^6]$ | |
|---|---|---|---|---|---|---|---|---|---|---|---|---|
| [5][1] | 1 | 1 | | | | | | | | | | $[1^5][1]$ |
| [41][1] | | 1 | 1 | 1 | | | | | | | | $[21^3][1]$ |
| [32][1] | | | 1 | | 1 | 1 | | | | | | $[2^21][1]$ |
| $[31^2][1]$ | | | | 1 | | 1 | 1 | | | | | $[31^2][1]$ |
| [4][2] | 1 | 1 | 1 | | | | | | | | | $[1^4][1^2]$ |
| [31][2] | | 1 | 1 | 1 | 1 | 1 | | | | | | $[21^2][1^2]$ |
| $[2^2][2]$ | | | 1 | | | 1 | | 1 | | | | $[2^2][1^2]$ |
| $[21^2][2]$ | | | | 1 | | 1 | 1 | | 1 | | | $[31][1^2]$ |
| $[1^4][2]$ | | | | | | | 1 | | | 1 | | $[4][1^2]$ |
| [3][3] | 1 | 1 | 1 | | 1 | | | | | | | $[1^3][1^3]$ |
| [3][21] | | 1 | 1 | 1 | | 1 | | | | | | $[1^3][21]$ |
| $[3][1^3]$ | | | | 1 | | | 1 | | | | | $[1^3][3]$ |
| [21][21] | | | 1 | 1 | 1 | 2 | 1 | 1 | 1 | | | [21][21] |
| | $[1^6]$ | $[21^4]$ | $[2^21^2]$ | $[31^3]$ | $[2^3]$ | [321] | $[41^2]$ | $[3^2]$ | [42] | [51] | [6] | |

*6.* $m+n=7$:

| | [7] | [61] | [52] | [51²] | [43] | [421] | [41³] | [3²1] | [32²] | [321²] | [31⁴] | [2³1] | [2²1³] | [21⁵] | [1⁷] | |
|---|---|---|---|---|---|---|---|---|---|---|---|---|---|---|---|---|
| [6][1] | 1 | 1 | | | | | | | | | | | | | | [1⁶][1] |
| [51][1] | | 1 | 1 | 1 | | | | | | | | | | | | [21⁴][1] |
| [42][1] | | | 1 | | 1 | 1 | | | | | | | | | | [2²1²][1] |
| [41²][1] | | | | 1 | | 1 | 1 | | | | | | | | | [31³][1] |
| [3²][1] | | | | | 1 | | | 1 | | | | | | | | [2³][1] |
| [321][1] | | | | | | 1 | | 1 | 1 | 1 | | | | | | [321][1] |
| [5][2] | 1 | 1 | 1 | | | | | | | | | | | | | [1⁵][1²] |
| [41][2] | | 1 | 1 | 1 | 1 | 1 | | | | | | | | | | [21³][1²] |
| [32][2] | | | 1 | | 1 | 1 | | 1 | 1 | | | | | | | [2²1][1²] |
| [31²][2] | | | | 1 | | 1 | 1 | 1 | | 1 | | | | | | [31²][1²] |
| [2²1][2] | | | | | | 1 | | | 1 | 1 | | 1 | | | | [32][1²] |
| [21³][2] | | | | | | | 1 | | | 1 | 1 | | 1 | | | [41][1²] |
| [1⁵][2] | | | | | | | | | | | 1 | | | 1 | | [5][1²] |
| [4][3] | 1 | 1 | 1 | | 1 | | | | | | | | | | | [1⁴][1³] |
| [31][3] | | 1 | 1 | 1 | 1 | 1 | | 1 | | | | | | | | [21²][1³] |
| [2²][3] | | | 1 | | | 1 | | | 1 | | | | | | | [2²][1³] |
| [21²][3] | | | | 1 | | 1 | 1 | | | 1 | | | | | | [31][1³] |
| [1⁴][3] | | | | | | | 1 | | | | 1 | | | | | [4][1³] |
| [4][21] | | 1 | 1 | 1 | | 1 | | | | | | | | | | [1⁴][21] |
| [31][21] | | | 1 | 1 | 1 | 2 | 1 | 1 | 1 | 1 | | | | | | [21²][21] |
| [2²][21] | | | | | 1 | 1 | | 1 | 1 | 1 | | 1 | | | | [2²][21] |
| | [1⁷] | [21⁵] | [2²1³] | [31⁴] | [2³1] | [321²] | [41³] | [32²] | [3²1] | [421] | [51²] | [43] | [52] | [61] | [7] | |

*7. $m+n=8$:*

| | [8] | [71] | [62] | $[61^2]$ | [53] | [521] | $[51^3]$ | $[4^2]$ | [431] | $[42^2]$ | $[421^2]$ | $[41^4]$ | $[3^22]$ | $[3^21^2]$ | $[32^21]$ | $[321^3]$ | $[31^5]$ | $[2^4]$ | $[2^31^2]$ | $[2^21^4]$ | $[21^6]$ | $[1^8]$ | |
|---|---|---|---|---|---|---|---|---|---|---|---|---|---|---|---|---|---|---|---|---|---|---|---|
| [7][1] | 1 | 1 | | | | | | | | | | | | | | | | | | | | | $[1^7][1]$ |
| [61][1] | | 1 | 1 | 1 | | | | | | | | | | | | | | | | | | | $[21^5][1]$ |
| [52][1] | | | 1 | | 1 | 1 | | | | | | | | | | | | | | | | | $[2^21^3][1]$ |
| $[51^2][1]$ | | | | 1 | | 1 | 1 | | | | | | | | | | | | | | | | $[31^4][1]$ |
| [43][1] | | | | | 1 | | | 1 | 1 | | | | | | | | | | | | | | $[2^31][1]$ |
| [421][1] | | | | | | 1 | | | 1 | 1 | 1 | | | | | | | | | | | | $[321^2][1]$ |
| $[41^3][1]$ | | | | | | | 1 | | | | 1 | 1 | | | | | | | | | | | $[41^3][1]$ |
| $[3^21][1]$ | | | | | | | | | 1 | | | | 1 | 1 | | | | | | | | | $[32^2][1]$ |
| [6][2] | 1 | 1 | 1 | | | | | | | | | | | | | | | | | | | | $[1^6][1^2]$ |
| [51][2] | | 1 | 1 | 1 | 1 | 1 | | | | | | | | | | | | | | | | | $[21^4][1^2]$ |
| [42][2] | | | 1 | | 1 | 1 | | 1 | 1 | 1 | | | | | | | | | | | | | $[2^21^2][1^2]$ |
| $[41^2][2]$ | | | | 1 | | 1 | 1 | | 1 | | 1 | | | | | | | | | | | | $[31^3][1^2]$ |
| $[3^2][2]$ | | | | | 1 | | | | 1 | | | | 1 | | | | | | | | | | $[2^3][1^2]$ |
| [321][2] | | | | | | 1 | | | 1 | 1 | 1 | | 1 | 1 | 1 | | | | | | | | $[321][1^2]$ |
| $[31^3][2]$ | | | | | | | 1 | | | | 1 | 1 | | 1 | | 1 | | | | | | | $[41^2][1^2]$ |
| $[2^3][2]$ | | | | | | | | | | 1 | | | | | 1 | | | 1 | | | | | $[3^2][1^2]$ |
| $[2^21^2][2]$ | | | | | | | | | | | 1 | | | | 1 | 1 | | | 1 | | | | $[42][1^2]$ |
| $[21^4][2]$ | | | | | | | | | | | | 1 | | | | 1 | 1 | | | 1 | | | $[51][1^2]$ |
| $[1^6][2]$ | | | | | | | | | | | | | | | | | 1 | | | | 1 | | $[6][1^2]$ |
| [5][3] | 1 | 1 | 1 | | 1 | | | | | | | | | | | | | | | | | | $[1^5][1^3]$ |
| [41][3] | | 1 | 1 | 1 | 1 | 1 | | 1 | 1 | | | | | | | | | | | | | | $[21^3][1^3]$ |
| [32][3] | | | 1 | | 1 | 1 | | | 1 | 1 | | | 1 | | | | | | | | | | $[2^21][1^3]$ |
| $[31^2][3]$ | | | | 1 | | 1 | 1 | | 1 | | 1 | | | 1 | | | | | | | | | $[31^2][1^3]$ |
| $[2^21][3]$ | | | | | | 1 | | | | 1 | 1 | | | | 1 | | | | | | | | $[32][1^3]$ |
| $[21^3][3]$ | | | | | | | 1 | | | | 1 | 1 | | | | 1 | | | | | | | $[41][1^3]$ |
| $[1^5][3]$ | | | | | | | | | | | | 1 | | | | | 1 | | | | | | $[5][1^3]$ |
| [5][21] | | 1 | 1 | 1 | | 1 | | | | | | | | | | | | | | | | | $[1^5][21]$ |
| [41][21] | | | 1 | 1 | 1 | 2 | 1 | | 1 | 1 | 1 | | | | | | | | | | | | $[21^3][21]$ |
| [32][21] | | | | | 1 | 1 | | 1 | 2 | 1 | 1 | | 1 | 1 | 1 | | | | | | | | $[2^21][21]$ |
| $[31^2][21]$ | | | | | | 1 | 1 | | 1 | 1 | 2 | 1 | 1 | 1 | 1 | 1 | | | | | | | $[31^2][21]$ |
| [4][4] | 1 | 1 | 1 | | 1 | | | 1 | | | | | | | | | | | | | | | $[1^4][1^4]$ |
| [31][4] | | 1 | 1 | 1 | 1 | 1 | | | 1 | | | | | | | | | | | | | | $[21^2][1^4]$ |
| $[2^2][4]$ | | | 1 | | | 1 | | | | 1 | | | | | | | | | | | | | $[2^2][1^4]$ |
| $[21^2][4]$ | | | | 1 | | 1 | 1 | | | | 1 | | | | | | | | | | | | $[31][1^4]$ |
| $[1^4][4]$ | | | | | | | 1 | | | | | 1 | | | | | | | | | | | $[4][1^4]$ |
| [31][31] | | | 1 | 1 | 1 | 2 | 1 | 1 | 2 | 1 | 1 | | 1 | 1 | | | | | | | | | $[21^2][21^2]$ |
| $[2^2][31]$ | | | | | 1 | 1 | | | 1 | 1 | 1 | | 1 | | 1 | | | | | | | | $[2^2][21^2]$ |
| $[21^2][31]$ | | | | | | 1 | 1 | | 1 | 1 | 2 | 1 | | 1 | 1 | 1 | | | | | | | $[31][21^2]$ |
| $[2^2][2^2]$ | | | | | | | | 1 | 1 | 1 | | | | 1 | 1 | | | 1 | | | | | $[2^2][2^2]$ |
| | $[1^8]$ | $[21^6]$ | $[2^21^4]$ | $[31^5]$ | $[2^31^2]$ | $[321^3]$ | $[41^4]$ | $[2^4]$ | $[32^21]$ | $[3^21^2]$ | $[421^2]$ | $[51^3]$ | $[3^22]$ | $[42^2]$ | [431] | [521] | $[61^2]$ | $[4^2]$ | [53] | [62] | [71] | [8] | |

*8. m+n=9:*

| | [9] | [81] | [72] | $[71^2]$ | [63] | [621] | $[61^3]$ | [54] | [531] | $[52^2]$ | $[521^2]$ | $[51^4]$ | $[4^21]$ | [432] | $[431^2]$ | $[42^21]$ | $[421^3]$ | $[41^5]$ | $[3^3]$ | $[3^221]$ | $[3^21^3]$ | $[32^3]$ | $[32^21^2]$ | $[321^4]$ | $[31^6]$ | $[2^41]$ | $[2^31^3]$ | $[2^21^5]$ | $[21^7]$ | $[1^9]$ | |
|---|---|---|---|---|---|---|---|---|---|---|---|---|---|---|---|---|---|---|---|---|---|---|---|---|---|---|---|---|---|---|---|
| [8][1] | 1 | 1 | | | | | | | | | | | | | | | | | | | | | | | | | | | | | $[1^8][1]$ |
| [71][1] | | 1 | 1 | 1 | | | | | | | | | | | | | | | | | | | | | | | | | | | $[21^6][1]$ |
| [62][1] | | | 1 | | 1 | 1 | | | | | | | | | | | | | | | | | | | | | | | | | $[2^21^4][1]$ |
| $[61^2][1]$ | | | | 1 | | 1 | 1 | | | | | | | | | | | | | | | | | | | | | | | | $[31^5][1]$ |
| [53][1] | | | | | 1 | | | 1 | 1 | | | | | | | | | | | | | | | | | | | | | | $[2^31^2][1]$ |
| [521][1] | | | | | | 1 | | | 1 | 1 | 1 | | | | | | | | | | | | | | | | | | | | $[321^3][1]$ |
| $[51^3][1]$ | | | | | | | 1 | | | | 1 | 1 | | | | | | | | | | | | | | | | | | | $[41^4][1]$ |
| $[4^2][1]$ | | | | | | | | 1 | | | | | 1 | | | | | | | | | | | | | | | | | | $[2^4][1]$ |
| [431][1] | | | | | | | | | 1 | | | | 1 | 1 | 1 | | | | | | | | | | | | | | | | $[32^21][1]$ |
| $[42^2][1]$ | | | | | | | | | | 1 | | | | 1 | | 1 | | | | | | | | | | | | | | | $[3^21^2][1]$ |
| $[421^2][1]$ | | | | | | | | | | | 1 | | | | 1 | 1 | 1 | | | | | | | | | | | | | | $[421^2][1]$ |
| $[3^22][1]$ | | | | | | | | | | | | | | 1 | | | | | 1 | 1 | | | | | | | | | | | $[3^22][1]$ |
| [7][2] | 1 | 1 | 1 | | | | | | | | | | | | | | | | | | | | | | | | | | | | $[1^7][1^2]$ |
| [61][2] | | 1 | 1 | 1 | 1 | 1 | | | | | | | | | | | | | | | | | | | | | | | | | $[21^5][1^2]$ |
| [52][2] | | | 1 | | 1 | 1 | | 1 | 1 | 1 | | | | | | | | | | | | | | | | | | | | | $[2^21^3][1^2]$ |
| $[51^2][2]$ | | | | 1 | | 1 | 1 | | 1 | | 1 | | | | | | | | | | | | | | | | | | | | $[31^4][1^2]$ |
| [43][2] | | | | | 1 | | | 1 | 1 | | | | 1 | 1 | | | | | | | | | | | | | | | | | $[2^31][1^2]$ |
| [421][2] | | | | | | 1 | | | 1 | 1 | 1 | | 1 | 1 | 1 | 1 | | | | | | | | | | | | | | | $[321^2][1^2]$ |
| $[41^3][2]$ | | | | | | | 1 | | | | 1 | 1 | | | 1 | | 1 | | | | | | | | | | | | | | $[41^3][1^2]$ |
| $[3^21][2]$ | | | | | | | | | 1 | | | | | 1 | 1 | | | | 1 | 1 | | | | | | | | | | | $[32^2][1^2]$ |
| $[32^2][2]$ | | | | | | | | | | 1 | | | | 1 | | 1 | | | | 1 | | 1 | | | | | | | | | $[3^21][1^2]$ |
| $[321^2][2]$ | | | | | | | | | | | 1 | | | | 1 | 1 | 1 | | | 1 | 1 | | 1 | | | | | | | | $[421][1^2]$ |
| $[31^4][2]$ | | | | | | | | | | | | 1 | | | | | 1 | 1 | | | 1 | | | 1 | | | | | | | $[51^2][1^2]$ |
| $[2^31][2]$ | | | | | | | | | | | | | | | | 1 | | | | | | 1 | 1 | | | 1 | | | | | $[43][1^2]$ |
| $[2^21^3][2]$ | | | | | | | | | | | | | | | | | 1 | | | | | | 1 | 1 | | | 1 | | | | $[52][1^2]$ |
| $[21^5][2]$ | | | | | | | | | | | | | | | | | | 1 | | | | | | 1 | 1 | | | 1 | | | $[61][1^2]$ |
| $[1^7][2]$ | | | | | | | | | | | | | | | | | | | | | | | | | 1 | | | | 1 | | $[7][1^2]$ |
| [6][3] | 1 | 1 | 1 | | 1 | | | | | | | | | | | | | | | | | | | | | | | | | | $[1^6][1^3]$ |
| [51][3] | | 1 | 1 | 1 | 1 | 1 | | 1 | 1 | | | | | | | | | | | | | | | | | | | | | | $[21^4][1^3]$ |
| [42][3] | | | 1 | | 1 | 1 | | 1 | 1 | 1 | | | 1 | 1 | | | | | | | | | | | | | | | | | $[2^21^2][1^3]$ |
| $[41^2][3]$ | | | | 1 | | 1 | 1 | | 1 | | 1 | | 1 | | 1 | | | | | | | | | | | | | | | | $[31^3][1^3]$ |
| | $[1^9]$ | $[21^7]$ | $[2^21^5]$ | $[31^6]$ | $[2^31^3]$ | $[321^4]$ | $[4,1^5]$ | $[2^41]$ | $[32^21^2]$ | $[3^21^3]$ | $[421^3]$ | $[51^4]$ | $[32^3]$ | $[3^221]$ | $[42^21]$ | $[431^2]$ | $[521^2]$ | $[61^3]$ | $[3^3]$ | [432] | $[52^2]$ | $[4^21]$ | [531] | [621] | $[71^2]$ | [54] | [63] | [72] | [81] | [9] | |

$m+n=9$ (continued):

| | $[9]$ | $[81]$ | $[72]$ | $[71^2]$ | $[63]$ | $[621]$ | $[61^3]$ | $[54]$ | $[531]$ | $[52^2]$ | $[521^2]$ | $[51^4]$ | $[4^21]$ | $[432]$ | $[431^2]$ | $[42^21]$ | $[421^3]$ | $[41^5]$ | $[3^3]$ | $[3^221]$ | $[3^21^3]$ | $[32^3]$ | $[32^21^2]$ | $[321^4]$ | $[31^6]$ | $[2^41]$ | $[2^31^3]$ | $[2^21^5]$ | $[21^7]$ | $[1^9]$ | |
|---|---|---|---|---|---|---|---|---|---|---|---|---|---|---|---|---|---|---|---|---|---|---|---|---|---|---|---|---|---|---|---|
| $[3^2][3]$ | | | | | 1 | | | | 1 | | | | | 1 | | | | | 1 | | | | | | | | | | | | $[2^3][1^3]$ |
| $[321][3]$ | | | | | | 1 | | | 1 | 1 | 1 | | | 1 | 1 | 1 | | | | 1 | | | | | | | | | | | $[321][1^3]$ |
| $[31^3][3]$ | | | | | | | 1 | | | | 1 | 1 | | | 1 | | 1 | | | | 1 | | | | | | | | | | $[41^2][1^3]$ |
| $[2^3][3]$ | | | | | | | | | | 1 | | | | | | 1 | | | | | | 1 | | | | | | | | | $[3^2][1^3]$ |
| $[2^21^2][3]$ | | | | | | | | | | | 1 | | | | | 1 | 1 | | | | | | 1 | | | | | | | | $[42][1^3]$ |
| $[21^4][3]$ | | | | | | | | | | | | 1 | | | | | 1 | 1 | | | | | | 1 | | | | | | | $[51][1^3]$ |
| $[1^6][3]$ | | | | | | | | | | | | | | | | | | 1 | | | | | | | 1 | | | | | | $[6][1^3]$ |
| $[6][21]$ | | 1 | 1 | 1 | | 1 | | | | | | | | | | | | | | | | | | | | | | | | | $[1^6][21]$ |
| $[51][21]$ | | | 1 | 1 | 1 | 2 | 1 | | 1 | 1 | 1 | | | | | | | | | | | | | | | | | | | | $[21^4][21]$ |
| $[42][21]$ | | | | | 1 | 1 | | 1 | 2 | 1 | 1 | | 1 | 1 | 1 | 1 | | | | | | | | | | | | | | | $[2^21^2][21]$ |
| $[41^2][21]$ | | | | | | 1 | 1 | | 1 | 1 | 2 | 1 | | 1 | 1 | 1 | 1 | | | | | | | | | | | | | | $[31^3][21]$ |
| $[3^2][21]$ | | | | | | | | 1 | 1 | | | | 1 | 1 | 1 | | | | | 1 | | | | | | | | | | | $[2^3][21]$ |
| $[321][21]$ | | | | | | | | | 1 | 1 | 1 | | 1 | 2 | 2 | 2 | 1 | | 1 | 2 | 1 | 1 | 1 | | | | | | | | $[321][21]$ |
| $[5][4]$ | 1 | 1 | 1 | | 1 | | | 1 | | | | | | | | | | | | | | | | | | | | | | | $[1^5][1^4]$ |
| $[41][4]$ | | 1 | 1 | 1 | 1 | 1 | | 1 | 1 | | | | 1 | | | | | | | | | | | | | | | | | | $[21^3][1^4]$ |
| $[32][4]$ | | | 1 | | 1 | 1 | | | 1 | 1 | | | | 1 | | | | | | | | | | | | | | | | | $[2^21][1^4]$ |
| $[31^2][4]$ | | | | 1 | | 1 | 1 | | 1 | | 1 | | | | 1 | | | | | | | | | | | | | | | | $[31^2][1^4]$ |
| $[2^21][4]$ | | | | | | 1 | | | | 1 | 1 | | | | | 1 | | | | | | | | | | | | | | | $[32][1^4]$ |
| $[21^3][4]$ | | | | | | | 1 | | | | 1 | 1 | | | | | 1 | | | | | | | | | | | | | | $[41][1^4]$ |
| $[1^5][4]$ | | | | | | | | | | | | 1 | | | | | | 1 | | | | | | | | | | | | | $[5][1^4]$ |
| $[5][31]$ | | 1 | 1 | 1 | 1 | 1 | | | 1 | | | | | | | | | | | | | | | | | | | | | | $[1^5][21^2]$ |
| $[41][31]$ | | | 1 | 1 | 1 | 2 | 1 | 1 | 2 | 1 | 1 | | 1 | 1 | 1 | | | | | | | | | | | | | | | | $[21^3][21^2]$ |
| $[32][31]$ | | | | | 1 | 1 | | 1 | 2 | 1 | 1 | | 1 | 2 | 1 | 1 | | | 1 | 1 | | | | | | | | | | | $[2^21][21^2]$ |
| $[31^2][31]$ | | | | | | 1 | 1 | | 1 | 1 | 2 | 1 | 1 | 1 | 2 | 1 | 1 | | | 1 | 1 | | | | | | | | | | $[31^2][21^2]$ |
| $[2^21][31]$ | | | | | | | | | 1 | 1 | 1 | | | 1 | 1 | 2 | 1 | | | 1 | | 1 | 1 | | | | | | | | $[32][21^2]$ |
| $[21^3][31]$ | | | | | | | | | | | 1 | 1 | | | 1 | 1 | 2 | 1 | | | 1 | | 1 | 1 | | | | | | | $[41][21^2]$ |
| $[1^5][31]$ | | | | | | | | | | | | | | | | | 1 | 1 | | | | | | 1 | 1 | | | | | | $[5][21^2]$ |
| $[5][2^2]$ | | | 1 | | | 1 | | | | 1 | | | | | | | | | | | | | | | | | | | | | $[1^5][2^2]$ |
| $[41][2^2]$ | | | | | 1 | 1 | | | 1 | 1 | 1 | | | 1 | | 1 | | | | | | | | | | | | | | | $[21^3][2^2]$ |
| $[32][2^2]$ | | | | | | | | 1 | 1 | 1 | | | 1 | 1 | 1 | 1 | | | | 1 | | 1 | | | | | | | | | $[2^21][2^2]$ |
| $[31^2][2^2]$ | | | | | | | | | 1 | | 1 | | | 1 | 1 | 1 | 1 | | 1 | 1 | | | 1 | | | | | | | | $[31^2][2^2]$ |
| | $[1^9]$ | $[21^7]$ | $[2^21^5]$ | $[31^6]$ | $[2^31^3]$ | $[321^4]$ | $[41^5]$ | $[2^41]$ | $[32^21^2]$ | $[3^21^3]$ | $[421^3]$ | $[51^4]$ | $[32^3]$ | $[3^221]$ | $[42^21]$ | $[431^2]$ | $[521^2]$ | $[61^3]$ | $[3^3]$ | $[432]$ | $[52^2]$ | $[4^21]$ | $[531]$ | $[621]$ | $[71^2]$ | $[54]$ | $[63]$ | $[72]$ | $[81]$ | $[9]$ | |

# I.H Character Tables of Wreath Products of Symmetric Groups

Here we show character tables of wreath products $S_m \operatorname{wr} S_n$, $m \cdot n \leqslant 10$, $m, n > 1$. We use the notation of Chapter 4, but for sake of simplicity we write $D^* \otimes D' \uparrow G$, $G := S_m \operatorname{wr} S_n$, instead of $\tilde{D}^* \otimes D' \uparrow G$, and $(\alpha; \beta)$ instead of $([\alpha]; [\beta])$.

The tables are due to F. Sänger.

$G = S_2 \operatorname{wr} S_2$

| Cycle-partition of representative | $[1^4]$ | $[1^2 2]$ | $[2^2]$ | $[2^2]$ | $[4]$ |
|---|---|---|---|---|---|
| Representative of conjugacy class | (e;1) | ((12),1;1) | ((12),(12);1) | (e;(12)) | ((12),1;(12)) |
| Class order | 1 | 2 | 1 | 2 | 2 |
| $(2;2)$ | 1 | 1 | 1 | 1 | 1 |
| $(2;1^2)$ | 1 | 1 | 1 | -1 | -1 |
| $(1^2;2)$ | 1 | -1 | 1 | 1 | -1 |
| $(1^2;1^2)$ | 1 | -1 | 1 | -1 | 1 |
| $([2]\#[1^2])\uparrow G$ | 2 | 0 | -2 | 0 | 0 |

$G = S_2 \text{ wr } S_3$

| Cycle-partition of representative | $[1^6]$ | $[1^4 2]$ | $[1^2 2^2]$ | $[1^2 2^2]$ | $[1^2 4]$ | $[2^3]$ | $[2^3]$ | $[24]$ | $[3^2]$ | $[6]$ |
|---|---|---|---|---|---|---|---|---|---|---|
| Representative of conjugacy class | (e;1) | ((12),1,1;1) | ((12),(12),1;1) | (e;(12)) | ((12),1,1;(12)) | ((12),(12),(12);1) | (1,1,(12);(12)) | ((12),1,(12);(12)) | (e;(123)) | ((12),1,1;(123)) |
| Class order | 1 | 3 | 3 | 6 | 6 | 1 | 6 | 6 | 8 | 8 |
| $(2;3)$ | 1 | 1 | 1 | 1 | 1 | 1 | 1 | 1 | 1 | 1 |
| $(2;21)$ | 2 | 2 | 2 | 0 | 0 | 2 | 0 | 0 | -1 | -1 |
| $(2;1^3)$ | 1 | 1 | 1 | -1 | -1 | 1 | -1 | -1 | 1 | 1 |
| $(1^2;3)$ | 1 | -1 | 1 | 1 | -1 | -1 | -1 | 1 | 1 | -1 |
| $(1^2;21)$ | 2 | -2 | 2 | 0 | 0 | -2 | 0 | 0 | -1 | 1 |
| $(1^2;1^3)$ | 1 | -1 | 1 | -1 | 1 | -1 | 1 | -1 | 1 | -1 |
| $(([2]\#[2]\#[1^2])\otimes[2]')\uparrow G$ | 3 | 1 | -1 | 1 | 1 | -3 | -1 | -1 | 0 | 0 |
| $(([2]\#[2]\#[1^2])\otimes[1^2]')\uparrow G$ | 3 | 1 | -1 | -1 | -1 | -3 | 1 | 1 | 0 | 0 |
| $(([2]\#[1^2]\#[1^2])\otimes[2]')\uparrow G$ | 3 | -1 | -1 | 1 | -1 | 3 | 1 | -1 | 0 | 0 |
| $(([2]\#[1^2]\#[1^2])\otimes[1^2]')\uparrow G$ | 3 | -1 | -1 | -1 | 1 | 3 | -1 | 1 | 0 | 0 |

$G = S_3 \text{ wr } S_2$

| Cycle-partition of representative | $[1^6]$ | $[1^4 2]$ | $[1^3 3]$ | $[1^2 2^2]$ | $[123]$ | $[2^3]$ | $[24]$ | $[3^2]$ | $[6]$ |
|---|---|---|---|---|---|---|---|---|---|
| Representative of conjugacy class | (e;1) | ((12),1;1) | ((123),1;1) | ((12),(12);1) | ((123),(12);1) | (e;(12)) | ((12),1;(12)) | ((123),(123);1) | ((123),1;(12)) |
| Class order | 1 | 6 | 4 | 9 | 12 | 6 | 18 | 4 | 12 |
| (3;2) | 1 | 1 | 1 | 1 | 1 | 1 | 1 | 1 | 1 |
| $(3;1^2)$ | 1 | 1 | 1 | 1 | 1 | -1 | -1 | 1 | -1 |
| (21;2) | 4 | 0 | -2 | 0 | 0 | 2 | 0 | 1 | -1 |
| $(21;1^2)$ | 4 | 0 | -2 | 0 | 0 | -2 | 0 | 1 | 1 |
| $(1^3;2)$ | 1 | -1 | 1 | 1 | -1 | 1 | -1 | 1 | 1 |
| $(1^3;1^2)$ | 1 | -1 | 1 | 1 | -1 | -1 | 1 | 1 | -1 |
| ([3]#[21])↑G | 4 | 2 | 1 | 0 | -1 | 0 | 0 | -2 | 0 |
| $([3]\#[1^3])$↑G | 2 | 0 | 2 | -2 | 0 | 0 | 0 | 2 | 0 |
| $([21]\#[1^3])$↑G | 4 | -2 | 1 | 0 | 1 | 0 | 0 | -2 | 0 |

$G = S_2 \,\mathrm{wr}\, S_4$

| Cycle-partition of representatitve | $[1^8]$ | $[1^6 2]$ | $[1^4 2^2]$ | $[1^4 2^2]$ | $[1^4 4]$ | $[1^2 2^3]$ | $[1^2 2^3]$ | $[1^2 24]$ | $[1^2 3^2]$ | $[1^2 6]$ | $[2^4]$ | $[2^4]$ | $[2^4]$ | $[2^2 4]$ | $[2^2 4]$ | $[23^2]$ | $[26]$ | $[4^2]$ | $[4^2]$ | $[8]$ |
|---|---|---|---|---|---|---|---|---|---|---|---|---|---|---|---|---|---|---|---|---|
| Representative of conjugacy class | (e;1) | ((12),1,1,1;1) | ((12),(12),1,1;1) | (e;(12)) | ((12),1,1,1;(12)) | ((12),(12),(12),1;1) | (1,1,(12),1;(12)) | ((12),1,(12),1;(12)) | (e;(123)) | ((12),1,1,1;(123)) | ((12),(12),(12),(12);1) | (1,1,(12),(12);(12)) | (e;(12)(34)) | ((12),1,(12),(12);(12)) | ((12),1,1,1;(12)(34)) | (1,1,1,(12);(123)) | ((12),1,1,(12);(123)) | ((12),1,(12),1;(12)(34) | (e;(1234)) | ((12),1,1,1;(1234)) |
| Class order | 1 | 4 | 6 | 12 | 12 | 4 | 24 | 24 | 32 | 32 | 1 | 12 | 12 | 12 | 23 | 32 | 32 | 12 | 48 | 48 |
| (2;4) | 1 | 1 | 1 | 1 | 1 | 1 | 1 | 1 | 1 | 1 | 1 | 1 | 1 | 1 | 1 | 1 | 1 | 1 | 1 | 1 |
| (2;31) | 3 | 3 | 3 | 1 | 1 | 3 | 1 | 1 | 0 | 0 | 3 | 1 | -1 | 1 | -1 | 0 | 0 | -1 | -1 | -1 |
| $(2;2^2)$ | 2 | 2 | 2 | 0 | 0 | 2 | 0 | 0 | -1 | -1 | 2 | 0 | 2 | 0 | 2 | -1 | -1 | 2 | 0 | 0 |
| $(2;21^2)$ | 3 | 3 | 3 | -1 | -1 | 3 | -1 | -1 | 0 | 0 | 3 | -1 | -1 | -1 | -1 | 0 | 0 | -1 | 1 | 1 |
| $(2;1^4)$ | 1 | 1 | 1 | -1 | -1 | 1 | -1 | -1 | 1 | 1 | 1 | -1 | 1 | -1 | 1 | 1 | 1 | 1 | -1 | -1 |
| $(1^2;4)$ | 1 | -1 | 1 | 1 | -1 | -1 | -1 | 1 | 1 | -1 | 1 | 1 | 1 | -1 | -1 | -1 | 1 | 1 | 1 | -1 |
| $(1^2;31)$ | 3 | -3 | 3 | 1 | -1 | -3 | -1 | 1 | 0 | 0 | 3 | 1 | -1 | -1 | 1 | 0 | 0 | -1 | -1 | 1 |
| $(1^2;2^2)$ | 2 | -2 | 2 | 0 | 0 | -2 | 0 | 0 | -1 | 1 | 2 | 0 | 2 | 0 | -2 | 1 | -1 | 2 | 0 | 0 |
| $(1^2;21^2)$ | 3 | -3 | 3 | -1 | 1 | -3 | 1 | -1 | 0 | 0 | 3 | -1 | -1 | 1 | 1 | 0 | 0 | -1 | 1 | -1 |
| $(1^2;1^4)$ | 1 | -1 | 1 | -1 | 1 | -1 | 1 | -1 | 1 | -1 | 1 | -1 | 1 | 1 | -1 | -1 | 1 | 1 | -1 | 1 |
| $(([2]\#[2]\#[2]\#[1^2])\otimes[3]')\uparrow G$ | 4 | 2 | 0 | 2 | 2 | -2 | 0 | 0 | 1 | 1 | -4 | -2 | 0 | -2 | 0 | -1 | -1 | 0 | 0 | 0 |
| $(([2]\#[2]\#[2]\#[1^2])\otimes[21]')\uparrow G$ | 8 | 4 | 0 | 0 | 0 | -4 | 0 | 0 | -1 | -1 | -8 | 0 | 0 | 0 | 0 | 1 | 1 | 0 | 0 | 0 |
| $(([2]\#[2]\#[2]\#[1^2])\otimes[1^3]')\uparrow G$ | 4 | 2 | 0 | -2 | -2 | -2 | 0 | 0 | 1 | 1 | -4 | 2 | 0 | 2 | 0 | -1 | -1 | 0 | 0 | 0 |
| $(([2]\#[2]\#[1^2]\#[1^2])\otimes([2]\#[2]\ )')\uparrow G$ | 6 | 0 | -2 | 2 | 0 | 0 | 0 | -2 | 0 | 0 | 6 | 2 | 2 | 0 | 0 | 0 | 0 | -2 | 0 | 0 |
| $(([2]\#[2]\#[1^2]\#[1^2])\otimes([2]\#[1^2])')\uparrow G$ | 6 | 0 | -2 | 0 | 2 | 0 | -2 | 0 | 0 | 0 | 6 | 0 | -2 | 2 | 0 | 0 | 0 | 2 | 0 | 0 |
| $(([2]\#[2]\#[1^2]\#[1^2])\otimes([1^2]\#[2])')\uparrow G$ | 6 | 0 | -2 | 0 | -2 | 0 | 2 | 0 | 0 | 0 | 6 | 0 | -2 | -2 | 0 | 0 | 0 | 2 | 0 | 0 |
| $(([2]\#[2]\#[1^2]\#[1^2])\otimes([1^2]\#[1^2])')\uparrow G$ | 6 | 0 | -2 | -2 | 0 | 0 | 0 | 2 | 0 | 0 | 6 | -2 | 2 | 0 | 0 | 0 | 0 | -2 | 0 | 0 |
| $(([2]\#[1^2]\#[1^2]\#[1^2])\otimes[3]')\uparrow G$ | 4 | -2 | 0 | 2 | -2 | 2 | 0 | 0 | 1 | -1 | -4 | -2 | 0 | 2 | 0 | 1 | -1 | 0 | 0 | 0 |
| $(([2]\#[1^2]\#[1^2]\#[1^2])\otimes[21]')\uparrow G$ | 8 | -4 | 0 | 0 | 0 | 4 | 0 | 0 | -1 | 1 | -8 | 0 | 0 | 0 | 0 | -1 | 1 | 0 | 0 | 0 |
| $(([2]\#[1^2]\#[1^2]\#[1^2])\otimes[1^3]')\uparrow G$ | 4 | -2 | 0 | -2 | 2 | 2 | 0 | 0 | 1 | -1 | -4 | 2 | 0 | -2 | 0 | 1 | -1 | 0 | 0 | 0 |

## $G = S_4 \, wr \, S_2$

| Cycle-partition of representative | [$1^8$] | [$1^6 2$] | [$1^5 3$] | [$1^4 2^2$] | [$1^4 2^2$] | [$1^4 4$] | [$1^3 2 3$] | [$1^2 2^3$] | [$1^2 2 4$] | [$1^2 3^2$] | [$1 2^2 3$] | [$1 3 4$] | [$2^4$] | [$2^4$] | [$2^2 4$] | [$2^2 4$] | [$2 6$] | [$4^2$] | [$4^2$] | [8] |
|---|---|---|---|---|---|---|---|---|---|---|---|---|---|---|---|---|---|---|---|---|
| **Representative of conjugacy class** | (e;1) | ((12),1;1) | ((123),1;1) | ((12)(34),1;1) | ((12),(12);1) | ((1234),1;1) | ((12),(123),;1) | ((12),(12)(34);1) | ((12),(1234),;1) | ((123),(123);1) | ((123),(12)(34);1) | ((123),(1234);1) | ((12)(34),(12)(34);1) | (e;(12)) | ((12)(34),(1234);1) | ((12),1;(12)) | ((123),1;(12)) | ((1234),(1234);1) | ((12)(34),1;(12)) | ((1234),1;(12)) |
| **Class order** | 1 | 12 | 16 | 6 | 36 | 12 | 96 | 36 | 72 | 64 | 48 | 96 | 9 | 24 | 36 | 144 | 192 | 36 | 72 | 144 |
| (4;2) | 1 | 1 | 1 | 1 | 1 | 1 | 1 | 1 | 1 | 1 | 1 | 1 | 1 | 1 | 1 | 1 | 1 | 1 | 1 | 1 |
| (4;$1^2$) | 1 | 1 | 1 | 1 | 1 | 1 | 1 | 1 | 1 | 1 | 1 | 1 | 1 | -1 | 1 | -1 | -1 | 1 | -1 | -1 |
| (31;2) | 9 | 3 | 0 | -3 | 1 | -3 | 0 | -1 | -1 | 0 | 0 | 0 | 1 | 3 | 1 | 1 | 0 | 1 | -1 | -1 |
| (31;$1^2$) | 9 | 3 | 0 | -3 | 1 | -3 | 0 | -1 | -1 | 0 | 0 | 0 | 1 | -3 | 1 | -1 | 0 | 1 | 1 | 1 |
| ($2^2$;2) | 4 | 0 | -2 | 4 | 0 | 0 | 0 | 0 | 0 | 1 | -2 | 0 | 4 | 2 | 0 | 0 | -1 | 0 | 2 | 0 |
| ($2^2$;$1^2$) | 4 | 0 | -2 | 4 | 0 | 0 | 0 | 0 | 0 | 1 | -2 | 0 | 4 | -2 | 0 | 0 | 1 | 0 | -2 | 0 |
| ($21^2$;2) | 9 | -3 | 0 | -3 | 1 | 3 | 0 | 1 | -1 | 0 | 0 | 0 | 1 | 3 | -1 | -1 | 0 | 1 | -1 | 1 |
| ($21^2$;$1^2$) | 9 | -3 | 0 | -3 | 1 | 3 | 0 | 1 | -1 | 0 | 0 | 0 | 1 | -3 | -1 | 1 | 0 | 1 | 1 | -1 |
| ($1^4$;2) | 1 | -1 | 1 | 1 | 1 | -1 | -1 | -1 | 1 | 1 | 1 | -1 | 1 | 1 | -1 | -1 | 1 | 1 | 1 | -1 |
| ($1^4$;$1^2$) | 1 | -1 | 1 | 1 | 1 | -1 | -1 | -1 | 1 | 1 | 1 | -1 | 1 | -1 | -1 | 1 | -1 | 1 | -1 | 1 |
| ([4]#[31])↑G | 6 | 4 | 3 | 2 | 2 | 2 | 1 | 0 | 0 | 0 | -1 | -1 | -2 | 0 | -2 | 0 | 0 | -2 | 0 | 0 |
| ([4]#[$2^2$])↑G | 4 | 2 | 1 | 4 | 0 | 2 | -1 | 2 | 0 | -2 | 1 | -1 | 4 | 0 | 2 | 0 | 0 | 0 | 0 | 0 |
| ([4]#[$21^2$])↑G | 6 | 2 | 3 | 2 | -2 | 4 | -1 | -2 | 0 | 0 | -1 | 1 | -2 | 0 | 0 | 0 | 0 | 2 | 0 | 0 |
| ([4]#[$1^4$])↑G | 2 | 0 | 2 | 2 | -2 | 0 | 0 | 0 | -2 | 2 | 2 | 0 | 2 | 0 | 0 | 0 | 0 | -2 | 0 | 0 |
| ([31]#[$2^2$])↑G | 12 | 2 | -3 | 4 | 0 | -2 | -1 | 2 | 0 | 0 | 1 | 1 | -4 | 0 | -2 | 0 | 0 | 0 | 0 | 0 |
| ([31]#[$21^2$])↑G | 18 | 0 | 0 | -6 | -2 | 0 | 0 | 0 | 2 | 0 | 0 | 0 | 2 | 0 | 0 | 0 | 0 | -2 | 0 | 0 |
| ([31]#[$1^4$])↑G | 6 | -2 | 3 | 2 | -2 | -4 | 1 | 2 | 0 | 0 | -1 | -1 | -2 | 0 | 0 | 0 | 0 | 2 | 0 | 0 |
| ([$2^2$]#[$21^2$])↑G | 12 | -2 | -3 | 4 | 0 | 2 | 1 | -2 | 0 | 0 | 1 | -1 | -4 | 0 | 2 | 0 | 0 | 0 | 0 | 0 |
| ([$2^2$]#[$1^4$])↑G | 4 | -2 | 1 | 4 | 0 | -2 | 1 | -2 | 0 | -2 | 1 | 1 | 4 | 0 | -2 | 0 | 0 | 0 | 0 | 0 |
| ([$21^2$]#[$1^4$])↑G | 6 | -4 | 3 | 2 | 2 | -2 | -1 | 0 | 0 | 0 | -1 | 1 | -2 | 0 | 2 | 0 | 0 | -2 | 0 | 0 |

$G = S_2 \,\mathrm{wr}\, S_5$

| Cycle-partition of representative | $[1^{10}]$ | $[1^8 2]$ | $[1^6 2^2]$ | $[1^6 2^2]$ | $[1^6 4]$ | $[1^4 2^3]$ | $[1^4 2^3]$ | $[1^4 24]$ | $[1^4 3^2]$ | $[1^4 6]$ | $[1^2 2^4]$ | $[1^2 2^4]$ | $[1^2 2^4]$ | $[1^2 2^2 4]$ | $[1^2 2^2 4]$ | $[1^2 2 3^2]$ | $[1^2 26]$ | $[1^2 4^2]$ | $[1^2 4^2]$ | $[1^2 8]$ |
|---|---|---|---|---|---|---|---|---|---|---|---|---|---|---|---|---|---|---|---|---|
| Representative of conjugacy class | (e;1) | ((12),1,1,1,1;1) | ((12),(12),1,1,1;1) | (e;(12)) | ((12),1,1,1,1;(12)) | ((12),(12),(12),1,1;1) | (1,1,(12),1,1;(12)) | ((12),1,(12),1,1;(12)) | (e;(123)) | ((12),1,1,1,1;(123)) | ((12),(12),(12),(12),1;1) | (1,1,(12),(12),1;(12)) | (e;(12)(34)) | ((12),1,(12),(12),1;(12)) | ((12),1,1,1,1;(12)(34)) | (1,1,1,(12),1;(123)) | ((12),1,1,(12),1;(123)) | ((12),1,(12),1,1;(12)(34)) | (e;(1234)) | ((12),1,1,1,1;(1234)) |
| Class order | 1 | 5 | 10 | 20 | 20 | 10 | 60 | 60 | 80 | 80 | 5 | 60 | 60 | 60 | 120 | 160 | 160 | 60 | 240 | 240 |
| (2;5) | 1 | 1 | 1 | 1 | 1 | 1 | 1 | 1 | 1 | 1 | 1 | 1 | 1 | 1 | 1 | 1 | 1 | 1 | 1 | 1 |
| (2;41) | 4 | 4 | 4 | 2 | 2 | 4 | 2 | 2 | 1 | 1 | 4 | 2 | 0 | 2 | 0 | 1 | 1 | 0 | 0 | 0 |
| (2;32) | 5 | 5 | 5 | 1 | 1 | 5 | 1 | 1 | -1 | -1 | 5 | 1 | 1 | 1 | 1 | -1 | -1 | 1 | -1 | -1 |
| $(2;31^2)$ | 6 | 6 | 6 | 0 | 0 | 6 | 0 | 0 | 0 | 0 | 6 | 0 | -2 | 0 | -2 | 0 | 0 | -2 | 0 | 0 |
| $(2;2^2 1)$ | 5 | 5 | 5 | -1 | -1 | 5 | -1 | -1 | -1 | -1 | 5 | -1 | 1 | -1 | 1 | -1 | -1 | 1 | 1 | 1 |
| $(2;21^3)$ | 4 | 4 | 4 | -2 | -2 | 4 | -2 | -2 | 1 | 1 | 4 | -2 | 0 | -2 | 0 | 1 | 1 | 0 | 0 | 0 |
| $(2;1^5)$ | 1 | 1 | 1 | -1 | -1 | 1 | -1 | -1 | 1 | 1 | 1 | -1 | 1 | -1 | 1 | 1 | 1 | 1 | -1 | -1 |
| $(1^2;5)$ | 1 | -1 | 1 | 1 | -1 | -1 | -1 | 1 | 1 | -1 | 1 | 1 | 1 | -1 | -1 | -1 | 1 | 1 | 1 | -1 |

| Cycle-partition of representative | $[2^5]$ | $[2^5]$ | $[2^5]$ | $[2^3 4]$ | $[2^3 4]$ | $[2^2 3^2]$ | $[2^2 3^2]$ | $[2^2 6]$ | $[2^2 6]$ | $[24^2]$ | $[24^2]$ | $[28]$ | $[3^2 4]$ | $[46]$ | $[5^2]$ | $[10]$ |
|---|---|---|---|---|---|---|---|---|---|---|---|---|---|---|---|---|
| Representative of conjugacy class | ((12),(12),(12),(12),(12);1) | (1,1,(12),(12),(12);(12)) | (1,1,1,1,(12);(12)(34)) | ((12),1,(12),(12),(12);(12)) | ((12),1,1,1,(12);(12)(34)) | (1,1,1,(12),(12);(123)) | (e;(12)(345)) | ((12),1,1,(12),(12);(123)) | (1,1,(12),1,1;(12)(345)) | ((12),1,(12),1,(12);(12)(34)) | (1,1,1,1,(12);(1234)) | ((12),1,1,1,(12);(1234)) | ((12),1,1,1,1;(12)(345)) | ((12),1,(12),1,1;(12)(345)) | (e;(12345)) | ((12),1,1,1,1;(12345)) |
| Class order | 1 | 20 | 60 | 20 | 120 | 80 | 160 | 80 | 160 | 60 | 240 | 240 | 160 | 160 | 384 | 384 |
| (2;5) | 1 | 1 | 1 | 1 | 1 | 1 | 1 | 1 | 1 | 1 | 1 | 1 | 1 | 1 | 1 | 1 |
| (2;41) | 4 | 2 | 0 | 2 | 0 | 1 | -1 | 1 | -1 | 0 | 0 | 0 | -1 | -1 | -1 | -1 |
| (2;32) | 5 | 1 | 1 | 1 | 1 | -1 | 1 | -1 | 1 | 1 | -1 | -1 | 1 | 1 | 0 | 0 |
| $(2;31^2)$ | 6 | 0 | -2 | 0 | -2 | 0 | 0 | 0 | 0 | -2 | 0 | 0 | 0 | 0 | 1 | 1 |
| $(2;2^2 1)$ | 5 | -1 | 1 | -1 | 1 | -1 | -1 | -1 | -1 | 1 | 1 | 1 | -1 | -1 | 0 | 0 |
| $(2;21^3)$ | 4 | -2 | 0 | -2 | 0 | 1 | 1 | 1 | 1 | 0 | 0 | 0 | 1 | 1 | -1 | -1 |
| $(2;1^5)$ | 1 | -1 | 1 | -1 | 1 | 1 | -1 | 1 | -1 | 1 | -1 | -1 | -1 | -1 | 1 | 1 |
| $(1^2;5)$ | -1 | -1 | -1 | 1 | 1 | 1 | 1 | -1 | -1 | -1 | -1 | 1 | -1 | 1 | 1 | -1 |

| | $[1^{10}]$ | $[1^82]$ | $[1^62^2]$ | $[1^62^2]$ | $[1^64]$ | $[1^42^3]$ | $[1^42^3]$ | $[1^424]$ | $[1^43^2]$ | $[1^46]$ | $[1^22^4]$ | $[1^22^4]$ | $[1^22^4]$ | $[1^22^24]$ | $[1^22^24]$ | $[1^223^2]$ | $[1^226]$ | $[1^24^2]$ | $[1^24^2]$ | $[1^28]$ | $[2^5]$ | $[2^5]$ | $[2^5]$ | $[2^34]$ | $[2^34]$ | $[2^23^2]$ | $[2^23^2]$ | $[2^26]$ | $[2^26]$ | $[24^2]$ | $[24^2]$ | $[28]$ | $[3^24]$ | $[46]$ | $[5^2]$ | $[10]$ |
|---|---|---|---|---|---|---|---|---|---|---|---|---|---|---|---|---|---|---|---|---|---|---|---|---|---|---|---|---|---|---|---|---|---|---|---|---|
| $(1^2;41)$ | 4 | -4 | 4 | 2 | -2 | -4 | -2 | 2 | 1 | -1 | 4 | 2 | 0 | -2 | 0 | -1 | 1 | 0 | 0 | 0 | -4 | -2 | 0 | 2 | 0 | 1 | -1 | -1 | 1 | 0 | 0 | 0 | 1 | -1 | -1 | 1 |
| $(1^2;32)$ | 5 | -5 | 5 | 1 | -1 | -5 | -1 | 1 | -1 | 1 | 5 | 1 | 1 | -1 | -1 | 1 | -1 | 1 | -1 | 1 | -5 | -1 | -1 | 1 | 1 | -1 | 1 | 1 | -1 | -1 | 1 | -1 | -1 | 1 | 0 | 0 |
| $(1^2;31^2)$ | [illegible] | -6 | 6 | 0 | 0 | -6 | 0 | 0 | 0 | 0 | 6 | 0 | -2 | 0 | 2 | 0 | 0 | -2 | 0 | 0 | -6 | 0 | 2 | 0 | -2 | 0 | 0 | 0 | 0 | 2 | 0 | 0 | 0 | 0 | 1 | -1 |
| $(1^2;2^21)$ | 5 | -5 | 5 | -1 | 1 | -5 | 1 | -1 | -1 | 1 | 5 | -1 | 1 | 1 | -1 | 1 | -1 | 1 | 1 | -1 | -5 | 1 | -1 | -1 | 1 | -1 | -1 | 1 | 1 | -1 | -1 | 1 | 1 | -1 | 0 | 0 |
| $(1^2;21^2)$ | 4 | -4 | 4 | -2 | 2 | -4 | 2 | -2 | 1 | -1 | 4 | -2 | 0 | 2 | 0 | -1 | 1 | 0 | 0 | 0 | -4 | 2 | 0 | -2 | 0 | 1 | 1 | -1 | -1 | 0 | 0 | 0 | -1 | 1 | -1 | 1 |
| $(1^2;1^5)$ | 1 | -1 | 1 | -1 | 1 | -1 | 1 | -1 | 1 | -1 | 1 | -1 | 1 | 1 | -1 | -1 | 1 | 1 | -1 | 1 | -1 | 1 | -1 | -1 | 1 | 1 | -1 | -1 | 1 | -1 | 1 | -1 | 1 | -1 | 1 | -1 |
| $(([2]\#[2]\#[2]\#[2]\#[1^2])\otimes[4]')\uparrow G$ | 5 | 3 | 1 | 3 | 3 | -1 | 1 | 1 | 2 | 2 | -3 | -1 | 1 | -1 | 1 | 0 | 0 | 1 | 1 | 1 | -5 | -3 | -1 | -3 | -1 | -2 | 0 | -2 | 0 | -1 | -1 | -1 | 0 | 0 | 0 | 0 |
| $(([2]\#[2]\#[2]\#[2]\#[1^2])\otimes[31]')\uparrow G$ | 15 | 9 | 3 | 3 | 3 | -3 | 1 | 1 | 0 | 0 | -9 | -1 | -1 | -1 | -1 | 0 | 0 | -1 | -1 | -1 | -15 | -3 | 1 | -3 | 1 | 0 | 0 | 0 | 0 | 1 | 1 | 1 | 0 | 0 | 0 | 0 |
| $(([2]\#[2]\#[2]\#[2]\#[1^2])\otimes[2^2]')\uparrow G$ | 10 | 6 | 2 | 0 | 0 | -2 | 0 | 0 | -2 | -2 | -6 | 0 | 2 | 0 | 2 | 0 | 0 | 2 | 0 | 0 | -10 | 0 | -2 | 0 | -2 | 2 | 0 | 2 | 0 | -2 | 0 | 0 | 0 | 0 | 0 | 0 |
| $(([2]\#[2]\#[2]\#[2]\#[1^2])\otimes[21^2]')\uparrow G$ | 15 | 9 | 3 | -3 | -3 | -3 | -1 | -1 | 0 | 0 | -9 | 1 | -1 | 1 | -1 | 0 | 0 | -1 | 1 | 1 | -15 | 3 | 1 | 3 | 1 | 0 | 0 | 0 | 0 | 1 | -1 | -1 | 0 | 0 | 0 | 0 |
| $(([2]\#[2]\#[2]\#[2]\#[1^2])\otimes[1^4]')\uparrow G$ | 5 | 3 | 1 | -3 | -3 | -1 | -1 | -1 | 2 | 2 | -3 | 1 | 1 | 1 | 1 | 0 | 0 | 1 | -1 | -1 | -5 | 3 | -1 | 3 | -1 | -2 | 0 | -2 | 0 | -1 | 1 | 1 | 0 | 0 | 0 | 0 |
| $(([2]\#[2]\#[2]\#[1^2]\#[1^2])\otimes([3]\#[2])')\uparrow G$ | 10 | 2 | -2 | 4 | 2 | -2 | 0 | -2 | 1 | 1 | 2 | 0 | 2 | -2 | 0 | -1 | -1 | -2 | 0 | 0 | 10 | 4 | 2 | 2 | 0 | 1 | 1 | 1 | 1 | -2 | 0 | 0 | -1 | -1 | 0 | 0 |
| $(([2]\#[2]\#[2]\#[1^2]\#[1^2])\otimes([21]\#[2])')\uparrow G$ | 20 | 4 | -4 | 2 | -2 | -4 | 2 | -2 | -1 | -1 | 4 | 2 | 0 | -2 | 0 | 1 | 1 | 0 | 0 | 0 | 20 | 2 | 0 | -2 | 0 | -1 | -1 | -1 | -1 | 0 | 0 | 0 | 1 | 1 | 0 | 0 |
| $(([2]\#[2]\#[2]\#[1^2]\#[1^2])\otimes([1^3]\#[2])')\uparrow G$ | 10 | 2 | -2 | -2 | -4 | -2 | 2 | 0 | 1 | 1 | 2 | 2 | -2 | 0 | 0 | -1 | -1 | 2 | 0 | 0 | 10 | -2 | -2 | -4 | 0 | 1 | 1 | 1 | 1 | 2 | 0 | 0 | -1 | -1 | 0 | 0 |
| $(([2]\#[2]\#[2]\#[1^2]\#[1^2])\otimes([3]\#[1^2])')\uparrow G$ | 10 | 2 | -2 | 2 | 4 | -2 | -2 | 0 | 1 | 1 | 2 | -2 | -2 | 0 | 0 | -1 | -1 | 2 | 0 | 0 | 10 | 2 | -2 | 4 | 0 | 1 | -1 | 1 | -1 | 2 | 0 | 0 | 1 | 1 | 0 | 0 |
| $(([2]\#[2]\#[2]\#[1^2]\#[1^2])\otimes([21]\#[1^2])')\uparrow G$ | 20 | 4 | -4 | -2 | 2 | -4 | -2 | 2 | -1 | -1 | 4 | -2 | 0 | 2 | 0 | 1 | 1 | 0 | 0 | 0 | 20 | -2 | 0 | 2 | 0 | -1 | 1 | -1 | 1 | 0 | 0 | 0 | -1 | -1 | 0 | 0 |
| $(([2]\#[2]\#[2]\#[1^2]\#[1^2])\otimes([1^3]\#[1^2])')\uparrow G$ | 10 | 2 | -2 | -4 | -2 | -2 | 0 | 2 | 1 | 1 | 2 | 0 | 2 | 2 | 0 | -1 | -1 | -2 | 0 | 0 | 10 | -4 | 2 | -2 | 0 | 1 | -1 | 1 | -1 | -2 | 0 | 0 | 1 | 1 | 0 | 0 |
| $(([2]\#[2]\#[1^2]\#[1^2]\#[1^2])\otimes([2]\#[3])')\uparrow G$ | 10 | -2 | -2 | 4 | -2 | 2 | 0 | -2 | 1 | -1 | 2 | 0 | 2 | 2 | 0 | 1 | -1 | -2 | 0 | 0 | -10 | -4 | -2 | 2 | 0 | 1 | 1 | -1 | -1 | 2 | 0 | 0 | 1 | -1 | 0 | 0 |
| $(([2]\#[2]\#[1^2]\#[1^2]\#[1^2])\otimes([1^2]\#[3])')\uparrow G$ | 10 | -2 | -2 | 2 | -4 | 2 | 2 | 0 | 1 | -1 | 2 | -2 | -2 | 0 | 0 | 1 | -1 | 2 | 0 | 0 | -10 | -2 | 2 | 4 | 0 | 1 | -1 | -1 | 1 | -2 | 0 | 0 | -1 | 1 | 0 | 0 |
| $(([2]\#[2]\#[1^2]\#[1^2]\#[1^2])\otimes([2]\#[21])')\uparrow G$ | 20 | -4 | -4 | 2 | 2 | 4 | -2 | -2 | -1 | 1 | 4 | 2 | 0 | 2 | 0 | -1 | 1 | 0 | 0 | 0 | -20 | -2 | 0 | -2 | 0 | -1 | -1 | 1 | 1 | 0 | 0 | 0 | -1 | 1 | 0 | 0 |
| $(([2]\#[2]\#[1^2]\#[1^2]\#[1^2])\otimes([1^2]\#[21])')\uparrow G$ | 20 | -4 | -4 | -2 | -2 | 4 | 2 | 2 | -1 | 1 | 4 | -2 | 0 | -2 | 0 | -1 | 1 | 0 | 0 | 0 | -20 | 2 | 0 | 2 | 0 | -1 | 1 | 1 | -1 | 0 | 0 | 0 | 1 | -1 | 0 | 0 |
| $(([2]\#[2]\#[1^2]\#[1^2]\#[1^2])\otimes([2]\#[1^3])')\uparrow G$ | 10 | -2 | -2 | -2 | 4 | 2 | -2 | 0 | 1 | -1 | 2 | 2 | -2 | 0 | 0 | 1 | -1 | 2 | 0 | 0 | -10 | 2 | 2 | -4 | 0 | 1 | 1 | -1 | -1 | -2 | 0 | 0 | 1 | -1 | 0 | 0 |
| $(([2]\#[2]\#[1^2]\#[1^2]\#[1^2])\otimes([1^2]\#[1^3])')\uparrow G$ | 10 | -2 | -2 | -4 | 2 | 2 | 0 | 2 | 1 | -1 | 2 | 0 | 2 | -2 | 0 | 1 | -1 | -2 | 0 | 0 | -10 | 4 | -2 | -2 | 0 | 1 | -1 | -1 | 1 | 2 | 0 | 0 | -1 | 1 | 0 | 0 |
| $(([2]\#[1^2]\#[1^2]\#[1^2]\#[1^2])\otimes[4]')\uparrow G$ | 5 | -3 | 1 | 3 | -3 | 1 | -1 | 1 | 2 | -2 | -3 | -1 | 1 | 1 | -1 | 0 | 0 | 1 | 1 | -1 | 5 | 3 | 1 | -3 | -1 | -2 | 0 | 2 | 0 | 1 | 1 | -1 | 0 | 0 | 0 | 0 |
| $(([2]\#[1^2]\#[1^2]\#[1^2]\#[1^2])\otimes[31]')\uparrow G$ | 15 | -9 | 3 | 3 | -3 | 3 | -1 | 1 | 0 | 0 | -9 | -1 | -1 | 1 | 1 | 0 | 0 | -1 | -1 | 1 | 15 | 3 | -1 | -3 | 1 | 0 | 0 | 0 | 0 | -1 | -1 | 1 | 0 | 0 | 0 | 0 |
| $(([2]\#[1^2]\#[1^2]\#[1^2]\#[1^2])\otimes[2^2]')\uparrow G$ | 10 | -6 | 2 | 0 | 0 | 2 | 0 | 0 | -2 | 2 | -6 | 0 | 2 | 0 | -2 | 0 | 0 | 2 | 0 | 0 | 10 | 0 | 2 | 0 | -2 | 2 | 0 | -2 | 0 | 2 | 0 | 0 | 0 | 0 | 0 | 0 |
| $(([2]\#[1^2]\#[1^2]\#[1^2]\#[1^2])\otimes[21^2]')\uparrow G$ | 15 | -9 | 3 | -3 | 3 | 3 | 1 | -1 | 0 | 0 | -9 | 1 | -1 | -1 | 1 | 0 | 0 | -1 | 1 | -1 | 15 | -3 | -1 | 3 | 1 | 0 | 0 | 0 | 0 | -1 | 1 | -1 | 0 | 0 | 0 | 0 |
| $(([2]\#[1^2]\#[1^2]\#[1^2]\#[1^2])\otimes[1^4]')\uparrow G$ | 5 | -3 | 1 | -3 | 3 | 1 | 1 | -1 | 2 | -2 | -3 | 1 | 1 | -1 | -1 | 0 | 0 | 1 | -1 | 1 | 5 | -3 | 1 | 3 | -1 | -2 | 0 | 2 | 0 | 1 | -1 | 1 | 0 | 0 | 0 | 0 |

$G = S_5 \,\mathrm{wr}\, S_2$

| Cycle-partition of representative | $[1^{10}]$ | $[1^8 2]$ | $[1^7 3]$ | $[1^6 2^2]$ | $[1^6 2^2]$ | $[1^6 4]$ | $[1^5 23]$ | $[1^5 23]$ | $[1^5 5]$ | $[1^4 2^3]$ | $[1^4 24]$ | $[1^4 3^2]$ |
|---|---|---|---|---|---|---|---|---|---|---|---|---|
| Representative of conjugacy class | (e;1) | ((12),1;1) | ((123),1;1) | ((12)(34),1;1) | ((12),(12);1) | ((1234),1;1) | ((12)(345),1;1) | ((12),(123);1) | ((12345),1;1) | ((12),(12)(34);1) | ((12),(1234);1) | ((123),(123);1) |
| Class order | 1 | 20 | 40 | 30 | 100 | 60 | 40 | 400 | 48 | 300 | 600 | 400 |
| (5;2) | 1 | 1 | 1 | 1 | 1 | 1 | 1 | 1 | 1 | 1 | 1 | 1 |
| $(5;1^2)$ | 1 | 1 | 1 | 1 | 1 | 1 | 1 | 1 | 1 | 1 | 1 | 1 |
| (41;2) | 16 | 8 | 4 | 0 | 4 | 0 | -4 | 2 | -4 | 0 | 0 | 1 |
| $(41;1^2)$ | 16 | 8 | 4 | 0 | 4 | 0 | -4 | 2 | -4 | 0 | 0 | 1 |
| (32;2) | 25 | 5 | -5 | 5 | 1 | -5 | 5 | -1 | 0 | 1 | -1 | 1 |
| $(32;1^2)$ | 25 | 5 | -5 | 5 | 1 | -5 | 5 | -1 | 0 | 1 | -1 | 1 |
| $(31^2;2)$ | 36 | 0 | 0 | -12 | 0 | 0 | 0 | 0 | 6 | 0 | 0 | 0 |
| $(31^2;1^2)$ | 36 | 0 | 0 | -12 | 0 | 0 | 0 | 0 | 6 | 0 | 0 | 0 |
| $(2^21;2)$ | 25 | -5 | -5 | 5 | 1 | 5 | -5 | 1 | 0 | -1 | -1 | 1 |
| $(2^21;1^2)$ | 25 | -5 | -5 | 5 | 1 | 5 | -5 | 1 | 0 | -1 | -1 | 1 |

| Cycle-partition of representative | $[1^3 2^2 3]$ | $[1^3 2^2 3]$ | $[1^3 25]$ | $[1^3 34]$ | $[1^2 2^4]$ | $[1^2 2^2 4]$ | $[1^2 2 3^2]$ | $[1^2 35]$ | $[1^2 4^2]$ | $[12^3 3]$ | $[12^2 5]$ | $[1234]$ |
|---|---|---|---|---|---|---|---|---|---|---|---|---|
| Representative of conjugacy class | ((12),(12)(345);1) | ((123),(12)(34);1) | ((12),(12345);1) | ((123),(1234);1) | ((12)(34),(12)(34);1) | ((1234),(12)(34);1) | ((123),(12)(345);1) | ((123),(12345);1) | ((1234),(1234);1) | ((12)(34),(12)(345);1) | ((12)(34),(12345);1) | ((1234),(12)(345);1) |
| Class order | 400 | 600 | 480 | 1200 | 225 | 900 | 800 | 960 | 900 | 600 | 720 | 1200 |
| (5;2) | 1 | 1 | 1 | 1 | 1 | 1 | 1 | 1 | 1 | 1 | 1 | 1 |
| $(5;1^2)$ | 1 | 1 | 1 | 1 | 1 | 1 | 1 | 1 | 1 | 1 | 1 | 1 |
| (41;2) | -2 | 0 | -2 | 0 | 0 | 0 | -1 | -1 | 0 | 0 | 0 | 0 |
| $(41;1^2)$ | -2 | 0 | -2 | 0 | 0 | 0 | -1 | -1 | 0 | 0 | 0 | 0 |
| (32;2) | 1 | -1 | 0 | 1 | 1 | -1 | -1 | 0 | 1 | 1 | 0 | -1 |
| $(32;1^2)$ | 1 | -1 | 0 | 1 | 1 | -1 | -1 | 0 | 1 | 1 | 0 | -1 |
| $(31^2;2)$ | 0 | 0 | 0 | 0 | 4 | 0 | 0 | 0 | 0 | 0 | -2 | 0 |
| $(31^2;1^2)$ | 0 | 0 | 0 | 0 | 4 | 0 | 0 | 0 | 0 | 0 | -2 | 0 |
| $(2^21;2)$ | 1 | -1 | 0 | -1 | 1 | 1 | 1 | 0 | 1 | -1 | 0 | -1 |
| $(2^21;1^2)$ | 1 | -1 | 0 | -1 | 1 | 1 | 1 | 0 | 1 | -1 | 0 | -1 |

| Cycle-partition of representative | $[145]$ | $[2^5]$ | $[2^3 4]$ | $[2^2 3^2]$ | $[2^2 6]$ | $[235]$ | $[24^2]$ | $[28]$ | $[46]$ | $[5^2]$ | $[10]$ |
|---|---|---|---|---|---|---|---|---|---|---|---|
| Representative of conjugacy class | ((1234),(12345);1) | (e;(12)) | ((12),1;(12)) | ((12)(345),(12)(345);1) | ((123),1;(12)) | ((12)(345),(12345);1) | ((12)(34),1;(12)) | ((1234),1;(12)) | ((12)(345),1;(12)) | ((12345),(12345);1) | ((12345),1;(12)) |
| Class order | 1440 | 120 | 1200 | 400 | 2400 | 960 | 1800 | 3600 | 2400 | 576 | 2880 |
| (5;2) | 1 | 1 | 1 | 1 | 1 | 1 | 1 | 1 | 1 | 1 | 1 |
| $(5;1^2)$ | 1 | -1 | -1 | 1 | -1 | 1 | -1 | -1 | -1 | 1 | -1 |
| (41;2) | 0 | 4 | 2 | 1 | 1 | 1 | 0 | 0 | -1 | 1 | -1 |
| $(41;1^2)$ | 0 | -4 | -2 | 1 | -1 | 1 | 0 | 0 | 1 | 1 | 1 |
| (32;2) | 0 | 5 | 1 | 1 | -1 | 0 | 1 | -1 | 1 | 0 | 0 |
| $(32;1^2)$ | 0 | -5 | -1 | 1 | 1 | 0 | -1 | 1 | -1 | 0 | 0 |
| $(31^2;2)$ | 0 | 6 | 0 | 0 | 0 | 0 | -2 | 0 | 0 | 1 | 1 |
| $(31^2;1^2)$ | 0 | -6 | 0 | 0 | 0 | 0 | 2 | 0 | 0 | 1 | -1 |
| $(2^21;2)$ | 0 | 5 | -1 | 1 | -1 | 0 | 1 | 1 | -1 | 0 | 0 |
| $(2^21;1^2)$ | 0 | -5 | 1 | 1 | 1 | 0 | -1 | -1 | 1 | 0 | 0 |

| | $[1^{10}]$ | $[1^8 2]$ | $[1^7 3]$ | $[1^6 2^2]$ | $[1^6 2^2]$ | $[1^6 4]$ | $[1^5 23]$ | $[1^5 23]$ | $[1^5 5]$ | $[1^4 2^3]$ | $[1^4 24]$ | $[1^4 3^2]$ | $[1^3 2^2 3]$ | $[1^3 2^2 3]$ | $[1^3 25]$ | $[1^3 34]$ | $[1^2 2^4]$ | $[1^2 2^2 4]$ | $[1^2 23^2]$ | $[1^2 35]$ | $[1^2 4^2]$ | $[12^3 3]$ | $[12^2 5]$ | $[1234]$ | $[145]$ | $[2^5]$ | $[2^3 4]$ | $[2^2 3^2]$ | $[2^2 6]$ | $[235]$ | $[24^2]$ | $[28]$ | $[46]$ | $[5^2]$ | $[10]$ |
|---|---|---|---|---|---|---|---|---|---|---|---|---|---|---|---|---|---|---|---|---|---|---|---|---|---|---|---|---|---|---|---|---|---|---|---|
| $(21^3;2)$ | 16 | -8 | 4 | 0 | 4 | 0 | 4 | -2 | -4 | 0 | 0 | 1 | -2 | 0 | 2 | 0 | 0 | 0 | 1 | -1 | 0 | 0 | 0 | 0 | 0 | 4 | -2 | 1 | 1 | -1 | 0 | 0 | 1 | 1 | -1 |
| $(21^3;1^2)$ | 16 | -8 | 4 | 0 | 4 | 0 | 4 | -2 | -4 | 0 | 0 | 1 | -2 | 0 | 2 | 0 | 0 | 0 | 1 | -1 | 0 | 0 | 0 | 0 | 0 | -4 | 2 | 1 | -1 | -1 | 0 | 0 | -1 | 1 | 1 |
| $(1^5;2)$ | 1 | -1 | 1 | 1 | 1 | -1 | -1 | -1 | 1 | -1 | 1 | 1 | 1 | 1 | -1 | -1 | 1 | -1 | -1 | 1 | 1 | -1 | 1 | 1 | -1 | 1 | -1 | 1 | 1 | -1 | 1 | -1 | -1 | 1 | 1 |
| $(1^5;1^2)$ | 1 | -1 | 1 | 1 | 1 | -1 | -1 | -1 | 1 | -1 | 1 | 1 | 1 | 1 | -1 | -1 | 1 | -1 | -1 | 1 | 1 | -1 | 1 | 1 | -1 | -1 | 1 | 1 | -1 | -1 | -1 | 1 | 1 | 1 | -1 |
| $([5]\#[41])\uparrow G$ | 8 | 6 | 5 | 4 | 4 | 4 | 3 | 3 | 3 | 2 | 2 | 2 | 1 | 1 | 1 | 1 | 0 | 0 | 0 | 0 | 0 | -1 | -1 | -1 | -1 | 0 | 0 | -2 | 0 | -2 | 0 | 0 | 0 | -2 | 0 |
| $([5]\#[32])\uparrow G$ | 10 | 6 | 4 | 6 | 2 | 4 | 6 | 0 | 5 | 2 | 0 | -2 | 2 | 0 | 1 | -2 | 2 | 0 | 0 | -1 | -2 | 2 | 1 | 0 | -1 | 0 | 0 | 2 | 0 | 1 | 0 | 0 | 0 | 0 | 0 |
| $([5]\#[31^2])\uparrow G$ | 12 | 6 | 6 | 4 | 0 | 6 | 6 | 0 | 7 | -2 | 0 | 0 | 0 | -2 | 1 | 0 | -4 | -2 | 0 | 1 | 0 | -2 | -1 | 0 | 1 | 0 | 0 | 0 | 0 | 1 | 0 | 0 | 0 | 2 | 0 |
| $([5]\#[2^2 1])\uparrow G$ | 10 | 4 | 4 | 6 | -2 | 6 | 4 | -2 | 5 | 0 | 0 | -2 | -2 | 0 | -1 | 0 | 2 | 2 | -2 | -1 | 2 | 0 | 1 | 0 | 1 | 0 | 0 | -2 | 0 | -1 | 0 | 0 | 0 | 0 | 0 |
| $([5]\#[21^3])\uparrow G$ | 8 | 2 | 5 | 4 | -4 | 4 | 5 | -1 | 3 | -2 | -2 | 2 | -1 | 1 | -3 | 1 | 0 | 0 | 2 | 0 | 0 | 1 | -1 | 1 | -1 | 0 | 0 | 2 | 0 | 0 | 0 | 0 | 0 | -2 | 0 |
| $([5]\#[1^5])\uparrow G$ | 2 | 0 | 2 | 2 | -2 | 0 | 0 | 0 | 2 | 0 | -2 | 2 | -2 | 2 | 0 | 0 | 2 | 0 | 0 | 2 | -2 | 0 | 2 | -2 | 0 | 0 | 0 | -2 | 0 | 0 | 0 | 0 | 0 | 2 | 0 |
| $([41]\#[32])\uparrow G$ | 40 | 14 | 1 | 4 | 4 | -4 | -1 | -1 | -5 | 2 | -2 | -2 | 1 | 1 | -1 | -1 | 0 | 0 | 2 | 1 | 0 | -1 | -1 | 1 | 1 | 0 | 0 | -2 | 0 | -1 | 0 | 0 | 0 | 0 | 0 |
| $([41]\#[31^2])\uparrow G$ | 48 | 12 | 6 | -8 | 0 | 0 | -6 | 0 | -2 | -4 | 0 | 0 | 0 | -2 | 2 | 0 | 0 | 0 | 0 | 1 | 0 | 2 | 2 | 0 | 0 | 0 | 0 | 0 | 0 | -1 | 0 | 0 | 0 | -2 | 0 |
| $([41]\#[2^2 1])\uparrow G$ | 40 | 5 | 1 | 4 | -4 | 4 | -9 | -3 | -5 | 2 | 2 | -2 | -1 | 1 | 1 | 1 | 0 | 0 | 0 | 1 | 0 | -1 | -1 | -1 | -1 | 0 | 0 | 2 | 0 | 1 | 0 | 0 | 0 | 0 | 0 |
| $([41]\#[21^3])\uparrow G$ | 32 | 0 | 8 | 0 | -8 | 0 | 0 | 0 | -8 | 0 | 0 | 2 | 4 | 0 | 0 | 0 | 0 | 0 | 0 | -2 | 0 | 0 | 0 | 0 | 0 | 0 | 0 | -2 | 0 | 0 | 0 | 0 | 0 | 2 | 0 |
| $([41]\#[1^5])\uparrow G$ | 8 | -2 | 5 | 4 | -4 | -4 | -5 | 1 | 3 | 2 | -2 | 2 | -1 | 1 | 3 | -1 | 0 | 0 | -2 | 0 | 0 | -1 | -1 | 1 | 1 | 0 | 0 | 2 | 0 | 0 | 0 | 0 | 0 | -2 | 0 |
| $([32]\#[31^2])\uparrow G$ | 60 | 6 | -6 | -4 | 0 | -6 | 6 | 0 | 5 | -2 | 0 | 0 | 0 | 2 | 1 | 0 | -4 | 2 | 0 | -1 | 0 | -2 | 1 | 0 | -1 | 0 | 0 | 0 | 0 | 1 | 0 | 0 | 0 | 0 | 0 |
| $([32]\#[2^2 1])\uparrow G$ | 50 | 0 | -10 | 10 | -2 | 0 | 0 | 0 | 0 | 0 | 2 | 2 | -2 | -2 | 0 | 0 | 2 | 0 | 0 | 0 | -2 | 0 | 0 | 2 | 0 | 0 | 0 | -2 | 0 | 0 | 0 | 0 | 0 | 0 | 0 |
| $([32]\#[21^3])\uparrow G$ | 40 | -6 | 1 | 4 | -4 | -4 | 9 | 3 | -5 | -2 | 2 | -2 | -1 | 1 | -1 | -1 | 0 | 0 | 0 | 1 | 0 | 1 | -1 | -1 | 1 | 0 | 0 | 2 | 0 | -1 | 0 | 0 | 0 | 0 | 0 |
| $([32]\#[1^5])\uparrow G$ | 10 | -4 | 4 | 6 | -2 | -6 | -4 | 2 | 5 | 0 | 0 | -2 | -2 | 0 | 1 | 0 | 2 | -2 | 2 | -1 | 2 | 0 | 1 | 0 | -1 | 0 | 0 | -2 | 0 | 1 | 0 | 0 | 0 | 0 | 0 |
| $([31^2]\#[2^2 1])\uparrow G$ | 60 | -6 | -6 | -4 | 0 | 6 | -6 | 0 | 5 | 2 | 0 | 0 | 0 | 2 | -1 | 0 | -4 | -2 | 0 | -1 | 0 | 2 | 1 | 0 | 1 | 0 | 0 | 0 | 0 | -1 | 0 | 0 | 0 | 0 | 0 |
| $([31^2]\#[21^3])\uparrow G$ | 48 | -12 | 6 | -8 | 0 | 0 | 6 | 0 | -2 | 4 | 0 | 0 | 0 | -2 | -2 | 0 | 0 | 0 | 0 | 1 | 0 | -2 | 2 | 0 | 0 | 0 | 0 | 0 | 0 | 1 | 0 | 0 | 0 | -2 | 0 |
| $([31^2]\#[1^5])\uparrow G$ | 12 | -5 | 6 | 4 | 0 | -6 | -6 | 0 | 7 | 2 | 0 | 0 | 0 | -2 | -1 | 0 | -4 | 2 | 0 | 1 | 0 | 2 | -1 | 0 | -1 | 0 | 0 | 0 | 0 | -1 | 0 | 0 | 0 | 2 | 0 |
| $([2^2 1]\#[21^3])\uparrow G$ | 40 | -14 | 1 | 4 | 4 | 4 | 1 | 1 | -5 | -2 | -2 | -2 | 1 | 1 | 1 | 1 | 0 | 0 | -2 | 1 | 0 | 1 | -1 | 1 | -1 | 0 | 0 | -2 | 0 | 1 | 0 | 0 | 0 | 0 | 0 |
| $([2^2 1]\#[1^5])\uparrow G$ | 10 | -6 | 4 | 6 | 2 | -4 | -6 | 0 | 5 | -2 | 0 | -2 | 2 | 0 | -1 | 2 | 2 | 0 | 0 | -1 | -2 | -2 | 1 | 0 | 1 | 0 | 0 | 2 | 0 | -1 | 0 | 0 | 0 | 0 | 0 |
| $([21^3]\#[1^5])\uparrow G$ | 8 | -6 | 5 | 4 | 4 | -4 | -3 | -3 | 3 | -2 | 2 | 2 | 1 | 1 | -1 | -1 | 0 | 0 | 0 | 0 | 0 | 1 | -1 | -1 | 1 | 0 | 0 | -2 | 0 | 2 | 0 | 0 | 0 | -2 | 0 |

## I.I Decompositions of Inner Tensor Products

This section contains tables of multiplicities

$$(Xi * Xj, Xk),$$

$Xi$, $Xj$, and $Xk$ being ordinary irreducible representations of $S_n$, $n \leqslant 8$. The ordering is the same as in Appendix I.A, and we use the fact that $(Xi * Xj, Xk) = (Xj * Xi, Xk)$, so that we need only show this multiplicity for the case when $j \geqslant i$.

The tables are due to A. Golembiowski.

Inner tensor products for $S_2$, $S_3$, $S_4$, and $S_5$

| | | X1 | X2 |
|---|---|---|---|
| X1*X1 | I | 1 | 0 |
| X1*X2 | I | 0 | 1 |
| X2*X2 | I | 1 | 0 |

| | | X1 | X2 | X3 |
|---|---|---|---|---|
| X1*X1 | I | 1 | 0 | 0 |
| X1*X2 | I | 0 | 1 | 0 |
| X1*X3 | I | 0 | 0 | 1 |
| X2*X2 | I | 1 | 1 | 1 |
| X2*X3 | I | 0 | 1 | 0 |
| X3*X3 | I | 1 | 0 | 0 |

| | | X1 | X2 | X3 | X4 | X5 |
|---|---|---|---|---|---|---|
| X1*X1 | I | 1 | 0 | 0 | 0 | 0 |
| X1*X2 | I | 0 | 1 | 0 | 0 | 0 |
| X1*X3 | I | 0 | 0 | 1 | 0 | 0 |
| X1*X4 | I | 0 | 0 | 0 | 1 | 0 |
| X1*X5 | I | 0 | 0 | 0 | 0 | 1 |
| X2*X2 | I | 1 | 1 | 1 | 1 | 0 |
| X2*X3 | I | 0 | 1 | 0 | 1 | 0 |
| X2*X4 | I | 0 | 1 | 1 | 1 | 1 |
| X2*X5 | I | 0 | 0 | 0 | 1 | 0 |
| X3*X3 | I | 1 | 0 | 1 | 0 | 1 |
| X3*X4 | I | 0 | 1 | 0 | 1 | 0 |
| X3*X5 | I | 0 | 0 | 1 | 0 | 0 |
| X4*X4 | I | 1 | 1 | 1 | 1 | 0 |
| X4*X5 | I | 0 | 1 | 0 | 0 | 0 |
| X5*X5 | I | 1 | 0 | 0 | 0 | 0 |

| | | X1 | X2 | X3 | X4 | X5 | X6 | X7 |
|---|---|---|---|---|---|---|---|---|
| X1*X1 | I | 1 | 0 | 0 | 0 | 0 | 0 | 0 |
| X1*X2 | I | 0 | 1 | 0 | 0 | 0 | 0 | 0 |
| X1*X3 | I | 0 | 0 | 1 | 0 | 0 | 0 | 0 |
| X1*X4 | I | 0 | 0 | 0 | 1 | 0 | 0 | 0 |
| X1*X5 | I | 0 | 0 | 0 | 0 | 1 | 0 | 0 |
| X1*X6 | I | 0 | 0 | 0 | 0 | 0 | 1 | 0 |
| X1*X7 | I | 0 | 0 | 0 | 0 | 0 | 0 | 1 |
| X2*X2 | I | 1 | 1 | 1 | 1 | 0 | 0 | 0 |
| X2*X3 | I | 0 | 1 | 1 | 1 | 1 | 0 | 0 |
| X2*X4 | I | 0 | 1 | 1 | 1 | 1 | 1 | 0 |
| X2*X5 | I | 0 | 0 | 1 | 1 | 1 | 1 | 0 |
| X2*X6 | I | 0 | 0 | 0 | 1 | 1 | 1 | 1 |
| X2*X7 | I | 0 | 0 | 0 | 0 | 0 | 1 | 0 |
| X3*X3 | I | 1 | 1 | 1 | 1 | 1 | 1 | 0 |
| X3*X4 | I | 0 | 1 | 1 | 2 | 1 | 1 | 0 |
| X3*X5 | I | 0 | 1 | 1 | 1 | 1 | 1 | 1 |
| X3*X6 | I | 0 | 0 | 1 | 1 | 1 | 1 | 0 |
| X3*X7 | I | 0 | 0 | 0 | 0 | 1 | 0 | 0 |
| X4*X4 | I | 1 | 1 | 2 | 1 | 2 | 1 | 1 |
| X4*X5 | I | 0 | 1 | 1 | 2 | 1 | 1 | 0 |
| X4*X6 | I | 0 | 1 | 1 | 1 | 1 | 1 | 0 |
| X4*X7 | I | 0 | 0 | 0 | 1 | 0 | 0 | 0 |
| X5*X5 | I | 1 | 1 | 1 | 1 | 1 | 1 | 0 |
| X5*X6 | I | 0 | 1 | 1 | 1 | 1 | 0 | 0 |
| X5*X7 | I | 0 | 0 | 1 | 0 | 0 | 0 | 0 |
| X6*X6 | I | 1 | 1 | 1 | 1 | 0 | 0 | 0 |
| X6*X7 | I | 0 | 1 | 0 | 0 | 0 | 0 | 0 |
| X7*X7 | I | 1 | 0 | 0 | 0 | 0 | 0 | 0 |

## Inner tensor products for $S_6$

| | | X1 | X2 | X3 | X4 | X5 | X6 | X7 | X8 | X9 | X10 | X11 |
|---|---|---|---|---|---|---|---|---|---|---|---|---|
| X1*X1 | I | 1 | 0 | 0 | 0 | 0 | 0 | 0 | 0 | 0 | 0 | 0 |
| X1*X2 | I | 0 | 1 | 0 | 0 | 0 | 0 | 0 | 0 | 0 | 0 | 0 |
| X1*X3 | I | 0 | 0 | 1 | 0 | 0 | 0 | 0 | 0 | 0 | 0 | 0 |
| X1*X4 | I | 0 | 0 | 0 | 1 | 0 | 0 | 0 | 0 | 0 | 0 | 0 |
| X1*X5 | I | 0 | 0 | 0 | 0 | 1 | 0 | 0 | 0 | 0 | 0 | 0 |
| X1*X6 | I | 0 | 0 | 0 | 0 | 0 | 1 | 0 | 0 | 0 | 0 | 0 |
| X1*X7 | I | 0 | 0 | 0 | 0 | 0 | 0 | 1 | 0 | 0 | 0 | 0 |
| X1*X8 | I | 0 | 0 | 0 | 0 | 0 | 0 | 0 | 1 | 0 | 0 | 0 |
| X1*X9 | I | 0 | 0 | 0 | 0 | 0 | 0 | 0 | 0 | 1 | 0 | 0 |
| X1*X10 | I | 0 | 0 | 0 | 0 | 0 | 0 | 0 | 0 | 0 | 1 | 0 |
| X1*X11 | I | 0 | 0 | 0 | 0 | 0 | 0 | 0 | 0 | 0 | 0 | 1 |
| X2*X2 | I | 1 | 1 | 1 | 1 | 0 | 0 | 0 | 0 | 0 | 0 | 0 |
| X2*X3 | I | 0 | 1 | 1 | 1 | 1 | 1 | 0 | 0 | 0 | 0 | 0 |
| X2*X4 | I | 0 | 1 | 1 | 1 | 0 | 1 | 1 | 0 | 0 | 0 | 0 |
| X2*X5 | I | 0 | 0 | 1 | 0 | 0 | 1 | 0 | 0 | 0 | 0 | 0 |
| X2*X6 | I | 0 | 0 | 1 | 1 | 1 | 2 | 1 | 1 | 1 | 0 | 0 |
| X2*X7 | I | 0 | 0 | 0 | 1 | 0 | 1 | 1 | 0 | 1 | 1 | 0 |
| X2*X8 | I | 0 | 0 | 0 | 0 | 0 | 1 | 0 | 0 | 1 | 0 | 0 |
| X2*X9 | I | 0 | 0 | 0 | 0 | 0 | 1 | 1 | 1 | 1 | 1 | 0 |
| X2*X10 | I | 0 | 0 | 0 | 0 | 0 | 0 | 1 | 0 | 1 | 1 | 1 |
| X2*X11 | I | 0 | 0 | 0 | 0 | 0 | 0 | 0 | 0 | 0 | 1 | 0 |
| X3*X3 | I | 1 | 1 | 2 | 1 | 0 | 2 | 1 | 1 | 0 | 0 | 0 |
| X3*X4 | I | 0 | 1 | 1 | 2 | 1 | 2 | 1 | 0 | 1 | 0 | 0 |
| X3*X5 | I | 0 | 1 | 0 | 1 | 1 | 1 | 0 | 0 | 1 | 0 | 0 |
| X3*X6 | I | 0 | 1 | 2 | 2 | 1 | 3 | 2 | 1 | 2 | 1 | 0 |
| X3*X7 | I | 0 | 0 | 1 | 1 | 0 | 2 | 2 | 1 | 1 | 1 | 0 |
| X3*X8 | I | 0 | 0 | 1 | 0 | 0 | 1 | 1 | 1 | 0 | 1 | 0 |
| X3*X9 | I | 0 | 0 | 0 | 1 | 1 | 2 | 1 | 0 | 2 | 1 | 1 |
| X3*X10 | I | 0 | 0 | 0 | 0 | 0 | 1 | 1 | 1 | 1 | 1 | 0 |
| X3*X11 | I | 0 | 0 | 0 | 0 | 0 | 0 | 0 | 0 | 1 | 0 | 0 |
| X4*X4 | I | 1 | 1 | 2 | 1 | 1 | 2 | 1 | 1 | 1 | 1 | 0 |
| X4*X5 | I | 0 | 0 | 1 | 1 | 0 | 1 | 1 | 1 | 0 | 0 | 0 |
| X4*X6 | I | 0 | 1 | 2 | 2 | 1 | 4 | 2 | 1 | 2 | 1 | 0 |
| X4*X7 | I | 0 | 1 | 1 | 1 | 1 | 2 | 1 | 1 | 2 | 1 | 1 |
| X4*X8 | I | 0 | 0 | 0 | 1 | 1 | 1 | 1 | 0 | 1 | 0 | 0 |
| X4*X9 | I | 0 | 0 | 1 | 1 | 0 | 2 | 2 | 1 | 1 | 1 | 0 |
| X4*X10 | I | 0 | 0 | 0 | 1 | 0 | 1 | 1 | 0 | 1 | 1 | 0 |
| X4*X11 | I | 0 | 0 | 0 | 0 | 0 | 0 | 1 | 0 | 0 | 0 | 0 |
| X5*X5 | I | 1 | 0 | 1 | 0 | 0 | 0 | 1 | 1 | 0 | 0 | 0 |
| X5*X6 | I | 0 | 1 | 1 | 1 | 0 | 2 | 1 | 0 | 1 | 1 | 0 |
| X5*X7 | I | 0 | 0 | 0 | 1 | 1 | 1 | 1 | 0 | 1 | 0 | 0 |
| X5*X8 | I | 0 | 0 | 0 | 1 | 1 | 0 | 0 | 0 | 1 | 0 | 1 |
| X5*X9 | I | 0 | 0 | 1 | 0 | 0 | 1 | 1 | 1 | 0 | 1 | 0 |
| X5*X10 | I | 0 | 0 | 0 | 0 | 0 | 1 | 0 | 0 | 1 | 0 | 0 |
| X5*X11 | I | 0 | 0 | 0 | 0 | 0 | 0 | 0 | 1 | 0 | 0 | 0 |
| X6*X6 | I | 1 | 2 | 3 | 4 | 2 | 5 | 4 | 2 | 3 | 2 | 1 |
| X6*X7 | I | 0 | 1 | 2 | 2 | 1 | 4 | 2 | 1 | 2 | 1 | 0 |
| X6*X8 | I | 0 | 1 | 1 | 1 | 0 | 2 | 1 | 0 | 1 | 1 | 0 |
| X6*X9 | I | 0 | 1 | 2 | 2 | 1 | 3 | 2 | 1 | 2 | 1 | 0 |
| X6*X10 | I | 0 | 0 | 1 | 1 | 1 | 2 | 1 | 1 | 1 | 0 | 0 |
| X6*X11 | I | 0 | 0 | 0 | 0 | 0 | 1 | 0 | 0 | 0 | 0 | 0 |
| X7*X7 | I | 1 | 1 | 2 | 1 | 1 | 2 | 1 | 1 | 1 | 1 | 0 |
| X7*X8 | I | 0 | 0 | 1 | 1 | 0 | 1 | 1 | 1 | 0 | 0 | 0 |
| X7*X9 | I | 0 | 1 | 1 | 2 | 1 | 2 | 1 | 0 | 1 | 0 | 0 |
| X7*X10 | I | 0 | 1 | 1 | 1 | 0 | 1 | 1 | 0 | 0 | 0 | 0 |
| X7*X11 | I | 0 | 0 | 0 | 1 | 0 | 0 | 0 | 0 | 0 | 0 | 0 |
| X8*X8 | I | 1 | 0 | 1 | 0 | 0 | 0 | 1 | 1 | 0 | 0 | 0 |
| X8*X9 | I | 0 | 1 | 0 | 1 | 1 | 1 | 0 | 0 | 1 | 0 | 0 |
| X8*X10 | I | 0 | 0 | 1 | 0 | 0 | 1 | 0 | 0 | 0 | 0 | 0 |
| X8*X11 | I | 0 | 0 | 0 | 0 | 1 | 0 | 0 | 0 | 0 | 0 | 0 |
| X9*X9 | I | 1 | 1 | 2 | 1 | 0 | 2 | 1 | 1 | 0 | 0 | 0 |
| X9*X10 | I | 0 | 1 | 1 | 1 | 1 | 1 | 0 | 0 | 0 | 0 | 0 |
| X9*X11 | I | 0 | 0 | 1 | 0 | 0 | 0 | 0 | 0 | 0 | 0 | 0 |
| X10*X10 | I | 1 | 1 | 1 | 1 | 0 | 0 | 0 | 0 | 0 | 0 | 0 |
| X10*X11 | I | 0 | 1 | 0 | 0 | 0 | 0 | 0 | 0 | 0 | 0 | 0 |
| X11*X11 | I | 1 | 0 | 0 | 0 | 0 | 0 | 0 | 0 | 0 | 0 | 0 |

## Inner tensor products for $S_7$

| | | X1 | X2 | X3 | X4 | X5 | X6 | X7 | X8 | X9 | X10 | X11 | X12 | X13 | X14 | X15 |
|---|---|---|---|---|---|---|---|---|---|---|---|---|---|---|---|---|
| X1*X1 | I | 1 | 0 | 0 | 0 | 0 | 0 | 0 | 0 | 0 | 0 | 0 | 0 | 0 | 0 | 0 |
| X1*X2 | I | 0 | 1 | 0 | 0 | 0 | 0 | 0 | 0 | 0 | 0 | 0 | 0 | 0 | 0 | 0 |
| X1*X3 | I | 0 | 0 | 1 | 0 | 0 | 0 | 0 | 0 | 0 | 0 | 0 | 0 | 0 | 0 | 0 |
| X1*X4 | I | 0 | 0 | 0 | 1 | 0 | 0 | 0 | 0 | 0 | 0 | 0 | 0 | 0 | 0 | 0 |
| X1*X5 | I | 0 | 0 | 0 | 0 | 1 | 0 | 0 | 0 | 0 | 0 | 0 | 0 | 0 | 0 | 0 |
| X1*X6 | I | 0 | 0 | 0 | 0 | 0 | 1 | 0 | 0 | 0 | 0 | 0 | 0 | 0 | 0 | 0 |
| X1*X7 | I | 0 | 0 | 0 | 0 | 0 | 0 | 1 | 0 | 0 | 0 | 0 | 0 | 0 | 0 | 0 |
| X1*X8 | I | 0 | 0 | 0 | 0 | 0 | 0 | 0 | 1 | 0 | 0 | 0 | 0 | 0 | 0 | 0 |
| X1*X9 | I | 0 | 0 | 0 | 0 | 0 | 0 | 0 | 0 | 1 | 0 | 0 | 0 | 0 | 0 | 0 |
| X1*X10 | I | 0 | 0 | 0 | 0 | 0 | 0 | 0 | 0 | 0 | 1 | 0 | 0 | 0 | 0 | 0 |
| X1*X11 | I | 0 | 0 | 0 | 0 | 0 | 0 | 0 | 0 | 0 | 0 | 1 | 0 | 0 | 0 | 0 |
| X1*X12 | I | 0 | 0 | 0 | 0 | 0 | 0 | 0 | 0 | 0 | 0 | 0 | 1 | 0 | 0 | 0 |
| X1*X13 | I | 0 | 0 | 0 | 0 | 0 | 0 | 0 | 0 | 0 | 0 | 0 | 0 | 1 | 0 | 0 |
| X1*X14 | I | 0 | 0 | 0 | 0 | 0 | 0 | 0 | 0 | 0 | 0 | 0 | 0 | 0 | 1 | 0 |
| X1*X15 | I | 0 | 0 | 0 | 0 | 0 | 0 | 0 | 0 | 0 | 0 | 0 | 0 | 0 | 0 | 1 |
| X2*X2 | I | 1 | 1 | 1 | 1 | 0 | 0 | 0 | 0 | 0 | 0 | 0 | 0 | 0 | 0 | 0 |
| X2*X3 | I | 0 | 1 | 1 | 1 | 1 | 1 | 0 | 0 | 0 | 0 | 0 | 0 | 0 | 0 | 0 |
| X2*X4 | I | 0 | 1 | 1 | 1 | 0 | 1 | 1 | 0 | 0 | 0 | 0 | 0 | 0 | 0 | 0 |
| X2*X5 | I | 0 | 0 | 1 | 0 | 1 | 1 | 0 | 1 | 0 | 0 | 0 | 0 | 0 | 0 | 0 |
| X2*X6 | I | 0 | 0 | 1 | 1 | 1 | 2 | 1 | 1 | 1 | 1 | 0 | 0 | 0 | 0 | 0 |
| X2*X7 | I | 0 | 0 | 0 | 1 | 0 | 1 | 1 | 0 | 0 | 1 | 1 | 0 | 0 | 0 | 0 |
| X2*X8 | I | 0 | 0 | 0 | 0 | 1 | 1 | 0 | 1 | 1 | 1 | 0 | 0 | 0 | 0 | 0 |
| X2*X9 | I | 0 | 0 | 0 | 0 | 0 | 1 | 0 | 1 | 1 | 1 | 0 | 1 | 0 | 0 | 0 |
| X2*X10 | I | 0 | 0 | 0 | 0 | 0 | 1 | 1 | 1 | 1 | 2 | 1 | 1 | 1 | 0 | 0 |
| X2*X11 | I | 0 | 0 | 0 | 0 | 0 | 0 | 1 | 0 | 0 | 1 | 1 | 0 | 1 | 1 | 0 |
| X2*X12 | I | 0 | 0 | 0 | 0 | 0 | 0 | 0 | 0 | 1 | 1 | 0 | 1 | 1 | 0 | 0 |
| X2*X13 | I | 0 | 0 | 0 | 0 | 0 | 0 | 0 | 0 | 0 | 1 | 1 | 1 | 1 | 1 | 0 |
| X2*X14 | I | 0 | 0 | 0 | 0 | 0 | 0 | 0 | 0 | 0 | 0 | 1 | 0 | 1 | 1 | 1 |
| X2*X15 | I | 0 | 0 | 0 | 0 | 0 | 0 | 0 | 0 | 0 | 0 | 0 | 0 | 0 | 1 | 0 |
| X3*X3 | I | 1 | 1 | 2 | 1 | 1 | 2 | 1 | 1 | 1 | 0 | 0 | 0 | 0 | 0 | 0 |
| X3*X4 | I | 0 | 1 | 1 | 2 | 1 | 2 | 1 | 1 | 0 | 1 | 0 | 0 | 0 | 0 | 0 |
| X3*X5 | I | 0 | 1 | 1 | 1 | 1 | 2 | 0 | 1 | 1 | 1 | 0 | 0 | 0 | 0 | 0 |
| X3*X6 | I | 0 | 1 | 2 | 2 | 2 | 4 | 2 | 2 | 2 | 3 | 1 | 1 | 0 | 0 | 0 |
| X3*X7 | I | 0 | 0 | 1 | 1 | 0 | 2 | 2 | 1 | 1 | 2 | 1 | 0 | 1 | 0 | 0 |
| X3*X8 | I | 0 | 0 | 1 | 1 | 1 | 2 | 1 | 2 | 1 | 2 | 0 | 1 | 1 | 0 | 0 |
| X3*X9 | I | 0 | 0 | 1 | 0 | 1 | 2 | 1 | 1 | 2 | 2 | 1 | 1 | 1 | 0 | 0 |
| X3*X10 | I | 0 | 0 | 0 | 1 | 1 | 3 | 2 | 2 | 2 | 4 | 2 | 2 | 2 | 1 | 0 |
| X3*X11 | I | 0 | 0 | 0 | 0 | 0 | 1 | 1 | 0 | 1 | 2 | 2 | 1 | 1 | 1 | 0 |
| X3*X12 | I | 0 | 0 | 0 | 0 | 0 | 1 | 0 | 1 | 1 | 2 | 1 | 1 | 1 | 1 | 0 |
| X3*X13 | I | 0 | 0 | 0 | 0 | 0 | 0 | 1 | 1 | 1 | 2 | 1 | 1 | 2 | 1 | 1 |
| X3*X14 | I | 0 | 0 | 0 | 0 | 0 | 0 | 0 | 0 | 0 | 1 | 1 | 1 | 1 | 1 | 0 |
| X3*X15 | I | 0 | 0 | 0 | 0 | 0 | 0 | 0 | 0 | 0 | 0 | 0 | 0 | 1 | 0 | 0 |
| X4*X4 | I | 1 | 1 | 2 | 1 | 1 | 2 | 1 | 0 | 1 | 1 | 1 | 0 | 0 | 0 | 0 |
| X4*X5 | I | 0 | 0 | 1 | 1 | 1 | 2 | 1 | 1 | 1 | 1 | 0 | 0 | 0 | 0 | 0 |
| X4*X6 | I | 0 | 1 | 2 | 2 | 2 | 4 | 2 | 3 | 2 | 3 | 1 | 1 | 1 | 0 | 0 |
| X4*X7 | I | 0 | 1 | 1 | 1 | 1 | 2 | 1 | 1 | 1 | 2 | 1 | 1 | 1 | 1 | 0 |
| X4*X8 | I | 0 | 0 | 1 | 0 | 1 | 3 | 1 | 1 | 2 | 2 | 1 | 1 | 0 | 0 | 0 |
| X4*X9 | I | 0 | 0 | 0 | 1 | 1 | 2 | 1 | 2 | 1 | 3 | 0 | 1 | 1 | 0 | 0 |
| X4*X10 | I | 0 | 0 | 1 | 1 | 1 | 3 | 2 | 2 | 3 | 4 | 2 | 2 | 2 | 1 | 0 |
| X4*X11 | I | 0 | 0 | 0 | 1 | 0 | 1 | 1 | 1 | 0 | 2 | 1 | 1 | 2 | 1 | 1 |
| X4*X12 | I | 0 | 0 | 0 | 0 | 0 | 1 | 1 | 1 | 1 | 2 | 1 | 1 | 1 | 0 | 0 |
| X4*X13 | I | 0 | 0 | 0 | 0 | 0 | 1 | 1 | 0 | 1 | 2 | 2 | 1 | 1 | 1 | 0 |
| X4*X14 | I | 0 | 0 | 0 | 0 | 0 | 0 | 1 | 0 | 0 | 1 | 1 | 0 | 1 | 1 | 0 |
| X4*X15 | I | 0 | 0 | 0 | 0 | 0 | 0 | 0 | 0 | 0 | 0 | 1 | 0 | 0 | 0 | 0 |
| X5*X5 | I | 1 | 1 | 1 | 1 | 1 | 1 | 1 | 1 | 1 | 1 | 0 | 1 | 0 | 0 | 0 |
| X5*X6 | I | 0 | 1 | 2 | 2 | 1 | 4 | 2 | 2 | 2 | 3 | 1 | 1 | 1 | 0 | 0 |
| X5*X7 | I | 0 | 0 | 0 | 1 | 1 | 2 | 2 | 1 | 1 | 2 | 1 | 1 | 0 | 0 | 0 |
| X5*X8 | I | 0 | 1 | 1 | 1 | 1 | 2 | 1 | 1 | 1 | 2 | 1 | 1 | 1 | 0 | 0 |
| X5*X9 | I | 0 | 0 | 1 | 1 | 1 | 2 | 1 | 1 | 1 | 2 | 1 | 1 | 1 | 1 | 0 |
| X5*X10 | I | 0 | 0 | 1 | 1 | 1 | 3 | 2 | 2 | 2 | 4 | 2 | 1 | 2 | 1 | 0 |
| X5*X11 | I | 0 | 0 | 0 | 0 | 0 | 1 | 1 | 1 | 1 | 2 | 1 | 1 | 1 | 0 | 0 |
| X5*X12 | I | 0 | 0 | 0 | 0 | 1 | 1 | 1 | 1 | 1 | 1 | 1 | 1 | 1 | 1 | 1 |
| X5*X13 | I | 0 | 0 | 0 | 0 | 0 | 1 | 0 | 1 | 1 | 2 | 1 | 1 | 1 | 1 | 0 |
| X5*X14 | I | 0 | 0 | 0 | 0 | 0 | 0 | 0 | 0 | 1 | 1 | 0 | 1 | 1 | 0 | 0 |
| X5*X15 | I | 0 | 0 | 0 | 0 | 0 | 0 | 0 | 0 | 0 | 0 | 0 | 1 | 0 | 0 | 0 |
| X6*X6 | I | 1 | 2 | 4 | 4 | 4 | 9 | 5 | 5 | 5 | 8 | 3 | 3 | 3 | 1 | 0 |
| X6*X7 | I | 0 | 1 | 2 | 2 | 2 | 5 | 2 | 3 | 3 | 5 | 2 | 2 | 2 | 1 | 0 |
| X6*X8 | I | 0 | 1 | 2 | 3 | 2 | 5 | 3 | 3 | 3 | 5 | 2 | 2 | 2 | 1 | 0 |
| X6*X9 | I | 0 | 1 | 2 | 2 | 2 | 5 | 3 | 3 | 3 | 5 | 3 | 2 | 2 | 1 | 0 |
| X6*X10 | I | 0 | 1 | 3 | 3 | 3 | 8 | 5 | 5 | 5 | 9 | 4 | 4 | 4 | 2 | 1 |
| X6*X11 | I | 0 | 0 | 1 | 1 | 1 | 3 | 2 | 2 | 3 | 4 | 2 | 2 | 2 | 1 | 0 |
| X6*X12 | I | 0 | 0 | 1 | 1 | 1 | 3 | 2 | 2 | 2 | 4 | 2 | 1 | 2 | 1 | 0 |
| X6*X13 | I | 0 | 0 | 0 | 1 | 1 | 3 | 2 | 2 | 2 | 4 | 2 | 2 | 2 | 1 | 0 |
| X6*X14 | I | 0 | 0 | 0 | 0 | 0 | 1 | 1 | 1 | 1 | 2 | 1 | 1 | 1 | 0 | 0 |
| X6*X15 | I | 0 | 0 | 0 | 0 | 0 | 0 | 0 | 0 | 0 | 1 | 0 | 0 | 0 | 0 | 0 |

| | | X1 | X2 | X3 | X4 | X5 | X6 | X7 | X8 | X9 | X10 | X11 | X12 | X13 | X14 | X15 |
|---|---|---|---|---|---|---|---|---|---|---|---|---|---|---|---|---|
| X7*X7 | I | 1 | 1 | 2 | 1 | 2 | 2 | 1 | 2 | 2 | 2 | 1 | 2 | 2 | 1 | 1 |
| X7*X8 | I | 0 | 0 | 1 | 1 | 1 | 3 | 2 | 2 | 2 | 3 | 1 | 1 | 1 | 0 | 0 |
| X7*X9 | I | 0 | 0 | 1 | 1 | 1 | 3 | 2 | 2 | 2 | 3 | 1 | 1 | 1 | 0 | 0 |
| X7*X10 | I | 0 | 1 | 2 | 2 | 2 | 5 | 2 | 3 | 3 | 5 | 2 | 2 | 2 | 1 | 0 |
| X7*X11 | I | 0 | 1 | 1 | 1 | 1 | 2 | 1 | 1 | 1 | 2 | 1 | 1 | 1 | 1 | 0 |
| X7*X12 | I | 0 | 0 | 0 | 1 | 1 | 2 | 2 | 1 | 1 | 2 | 1 | 1 | 0 | 0 | 0 |
| X7*X13 | I | 0 | 0 | 1 | 1 | 0 | 2 | 2 | 1 | 1 | 2 | 1 | 0 | 1 | 0 | 0 |
| X7*X14 | I | 0 | 0 | 0 | 1 | 0 | 1 | 1 | 0 | 0 | 1 | 1 | 0 | 0 | 0 | 0 |
| X7*X15 | I | 0 | 0 | 0 | 0 | 0 | 0 | 1 | 0 | 0 | 0 | 0 | 0 | 0 | 0 | 0 |
| X8*X8 | I | 1 | 1 | 2 | 1 | 1 | 3 | 2 | 1 | 2 | 3 | 2 | 1 | 1 | 1 | 0 |
| X8*X9 | I | 0 | 1 | 1 | 2 | 1 | 3 | 2 | 2 | 1 | 3 | 1 | 1 | 2 | 1 | 1 |
| X8*X10 | I | 0 | 1 | 2 | 2 | 2 | 5 | 3 | 3 | 3 | 5 | 3 | 2 | 2 | 1 | 0 |
| X8*X11 | I | 0 | 0 | 0 | 1 | 1 | 2 | 1 | 2 | 1 | 3 | 0 | 1 | 1 | 0 | 0 |
| X8*X12 | I | 0 | 0 | 1 | 1 | 1 | 2 | 1 | 1 | 1 | 2 | 1 | 1 | 1 | 1 | 0 |
| X8*X13 | I | 0 | 0 | 1 | 0 | 1 | 2 | 1 | 1 | 2 | 2 | 1 | 1 | 1 | 0 | 0 |
| X8*X14 | I | 0 | 0 | 0 | 0 | 0 | 1 | 0 | 1 | 1 | 1 | 0 | 1 | 0 | 0 | 0 |
| X8*X15 | I | 0 | 0 | 0 | 0 | 0 | 0 | 0 | 0 | 1 | 0 | 0 | 0 | 0 | 0 | 0 |
| X9*X9 | I | 1 | 1 | 2 | 1 | 1 | 3 | 2 | 1 | 2 | 3 | 2 | 1 | 1 | 1 | 0 |
| X9*X10 | I | 0 | 1 | 2 | 3 | 2 | 5 | 3 | 3 | 3 | 5 | 2 | 2 | 2 | 1 | 0 |
| X9*X11 | I | 0 | 0 | 1 | 0 | 1 | 3 | 1 | 1 | 2 | 2 | 1 | 1 | 0 | 0 | 0 |
| X9*X12 | I | 0 | 1 | 1 | 1 | 1 | 2 | 1 | 1 | 1 | 2 | 1 | 1 | 1 | 0 | 0 |
| X9*X13 | I | 0 | 0 | 1 | 1 | 1 | 2 | 1 | 2 | 1 | 2 | 0 | 1 | 1 | 0 | 0 |
| X9*X14 | I | 0 | 0 | 0 | 0 | 1 | 1 | 0 | 1 | 1 | 1 | 0 | 0 | 0 | 0 | 0 |
| X9*X15 | I | 0 | 0 | 0 | 0 | 0 | 0 | 0 | 1 | 0 | 0 | 0 | 0 | 0 | 0 | 0 |
| X10*X10 | I | 1 | 2 | 4 | 4 | 4 | 9 | 5 | 5 | 5 | 8 | 3 | 3 | 3 | 1 | 0 |
| X10*X11 | I | 0 | 1 | 2 | 2 | 2 | 4 | 2 | 3 | 2 | 3 | 1 | 1 | 1 | 0 | 0 |
| X10*X12 | I | 0 | 1 | 2 | 2 | 1 | 4 | 2 | 2 | 2 | 3 | 1 | 1 | 1 | 0 | 0 |
| X10*X13 | I | 0 | 1 | 2 | 2 | 2 | 4 | 2 | 2 | 2 | 3 | 1 | 1 | 0 | 0 | 0 |
| X10*X14 | I | 0 | 0 | 1 | 1 | 1 | 2 | 1 | 1 | 1 | 1 | 0 | 0 | 0 | 0 | 0 |
| X10*X15 | I | 0 | 0 | 0 | 0 | 0 | 1 | 0 | 0 | 0 | 0 | 0 | 0 | 0 | 0 | 0 |
| X11*X11 | I | 1 | 1 | 2 | 1 | 1 | 2 | 1 | 0 | 1 | 1 | 1 | 0 | 0 | 0 | 0 |
| X11*X12 | I | 0 | 0 | 1 | 1 | 1 | 2 | 1 | 1 | 1 | 1 | 0 | 0 | 0 | 0 | 0 |
| X11*X13 | 1 | 0 | 1 | 1 | 2 | 1 | 2 | 1 | 1 | 0 | 1 | 0 | 0 | 0 | 0 | 0 |
| X11*X14 | I | 0 | 1 | 1 | 1 | 0 | 1 | 1 | 0 | 0 | 0 | 0 | 0 | 0 | 0 | 0 |
| X11*X15 | I | 0 | 0 | 0 | 1 | 0 | 0 | 0 | 0 | 0 | 0 | 0 | 0 | 0 | 0 | 0 |
| X12*X12 | I | 1 | 1 | 1 | 1 | 1 | 1 | 1 | 1 | 1 | 1 | 0 | 1 | 0 | 0 | 0 |
| X12*X13 | 1 | 0 | 1 | 1 | 1 | 1 | 2 | 0 | 1 | 1 | 1 | 0 | 0 | 0 | 0 | 0 |
| X12*X14 | I | 0 | 0 | 1 | 0 | 1 | 1 | 0 | 1 | 0 | 0 | 0 | 0 | 0 | 0 | 0 |
| X12*X15 | I | 0 | 0 | 0 | 0 | 1 | 0 | 0 | 0 | 0 | 0 | 0 | 0 | 0 | 0 | 0 |
| X13*X13 | I | 1 | 1 | 2 | 1 | 1 | 2 | 1 | 1 | 1 | 0 | 0 | 0 | 0 | 0 | 0 |
| X13*X14 | I | 0 | 1 | 1 | 1 | 1 | 1 | 0 | 0 | 0 | 0 | 0 | 0 | 0 | 0 | 0 |
| X13*X15 | I | 0 | 0 | 1 | 0 | 0 | 0 | 0 | 0 | 0 | 0 | 0 | 0 | 0 | 0 | 0 |
| X14*X14 | I | 1 | 1 | 1 | 1 | 0 | 0 | 0 | 0 | 0 | 0 | 0 | 0 | 0 | 0 | 0 |
| X14*X15 | I | 0 | 1 | 0 | 0 | 0 | 0 | 0 | 0 | 0 | 0 | 0 | 0 | 0 | 0 | 0 |
| X15*X15 | I | 1 | 0 | 0 | 0 | 0 | 0 | 0 | 0 | 0 | 0 | 0 | 0 | 0 | 0 | 0 |

# Inner tensor products for $S_8$

| | | X1 | X2 | X3 | X4 | X5 | X6 | X7 | X8 | X9 | X10 | X11 | X12 | X13 | X14 | X15 | X16 | X17 | X18 | X19 | X20 | X21 | X22 |
|---|---|---|---|---|---|---|---|---|---|---|---|---|---|---|---|---|---|---|---|---|---|---|---|
| X1*X1 | I | 1 | 0 | 0 | 0 | 0 | 0 | 0 | 0 | 0 | 0 | 0 | 0 | 0 | 0 | 0 | 0 | 0 | 0 | 0 | 0 | 0 | 0 |
| X1*X2 | I | 0 | 1 | 0 | 0 | 0 | 0 | 0 | 0 | 0 | 0 | 0 | 0 | 0 | 0 | 0 | 0 | 0 | 0 | 0 | 0 | 0 | 0 |
| X1*X3 | I | 0 | 0 | 1 | 0 | 0 | 0 | 0 | 0 | 0 | 0 | 0 | 0 | 0 | 0 | 0 | 0 | 0 | 0 | 0 | 0 | 0 | 0 |
| X1*X4 | I | 0 | 0 | 0 | 1 | 0 | 0 | 0 | 0 | 0 | 0 | 0 | 0 | 0 | 0 | 0 | 0 | 0 | 0 | 0 | 0 | 0 | 0 |
| X1*X5 | I | 0 | 0 | 0 | 0 | 1 | 0 | 0 | 0 | 0 | 0 | 0 | 0 | 0 | 0 | 0 | 0 | 0 | 0 | 0 | 0 | 0 | 0 |
| X1*X6 | I | 0 | 0 | 0 | 0 | 0 | 1 | 0 | 0 | 0 | 0 | 0 | 0 | 0 | 0 | 0 | 0 | 0 | 0 | 0 | 0 | 0 | 0 |
| X1*X7 | I | 0 | 0 | 0 | 0 | 0 | 0 | 1 | 0 | 0 | 0 | 0 | 0 | 0 | 0 | 0 | 0 | 0 | 0 | 0 | 0 | 0 | 0 |
| X1*X8 | I | 0 | 0 | 0 | 0 | 0 | 0 | 0 | 1 | 0 | 0 | 0 | 0 | 0 | 0 | 0 | 0 | 0 | 0 | 0 | 0 | 0 | 0 |
| X1*X9 | I | 0 | 0 | 0 | 0 | 0 | 0 | 0 | 0 | 1 | 0 | 0 | 0 | 0 | 0 | 0 | 0 | 0 | 0 | 0 | 0 | 0 | 0 |
| X1*X10 | I | 0 | 0 | 0 | 0 | 0 | 0 | 0 | 0 | 0 | 1 | 0 | 0 | 0 | 0 | 0 | 0 | 0 | 0 | 0 | 0 | 0 | 0 |
| X1*X11 | I | 0 | 0 | 0 | 0 | 0 | 0 | 0 | 0 | 0 | 0 | 1 | 0 | 0 | 0 | 0 | 0 | 0 | 0 | 0 | 0 | 0 | 0 |
| X1*X12 | I | 0 | 0 | 0 | 0 | 0 | 0 | 0 | 0 | 0 | 0 | 0 | 1 | 0 | 0 | 0 | 0 | 0 | 0 | 0 | 0 | 0 | 0 |
| X1*X13 | I | 0 | 0 | 0 | 0 | 0 | 0 | 0 | 0 | 0 | 0 | 0 | 0 | 1 | 0 | 0 | 0 | 0 | 0 | 0 | 0 | 0 | 0 |
| X1*X14 | I | 0 | 0 | 0 | 0 | 0 | 0 | 0 | 0 | 0 | 0 | 0 | 0 | 0 | 1 | 0 | 0 | 0 | 0 | 0 | 0 | 0 | 0 |
| X1*X15 | I | 0 | 0 | 0 | 0 | 0 | 0 | 0 | 0 | 0 | 0 | 0 | 0 | 0 | 0 | 1 | 0 | 0 | 0 | 0 | 0 | 0 | 0 |
| X1*X16 | I | 0 | 0 | 0 | 0 | 0 | 0 | 0 | 0 | 0 | 0 | 0 | 0 | 0 | 0 | 0 | 1 | 0 | 0 | 0 | 0 | 0 | 0 |
| X1*X17 | I | 0 | 0 | 0 | 0 | 0 | 0 | 0 | 0 | 0 | 0 | 0 | 0 | 0 | 0 | 0 | 0 | 1 | 0 | 0 | 0 | 0 | 0 |
| X1*X18 | I | 0 | 0 | 0 | 0 | 0 | 0 | 0 | 0 | 0 | 0 | 0 | 0 | 0 | 0 | 0 | 0 | 0 | 1 | 0 | 0 | 0 | 0 |
| X1*X19 | I | 0 | 0 | 0 | 0 | 0 | 0 | 0 | 0 | 0 | 0 | 0 | 0 | 0 | 0 | 0 | 0 | 0 | 0 | 1 | 0 | 0 | 0 |
| X1*X20 | I | 0 | 0 | 0 | 0 | 0 | 0 | 0 | 0 | 0 | 0 | 0 | 0 | 0 | 0 | 0 | 0 | 0 | 0 | 0 | 1 | 0 | 0 |
| X1*X21 | I | 0 | 0 | 0 | 0 | 0 | 0 | 0 | 0 | 0 | 0 | 0 | 0 | 0 | 0 | 0 | 0 | 0 | 0 | 0 | 0 | 1 | 0 |
| X1*X22 | I | 0 | 0 | 0 | 0 | 0 | 0 | 0 | 0 | 0 | 0 | 0 | 0 | 0 | 0 | 0 | 0 | 0 | 0 | 0 | 0 | 0 | 1 |
| X2*X2 | I | 1 | 1 | 1 | 1 | 0 | 0 | 0 | 0 | 0 | 0 | 0 | 0 | 0 | 0 | 0 | 0 | 0 | 0 | 0 | 0 | 0 | 0 |
| X2*X3 | I | 0 | 1 | 1 | 1 | 1 | 1 | 0 | 0 | 0 | 0 | 0 | 0 | 0 | 0 | 0 | 0 | 0 | 0 | 0 | 0 | 0 | 0 |
| X2*X4 | I | 0 | 1 | 1 | 1 | 0 | 1 | 1 | 0 | 0 | 0 | 0 | 0 | 0 | 0 | 0 | 0 | 0 | 0 | 0 | 0 | 0 | 0 |
| X2*X5 | I | 0 | 0 | 1 | 0 | 1 | 1 | 0 | 1 | 1 | 0 | 0 | 0 | 0 | 0 | 0 | 0 | 0 | 0 | 0 | 0 | 0 | 0 |
| X2*X6 | I | 0 | 0 | 1 | 1 | 1 | 2 | 1 | 0 | 1 | 1 | 1 | 0 | 0 | 0 | 0 | 0 | 0 | 0 | 0 | 0 | 0 | 0 |
| X2*X7 | I | 0 | 0 | 0 | 1 | 0 | 1 | 1 | 0 | 0 | 0 | 1 | 1 | 0 | 0 | 0 | 0 | 0 | 0 | 0 | 0 | 0 | 0 |
| X2*X8 | I | 0 | 0 | 0 | 0 | 1 | 0 | 0 | 0 | 1 | 0 | 0 | 0 | 0 | 0 | 0 | 0 | 0 | 0 | 0 | 0 | 0 | 0 |
| X2*X9 | I | 0 | 0 | 0 | 0 | 1 | 1 | 0 | 1 | 2 | 1 | 1 | 0 | 1 | 1 | 0 | 0 | 0 | 0 | 0 | 0 | 0 | 0 |
| X2*X10 | I | 0 | 0 | 0 | 0 | 0 | 1 | 0 | 0 | 1 | 1 | 1 | 0 | 1 | 0 | 1 | 0 | 0 | 0 | 0 | 0 | 0 | 0 |
| X2*X11 | I | 0 | 0 | 0 | 0 | 0 | 1 | 1 | 0 | 1 | 1 | 2 | 1 | 0 | 1 | 1 | 1 | 0 | 0 | 0 | 0 | 0 | 0 |
| X2*X12 | I | 0 | 0 | 0 | 0 | 0 | 0 | 1 | 0 | 0 | 0 | 1 | 1 | 0 | 0 | 0 | 1 | 1 | 0 | 0 | 0 | 0 | 0 |
| X2*X13 | I | 0 | 0 | 0 | 0 | 0 | 0 | 0 | 0 | 1 | 1 | 0 | 0 | 1 | 1 | 1 | 0 | 0 | 0 | 0 | 0 | 0 | 0 |
| X2*X14 | I | 0 | 0 | 0 | 0 | 0 | 0 | 0 | 0 | 1 | 0 | 1 | 0 | 1 | 1 | 1 | 1 | 0 | 0 | 0 | 0 | 0 | 0 |
| X2*X15 | I | 0 | 0 | 0 | 0 | 0 | 0 | 0 | 0 | 0 | 1 | 1 | 0 | 1 | 1 | 2 | 1 | 0 | 1 | 1 | 0 | 0 | 0 |
| X2*X16 | I | 0 | 0 | 0 | 0 | 0 | 0 | 0 | 0 | 0 | 0 | 1 | 1 | 0 | 1 | 1 | 2 | 1 | 0 | 1 | 1 | 0 | 0 |
| X2*X17 | I | 0 | 0 | 0 | 0 | 0 | 0 | 0 | 0 | 0 | 0 | 0 | 1 | 0 | 0 | 0 | 1 | 1 | 0 | 0 | 1 | 1 | 0 |
| X2*X18 | I | 0 | 0 | 0 | 0 | 0 | 0 | 0 | 0 | 0 | 0 | 0 | 0 | 0 | 0 | 1 | 0 | 0 | 0 | 1 | 0 | 0 | 0 |
| X2*X19 | I | 0 | 0 | 0 | 0 | 0 | 0 | 0 | 0 | 0 | 0 | 0 | 0 | 0 | 0 | 1 | 1 | 0 | 1 | 1 | 1 | 0 | 0 |
| X2*X20 | I | 0 | 0 | 0 | 0 | 0 | 0 | 0 | 0 | 0 | 0 | 0 | 0 | 0 | 0 | 0 | 1 | 1 | 0 | 1 | 1 | 1 | 0 |
| X2*X21 | I | 0 | 0 | 0 | 0 | 0 | 0 | 0 | 0 | 0 | 0 | 0 | 0 | 0 | 0 | 0 | 0 | 1 | 0 | 0 | 1 | 1 | 1 |
| X2*X22 | I | 0 | 0 | 0 | 0 | 0 | 0 | 0 | 0 | 0 | 0 | 0 | 0 | 0 | 0 | 0 | 0 | 0 | 0 | 0 | 0 | 1 | 0 |
| X3*X3 | I | 1 | 1 | 2 | 1 | 1 | 2 | 1 | 1 | 1 | 1 | 0 | 0 | 0 | 0 | 0 | 0 | 0 | 0 | 0 | 0 | 0 | 0 |
| X3*X4 | I | 0 | 1 | 1 | 2 | 1 | 2 | 1 | 0 | 1 | 0 | 1 | 0 | 0 | 0 | 0 | 0 | 0 | 0 | 0 | 0 | 0 | 0 |
| X3*X5 | I | 0 | 1 | 1 | 1 | 2 | 2 | 0 | 0 | 2 | 1 | 1 | 0 | 1 | 0 | 0 | 0 | 0 | 0 | 0 | 0 | 0 | 0 |
| X3*X6 | I | 0 | 1 | 2 | 2 | 2 | 4 | 2 | 1 | 3 | 2 | 3 | 1 | 1 | 1 | 1 | 0 | 0 | 0 | 0 | 0 | 0 | 0 |
| X3*X7 | I | 0 | 0 | 1 | 1 | 0 | 2 | 2 | 0 | 1 | 1 | 2 | 1 | 0 | 1 | 0 | 1 | 0 | 0 | 0 | 0 | 0 | 0 |
| X3*X8 | I | 0 | 0 | 1 | 0 | 0 | 1 | 0 | 1 | 1 | 1 | 0 | 0 | 0 | 1 | 0 | 0 | 0 | 0 | 0 | 0 | 0 | 0 |
| X3*X9 | I | 0 | 0 | 1 | 1 | 2 | 3 | 1 | 1 | 4 | 2 | 3 | 0 | 2 | 2 | 2 | 1 | 0 | 0 | 0 | 0 | 0 | 0 |
| X3*X10 | I | 0 | 0 | 1 | 0 | 1 | 2 | 1 | 1 | 2 | 3 | 2 | 1 | 1 | 2 | 2 | 1 | 0 | 1 | 0 | 0 | 0 | 0 |
| X3*X11 | I | 0 | 0 | 0 | 1 | 1 | 3 | 2 | 0 | 3 | 2 | 5 | 2 | 2 | 2 | 3 | 3 | 1 | 0 | 1 | 0 | 0 | 0 |
| X3*X12 | I | 0 | 0 | 0 | 0 | 0 | 1 | 1 | 0 | 0 | 1 | 2 | 2 | 0 | 1 | 1 | 2 | 1 | 0 | 0 | 1 | 0 | 0 |
| X3*X13 | I | 0 | 0 | 0 | 0 | 1 | 1 | 0 | 0 | 2 | 1 | 2 | 0 | 2 | 1 | 2 | 1 | 0 | 0 | 1 | 0 | 0 | 0 |
| X3*X14 | I | 0 | 0 | 0 | 0 | 0 | 1 | 1 | 1 | 2 | 2 | 2 | 1 | 1 | 3 | 2 | 2 | 0 | 1 | 1 | 1 | 0 | 0 |
| X3*X15 | I | 0 | 0 | 0 | 0 | 0 | 1 | 0 | 0 | 2 | 2 | 3 | 1 | 2 | 2 | 4 | 3 | 1 | 1 | 2 | 1 | 0 | 0 |
| X3*X16 | I | 0 | 0 | 0 | 0 | 0 | 0 | 1 | 0 | 1 | 1 | 3 | 2 | 1 | 2 | 3 | 4 | 2 | 1 | 2 | 2 | 1 | 0 |
| X3*X17 | I | 0 | 0 | 0 | 0 | 0 | 0 | 0 | 0 | 0 | 0 | 1 | 1 | 0 | 0 | 1 | 2 | 2 | 0 | 1 | 1 | 1 | 0 |
| X3*X18 | I | 0 | 0 | 0 | 0 | 0 | 0 | 0 | 0 | 0 | 1 | 0 | 0 | 0 | 1 | 1 | 1 | 0 | 1 | 0 | 1 | 0 | 0 |
| X3*X19 | I | 0 | 0 | 0 | 0 | 0 | 0 | 0 | 0 | 0 | 0 | 1 | 0 | 1 | 1 | 2 | 2 | 1 | 0 | 2 | 1 | 1 | 0 |
| X3*X20 | I | 0 | 0 | 0 | 0 | 0 | 0 | 0 | 0 | 0 | 0 | 0 | 1 | 0 | 1 | 1 | 2 | 1 | 1 | 1 | 2 | 1 | 1 |
| X3*X21 | I | 0 | 0 | 0 | 0 | 0 | 0 | 0 | 0 | 0 | 0 | 0 | 0 | 0 | 0 | 0 | 1 | 1 | 0 | 1 | 1 | 1 | 0 |
| X3*X22 | I | 0 | 0 | 0 | 0 | 0 | 0 | 0 | 0 | 0 | 0 | 0 | 0 | 0 | 0 | 0 | 0 | 0 | 0 | 0 | 1 | 0 | 0 |
| X4*X4 | I | 1 | 1 | 2 | 1 | 1 | 2 | 1 | 0 | 0 | 1 | 1 | 1 | 0 | 0 | 0 | 0 | 0 | 0 | 0 | 0 | 0 | 0 |
| X4*X5 | I | 0 | 0 | 1 | 1 | 1 | 2 | 1 | 1 | 2 | 1 | 1 | 0 | 0 | 1 | 0 | 0 | 0 | 0 | 0 | 0 | 0 | 0 |
| X4*X6 | I | 0 | 1 | 2 | 2 | 2 | 4 | 2 | 1 | 3 | 2 | 3 | 1 | 1 | 1 | 1 | 1 | 0 | 0 | 0 | 0 | 0 | 0 |
| X4*X7 | I | 0 | 1 | 1 | 1 | 1 | 2 | 1 | 0 | 1 | 1 | 2 | 1 | 0 | 0 | 1 | 1 | 1 | 0 | 0 | 0 | 0 | 0 |
| X4*X8 | I | 0 | 0 | 0 | 0 | 1 | 1 | 0 | 0 | 1 | 0 | 1 | 0 | 1 | 0 | 0 | 0 | 0 | 0 | 0 | 0 | 0 | 0 |
| X4*X9 | I | 0 | 0 | 1 | 0 | 2 | 3 | 1 | 1 | 4 | 3 | 3 | 1 | 2 | 2 | 2 | 1 | 0 | 0 | 0 | 0 | 0 | 0 |
| X4*X10 | I | 0 | 0 | 0 | 1 | 1 | 2 | 1 | 0 | 3 | 1 | 3 | 0 | 2 | 2 | 2 | 1 | 0 | 0 | 1 | 0 | 0 | 0 |
| X4*X11 | I | 0 | 0 | 1 | 1 | 1 | 3 | 2 | 1 | 3 | 3 | 4 | 2 | 2 | 3 | 3 | 3 | 1 | 1 | 1 | 1 | 0 | 0 |
| X4*X12 | I | 0 | 0 | 0 | 1 | 0 | 1 | 1 | 0 | 1 | 0 | 2 | 1 | 0 | 1 | 1 | 2 | 1 | 0 | 1 | 1 | 1 | 0 |
| X4*X13 | I | 0 | 0 | 0 | 0 | 0 | 1 | 0 | 1 | 2 | 2 | 2 | 0 | 1 | 2 | 2 | 1 | 0 | 1 | 0 | 0 | 0 | 0 |
| X4*X14 | I | 0 | 0 | 0 | 0 | 1 | 1 | 0 | 0 | 2 | 2 | 3 | 1 | 2 | 1 | 3 | 2 | 1 | 0 | 1 | 0 | 0 | 0 |
| X4*X15 | I | 0 | 0 | 0 | 0 | 0 | 1 | 1 | 0 | 2 | 2 | 3 | 1 | 2 | 3 | 4 | 3 | 0 | 1 | 2 | 1 | 0 | 0 |
| X4*X16 | I | 0 | 0 | 0 | 0 | 0 | 1 | 1 | 0 | 1 | 1 | 3 | 2 | 1 | 2 | 3 | 4 | 2 | 1 | 2 | 2 | 1 | 0 |
| X4*X17 | I | 0 | 0 | 0 | 0 | 0 | 0 | 1 | 0 | 0 | 0 | 1 | 1 | 0 | 1 | 0 | 2 | 1 | 0 | 1 | 2 | 1 | 1 |
| X4*X18 | I | 0 | 0 | 0 | 0 | 0 | 0 | 0 | 0 | 0 | 0 | 1 | 0 | 1 | 0 | 1 | 1 | 0 | 0 | 1 | 0 | 0 | 0 |
| X4*X19 | I | 0 | 0 | 0 | 0 | 0 | 0 | 0 | 0 | 0 | 1 | 1 | 1 | 0 | 1 | 2 | 2 | 1 | 1 | 1 | 1 | 0 | 0 |
| X4*X20 | I | 0 | 0 | 0 | 0 | 0 | 0 | 0 | 0 | 0 | 0 | 1 | 1 | 0 | 0 | 1 | 2 | 2 | 0 | 1 | 1 | 1 | 0 |
| X4*X21 | I | 0 | 0 | 0 | 0 | 0 | 0 | 0 | 0 | 0 | 0 | 0 | 1 | 0 | 0 | 0 | 1 | 1 | 0 | 0 | 1 | 1 | 0 |
| X4*X22 | I | 0 | 0 | 0 | 0 | 0 | 0 | 0 | 0 | 0 | 0 | 0 | 0 | 0 | 0 | 0 | 0 | 1 | 0 | 0 | 0 | 0 | 0 |

| | | X1 | X2 | X3 | X4 | X5 | X6 | X7 | X8 | X9 | X10 | X11 | X12 | X13 | X14 | X15 | X16 | X17 | X18 | X19 | X20 | X21 | X22 |
|---|---|---|---|---|---|---|---|---|---|---|---|---|---|---|---|---|---|---|---|---|---|---|---|
| X5*X5 | I | 1 | 1 | 2 | 1 | 1 | 2 | 1 | 1 | 2 | 2 | 1 | 0 | 1 | 1 | 1 | 0 | 0 | 0 | 0 | 0 | 0 | 0 |
| X5*X6 | I | 0 | 1 | 2 | 2 | 2 | 5 | 2 | 1 | 4 | 3 | 4 | 1 | 2 | 2 | 2 | 1 | 0 | 0 | 0 | 0 | 0 | 0 |
| X5*X7 | I | 0 | 0 | 0 | 1 | 1 | 2 | 2 | 0 | 2 | 1 | 3 | 1 | 1 | 1 | 1 | 1 | 0 | 0 | 0 | 0 | 0 | 0 |
| X5*X8 | I | 0 | 1 | 0 | 1 | 1 | 1 | 0 | 0 | 1 | 0 | 1 | 0 | 1 | 0 | 1 | 0 | 0 | 0 | 0 | 0 | 0 | 0 |
| X5*X9 | I | 0 | 1 | 2 | 2 | 2 | 4 | 2 | 1 | 4 | 3 | 4 | 1 | 2 | 3 | 3 | 2 | 0 | 1 | 1 | 0 | 0 | 0 |
| X5*X10 | I | 0 | 0 | 1 | 1 | 2 | 3 | 1 | 0 | 3 | 2 | 4 | 1 | 2 | 1 | 3 | 2 | 1 | 0 | 1 | 0 | 0 | 0 |
| X5*X11 | I | 0 | 0 | 1 | 1 | 1 | 4 | 3 | 1 | 4 | 4 | 6 | 3 | 2 | 4 | 4 | 4 | 1 | 1 | 1 | 1 | 0 | 0 |
| X5*X12 | I | 0 | 0 | 0 | 0 | 0 | 1 | 1 | 0 | 1 | 1 | 3 | 2 | 1 | 1 | 2 | 2 | 1 | 0 | 1 | 0 | 0 | 0 |
| X5*X13 | I | 0 | 0 | 1 | 0 | 1 | 2 | 1 | 1 | 2 | 2 | 2 | 1 | 1 | 2 | 2 | 2 | 0 | 1 | 1 | 1 | 0 | 0 |
| X5*X14 | I | 0 | 0 | 0 | 1 | 1 | 2 | 1 | 0 | 3 | 1 | 4 | 1 | 2 | 2 | 3 | 3 | 1 | 0 | 2 | 1 | 0 | 0 |
| X5*X15 | I | 0 | 0 | 0 | 0 | 1 | 2 | 1 | 1 | 3 | 3 | 4 | 2 | 2 | 3 | 4 | 4 | 2 | 1 | 2 | 2 | 1 | 0 |
| X5*X16 | I | 0 | 0 | 0 | 0 | 0 | 1 | 1 | 0 | 2 | 2 | 4 | 2 | 2 | 3 | 4 | 5 | 2 | 1 | 2 | 2 | 1 | 0 |
| X5*X17 | I | 0 | 0 | 0 | 0 | 0 | 0 | 0 | 0 | 0 | 1 | 1 | 1 | 0 | 1 | 2 | 2 | 1 | 1 | 1 | 1 | 0 | 0 |
| X5*X18 | I | 0 | 0 | 0 | 0 | 0 | 0 | 0 | 0 | 1 | 0 | 1 | 0 | 1 | 0 | 1 | 1 | 1 | 0 | 1 | 0 | 1 | 0 |
| X5*X19 | I | 0 | 0 | 0 | 0 | 0 | 0 | 0 | 0 | 1 | 1 | 1 | 1 | 1 | 2 | 2 | 2 | 1 | 1 | 1 | 2 | 1 | 1 |
| X5*X20 | I | 0 | 0 | 0 | 0 | 0 | 0 | 0 | 0 | 0 | 0 | 1 | 0 | 1 | 1 | 2 | 2 | 1 | 0 | 2 | 1 | 1 | 0 |
| X5*X21 | I | 0 | 0 | 0 | 0 | 0 | 0 | 0 | 0 | 0 | 0 | 0 | 0 | 0 | 0 | 1 | 1 | 0 | 1 | 1 | 1 | 0 | 0 |
| X5*X22 | I | 0 | 0 | 0 | 0 | 0 | 0 | 0 | 0 | 0 | 0 | 0 | 0 | 0 | 0 | 0 | 0 | 0 | 0 | 1 | 0 | 0 | 0 |
| X6*X6 | I | 1 | 2 | 4 | 4 | 5 | 9 | 5 | 2 | 9 | 6 | 9 | 3 | 4 | 5 | 5 | 4 | 1 | 1 | 1 | 0 | 0 | 0 |
| X6*X7 | I | 0 | 1 | 2 | 2 | 2 | 5 | 2 | 1 | 4 | 3 | 5 | 2 | 2 | 3 | 3 | 3 | 1 | 0 | 1 | 1 | 0 | 0 |
| X6*X8 | I | 0 | 0 | 1 | 1 | 1 | 2 | 1 | 0 | 2 | 2 | 2 | 0 | 1 | 1 | 1 | 1 | 0 | 0 | 0 | 0 | 0 | 0 |
| X6*X9 | I | 0 | 1 | 3 | 3 | 4 | 9 | 4 | 2 | 9 | 7 | 10 | 3 | 5 | 6 | 7 | 5 | 1 | 1 | 2 | 1 | 0 | 0 |
| X6*X10 | I | 0 | 1 | 2 | 2 | 3 | 6 | 3 | 2 | 7 | 5 | 8 | 3 | 4 | 5 | 6 | 5 | 1 | 1 | 2 | 1 | 0 | 0 |
| X6*X11 | I | 0 | 1 | 3 | 3 | 4 | 9 | 5 | 2 | 10 | 8 | 13 | 5 | 6 | 8 | 10 | 9 | 3 | 2 | 4 | 3 | 1 | 0 |
| X6*X12 | I | 0 | 0 | 1 | 1 | 1 | 3 | 2 | 0 | 3 | 3 | 5 | 2 | 2 | 3 | 4 | 5 | 2 | 1 | 2 | 2 | 1 | 0 |
| X6*X13 | I | 0 | 0 | 1 | 1 | 2 | 4 | 2 | 1 | 5 | 4 | 6 | 2 | 3 | 4 | 5 | 4 | 1 | 1 | 2 | 1 | 0 | 0 |
| X6*X14 | I | 0 | 0 | 1 | 1 | 2 | 5 | 3 | 1 | 6 | 5 | 8 | 3 | 4 | 5 | 7 | 6 | 2 | 2 | 3 | 2 | 1 | 0 |
| X6*X15 | I | 0 | 0 | 1 | 1 | 2 | 5 | 3 | 1 | 7 | 6 | 10 | 4 | 5 | 7 | 9 | 9 | 3 | 2 | 4 | 3 | 1 | 0 |
| X6*X16 | I | 0 | 0 | 0 | 1 | 1 | 4 | 3 | 1 | 5 | 5 | 9 | 5 | 4 | 6 | 9 | 9 | 4 | 2 | 5 | 4 | 2 | 1 |
| X6*X17 | I | 0 | 0 | 0 | 0 | 0 | 1 | 1 | 0 | 1 | 1 | 3 | 2 | 1 | 2 | 3 | 4 | 2 | 1 | 2 | 2 | 1 | 0 |
| X6*X18 | I | 0 | 0 | 0 | 0 | 0 | 1 | 0 | 0 | 1 | 1 | 2 | 1 | 1 | 2 | 2 | 2 | 1 | 0 | 1 | 1 | 0 | 0 |
| X6*X19 | I | 0 | 0 | 0 | 0 | 0 | 1 | 1 | 0 | 2 | 2 | 4 | 2 | 2 | 3 | 4 | 5 | 2 | 1 | 2 | 2 | 1 | 0 |
| X6*X20 | I | 0 | 0 | 0 | 0 | 0 | 0 | 1 | 0 | 1 | 1 | 3 | 2 | 1 | 2 | 3 | 4 | 2 | 1 | 2 | 2 | 1 | 0 |
| X6*X21 | I | 0 | 0 | 0 | 0 | 0 | 0 | 0 | 0 | 0 | 0 | 1 | 1 | 0 | 1 | 1 | 2 | 1 | 0 | 1 | 1 | 0 | 0 |
| X6*X22 | I | 0 | 0 | 0 | 0 | 0 | 0 | 0 | 0 | 0 | 0 | 0 | 0 | 0 | 0 | 0 | 1 | 0 | 0 | 0 | 0 | 0 | 0 |
| X7*X7 | I | 1 | 1 | 2 | 1 | 2 | 2 | 1 | 1 | 2 | 2 | 2 | 1 | 1 | 1 | 2 | 2 | 1 | 1 | 1 | 1 | 1 | 0 |
| X7*X8 | I | 0 | 0 | 0 | 0 | 0 | 1 | 1 | 1 | 1 | 1 | 1 | 1 | 0 | 1 | 1 | 0 | 0 | 0 | 0 | 0 | 0 | 0 |
| X7*X9 | I | 0 | 0 | 1 | 1 | 2 | 4 | 2 | 1 | 5 | 4 | 6 | 2 | 3 | 3 | 4 | 3 | 1 | 1 | 1 | 0 | 0 | 0 |
| X7*X10 | I | 0 | 0 | 1 | 1 | 1 | 3 | 2 | 1 | 4 | 3 | 4 | 1 | 2 | 4 | 3 | 3 | 0 | 1 | 1 | 1 | 0 | 0 |
| X7*X11 | I | 0 | 1 | 2 | 2 | 3 | 5 | 2 | 1 | 6 | 4 | 6 | 2 | 4 | 4 | 6 | 5 | 2 | 1 | 3 | 2 | 1 | 0 |
| X7*X12 | I | 0 | 1 | 1 | 1 | 1 | 2 | 1 | 1 | 2 | 1 | 2 | 1 | 1 | 2 | 2 | 2 | 1 | 1 | 2 | 2 | 1 | 1 |
| X7*X13 | I | 0 | 0 | 0 | 0 | 1 | 2 | 1 | 0 | 3 | 2 | 4 | 1 | 2 | 2 | 3 | 2 | 0 | 0 | 1 | 0 | 0 | 0 |
| X7*X14 | I | 0 | 0 | 1 | 0 | 1 | 3 | 1 | 1 | 3 | 4 | 4 | 2 | 2 | 3 | 4 | 3 | 1 | 1 | 1 | 1 | 0 | 0 |
| X7*X15 | I | 0 | 0 | 0 | 1 | 1 | 3 | 2 | 1 | 4 | 3 | 0 | 2 | 3 | 4 | 5 | 4 | 1 | 1 | 2 | 1 | 0 | 0 |
| X7*X16 | I | 0 | 0 | 1 | 1 | 1 | 3 | 2 | 0 | 3 | 3 | 5 | 2 | 2 | 3 | 4 | 5 | 2 | 1 | 2 | 2 | 1 | 0 |
| X7*X17 | I | 0 | 0 | 0 | 1 | 0 | 1 | 1 | 0 | 1 | 0 | 2 | 1 | 0 | 1 | 1 | 2 | 1 | 0 | 1 | 1 | 1 | 0 |
| X7*X18 | I | 0 | 0 | 0 | 0 | 0 | 0 | 1 | 0 | 1 | 1 | 1 | 1 | 0 | 1 | 1 | 1 | 0 | 1 | 0 | 0 | 0 | 0 |
| X7*X19 | I | 0 | 0 | 0 | 0 | 0 | 1 | 1 | 0 | 1 | 1 | 3 | 2 | 1 | 1 | 2 | 2 | 1 | 0 | 1 | 0 | 0 | 0 |
| X7*X20 | I | 0 | 0 | 0 | 0 | 0 | 1 | 1 | 0 | 0 | 1 | 2 | 2 | 0 | 1 | 1 | 2 | 1 | 0 | 0 | 1 | 0 | 0 |
| X7*X21 | I | 0 | 0 | 0 | 0 | 0 | 0 | 1 | 0 | 0 | 0 | 1 | 1 | 0 | 0 | 0 | 1 | 1 | 0 | 0 | 0 | 0 | 0 |
| X7*X22 | I | 0 | 0 | 0 | 0 | 0 | 0 | 0 | 0 | 0 | 0 | 0 | 1 | 0 | 0 | 0 | 0 | 0 | 0 | 0 | 0 | 0 | 0 |
| X8*X8 | I | 1 | 0 | 1 | 0 | 0 | 0 | 1 | 1 | 0 | 1 | 0 | 0 | 0 | 1 | 0 | 0 | 0 | 1 | 0 | 0 | 0 | 0 |
| X8*X9 | I | 0 | 1 | 1 | 1 | 1 | 2 | 1 | 0 | 2 | 1 | 2 | 1 | 1 | 1 | 2 | 1 | 0 | 0 | 1 | 0 | 0 | 0 |
| X8*X10 | I | 0 | 0 | 1 | 0 | 0 | 2 | 1 | 1 | 1 | 2 | 1 | 1 | 0 | 2 | 1 | 1 | 0 | 1 | 0 | 1 | 0 | 0 |
| X8*X11 | I | 0 | 0 | 0 | 1 | 1 | 2 | 1 | 0 | 2 | 1 | 4 | 1 | 2 | 1 | 2 | 2 | 1 | 0 | 1 | 0 | 0 | 0 |
| X8*X12 | I | 0 | 0 | 0 | 0 | 0 | 0 | 1 | 0 | 1 | 1 | 1 | 1 | 0 | 1 | 1 | 1 | 0 | 1 | 0 | 0 | 0 | 0 |
| X8*X13 | I | 0 | 0 | 0 | 1 | 1 | 1 | 0 | 0 | 1 | 0 | 2 | 0 | 1 | 0 | 1 | 1 | 1 | 0 | 1 | 0 | 0 | 0 |
| X8*X14 | I | 0 | 0 | 1 | 0 | 0 | 1 | 1 | 1 | 1 | 2 | 1 | 1 | 0 | 2 | 1 | 2 | 0 | 1 | 0 | 1 | 0 | 0 |
| X8*X15 | I | 0 | 0 | 0 | 0 | 1 | 1 | 1 | 0 | 2 | 1 | 2 | 1 | 1 | 1 | 2 | 2 | 1 | 0 | 1 | 1 | 1 | 0 |
| X8*X16 | I | 0 | 0 | 0 | 0 | 0 | 1 | 0 | 0 | 1 | 1 | 2 | 1 | 1 | 2 | 2 | 2 | 1 | 0 | 1 | 1 | 0 | 0 |
| X8*X17 | I | 0 | 0 | 0 | 0 | 0 | 0 | 0 | 0 | 0 | 0 | 1 | 0 | 1 | 0 | 1 | 1 | 0 | 0 | 1 | 0 | 0 | 0 |
| X8*X18 | I | 0 | 0 | 0 | 0 | 0 | 0 | 0 | 1 | 0 | 1 | 0 | 1 | 0 | 1 | 0 | 0 | 0 | 1 | 0 | 1 | 0 | 1 |
| X8*X19 | I | 0 | 0 | 0 | 0 | 0 | 0 | 0 | 0 | 1 | 0 | 1 | 0 | 1 | 0 | 1 | 1 | 1 | 0 | 1 | 0 | 1 | 0 |
| X8*X20 | I | 0 | 0 | 0 | 0 | 0 | 0 | 0 | 0 | 0 | 1 | 0 | 0 | 0 | 1 | 1 | 1 | 0 | 1 | 0 | 1 | 0 | 0 |
| X8*X21 | I | 0 | 0 | 0 | 0 | 0 | 0 | 0 | 0 | 0 | 0 | 0 | 0 | 0 | 0 | 1 | 0 | 0 | 0 | 1 | 0 | 0 | 0 |
| X8*X22 | I | 0 | 0 | 0 | 0 | 0 | 0 | 0 | 0 | 0 | 0 | 0 | 0 | 0 | 0 | 0 | 0 | 0 | 1 | 0 | 0 | 0 | 0 |
| X9*X9 | I | 1 | 2 | 4 | 4 | 4 | 9 | 5 | 2 | 8 | 7 | 11 | 4 | 5 | 6 | 8 | 7 | 2 | 2 | 3 | 2 | 0 | 0 |
| X9*X10 | I | 0 | 1 | 2 | 3 | 3 | 7 | 4 | 1 | 7 | 5 | 9 | 3 | 4 | 5 | 6 | 6 | 2 | 1 | 3 | 2 | 1 | 0 |
| X9*X11 | I | 0 | 1 | 3 | 3 | 4 | 10 | 6 | 2 | 11 | 9 | 14 | 6 | 6 | 9 | 11 | 10 | 3 | 2 | 4 | 3 | 1 | 0 |
| X9*X12 | I | 0 | 0 | 0 | 1 | 1 | 3 | 2 | 1 | 4 | 3 | 6 | 2 | 3 | 4 | 5 | 4 | 1 | 1 | 2 | 1 | 0 | 0 |
| X9*X13 | I | 0 | 1 | 2 | 2 | 2 | 5 | 3 | 1 | 5 | 4 | 6 | 3 | 2 | 4 | 5 | 5 | 2 | 1 | 2 | 2 | 1 | 0 |
| X9*X14 | I | 0 | 1 | 2 | 2 | 3 | 6 | 3 | 1 | 6 | 5 | 9 | 4 | 4 | 5 | 7 | 7 | 3 | 1 | 3 | 2 | 1 | 0 |
| X9*X15 | I | 0 | 0 | 2 | 2 | 3 | 7 | 4 | 2 | 8 | 6 | 11 | 5 | 5 | 7 | 8 | 9 | 4 | 2 | 4 | 4 | 2 | 1 |
| X9*X16 | I | 0 | 0 | 1 | 1 | 2 | 5 | 3 | 1 | 7 | 6 | 10 | 4 | 5 | 7 | 9 | 9 | 3 | 2 | 4 | 3 | 1 | 0 |
| X9*X17 | I | 0 | 0 | 0 | 0 | 0 | 1 | 1 | 0 | 2 | 2 | 3 | 1 | 2 | 3 | 4 | 3 | 0 | 1 | 2 | 1 | 0 | 0 |
| X9*X18 | I | 0 | 0 | 0 | 0 | 1 | 1 | 1 | 0 | 2 | 1 | 2 | 1 | 1 | 1 | 2 | 2 | 1 | 0 | 1 | 1 | 1 | 0 |
| X9*X19 | I | 0 | 0 | 0 | 0 | 1 | 2 | 1 | 1 | 3 | 3 | 4 | 2 | 2 | 3 | 4 | 4 | 2 | 1 | 2 | 2 | 1 | 0 |
| X9*X20 | I | 0 | 0 | 0 | 0 | 0 | 1 | 0 | 0 | 2 | 2 | 3 | 1 | 2 | 2 | 4 | 3 | 1 | 1 | 2 | 1 | 0 | 0 |
| X9*X21 | I | 0 | 0 | 0 | 0 | 0 | 0 | 0 | 0 | 0 | 1 | 1 | 0 | 1 | 1 | 2 | 1 | 0 | 1 | 1 | 0 | 0 | 0 |
| X9*X22 | I | 0 | 0 | 0 | 0 | 0 | 0 | 0 | 0 | 0 | 0 | 0 | 0 | 0 | 0 | 1 | 0 | 0 | 0 | 0 | 0 | 0 | 0 |
| X10*X10 | I | 1 | 1 | 3 | 1 | 2 | 5 | 3 | 2 | 5 | 6 | 6 | 4 | 2 | 5 | 5 | 5 | 2 | 2 | 1 | 2 | 0 | 0 |
| X10*X11 | I | 0 | 1 | 2 | 3 | 4 | 8 | 4 | 1 | 9 | 6 | 12 | 4 | 6 | 6 | 9 | 8 | 3 | 1 | 4 | 2 | 1 | 0 |
| X10*X12 | I | 0 | 0 | 1 | 0 | 1 | 3 | 1 | 1 | 3 | 4 | 4 | 2 | 2 | 3 | 4 | 3 | 1 | 1 | 1 | 1 | 0 | 0 |
| X10*X13 | I | 0 | 1 | 1 | 2 | 2 | 4 | 2 | 0 | 4 | 2 | 6 | 2 | 3 | 2 | 4 | 4 | 2 | 0 | 2 | 1 | 1 | 0 |
| X10*X14 | I | 0 | 0 | 2 | 2 | 1 | 5 | 4 | 2 | 5 | 5 | 6 | 3 | 2 | 6 | 5 | 5 | 1 | 2 | 2 | 3 | 1 | 1 |
| X10*X15 | I | 0 | 1 | 2 | 2 | 3 | 6 | 3 | 1 | 6 | 5 | 9 | 4 | 4 | 5 | 7 | 7 | 3 | 1 | 3 | 2 | 1 | 0 |
| X10*X16 | I | 0 | 0 | 1 | 1 | 2 | 5 | 3 | 1 | 6 | 5 | 8 | 3 | 4 | 5 | 7 | 6 | 2 | 2 | 3 | 2 | 1 | 0 |
| X10*X17 | I | 0 | 0 | 0 | 0 | 1 | 1 | 0 | 0 | 2 | 2 | 3 | 1 | 2 | 1 | 3 | 2 | 1 | 0 | 1 | 0 | 0 | 0 |
| X10*X18 | I | 0 | 0 | 1 | 0 | 0 | 1 | 1 | 1 | 1 | 2 | 1 | 1 | 0 | 2 | 1 | 2 | 0 | 1 | 0 | 1 | 0 | 0 |
| X10*X19 | I | 0 | 0 | 0 | 1 | 1 | 2 | 1 | 0 | 3 | 1 | 4 | 1 | 2 | 2 | 3 | 3 | 1 | 0 | 2 | 1 | 0 | 0 |
| X10*X20 | I | 0 | 0 | 0 | 0 | 0 | 1 | 1 | 1 | 2 | 2 | 2 | 1 | 1 | 3 | 2 | 2 | 0 | 1 | 1 | 1 | 0 | 0 |
| X10*X21 | I | 0 | 0 | 0 | 0 | 0 | 0 | 0 | 0 | 1 | 0 | 1 | 0 | 1 | 1 | 1 | 1 | 0 | 0 | 0 | 0 | 0 | 0 |
| X10*X22 | I | 0 | 0 | 0 | 0 | 0 | 0 | 0 | 0 | 0 | 0 | 0 | 0 | 0 | 1 | 0 | 0 | 0 | 0 | 0 | 0 | 0 | 0 |

| | | X1 | X2 | X3 | X4 | X5 | X6 | X7 | X8 | X9 | X10 | X11 | X12 | X13 | X14 | X15 | X16 | X17 | X18 | X19 | X20 | X21 | X22 |
|---|---|---|---|---|---|---|---|---|---|---|---|---|---|---|---|---|---|---|---|---|---|---|---|
| X11*X11 | I | 1 | 2 | 5 | 4 | 6 | 13 | 6 | 4 | 14 | 12 | 17 | 6 | 8 | 12 | 14 | 13 | 4 | 4 | 6 | 5 | 2 | 1 |
| X11*X12 | I | 0 | 1 | 2 | 2 | 3 | 5 | 2 | 1 | 6 | 4 | 6 | 2 | 4 | 4 | 6 | 5 | 2 | 1 | 3 | 2 | 1 | 0 |
| X11*X13 | I | 0 | 0 | 2 | 2 | 2 | 6 | 4 | 2 | 6 | 6 | 8 | 4 | 4 | 6 | 6 | 6 | 2 | 2 | 2 | 2 | 0 | 0 |
| X11*X14 | I | 0 | 1 | 2 | 3 | 4 | 8 | 4 | 1 | 9 | 6 | 12 | 4 | 6 | 6 | 9 | 6 | 3 | 1 | 4 | 2 | 1 | 0 |
| X11*X15 | I | 0 | 1 | 3 | 3 | 4 | 10 | 6 | 2 | 11 | 9 | 14 | 6 | 6 | 9 | 11 | 10 | 3 | 2 | 4 | 3 | 1 | 0 |
| X11*X16 | I | 0 | 1 | 3 | 3 | 4 | 9 | 5 | 2 | 10 | 8 | 13 | 5 | 6 | 8 | 10 | 9 | 3 | 2 | 4 | 3 | 1 | 0 |
| X11*X17 | I | 0 | 0 | 1 | 1 | 1 | 3 | 2 | 1 | 3 | 3 | 4 | 2 | 2 | 3 | 3 | 3 | 1 | 1 | 1 | 1 | 0 | 0 |
| X11*X18 | I | 0 | 0 | 0 | 1 | 1 | 2 | 1 | 0 | 2 | 1 | 4 | 1 | 2 | 1 | 2 | 2 | 1 | 0 | 1 | 0 | 0 | 0 |
| X11*X19 | I | 0 | 0 | 1 | 1 | 1 | 4 | 3 | 1 | 4 | 4 | 6 | 3 | 2 | 4 | 4 | 4 | 1 | 1 | 1 | 1 | 0 | 0 |
| X11*X20 | I | 0 | 0 | 0 | 1 | 1 | 3 | 2 | 0 | 3 | 2 | 5 | 2 | 2 | 2 | 3 | 3 | 1 | 0 | 1 | 0 | 0 | 0 |
| X11*X21 | I | 0 | 0 | 0 | 0 | 0 | 1 | 1 | 0 | 1 | 1 | 2 | 1 | 0 | 1 | 1 | 1 | 0 | 0 | 0 | 0 | 0 | 0 |
| X11*X22 | I | 0 | 0 | 0 | 0 | 0 | 0 | 0 | 0 | 0 | 0 | 1 | 0 | 0 | 0 | 0 | 0 | 0 | 0 | 0 | 0 | 0 | 0 |
| X12*X12 | I | 1 | 1 | 2 | 1 | 2 | 2 | 1 | 1 | 2 | 2 | 2 | 1 | 1 | 1 | 2 | 2 | 1 | 1 | 1 | 1 | 1 | 0 |
| X12*X13 | I | 0 | 0 | 0 | 0 | 1 | 2 | 1 | 0 | 3 | 2 | 4 | 1 | 2 | 2 | 3 | 2 | 0 | 0 | 1 | 0 | 0 | 0 |
| X12*X14 | I | 0 | 0 | 1 | 1 | 1 | 3 | 2 | 1 | 4 | 3 | 4 | 1 | 2 | 4 | 3 | 3 | 0 | 1 | 1 | 1 | 0 | 0 |
| X12*X15 | I | 0 | 0 | 1 | 1 | 2 | 4 | 2 | 1 | 5 | 4 | 6 | 2 | 3 | 3 | 4 | 3 | 1 | 1 | 1 | 0 | 0 | 0 |
| X12*X16 | I | 0 | 1 | 2 | 2 | 2 | 5 | 2 | 1 | 4 | 3 | 5 | 2 | 2 | 3 | 3 | 3 | 1 | 0 | 1 | 1 | 0 | 0 |
| X12*X17 | I | 0 | 1 | 1 | 1 | 1 | 2 | 1 | 0 | 1 | 1 | 2 | 1 | 0 | 0 | 1 | 1 | 1 | 0 | 0 | 0 | 0 | 0 |
| X12*X18 | I | 0 | 0 | 0 | 0 | 0 | 1 | 1 | 1 | 1 | 1 | 1 | 1 | 0 | 1 | 1 | 0 | 0 | 0 | 0 | 0 | 0 | 0 |
| X12*X19 | I | 0 | 0 | 0 | 1 | 1 | 2 | 2 | 0 | 2 | 1 | 3 | 1 | 1 | 1 | 1 | 1 | 0 | 0 | 0 | 0 | 0 | 0 |
| X12*X20 | I | 0 | 0 | 1 | 1 | 0 | 2 | 2 | 0 | 1 | 1 | 2 | 1 | 0 | 1 | 0 | 1 | 0 | 0 | 0 | 0 | 0 | 0 |
| X12*X21 | I | 0 | 0 | 0 | 1 | 0 | 1 | 1 | 0 | 0 | 0 | 1 | 1 | 0 | 0 | 0 | 0 | 0 | 0 | 0 | 0 | 0 | 0 |
| X12*X22 | I | 0 | 0 | 0 | 0 | 0 | 0 | 1 | 0 | 0 | 0 | 0 | 0 | 0 | 0 | 0 | 0 | 0 | 0 | 0 | 0 | 0 | 0 |
| X13*X13 | I | 1 | 1 | 2 | 1 | 1 | 3 | 2 | 1 | 2 | 3 | 4 | 2 | 1 | 3 | 2 | 3 | 1 | 1 | 1 | 2 | 1 | 1 |
| X13*X14 | I | 0 | 1 | 1 | 2 | 2 | 4 | 2 | 0 | 4 | 2 | 6 | 2 | 3 | 2 | 4 | 4 | 2 | 0 | 2 | 1 | 1 | 0 |
| X13*X15 | I | 0 | 1 | 2 | 2 | 2 | 5 | 3 | 1 | 5 | 4 | 6 | 3 | 2 | 4 | 5 | 5 | 2 | 1 | 2 | 2 | 1 | 0 |
| X13*X16 | I | 0 | 0 | 1 | 1 | 2 | 4 | 2 | 1 | 5 | 4 | 6 | 2 | 3 | 4 | 5 | 4 | 1 | 1 | 2 | 1 | 0 | 0 |
| X13*X17 | I | 0 | 0 | 0 | 0 | 0 | 1 | 0 | 1 | 2 | 2 | 2 | 0 | 1 | 2 | 2 | 1 | 0 | 1 | 0 | 0 | 0 | 0 |
| X13*X18 | I | 0 | 0 | 0 | 1 | 1 | 1 | 0 | 0 | 1 | 0 | 2 | 0 | 1 | 0 | 1 | 1 | 1 | 0 | 1 | 0 | 0 | 0 |
| X13*X19 | I | 0 | 0 | 1 | 0 | 1 | 2 | 1 | 1 | 2 | 2 | 2 | 1 | 1 | 2 | 2 | 2 | 0 | 1 | 1 | 1 | 0 | 0 |
| X13*X20 | I | 0 | 0 | 0 | 0 | 1 | 1 | 0 | 0 | 2 | 1 | 2 | 0 | 2 | 1 | 2 | 1 | 0 | 0 | 1 | 0 | 0 | 0 |
| X13*X21 | I | 0 | 0 | 0 | 0 | 0 | 0 | 0 | 0 | 1 | 1 | 0 | 0 | 1 | 1 | 1 | 0 | 0 | 0 | 0 | 0 | 0 | 0 |
| X13*X22 | I | 0 | 0 | 0 | 0 | 0 | 0 | 0 | 0 | 0 | 0 | 0 | 0 | 1 | 0 | 0 | 0 | 0 | 0 | 0 | 0 | 0 | 0 |
| X14*X14 | I | 1 | 1 | 3 | 1 | 2 | 5 | 3 | 2 | 5 | 6 | 6 | 4 | 2 | 5 | 5 | 5 | 2 | 2 | 1 | 2 | 0 | 0 |
| X14*X15 | I | 0 | 1 | 2 | 3 | 3 | 7 | 4 | 1 | 7 | 5 | 9 | 3 | 4 | 5 | 6 | 6 | 2 | 1 | 3 | 2 | 1 | 0 |
| X14*X16 | I | 0 | 1 | 2 | 2 | 3 | 6 | 3 | 2 | 7 | 5 | 8 | 3 | 4 | 5 | 6 | 5 | 1 | 1 | 2 | 1 | 0 | 0 |
| X14*X17 | I | 0 | 0 | 0 | 1 | 1 | 2 | 1 | 0 | 3 | 1 | 3 | 0 | 2 | 2 | 2 | 1 | 0 | 0 | 1 | 0 | 0 | 0 |
| X14*X18 | I | 0 | 0 | 1 | 0 | 0 | 2 | 1 | 1 | 1 | 2 | 1 | 1 | 0 | 2 | 1 | 1 | 0 | 1 | 0 | 1 | 0 | 0 |
| X14*X19 | I | 0 | 0 | 1 | 1 | 2 | 3 | 1 | 0 | 3 | 2 | 4 | 1 | 2 | 1 | 3 | 2 | 1 | 0 | 1 | 0 | 0 | 0 |
| X14*X20 | I | 0 | 0 | 1 | 0 | 1 | 2 | 1 | 1 | 2 | 3 | 2 | 1 | 1 | 2 | 2 | 1 | 0 | 1 | 0 | 0 | 0 | 0 |
| X14*X21 | I | 0 | 0 | 0 | 0 | 0 | 1 | 0 | 0 | 1 | 1 | 1 | 0 | 1 | 0 | 1 | 0 | 0 | 0 | 0 | 0 | 0 | 0 |
| X14*X22 | I | 0 | 0 | 0 | 0 | 0 | 0 | 0 | 0 | 0 | 1 | 0 | 0 | 0 | 0 | 0 | 0 | 0 | 0 | 0 | 0 | 0 | 0 |
| X15*X15 | I | 1 | 2 | 4 | 4 | 4 | 9 | 5 | 2 | 8 | 7 | 11 | 4 | 5 | 6 | 8 | 7 | 2 | 2 | 3 | 2 | 0 | 0 |
| X15*X16 | I | 0 | 1 | 3 | 3 | 4 | 9 | 4 | 2 | 9 | 7 | 10 | 3 | 5 | 6 | 7 | 5 | 1 | 1 | 2 | 1 | 0 | 0 |
| X15*X17 | I | 0 | 0 | 1 | 0 | 2 | 3 | 1 | 1 | 4 | 3 | 3 | 1 | 2 | 2 | 2 | 1 | 0 | 0 | 0 | 0 | 0 | 0 |
| X15*X18 | I | 0 | 1 | 1 | 1 | 1 | 2 | 1 | 0 | 2 | 1 | 2 | 1 | 1 | 1 | 2 | 1 | 0 | 0 | 1 | 0 | 0 | 0 |
| X15*X19 | I | 0 | 1 | 2 | 2 | 2 | 4 | 2 | 1 | 4 | 3 | 4 | 1 | 2 | 3 | 3 | 2 | 0 | 1 | 1 | 0 | 0 | 0 |
| X15*X20 | I | 0 | 0 | 1 | 1 | 2 | 3 | 1 | 1 | 4 | 2 | 3 | 0 | 2 | 2 | 2 | 1 | 0 | 0 | 0 | 0 | 0 | 0 |
| X15*X21 | I | 0 | 0 | 0 | 0 | 1 | 1 | 0 | 1 | 2 | 1 | 1 | 0 | 1 | 1 | 0 | 0 | 0 | 0 | 0 | 0 | 0 | 0 |
| X15*X22 | I | 0 | 0 | 0 | 0 | 0 | 0 | 0 | 0 | 1 | 0 | 0 | 0 | 0 | 0 | 0 | 0 | 0 | 0 | 0 | 0 | 0 | 0 |
| X16*X16 | I | 1 | 2 | 4 | 4 | 5 | 9 | 5 | 2 | 9 | 6 | 9 | 3 | 4 | 5 | 5 | 4 | 1 | 1 | 1 | 0 | 0 | 0 |
| X16*X17 | I | 0 | 1 | 2 | 2 | 2 | 4 | 2 | 1 | 3 | 2 | 3 | 1 | 1 | 1 | 1 | 1 | 0 | 0 | 0 | 0 | 0 | 0 |
| X16*X18 | I | 0 | 0 | 1 | 1 | 1 | 2 | 1 | 0 | 2 | 2 | 2 | 0 | 1 | 1 | 1 | 1 | 0 | 0 | 0 | 0 | 0 | 0 |
| X16*X19 | I | 0 | 1 | 2 | 2 | 2 | 5 | 2 | 1 | 4 | 3 | 4 | 1 | 2 | 2 | 2 | 1 | 0 | 0 | 0 | 0 | 0 | 0 |
| X16*X20 | I | 0 | 1 | 2 | 2 | 2 | 4 | 2 | 1 | 3 | 2 | 3 | 1 | 1 | 1 | 1 | 0 | 0 | 0 | 0 | 0 | 0 | 0 |
| X16*X21 | I | 0 | 0 | 1 | 1 | 1 | 2 | 1 | 0 | 1 | 1 | 1 | 0 | 0 | 0 | 0 | 0 | 0 | 0 | 0 | 0 | 0 | 0 |
| X16*X22 | I | 0 | 0 | 0 | 0 | 0 | 1 | 0 | 0 | 0 | 0 | 0 | 0 | 0 | 0 | 0 | 0 | 0 | 0 | 0 | 0 | 0 | 0 |
| X17*X17 | I | 1 | 1 | 2 | 1 | 1 | 2 | 1 | 0 | 0 | 1 | 1 | 1 | 0 | 0 | 0 | 0 | 0 | 0 | 0 | 0 | 0 | 0 |
| X17*X18 | I | 0 | 0 | 0 | 0 | 1 | 1 | 0 | 0 | 1 | 0 | 1 | 0 | 1 | 0 | 0 | 0 | 0 | 0 | 0 | 0 | 0 | 0 |
| X17*X19 | I | 0 | 0 | 1 | 1 | 1 | 2 | 1 | 1 | 2 | 1 | 1 | 0 | 0 | 1 | 0 | 0 | 0 | 0 | 0 | 0 | 0 | 0 |
| X17*X20 | I | 0 | 1 | 1 | 2 | 1 | 2 | 1 | 0 | 1 | 0 | 1 | 0 | 0 | 0 | 0 | 0 | 0 | 0 | 0 | 0 | 0 | 0 |
| X17*X21 | I | 0 | 1 | 1 | 1 | 0 | 1 | 1 | 0 | 0 | 0 | 0 | 0 | 0 | 0 | 0 | 0 | 0 | 0 | 0 | 0 | 0 | 0 |
| X17*X22 | I | 0 | 0 | 0 | 1 | 0 | 0 | 0 | 0 | 0 | 0 | 0 | 0 | 0 | 0 | 0 | 0 | 0 | 0 | 0 | 0 | 0 | 0 |
| X18*X18 | I | 1 | 0 | 1 | 0 | 0 | 0 | 1 | 1 | 0 | 1 | 0 | 0 | 0 | 1 | 0 | 0 | 0 | 1 | 0 | 0 | 0 | 0 |
| X18*X19 | I | 0 | 1 | 0 | 1 | 1 | 1 | 0 | 0 | 1 | 0 | 1 | 0 | 1 | 0 | 1 | 0 | 0 | 0 | 0 | 0 | 0 | 0 |
| X18*X20 | I | 0 | 0 | 1 | 0 | 0 | 1 | 0 | 1 | 1 | 1 | 0 | 0 | 0 | 1 | 0 | 0 | 0 | 0 | 0 | 0 | 0 | 0 |
| X18*X21 | I | 0 | 0 | 0 | 0 | 1 | 0 | 0 | 0 | 1 | 0 | 0 | 0 | 0 | 0 | 0 | 0 | 0 | 0 | 0 | 0 | 0 | 0 |
| X18*X22 | I | 0 | 0 | 0 | 0 | 0 | 0 | 0 | 1 | 0 | 0 | 0 | 0 | 0 | 0 | 0 | 0 | 0 | 0 | 0 | 0 | 0 | 0 |
| X19*X19 | I | 1 | 1 | 2 | 1 | 1 | 2 | 1 | 1 | 2 | 2 | 1 | 0 | 1 | 1 | 1 | 0 | 0 | 0 | 0 | 0 | 0 | 0 |
| X19*X20 | I | 0 | 1 | 1 | 1 | 2 | 2 | 0 | 0 | 2 | 1 | 1 | 0 | 1 | 0 | 0 | 0 | 0 | 0 | 0 | 0 | 0 | 0 |
| X19*X21 | I | 0 | 0 | 1 | 0 | 1 | 1 | 0 | 1 | 1 | 0 | 0 | 0 | 0 | 0 | 0 | 0 | 0 | 0 | 0 | 0 | 0 | 0 |
| X19*X22 | I | 0 | 0 | 0 | 0 | 1 | 0 | 0 | 0 | 0 | 0 | 0 | 0 | 0 | 0 | 0 | 0 | 0 | 0 | 0 | 0 | 0 | 0 |
| X20*X20 | I | 1 | 1 | 2 | 1 | 1 | 2 | 1 | 1 | 1 | 1 | 0 | 0 | 0 | 0 | 0 | 0 | 0 | 0 | 0 | 0 | 0 | 0 |
| X20*X21 | I | 0 | 1 | 1 | 1 | 1 | 1 | 0 | 0 | 0 | 0 | 0 | 0 | 0 | 0 | 0 | 0 | 0 | 0 | 0 | 0 | 0 | 0 |
| X20*X22 | I | 0 | 0 | 1 | 0 | 0 | 0 | 0 | 0 | 0 | 0 | 0 | 0 | 0 | 0 | 0 | 0 | 0 | 0 | 0 | 0 | 0 | 0 |
| X21*X21 | I | 1 | 1 | 1 | 1 | 0 | 0 | 0 | 0 | 0 | 0 | 0 | 0 | 0 | 0 | 0 | 0 | 0 | 0 | 0 | 0 | 0 | 0 |
| X21*X22 | I | 0 | 1 | 0 | 0 | 0 | 0 | 0 | 0 | 0 | 0 | 0 | 0 | 0 | 0 | 0 | 0 | 0 | 0 | 0 | 0 | 0 | 0 |
| X22*X22 | I | 1 | 0 | 0 | 0 | 0 | 0 | 0 | 0 | 0 | 0 | 0 | 0 | 0 | 0 | 0 | 0 | 0 | 0 | 0 | 0 | 0 | 0 |

APPENDIX II

# *Notes and References*

The intention of this appendix is to provide a guide to the literature on representations of symmetric groups and related series of groups. Since the literature on this subject is so vast, we can by no means claim to be complete, but we try to help the reader find further interesting results and problems.

We emphasize the references to results published after the appearance of the book by G. de B. Robinson in 1961, for that book contains a nearly complete bibliography up to the date of its publication. We also point out the bibliography given in Boerner [1967], where the literature until 1967 is gathered.

## II.A Books and Lecture Notes

In the first place we list the books and the lecture notes which are mainly concerned with representation theory of symmetric groups and closely related topics and give a short comment concerning their content.

The first book which contained most of the results known at the time on the representation theory of $S_n$ was Murnaghan [1938c], which deals mainly with characters and their evaluation.

Then came Rutherford's book [1948], which was the first to be totally devoted to $S_n$. This book covers the main results of Young on the ordinary irreducible representations of $S_n$, in particular their matrix forms. It is easy to read, in contrast to Young's original papers, which can be found in the Mathematical Expositions series of the University of Toronto Press (No. 21, edited by G. de B. Robinson [1977]).

When Robinson's book appeared in 1961, it was the first to lay the main emphasis on the modular representation theory of $S_n$. Until the present

ENCYCLOPEDIA OF MATHEMATICS and Its Applications, Gian-Carlo Rota (ed.). Vol. 16: G. D. James and A. Kerber, The Representation Theory of the Symmetric Group
ISBN 0-201-13515-9

volume, it was the only book dealing with the modular theory. Robinson's book shows the level of knowledge in 1961.

Besides these books, the lecture notes of Kerber [1971b] cover large parts of the theory. These notes are mainly concerned with the evaluation of decomposition numbers and generalized decomposition numbers of $S_n$ as well as with the irreducible matrix representations of wreath products.

In the lecture notes of Knutson [1973] and Hoffman [1979] the emphasis lies on the isomorphism between a ring constructed by composing in a certain way all the representation rings of the symmetric groups $S_n$, and the ring of all symmetric polynomials in an infinite number of variables. The main technical tool is the notion of $\lambda$-ring.

The second volume of lecture notes by Kerber [1975a] is mainly concerned with characters of wreath products and applications to representation theory and combinatorics.

In the lecture notes of James [1978e] the approach is characteristic-free and relies heavily on Specht modules.

Besides these books and lecture notes which are devoted to the representation theory of symmetric groups, there are many books where large sections contain results on $S_n$, while the main aim is the representation theory of the classical groups.

One of these books needs special attention: Littlewood [1940]. This book is the standard reference for the approach to representations of $S_n$ along Schur functions which are generating functions for the values of the ordinary irreducible characters of $S_n$. In connection with this Macdonald's book [1979] is important.

The most important books dealing with the representation theory of classical groups in connection with $S_n$ are those of Weyl [1939], Boerner ([1955], 2nd. ed. [1967] and [1967], 2nd. ed. [1970]) and Hamermesh [1962]. The books of Boerner and Hamermesh are concerned with the needs of physicists.

## II.B Comments on the Chapters

### *Comments on Chapter 1*

*1.1* The fact that the elements of $A_n$ are commutators is shown in Ore [1951]. There are very many papers which consider sets of generators of symmetric and alternating groups. The interested reader is referred to Piccard [1946], [1948h] and [1957].

*1.2* The book we mentioned by Andrews [1976] on partitions covers nearly everything known on the subject; in particular it gives many results on the enumeration of partitions.

The ambivalency of alternating groups is discussed in Berggren [1969]. The corresponding results on wreath products can be found in Kerber [1970b].

*1.3* The approach we adopt using intertwining numbers was suggested first in the lecture notes of Coleman [1966]. A precursor was Burrow [1954], and further information appears in Munkholm [1968]. These last two papers give other applications of the method of using representations induced from pairs of subgroups.

*1.4* The dominance order $\trianglelefteq$ on the set of partitions $\alpha \vdash n$, which is the main subject of this section, seems to have been introduced first by Grace and Young [1903] and independently by Muirhead [1903].

The Gale-Ryser theorem which characterized $\alpha \trianglelefteq \beta$ by the existence of certain 0-1 matrices can be found in Ryser's book [1963], for example. The characterization of $\alpha \trianglelefteq \beta$ in terms of representations of $S_n$ (in fact by idempotents) is due to Ruch and Schönhofer [1970]. The importance of dominance for the representations was also noticed by Mayer [1971] and [1975a].

Interesting papers concerning the lattice of partitions under the dominance order are Brylawski [1973], Ruch and Gutman, and Thürlings [1977]. The first of these papers deals with the Moebius function of the lattice, the second with partitions which represent edge degree sequences, for which topic the reader is also referred to the book of Berge [1976].

We suggest that the reader who is interested in the Ruch-Schönhofer theorem should consult at least some of the numerous papers on the enumeration of nonnegative integral matrices with prescribed row and column sums and try to rephrase the results given there in terms of intertwining numbers of representations of symmetric groups.

*1.5* Tableaux and diagrams were considered in combinatorics quite a while before Frobenius started developing representation theory. That they are useful for representations of symmetric groups was noticed first by Young, who came to them from invariant theory. The reader should have a look at his collected works.

Many of Young's proofs were simplified by D. E. Rutherford (see his *Substitutional Analysis*). A considerable contribution was also made by J. von Neumann: see van der Waerden's remarks in his *Algebra II*.

*Comments on Chapter 2*

*2.1* As we have already mentioned, the chosen approach to the representation theory of $S_n$ by way of intertwining numbers was suggested by Coleman. It has the advantage that it is (via Mackey's theorem) very lucid under the aspect of general representation theory of finite groups. Furthermore it clearly shows the meaning of numbers which need to be evaluated in order to get the character table, say.

*2.2* The papers of Snapper [1971] and Liebler and Vitale [1973] made this part of the theory much easier than it was before. In particular Snapper was the first to point to the significance of the Gale-Ryser theorem in this context.

The method using $M_n$ and $\Xi_n$ in order to evaluate the character table of $S_n$ was used by Murnaghan (see his book and his papers). It is also discussed in Fox [1967] and Kerber [1969b].

*2.3* The main tool of this section is the determinantal form 2.3.4 of $[\alpha]$, which is a consequence of the quite old so-called Jacobi-Trudi equation (see the remark on p. 88 of Littlewood's book [1940]). Its derivation in connection with the proof of the Murnaghan-Nakayama rule uses ideas of M. Klemm and B. Wagner (unpublished). This determinantal form is in fact equivalent to a formula given by Frobenius [1900].

The characters $\chi^{\lambda/\mu}$ (or rather, the corresponding skew diagrams) were introduced in Young's eighth paper on quantitative substitutional analysis [QSA VIII, 1934]. Further results were found by Robinson [1947a].

The result which is called Young's rule was stated in the form of a substitutional equation in Young's QSA VII [1934]. The proof given there uses raising operators and is unsatisfactory. The equation in question is also discussed in Robinson [1935b].

The papers where hooks were introduced are by Nakayama [1940a, 1940b] and the very useful expression 2.3.21 of the dimensions in terms of hook lengths is due to Frame, Robinson, and Thrall [1954].

*2.4* See the comment on Section 2.3.

*2.5* The main result of this section is in Frobenius [1901].

The present proof of 2.5.13 uses ideas of B. Wagner (unpublished). Further interesting results on the characters of $A_n$ can be found in Specht [1938].

*2.6* The first proof that $S_n$ is characterized by its character table was given by Nagao [1957]. His proof is more elementary than the elegant one presented here, which is due to H. Pahlings. Nagao's method is to show that certain elements of $C^{(2,1^{n-2})}$ satisfy the same relations as the transpositions $(i, i+1) \in C^{(2,1^{n-2})}$ do. But the calculations are very long winded. The proof of the corresponding result on $A_n$ was given by Oyama [1964]. Here the necessary calculations are even longer. A shorter proof for $A_n$ was presented by Higman [1971]. Each of these proofs needs a separate treatment of $S_n$ and $A_n$. A very short proof of both these results can be given by an application (to $[n-1, 1]$) of a deep characterization theorem of Huffman [1975].

It should also be mentioned that the groups $C_p \operatorname{wr} S_n$, $C_p$ cyclic of prime order $p$, $n \geqslant 3$, are characterized by their character table; see Yokonuma [1965]. Thus the Weyl groups of types $A_n$ and $B_n$ are characterized in this way. For the Weyl groups of type $D_n$ this is true only if $n \not\equiv 0 \bmod 4$. For this last and further interesting results see Pahlings [1976a, 1976b].

*2.7* Cores of partitions or diagrams were introduced in the papers of T. Nakayama (mentioned already in the comments on Section 2.3). The $\beta$-numbers occur in the papers of H. K. Farahat, and the visualization by an (unnumbered) abacus is due to G. James. The idea of numbering such a scheme is due to B. Wagner. The important concept of $q$-quotient or star $q$-diagram is due to Robinson [1948] and Littlewood [1951]. The equivalence of the two concepts was proved by Farahat [1952, 1953].

*2.8* The main result of this section is the Littlewood-Richardson rule stated in 2.8.13 and rephrased in 2.8.14. The so-called Young's rule is a particular case of it. This Littlewood-Richardson rule may be considered to be the most important result of ordinary representation theory of $S_n$. There are very few really satisfactory proofs of it; the present one follows the lines of Wagner [1975/76]. Thanks to the work of M. Schützenberger, this result, together with related topics, nowadays forms an important section of combinatorics; see for example the review article by Schützenberger [1977] and the references mentioned therein.

*2.9* Many papers have been published on the decomposition of inner tensor products of ordinary irreducible characters of $S_n$ (see e.g. the additional references appearing in Littlewood, Murnaghan, and Robinson). The results known are still far from being complete; it can be shown that this problem is much more difficult than the Littlewood-Richardson rule.

2.9.5 is due to Robinson and Taulbee [1954].

2.9.16 and 2.9.17 stem from a paper of Klemm [1977b]

*Comments on Chapter 3*

This chapter does not need much comment. The matrix form of the irreducible ordinary representations of $S_n$ is due to Young; see his collected works and also Thrall [1941]. But the standard reference is D. E. Rutherford's book *Substitutional Analysis* [1948]. The present description follows its lines and makes use of some simplifications by H. Boerner (see the second edition of his book [1970]; the first edition gives the so-called natural form, which is rational integral). An apparently different integral form of the representing matrices can be found in Kazhdan and Lusztig [1979].

*Comments on Chapter 4*

*4.1* Wreath products were first considered by Specht [1932 and 1933]. (He followed a hint given by Schur.) Another important reference is Ore [1942].

*4.2* See Specht's papers mentioned above and part I of Kerber's lecture notes [1971b]. For additional results concerning the splitting of conjugacy classes and representations over particular subgroups of index 2 see Celik, Kerber, and Pahlings [1975] and Mayer [1975b].

*4.3* The matrix representations of $G \operatorname{wr} H$ over $\mathbb{C}$ were described by Specht. The present characteristic-free approach is due to Kerber [1968a].

*4.4* The first examples of wreath products can be found in Cauchy [1844], Netto [1882], and Radzig [1895], where wreath products arise in connection with the construction of Sylow subgroups of symmetric groups.

The first representation-theoretical consideration of particular wreath products was given by Young in QSA V [1930]. Here one can find the ordinary representation theory of hyperoctahedral groups $S_2 \operatorname{wr} S_n$. See also Mayer [1975c].

After Specht's consideration of the general case the particular case $C_m \operatorname{wr} S_n$ was examined in detail by Osima ([1954b], [1956b]) and Puttaswamaiah ([1963] and [1969]). Wreath products $S_m \operatorname{wr} S_n$ were considered by Kerber [1969a].

Certain wreath products have bases of their character rings which consist of permutation characters: see Geissinger and Kinch [1978] and Kerber and Tappe [1976].

Results concerning wreath products being $M$-groups are due to Dade (see Huppert I [1967], V, §18), Seitz [1969], and Kerber [1970a].

*Comments on Chapter 5*

*5.1* The combinatorial theory of enumeration, which may be considered as the theory of the Cauchy-Frobenius lemma (see P. M. Neumann's paper [1979]) and its generalizations is (after Frobenius and Burnside) due to the papers of Redfield [1927] and Pólya [1937]. The standard reference for its main applications in enumerative graph theory is the book of Harary and Palmer [1973].

The exponentiation-group enumeration theorem in weighted form is due to Lehmann [1976], whose approach is slightly different from the one adopted here.

The reader is also referred to the following papers, which give a description from the representation-theoretical point of view: Foulkes [1963, 1966], Kerber [1974, 1975b, 1975c, 1975d, 1976], and Kerber and Lehmann [1977].

*5.2* The construction in this section is due to Schur's famous dissertation [1901]. Further important papers on this subject are Weyl [1925, 1929, 1937] and van der Waerden [1931].

The modular case was considered first by Thrall [1942b and 1944] but more recent papers on the modular or characteristic-free case are those by Carter and Lusztig [1974], Clausen [1979, 1980a, 1980b], and James [1980]. See also Chapter 8.

Symmetry classes play an important role in multilinear algebra, and the reader should consult Marcus [1973, 1975]. In particular the question of irreducibility of symmetry classes is interesting. By the results given it is equivalent to the question for monomial representations of $S_n$. The interested reader may consult Clausen [1974] and Djokovic and Malzan [1975].

The given proof of 5.2.14 is due to B. Wagner (unpublished).

The proof of the Amitsur-Levitzki theorem is more or less the one due to Kostant [1958].

The results 5.2.35 ff. can be found in Esper [1976a].

*5.3* This construction was introduced (and applied) by Esper [1976a], Esper and Kerber [1976, 1977], and Kerber [1980].

For the divisibility questions discussed in this section the reader should compare Frobenius [1907], Solomon [1962, 1969], Isaacs [1970], End [1971], Ritter [1974], and Kerber and Wagner [1980]. Further references can be found in Finkelstein's very useful review article [1978].

Readers interested in representations of the first, second, and third kind or simple-phase ones are referred to van Zanten and de Vries [1973b, 1974, 1975].

*5.4* Plethysms were introduced by Littlewood [1936c]. Various other papers are concerned with the decomposition of plethysms of ordinary irreducible representations of symmetric groups; see for example, Duncan [1952a, 1952b, 1954], Foulkes [1949a, 1951, 1954], and the papers of Ibrahim.

*5.5* The relevant papers are by Frobenius [1904], Tsuzuku [1961], Foulkes [1970], Kerber [1978], and Klemm and Wagner [1979].

For further reading and for applications to permutation group theory we recommend Klemm [1976a, 1976b and 1977b] and Saxl [1975a and 1975b].

At the end of this section (see 5.5.48) we mentioned the character polynomials. This is a very important topic, since each such polynomial yields by the corresponding polynomial function an infinite number of ordinary irreducible characters of symmetric groups. It is in fact possible to do a great deal of the representation theory of $S_n$ by examining such polynomials. This idea is due to Frobenius. It was used extensively by Murnaghan, and later on by Specht; a recent treatment is given by B. Wagner [1979].

*Comments on Chapter 6*

*6.1* Standard references for the modular representation theory of finite groups are Curtis and Reiner [1962] and Puttaswamaiah and Dixon [1977].

The analysis of diagrams or partitions used in this and the next section to evaluate the total number of irreducible $p$-modular representations (6.1.12) and the number of such representations which belong to a given $p$-block (6.2.2) is due to Osima [1956a].

Nakayama made his conjecture (6.1.21) in his paper [1940b]. He showed there that it is true for each $n<2p$. These cases were also studied by Thrall and Nesbitt [1942]. The first proof of Nakayama's conjecture was given by Brauer [1947] and Robinson [1947b]. Their method of proof uses properties of defect groups and a particular property of the ordinary characters of symmetric groups, namely 2.7.25. Later Nakayama and Osima [1951] published an alternative proof which used the methods of Nakayama's two papers. A further proof which lays more emphasis on the defect groups is due to Osima [1955]. Quite different from all these proofs is the one by Farahat and Higman [1959]. Here the main tool is a set of generators for the center of the integral group ring of $S_n$. A proof (using $S$-functions) of one half of the conjecture can be found in Farahat [1956].

We here follow the lines of Meier and Tappe [1976] whose proof is the most elegant one. (But see also Section 7.3.) It uses Brauer's second main theorem as well as the generalized decomposition numbers of symmetric groups, which were first considered by Kerber [1966]. See also Osima [1968, 1971].

*6.2* Staal [1950] realized the significance of the star $p$-diagram or $p$-quotient for the $p$-modular representation theory of $S_n$.

The fact that both the numbers of irreducible ordinary and modular representations in a $p$-block of $S_n$ depend only on the defect of the $p$-block, which is in turn determined uniquely by the number of removable $p$-hooks of the corresponding diagrams, was shown by Robinson [1951]. Our derivation of these results follows Osima [1954a]. We use this approach to demonstrate the usefulness of Osima's $u$-numbers, which themselves merit further investigation. An entirely different method is given in James [1978c].

*6.3* The methods of $r$-inducing and $r$-restricting are powerful tools in the determination of decomposition matrices of symmetric groups. Further examples of their application can be found in the lecture notes by A. Kerber [1971b], mentioned above.

The first complete proof of the "wedge shape" (6.3.1) of the decomposition matrix was given by Farahat, Müller, and Peel [1976], but their result did not include the 1's down the diagonal. The proof of 6.3.50 is based on the ideas of James [1979d]. It is noteworthy that the entirely different method appearing in Chapter 7 was discovered before a construction of 6.3.1 relying solely on $r$-inducing was found.

There is, as yet, no known module-theoretic interpretation of the process (6.3.48) whereby a diagram is transformed into a $p$-regular diagram. In the paper of Littlewood [1951] mentioned after 6.3.68, Littlewood proves that a set of relations, each one of which involves only partitions with the same $p$-core, gives all the relations between ordinary characters restricted to $p$-regular classes. He claims that the Nakayama conjecture follows, but this appears to be incorrect.

*Comments on Chapter 7*

*7.1* Specht introduced his modules in 1935. Our notation differs from Specht's, but the translation is trivial. The source for this section is James [1976a], where the modular irreducible representations of $S_n$ were first constructed. An interesting alternative approach, in terms of letter place algebras, appears in Clausen [1980a].

*7.2* The Garnir relations were given in Garnir's paper [1950].

Our construction of the standard basis follows that of Peel [1975a].

*7.3* G. E. Murphy's results involving the element $R_n$ are not yet published.

The evaluation of the determinant of the Gram matrix appears in the paper of James and Murphy [1979], where the calculations for 7.3.20 may be found.

Theorem 7.3.23 was conjectured (in an alternative form) by R. W. Carter. One implication was proved by James [1978d]; the other implication is the one which follows at once as soon as the formula for the determinant of the Gram matrix is in the form 7.3.20.

*Comments on Chapter 8*

*8.1* The idea of studying Weyl modules and Specht modules simultaneously embedded in tensor space is due to Carter and Lusztig [1974] and we are guided by their approach.

*8.2* The algebra which we call $\overline{U}_F$ is denoted by $\mathcal{U}_A$ in Carter and Lusztig's paper [1974]. For further information, the reader may consult Kostant [1966].

*8.3* The ideas here stem from Wong [1972], but some of the results are new.

*8.4* The majority of the results in this section appear in Carter and Lusztig's paper [1974] cited above. An alternative proof of Theorem 8.4.2 appears in James [1977a], and applications of the result may be found in the lecture notes by James [1978e] already mentioned.

## II.C Suggestions for Further Reading

An important part of the representation theory of symmetric groups, namely the theory of their projective representations, remains completely untouched in this book. The standard reference for this is the classical paper of Schur [1910], together with the publications of A. O. Morris and his coworkers (in particular E. W. Read), which are listed among the references.

Another important subject is the very strong connection between ordinary representation theory of symmetric groups and ordinary invariant theory. The first publication which made this clear was the book of Grace and Young [1903]. See also Schur and Grunsky [1968] or the modern treatment by Dieudonné and Carrell [1970].

Recently the development of characteristic-free invariant theory began, and as with ordinary theory, it happened with the characteristic-free approach that the tools of the theory turned out to be very useful for a characteristic-free approach to representation theory of symmetric and general linear groups (which is different from the one given here in Chapters 7 and 8). The reader is referred to Désarménien, Kung, and Rota [1978] and Clausen [1979, 1980a, 1980b].

The reader who is particularly interested in the general linear group should read the recent papers of J. Towber and the excellent lecture notes by J. A. Green [1980].

## II.D References

This last part of Appendix II contains references in alphabetical order. We have tried to give a more or less complete collection of references concerning the symmetric group, its representation theory, and the topics covered in this book.

Abramsky, Y. J., Jahn, H. A., and King, R. C, "Frobenius symbols and the groups $S_s$, GL($n$), O($n$) and Sp($n$)", *Can. J. Math.* **25** (1973), 941–959; ZB 235 #20043; MR 48 #4135.

Aitken, A. C., "On induced permutation matrices and the symmetric group", *Proc. Edinburgh Math. Soc.* (2) **5** (1936), 1–13; ZB 15, p. 248.

Aitken, A. C., "On compound permutation matrices", *Proc. Edinburgh Math. Soc.* (2) **7** (1946), 196–203; MR 8, p. 310.

Aizenberg, N. N., "On the representations of the wreath product of finite groups", *Ukrain. Mat. Z.* **13** (1961), 5–12; ZB 109, p. 15; MR 25, #2134.

Aizenberg, N. N., and Leticevskii, A. A., "Über die Berechnung der Darstellungen eines Kranzprodukts endlicher Gruppen auf elektronischen Digitalrechnern", *Kibernetika* (*Kiev*), **1965**, No. 3, 63–71; ZB 203, p. 30.

Amato, V., "Sul rango del gruppo totale delle sostituzioni sopra n elementi", *Note Escrcit. Mat.* **6** (1931), 75–81; ZB 4, p. 51.

Amato, V., "Sul gruppo totale di sostituzioni su $n$ lettre", *Acta Catania* **20** (1934), No. 11, 1–21; ZB 10, p. 393.

Amato, V., "Sul rango del gruppo totale di sostituzioni sopra $n$ lettere", *Rend. Circ. Mat. Palermo* **61** (1937), 83–86; ZB 18, p. 145.

Andrews, G. E., *The Theory of Partitions*, Encyclopedia of Math. Appl., Vol. 2, Addison-Wesley, 1976.

Astie, A., "Caracterisation des sous-groupes d'ordre impair et de rang minimun du groupe symétrique $S_n$", *C. R. Acad. Sci., Paris, Ser. A-B* **281** (1975), No. 19, A783–A785; MR 53, #3070.

Backhouse, N. B., and Gard, P., "Symmetrized powers of point group representations", *J. Phys. A: Math. Nucl. Gen.* **7** (1974), 1239–1250.

Ball, R. W., "Maximal subgroups of symmetric groups", *Trans. Amer. Math. Soc.* **121** (1966), 393–407; ZB 136, p. 278; MR 34, #2672.

Bannai, E., "Multiply transitive permutation representations of finite symmetric groups", *J. Fac. Sci. Univ. Tokyo Sect. I* **16** (1969), 287–296; ZB 197, p. 22, MR 41, #5518.

Bannai, E., "Maximal subgroups of low rank of finite symmetric and alternating groups", *J. Fac. Sci. Univ. Tokyo, Sect. I A* **18** (1972), 475–486; ZB 262, #20003; MR 50, #10027.

Barbosa, R. M., "Cycle index of symmetric group—direct proofs and determinant form", *Naturalia* **4** (1978), 55–60.

Bateman, J. M., Richard, E., Phillips, R. E., and Sonneborn, L. M., "Wreath products and representations of degree one or two", *Trans. Amer. Math. Soc.* **181** (1973), 143–153.

Bauer, M., "Über die alternierende Gruppe", *Mat. fiz. Lap.* **39**, (1932), 25–26; ZB 7, p. 52.

Bauer, M., "Über die alternierende Gruppe", *Acta Szeged* **6**, (1934), 222–223; ZB 8, p. 338.

Bayar, E., "Eine neue Einführung in die Darstellungstheorie symmetrischer Gruppen", *Mitt. Math. Sem. Giessen* **81** (1969), 1–45; ZB 197, p. 304; MR 39, #1573.

Bayar, E., "Über eine spezielle Darstellung der symmetrischen Gruppen", *Commun. Fac. Sci. Univ. Ankara, Ser. A* **21** (1971) 1–12 (1972); ZB 255, #20009; MR 47, #5089.

Bayramow, R. A., "Über den Untergruppenverband der symmetrischen Gruppe, den Unterhalbgruppenverband der symmetrischen Halbgruppe und den Unteralgebrenverband der Postschen Algebra", *Semigroup Forum* **2** (1971), 271–280.

Bays, S., "Les répartitions imprimitives des $n$-uples dans le groupe symétrique de degré $n$", *Comment. Math. Helv.* **26** (1952), 68–77; ZB 46, p. 249; MR 13, p. 817.

Beke, E., "Über die Einfachheit der alternierenden Gruppe", *Math. Ann.* **49** (1897), 581–582.

Bender, H. A., "A new method for the determination of the group of isomorphisms of the symmetric group of degree $N$", *Amer. Math. Monthly* **31** (1924), 287–289.

Benson, C. T., and Curtis, C. W., "On the degrees and rationality of certain characters of finite Chevalley groups", *Trans. Amer. Math. Soc.* **165** (1972), 251–273; ZB 246, #20008; MR 46, #3608.

Benson, C. T., and Curtis, C. W., "On the degrees and rationality of certain characters of finite Chevalley groups", Representation theory, finite groups, related topics, *Proc. Sympos. Pure Math.* **21** (1970), 1–5; ZB 268, #20029.

Benson, C. T., and Grove, L. C., *Finite Reflection Groups*, Bogden and Quigley, New York, 1971.

Berge, C., *Graphs and Hypergraphs*, North Holland Publishing Company, Amsterdam, 1976 ix+528 pp. ZB 311 #05101; MR 52 #5453.

Berggren, J. L., "Finite groups in which every element is conjugate to its inverse.", *Pacific J. Math* **28** (1969), 289–293; ZB 172, p. 31; MR 39, #1539.

Bertin, M.-J., "Anneau des invariants du groupe alterné en charactéristique 2", *Bull. Sci. Math.* (2) **94** (1970), 65–72.

Bertram, E., "Even permutations as a product of two conjugate cycles", *J. Combinat. Theory, Ser. A* **12** (1972), 368–380; ZB 238, #20004; MR 45, #6905.

Best, M. R., "The distribution of some variables on symmetric groups", *Nederl. Akad. Wetensch. Proc. Ser. A* **73**=*Indag. Math.* **32** (1970), 385–402; ZB 208, p. 33; MR 45, #2003.

Biedenharn, L. C., and Louck, J. D., "Representations of the symmetric group as special cases of the boson polynomials in U($n$)", The permutation group in physics and chemistry, Symp. Bielefeld 1978, *Lect. Notes Chem.* **12** (1979), 121–147.

Binder, G. J., "The bases of the symmetric group", *Izv. Vyss. Ucebn. Zaved. Mat.* **1968**, No. 11/78, 19–25; ZB 177, p. 37; MR 40, #1461.

Binder, G. J., "The two-element bases of the symmetric group", *Izv. Vyss. Ucebn. Zaved. Mat.* **1970**, No. 1(92), 9–11; ZB 218, p. 133; MR 41, #1850.

Binder, G. J., "Certain complete sets of complementers of the symmetric and alternating group of $n$-th degree", *Mat. Zametki* **7** (1970), 173–180; ZB 218, p. 134; MR 41, #1851.

Binder, G. J., "Some complete sets of complementary elements of the symmetric and alternating group of $n$-th degree", *Math. Notes* **7** (1970), 105–109; ZB 239, #20001.

Bivins, R. L., Metroplis, N., Stein, P. R., and Wells, M. B., "Characters of $S_n$ for $n=15,16$", *Math. Tables and other aids to computing* **8** (1954), 212–216.

Blackburn, N., "The extension theory of the symmetric and alternating groups", *Math. Z.* **117** (1970), 191–206; ZB 205, p. 324; MR 44, #4087.

Blichfeldt, H. F., "A theorem concerning the invariants of linear homogeneous groups, with some applications to substitution-groups", *Trans. Amer. Math. Soc.* **5** (1904), 461–466.

Blum, J., "Enumeration of the square permutations in $S_n$", *J. Combinat. Theory (A)* **17** (1974), 156–161; ZB 288, #05005.

Boccara, G., "Sur certaines relations d'ordre dans les groupes symétriques finis ou infinis", in *Permutations* (Actes Colloq., Univ. René-Descartes, Paris, *1972*), Gauthier-Villars, Paris, 1974, pp. 167–185; MR 50, #10028.

Boerner, H., "Über die rationalen Darstellungen der allgemeinen linearen Gruppe.", *Arch. Math* **1** (1948), 52–55; ZB 34, p. 162; MR 10, p. 99.

Boerner, H., *Darstellungen von Gruppen mit Berücksichtigung der Bedürfnisse der modernen Physik*, Springer-Verlag, 1955, 2nd ed. 1967.

Boerner, H., "Darstellungstheorie der endlichen Gruppen", in *Enzyklopädie der Math. Wiss. I* 1, 15, Teubner-Verlag, 1967, pp. 1–80; ZB 153, p. 351; MR 37, #2872.

Boerner, H., *Representations of Groups*, North-Holland Publishing Company, 1967, 2nd ed. 1970.

Boorman, E. H., "$S$-operations in representation theory", *Trans. Amer. Math. Soc.* **205** (1975), 127–149; MR 51, #678.

Bovey, J., and Williamson, A., "The probability of generating the symmetric group", *Bull. London Math. Soc.* **10** (1978), 91–96.

Bovey, J. D., "An approximate probability distribution for the order of elements of the symmetric group", *Bull. London Math. Soc.* **12** (1980), 41–46.

Brandt, A. J., "The free Lie ring and Lie representations of the full linear group", *Trans. Amer. Math. Soc.* **56** (1944), 528–536; MR 6, p. 146.

Brauer, R., "On a conjecture by Nakayama", *Trans. Roy. Soc. Canada Sect. III (3)* **41** (1947), 11–19; ZB 29, p. 199; MR 10, p. 678.

Brauer, R., "Symmetrische Funktionen, Invarianten von linearen Gruppen endlicher Ordnung", *Math. J. Okayama* **21** (1979), 91–113.

Brender, M., "Spherical functions on the symmetric groups", *J. Algebra* **42** (1976), 302–314; MR 54, #12881.

Brenner, J. L., "Covering theorems for finite nonabelian simple groups. IX. How the square of a class with two nontrivial orbits in $S_n$ covers $A_n$", *Ars Combinatoria* **4** (1977), 151–176; MR 58, #28162a.

Brenner, J. L, "Corrigendum: Covering theorems for finite nonabelian simple groups IX. How the square of a class with two nontrivial orbits in $S_n$ covers $A_n$", *Ars Combinatoria* **6** (1978), 315; MR 58, #28162b.

Bröcker, T., "Homologie symmetrischer und alternierender Gruppen.", *Inventiones Math.* **2** (1967), 222–237; ZB 144, p. 260.

de Bruijn, N. G., "Pólya's theory of counting", in E. F. Beckenbach (ed.), *Applied Combinatorial Mathematics*, Wiley and Sons, 1964, pp. 144–184.

de Bruijn, N. G., "A survey of generalizations of Pólya's enumeration theorem", *Nieuw Archief voor Wiskunde (2)* **XIX** (1971), 89–112.

Brylawski, T., "The lattice of integer partitions", *Discrete Math.* **6** (1973), 201–219; MR 48, #3752.

Burge, W. H., "Four correspondences between graphs and generalized Young tableaux", *J. Comb. Theory (A)* **17** (1974), 12–30.

Burnside, W., "The determination of all groups of rational linear substitution of finite order which contain the symmetric group in the variables", *Proc. London Math. Soc. (2)* **10**, (1911) 284–308.

Burroughs, J., "Operations in Grothendieck rings and the symmetric group", *Can. J. Math.* **26** (1974), 543–550; MR 49, #5158.

Burrow, M. D., "A generalization of the Young diagram", *Can. J. Math.* **6** (1954), 498–508; ZB 56, p. 26; MR 16, p. 333.

Burrow, M., *Representation Theory of Finite Groups*, Academic Press, 1965.

Butler, P. H., and King, R. C., "Branching rules for $U(N) \supset U(M)$ and the evaluation of outer plethysms", *J. Math. Phys.* **14** (1973), 741–745; ZB 262, #20048.

Butler, P. H., and King, R. C., "The symmetric group: Characters, products and plethysms", *J. Math. Phys.* **14** (1973), 1176–1183.

Butler, P. H., and King, R. C., "Symmetrized Kronecker products of group representations", *Can. J. Math.* **26** (1974), 328–339.

Calame, A., "Les relations caractéristiques des bases du groupe symétrique", Thèse, Université de Neuchatel, 1955, 101 pp.; ZB 67, p. 260; MR 17, p. 940.

Calame, A., and Piccard, S., "Les relations caractéristiques des bases du groupe symétrique", *C. R. Acad. Sci. Paris* **240** (1955), 2477–2478; ZB 67, p. 260; MR 17, p. 941.

Cardenas, H., "The cohomology algebra of a symmetric group", *An. Inst. Mat. Univ. Nac. Autonoma Mexico* **4** (1964), 3–31; MR 34, #1383.

Cardenas, H., "The cohomology algebra of the symmetric group of order $p^2$", *Bol. Soc. Mat. Mexicana (2)* **10** (1965), 1–30; MR 35, #4308.

Cardenas, H., and Lluis, E., "The normalizer of the Sylow $p$-group of the symmetric $S_{p^2}$", *An. Inst. Mat. Univ. Nac. Autonoma Mexico* **2** (1962), 1–7; MR 29, #4802.

Cardenas, H., and Lluis, E., "The normalizer of the Sylow $p$-group of the symmetric group $S_{p^n}$", *Bol. Soc. Math. Mexicana (2)* **9** (1964), 1–6; MR 31, #5898.

Carter, R. W., and Lusztig, G., "On the modular representations of the general linear and symmetric groups", *Math. Z.* **136** (1974), 193–242.

Cauchy, A., *Exercises d'analyse et de physique mathématique*, Vol. III, Paris. 1844.

Celik, Ö., "Über die Normalteiler von $G \operatorname{wr} S_n$, insbesondere $S_2 \operatorname{wr} S_n$, vom Index 2", *Revue Fac. Sci. Univ. Istanbul, Ser. A* **40** (1975), 129–166 (1978); ZB 372, #20015.

Celik, Ö., Kerber, A., and Pahlings, H., "Zur Darstellungstheorie gewisser Verallgemeinerungen der Serien von Weyl-Gruppen", *Mitt. Math. Sem. Giessen* **119** (1975), 15–90; ZB 342, #20004; MR 54, #7600.

Chan, G. H., "A characterization of minimal $(k)$-groups of degree $n \leqslant 3k$", *Linear and Multilinear Algebra* **4** (1977), 285–305; ZB 348, #20002.

Chan, G. H., "On the triviality of a symmetry class of tensors", *Linear and Multilinear Algebra* **6** (1978), 73–82.

Chan, G. -H., "$(k)$-characters and the triviality of symmetry classes", *Linear Algebra and Its Appl.* **25** (1979), 139–149.

Chernysh, V. V., "A graphic method for finding the Clebsch-Gordan series for a symmetric group", *Izv. Vyss. Ucebn. Zaved. Mat.* **22** (1978), 89–92; *Soviet Mathematics* **22** (1978), 73–76; MR 80, #20007.

Chowla, S., Herstein, I. N., and Moore, W. K., "On recursions connected with symmetric groups I", *Can. J. Math.* **3** (1951), 328–334; ZB 43, p. 259; MR 13, p. 10.

Chowla, S., Herstein, I. N., and Scott, W. R., "The solutions of $x^d = 1$ in symmetric groups", *Norske Vid. Selsk. Forh., Trondheim* **25** (1952), 29–31; MR 14, p. 947.

Chung, J. H., "Modular representations of the symmetric group", *Can. J. Math* **3** (1951), 309–327; MR 13, p. 106.

Clausen, M., "Zentralisatorenverbände von Moduln über halbeinfachen Gruppenalgebren; zur Theorie der Symmetrieklassen in Tensorräumen", Diplomarbeit, Giessen, 1974.

Clausen, M., "Letter place algebras and a characteristic-free approach to the representation theory of the general linear and symmetric groups, I", *Advances in Math.* **33** (1979), 161–191.

Clausen, M., "Letter-Place-Algebren und ein charakteristik-freier Zugang zur Darstellungstheorie symmetrischer und voller linearer Gruppen", *Bayreuther Mathematische Schriften* **4** (1980a), 151 pp.

Clausen, M., "Letter place algebras and a characteristic-free approach to the representation theory of the general linear and symmetric groups, II", *Advances in Math.*, **38** (1980b), 152–177.

Clifford, A. H., "Representations induced in an invariant subgroup", *Ann. of Math.* **38** (1937), 533–550.

Coleman, A. J., "Induced representations with applications to $S_n$ and $GL(n)$", Queen's Papers in Pure and Applied Mathematics, No. 4, Queen's University, Kingston, Ontario, 1966; ZB 141, p. 25; MR 34, #2718.

Comét, S., "Une propriété des déterminants et son application au calcul des caractéres des groupes symétriques", *Kungl. Fysiografiska Sällskapets i Lund Förhandlingar* (*Proc. Roy. Physiog. Soc. Lund*) **14** No. 7, (1945), 84–94; MR 7, p. 113.

Comét, S., "Über die Anwendung von Binärmodellen in der Theorie der Charaktere der symmetrischen Gruppen", *Numer. Math.* **1** (1959), 10–109; ZB 100, p. 30; MR 21, #692.

Comét, S., "Improved methods to calculate the characters of the symmetric group", *Math. Comput.* **14** (1960), 104–117; ZB 103, p. 264; MR 22, #10212.

de Concini, C., "Symplectic standard tableaux" *Advances in Math.* **34** (1979), 1–27.

de Concini, C., Eisenbud, D., and Procesi, C., "Young diagrams and determinantal varieties", *Inventiones Math.* **56** (1980), 129–165.

de Concini, C., and Procesi, C., "A characteristic free approach to invariant theory", *Advances in Math.* **21** (1976), 330–354.

de Concini, C., and Strickland, E., "Traceless tensors and the symmetric group", *J. Algebra* **61**, (1979) 112–128.

Cooper, C. D. H., "Maximal $\bar{p}$-subgroups of the symmetric groups", *Math. Z.* **123** (1971), 285–289; ZB 224, #20001.

Coppersmith, D., and Grossman, E., "Generators for certain alternating groups with applications to cryptography", *SIAM J. Appl. Math.* **29** (1975), 624–627.

Coxeter, H. S. M., "An abstract definition for the alternating group in terms of two generators", *J. London Math. Soc.* **11** (1936), 150–156; ZB 14, p. 53.

Craig, W., "A presentation of the symmetric group based on unique irredundant nondecreasing factorization", *J. Pure Appl. Algebra* **13** (1978) 165–168; ZB 395, #20018.

Crouch, R. B., "Monomial groups", *Trans. Amer. Math. Soc.* **80** (1955), 187–215; ZB 65, p. 259.

Cummings, L. J., and Robinson, R. W., "Linear symmetry classes", *Can. J. Math.* **28** (1976), 1311–1319.

Curtis, C. W., "Corrections and additions to 'On the degrees and rationality of certain characters of finite Chevalley groups'", *Trans. Amer. Math. Soc.* **202** (1975), 405–406.

Curtis, C. W., and Reiner, I., *Representation Theory of Finite Groups and Associative Algebras*, Pure and Applied Mathematics, Vol. XI, John Wiley & Sons, New York–London–Sydney, 1962, xiv+689 pp.

Czyzo, E., "On the determination of Hall polynomials", *Bull. Acad. Polon. Sci. Ser. Sci. Math. Astron. Phys.* **23** (1975), 739–745; ZB 314, #20022.

Dagger, S. W., "A class of irreducible characters for certain classical groups", *J. London Math. Soc.* (2) **2** (1970), 513–520.

Davies, J. W., and Morris, A. O., "The Schur multiplier of the generalized symmetric group", *J. London Math. Soc.* (2) **8** (1974), 615–620; MR 50, #485.

Dénes, J., Erdös, P., and Turan, P., "On some statistical properties of the alternating group of degree *n*", *Enseignement Math.* (2) **15** (1969), 89–99; MR 40, #214.

Désarménien, J., "An algorithm for the Rota straightening formula", *Discrete Math.* **30** (1980), 51–68.

Désarménien, J., Kung, J. P. S., and Rota, G.-C., "Invariant theory, Young bitableaux, and combinatorics", *Advances in Math.* **27** (1978), 63–92; MR 58, #5737.

Dey, I. M. S., and Wiegold, J., "Generators for alternating and symmetric groups", *J. Austral. Math. Soc.* **12** (1971), 63–68; ZB 213, p. 302.

Dickson, L. E., "Representations of the general symmetric group as linear groups in finite and infinite fields", *Trans. Amer. Math. Soc.* **9** (1908), 121–148.

Dieudonné, J. A., and Carrell, J. B., "Invariant theory, old and new", *Advances in Math.* **4** (1970), 2–80; Academic Press, 1971; ZB 196, p. 58; MR 43, #4828.

Dinkines, F., "Semi-automorphisms of symmetric and alternating groups.", *Proc. Amer. Math. Soc.* **2** (1951), 478–486; MR 12, p. 801; ZB 42, p. 254.

Dixon, J. D., "The probability of generating the symmetric group", *Math. Z.* **110** (1969), 199–205; ZB 176, p. 299.

Dixon, J. D., "Maximal abelian subgroups of the symmetric group", *Can. J. Math.* **23** (1971), 426–438; MR 43, #7496.

Djoković, D. Z., and Malzan, J., "Imprimitive irreducible complex characters of the symmetric group", *Math. Z.* **138** (1974), 219–224.

Djoković, D. Z., and Malzan, J., "Monomial irreducible characters of the symmetric group", *J. Algebra* **35** (1975), 153–158; MR 51, #5737.

Djoković, D. Z., and Malzan, J., "Imprimitive, irreducible complex characters of the alternating group", *Can. J. Math.* **28** (1976), 1199–1204; MR 54, #7601.

Doubilet, P., "An inversion formula involving partitions", *Bull. Am. Math. Soc.* **79** (1973), 177–179.

Doubilet, P., Rota, G.-C., and Stein, J., "On the foundations of combinatorial theory IX: Combinatorial methods in invariant theory", *Stud. Appl. Math.* **53** (1974), 185–216; MR 58, #16736.

Dress, A., "Eine Bemerkung zur Ruch-Schönhoferschen Halbordnung von Young-Diagrammen", *Match* **7** (1979), 317–325.

Duncan, D. G., "Note on a formula by Todd", *J. London Math. Soc.* **27** (1952a), 235–236; MR 13, p. 910.

Duncan, D. G., "On D. E. Littlewood's algebra of *S*-functions", *Can. J. Math.* **4** (1952b), 504–512; ZB 48, p. 11; MR 14, p. 443.

Duncan, D. G., "Note on the algebra of *S*-functions", *Can. J. Math.* **6** (1954), 509–510; ZB 56, p. 17; MR 16, p. 328.

Durbin, J. R., "On locally compact wreath products", *Pacific J. Math.* **57** (1975), 99–107.

Durbin, J. R., "Spherical functions on wreath products", *J. Pure Appl. Algebra* **10** (1977), 127–134.

Durbin, J. R., and Farmer, K. B., "On projective representations of finite wreath products", *Math. Comp.* **31** (1977), 527–535; ZB 357, #20005.

Dussaud, R., "Formules de Gamba et représentations linéaires du groupe symétrique", *Publ. Cent. Rech. Math. Pures. Ser. I* **13** (1978), 35–38; ZB 404, #20007.

Dye, R. H., "Symmetric groups as maximal subgroups of orthogonal and symplectic groups", *J. London Math. Soc.* **20** (1979), 227–237.

Edwards, S. A., "Gel' fand bases and the permutation representations of the symmetric group associated with the subgroups $S_{\lambda_1}\times\cdots\times S_{\lambda_n}$", *J. Physics A* **13** (1980), 1563–1573.

End, W., "Über einen Satz von Frobenius und Solomon", *Archiv der Math.* **22** (1971), 241–245; ZB 224 #20032.

Esper, N., "Tables of reductions of symmetrized inner products ('inner plethysms') of ordinary irreducible representations of symmetric groups", *Math. Comput.* **29** (1975), 1150–1151; ZB 314, #20014.

Esper, N., "Die Symmetrisierungsmatrix", Dissertation, Aachen, 1976a.

Esper, N., "On the decomposition of symmetrized inner products of ordinary representations of symmetric groups", *Mitt. Math. Sem. Giessen* **121** (1976b), 169–179; MR 58, #867.

Esper, N., and Kerber, A., "Permutrization of representations with applications to simple-phase groups and the enumeration of roots in groups and of conjugacy classes which are invariant under power maps", *Mitt. Math. Sem. Univ. Giessen* **121** (1976), 181–200; ZB 358, #20008; MR 57, #12665.

Esper, N., and Kerber, A., "Permutrization of representations", in *Combinatoire et representation du groupe symetrique* (Actes Table Ronde C.N.R.S. Univ. Louis-Pasteur Strasbourg, Strasbourg 1976), Lecture Notes in Math., Vol. **579**, Springer, Berlin, 1977, pp. 267–280; ZB 358, #20007; MR 57, #6163.

Evens, L., and Kahn, D. S., "Chern classes of certain representations of symmetric groups", *Trans. Amer. Math. Soc.* **245** (1978), 309–330; ZB 402, #20009.

Fakirov, D., "Bemerkungen über die vollständige Reduzibilität einer beliebigen endlichdimensionalen, mittels des Youngschen Schemas konstruierten irreduziblen Darstellung der vollständigen linearen Gruppe $G_n$ beim Übergang zu ihrer Untergruppe $G_{n-1}$", *C. R. Acad. Bulgare Sci.* **21** (1968), 207–210; ZB 241, #20030.

Farahat, H. K., "On *p*-quotients and star diagrams of the symmetric group", *Proc. Cambridge Philos. Soc.* **48** (1952), 737–740; MR 14, p. 351.

Farahat, H. K., "On *p*-quotients and star diagrams of the symmetric group", *Proc. Cambridge Philos. Soc.* **49** (1953), 157–160; ZB 50, p. 25; MR 14, p. 723.

Farahat, H. K., "On the representations of the symmetric group", *Proc. London Math. Soc.* (*3*) **4** (1954), 303–316; MR 16, p. 11.

Farahat, H. K., "On the blocks of characters of symmetric groups", *Proc. London Math. Soc.* (*3*) **6** (1956), 501–517; MR 19, p. 634.

Farahat, H. K., "On Schur functions", *Proc. London Math. Soc.* (*3*) **8** (1958), 621–630; ZB 85, p. 16; MR 20, #5243.

Farahat, H. K., "The symmetric group as metric space", *J. London Math. Soc.* **35** (1960), 215–220; MR 22, #2645.

Farahat, H. K., "On the natural representation of the symmetric groups", *Proc. Glasgow Math. Assoc.* **5** (1962), 121–136; ZB 107, p. 261; MR 25, #3097.

Farahat, H. K., "Note on a paper of Tsuzuku", *Proc. Glasgow Math. Assoc.* **6** (1964), 196–197; ZB 139, p. 18.

Farahat, H. K., "Induction algebra and group systems", workshop on symmetric groups and related topics, Research Paper 293, The University of Calgary, 1975.

Farahat, H. K., and Higman, G., "The centres of symmetric group rings", *Proc. Roy. Soc. London Ser. A* **250** (1959), 212–221; ZB 84, p. 30; MR 21, #2697.

Farahat, H. K., Kerber, A., and Peel, M. H., "Modular representation theory of the symmetric groups", Research paper No. 131, The University of Calgary, 1971.

Farahat, H. K., Müller, W., and Peel, M. H., "The modular characters of the symmetric groups", *J. Algebra* **40** (1976), 354–363; MR 54, #392.

Farell, R. H., *Techniques of Multivariant Calculation*, Lecture Notes in Math., Vol. 520, Springer-Verlag, 1976, x+337 pp.

Farmer, K. B., "Representation of wreath products", Ph.D. thesis, The Univ. of Texas at Austin, 1974; Diss. Abstract Int. 35 (1974/75) 4033-B.

Farmer, K. B., "On projective representations of finite wreath products", in *Proc. Conf. Finite Groups*, Park City, Utah, 1975, pp. 357–363; ZB 342, #20003; MR 53, #8222.

Feit, W., "The degree formula for the skew-representations of the symmetric group", *Proc. Amer. Math. Soc.* **4** (1953), 740–744; MR 15, p. 287.

Festraets, A., "Sur les involutions du groupe alterné d'un ensemble fini", *Acad. Roy. Belg. Bull. Cl. Sci.* (*5*) **50** (1964), 287–293; MR 30, #3127.

Findlay, W., "The Sylow subgroups of the symmetric group", *Trans. Amer. Math. Soc.* **5** (1904), 263–278.

Finkelstein, H., "Solving equations in groups: A survey of Frobenius' theorem", *Period. Math. Hung.* **9** (1978), 187–204.

Finkelstein, H., and Mandelberg, K. I., "On solutions of 'equations in symmetric groups'", *J. Comb. Theory (A)* **25** (1978), 142–152; ZB 402, #20002.

Fischer, B., "A characterization of the symmetric groups on 4 and 5 letters", *J. Algebra* **3** (1966), 88–98; ZB 139, p. 250; MR 33, #2708.

Fischer, B., "Eine Kennzeichnung der symmetrischen Gruppen von Grade 6 and 7", *Math. Z.* **95** (1967), 288–298; ZB 139, p. 251; MR 34, #5909.

Foata, D., "Une propriété du vidage-remplissage des tableaux de Young", in *Combinatoire et representation du groupe symetrique* (Actes Table Ronde C.N.R.S. Univ. Louis-Pasteuer Strasbourg, Strasbourg, 1976), Lecture Notes in Math. Vol. 579, Springer, Berlin, 1977, pp. 121–135; ZB 398, #05012; MR 57, #6173.

Foata, D., "A matrix-analog for Viennot's construction of the Robinson correspondence", *Linear and Multilinear Algebra* **7** (1979), 281–298.

Foata, D., and Schützenberger, M., "Major index and inversion number of permutations", *Math. Nachrichten* **83** (1978), 143–159.

Foata, D., and Strehl, V., "Rearrangements of the symmetric group and enumerative properties of the tangent and secant numbers", *Math. Z.* **137** (1974), 257–264.

Fomin, A. N., "Über einfache Untergruppen symmetrischer Gruppen, die endliche Potenzen substituieren", *Algebra Logika* **11** (1972), 724–730; ZB 275, #20002.

Fomin, S. V., Finite, partially ordered sets and Young diagrams (Russian), Dokl. Akad. Nauk SSSR 243 (1978), **5** 1144–1147.

Foulkes, H. O., "Differential operators associated with $S$-functions", *J. London Math. Soc.* **24** (1949a), 136–143; ZB 37, p. 9; MR 11, p. 4.

Foulkes, H. O., "A note on $S$-functions", *Quart. J. Math (Oxford Ser.)* **20** (1949b), 150–152; ZB 33, p. 150.

Foulkes, H. O., "Concomitants of the quintic and sextic up to degree four in the coefficients of the ground form", *J. London Math. Soc.* **25** (1950), 205–209; ZB 37, p. 149.

Foulkes, H. O., "Modified bialternants and symmetric function identities", *J. Lond. Math. Soc.* **25** (1950), 268–275.

Foulkes, H. O., "The new multiplication of $S$-functions", *J. London Math. Soc.* **26** (1951), 132–139; ZB 42, p. 251; MR 12, p. 666.

Foulkes, H. O., "Monomial symmetric functions $S$-functions and group characters", *Proc. London Math. Soc. (3)* **2** (1952), 45–59; MR 13, p. 820.

Foulkes, H. O., "Matrix differentiation of $S$-functions", *Proc. Edinburgh Math. Soc. (2)* **10** (1953), 5–10; ZB 50, p. 15.

Foulkes, H. O., "Plethysm of $S$-functions", *Philos. Trans. Roy Soc. London Ser. A* **246** (1954), 555–591; MR 15, p. 926.

Foulkes, H. O., "The analysis of the characters of the Lie representation of the general linear group", *Proc. Amer. Math. Soc.* **10** (1959), 497–501; MR 21, #3497.

Foulkes, H. O., "On Redfield's group reduction functions", *Can. J. Math.* **15** (1963), 272–284; ZB 113, p. 256; MR 26, #6239.

Foulkes, H. O., "On Redfield's range-correspondences", *Can. J. Math.* **18** (1966), 1060–1971; ZB 149, p. 275; MR 34, #87.

Foulkes, H. O., "Linear graphs and $S$-functions", paper read to the Conf. on Comb. Math. and its Appl., Oxford, 1969.

Foulkes, H. O., "Group transitivity and a multiplicative function of a partition," *J. Combinat. Theory* **9** (1970), 261–266; ZB 213, p. 302; MR 43, #2061.

Foulkes, H. O., "A survey of some combinatorial aspects of symmetric functions", in *Permutations*, actes du colloque Paris, 1972, pp. 79–92.

Foulkes, H. O., "Characters of symmetric groups induced by characters of cyclic subgroups", in *Combinatorics* (Proc. Conf. Comb. Math., Math. Inst. Oxford, 1972), Inst. Math. Appl., Southend-on-Sea, 1972, pp. 141–154; MR 49, #7346.

Foulkes, H. O., "On a combinatorial lemma relating to standard tableaux", *J. Comb. Theory* (*A*) **17** (1974), 258.

Foulkes, H. O., "Paths in ordered structures of partitions", *Discrete Math.* **9** (1974), 365–374.

Foulkes, H. O., "Enumeration of permutations with prescribed up-down and inversion sequences", *Discrete Math.* **15** (1976), 235–252.

Foulkes, H. O., "Tangent and secant numbers and representations of symmetric groups", *Discrete Math.* **15** (1976), 311–324; ZB 342, #20005.

Foulkes, H. O., "Recurrences for characters of the symmetric group", *Discrete Math.* **21** (1978), 137–144; ZB 394, #20008.

Foulkes, H. O., "Eulerian numbers, Newcomb's problem and representations of symmetric groups", *Discrete Math.* **30** (1980), 3–49.

Fox, R. F., "A simple new method for calculating the characters of the symmetric groups", *J. Combinat. Theory* **2** (1967), 186–212; ZB 153, p. 38; MR 34, #7678.

Frame, J. S., "On the decomposition of transitive permutation groups generated by the symmetric group", *Proc. Nat. Acad. Sci. U.S.A.* **26** (1940), 132–139; ZB 23, p. 13; MR 1, p. 161.

Frame, J. S., "Orthogonal group matrices of hyperoctahedral groups", *Nagoya Math. J.* **27** (1966), 585–590; ZB 145, p. 29; MR 33, #5748.

Frame, J. S., and Robinson, G. de B., "On a theorem of Osima and Nagao", *Can. J. Math.* **6** (1954), 125–127; ZB 55, p. 255; MR 15, p. 682.

Frame, J. S., Robinson, G. de B., and Thrall, R. M., "The hook graphs of the symmetric group", *Can. J. Math.* **6** (1954), 316–324; MR 15, p. 931.

Freudenthal, H., "Une interprétation géométrique des automorphismes exterieurs du groupe symétrique $S_6$", *Rend. Sem. Mat. Fis. Milano* **42** (1972), 47–56; ZB 335, #20025; MR 49, #2908.

Frobenius, G., "Über die Congruenz nach einem aus zwei endlichen Gruppen gebildeten Doppelmodul", *Journal für die reine und angew. Mathematik* **101** (1887), 273–299.

Frobenius, G., "Über die Charaktere der symmetrischen Gruppe", *Berl. Ber.* **1900**, 516–534.

Frobenius, G., "Über die Charaktere der alternierenden Gruppe", *Berl. Ber.* **1901**, 303–315.

Frobenius, G., "Über die charakteristischen Einheiten der symmetrischen Gruppe", *Berl. Ber.* **1903**, 328–358.

Frobenius, G., "Über die Charaktere der mehrfach transitiven Gruppen", *Berl. Ber.* **1904**, 558–571.

Frobenius, G., "Über einen Fundamentalsatz der Gruppentheorie II", *Berl. Ber* **1907**, 428–437.

Frucht, R., "Kronen von Gruppen und ihre Untergruppen mit einer Anwendung auf Determinanten", *Revista Un. mat. Argentina* **8**, (1942) 42–69; Un. mat. Argentina Publ. No. 24, 1942, 30 pp.; ZB 61, p. 33.

Frucht, R., "The subgroups of the complete monomial groups of degree 2", *Univ. Nac. Tucumán. Revista A* **4** (1944), 47–54; MR 7, p. 5.

Gabriel, J. R., "On the construction of irreducible representations of the symmetric group", *Proc. Cambridge Phil. Soc.* **57** (1961), 330–340; ZB 96, p. 252; MR 22, #11053.

Gamba, A., "Sui caratteri delle rappresentazioni del gruppo simmetrico", *Atti. Acad. Naz. Lincei Rend Cl. Sci. Fis. Mat. Nat.* (*8*) **12** (1952), 167–169; MR 14, p. 16.

Gamba, A., and Radicati, L. A., "Sopra un teorema per la riduzione di talune rappresentazioni del gruppo simmetrico", *Atti. Acad. Naz. Lincei Rend. Cl. Sci. Fis. Mat. Nat.* (*8*) **14** (1953), 632–634; ZB 51, p. 259; MR 15, p. 504.

Gansner, E. R., "A characterization of permutations via skew-hooks", *J. Combinat. Theory* (*A*) **23** (1977), 176–179; ZB 363, #05014; MR 57, #9558.

Gard, P., "Symmetrized $n$-th powers of induced representations", *J. Phys. A* **6** (1973), 1807–1828; ZB 272, #20009.

Gard, P., and Backhouse, N. B., "The reduction of symmetrized powers of corepresentations of magnetic groups", *J. Physics A* **8** (1975), 450–458; ZB 299, #20004.

Gardiner, A., "Generators of period two for alternating groups", *Arch. der Math.* **24** (1973), 472–474; ZB 288, #20001.

Garnir, H., "Théorie de la représentation linéaire des groupes symétriques", *Mém. Soc. Roy. Sci. Liége* (*4*) **10**, No. 2 (1950), 5–100; ZB 45, p. 158; MR 12, p. 77.

Garnir, H. G., "Théorie de la représentation linéaire des groupes alternés", *Acad. Roy. Belg. Cl. Sci. Mém. Coll. in 8°* (*2*) **26** (1951), No. 1615, 22 pp.; MR 13, p. 722.

Garsia, A. M., and Gessel, I., "Permutation statistics and partitions", *Advances in Math.* **31** (1979), 288–305.

Garsia, A. M., and Remmel, J., "On the raising operators of Alfred Young", Relations between combinatorics and other parts of mathematics, Proc. Symp. Pure Math. Amer. Math. Soc., Columbus, Ohio, 1978, *Proc. Symp. Pure Math.* **34** (1979), 181–198; ZB 426, #20009.

Geissinger, L., and Kinch, D., "Representations of the hyperoctahedral groups", *J. Algebra* **53** (1978), 1–20; ZB 379, #20007; MR 58, #11092.

Gel'fand, A. V., "Defining relations of the symmetric and alternating groups that correspond to some of their factorizations into cyclic subgroups", *Perm. Gos. Uni. Ucen. Zap.*, No. 343 (1975), 3–28, 103; MR 57, #3223.

Gel'fand, S. I., "Representations of the full linear group over a finite field", *Math. USSR Sbornik* **12**, No. 1, (1970), 13–39.

Golomb, S. W., and Hales, A. W., "On enumerative equivalence of group elements", *J. Comb. Theory* **5** (1968), 308–312.

Gomi, K., "On maximal $p$-local subgroups of $S_n$ and $A_n$", *J. Fac. Sci. Univ. Tokyo Sect. I A* **19** (1972), 215–229.

Gould, M., and James, H. H., "Automorphism groups retracting onto symmetric groups", *Pacific J. Math.* **81** (1979), 93–100.

Grace, J. H., and Young, A., *The Algebra of Invariants*, Chelsea Publishing Company, Bronx, N.Y., 1903, vii+384 pp.; Reprinted 1978.

Grassl, R. M., and Hillman, A. P., "Functions on tableau frames", *Discrete Math.* **25** (1979), 245–255; ZB 402, #05006.

Gray, A. B., Jr., "Normal subgroups of monomial groups", *Pacific J. Math.* **12** (1962), 527–532; ZB 109, p. 21; MR 26, #2514.

Green, J. A., "The characters of the finite general linear groups", *Trans. Amer. Math. Soc.* **80** (1955), 402–447; MR 17, p. 345.

Green, J. A., "Les polynomes de Hall et les caractéres des groupes GL$(n, q)$", in *Colloque d'Algébre supérieure*, Bruxelles du 19 au 22 déc. 1956, Centre Belge Rech. Math., pp. 207–215; ZB 86, p. 25.

Green, J. A., *Polynomial Representations of* $GL_n$, Lecture Notes in Math. Vol. 830, Springer-Verlag, 1980.

Greene, C., Nijenhuis, A., and Wilf, H. S., "A probabilistic proof of a formula for the number of Young tableaux of a given shape", *Advances in Math.* **31** (1979), 104–109; ZB 398, #05008; MR 80b, #05016.

Guenoche, A., "Enumeration des tableaux Standards", *Discrete Math.* **25** (1979), 257–267; ZB 402, #05005.

Gündüzalp, Y., "Über die gewöhnlichen irreduziblen Charactere der symmetrischen Gruppe", *Mitt. Math. Sem. Giessen* **81** (1969), i+53 pp.; MR 39, #2891.

Gurevic, G. B., "Some properties of Young symmetrizers", in *Questions of Differential and Non-Euclidean Geometry*, Moskov, Gos. Ped. Inst., Moscow, 1972, pp. 119–145; MR 57, #6174.

Hadwiger, H., "Bemerkungen über eine spezielle Basis für die symmetrische und alternierende Gruppe", *Tohoku Math. J.* **49** (1942), 87–89; ZB 61, p. 31; MR 7, p. 371.

Halberstadt, E., "Sur certains sous-groupes des groupes symétriques", *C. R. Acad. Sci., Paris, Sér. A* **279** (1974), 733–735; ZB 295, #20003.

Halberstadt, E., "On certain maximal subgroups of symmetric or alternating groups", *Math. Z.* **151** (1976), 117–125; ZB 336, #20001.

Hall, P., "The algebra of partitions" in *Proc. Fourth Canad. Math. Congr. Banff 1957,* 1959, pp. 147–159; ZB 122, p. 34.

Halperin, I., "Odd and even permutations", *Canadian Math. Bull.* **3** (1960), 185–187.

Hamermesh, M., *Group Theory and Its Application to Physical Problems*, Addison-Wesley, 1962; MR 25, #132.

Hamernik, W., "Specht modules and the radical of the group ring over the symmetric group $S_p$", *Comm. Algebra* **4** (1976), 435–457; MR 53, #8210.

Hannabuss, K. C., "Symmetrized tensor products of induced representations", *J. Physics A* **9** (1976), 325–334; ZB 317, #22003.

Harary, F., and Palmer, E. M., *Graphical Enumeration*, Academic Press, New York–London, 1973, xiv+271 pp; ZB 266, #05108.

Held, D., "A characterization of the alternating groups of degrees eight and nine", *J. Algebra* **7** (1967), 218–237; MR 36, #1530.

Held, D., "Eine Kennzeichnung der Mathieu-Gruppe $M_{22}$ und der alternierenden Gruppe $A_{10}$", *J. Algebra* **8** (1968), 436–449; ZB 157, p. 355; MR 37, #5283.

Herman, J.E., and Chung, F. R. K., "Some results on hook lengths", *Discrete Math.* **20** (1977), 33–40.

Herrera, R. B., "The number of elements of given period in finite symmetric groups", *Amer. Math. Monthly* **64** (1957), 488–490.

Higman, G., "Representation of general linear groups and varieties of $p$-groups", in *Proc. Internat. Conf. Theory Groups*, Canberra, 1965, pp. 167–173; ZB 242, #20040.

Higman, G., "Construction of simple groups from character tables", in *Finite Simple Groups* (Proc. Instructional Conf. Oxford, 1969), Academic Press, London, 1971, pp. 205–214; MR 49, #7349.

Hillel, J., "Algebras of symmetry classes of tensors and their underlying permutation groups", *J. Algebra* **23** (1972), 215–227.

Hillel, J., "Dimensionality of algebras of symmetry classes of tensors", *Linear and Multilinear Algebra* **4** (1976), 41–44.

Hiller, H. L., "Flag manifolds and the hook formula", *Math. Z.* **170** (1980), 105–107; ZB 424, #05006.

Hillman, A. P., and Grassl, R. M., "Reverse plane partitions and tableau hook numbers", *J. Comb. Theory* **21** (1976), 216–221.

Hoffman, P., "Products of representations of wreath products", *Discrete Math.* **23**, (1978), 37–48; ZB 387, #20007.

Hoffman, P., *τ-Rings and Wreath Product Representations*, Lecture Notes in Math., Vol. 746, Springer, 1979.

Huffman, W. C., "Linear groups containing an element with an eigenspace of codimension two", *J. of Algebra* **34** (1975), 260–287; ZB 302, #20037; MR 53, #5762.

Hunter, D. B., "On a result in Young's quantitative substitutional analysis", *Proc. Edinburgh Math. Soc. (2)* **15** (1967), 257–262; ZB 178, p. 26; MR 37, #4179.

Hunter, D. B., "Analysis of the outer product of symmetric group representations", *BIT, Nordisk. Tidskr. Inform.-Behandl.* **10** (1970), 106–114; ZB 197, p. 303.

Huppert, B., *Endliche Gruppen I*, Die Grundlehren der Mathematischen Wissenschaften, Band 134, Springer-Verlag, Berlin–New York, 1967, xii+793 pp.; ZB 217, p. 72, MR 37 #302.

Huynh, M., "Modular invariant theory and cohomology algebras of symmetric groups", *J. Fac. Sci. Univ. Tokyo Sect. IA Math.* **22** (1975), 319–369; MR 54, #10440.

Ibrahim, E. M., "The plethysm of $S$-functions" *Quart. J. Math., Oxford Ser. (2)* **3** (1952), 50–55; ZB 46, p. 15; MR 14, p. 243.

Ibrahim, E. M., "On a theorem by Murnaghan", *Proc. Nat. Acad. Sci. U.S.A.* **40** (1954), 306–309.

Ibrahim, E. M., "Tables for the plethysm of $S$-function of degree 10 and 12", *Proc. Math. Phys. Soc. Egypt* **5**, No. 2 (1954), 85–86 (2 plates); MR 17, p. 1182.

Ibrahim, E. M., "On D. E. Littlewood's algebra of $S$-function", *Proc. Amer. Math. Soc.* **7** (1956), 199–202; MR 17, p. 1182.

Ibrahim, E. M., "Note on a paper of Murnaghan", *Proc. Amer. Math. Soc.* **7** (1956), 1000–1001.

Ibrahim, E. M., $S$-functional plethysms of degrees 14 and 15, *Proc. Math. Phys. Soc. Egypt* **22**, (1959), 137–142; ZB 88, p. 27; MR 21, #7259.

Ibrahim, E. M., "A formula for $\{\lambda\}\otimes\{\mu\}$", *Proc. Math. Phys. Soc. U.A.R.* **30**, (1966), 73–79; ZB 325, #20045; MR 41, #6998.

Ibrahim, E. M., "On the Kronecker product of the irreducible representation of the symmetric group", *J. Natur. Sci. and Math.* **8** (1968), 243–250; ZB 213, p. 36; MR 38, #5949.

Ibrahim, E. M., "The symmetric and skew-symmetric concomitants of 2 ground forms of type $\{\lambda\}$", *Proc. Math. Phys. Soc, U.A.R. (Egypt)* **32** (1968), 71–73; ZB 214, p. 50; MR 44, #1745.

Ibrahim, E. M., "The inner product of $S$-functions", *Časopis Pěst. Math.* **95** (1970), 360–366; ZB 213, p. 36; MR 43, #7523.

Ibrahim, E. M., "A note on the principal parts of the products of $S$-functions", *Proc. Math. Phys. Soc. A.R.E.* **34** (1970), 41–42; ZB 327, #15014; MR 50, #10047.

Ingram, R. E., "Some characters of the symmetric group", *Proc. Amer. Math. Soc.* **1** (1950), 358–369; MR 12, p. 157.

Isaacs, I. M., "Systems of equations and generalized characters in groups", *Can. J. Math.* **22** (1970) 1040–1046.

Ito, N., "A theorem on the alternating group $A_n$ ($n \geqslant 5$)", *Math. Japonicae* **2** (1951), 59–60; ZB 44, p. 15; MR 13, p. 621.

Jacobsthal, E., "Sur le nombre d'elements du groupe symétrique $S_n$ dont l'ordre est un nombre premier", *Norske Vid. Selsk. Forh.* **21**, No. 12 (1949), 49–51; ZB 40, p. 151.

Jacques, A., "Nombre de cycles d'une permutation et caractères du groupe symétrique", in *Permutations* (Acte Colloq. Univ. René-Descartes, Paris, 1972), Gauthier-Villars, Paris, 1974, pp. 93–96; MR 50, #4713.

Jahn, H. A., "Hassitt-type Young operator expansions I: An orthogonal transformation between (a) the Young operators of the symmetric group $S_{A+1}$ and (b) the two-sided products of the Young operators of $S_A$ with the transposition $P_{A,A+1}$", *Philos. Trans. Roy. Soc. London Ser. A* **253** (1960), 27–53; ZB 94, p. 14.

James, G. D., "The irreducible representations of the symmetric groups", *Bull. London Math. Soc.* **8** (1976a), 229–232; MR 54, #5329.

James, G. D., "Representations of the symmetric groups over the field of order 2", *J. Algebra* **38** (1976b), 280–308; MR 53, #595.

James, G. D., "On the decomposition matrices of the symmetric groups I", *J. Algebra* **43** (1976c), 42–44.; MR 55, #3057a .

James, G. D., "On the decomposition matrices of the symmetric group II", *J. Algebra* **43** (1976d), 45–54; MR 55, #3057b.

James, G. D., "A characteristic-free approach to the representation theory of $\mathfrak{S}_n$", *J. Algebra* **46** (1977a), 430–450; MR 55, #12805.

James, G. D., "The module orthogonal to the Specht module", *J. Algebra* **46** (1977b), 451–456; MR 56, #460.

James, G. D., "Some counterexamples in the theory of Specht modules", *J. Algebra* **46** (1977c), 457–461; MR 56, #461.

James, G. D., "A note on the decomposition matrices of $S_{12}$ and $S_{13}$ for the prime 3", *J. Algebra* **53** (1978a), 410–411; MR 80a, #20005.

James, G. D., "A note on the *T*-ideal generated by $s_3[x_1, x_2, x_3]$", *Israel J. Math.* **29** (1978b), 105–112; MR 58, #10869.

James, G. D., "Some combinatorial results involving Young diagrams", *Math. Proc. Cambridge Philos. Soc.* **83**, No. 1 (1978c), 1–10; MR 57, #3233.

James, G. D., "On a conjecture of Carter concerning irreducible Specht modules", *Math. Proc. Cambridge Philos. Soc.* **83** (1978d), 11–17; MR 57, #3234.

James, G. D., *The Representation Theory of the Symmetric Groups*, Lecture Notes in Math., Vol. 682, Springer-Verlag, 1978e, v + 156pp.; MR 80g, #20019.

James, G. D., "The decomposition of tensors over fields of prime characteristic", *Math Z.* **172** (1980), 161–178.

James, G. D., and Murphy, G. E., "The determinant of the Gram matrix for a Specht module", *J. Algebra* **59** (1979), 222–235.

James, G. D., and Peel, M. H., "Specht series for skew representations of symmetric groups", *J. Algebra* **56** (1979), 343–364; ZB 398, #20016; MR 80h, #20021.

Johnson, D., "Representations of the symmetric group", Thesis, University of Toronto, 1958.

Johnson, N. W., "A geometric model for the generalized symmetric group", *Canad. Math. Bull.* **3** (1960), 133–142; MR 23, #1733.

Jones, R. H., "Modular spin representations of the symmetric group", *Proc. Cambridge Philos. Soc.* **69** (1971), 365–372.

Jordan, P., "Der Zusammenhang der symmetrischen und linearen Gruppen und das Mehrkörperproblem", *Z. f. Physik* **94** (1935), 531–535.

Jucys, A. A., "On the Young operators of symmetric groups", *Litovsk. Fiz. Sb.* **6** (1966), 163–180; MR 34, #2725.

Jucys, A. A., "Representations of the symmetric groups and the Clebsch-Gordan coefficients of unitary groups", *Litovsk. Mat. Sb.* **8** (1968), 597–609; ZB 214, p. 47.

Jucys, A. A., "Symmetric polynomials and the center of the symmetric group ring", *Rep. Math. Phys.* **5** (1974), 107–112; ZB 288, #20014; MR 54, #7597.

Jucys, A. A., "The number of symmetric non-negative integral matrices with prescribed row sums", *Litovsk. Mat. Sb.* **17** (1977), 205–208; ZB 361, #05006.

Jucys, A. A., "Tournaments and generalized Young tableaux" Mat. Zametki **27** (1980), 353–359.

Kahn, D. S., and Priddy, S. B., "On the transfer in the homology of symmetric groups", *Math. Proc. Cambridge Philos. Soc.* **83** (1978), 91–101.

Kaloujnine, L., "Sur les *p*-groupes de Sylow du groupe symetrique du degré $p^m$", *C. R. Acad. Sci. Paris* **221** (1945), 222–224; ZB 61, p. 33; MR 7, p. 239.

Kaloujnine, L., "Sur les $p$-groupes de Sylow du groupe symétrique du degré $p^m$. (Suite antrale ascendante et descendante.)", *C. R. Acad. Sci. Paris* **223** (1946), 703–705; MR 8, p. 251.

Kaloujnine, L., "La structure du $p$-groupe de Sylow du groupe symétrique du degré $p^2$", *C. R. Acad. Sci. Paris* **222** (1946), 1424–1425; ZB 61, p. 33; MR 8, p. 13.

Kaloujnine, L., "Sur les $p$-groupes de Sylow du groupe symétrique de degré $p^m$. (Sousgroupes caractéristiques, sousgroupes parallelotopiques)", *C. R. Acad. Sci. Paris* **224** (1947), 253–255; ZB 30, p. 106; MR 8, p. 367.

Kaloujnine, L., "La structure des $p$-groupes de Sylow des groupes symétriques finis", *Ann. Sci. Ecole Norm. Sup.* (*3*) **65** (1948), 239–276; ZB 34, p. 305; MR 10, p. 505.

Kaloujnine, L., "Über eine Verallgemeinerung der $p$-Sylowgruppen symmetrischer Gruppen", *Acta Math. Acad. Sci. Hungar.* **2** (1951), 197–221; ZB 44, p. 256; MR 14, p. 617.

Kaloujnine, L., "Sur la structure de $p$-groupe de Sylow des groups symétriques finis et de quelques generalisations infinies des ces groupes.", Seminaire Bourbaki 1948/1949, No. 5, 1959, 3 pp.; ZB 100, p. 257.

Kaloujnine, L., and Klin, M. H., "Certain maximal subgroups of symmetric and alternating groups", *Math. Sb.* (*N.S.*) **87**, No. 129 (1972), 95–121; ZB 238, #20009; MR 45, #6906.

Kaloujnine, L., and Klin, M. H., "Some numerical invariants of permutation groups", *Latviisk. Mat. Ezegodnik Vyp.* **18** (1976), 81–99, 222; MR 57, #3221.

Kaloujnine, L., and Krasner, M., "Le produit complet des groupes de permutations et le probleme d'extension des groupes", *C. R. Acad. Sci. Paris* **227** (1948), 806–808; ZB 38, p. 162; MR 10, p. 351.

Kazhdan, D., and Lusztig, G., "Representations of Coxeter groups and Hecke algebras", *Inventiones Math.* **53** (1979), 165–184.

Kemer, A. R., "$T$-Ideals with power growth of the codimensions are Specht", *Siber. Math. J.* **19** (1978), 37–48.

Kerber, A., "Zur modularen Darstellungstheorie symmetrischer und alternierender Gruppen", *Mitt. Math. Sem. Giessen* **68** (1966), iii+80 pp.; ZB 139, p. 251; MR 33, #5753.

Kerber, A., "Zur Darstellungstheorie von Kranzprodukten", *Can. J. Math.* **20** (1968a), 665–672; ZB 157, p. 65; MR 38, #2223.

Kerber, A., "Zur modularen Darstellungstheorie symmetrischer und alternierender Gruppen II", *Archiv Math.* **19** (1968b), 588–594; ZB 175, p. 20; MR 39, #5725.

Kerber, A., "Zur Darstellungstheorie von Symmetrien symmetrischer Gruppen", *Mitt. Math. Sem. Giessen* **80** (1969a), 1–27; ZB 167, p. 299; MR 40, #4385.

Kerber, A., "On a paper of Fox about a method for calculating the ordinary irreducible characters of symmetric groups", *J. Combinatorial Theory* **6** (1969b), 90–93; ZB 165, p. 342; MR 38, #4579.

Kerber, A., "Zur Theorie der $M$-Gruppen", *Math. Z.* **115** (1970a), 4–6; ZB 186, p. 317; MR 41, #3628.

Kerber, A., "Zu einer Arbeit von J. L. Berggren über ambivalente Gruppen", *Pacific J. Math.* **33** (1970b), 669–675; ZB 182, p. 359; MR 42, #3195.

Kerber, A., "Zur modularen Darstellungstheorie symmetrischer und alternierender Gruppen III", *Math. J. Okayama* **15** (1971a), 25–33; ZB 241, #20012.

Kerber, A., *Representations of Permutation Groups I*, Lecture Notes in Math., Vol. 240, Springer-Verlag, 1971b; ZB 232, #20014; MR 48, #4098.

Kerber, A., "Der Zykelindex der Exponentialgruppe", *Mitt. Math. Sem. Giessen* **98** (1973), 5–20; ZB 258, #20045.

Kerber, A., "Symmetrization of representations", in *Proc. 2nd. Ind. Coll. on Group Theoretical Methods in Physics*, June 25–29, 1973, University of Nijmegen, pp. B146–151.

Kerber, A., "On the representation theory of finite permutation groups", in *Lecture Notes on the Representation Theory of Finite Groups* (Sympos. Univ. Istanbul, Istanbul, 1973), Math. Res. Inst., Univ. Istanbul, Istanbul, 1974, pp. 29–53; ZB 289, #20007; MR 52, #3308.

Kerber, A., *Representations of Permutation Groups II*, Lecture Notes in Math., Vol. 495, Springer-Verlag, 1975a; ZB 318, #20005; MR 53, #13376.

Kerber, A., "Characters of wreath products and some applications to representation theory and combinatorics", *Discrete Math.* **13** (1975b), 13–30.

Kerber, A., "The enumeration theories of Redfield, Pólya and de Bruijn and some related topics", Workshop on Symm. Groups and Related Topics, July–Aug. 1975, Part I, Univ. of Calgary, Research Paper #290, (1975c).

Kerber, A., "On graphs and their enumeration I", *MATCH* **1** (1975d), 5–10.

Kerber, A., "On graphs and their enumeration II", *MATCH* **2** (1976), 17–34; ZB 403, #05051.

Kerber, A., "A matrix of combinatorial numbers related to the symmetric groups", *Discrete Math.* **22** (1978), 319–321; ZB 377, #20004.

Kerber, A., "Counting isomers and such." The permutation group in physics and chemistry, Symp. Bielefeld 1978, Lect. Notes Geom. **12** (1979), 1–18; ZB 426, #05031.

Kerber, A., "On certain connections between the representation theory of the symmetric group and the representation theory of an arbitrary finite group", in *Lecture Notes Pure Appl. Math.*, Vol. 57, M. Dekkcr, 1980, pp. 137–152.

Kerber, A., and Lehmann, W., "On graphs and their enumeration III", *MATCH* **3** (1977), 67–86.

Kerber, A., and Peel, M. H., "On the decomposition numbers of symmetric and alternating groups", *Mitt. Math. Sem. Univ. Giessen* **91** (1971), 45–81; ZB 219, #20007; MR 45, #5206.

Kerber, A., and Tappe, J., "On permutation characters of wreath products", *Discrete Math.* **15** (1976), 151–161; ZB 335, #20021; MR 53, #13373.

Kerber, A., and Wagner, B., "Gleichungen in endlichen Gruppen", *Archiv d. Math.* **35** (1980), 252–262.

Kervaire, M., "Operations d'Adams on theorie des representations lineaires des groupes finis", *Enseignement Math., II. Ser.* **22** (1976), 1–28; ZB 335, #20003; MR 54, #5325.

King, R. C., "The dimensions of irreducible representations of linear groups", *Can. J. Math.* **22** (1970), 436–448; MR 41, #5517.

King, R. C., "Generalized Young tableaux and the general linear groups", *J. Math. Phys.* **11** (1970), 280–293; ZB 199, p. 346.

King, R. C., "Branching rules for $\mathrm{GL}(N) \supset \Sigma_m$ and the evaluation of inner plethysms", *J. Math. Phys.* **15** (1974), 258–267.

King, R. C., "Generalised Vandermonde determinants and Schur functions", *Proc. Amer. Math. Soc.* **48** (1975), 53–56.

King, R. C., and Plunkett, S. P. O., "The evaluation of weight multiplicities using characters and $S$-functions", *J. Phys. A: Math. Gen.* **9** (1976), 863–887.

Klaiber, B., "Fortsetzbarkeit und Korrespondenz von Darstellungen", Dissertation, Mainz, 1969.

Klass, M. J., "A generalization of Burnside's combinatorial lemma", *J. Comb. Theory (A)* **20** (1976), 273–278.

Klein, T., "The multiplication of Schur-functions and extensions of $p$-modules", *J. London Math. Soc.* **43** (1968), 280–284.

Klein, T., "The Hall polynomial", *J. Algebra* **12** (1969), 61–78; MR 38, #4557.

Klemm, M., "Permutationscharaktere und irreduzible Charaktere der symmetrischen Gruppen", *Mitt. Math. Sem. Giessen* **121** (1976a), 59–64; ZB 361, #20016; MR 58, #5888.

Klemm, M., "Über Charaktere mehrfach transitiver Gruppen", *Archiv d. Math.* **27** (1976b), 241–248; ZB 328, #20002.

Klemm, M., "Charaktere mehrfach transitiver Permutationsgruppen", in *Combinatoire et représentation du groupe symétrique* (Actes Table Ronde C.N.R.S. Univ. Louis-Pasteur Strasbourg, Strasbourg, 1976), Lecture Notes in Math., Vol. 579, Springer, Berlin, 1977a, pp. 281–286; MR 57, #9819.

Klemm, M., "Tensorprodukte von Charakteren der symmetrischen Gruppe", *Archiv d. Math.* **28** (1977b), 455–459; ZB 362, #20008; MR 56, #15753.

Klemm, M., and Wagner, B., "A matrix of combinatorial numbers related to the symmetric groups", *Discrete Math.* **28** (1979), 173–177.

Klevacek, V. I., "Über Kriterien der Vollständigkeit in einer symmetrischen Gruppe endlichen Grades", *Kibernetika (Kiev)* **1975**, Nos. 22–25; ZB 314, #20004.

Knörr, R., "Uniserial blocks of symmetric and alternating groups", *Arch. d. Math.* **27** (1976), 584–587.

Knuth, D. E., "Permutations, matrices and generalized Young tableaux", *Pacific J. Math.* **34** (1970), 709–727; ZB 185, p. 32.

Knutson, D., *λ-Rings and the Representation Theory of the Symmetric Group*, Lecture Notes in Math., Vol. 308, Springer-Verlag, 1973; ZB 272, #20008; MR 51, #679.

Kodama, T., and Koichi, Y., "Some properties of characters of the symmetric groups", *Mem. Fac. Sci. Kyusyu Univ. Ser. A* **12** (1958), 104–112; MR 21, #3498.

Kondo, K., "Table of characters of the symmetric group of degree 14", *Proc. Phys. Math. Soc. Japan (3)* **22** (1940), 585–593.

Kondo, K., "Über die Zerlegung der Charaktere der alternierenden Gruppe", *Proc. Imp. Acad. Tokyo* **16** (1940), 131–135; ZB 23, p. 209; MR 2, p. 3.

Kondo, T., "On the alternating groups I", *J. Fac. Sci. Univ. Tokyo* **15** (1968), 87–97; ZB 196, p. 47; MR 39, #1534.

Kondo, T., "A characterization of the alternating group of degree eleven", *Illinois J. Math.* **13** (1969), 528–541; ZB 194, p. 38; MR 40, #225.

Kondo, T., "On the alternating groups II", *J. Math. Soc. Japan* **21** (1969), 116–139; ZB 196, p. 47; MR 39, #2856.

Kondo, T., "On the alternating groups III", *J. of Algebra* **14** (1970), 35–69; ZB 205, p. 35; MR 41, #5478.

Kondo, T., "On the alternating groups IV", *J. Math. Soc. Japan* **23** (1971), 527–547; MR 44, #4085.

König, J., *Über die alternierende Gruppe*, Buda-Pest, Akademisitry von Februar 1884.

Kostant, B., "A theorem of Frobenius, a theorem of Amitsur-Levitski and cohomology theory", *J. Math. Mech.* **7** (1958), 237–264; ZB 87, p. 257; MR 19, p. 1153.

Kostant, B., "Groups over $\mathbb{Z}$", in *Proceedings of Symposia in Pure Mathematics, Vol. IX (Boulder 1965), Algebraic Groups and Discontinuous Subgroups*, Amer. Math. Soc., 1966, pp. 90–98.

Kraljevic, H., "Irreducibility of the principal series representations of the group $GL(n, R)$", *Glasnik Matematicki Ser. III* **7** (1972), 49–51.

Kramer, P., "Recoupling coefficients of the symmetric group for shell and cluster model configurations", *Z. f. Physik* **216** (1968), 68–83.

Krasner, M., "Une nouvelle presentation de la theorie des groupes de permutations et ses applications a la theorie de Galois et de produit d'entrelacement ('wreath product') de groupes", *Math. Balkanica* **3** (1973), 229–280; MR 51, #3273.

Kreweras, G., "Classification des permutations suivant certaines propriétés ordinales de leur représentation place", in *Permutations*, actes du colloque Paris, 1972, pp. 97–115.

Krull, W., "Die $\langle p'_1, \ldots, p'_r \rangle$-Sylowgruppen der symmetrischen Gruppe $\mathfrak{S}^{(n)}$", *Arch. Math.* **20** (1969), 453–458; ZB 202, p. 23; MR 41, #5492.

Kuyk, W., "Certain representations of the wreath product and of a certain type of its subgroups", Math. Centrum Amsterdam Afd. Zuivere Wisk., ZW-013, 1964, 10 pp.; ZB 138, p. 23; MR 35, 1, #2969.

Lam, T. V., "Young diagrams, Schur functions, the Gale-Ryser theorem and a conjecture of Snapper", *J. Pure Appl. Algebra* **10** (1977/78), 81–94; MR 57, #12671.

Lascoux, A., "Tableaux de Young gauches", Semin. Théor. Nombres 1974–1975, Univ. Bordeaux Exposé No. 4, 8 pp.; ZB 321, #05024.

Lascoux, A., "Calcul de Schur et extensions grassmanniennes des λ-anneaux", in *Combinatoire et representation du groupe symetrique* (Actes Table Ronde CNRS, Univ. Louis-Pasteur Strasbourg, Strasbourg 1976), Lecture Notes in Math., Vol. 579, Springer-Verlag, Berlin, 1977, pp. 182–216; MR 58, #268.

Lascoux, A., "Produit de Kronecker des représentations du groupe symetrique", in *Sém. d'Algèbre Paul Dubreil et Marie-Paule Malliavin*, *Proc.* (Paris, 1979), Lecture Notes in Math., Vol. 795, Springer-Verlag, 1980, pp. 319–329.

Lascoux, A. and Schützenberger, M. -P., "Sur une conjecture de H. O. Foulkes", *C. R. Acad. Sci. Paris Ser. A-B* **286**, No. 7 (1978), A323–A324; MR 57, #12672.

Lascoux, A. and Schützenberger M. P., "Croissance des polynomes de Foulkes-Green", *C. R. Acad. Sci., Paris, Ser. A* **288**, 95–98 (1979); ZB 398, #20052.

Lehmann, W., "Ein vereinheitlichender Ansatz für die Redfield–Pólya–de Bruijnsche Abzähltheorie", Dissertation, Aachen, 1976.

Leron, U., "Young subgroups and polynomial identities", *J. Algebra* **49** (1977), 304–314 ZB 371, #20016; MR 57, #6151.

Leron, U., and Moran, G., "The full cycles in a Young coset", *J. Comb. Theory (A)* **27** (1979), 218–221.

Lesieur, L. "Sur la multiplication des fonctions caracteristiques de Schur", *C. R. Acad. Sci. Paris* **225** (1947), 848–850; MR 9, p. 268.

Lévy-Bruhl, J., "Opérateurs d'antisymétriseurs", Algébre et Théorie des Nombres, Sém. P. Dubreil, M.-L. Dubreil-Jacotin et C. Pisot **13** (1959/60), No. 11, 20 pp. (1961); ZB 114, p. 20.

Lewin, M., "Generating the alternating group by cyclic triples", *Discrete Math.* **11** (1975), 187–189; ZB 299, #05004.

Liber, A. E., "On symmetric generalized groups", *Mat. Sbornik (N. S.)* **33 (75)**, 531–544 (1953); ZB 52, p. 17; MR 15, p. 502.

Liebeck, H., "Even and odd permutations", *Amer. Math. Monthly* **76** (1969), 668.

Liebler, R. A., and Vitale, M. R., "Ordering the partition characters of the symmetric group", *J. Algebra* **25** (1973), 487–489; ZB 274, #20016; MR 47, #5091.

Littlewood, D. E., "Group characters and the structure of groups", *Proc. London Math. Soc. (2)* **39** (1935), 150–199; ZB 11, p. 250.

Littlewood, D. E., "Some properties of *S*-functions", *Proc. London Math. Soc. (2)* **40** (1936a), 49–70.

Littlewood, D. E., "On induced and compound matrices", *Proc. London Math. Soc. (2)* **40** (1936b), 370–381.

Littlewood, D. E., "Polynomial concomitants and invariant matrices", *J. London Math. Soc.* **11** (1936c), 49–55.

Littlewood, D. E., "The construction of invariant matrices", *Proc. London Math. Soc.* (2) **43** (1937) 226–240; ZB 16, p. 394.

Littlewood, D. E., *The Theory of Group Characters and Matrix Representations of Groups*, Oxford University Press, 1940; MR 2, p. 3.

Littlewood, D. E., "On invariant theory under restricted groups", *Trans. London Roy. Soc. A* **239** (1943), 387–417.

Littlewood, D. E., "Invariant theory, tensors and group characters", *Philos. Trans. Roy. Soc. London Ser. A* **239** (1944), 305–365; MR 6, p. 41.

Littlewood, D. E., *A University Algebra*, W. Heinemann Ltd., Melbourne, 1950.

Littlewood, D. E., "Modular representations of symmetric groups", *Proc. Roy. Soc. London Ser. A* **209** (1951), 333–353; ZB 44, p. 257; MR 14, p. 243.

Littlewood, D. E., "On the Poincaré polynomials of the classical groups", *J. London Math. Soc.* **28** (1953), 494–500; ZB 51, p. 18; MR 15, p. 198.

Littlewood, D. E., "The Kronecker product of symmetric group representations", *J. London Math. Soc.* **31** (1956), 89–93; MR 17, p. 583.

Littlewood, D. E., "The characters and representations of imprimitive groups", *Proc. London Math. Soc.* (3) **6** (1956), 251–266; MR 17, p. 1182.

Littlewood, D. E., "Plethysm and the inner product of $S$-functions", *J. London Math. Soc.* **32** (1957), 18–22; MR 18, p. 640.

Littlewood, D. E., "The inner plethysm of $S$-functions", *Canad. J. Math.* **10** (1958), 1–16; ZB 79, p. 36; MR 20, #1714.

Littlewood, D. E., "Products and plethysms of characters with orthogonal symplectic and symmetric groups", *Canad. J. Math.* **10** (1958), 17–32; ZB 79, p. 36; MR 20, #1715.

Littlewood, D. E., "On certain symmetric functions", *Proc. London Math. Soc.* (3) **11** (1961), 485–498; MR 24 #173.

Littlewood, D. E., and Richardson, A. R., "Immanants of some special matrices", *Quart. J. Math.* (*Oxford Series*) **5** (1934), 269–282; ZB 10, p. 252.

Littlewood, D. E., and Richardson, A. R., "Group characters and algebra", *Philos. Trans. Roy. Soc. London A* **233** (1934), 99–142; ZB 9, p. 202.

Littlewood, D. E., and Richardson, A. R., "Some special $S$-functions and $q$-series", *Quart. J. Math.* (*Oxford series*) **8** (1935), 184–198.

Livingstone, D., "Proof of a theorem discovered by Murnaghan", *Proc. Nat. Acad. Sci. U.S.A.* **43** (1957), 618–619; MR 20, #1713.

Livingstone, D. and Wagner, A., "Transitivity of finite permutation groups on unordered sets", *Math. Z.* **90** (1965), 393–403.

Lorimer, P. J., "The outer automorphisms of $S_6$", *Amer. Math. Monthly* **73** (1966), 642–643; ZB 136, p. 282.

Lugowski, H., "Gruppenring und Kommutatoralgebra nach H. Weyl", *Wis. Z. Päd. Hochsch. Potsdam Math. -Natur.* **R5** (1960), 207–213; MR 24, #1931.

Lusztig, G. "On the discrete representations of the general linear groups over a finite field", *Bull. Amer. Math. Soc.* **79** (1973), 550–554.

Lusztig, G., *The Discrete Series of* $GL_n$ *over a Finite Field*, Annals of Mathematics Studies, No. 81, Princeton University Press, 1974, v+99 pp.

Lusztig, G., "On the Green polynomials of classical groups", *Proc. London Math. Soc.* (3) **33** (1976), 443–475; MR 54, #12917.

Lusztig, G., "Irreducible representations of finite classical groups", *Inventiones Math.* **43** (1977), 125–175.

MacAogain, E., "Decomposition matrices of symmetric and alternating Groups", Trinity College, Dublin, Research Notes TCD 1976-10.

Macdonald, I. G., "On the degrees of the irreducible representations of symmetric groups", *Bull. London Math. Soc.* **3** (1971), 189–192; ZB 219, #20008; MR 44, #6865.

Macdonald, I. G., "On the degrees of the irreducible representations of finite Coxeter groups", *J. London Math. Soc.* **6** (1973), 298–300.

Macdonald, I. G., *Symmetric Functions and Hall Polynomials*, Oxford University Press, 1979.

Macdonald, I. G., "Polynomial functors and wreath products", *J. Pure Appl. Algebra* **18** (1980), 173–204.

Mackey, G. W., "Symmetric and antisymmetric Kronecker squares and intertwinding numbers of induced representations of finite groups", *Amer. J. Math.* **75** (1953), 387–405.

MacMahon, P. A., *Combinatorial Analysis*, Vols. 1,2, Cambridge University Press, Cambridge, 1915; reprinted, Chelsea, New York, 1960.

Maillet, E., "Sur les isomorphes holoédriques et transitifs des groupes symetriques on alternés, *J. Math. Pures Appl. Ser. (2)* **1** (1895), 5–34.

Makar, R. H., "The irreducible representation of the symmetric groups of degrees 3, 4, and 5" *Proc. Math. Phys. Soc. Egypt.* **3** (1948), 13–21; MR 11, p. 77.

Makar, R. H., "On the analysis of the Kronecker product of irreducible representations of the symmetric group" *Proc. Edinb. Math. Soc. (2)* **8** (1949), 133–137; ZB 37, p. 152; MR 12, p. 10.

Makar, R. H., "On the analysis of the representations of the linear group of dimension 2", *Nederl. Akad. Wet., Proc. Ser. A* **61** (1958), 475–479; ZB 83, p. 20.

Makar, R. H., and Missiha, S. A., "The coefficient of the $S$-function $\{nm-k-r, k, r\}$, $k \leqslant m$, in the analysis of $\{m\}\otimes\{\nu\}$, where $(\nu)$ is any partition of $n$, and $n=5$ or 6. I, II", *Nederl. Akad. Wet. Proc. Ser. A* **61**; *Indag. Math.* **20** (1958), 77–93; ZB 80, p. 22; MR 20, #3219.

Malzan, J., "Real finite groups", Thesis, University of Toronto, 1969.

Malzan, J., "Vertex subgroups of irreducible representations of solvable groups", *Canad. J. Math.* **23** (1971), 12–21; ZB 218, #20040; MR 42, #4647.

Malzan, J., "Matrix groups of the second kind", *Illinois J. Math.* **16** (1972), 154–157.

Mann, B. M., "The cohomology of the symmetric groups", *Trans. Amer. Math. Soc.* **242** (1978), 157–184.

Marcus, M., "The use of multilinear algebra for proving matrix inequalities", in *Proc. of Conference on Matrix Theory* (ed. by H. Schneider), University of Wisconsin Press, Madison, 1964.

Marcus, M., *Finite Dimensional Multilinear Algebra*, Parts I, II, Marcel Dekker, New York, 1973, 1975.

Marcus, M., and Andresen, P., "The numerical radius of exterior powers", *Linear Algebra and its Appl.* **16** (1977), 131–151.

Marcus, M. and Katz, S. M., "Matrices of Schur functions", *Duke Math. J.* **36** (1969), 343–352.

Marzec, W., "Die Charaktere symmetrischer und antisymmetr. $n$-ter Potenzen einer Gruppendarst.", *Zeszyty nank Politechn. Poznań* **12**, Mat. 1, 43–60 (1962) (English summary).

Matthews, J. A. J., and Robinson, G. de B., "On the dimension of an irreducible tensor representation of the general linear group GL($d$)" *Canad. Math. Bull.* **13** (1970), 389–390; ZB 226, #20010.

Mayer, S. J., "On the irreducible characters of the Weyl groups", Dissertation, University of Warwick, 1971.

Mayer, S. J., "On the irreducible characters of the symmetric group", *Advances in Math.* **15** (1975a), 127–132; ZB 351, #20005.

Mayer, S. J., "On the characters of the Weyl group of type $D$", *Math. Proc. Camb. Philos. Soc.* **77** (1975b), 259–264; ZB 296, #20003; MR 51, #733.

Mayer, S. J., "On the characters of the Weyl group of type $C$", *J. Algebra* **33** (1975c), 59–67; ZB 296, #20004.

McConnell, J., "Reduction of representation of the general linear group using Lie algebras", *SIAM J. Appl. Math.* **25** (1973), 287–299; ZB 267, #20034.

McConnell, J., "Note on multiplication theorems for Schur functions", in *Combinatoire et representation du groupe symétrique* (Actes Table Ronde *C.N.R.S.*, Univ. Louis-Pasteur Strasbourg, Strasbourg; 1976), Lecture Notes in Math., Vol. 579, Springer, Berlin, 1977, pp. 252–257; MR 57, #6175.

McConnell, J. R., and Newell, M. J., "Expansion of symmetric products in series of Schur functions", *Proc. Roy. Irish Acad., Sect. A* **73** (1973), 255–274; ZB 257, #05004.

McIntosh, H. V., "Symmetry-adapted functions belonging to the symmetric groups", *J. Math Phys.* **1** (1960), 453–460; ZB 100, p. 258.

McKay, J., "Irreducible representations of odd degree", *J. Algebra* **20** (1972), 416–418; MR 44, #4111.

McKay, J., "The largest degrees of irreducible characters of the symmetric group", *Math. Comput.* **30** (1976), 624–631; MR 53, #8216.

Mead, C. A., *Permutation Groups, Symmetry and Chirality in Molecules*, Topics in Current Chemistry, Vol. 49, Springer-Verlag, 1974.

Mead, A., Schönhofer, A., and Ruch, E., "Theory of chirality functions, generalized for molecules with chiral ligands", *Theoret. Chim. Acta* **29** (1973), 269–304.

Meier, N., and Tappe, J., "Ein neuer Beweis der Nakayama-Vermutung über die Blockstruktur symmetrischer Gruppen", *Bull. London Math. Soc.* **8** (1976), 34–37; ZB 332, #20007; MR 53, #10907.

Men, B. A. Cherepanov, V. L., and Men, A. N., "Group theoretical methods for determining permitted terms of the electronic states of complexes in crystals. The use of plethysm for classification of the permitted terms of impurity complexes in crystal", *International J. Quantum Chem.* **7** (1973), 739–743.

Merris, R., "Inequalities for matrix functions", *J. Algebra* **22** (1972), 451–460.

Merris, R., "On characters of subgroups", *J. Res. Nat. Bur. Standards, Sect. B* **78** (1974), 35–38; ZB 282, #20004.

Merris, R., "Two problems involving Schur functions", *Linear Algebra and Its Appl.* **10** (1975), 155–162.

Merris, R., "The irreducibility of $K(A)$", *Linear and Multilinear Algebra* **2** (1975), 299–303.

Merris, R., "Relations among generalized matrix functions", *Pacific J. Math.* **62** (1976), 153–161.

Merris, R., "The Kronecker power of a permutation", *J. Res. Nat. Bur. Standards Sect. B* **80** (1976), 265–268; ZB 352, #20011; MR 55, #476.

Merris, R., "Nonzero decomposable symmetrized tensors", *Linear Algebra and Its Appl.* **17** (1977), 287–292.

Merris, R., "On vanishing decomposable symmetrized tensors", *Linear and Multilinear Algebra* **5** (1977), 79–86.

Merris, R., "Generalized matrix functions: A research problem", *Linear and Multilinear Algebra* **8** (1979), 83–86.

Merris, R., "Recent advances in symmetry classes of tensors", *Linear and Multilinear Algebra* **7** (1979), 317–328.

Merris, R., and Pierce, S., "A class of representations of the full linear group", *J. Algebra* **17** (1971), 346–351; ZB 217, p. 79; MR 43, #374.

Merris, R., and Pierce, S., "The Bell numbers and r-fold transitivity", *J. Comb. Theory* **12** (1972), 155–157.

Merris, R., and Rushid, M. A., "The dimension of certain symmetry classes of tensors", *Linear and Multilinear Algebra* **2** (1974), 245–248.

Merris, R., and Watkins, W., "Character induced subgroups", *J. Res. Nat. Bur. Standards, Sect. B* **77** (1973), 93–99; ZB 277, #20007.

Michel, H., "Über Strömungen in Gruppen, speziell in symmetrischen Gruppen", *Math. Z.* **105** (1968), 141–149; ZB 155, p. 50; MR 37, #2847.

Mignotte, M., "Etude arithmetique des nombres de Young", *Acta Math. Acad. Sci. Hung* **29** (1977), 1–22.

Miller, G. A., "Possible orders of two generators of the alternating and of the symmetric group", *Trans. Amer. Math. Soc.* **30** (1928), 24–32.

Molcanov, V. F., "On matrix elements of irreducible representations of the symmetric group", *Vestnik Moskov, Univ. Ser. I Mat. Meh.* **21**, No. 1 (1966), 52–57; ZB 207, p. 334; MR 33, #5754.

Moran, G., "Reflection classes whose cubes cover the alternating group", *J. Comb. Theory A* **21** (1976), 1–19; MR 54, #2769.

Moran, G., "The bireflections of a permutation", *Discrete Math* **15** (1976), 55–62; ZB 335, #05005.

Moran, G., "The product of two reflection classes of the symmetric group", *Discrete Math.* **15** (1976), 63–77; ZB 338, #05001; MR 54, #423.

Moran, G., "Some coefficients in the center of the group algebra of the symmetric group", *Discrete Math.* **21** (1978), 75–81.

Morris, A. O., "The spin characters of the symmetric group", *Quart. J. Math. Oxford Ser. (2)* **13** (1962), 241–246; ZB 112, p. 23; MR 26, #2524.

Morris, A. O., "The spin representation of the symmetric group", *Proc. London Math. Soc. (3)* **12** (1962), 55–76; ZB 104, p. 252; MR 25, #133.

Morris, A. O., "On *Q*-functions", *J. London Math. Soc.* **37** (1962), 445–455; ZB 112, p. 23; MR 26, #239.

Morris, A. O., "The spin characters of the symmetric groups II", *Quart. J. Math. Oxford Ser. (2)* **14** (1963), 247–253; ZB 131, p. 264; MR 27, #4865.

Morris, A. O., "The multiplication of Hall functions", *Proc. London Math. Soc. (3)* **13** (1963), 733–742; ZB 125, p. 17; MR 27, #3712.

Morris, A. O., "The characters of the group GL $(n, q)$", *Math. Z.* **81** (1963), 112–123; ZB 118, p. 38; MR 27, #3711.

Morris, A. O., "A note on the multiplication of Hall functions", *J. London Math. Soc.* **39** (1964), 481–488; ZB 125, p. 17; MR 29, #5897.

Morris, A. O., "The spin representation of the symmetric group", *Canad. J. Math.* **17** (1965), 543–549; ZB 135, p. 56; MR 31, #240.

Morris, A. O., "On an algebra of symmetric functions", *Quart. J. Math. Oxford Ser. (2)* **16** (1965), 53–64; MR 40, #252.

Morris, A. O., "Generalizations of the Cauchy and Schur Identities", *J. Comb. Theory (A)* **11** (1971), 163–169.

Morris, A. O., "A survey on Hall-Littlewood functions and their applications to representation theory", in *Combinatoire et representation du groupe symétrique* (Actes Table Ronde CNRS, Univ. Louis-Pasteur Strasbourg, Strasbourg, 1976), Lecture Notes in Math., Vol. 579, Springer, Berlin, 1977, pp. 136–154; MR 57, #12247.

Morris, A. O., "The projective characters of the symmetric group, an alternative proof", *J. London Math. Soc. (2)* **19** (1979), 57–58.

Morris, P. A., "Computational problems in *S*-function theory", *Math. Comp.* **27** (1973), 965–971; MR 49, #2900.

Morris, P. A., "Applications of graph theory to $S$-function theory", *J. London Math. Soc.* **8** (1974), 63–72.

Morris, P. A., and van Rees, G. H. J., "On computing plethysms of Schur-functions", *MATCH* **3** (1977), 51–66; ZB 381, #05005; MR 57, #6176.

Moser, L., and Wyman, M., "On solutions of $x^d = 1$ in symmetric groups", *Canad. J. Math.* **7** (1955), 159–168; ZB 64, p. 26.

Muirhead, R. F., "Some methods applicable to identities and inequalities of symmetric algebraic functions of $n$ letters", *Proc. Edinburgh Math. Soc.* **21** (1903), 144–157.

Müller, W., "Darstellungstheoretische Eigenschaften der symmetrischen Gruppe", Dissertation, München, 1969.

Müller, W., "Über die Grösse des Radikals in Gruppenringen über der symmetrischen Gruppe", *Manuscripta Math.* **4** (1971), 39–60; ZB 213, p. 310; MR 44 #332.

Mullineux, G., "Representations of $A_n$ in characteristic 2 with the 3-cycles fixed-point-free", *Quarterly J. Math. Oxford* **29** (1978), 199–212; MR 58, #868.

Mullineux, G., "A characterization of $A_n$ by centralizers of short involutions", *Quarterly J. Math. Oxford* **29** (1978), 213–220.

Mullineux, G., "Bijections of $p$-regular partitions and $p$-modular irreducibles of the symmetric groups", *J. London Math. Soc. (2)* **20** (1979), 60–66.

Mullineux, G., "On the $p$-cores of $p$-regular diagrams", *J. London Math. Soc.* **20** (1979), 222–226.

Munkholm, H. J., "Induced monomial representations, Young elements, and metacyclic groups", *Proc. Amer. Math. Soc.* **19** (1968), 453–458; ZB 155, p. 55; MR 36, #6513.

Murnaghan, F. D., "On the representations of the symmetric group", *Amer. J. Math.* **59** (1937), 437–488; ZB 17, p. 155.

Murnaghan, F. D., "The characters of the symmetric groups", *Amer. J. Math.* **59** (1937), 739–753; ZB 17, p. 391.

Murnaghan, F. D., "The irreducible representations of the symmetric group", *Proc. Nat. Acad. Sci. U.S.A.* **23** (1937), 277–280; ZB 17, p. 155.

Murnaghan, F. D., "On the direct product of the irreducible representations of the symmetric group", *Proc. Nat. Acad. Sci.* **23** (1937), 488–490; ZB 17, p. 197.

Murnaghan, F. D., "The analysis of the direct product of irreducible representations of the symmetric groups", *Amer. J. Math.* **60** (1938a), 44–65; ZB 18, p. 297.

Murnaghan, F. D., "The analysis of the Kronecker product of irreducible representations of the symmetric group", *Amer. J. Math.* **60** (1938b), 761–784.

Murnaghan, F. D., *The Theory of Group Representations*, The John Hopkins Press, 1938c; MR 31, #258, Dover Publ., New York, 1963.

Murnaghan, F. D., "On the multiplication of $S$-functions", *Proc. Nat. Acad. Sci. U.S.A.* **36** (1950), 476–479; MR 12, p. 479.

Murnaghan, F. D., "On the analysis of representations of the linear group", *Proc. Nat. Acad. Sci. U.S.A.* **37** (1951), 51–55; ZB 42, p. 24; MR 12, p. 588.

Murnaghan, F. D., "The characters of the symmetric group", *Proc. Nat. Acad. Sci. U.S.A.* **37** (1951), 55–58; MR 12, p. 587.

Murnaghan, F. D., "A generalization of Hermite's law of reciprocity", *Proc. Nat. Acad. Sci. U.S.A.* **37** (1951), 439–441; ZB 43, p. 19.

Murnaghan, F. D., "The analysis of representations of the linear group", *Anais Acad. Brasil. Ci.* **23** (1951), 1–19; ZB 43, p. 260; MR 13, p. 204.

Murnaghan, F. D., "The characters of the symmetric group", *Anais Acad. Brasil. Ci.* **23** (1951), 141–154; ZB 45, p. 157; MR 14, p. 843.

Murnaghan, F. D., "A generalization of Hermite's law of reciprocity", *Anais Acad. Brasil Ci.* **23** (1951), 347–368; ZB 44, p. 258.

Murnaghan, F. D., "On the Poincaré-polynomial of the full linear group", *Proc. Nat. Acad. Sci. U.S.A.* **38** (1952), 606–608.

Murnaghan, F. D., "On the Poincaré-polynomials of the classical groups", *Proc. Nat. Acad. Sci. U.S.A.* **38** (1952), 608–611.

Murnaghan, F. D., "On the multiplication of representations of the linear group", *Proc. Nat. Acad. Sci. U.S.A.* **38** (1952), 738–741; ZB 49, p. 301; MR 14, p. 244.

Murnaghan, F. D., "On the invariant theory of the classical groups", *Proc. Nat. Acad. Sci. U.S.A.* **38** (1952), 966–973; ZB 48, p. 255; MR 14, p. 447.

Murnaghan, F. D., "On the decomposition of tensors by contraction", *Proc. Nat. Acad. Sci.* **38** (1952), 973–979; MR 14, p. 533.

Murnaghan, F. D., "On the Poincare-polynomials of the classical groups—Addendum", *Proc. Nat. Acad. Sci. U.S.A.* **39** (1953), 48; MR 14, p. 619.

Murnaghan, F. D., "On thc analysis of $\{m\}\otimes\{1^k\}$ and $\{m\}\otimes\{k\}$", *Proc. Nat. Acad. Sci. U.S.A.* **40** (1954), 721–723; ZB 56, p. 257.

Murnaghan, F. D., "On the characters of the symmetric group", *Proc. Nat. Acad. Sci. U.S.A.* **41** (1955), 396–398; MR 16, p. 996.

Murnaghan, F. C., "On the generation of the irreducible representations of the symmetric groups", *Proc. Nat. Acad. Sci. U.S.A.* **41** (1955), 514–515; MR 17, p. 12.

Murnaghan, F. D., "On the analysis of the Kronecker product of irreducible representations of $S_n$", *Proc. Nat. Acad. Sci. U.S.A.* **41** (1955), 515–518; MR 17, p. 12.

Murnaghan, F. D., "On the irreducible representations of the symmetric group", *Proc. Nat. Acad. Sci. U.S.A.* **41** (1955), 1096–1103; MR 17, p. 583.

Murnaghan, F. D., "On the Kronecker product of irreducible representations of the symmetric group", *Proc. Nat. Acad. Sci. U.S.A.* **42** (1956), 95–98; MR 17, p. 710.

Murnaghan, F. D., "The characters of the symmetric group", *Proc. Nat. Acad. Sci. U.S.A.* **68** (1971), 399–401; ZB 209, p. 333.

Murnaghan, F. D., "Powers of representations of the rotation group (their symmetric, alternating, and other parts)", *Proc. Nat. Acad. Sci. U.S.A.* **69** (1972), 1181–1184; ZB 241, #20031; MR 45, #6937.

Murphy, G., "On decomposability of some Specht modules for symmetric groups", *J. Algebra* **66** (1980), 156–168.

Nagao, H., "Note on the modular representations of symmetric groups", *Canad. J. Math.* **5** (1953), 356–363; MR 14, p. 1061.

Nagao, H., "On the groups with the same table of characters as symmetric groups", *J. Inst. Polytech. Osaka City Univ. Serv. A* **8** (1957), 1–8; ZB 77, p. 32; MR 19, p. 387.

Nakaoka, M., "Note on cohomology algebras of symmetric groups", *J. Math. Osaka City Univ. Ser. A* **13** (1962), 45–55; ZB 134, p. 267.

Nakayama, T., "On some modular properties of irreducible representations of a symmetric group I", *Jap. J. Math.* **17** (1940a), 165–184; ZB 61, p. 40; MR 3, p. 195.

Nakayama, T., "On some modular properties of irreducible representations of a symmetric group II", *Jap. J. Math.* **17** (1940b), 411–423; ZB 61, p. 40; MR 3, p. 196.

Nakayama, T., and Osima, M., "Note on blocks of symmetric groups", *Nagoya Math. J.* **2** (1951), 111–117; ZB 42, p. 24; MR 12, p. 672.

Nathanson, M. B., "On the greatest order of an element of the symmetric group", *Amer. Math. Monthly* **79** (1972), 500–501; MR 45, #6909.

Nazarova, L. A., "Unimodular representations of the alternating group of degree 4", *Ukrain. Mat. Ž.* **15** (1963), 437–444; ZB 119, p. 30; MR 281, #2148.

Nazarova, L. A., and Roiter, A. V., "Integral representations of a symmetric group of third degree", *Ukrain. Mat. Ž.* **14** (1962), 271–288; ZB 145, p. 30; MR 26, #6273.

Nenov, N. D., "The representation of the abstract wreath product of groups by permutations", *Annuaire Univ. Sofia Fac. Math. Méc.* **67** (1972/73), 565–572; ZB 359, #20023; MR 54, #12915.

Netto, E., *Substitutionentheorie und ihre Anwerdung auf die Algebra*, Teubner-Verlag, Leipzig, 1882.

Neumann, M., and Visa, A., "La représentation des substitutions de $S_4$ et $A_5$ par les rotations des corps réguliers", *Lucrarile sti. Inst. Ped. Timisoara Mat. -Fiz.* **1959**, 103–110; ZB 128, p. 254.

Neumann, P. M., "A lemma that is not Burnside's", *Math. Scientist* **4** (1979), 133–141.

Newell, M. J., "A theorem on the plethysm of *S*-functions", *Quart. J. Math. Oxford Ser. II* **2** (1951), 161–166; ZB 43, p. 260.

Newell, M. J., "On the quotients of alternants and the symmetric group", *Proc. London Math. Soc. (2)* **53** (1951), 345–355; ZB 54, p. 12; MR 13, p. 10.

Newell, M. J., "On the multiplication of *S*-functions", *Proc. London Math. Soc. (2)* **53** (1951), 356–362; ZB 54, p. 12.

Nicolas, J.-L., "Sur l'ordre maximum d'un elément dans le groupe $S_n$ des permutations", *Acta Arith.* **14** (1967), 315–332; MR 37, #6361.

Nicolas, J.-L., "Ordre maximal d'un element du groupe $S_n$ des permutations, et 'highly composite numbers' (Thèse)", *Bull. Soc. Math. France* **97** (1969), 129–191; MR 40, #7340.

Nielson, J., "Die symmetrische und die alternierende Gruppe", *Math. Tidsskr. B* **1940**, 7–18; ZB 23, p. 209; MR 2, p. 211.

Nilov, G. N., "The number of cyclic subgroups of arbitrary order of the symmetric group", *Kabardino-Balkarsk. Gos. Univ. Učen. Zap. Ser. Fiz. -Mat.* **19** (1963), 249–250(a), 250–251(b); MR 32, #1248.

Nilov, G. N., "The number of subgroups of order four of a symmetric group" (Proceedings of the Annual Scientific Conference, Nal'chik, 1965), *Kabardino-Balkarsk. Gos. Univ. Učen. Zap.* **24** (1965), 191–193; MR 35, #6750.

Noda, R., and Yamaki, H., "A characterization of the alternating groups of degrees six and seven", *Osaka J. Math.* **7** (1970), 313–319.

Oberschelp, W., "Kombinatorische Anzahlbestimmungen in Relationen", *Math. Ann.* **174** (1967), 53–78.

Ohmori, Z., "On the Schur indices of GL $(n, q)$ and SL $(2n+1, q)$", *J. Math. Soc. Japan* **29** (1977), 693–707.

Oliveira, G. N., and Dias da Silva, J. A., "Conditions for equality of decomposable symmetric tensors II", *Linear Algebra Appl.* **28** (1979), 161–176.

Olsson, J. B., "On the blocks of GL $(n, q)$ I", *Trans. Amer. Math. Soc.* **222** (1976), 143–156.

Olsson, J. B., "McKay numbers and heights of characters", *Math. Scand.* **38** (1976), 25–42.

Olsson, J. B., and Regev, A., "An application of representation theory to PI-algebras", *Proc. Amer. Math. Soc.* **55** (1976), 253–257.

Olsson, J. B., and Regev, A., "Colength sequence of some *T*-ideals", *J. Algebra* **38** (1976), 100–111.

Olsson, J. B., and Regev, A., "On the *T*-ideal generated by a standard identity", *Israel J. Math.* **26** (1977), 97–104.

Ore, O., "Theory of monomial groups", *Trans. Amer. Math. Soc.* **51** (1942), 15–64; ZB 28, p. 3; MR 3, p. 197.

Ore, O., "Some remarks on commutators", *Proc. Amer. Math. Soc.* **2** (1951), 307–314; ZB 43, p. 24; MR 12, p. 671.

Osima, M., "On some character relations of symmetric groups", *Math. J. Okayama Univ.* **1** (1952), 63–68; MR 14, p. 243.

Osima, M., "On the irreducible representations of the symmetric group", *Canad. J. Math.* **4** (1952), 381–384; MR 13, p. 911.

Osima, M., "Some remarks on the characters of the symmetric group", *Canad. J. Math.* **5** (1953), 336–343; ZB 52, p. 23; MR 15, p. 100.

Osima, M., "Some remarks on the characters of the symmetric group II", *Canad. J. Math.* **6** (1954a), 511–521; ZB 58, p. 260; MR 16, p. 566.

Osima, M., "On the representations of the generalized symmetric group", *Math J. Okayama Univ.* **4** (1954b), 39–56; ZB 58, p. 21; MR 16, p. 794.

Osima, M., "On blocks of characters of the symmetric group", *Proc. Japan Acad.* **31** (1955), 131–134; MR 17, p. 941.

Osima, M., "Note on a paper by J. S. Frame and G. de B. Robinson", *Math. J. Okayama* **6** (1956a), 77–79; ZB 72, p. 260; MR 18, p. 560.

Osima, M., "On the representations of the generalized symmetric group II", *Math. J. Okayama Univ.* **6** (1956b), 81–97; ZB 72, p. 260; MR 18, p. 716.

Osima, M., "On the generalized decomposition numbers of the symmetric group", *J. Math. Soc. Japan* **20** (1968), 289–296; ZB 182, p. 351; MR 36, #6515.

Osima, M., "On the generalized decomposition numbers of the alternating group", *Proc. Japan Acad.* **47** (1971), 757–760; ZB 246, # 20006; MR 46, #3603.

Oyama, T., "On the groups with the same table of characters as alternating groups", Osaka J. Math **1** (1964), 91–101; ZB 139, p. 19; MR 29, #3529.

Pahlings, H., "Irreducible odd representations of wreath products", *J. London Math. Soc. Ser. II* **12** (1975), 45–48; ZB 314, #20013.

Pahlings, H., "Characterization of groups by their character tables I", *Communications in Algebra* **4** (1976a), 111–153; ZB 336, #20004; MR 53, #8217.

Pahlings, H., "Characterization of groups by their character tables II", *Communications in Algebra* **4** (1976b), 155–178; ZB 336, #20005; MR 53, #8217.

Parkinson, C., "Ambivalence in alternating symmetric groups", *Amer. Math. Monthly* **80** (1973), 190–192.

Patera, J., and Sharp, R. T., "Generating functions for plethysms of finite and continuous groups", *J. Phys. A: Math. Gen.* **13** (1980), 397–416.

Patterson, C. W., and Harter, W. G., "Canonical symmetrization for unitary bases. I. Canonical Weyl bases", *J. Math. Phys.* **17** (1976), 1125–1136.

Patterson, C. W., and Harter, W. G., "Canonical symmetrization for the unitary bases. II. Boson and fermion bases", *J. Math. Phys.* **17** (1976), 1137–1142.

Pavlov, A. I., "On the equation $x^k=a$ in the symmetric group", *Dokl. Akad. Nauk SSSR* **240** (1978); *Soviet Math. Dokl.* **19** (1978), 644–646.

Peel, M. H., "On the second natural representation of the symmetric groups", *Glasgow Math. J.* **10** (1969), 25–37; ZB 175, p. 303; MR 39, #2892.

Peel, M. H., "Hook representations of the symmetric groups", *Glasgow Math. J.* **12** (1971), 136–149; ZB 235, #20012; MR 46, #7363.

Peel, M. H., "Specht modules and the symmetric groups", *J. Algebra* **36** (1975a), 88–97; ZB 313, #20005.

Peel, M. H., "Modular representations of the symmetric groups", Univ. of Calgary Research Paper No. 292, 1975b.

Peterson, G. L., "Automorphisms of the integral group ring of $S_n$", *Proc. Amer. Math. Soc.* **59** (1976), 14–18; ZB 352, #20004.

Piccard, S., "Sur les bases du groupe symétrique et du groupe alternant", *Comment. Math. Helv.* **11** (1938), 1–8; ZB 19, p. 396.

Piccard, S., "Sur les bases du groupe symétrique", *Časopis Mat. Fysik*, Praha **68** (1939), 15–30; ZB 21, p. 106.

Piccard, S., "Sur les bases du groupe symétrique et du groupe alternant", *Math. Ann.* **116** (1939), 752–767; ZB 22, p. 10.

Piccard, S., "Sur les bases du groupe symétrique et du groupe alternant", *Wiadom. mat. Warzawa* **47** (1939), 141–179; ZB 21, p. 106.

Piccard, S., "Quelques propositions concernant les bases du groupe symétrique et du groupe alternant", *Comment. Math. Helv.* **12** (1940), 130–148; ZB 22, p. 312; MR 1, p. 161.

Piccard, S., "Sur les bases du groupe symétrique", *Mathematica Timisoara* **17** (1941), 147–166; ZB 26, p. 56; MR 4, p. 1.

Piccard, S., "Quelques propositions concernant les bases du groupe symétrique et du groupe alterné", *Enseignement Math.* **38** (1942), 276–286; ZB 26, p. 386; MR 4, p. 133.

Piccard, S., "Sur les bases du groupe symétrique et du groupe alternant", *Ann. Soc. Polon. Math.* **18** (1945), 25–46; ZB 61, p. 32; MR 8, p. 310.

Piccard, S., "Sur les bases du groupe symétrique et les couples de substitutions qui engendrent un groupe régulier", in *Mém. Univ. Neuchatel*, Vol. 19, Librairie Vuibert, Paris, 1946; ZB 61, p. 32; MR 8, p. 13.

Piccard, S., "Sur les bases du groupe symétrique d'ordre 7!", *C. R. Acad. Sci. Paris* **225** (1947), 1246–1247; ZB 30, p. 106; MR 9, p. 224.

Piccard, S., "Note sur les bases du groupe symétrique", *Comment. Math. Helv.* **21** (1948a), 142–149; ZB 30, p. 292; MR 9, p. 491.

Piccard, S., "Les systèmes de substitutions qui engendrent le groupe symétrique ou le groupe alterné", *Ann. Univ. Lyon Sect. A (3)* **11** (1948b), 21–29; MR 10, p. 351.

Piccard, S., "Sur les bases du groupe symétrique d'ordre 7!", *C. R. Acad. Sci. Paris* **226** (1948c), 42–43; ZB 30, p. 106; MR 9, p. 224.

Piccard, S., "Les bases du groupe symétrique et du groupe alterné dont l'une des substitutions est formée de deux transpositions", *C. R. Acad. Sci. Paris* **226** (1948d), 146–148; ZB 30, p. 292; MR 9, p. 409.

Piccard, S., "Un théorème concernant le nombre des bases d'un sous-groupe transitif et primitif, à base du second ordre, du groupe symétrique", *C. R. Acad. Sci. Paris* **227** (1948e), 254–256; ZB 33, p. 347; MR 10, p. 8.

Piccard, S., "Un théorème concernant le nombre des bases d'un sous-groupe transitif et primitif, à base du second ordre, du groupe symétrique", *C. R. Acad. Sci. Paris* **227** (1948f), 745–747; ZB 33, p. 347; MR 10, p. 180.

Piccard, S., "Relations caractéristiques des bases du groupe symétrique", *Math. Timisoara* **23** (1948g), 88–100; ZB 31, p. 5; MR 10, p. 180.

Piccard, S., *Sur les Bases du Groupe Symétrique* II, Librairie Vuibert, Paris, 1948h; ZB 31, p. 107; MR 10, p. 281.

Piccard, S., "Les groupes engendrés par un systéme connexe de cycles d'ordre sept et les bases des groupes symétrique et alterné de degré $n > 10$ dont l'une des substitutions est un cycle du septième ordre", *Comment. Math. Helv.* **24** (1950), 4–17; ZB 36, p. 295; MR 11, p. 712.

Piccard, S., "Les bases du groupe symétrique dont l'une des substitutions est un cycle du sixième ordre", *Comment. Math. Helv.* **25** (1951), 91–130; ZB 42, p. 254; MR 13, p. 104.

Piccard, S., "Les relations caractéristiques des bases du second ordre du groupe symétrique", *C. R. Acad. Sci. Paris* **240** (1955), 1751–1754; ZB 64, p. 25; MR 16, p. 994.

Piccard, S., *Sur les bases des groupes d'ordre fini*, Mémoires de l'Université de Neuchatel, Tome 25, Secrétariat de l'Université, Neuchatel, 1957; ZB 77, p. 31; MR 20, #902.

Pierce, S., "A class of representations of the full linear group II", *Trans. Amer. Math. Soc.* **173** (1972), 251–262; ZB 292, #15013.

Playtis, A. S., Sehgal, S., and Zassenhaus, H., "Equidistributed permutation groups", *Comm. Algebra* **6** (1978), 35–57; ZB 372, #20003; MR 57, #6152.

Plunkett, S. P. O., "On the plethysm of $S$-functions", *Can. J. Math.* **24** (1972), 541–552; MR 45, #3596.

Pogorelov, B. A., "Maximal subgroups of symmetric groups that are defined on projective spaces over finite fields", *Math. Zametki* **16** (1974), 91–100; MR 50, #10030.

Pollak, G., "A new proof of the simplicity of the alternating group", *Acta Sci. Math. Szeged.* **16** (1955), 63–64; ZB 64, p. 253; MR 16, p. 994.

Pólya, G., "Kombinatorische Anzahlbestimmungen für Gruppen, Graphen und chemische Verbindungen", *Acta. Sci. Math.* **68** (1937), 145–254.

Puttaswamaiah, B. M., "Group representations (alternating and generalized symmetric groups)", Thesis, University of Toronto, 1963, vi+204 pp.

Puttaswamaiah, B. M., "Unitary representations of generalized symmetric groups", *Can. J. Math.* **21** (1969), 28–38; ZB 169, p. 345; MR 38, #5950.

Puttaswamaiah, B. M., and Dixon, J. D., *Modular Representations of Finite Groups*, Academic Press, New York–San Francisco–London, 1977 vii+242 pp.

Puttaswamaiah, B. M., and Robinson, G. de B., "Induced representations and alternating groups", *Canad. J. Math.* **16** (1964), 587–601; MR 29, #4814.

Rabinovic, E. B., "Composition factors of the infinite symmetric group" (Russian), *Dokl. Akad. Nauk BSSR* **20** No. 7 (1976), 593–596, 667; MR 58, #28183.

Radzig, A., "Die Anwendugen des Sylowschen Satzes auf die symmetrische und die alternierende Gruppe", Dissertation, Berlin, 1895.

Ranganathan, N. R., and Prakash, J. S., "Molien Function for a Symmetric Group" (Proc. Conf. Group Theoretical Methods in Physics, Austin, 1978), *Lecture Notes in Physics* **94** (1978), 448–449.

Ranganathan, N. R., and Prakash, J. S., "Two algorithms for symmetric groups", *J. Phys. A.: Math. Gen.* **13** (1980), 2653–2658.

Rasala, R., "On the minimal degrees of characters of $S_n$", *J. Algebra* **45** (1977), 132–181; ZB 348, #20009; MR 55, #477.

Read, E. W., "On projective representations of the finite reflection groups of type $B_l$ and $D_l$", *J. London Math. Soc.* **10** (1975), 129–142; MR 51, #3289.

Read, E. W., "On the Schur multiplier of a wreath product", *Illinois J. Math.* **20** (1976), 456–466; MR 54, #2818.

Read, E. W., "On the Schur multipliers of the finite imprimitive unitary reflection groups $G(m, p, n)$", *J. London Math. Soc.* **13** (1976), 150–154; MR 53, #3094.

Read, E. W., "The $\alpha$-regular classes of the generalized symmetric group", *Glasgow Math. J.* **17** (1976), 144–150; MR 54, #393.

Read, E. W., "The projective representations of the generalized symmetric group", *J. Algebra* **46** (1977), 102–133; ZB 357, #20007.

Read, E. W., "On the projective characters of the symmetric group", *J. London Math. Soc. Ser. II* **15** (1977), 456–464; ZB 379, #20008; MR 58, #22269.

Read, R. C., "The use of $S$-functions in combinatorial analysis", *Can. J. Math.* **20** (1968), 808–841.

Rédei, L., "Die Einfachheit der alternierenden Gruppe", *Monatsheft Math.* **55** (1951), 328–329; ZB 44, p. 15; MR 13, p. 528.

Redfield, J. H., "The theory of group reduced distributions", *Amer. J. Math.* **49** (1927), 433–455.

Regev, A., "The $T$-ideal generated by the standard identity $s_3[x_1, x_2, x_3]$", *Israel J. Math.* **26** (1977), 105–125.

Regev, A., "The representations of $S_n$ and explicit identities for P.I. algebras", *J. Algebra* **51** (1978), 25–40.

Regev, A., "Algebras satisfying a Capelli identity", *Israel J. Math.* **33** (1979), 149–154.

Regev, A., "The Kronecker product of $S_n$-characters and an $A \otimes B$ theorem for Capelli identities", *J. Algebra* **66** (1980), 505–510.

Rigby, J. F., "Monomial groups with respect to a basic Abelian group", *Proc. London Math. Soc.* **10**, No. 3 (1960), 239–252; ZB 91, p. 28; MR 22, #5664.

Ritter, J., "Bemerkungen zu den Frobeniusoperatoren $\psi^n$ des Charakterringes $R_kG$ einer endlichen Gruppe $G$", *Journal of Algebra* **30** (1974), 355–367; ZB 284, #20011.

Robinson, G. de B., "A geometrical study of the alternating and symmetric groups", *Proc. Cambridge Philos. Soc.* **25** (1929), 168–174.

Robinson, G. de B., "A geometrical study of the hyperoctahedral group", *Proc. Cambridge Philos. Soc.* **26** (1930), 94–98.

Robinson, G. de B., "On the geometry of the linear representations of the symmetric groups", *Proc. London Math. Soc. (2)* **38** (1935a), 402–413; ZB 11, p. 393.

Robinson, G. de B., "Note on an equation of quantitative substitutional analysis", *Proc. London Math. Soc. (2)* **38** (1935b), 414–416.

Robinson, G. de B., "On the fundamental region of an orthogonal representation of a finite group", *Proc. London Math. Soc. (2)* **43** (1937), 289–301; ZB 17, p. 6.

Robinson, G. de B., "On the representations of the symmetric groups", *Amer. J. Math.* **60** (1938), 745–760; ZB 19, p. 251.

Robinson, G. de B., "On the representations of the symmetric group II", *Amer. J. Math.* **69** (1947a), 286–298; ZB 36, p. 154; MR 8, p. 563.

Robinson, G. de B., "On a conjecture by Nakayama", *Trans. Roy. Soc. Can. Sect. III (3)* **41** (1947b), 20–25; ZB 29, p. 199.

Robinson, G. de B., "On the representations of the symmetric group III", *Amer. J. Math.* **70** (1948), 277–294; ZB 36, p. 155; MR 10, p. 678.

Robinson, G. de B., "On the disjoint product of irreducible representations of the symmetric group", *Canad. J. Math.* **1** (1949), 166–175; ZB 36, p. 155; MR 10, p. 504.

Robinson, G. de B., "Induced representations and invariants", *Canad. J. Math.* **2** (1950), 334–343; ZB 39, p. 20; MR 12, p. 74.

Robinson, G. de B., "On the modular representations of the symmetric group", *Proc. Nat. Acad. Sci. U.S.A.* **37** (1951), 694–696; ZB 44, p. 257; MR 13, p. 530.

Robinson, G. de B., "On a conjecture by J. H. Chung", *Canad. J. Math.* **4** (1952), 373–380; ZB 46, p. 250; MR 14, p. 243.

Robinson, G. de B., "On the modular representations of the symmetric group II", *Proc. Nat. Acad. Sci. U.S.A.* **38** (1952), 129–133; ZB 46, p. 250; MR 14, p. 243.

Robinson, G. de B., "On the modular representations of the symmetric group III", *Proc. Nat. Acad. Sci. U.S.A.* **38** (1952), 424–426; ZB 46, p. 251; MR 14, p. 244.

Robinson, G. de B., "On the modular representations of the symmetric group IV", *Canad. J. Math.* **6** (1954), 486–497; MR 16, p. 333.

Robinson, G. de B., "On the modular representation of the symmetric group V" *Canad. J. Math.* **7** (1955), 391–400; MR 17, p. 12.

Robinson, G. de B., "The degree of an irreducible representation of $S_n$", *Proc. Nat. Acad. Sci. U.S.A.* **42** (1956), 357–359; MR 18, p. 13.

Robinson, G. de B., "A remark by Philip Hall", *Canad. Math. Bull.* **1** (1958), 21–23; ZB 86, p. 24; MR 20, #2383.

Robinson, G. de B., "Group representations", *Trans. Roy. Soc. Canada Sect. III (3)* **54** (1960), 1–8; ZB 96, p. 20; MR 24, #1324.

Robinson, G. de B., *Representation Theory of the Symmetric Group* Mathematical Expositions No. 12, University of Toronto Press, Toronto, 1961; ZB 102, p. 20; MR 23, #3182.

Robinson, G. de B., "Modular representations of $S_n$", *Canad. J. Math.* **16** (1964), 191–203; ZB 117, p. 271; MR 28, #2161.

Robinson, G. de B., "Geometry of group representations", *Nagoya Math. J.* **27** (1966), 509–513; ZB 145, p. 30; MR 33, #5751.

Robinson, G. de B., "Note on a theorem of Livingstone and Wagner", *Math. Z.* **102** (1967), 351–352; ZB 149, p. 271.

Robinson, G. de B., "The algebras of representations and classes of finite groups", *J. Math. Phys.* **12** (1971), 2212–2215; ZB 235, #20013; MR 44, #4113.

Robinson, G. de B., "Tensor product representations", *J. of Algebra* **20** (1972), 118–123; ZB 225, #20007; MR 44, #4114.

Robinson, G. de B., "The dual of Frobenius' reciprocity theorem", *Canad. J. Math.* **25** (1973), 1051–1059.

Robinson, G. de B., "Restricting and inducing on inner products of representations of finite groups", *Canad. J. Math.* **27** (1975), 1349–1354; MR 53, #3089.

Robinson, G. de B. (ed.), *The Papers of A. Young (1873–1940)*, University of Toronto Press, Mathematical Expositions No. 21, Toronto, 1977.

Robinson, G. de B., and Taulbee, O. E., "The reduction of the inner product of two irreducible representations of $S_n$", *Proc. Nat. Acad. Sci. U.S.A.* **40** (1954), 723–726; ZB 56, p. 257; MR 16, p. 110.

Robinson, G. de B., and Taulbee, O. E., "On the modular representations of the symmetric groups VI", *Proc. Nat. Acad. Sci. U.S.A.* **41** (1955), 596–598; ZB 66, p. 19; MR 17, p. 126.

Robinson, G. de B., and Thrall, R. M., "The content of a Young diagram", *Michigan Math. J.* **2** (1954), 157–167.

Roe, E. D., "On the coefficients in the product of an alternant and a symmetric function", *Trans. Amer. Math. Soc.* **5** (1904), 193–213.

Roe, E. D., "On the coefficients in the quotient of two alternants", *Trans. Amer. Math. Soc.* **6** (1905), 63–74.

Röel, R. W. J., "Invariance groups of Young operators; Pauling numbers" (Group Theor. Meth. Phys., 4th int. Coll., Nijmegen, 1975), *Lect. Notes Phys.* **50** (1976), 376–385; ZB 364, #20003.

Rothaus, O., and Thompson, J. G., "A combinatorial problem in the symmetric group", *Pacific J. Math.* **18** (1966), 175–178; ZB 145, p. 29; MR 33, #4130.

Ruch, E., "The diagram lattice as structural principle", *Theoret. Chim. Acta (Berl.)* **38** (1975), 167–183.

Ruch, E., and Gutman, I., "The branching extent of graphs", to appear.

Ruch, E., and Schönhofer, A., "Theorie der Chiralitätsfunktionen", *Theor. Chim. Acta (Berl.)* **19** (1970), 225–287.

Rudvalis, A., and Snapper, E., "Permutation representations of finite groups", mimeographed notes (unpublished).

Rudvalis, A., and Snapper, E., "Numerical polynomials for arbitrary characters", *J. Comb. Theory* **10** (1977), 145–159.

Rutherford, D. E., "On the relations between the numbers of standard tableaux", *Proc. Edinburgh Math. Soc. (2)* **7** (1942), 51–54; MR 4, p. 133.

Rutherford, D. E., "On substitutional equations", *Proc. Roy. Soc. Edinburgh, Sect. A* **62** (1944), 117–126; MR 7, p. 112.

Rutherford, D. E., *Substitutional Analysis*, University Press, Edinburgh, 1948; ZB 38, p. 16; MR 10, p. 280.

Ryser, H. J., *Combinatorial Mathematics*, The Carus Mathematical Monographs, 1963, xi + 154 pp.

Sagan, B., "An analog of Schensted's algorithm for shifted Young tableaux", *J. Comb. Theory (A)* **27** (1979), 10–18.

El Samra, N., and King, R. C., "Reduced Determinantal forms for characters of the classical Lie groups", *J. Physics A: Math. Gen.* **12** (1979), 2305–2315.

El Samra, N., and King, R. C., "Dimensions of irreducible representations of the classical Lie groups", *J. Physics A: Math. Gen.* **12** (1979), 2317–2328.

Sänger, F., "Einige Charakterentafeln von Symmetrien symmetrischer Gruppen", *Mitt. Math. Sem. Giessen* **98** (1973), 21–38; ZB 267, #20006.

Sänger, F., "Plethysmen von irreduziblen Darstellugen symmetrischer Gruppen" Dissertation, Aachen, 1980.

Sass, H., "Eine abstrakte Definition gewisser alternierender Gruppen", *Math. Z.* **128** (1972), 109–113.

Sato, M., "On formal fractions associated with the symmetric groups", *J. Comb. Theory (A)* **20** (1976), 124–131.

Saxl, J., "Characters and generosity of permutation groups", *Proc. Amer. Math. Soc.* **47** (1975a), 73–76; ZB 293, #20008; MR 50, #10031.

Saxl, J., "Characters of multiply transitive permutation groups", *J. Algebra* **34** (1975b), 528–539; ZB 323, #20004; MR 56, #8680.

Saxl, J., "Restrictions of characters, generosity, interchange and coloured graphs", in *Combinatoire et représentation du groupe symétrique* (Actes Table Ronde C.N.R.S., Univ. Louis-Pasteur Strasbourg, Strasbourg, 1976), Lecture Notes in Math., Vol. 579, Springer, Berlin, 1977, pp. 258–266; MR 57, #12658.

Schensted, C., "Longest increasing and decreasing subsequences", *Canad. J. Math.* **13** (1961), 179–191.

Schindler, S., and Mirman, R., "The Clebsch-Gordan decomposition and the coefficients for the symmetric group", *J. Math. Phys.* **18** (1977), 1688–1696, 1697–1704.

Schur, I., "Über eine Klasse von Matrizen, die sich einer gegebenen Matrix zuordnen lassen", Dissertation, Berlin, 1901.

Schur, I., "Über die Darstellung der symmetrischen Gruppe durch lineare homogene Substitutionen", *Berl. Ber.* **1908**, 664–678.

Schur, I., "Über die Darstellung der symmetrischen und der alternierenden Gruppe durch gebrochene lineare Substitutionen", *J. Math.* **139** (1910), 155–250.

Schur, I., "Über die rationalen Darstellungen der allgemeinen linearen Gruppe", *Berl. Ber.* **1927**, 58–75.

Schur, I., "Über die reellen Kollineationsgruppen, die der symmetrischen oder der alternierenden Gruppe isomorph sind", *J. Math.* **158** (1927), 63–79.

Schur, I., and Grunsky, H., *Vorlesungen über Invariantentheorie*, Springer-Verlag, 1968; ZB 159, p. 37; MR 37 #5248.

Schützenberger, M. P., "Quelques remarques sur une construction de Schensted", *Math. Scand.* **12** (1963), 117–128.

Schützenberger, M. P., "Sur un théorème de G. de B. Robinson", *C. R. Acad. Sci. Paris Ser. A-B* **272** (1971), A420–A421; MR 45, #3595.

Schützenberger, M. P., "Sur une construction de Gilbert de B. Robinson", Semin. P. Dubreil, 25 année 1971/72, Algebre, Fasc. 1, 2, Expose 8 (1973), 4 pp.; ZB 386, #05008.

Schützenberger, M. P., "Evacuations", in *Colloquio Internazionale sulle Teorie Combinatorie (Rome, 1973)*, Tomo I, Atti dei Convegni Lincei, No. 17, Accad. Naz. Lincei, Rome, 1976, pp. 257–264; ZB 377, #05015; MR 57, #16393.

Schützenberger, M. P., "La correspondance de Robinson", Combinatoire et représentation du groupe symétrique. (Actes Table Ronde C.N.R.S. Strasbourg 1976), Lecture Notes in Math., Vol. 579, (1977), 59–113; ZB 398, #05011.

Schwenk, A. J., "An asymptotic evaluation of the cycle index of a symmetric group", *Discrete Math.* **18** (1977), 71–78.

Segal, I. E., "The automorphisms of the symmetric group", *Bull. Amer. Math. Soc.* **46** (1940), 565; ZB 61, p. 33; MR 2, p. 1.

de Séguier, J., "Sur la représentation linéaire homogène des groupes symétriques et alternés", *J. de Math.* **6** (1910), 387–436.

de Séguier, J., "Sur la répresentation linéaire homogène des groupes symétriques et alternés", *J. de Math. (6)* **7**, (1911), 113–121.

Seitz, G., "M-groups and the supersolvable residual", *Math. Z.* **110** (1969), 101–122; ZB 214 p. 43; MR 40 #1500.

Servedio, F. J., "The GL $(V)$-module structure of $S^t(S^r(V))$, symmetric tensors of degree $t$ on symmetric tensors of degree $r$ on $V$", *Communications in Algebra* **8** (1980), 1387–1401.

Siebeneicher, Ch., "λ-Ringstrunkturen auf dem Burnsidering der Permutationsdarstellugen einer endlichen Gruppe", *Math. Z.* **146** (1976), 223–238; ZB 313, #20006.

Silcock, H. L., "Generalized wreath products and the lattice of normal subgroups of a group", *Algebra Universalis* **7** (1977), 361–372.

Slepian, D., "On the symmetrized Kronecker power of a matrix and extensions of Mehler's formula for hermite polynomials", *SIAM J. Math. Analysis* **3** (1972), 606–616.

Smith, P. R., and Wybourne, B. G., "Selection rules and the decomposition of the Kronecker square of irreducible representations", *J. Math. Phys.* **8** (1968), 2434–2440; ZB 155, p. 58.

Smith, P. R., and Wybourne, B. G., "Plethysm and the theory of complex spectra", *J. Math. Phys.* **9** (1968), 1040–1051.

Smith, R. K., "Integral Representations of the Symmetric Group $S_n$", M. Sc. thesis, Ames Iowa, 1978.

Snapper, E., "The polynomial of a permutation representation", *J. Comb. Theory* **5** (1968), 105–114.

Snapper, E., "Group characters and nonnegative integral matrices", *Journal of Algebra* **19** (1971), 520–535; ZB 226, #20008.

Snapper, E., "Characteristic polynomials of a permutation representation", *J. Comb. Theory (A)* **26** (1979), 65–81.

Solomon, L., "On Schur's index and the solutions of $G^n = 1$ in a finite group", *Math. Z.* **78** (1962) 122–125; MR 25, #2125.

Solomon, L., "The solution of equations in groups", Archiv. der Math. **20** (1969), 241–247.

Solomon, L., "Partition identities and invariants of finite groups", *J. Comb. Theory Ser. A* **23** (1977), 148–175; MR 56, #8677.

Specht, W., "Eine Verallgemeinerung der symmetrischen Gruppe", *Schriften Berlin* **1** (1932), 1–32; ZB 4, p. 338.

Specht, W., "Eine Verallgemeinerung der Permutationsgruppen", *Math. Z.* **37** (1933a), 321–341; ZB 7, p. 149.

Specht, W., “Die irreduziblen Darstellungen der symmetrischen Gruppe”, *Jahresbericht D.M.V.* **42** (1933b), 124–126.

Specht, W., “Die irreduziblen Darstellungen der symmetrischen Gruppe”, *Math. Z.* **39** (1935), 696–711; ZB 11, p. 103.

Specht, W., “Darstellungstheorie der Hyperoktaedergruppe”, *Math. Z.* **42** (1937a), 629–640; ZB 17, p. 6.

Specht, W., “Zur Darstellungstheorie der symmetrischen Gruppe”, *Math. Z.* **42** (1937b), 774–779; ZB 17, p. 6.

Specht, W., “Darstellungstheorie der alternierenden Gruppe”, *Math. Z.* **43** (1938), 553–572; ZB 18, p. 204; MR 2, p. 126.

Specht, W., “Beiträge zur Darstellungstheorie der allgemeinen linearen Gruppe”, *Math. Z.* **51** (1948), 377–403; ZB 30, p. 340; MR 10, p. 352.

Specht, W., “Die Charaktere der symmetrischen Gruppe”, *Math. Z.* **73** (1960), 312–329; ZB 96, p. 19; MR 22, #4786.

Springer, T. A., “Trigonometric sums, Green functions of finite groups and representations of Weyl groups”, *Invent. Math.* **36** (1976), 173–207.

Springer, T. A., “Darstellungen symmetrischer Gruppen”, *Nederl. Akad. Wet., Verslag Afd. Natuurk* **87**, 25–27 (1978); ZB 379, #20010; MR 58, #5890.

Springer, T. A., “A construction of representations of Weyl groups”, *Inventiones Math.* **44** (1978), 279–293.

Srinivasan, B., “Green Polynomials of finite classical groups”, *Communications in Algebra* **5** (1977), 1241–1258.

Staal, R. A., “Star diagrams and the symmetric group”, *Canad. J. Math.* **2** (1950), 79–92; ZB 36, p. 155; MR 11, p. 415.

Stams, W., “MacMahon-Spiel und symmetrische Gruppe $S_6$”, *Deutsche Math.* **2** (1937), 691–697; ZB 18, p. 11.

Stanley, R. P., “Theory and application of plane partitions: Part 1”, *Studies in Applied Mathematics* **50** (1971), 167–188; ZB 225, #05011.

Stanley, R. P., “Theory and application of plane partitions: Part 2”, *Studies in Applied Math.* **50** (1971), 259–279; ZB 225, #05012.

Steinberg, R., “A geometric approach to the representations of the full linear group over a Galois field”, *Trans. Amer. Math. Soc.* **71** (1951), 274–282.

Stockhofe, D., “Die Zerlegungsmatrizen der Symmetrischen Gruppen $S_{12}$ und $S_{13}$ zur Primzahl 2”, *Communications in Algebra* **7** (1979), 39–45; ZB 394, # 20007; MR 58 #22270.

Stojakovic, M., “Sur une relation d'ordre dans le groupe symétrique”, *Univ. Beogradu. Godisnjak Filozof. Fak. Novom Sadu* **1** (1956), 281–292; MR 20, #2377.

Stojakovic, M., “Sur une relation d'ordre dans le groupe symétrique”, *Acad. Serbe Sci. Publ. Inst. Math.* **10** (1956), 71–78; ZB 71, p. 254; MR 18, p. 559.

Suetuna, Z., “Über die sich selbst assoziierten Charaktere der symmetrischen Gruppe”, *J. Reine Angew. Math.* **183** (1941), 92–97; ZB 24, p. 149; MR 4, p. 2.

Sullivan, J. J., “$N$-th rank tensor representations of $U(n)$ symmetry adapted to subgroups of the symmetric group $S_N$”, *J. Math. Phys.* **14** (1973), 387–395.

Sullivan, J. J., “Recoupling coefficients of the symmetric group involving outer plethysms”, *J. Math. Phys.* **19** (1978), 1674–1680; ZB 389, #20039.

Sullivan, J. J., “Generalized back coupling rules for the Racah algebra of GL$n$”, *J. Math. Phys.* **21** (1980), 227–233.

Suprunenko, D., “Über nilpotente transitive Untergruppen der symmetrischen Gruppe”, *Dokl. Akad. Nauk SSSR (N.S.)* **99** (1954), 23–25; ZB 56, p. 23.

Suprunenko, D. A., and Metel'skii, N. N., "Das Problem der Zuordnungen und der Minimierung einer Summe linearer Formen auf einer symmetrischen Gruppe", *Kibernetika, Kiev* **1973**, No. 3, 64–68; ZB 267, #200001.

Szalay, M., "A note on the dimensions of representations of $S_n$", in *Topics in Number Theory* (Proc. Coll. Debrecen, 1974), 383–388.

Szalay, M., and Turau, P., "On some problems of statistical theory of partitions with applications to characters of the symmetric group I, II", *Acta. Math. Acad. Sci. Hung.* **29** (1977), 361–379, 381–392.

Tappe, J., "Die Blockstruktur von Kranzprodukten der Form $G \operatorname{wr} S_n$ und $G \operatorname{wr} C_q$", Dissertation, Aachen, 1974.

Tappe, J., "Zur modularen Darstellungstheorie von Kranzprodukten", *Mitt. Math. Sem. Giessen* **112** (1974), 1–18.

Tappe, J., "Blocks and defect groups of monomial groups", *Manuscripta Math.* **17** (1975), 227–252; ZB 321, #20009.

Tappe, J., "Über die Charaktere und die Zerlegungszahlen der Weylgruppen $B_n$ und $D_n$", *Mitt. Math. Sem. Giessen* **121** (1976), 39–57; ZB 361, # 20017; MR 57, #12673.

Tappe, J., "Brauer correspondence and characters of height zero of monomial groups", *Manuscripta Math.* **18** (1976), 399–416; ZB 338, #20015; MR 53, #8221.

Tappe, J., "On the blocks of certain subgroups of monomial groups", *Arch. der Math.* **27** (1977), 25–33; ZB 361, #20018.

Taulbee, O. E., "Modular representations of $\mathfrak{S}_n$", Thesis, Michigan State University, 1957.

Taylor, M. J., "The locally free classgroup of the symmetric group", *Illinois J. Math.* **23** (1979), 687–702.

Taylor, T., "Representation theory of the symmetric and hyperoctahedral groups", Dissertation, Aberystwyth.

Taylor, T., "Representations of Weyl groups", Thesis, Aberystwyth, 1973.

Thomas, G. P., "Baxter algebras and Schur-functions", Dissertation, University of Wales, 1974.

Thomas, G. P., "A combinatorial interpretation of the wreath product of Schur functions", *Canad. J. Math.* **28** (1976), 879–884; MR 54, #108.

Thomas, G. P., "A generalization of a construction due to Robinson", *Canad. J. Math.* **28** (1976), 665–672.

Thomas, G. P., "Frames, Young tableaux, and Baxter sequences.", *Advances Math.* **26** (1977), 275–289; MR 58, #264.

Thomas, G. P., "Further results on Baxter sequences and generalized Schur functions", in *Combinatoire et représentation du groupe symétrique* (Actes Table Ronde CNRS, Univ. Louis-Pasteur Strasbourg, Strasbourg, 1976), Lecture Notes in Math., Vol. 579, Springer, Berlin, 1977, pp. 155–167; MR 57, #12243.

Thomas, G. P., "On a construction of Schützenberger", *Discrete Math.* **17** (1977), 107–118, MR 56, #8383.

Thomas, G. P., "On Schensted's construction and the multiplication of Schur-functions", *Advances Math.* **30** (1978), 8–32.

Thomas, G., "A note on Young's raising operator", *C. R. Math. Rep. Acad. Sci. Canada* **II** (1980), 35–36.

Thompson, J. G., "Hall subgroups of the symmetric groups", *J. Comb. Theory* **1** (1966), 271–179; ZB 144, p. 261; MR 33, #5711.

Thrall, R. M., "Young's seminormal representation of the symmetric group", *Duke Math. J.* **8** (1941), 611–624; ZB 61, p. 41; MR 3, p. 195.

Thrall, R. M., "On symmetrized Kronecker powers and the structure of the free Lie ring", *Amer. J. Math.* **64** (1942a), 371–388; ZB 61, p. 42.

Thrall, R. M., "On the decomposition of modular tensors I", *Ann. of Math. (2)* **43** (1942b), 671–684; ZB 61, p. 41; MR 4, p. 134.

Thrall, R. M., "On the decomposition of modular tensors II", *Ann. of Math. (2)* **45** (1944), 639–657; ZB 61, p. 41; MR 6, p. 146.

Thrall, R. M., and Nesbitt, C. J., "On the modular representations of symmetric groups", *Ann. of Math. (2)* **43** (1942), 656–670; ZB 61, p. 41; MR 4, p. 134.

Thrall, R. M., and Robinson, G. de B., "Supplement to a paper by G. de B. Robinson", *Amer. J. Math.* **73** (1951), 721–724; ZB 43, p. 260; MR 13, p. 205.

Thürlings, K.-J., "Der Diagrammverband und dessen Anwendung", Diplomarbeit RWTH, Aachen, 1977.

Todd, J. A., "A note on the algebra of $S$-functions", *Proc. Cambridge Philos. Soc.* **45** (1949), 328–334.

Todd, J. A., "Note on a paper by Robinson", *Canad. J. Math.* **2** (1950), 331–333; ZB 39, p. 20.

Tompkins, D. R., "Decomposition of tensors of the classical groups", *J. Math. Phys.* **8** (1967), 1502–1514; ZB 178, p. 25.

Tompkins, D. R., "A modified Frobenius equation for characters and branching laws of the permutation groups", *Nuovo Cimento, X. Ser.*, **B56** (1968), 316–322; ZB 164, p. 26; MR 38, #252.

Towber, J., "Two new functors from modules to algebras", *J. of Algebra* **48** (1977), 80–104.

Towber, J., "Young symmetry, the flag manifold, and representations of GL($n$)", *J. of Algebra* **61** (1979), 414–462.

Tsuzuku, T., "On multiple transitivity of permutation groups", *Nagoya Math. J.* **18** (1961), 93–109; ZB 109, p. 15; MR 23, #1732.

Turkin, W. K., "Über Herstellung und Anordnungen der monomialen Darstellungen endlicher Gruppen", *Math. Ann.* **111** (1935), 743–747; ZB 12, p. 249.

Tyskevic, R. I., and Amidi, Z. A., "Permutation groups and invariant relations", *Vescı Akad. Navuk BSSR Ser. Fız. -Mat. Navuk* **1973**, No. 4, 17–27, 135; MR 49, #7341.

Usher, A. O., "Plethysm of $S$-functions", *Canad. J. Math.* **28** (1976), 440–445; ZB 361, #20020; MR 53, #3095.

Ustimenko-Bakumovskii, V. A., "The lattice of supergroups of an induced symmetric group", *Soviet Math. Doklady* **18** (1977), 1433–1437 translation from *Doklady Akad. Nauk SSSR* **237** (1977), 276–279; ZB 396; #20002.

Vapné, Ju. E., "The criterion for the representability of wreath products of groups by matrices", *Dokl. Akad. Nauk SSSR* **195** (1970), 13–16; *Soviet Math. Dokl.* **11** (1970), 1396–1399; ZB 218, # 20032.

Venkatarayudu, T., "The characters of the classes $(n\text{-}k, k)$ of the symmetric group of degree $n$", *J. Indian Math. Soc. (N.S.)* **7** (1943), 42–45; ZB 61, p. 39; MR 5, p. 58.

Venkatarayudu, T., "The character table of a subgroup of the symmetric group of degree 8", *Proc. Indian Acad. Sci. Sect. A* **17** (1943), 79–82; MR 4, p. 267.

Venkatarayudu, T., "Characters of the classes of the form $(n_1, n_2, n_3)$ in symmetric groups", *Proc. Indian Acad. Sci. Sect. A* **22** (1945), 42–45; ZB 61, p. 39; MR 7, p. 113.

Veršik, A. M., and Kerov, S. V., "Asymptotics of the Plancherel measure of the symmetric group and the limiting form of Young-tables", *Dokl. Akad. Nauk SSSR* **233** (1977), 1024–1027; *Soviet Math. Dokl.* **18** (1977), 527–531.

Veršik, A. M., and Šmidt, A. A., "Symmetric groups of high degree", *Dokl. Akad. Nauk SSSR* **206** (1972), 269–272; *Soviet Math. Dokl.* **13** (1972), 1190–1194; MR 47, #4300.

Viennot, G., "Une forme géométrique de la correspondance de Robinson-Schensted", *Combinatoire et représentation du groupe symétrique*. (Actes Tables Ronde C.N.R.S. Strasbourg, 1976), Lecture notes in Math., Vol. 579, 29–58 (1977); ZB 389, #05016; MR 57, #9827.

van der Waerden, B., "Der Zusammenhang zwischen den Darstellungen der symmetrischen und der linearen Gruppen", *Math. Ann.* **104** (1931), 92–95, 800.

van der Waerden, B., *Algebra II*, Die Grundlehren der Mathematischen Wissenschaften in Einzeldarstellungen, Vol. 34, 4th edition, x+300 pp.; MR 38, #1968.

Wagner, A., "The faithful linear representations of least degree of $S_n$ and $A_n$ over a field of characteristic 2", *Math. Z.* **151** (1976), 127–137; ZB 321, #20008.

Wagner, A., "An observation on the degrees of projective representations of the symmetric and alternating group over an arbitrary field", *Archiv der Math.* **29** (1977), 583–589.

Wagner, A., "The faithful linear representations of least degree of $S_n$ and $A_n$ over a field of odd characteristic", *Math. Z.* **154** (1977), 103–114; ZB 336, #20008, MR 55, #10555.

Wagner, B., "Symmetrische Polynome und Darstellungen der symmetrischen Gruppen", Diplomarbeit, Aachen, 1975/76.

Wagner, B., "Charaktere symmetrischer und monomialer Gruppen als Polynomfunktionen", *Beyreuther Math. Schriften* **2** (1979), xii+111 pp.

Wales, D. B., "Some projective representations of $S_n$", *J. Algebra* **61** (1979), 37–57.

Wallace, A. H., "Invariant matrices and the Gordon-Capelli series", *Proc. London Math. Soc. (3)* **2** (1952), 98–127.

Wallace, A. H., "Generalized Young tabeaux", *Proc. Edinb. Math. Soc. (2)* **9** (1953), 35–43.

Watkins, M. E., "Graphical regular representations of alternating, symmetric, and miscellaneous small groups", *Aequationes Math.* **11** (1974), 40–50.

Weir, A. J., "The Sylow subgroups of the symmetric groups", *Proc. Amer. Math. Soc.* **6** (1955), 534–541; ZB 65, p. 256; MR 17, p. 235.

Weisner, L., "On the Sylow subgroups of the symmetric and alternating groups", *Amer. J. Math.* **47** (1925), 121–124.

Wells, C., "Some applications of the wreath product construction", *Amer. Math. Monthly* **83** (1976), 317–338; MR 53, # 8309.

Weyl, H., "Das gruppentheoretische Fundament der Tensorrechnung", *Göttinger Nachrichten* **1924**, 218–224.

Weyl, H., "Theorie der Darstellung kontinuierlicher halbeinfacher Gruppen durch lineare Transformationen", *Math. Z.* **23** (1925), 271–309.

Weyl, H., "Der Zusammenhang zwischen der symmetrischen und der linearen Gruppe", *Annals of Math. (2)* **30** (1929), 499–516.

Weyl, H., "Commutator algebra of a finite group of collineations", *Duke Math. J.* **3** (1937), 200–212.

Weyl, H., *The Classical Groups. Their Invariants and Representations*, Princeton University Press, 1939; MR 1, p. 42.

White, D. E., "A de Bruijn-type formula for enumerating patterns of partitions", *Aequationes Math.* **16** (1977), 59–64; ZB 381, #05002.

White, D. E., "Multilinear techniques in Pólya enumeration theory", *Linear and Multilinear Algebra* **7** (1979), 299–315.

White, D. E., "Monotonicity and unimodality of the pattern inventory", *Advances in Math.* **38** (1980), 101–108.

White, D. E., "A Pólya interpretation of the Schur function", *J. Comb. Theory (A)* **28** (1980), 272–281.

Williams, A. G., "Characteristics of $G\mathrm{wr}S_n$", *Math. Proc. Camb. Philos. Soc.* **79** (1976), 433–441; ZB 338, #20011; MR 53, #8225.

Williamson, S. G., "Symmetry operators of Kranz products", *J. Comb. Theory* **11** (1971), 122–138.

Woltermann, M., and Sehgal, S., "Equidistributed $\frac{3}{2}$-transitive solvable permutation groups", *Communications in Algebra* **7** (1979), 1599–1643.

Woltermann, M., and Sehgal, S., "Equidistributed $\frac{3}{2}$-transitive solvable permutation groups II", *Communications in Algebra* **7** (1979), 1645–1672.

Wong, W. J., "A characterization of the alternating group of degree 8", *Proc. London Math. Soc. (3)* **13** (1963), 359–383; ZB 112, p. 261; MR 26, #5041.

Wong, W. J., "Irreducible modular representations of finite Chevalley groups", *J. Algebra* **20** (1972), 355–367.

Xu Cheng-hao, "The commutators of the alternating group", *Sci. Sinica* **14** (1965), 339–342; ZB 152, p. 4; MR 32, #1241.

Yamaki, H., "A characterization of the alternating groups of degrees 12, 13, 14, 15", *J. Math. Soc. Japan* **20** (1968), 673–694; ZB 167, p. 23; MR 38, #1157.

Yamaki, H., "A characterization of the simple groups $A_7$ and $M_{11}$", *J. Math. Soc. Japan* **23** (1971), 130–136; ZB 206, p. 310.

Yamanouchi, T., "On the construction of unitary irreducible representations of the symmetric group", *Proc. Phys. Math. Soc. Jap. (3)* **19** (1937), 436–450; ZB 16, p. 293.

Yamanouchi, T., "Tables useful for construction of irreducible representation matrices of symmetric group", *J. Phys. Soc. Japan* **3** (1948), 245–253; MR 12, p. 479.

Yokonuma, T., "On a property of some generalized symmetric groups", *J. Fac. Sci. Tokyo* **12** (1965), 193–211; ZB 139, p. 20; MR 32, 2, #7650.

Young, A., "On quantitative substitutional analysis", *Proc. London Math. Soc.* **33** (1901), 97–146.

Young, A., "On quantitative substitutional analysis (second paper)", *Proc. London Math. Soc.* **34** (1902), 361–397.

Young, A., "On quantitative substitutional analysis", *J. London Math. Soc.* **3** (1928), 14–19.

Young, A., "On quantitative substitutional analysis (third paper)", *Proc. London Math. Soc.* **28** (1928), 255–292.

Young, A., "On quantitative substitutional analysis (fourth paper)", *Proc. London Math. Soc. (2)* **31** (1930), 253–272.

Young, A., "On quantitative substitutional analysis (fifth paper)", *Proc. London Math. Soc. (2)* **31** (1930), 273–288.

Young, A., "On quantitative substitutional analysis (sixth paper)", *Proc. London Math. Soc. (2)* **34** (1932), 196–230; ZB 5, p. 97.

Young, A., "On quantitative substitutional analysis (seventh paper)", *Proc. London Math. Soc. (2)* **36** (1934), 304–368.

Young, A., "On quantitative substitutional analysis (eighth paper)", *Proc. London Math. Soc. (2)* **37** (1934), 441–495; ZB 9, p. 301.

Young, A., "The application of substitutional analysis to invariants", *Phil. Trans. Roy. Soc. London A* **234** (1935), 79–114.

Young, A., "On quantitative substitutional analysis (ninth paper)", *Proc. London Math. Soc. (2)* **54** (1952), 219–253.

Zacher, G., "The lattice of subgroups of the symmetric group" (Italian, English summary), *Rend. Istit. Mat. Univ. Trieste* **9**, No. 1–2 (1977), 122–126.

van Zanten, A. J., and de Vries, E., "Tensor operators of finite groups", *J. Math. Physics* **14** (1973a), 1423–1429; MR 49, # 399.

van Zanten, A. J., and de Vries, E., "On the number of roots of the equation $x^n = 1$ in finite groups and related properties", *J. Algebra* **25** (1973b), 475–486; ZB 258, #20006; MR 47, #3509.

van Zanten, A. J., and de Vries, E., "On the number of classes of a finite group invariant for certain substitutions", *Canad. J. Math.* **26** (1974), 1090–1097; ZB 304, #20004.

van Zanten, A. J., and de Vries, E., "Criteria for groups with representations of the second kind and for simple phase groups", *Canad.. J. Math.* **27** (1975), 528–544; ZB 338, #20010; MR 52, #555.

Zavrid, G. P., "Nilpotente Untergruppen einer alternierenden Gruppe", *Izv. Akad. Nauk BSSR, Ser. Fiz. -Mat. Nauk* **1965**, No. 2, 110–112; ZB 154, p. 20.

Zia-ud-Din, M., "The characters of the symmetric group of order 11!", *Proc. London Math. Soc. (2)* **39** (1935), 200–204.

Zia-ud-Din, M., "The characters of the symmetric group of degrees 12 and 13", *Proc. London Math. Soc. (2)* **42** (1936), 340–355.

# Index

Let $(\alpha \in \mathbb{Z}^n)$ $\quad \varphi(\alpha) = \dfrac{n!\,\Delta(\alpha)}{\alpha!} = \dfrac{n!}{\alpha!} \prod_{i<j} (\alpha_i - \alpha_j)$

$(= 0 \text{ if any } \alpha_i < 0)$ $\quad [\varphi(\beta) = d(\beta)]$

$s \in \lambda$:

$$f_\lambda(s) = \varphi(\xi) - \sum_{i=1}^{\lambda'_k} \varphi(\xi + h(s)(\varepsilon_i - \varepsilon_j))$$

where $\xi = \lambda + \delta$, $\quad s = (j, k) \in \lambda$

Then

$$\boxed{\det \lambda = \prod_{s \in \lambda} h(s)^{f_\lambda(s)}.} \qquad (7.3.20)$$

Hence
$$a_n - (A+\varepsilon) b_n < a_{n+1} - (A+\varepsilon) b_{n+1}$$

and
$$a_n - (A-\varepsilon) b_n > a_{n+1} - (A-\varepsilon) b_{n+1}$$

Consider the sequence $\left(a_n - (A+\varepsilon) b_n\right)_{n>N}$. Increasing and $\to 0$, hence terms are $< 0$. Hence
$$a_n < (A+\varepsilon) b_n \quad \text{for } n > N$$

ie
$$\frac{a_n}{b_n} < A+\varepsilon.$$

Similarly
$$\frac{a_n}{b_n} > A-\varepsilon. \;|$$